MW00633111

RENOVATION 5TH EDITION

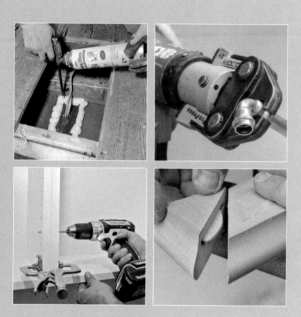

RENOVATION 5TH EDITION

Michael Litchfield & Chip Harley

The Taunton Press

Text © 2019 by Michael W. Litchfield and George Harley

Photographs © 2019 by Michael W. Litchfield and Ken Gutmaker, except where noted.

Illustrations by Vincent Babak, © 2019 by The Taunton Press, Inc.

All rights reserved.

The Taunton Press
Inspiration for hands-on living®

The Taunton Press, Inc., 63 South Main Street, Newtown, CT 06470-2344

e-mail: tp@taunton.com

Editors:	Carolyn Mandarano, Peter Chapman
Copy editor:	Diane Sinitsky
Indexer:	Cathy Goddard
Cover design:	Jean-Marc Troadec
Interior design:	Carol Petro
Layout:	Laura Lind Design, Barbara Cottingham, Lynne Phillips
Illustrator:	Vincent Babak
Photographers:	Michael W. Litchfield and Ken Gutmaker, except where noted
Cover photo:	Rodney Diaz, *Fine Homebuilding*

Library of Congress Cataloging-in-Publication Data

Names: Litchfield, Michael W., author. | Harley, Chip, author.
Title: Renovation / Michael W. Litchfield and Chip Harley.
Description: 5th edition, Completely revised and updated. | Newtown, CT :
   Taunton Press, Inc., [2019]
Identifiers: LCCN 2018049603| ISBN 9781631869594 (hardcover) | ISBN
   9781631869600 (paperback)
Subjects: LCSH: Dwellings--Remodeling. | Dwellings--Conservation and
   restoration. | Dwellings--Maintenance and repair.
Classification: LCC TH4816 .L57 2019 | DDC 643/.7--dc23
LC record available at https://lccn.loc.gov/2018049603

Printed in the United States of America
10 9 8 7 6 5 4 3 2

The following manufacturers/names appearing in *Renovation* are trademarks: AccuVent®, Acousto-Plumb®, AdvanTech®, Ahrens® Chamber-Tech 2000™, Airkrete®, AlumaFlex™, American Aldes®, Aqua Mix® Sealer's Choice®, Ardex Feather Finish®, Autodesk AudoCAD®, Azek®, Bahco®, Basement Systems®, BaySeal®, Bazooka®, Benjamin Moore® Natura®, B-I-N®, Bituthene®, Black & Decker® Workmate®, Bondo®, Bostick®, Build It® Green, Caesarstone®, Carborundum®, Carlisle Wide Plank Floor®, Cedar Breather®, CertainTeed Winterguard™, Channellock®, ChoiceTrim®, Citistrip®, Clear Lam®, Climatek™, ColorSnap®, COPALUM®, Corian®, Crack-Pac®, Crescent®, DAP®, DELTA®, DensGlass®, DensShield®, DeWalt®, Diablo®, DMF Lighting®, Dow®, Dremel®, Dropbox™, Duct Blaster®, Dumpster®, DuPont™, Durabond® 90, Durapalm®, Durock®, EcoSeal™, Elmer's® Glue-All®, EnergyComplete™, FasCap™, Fastrip®, Federal Pacific®, Festool®, Fix-It-All™, Flood® Floetrol®, Flood® Penetrol®, Formica®, Fortifiber®, Fypon®, GacoFlex®, GAF Weather Watch®, Gammon Reel®, Genie®, GenieClip®, Genie Lift™, Goof Off®, Google Maps™, Google SketchUp™, Gorilla Glue®, GreenHomes® America, Greenlee® Nail Eater®, Grip-Lok®, GRK Fasteners™, Habitat for Humanity®, Hackzall®, Hafele®, HardieBacker®, HardieTrim®, Henry 208 Wet Patch®, Hole-Hawg®, Homax®, Home Depot®, Home Slicker®, HomerWood®, IKEA®, iPad®, Kilz®, Kleer®, Kohler®, Krazy Glue®, LaHabra®, Lauzon®, layfastSBS®, LeadCheck™, LevelQuik®, Lift Off®, Lincoln Logs®, Lion Miter-Trimmer®, Lutron®, Maestro Wireless®, Magic Trowel®, Makita®, Malco® TurboShear™, Mapelath™, Marmoleum®, Marvin Tilt Pac®, Masonite®, Mastercut®, Medex®, Medite®, Metabo®, Microllam®, Milwaukee®, Miracle Sealants® 511 Porous Plus®, MiraTEC®, Mohawk®, Moistop Corner Shield®, Multi-Strip™, Murphy® Oil Soap, Mylar®, Natural Cork®, Nibco®, OSI® Pro-Series®, Paint Shaver®, Panasonic™, PaperTiger®, Parabond®, Parallam®, Peacekeeper®, Peel Away®, Perform Guard®, Pico® Wireless, Plaster-Weld®, PL Turbo™, Plyboo®, Polyken® Foilastic®, Power Lite®, ProTrim®, Quikrete®, Q-Lon®, Ram Board®, Ready-Strip®, ReStore®, Rev-a-Shelf®, Rhodia®, Rock-On®, Romex®, Roto-Split®, R-Wrap®, Safecoat®, Sawzall®, Senco® DuraSpin®, Shaw® Floors, Sheetrock®, Sherwin-Williams®, SierraPine®, Silestone®, Simplex®, Simpson Crack-Pac®, Simpson Strong-Tie®, Skilsaw®, Skype™, Smart Strip®, Snap-On®, Sto®, Structurwood®, Sturd-I-Floor®, Super Glue®, Super Jumbo Tex®, Suppress®, Surewall®, Surform®, Swanson® Speed® Square, TEC™, Teco®, Teflon®, Tergitol™, Thorobond®, Thoroseal®, 3M® Fastbond®, TileLab®, TimberStrand®, Titebond®, TopNotch®, Torx®, Triton™, TruExterior®, True Value®, TrusJoist®, TurboShear®, Tyvek®, Ultraflex®, U-Sand®, Velux®, Vise-Grip®, Vulkem® 116, Wall-Nuts™, Watco®, Weld-Crete®, Weldwood®, Weyerhaeuser®, Wire-Nut®, Wonder Bar®, WonderBoard®, W. R. Grace Ice & Water Shield®, Zinsco®, Zinsser® DIF™, Zinsser® Jomax®, ZipWall®, Zircon® StudSensor™, Zodiaq®

"Many hands make work light."

This book is dedicated to the many
hands that have helped ours along the way.

Acknowledgments

ONE SPRING DAY IN 1972, I bought my first home, an old Vermont schoolhouse built in 1826 that was standing at all because of its massive oak post-and-beam frame. Its roof was shot, its foundation was caving in, and its new owner had more enthusiasm than sense. (Or money. The old place cost $6,000, and I had to borrow most of that.) Once I realized what I had gotten into—and that no books were complete or clear enough to guide me—I started peppering local builders with questions, and they, good souls, shared what they knew.

More than 45 years later, I'm still at it.

In intervening years I had the good fortune to help launch *Fine Homebuilding* magazine and so got access to talented designers, builders, and tradespeople across North America. The first edition of *Renovation* was published in 1982, not long after *Fine Homebuilding* hit the stands, and the information-gathering model was the same for both: Find the best people, go to their job sites, and work as hard as they do, to photograph and record the process.

About the same time, Chip Harley got his California contractor's license, after learning the business from the bottom up. Chip was drawn to tools and building as far back as he can remember—as were his brothers Duffy and Tom—so in 1984, they launched a construction company in Berkeley. Over the years, Chip and his crews have remodeled more homes than they can remember, dealing with common and unique problems along the way.

In 2002, when I was about to revise the third edition of this book, I went looking for a technical editor and found Chip. Three minutes into the conversation, it was apparent that he had an encyclopedic knowledge of building and could explain it all simply. It's a rare gift. I have called him on-site scores of times, and though I could hear saws whining, hammers thudding, and guys yelling in the background, Chip always answered my questions calmly and in great detail.

Chip's knowledge is remarkable, but his generosity is typical of many of the builders I've met over the years—people who are good at what they do love to talk about it.

All told, more than 500 people have helped me create or update the information in these editions. So once again, I have a lot of people to thank.

Sally and George Kiskaddon of Builders Booksource in Berkeley, CA, have been champions of all five editions of *Renovation*. Their independent bookstore has perhaps the widest and deepest selection of design, construction, and code books anywhere, and they will gladly ship orders.

Among architects and designers, I would like to thank Glenn Jarvis, Robin Pennell, and Jon Larson of Jarvis Architects, Oakland, CA, for generously sharing projects, contacts, and insights; Russell Hamlet of Studio Hamlet on Bainbridge Island, WA, for always exceeding my expectations; Fran Halperin of Halperin & Christ, San Rafael, CA, for her exquisite taste; Jerri Holan, FAIA, Oakland, CA; Stephen Shoup of BuildingLab, Emeryville, CA, for his friendship and great insights into green building practices; thanks also to BuildingLab's Hide Kawato for many kindnesses. Revising this edition also gave me the chance to reconnect with old friends Carolynn and Gregory Schipa of The Weather Hill Company of Warren, VT.

Scores of builders are listed below. But special thanks to those contractors whom I relied on again and again: Dave Carley of Bainbridge Island, WA; Eric Christ of Halperin & Christ, San Rafael, CA; Dean Rutherford of Rutherford & Singelstadt, Richmond, CA, and his wife Marty, both great souls.

These people contributed significantly to individual chapters: **Reading a House:** Roger Robinson of First Rate Property Inspection in Oakland, CA, and co-author of *House Check: Finding and Fixing Common House Problems* (The Taunton Press, 2003); Brian Cogley of Cogley Property Inspections, Oakland, CA. **Planning Your Renovation:** Taya and Stephen Shoup, Carolynn and Gregory Schipa, Fran Halperin, Sharon Low, Sandra Almador Mora, David Glaser, Russell Hamlet, Dave Carley, Robin Supplee, Mike Derzon, Beth and Gary Sumner. **Tools, Building Materials:** Warren White and Dave Yungert of Truitt & White Lumber in Berkeley, CA, once again let us cherry-pick and photograph their latest releases. **Roofs:** The insulation crew of Abril Roofing:

Jose Bustos, Francisco Padilla, Francisco Vargas, Raul Foyfan. **Doors, Windows, and Skylights:** Bill Essert of Wooden Window in Oakland, CA, and Manuel Morgado of Truitt & White were extremely gracious with their time and contacts. The talented crews doing the work: Ray DeMuro and Joe Gonzalves; Franco Arellano, Claudio Matute, and Ioane Serguetasi of Wooden Window; J. Luis Perez of Brooks Weitzman; Tim Thorvick and Eli Chavez of BuildingLab. And Gary Schroeder, at the top of his game, in skylights. **Exteriors:** Les Williams of L.A. Williams Copper Gutters, Oliver Govers, and my old friends from Holland & Harley, Jesus Beltran and Heberto Mendez. **Structural Carpentry:** Rick Collins and his wife, Nicole, principals of Trillium Dell Timberworks of Galesburg, IL; Rick's masterful big-timber restorations are inspiring. Mike Guertin, John Michael Davis, Bob's Iron in Oakland, CA. **Masonry:** Chad Corley of The Quikrete Companies and Sally McKnight of the Irish Sweep, Alameda, CA. **Foundations and Concrete:** Mike Massoumi of Impact Construction, Albany, CA. **Electrical Wiring:** A thousand thanks (one for each question) to Michael McAlister, my co-author on three editions of *Wiring Complete* (The Taunton Press). Also, Troy Mittone of Oakland Electric. **Plumbing:** Martin Kelleher of Kelleher Plumbing for the latest on PEX plumbing systems. Albert Nahman, Mario Hernandez, and Cesar Enrique Ruano of Albert Nahman Plumbing, Oakland, CA. Thanks also to Ron Kyle. **Kitchens and Baths:** Catherine Moncrieff & Stuart Brotman of Bowman's Gate Design, Marciel Dornelio, Steve Kirby, and Les Baker of Baker Granite & Marble. **Energy Conservation and Air Quality:** I am very grateful for the help of Scott Gibson, who, along with David Johnston, wrote *Green from the Ground Up* (The Taunton Press, 2008); Donn Davy of Green Home Savvy; Mike Rogers of GreenHome America. Also Don Cain, Edward Joachin, and Rolando Ceja Valadez of SDI Insulation. **Finish Surfaces:** Steve Marshall of The Little House for the soundproofing sequence and his crew: Jose Rodriquez, Javier Santiago, Roberto Padilla Lopez, Benjamin Santiago. **Tiling** seems to attract colorful characters, including Dan Blume, Josue "Blanco" Zazueta,

and Victor Estrada Martinez. **Finish Carpentry:** Chris Rogers of BuildingLab is a great talent, as are restoration carpenters Jim Spaulding of B.C.E. in Alameda, CA, and John Michael Davis of New Orleans, LA. Thanks for all your help. **Painting:** Greg Scillitani and his capable crew, Greg and Scott. **Wallpapering:** Peter Bridgman is considered to be one of the finest installers of wall coverings in the San Francisco, CA, area. **Flooring:** Pawel Bajer of Tegan Flooring. Also to Mark Clarine and Jeff Molina of Floor Dimensions, shown installing resilient flooring and carpeting, respectively.

Thanks to the good people at The Taunton Press who helped me in myriad ways or who will, in the course of things, produce this book. At *Fine Homebuilding* and *Green Building Advisor*, thanks to Brian Pontolilo, Chuck Miller, Dan Morrison, Rob Wotzak, Martin Holladay, Joe Lstiburek, Andy Engel, Mike Guertin, and others whose insights or photos I drew on. On the book side, thanks to Lynne Phillips, Barbara Cottingham, and Sharon Zagata.

I am delighted that illustrator Vincent Babak returned for the fifth edition and very grateful that Muffy Kibbey of Berkeley, CA, Art Grice of Seattle, WA, David Milne of Toronto, Ont., Scott Hargis of Oakland, CA, and David Simone took the beautiful house portraits in chapter 2. On the copy side, Scott Gibson made this a much better book. Carolyn Mandarano shepherded this edition through production and was a treat to work with. Finally, in 40-some years of book writing, I have never had an executive editor who was the equal of Peter Chapman. If I ever get caught in a hurricane at sea, I want Peter at the helm.

And so on to the home front. Chip says of his wife, Allyson Page, "I sometimes borrow her strength to keep going. My world revolves around her and our daughters, Ronnie and Lucia." In the same vein, I would like to express my great joy, deep love and unbelievable good fortune that Jeannie is part of my life, and that the world's best dog was part of the bargain.

—Mike Litchfield
Point Reyes, CA

Table of Contents

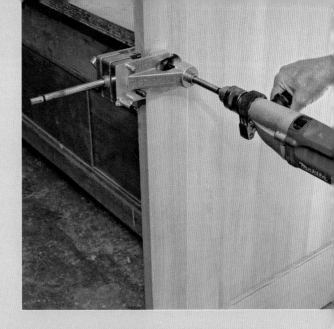

INTRODUCTION   **4**

CHAPTER 1:
Reading a House   **6**

The Roof   9
House Exterior   12
Interiors   14
Mechanical Systems   18
Environmental Hazards   21

CHAPTER 2:
Planning Your Renovation   **23**

A Home for the Long Run   23
Clarifying Your Goals   24
Documenting What's There   26
Getting Help with Your Renovation   29
Developing a Renovation Plan, and Getting It Drawn   31
Making the Most of Your Renovation   35
Creating Lasting Value: Five Case Histories   36

CHAPTER 3:
Tools   **51**

Tool Safety   51
Tools to Own   53
Tools to Rent   64

CHAPTER 4:
Building Materials   **68**

Standard Lumber   68
Salvage Lumber   71
Engineered Lumber   71
Framing with Steel   74
Structural and Nonstructural Panels   75
Metal Connectors   78
Bolts   84
Adhesives, Caulks, and Sealants   86

CHAPTER 5:
Roofs   **90**

Roof Safety and Equipment   90
Preparing to Install a New Roof   92
Roof Flashing   96
Asphalt Shingles   101
Wood Shingles and Shakes   110
A Medley of Roofing Types   113

CHAPTER 6:

Doors, Windows, and Skylights 116

Ordering Doors: An Overview	116
Installing an Interior Door	119
Installing an Exterior Door	124
Installing French Doors	131
Installing Hardware	133
Weatherstripping Door Frames	136
Keeping the Weather Away: A Primer on Flashing	139
Choosing Windows	142
Refurbishing Old Double-Hung Windows	156
Installing a Skylight	159

CHAPTER 7:

Exteriors 166

Water-Resistive Barriers	168
Exterior Trim	170
Siding	176
Fiber-Cement Siding	183
Stucco Repairs	185
Rain Gutters	190

CHAPTER 8:

Structural Carpentry 194

Understanding Structure	195
Framing Walls	195
Demolition	206
Reinforcing and Repairing the Structure	211
Stair Repairs	217
Structural Remodeling	222

CHAPTER 9:

Masonry 234

Terms, Tools, and Tips	234
Working with Brick	237
Chimneys	244

CHAPTER 10:

Foundations and Concrete 258

An Overview	258
Jacking and Shoring	263
Minor Repairs and Upgrades	269
Replacing a Shallow Foundation	272
Concrete Work	279
A Mat Slab Foundation	282
Damp Basement Solutions	284

CHAPTER 11:
Electrical Wiring 286

Understanding Electricity   286
Electricity in Your Home   288
Planning an Electrical Remodel   294
Tools and Materials   304
Rough-in Wiring   312
Wiring Receptacles   327
Wiring Switches   332
Fixture Wiring   338
Portfolio of Wiring Schematics   343

CHAPTER 12:
Plumbing 346

An Overview of Plumbing Systems   346
Planning   349
Tools   352
Copper Water-Supply Pipe   354
PEX Supply Pipes   358
PEX Pros and Cons   358
Galvanized Steel Pipe   362
CPVC Supply   362
DWV Materials   362
Venting Options   366
Roughing-in DWV Pipes   370
Roughing-in Supply Pipes   376
Installing Fixtures   377
Replacing a Water Heater   383
Installing a Tankless Water Heater 385

CHAPTER 13:
Kitchens and Baths 389

Kitchen Planning   389
Installing Cabinets   394
A Personal Take on IKEA Cabinets   400
Countertops for Kitchens and Baths   403
Kitchen Sinks   408
Bathroom Planning   412

CHAPTER 14:
Energy Conservation and Air Quality 416

A Nine-Step Energy Retrofit   417
Getting an Energy Audit   419
Air-Sealing   423
Increasing Controlled Ventilation   432
Controlling Moisture and Mold   435
Choosing Insulation Wisely   437
Installing Insulation   443

CHAPTER 15:
Finish Surfaces 454

Drywall   454
Plastering   475
Soundproofing   479

CHAPTER 16:

Tiling 484

Choosing Tile	484
Tools	486
Materials	489
Getting Ready to Tile	493
Installing Setting Beds	496
Tile Estimation and Layout	500
Tiling a Floor	502
Countertops	505
Tub Surround	508

CHAPTER 17:

Finish Carpentry 512

Tools	512
Materials	517
Basic Skills	519
Casing a Door	523
Casing a Window	528
Baseboard and Crown Molding	530

CHAPTER 18:

Painting 538

Essential Prep Work	538
Choosing Paint	538
Tools and Equipment	541
Painting Basics	544
Spray Painting	546
Lead-Paint Safety	548
Painting the Interior	551
Stripping and Refinishing Interior Trim and Wood Paneling	558
Painting the Exterior	563

CHAPTER 19:

Wallpapering 570

Selecting Materials	570
Ordering Wallcovering	573
Equipment	575
Preparing Surfaces	576
Laying Out the Work	581
Basic Papering Techniques	581
Complex and Special-Care Areas	586

CHAPTER 20:

Flooring 590

Flooring Choices	591
Refinishing Wood Floors	595
Installing Strip Flooring	605
"Floating" an Engineered Wood Floor	610
Resilient Flooring	614
Wall-to-Wall Carpeting	618

GLOSSARY OF BUILDING TERMS	626
INDEX	633
CREDITS	647

Introduction

I HAVE BEEN WRITING OR REVISING this book through four decades. I started the first edition in 1978, when Jimmy Carter was president and most computers were the size of closets. After gutting and rebuilding *Renovation* in 2012, I was sure the fourth edition would stand forever, but the upheaval in housing in the last decade convinced me otherwise. More than ever, we need to plan carefully, spend wisely, and build durably. So this fifth edition has been revised to help builders and homeowners add value, maximize space and functionality, conserve resources, and create comfortable homes that can accommodate the inevitable changes that life brings.

As in earlier editions, *R5* draws on innumerable conversations with carpenters, electricians, engineers, energy auditors, plumbers, painters, masons, architects, and other building professionals. All told, the photos in *R5* were drawn from roughly 40,000 taken over the years, mostly on job sites across North America.

That last point—"on job sites"—is what distinguishes this book from others in the field, and it's what should prove most useful to you when you're in the thick of a renovation. *R5* tells you which sawblade to use, the size and spacing of nails, when to tear out and when to make do, and how to lay out and prep a job so it goes smoothly. Because this book contains thousands of tips and techniques from contractors who had schedules and budgets to meet, it also will save you time and money. In other words, the methods in this book are field-tested. Supported by lifetimes of practical experience, you can proceed confidently.

This book is as much concerned with *what* and *why* as it is with *how*. Thus, for every topic—from foundations to finish flooring—you'll find the tools and materials you'll need, the problems you may encounter, and workable solutions to see you through. Because the information in each chapter follows the sequence of an actual renovation, you'll know what to anticipate at every stage. Equally important, *R5's* often-ingenious solutions will help you deal with the unexpected situations that are a part of every renovation.

PLAY IT Safe

Please heed all safety warnings: They are there for your protection. The publisher and I have made every effort to describe safe construction procedures in a clear and straightforward manner. But because of the differences in skill and experience of each reader and because of variations in materials, site conditions, and the like, neither I nor the publisher can assume responsibility for results with particular projects.

How to Use This Book

Read the opening remarks in a chapter before reading up on specific tasks. That is, the information in each chapter tends to be cumulative. The first few paragraphs often introduce important terms and concepts. Thereafter, you'll find tools and techniques presented more or less chronologically, in the order you'd need them in a renovation.

Although new terms are defined early in each chapter and later in context, you may come across terms whose definitions you skipped earlier. If you need a definition, consult the glossary or the index.

An in-depth review of tools and materials is beyond the scope of this book. If you want more information on either, consider browsing the Internet. Although I do mention specific brand names and occasional Internet addresses, please consider them reference points for research and not product endorsements. Most of the brand names are those I encountered on job sites or were praised by a builder whose opinions I value.

Maybe it's always been so, but research has become a big part of renovation. So supplement your reading and Internet searches by talking to neighbors, local contractors, and building-material suppliers. Experience is always the best teacher—even if it's someone else's experience. A friend or neighbor who's been through a renovation may be able to recommend reliable builders and suppliers and may also be a calm voice when you need one most. So go to it. As Aristotle once said (though not to me directly), "Courage is first among human virtues, for without it, we're unlikely to practice many of the others."

Reading a House

With a little practice, you can train your eye to see a house's potential and pitfalls.

Every house has stories to tell. If you know where to look, you can see how skillfully the house was built or remodeled, how well it has weathered the elements, and how carefully its owners have taken care of it.

This chapter explains how to read a house's sometimes subtle history, how to spot problems, and how to determine what caused them. Whether you're a homeowner, a house shopper, or a renovation contractor, look closely and search for patterns. You may be surprised to discover how many areas need attention, whether for safety, updating, appearance, or preventive maintenance. A careful inspection can be your guide to future renovations.

If you're house shopping, your inspection may reveal conditions serious enough to dissuade you from buying. Or, if you decide to buy, those problems may give you leverage in negotiating a lower price. Remember, most aspects of purchase agreements are negotiable. If you're a remodeling contractor, this chapter may be helpful in assessing systems you are less familiar with, and subsequent chapters will specify techniques and materials that can make your renovation projects more time- and cost-effective.

Finally, think of this chapter as a gateway to solutions throughout the book. Many of the problems described in this chapter are followed by page numbers or chapter numbers directing you to further explanations or possible solutions. If you don't find specific cross-references to topics you'd like to learn more about, consult the book's index.

TRAINING YOUR EYE

The house on the facing page says much to a trained eye. This rural Victorian dates from about 1890 and though it's plain compared to its city cousins, it was built by skilled hands. Inlet corner trim, triangular pediments over the windows, and ornamental brackets were typical architectural touches of that era, and for a country house, it's proportions are rather stately.

The porch, however, was a clumsy afterthought, lacking the sophistication of the original structure. The porch posts are 4x4s without any molding or shaping, the balusters are just 2x2s, the roof would be more appropriate to a wood shed, and both ends of the porch roof sag because its post loads are inadequately supported. And the stairs are shot.

Still, the house has great bones. Seasoned carpenters were needed to frame the converging planes of the roof, so there's probably good workmanship throughout the house, which is old

For Nesters: Keeping Emotions in Check

When shopping for a house, it's hard to keep emotions in check. Unless you're buying a property solely as an investment, you're probably looking for a nest. If you're like most of us, you'll imagine yourself living there, surrounded by friends and family. Those warm feelings are all understandably human but probably not the best frame of mind for making one of the biggest financial decisions of your life. By all means, listen to your feelings; just don't lead with them.

Look at a lot of houses. Read this chapter to get an overview of house systems and learn building lingo. Then scrutinize every house you enter—whether it's for sale or not. Be cold-eyed: Look beyond the lace curtains and the fresh paint. Look for problems and try to figure out what's causing them. Then, when you begin shopping for real and find that certain place that wins your heart, you won't lose your head.

Also, if you like a house, check out the neighborhood and talk to neighbors to see what they are like. Ask about traffic, schools, shopping, city services, and crime. This will help you imagine what living there will be like.

enough that its 2x4s are probably full-size 2x4s. The walls are plumb; the roof ridges don't sag; and despite its weathered appearance, the ship-lap siding is largely intact. As important, the roofing is recent. Despite the grass clogging the porch gutter, someone has been keeping an eye on the old homestead.

Before making an offer, you'd want to do some digging. From the street, there's no sign of gutters or downspouts on the main house, so it would be crucial to inspect the joists and mud-sills for signs of water damage and insect infestation. The foundation is unreinforced brick, so it would need to be replaced. Given the missing brackets and boarded-up windows, vandals may have made off with some of the interior trim.

All in all, though, this house is an exciting prospect and certainly worth a closer look.

GATHERING INFORMATION

If you feel strongly about a house, start by asking the real estate agent or owner for a recent termite report and a disclosure statement, and read them closely. Most states require such disclosures from owners; if you are working with an agent, such statements are probably mandatory. Disclosure statements describe (1) things not originally built with a permit or not built according to code, (2) code violations recently observed by an inspector, and (3) other conditions the homeowner knows need fixing. Armed with this information, you can begin looking for unreported problems, which always exist.

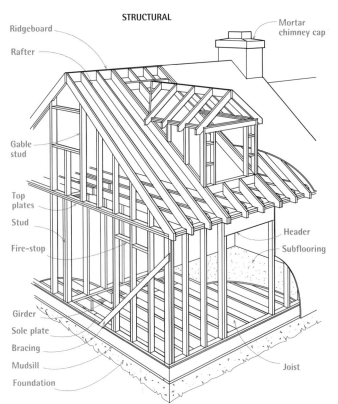

STRUCTURAL

Ridgeboard
Rafter
Mortar chimney cap
Gable stud
Top plates
Stud
Fire-stop
Header
Subflooring
Girder
Sole plate
Bracing
Mudsill
Joist
Foundation

These drawings contain most
of the building terms used in this chapter.
For additional terms, consult the index, the glossary,
and pertinent chapters.

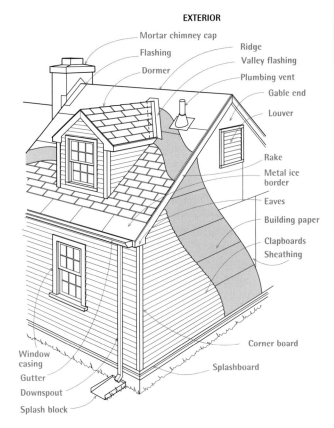

EXTERIOR

Mortar chimney cap
Flashing
Dormer
Ridge
Valley flashing
Plumbing vent
Gable end
Louver
Rake
Metal ice border
Eaves
Building paper
Clapboards
Sheathing
Corner board
Window casing
Gutter
Splashboard
Downspout
Splash block

Because of competitive bidding, buying a house can be nerve-wracking. But you can reduce some of the pressure by making your own preliminary inspection. This will enable you to delay paying for a professional inspector until you're sure it's a house you should seriously consider. You'll be able to red-flag special concerns for the inspector. And, as a bonus, after conducting your own inspection, you'll better understand the inspector's report.

When you're ready, the opinion of an accredited house inspector can be invaluable. Typically, house inspections take two to three hours and yield a detailed field report that can run 20 pages or more. A report also may recommend additional inspections, if warranted, by structural engineers, HVAC (heating, ventilation, and air-conditioning) specialists, and the like.

GETTING READY TO INSPECT

Before crawling around attics or basements, get a tetanus shot. Wear old clothes or coveralls and a respirator mask for dusty spaces. Use a cellphone camera to record what you see or want to ask about later. Other useful tools: electrical tester (p. 294), flashlight, penknife, small spirit-level to check plumb and level, and binoculars so you can scan the roof. For more extensive inspections, get a copy of *The Complete Guide to Home Inspection* (The Taunton Press, 2015).

You're less likely to be distracted if you conduct your inspection alone. You're after facts, not the opinions of an owner or real estate agent who's eager to sell. If you're working with a partner, discuss your intentions with him or her beforehand; you may want to divvy up the inspection tasks. Likewise, if your real estate agent will be present, make it clear that time is tight and you need to focus.

Begin outside, scrutinizing the house methodically from top to bottom. As you see flaws and suspect areas, record them on a sketch of the building. Then go inside and repeat the process, as suggested in this chapter. Finally, as you inspect, look for patterns in what you observe: If there's water damage at the top of an interior wall or near a window, look outside for worn roofing, missing flashing, and the like. Water will be the cause of many, if not most, problems.

Finding a Home with a Future

None of us has a crystal ball. But we can anticipate and, to some degree, plan for a number of life changes. Your home should be flexible enough to accommodate them. If, for example, you're in your 30s and plan to have kids, the house should have enough rooms—or a large enough lot for an addition. If you long to work at home, is there a garage you can convert into an office? For folks in their 50s, the future may entail kids leaving home soon, caring for an elderly parent, or planning for a more secure retirement.

Scenarios can get quite specific: If an adult child returns home, will you need to add soundproofing and a separate entrance? Would a flight of stairs be problematic for an elderly parent? Is there enough room to create an in-law unit to rent out—or for you to move into? Choose a home carefully because you may live in it a long time.

Staying put and planning for the future is a big departure from the frothy years of the housing bubble. But living within your means, conserving resources, and valuing family, friends, and neighbors can also be part of a rich life.

When evaluating a house's potential, consider all of its assets, even those that need a lot of work because they may be diamonds in the rough. Here, a dilapidated garage became an elegant guest bedroom suite.

The Roof

Because water is usually a house's main enemy, spend time examining the roof—the first line of defense against rain, snow, and ice. Few homeowners would allow a prospective buyer on a sloped roof, whether fearing roof damage or a fall. Even if you're sure-footed and could obtain permission, it's probably wiser to stay on the ground. Use your binoculars to take a closer look, unless, of course, you'll be inspecting a flat roof.

ROOF CHECKPOINTS

Sight along the ridge to see if it's straight. If the ridge sags in the middle, suspect too many layers of roofing or undersize rafters. If the roof sags between rafters, the roof sheathing may be too thin and should be replaced during the next reroofing.

Next, look for flashing at the bases of chimneys and plumbing vents. These objects can dam water and allow it to leak through the roof. If flashing is absent, rusty, or otherwise deteriorated, there's a good chance of water damage.

The valleys between roof sections should be flashed because they carry a lot of water. Thus, where roof planes converge, you should see either metal flashing down the valley (an open valley) or interwoven shingles (a closed valley).

Drip-edge is specialized flashing that should protrude from beneath the lowest courses of roofing. It allows water to drip clear of the roof. Older homes lacking drip-edges often suffer water damage because water soaks backward

Eaves Flashing

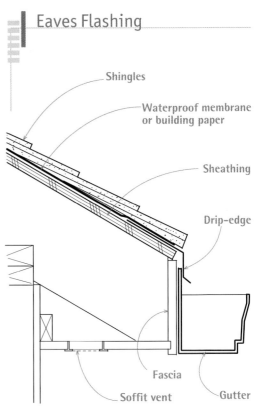

Shingles

Waterproof membrane or building paper

Sheathing

Drip-edge

Fascia

Soffit vent

Gutter

under sheathing onto the tops of walls (see "Eaves Flashing" above).

Wherever roofs adjoin walls or dormer walls, look for roof-to-wall and step flashings, as shown in "Flashing a Shed Roof That Abuts a House Wall" on p. 101. At brick chimneys, consider whether saddle and step flashings are properly counterflashed, as shown in "Chimney Flashing" on p. 100.

ASPHALT SHINGLES

If the granular surface of asphalt shingles is worn and if shingles are cupped and dog-eared, it's time for a new roof. Ditto if gutters contain a significant amount of gravel washed off the surface of the shingles.

If the roofing is lumpy and uneven, there are probably two or more layers of shingles that weren't well installed. They'll need to be stripped to the sheathing before reroofing.

Do you see odd-colored shingles? If so, they are probably patches over old leaks. Or if the roof is relatively new and shingles are worn in only one area, perhaps one bundle of shingles was defective.

A roof less than five years old with a large number of loose or missing shingles indicates that the installer's power nailer was set too deep and drove the nails through the shingles. In this case, that roofing needs to be replaced.

WOOD SHINGLES AND SHAKES

Shingle wear will almost always be greater on a south-facing roof because that side gets the most sun. If shingle ends are cupping and splitting, plan to reroof soon.

Mossy shingles or shakes are common in wet climates and on shady roof sections. Although moss-covered shingles can be relatively sound, moss retains water and will induce rot.

If the house is in a fire-risk area, insurers may refuse to give a policy on a wood-shingle or wood-shake roof. In this case, replace it with noncombustible roofing.

SLATE SHINGLES

Do not attempt to walk on slate roofs. Even when dry, they're slippery. Slate also is brittle and breaks easily.

Off-color areas may indicate replacement shingles in an area damaged by a falling tree limb. Later, when you are in the attic, check for water stains on supporting lumber.

If you see rust-colored streaks or cockeyed slates, the installer may have used nails that weren't galvanized, which by now have rusted through. Although it's possible to remove and renail slate, the job requires a costly specialist. However, if many nails have rusted through, the roof is dangerous, and the slate should be removed.

ROOF TILES

Stay off tile roofs. Even when the slope isn't steep, your weight could damage the tiles. Inspect them from an extension ladder or with binoculars.

Look for odd-colored tiles from earlier repairs. Obviously cracked or broken tiles can be replaced, but the job is costly. Under the roof, check for water stains.

Look closely at the ridge. Sagging ridge and rafters suggest too much weight on the framing. It's a big expense to remove the tiles, bolster or replace sagging rafters, replace sheathing, and then replace tiles.

METAL ROOFING

Stay off metal roofs. They are slippery, whether wet or dry.

Even though a rusty roof may not look great, it could be watertight with a lot of years left in it. Roofs with superficial rust can be sanded and repainted. Check for evidence of leaks in the attic.

Roofing panels should be nailed at the high point of metal folds. If you see many nails in the roofing channels themselves—where the water runs—the installation was inept, and you'll need a new roof.

BUILT-UP ROOFING

On older built-up roofs, there were alternating layers of heavy building paper and hot tar, covered with light-colored gravel to reflect sunlight and protect the surface from ultraviolet (UV) damage. More recently, modified bitumen (MB) has largely replaced hot tar and paper. MB roofs typically have cap membranes "torched on" (heated with a propane flame) to fuse them to fiberglass-reinforced interplies, or base coats.

Blisters in built-up roofs are usually caused by trapped water. Individual blisters can be patched with three-course patches (see chapter 5), but if blisters are widespread, it's time to reroof.

Foot traffic can abrade and puncture flat roofs. If you find no evidence of water damage below, you can spot-patch abused areas, lay

down new gravel, lock the door to the roof, and consider yourself lucky.

Most leaks occur at turn-ups, where the flat roof joins walls, parapets, and other vertical surfaces. If the turn-up surfaces are cracked, split, sagging, or unpainted, water may have gotten in and done damage. A large amount of tar at the base of walls may indicate inadequate flashing and frequent repairs.

Cracking or blistering around downspout outlets and internal drains indicates inadequate maintenance. Are there wire baskets in the openings? Are openings free of debris? If you have doubts, flush the outlets with a hose to see how well they drain.

Is the flashing around plumbing vents sound? This is not a major repair, but it can indicate general neglect.

CHIMNEYS

Although the homeowner probably won't let you on the roof, chimneys need closer inspection than binoculars allow. Even if the chimney looks good through binoculars, make the purchase agreement contingent on a professional chimney inspection. Here's what a pro will look for:

▶ **Are mortar joints solid or crumbly?** Repointing mortar isn't a big job unless it's badly eroded—in which case, the chimney may be unsound and may need to be rebuilt or removed.

▶ **Is there a sloping mortar chimney cap or crown at the top of the chimney to shed water?** A mortar cap is easy enough to repair, but a cracked or missing cap suggests a lack of general maintenance and possible water and ice damage inside the chimney. Concrete-based caps in snow country should overhang bricks by at least 1 in. (see "Overhanging Chimney Crown" on p. 246).

▶ **If there is a prefabricated metal or concrete cap elevated above the chimney top** to keep precipitation out, it, too, must be crack-free and well attached. Elevated caps may interfere with fireplace draft, so look for smoke marks above the fireplace opening.

▶ **Is chimney flashing at the roof intact?** Tired flashing can be replaced when it's time to reroof, but missing or degraded flashing may mean water damage and rotted framing in the attic.

▶ **To be safe, chimneys must have an intact liner** (usually flue tile in older houses). If the flue is only brick and mortar instead of flue tile or if there's creosote running down the outside of the chimney, the chimney is unsafe. Any

This old brick chimney has four strikes against it: crumbling mortar, failed base flashing, no flue lining, and a wind–buffeted TV antenna that stresses mortar joints.

cracks in a flue—or no flue at all—can allow superheated gases to escape and ignite the adjacent framing.

House Exterior

After inspecting the roof, examine the gutters and eaves. The overhanging eaves are actually a transition from roof to wall and are composed of several building materials.

GUTTERS AND DOWNSPOUTS

Eyeball the trim at the eaves. Is it intact or splitting? Do you see stains or discoloration? Water damage along the eaves usually is caused by clogged gutters or missing drip-edge flashing on the roof and, less often, by roof leaks.

You'll need to replace metal gutters that have rusted through or that leak widely. But this is a moderate expense, unless you replace them with copper gutters. If the deteriorated gutters are wood, they will usually be nailed directly to fascia trim or, less often, built into a cornice. By the time they fail, they may have allowed a lot of rot to occur behind them. Probe to see how much. Wood gutters are the most expensive to replace, so consider alternatives.

Stained siding behind downspouts may have been caused by clogs or gutter seams that rusted through. Or if the downspouts and gutters are new, upper downspout sections may have been mistakenly slipped over lower ones, rather than into them.

If rainwater gushes over the gutter, either the downspout screen, the downspout itself, or the ground drainpipe is clogged. Aboveground clogs are easy to fix. Belowground clogs require either reaming tools or digging, still only moderate expenses. But if the drainpipe is clogged with tree roots, you may need to replace it.

Leaking pipes and vegetation too close to the house were factors in causing this siding to rot.

SIDING

Wood siding will deteriorate if it's not well maintained, especially along south-facing walls, which get the most sun. A certain amount of weathering is normal. Cracked and worn shingles or clapboards can be replaced, but if the deterioration is widespread, it makes sense to re-side the whole wall.

If you see widespread vertical black-brown stains on siding that's otherwise in good shape, installers likely failed to use galvanized or corrosion-resistant nails. Fixing the problem may mean sinking, filling, and priming the nail heads—a tedious undertaking.

Discoloration of siding along the base of the wall could be caused by any of several factors: (1) siding that's too close to the ground, (2) nearby plants that keep the siding damp, (3) splashback from roof runoff, or (4) a badly positioned lawn sprinkler. All are easy fixes themselves, but moisture may already have caused the underlying framing to rot.

Imitation wood siding that delaminates or sprouts fungus is probably exterior hardboard that's been discontinued because of class-action lawsuits. Replace the siding.

Chronically peeling paint on the exterior walls of bathrooms and kitchens is usually caused by excessive room moisture migrating outdoors. The remedy is usually to add a vent fan.

Stucco is strong and relatively maintenance-free, but it will crack if the building shifts (see "Tattletale Cracks" at right). Loose or bulging stucco has separated from the lath behind it. This is a modest repair, unless the problem is widespread. If you see extensive patching, be suspicious. Loose, crumbling stucco is common if the base of a wall is close to the soil or in contact with it. This repair usually requires cutting back the loose stucco and adding a weep screed (see p. 185) so water can drip free.

Newer homes with rot in the walls may have incorrectly installed "synthetic stucco," or EIFS (exterior finish insulation system); this is especially a problem in the Southeast. Have an EIFS specialist certify that the house is sound; this is a headache you don't need.

Brick is strong but its joints may crack if the foundation moves, whether from settling, frost heave, or earth tremors. If a brick veneer half-wall is pulling away from an exterior wall or if a full-story brick facade is bowing outward, the metal ties holding the brick to the wall sheathing may have rusted out. Repairs will be expensive. This condition may also signal water damage and foundation problems. Eroded mortar joints can be repointed with mortar if the bricks are sound, however.

Tattletale CRACKS

Diagonal cracks running out from the upper corners of windows or doors may telegraph big trouble. Building loads often concentrate on a header—a load-bearing member over a door or window opening—and diagonal cracking may be a sign the header is not adequately supported. That is, the house's framing or foundation may be shifting. Get a structural engineer's opinion.

Stepped cracks in mortar joints above doors and windows usually are caused by rusting steel lintels. Support the bricks above the opening before replacing the lintel.

Vinyl or metal siding doesn't require much maintenance and protects the structure if it's properly installed. However, if there are gaps between sections or where siding abuts trim, suspect a sloppy installation and hidden rot.

WINDOWS AND DOORS

Is there flashing over doors and windows? If not, suspect water damage behind.

Examine windows and doors very carefully. Are frames solid? Deteriorated window sashes should be replaced, which can be expensive. Also inspect doorsills and windowsills, which are rot-prone if water collects there.

Carefully inspect doors for fit and function. Look for signs of warping, sagging, or separation between rails and stiles. Examine the jambs of exterior doors for damage from abuse, hardware changes, or even forced entry.

In areas with heavy rainfall and expansive clay soils, it's best to connect downspouts to solid drainpipes that carry roof runoff away from the foundation. Flexible plastic drain extensions that run along the surface are also effective, though unsightly.

▌▌▌▌

Doors and windows badly out of square suggest a house that has shifted and may still be shifting. This can result from poor drainage and an inadequately sized foundation.

Swollen or rotted basement windows will need to be replaced with durable all-vinyl units. But first you'll need to attend to drainage problems that have allowed water to collect.

AROUND THE HOUSE

Walk around the house. Although you will be able to see more of the foundation inside, damp basements and cracked foundations are often caused by faulty drainage outside.

Does the ground slope away from the base of the house? Or would runoff from the roof collect next to the building? Is the soil damp or compacted next to the house? Although drainage may seem a minor factor, faulty drainage can cause wet basements and even foundation failure.

Where do downspouts empty? Is runoff carried away from the house by drainpipes or, at the very least, are there splash blocks beneath downspouts to direct water away from the foundation?

Note the positions of foundation cracks, especially cracks greater than ¼ in. wide. Also look for signs of foundation settling or leaning around downspouts, water sources, and areas on the uphill side of the house.

Inspect chimney bases closely, both where additions join the main house and where loads concentrate on foundation bearing points. If there's cracking where the chimney base joins the main house foundation or if the chimney base is tilting, it may be undersize and need replacing.

Do bushes or dirt touch the siding? If so, prod the siding and splashboard (also called a water table), using the awl on your pocketknife. If the wood is soft, it may be retaining a lot of moisture. Dirt can also be an avenue for termites, so look for the telltale dirt tubes that termites construct, further discussed on p. 215.

If you see tar or roof cement slathered around chimney bases, plumbing stacks, roof valleys, and other roof joints, assume leaks may have occurred there, either because flashing failed or was never installed. Tar can be a functional, though ugly, short-term fix, but be sure to replace old flashing when installing a new roof.

▌▌▌▌

Bricks AND EARTHQUAKES

In earthquake country, brick chimneys may break off at the roofline, fall through the roof, and harm inhabitants. A structural engineer can analyze the chimney and suggest remedies, such as seismic steel bracing of the chimney to the roof or removing the chimney above the roof and replacing it with metal flue pipe.

Interiors

Armed with your outdoor observations, go inside. Start with attics and basements, which are prime places to look for signs of water damage, especially if they're unfinished.

IN THE ATTIC

Review the notations you made about the roof, chimney, and eaves. Now look for outdoor–indoor relationships such as missing flashing and stains on the underside of roof sheathing.

Water damage is often visible in the attics of older homes.

Water stains around plumbing vents, dormers, and chimneys are more likely caused by failed or missing flashing. If the wood is damp after a rain, the leaks are active.

Dark brown stains around the chimney that smell of creosote are probably caused by cracked flue tile, which allowed caustic creosote compounds to work their way through mortar joints. Such a chimney is unsafe to use and must be either relined or replaced.

Cracks in the chimney's mortar joints may be caused by an undersize or shifting chimney footing—another major cause of flue failure.

Ventilation and insulation are vital to a healthy house but are often misunderstood. Attics that lack adequate ventilation will be excessively hot in the summer. In winter in cold climates, inadequately ventilated attics allow rising water vapor from the living area to collect as frost on the underside of roof sheathing. The frost eventually melts, soaks the sheathing, drips onto the attic floor, and perhaps soaks through top-floor ceilings.

Also in winter, inadequately ventilated attics can trap warm air from below, which warms the roof, causing snow to melt and run into unwarmed overhangs where it refreezes, resulting in ice dams that can damage roofing and leak behind the siding. Adding soffit, gable, and ridge vents sometimes alleviates these problems.

Discolored rafters along the roof–wall joint and delaminated roof sheathing, coupled with stains at the top of interior walls below, are usually caused by warm, moist air migrating upward from living spaces. Often, that air escapes via holes in the ceiling—whether recessed light fixtures, plumbing and wiring chases, or gaps around the chimney. Once you've air-sealed these holes, then add insulation to the attic floor, upgrade bath and kitchen vent fans, and otherwise reduce excess moisture in living spaces.

Here, rafter rot, mildewed ceilings, and delaminating roof sheathing were largely caused by inadequate ventilation.

Air-sealing, types of attic insulation, and methods of installing insulation are addressed in chapter 14.

Water-stained attic floors and discolored attic insulation often show roof leaks clearly. If moisture is making its way up from living spaces, the insulation's underside may be moldy. Pull up affected sections and see if water has collected there and caused damage.

Structural condition may be an issue if a roof was overloaded.

If the roof sheathing is bellying (sagging) between the rafters, it's probably too thin and should be replaced with thicker plywood when you replace the roofing.

Are the rafters and ridge sagging? If so, you should hire a seasoned contractor or a structural engineer to see what the remedy might be. If this condition resulted from too many layers of roofing, stripping what's on the roof may be all that's needed. However, deformed framing may need replacing or additional support, as shown in "Reinforcing a Roof" at right.

Roof trusses must not be cut or modified or their structural integrity will be compromised. If you note cuts made to accommodate air ducts and the like, the truss system should be inspected by an engineer.

WALLS

Most homebuyers repaint or repaper walls to suit their tastes. Instead of concerning yourself with paint colors or wallpaper patterns, focus on surfaces that suggest underlying problems that may require remedies.

After consulting the notes you recorded outside, study wall–ceiling joints, which also can tell

stories. Water stains on interior outside walls, especially above windows, may have multiple causes: missing flashing, gaps between siding and exterior trim, and leaks in the gutters or roof.

Crumbling drywall or plaster and extensive mold at the top of walls may be caused by exterior leaks or, just as likely, by excessive moisture in the living areas. If the problems are severe, mold may be growing on framing inside the walls. In extreme cases, after correcting the sources of moisture, you may need to tear out drywall or plaster and replace studs and plates.

Large diagonal cracks in drywall or plaster at the corners of doors and windows may correspond with cracks on the house exterior. Such cracks suggest structural shifting and foundation distress.

Door and window trim that tilts toward a common low point suggests failure in the substructure (girder, post, or pad) or in the foundation itself.

FLOORS AND STAIRS

Squeaky floors may take two minutes to repair or, if the cause is elusive, two days. Isolated floor squeaks are annoying but rarely a sign of anything serious.

Excessively springy floors suggest subflooring or underlayment that's not strong enough to span the distance between joists. Additional layers of subflooring or flooring should firm things up.

PRO TIP

In the attic, walk on the ceiling joists or use plank walkways across them. It's unsafe to step anywhere else. Wear a hard hat so roofing nails above don't stab your head—another good reason for that tetanus shot. You'll likely get dirty crawling around, so dress down.

PRO TIP

People selling a house often spiff it up with a new coat of paint—except inside the closets. So it's smart to inspect clothes closets if they're not too clogged with clothes. That's also an area you'll often find exposed lead paint in older homes. If you test for lead paint, be sure that's one area you investigate.

Reinforcing a Roof

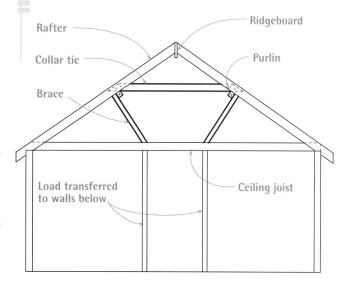

Tired roof framing sometimes needs additional support to keep rafters from sagging or spreading.

Deteriorating flooring near an exterior door-sill suggests that rain or snow has soaked the sill and subflooring. You'll need to replace damaged materials and install an all-weather doorsill.

Widespread cracking in tile floors may result from insufficient adhesive between tile and sub-flooring or an underlayment that's too thin and thus not rigid enough. Repairing it may mean tearing out the tile and perhaps the underlayment before replacing both.

Cracked stair treads may be worn, undersize, or inadequately supported by the carriages underneath. If the problem is widespread, remove the treads and examine the substructure.

Stairs sloping badly to one side with cracked walls along the stair suggest a stair carriage that is pulling loose from wall mountings or other framing members. If the underside of the stairs is not accessible, repairing the problem can be complex and costly.

Handrails and newel posts that wobble excessively should be resecured. If balusters are missing, it can be costly to have replacements hand-turned to match. As a cost-saving alternative, you might be able to find replacements at a salvage yard.

KITCHENS AND BATHROOMS

If there are stained, springy floors around the base of a toilet, the subfloor and the joists below may be water damaged. The cause of the leak may be simple—just a worn-out wax ring gasket under the toilet. But if damage is significant, you may need to pull up flooring and replace it.

Old resilient flooring that's worn around the edges is common and needs to be replaced. Damaged linoleum or vinyl around the base of cabinets, shower stalls, tubs, and other fixtures often foretells water damage below. If there's an unfinished basement beneath those fixtures, look there for damage. Otherwise, look for water damage in finished ceilings below.

If tiled tub enclosures are in poor repair, test the firmness of the substrate behind by pushing with the heel of your hand. If the walls flex, the tile may be installed over ordinary drywall, which deteriorates if it absorbs water. One remedy is tearing out the tile and drywall and installing a cement-based backer board before retiling. Check tub–wall joints closely. They must be well caulked to forestall leaks.

Use your pocketknife awl to prod gently for damage under lavatory and kitchen sink cabinets. Rusted-out drainpipes or leaking supply-pipe connections are easily replaced, but extensive water damage to the subfloor or floor joists can be a major repair.

The water damage and rot beneath this toilet could have been avoided by replacing a $2 wax gasket.

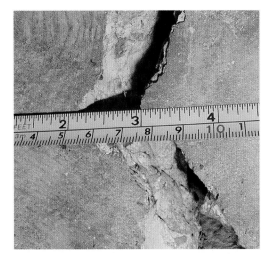

Any crack that runs through a foundation is serious.

The base of this wall was kicked out by a rotating foundation.

If bathrooms or kitchens smell musty or are mildewed, especially at the top of walls, there's inadequate ventilation. Scrubbing walls and adding vent fans will probably cure the problem, unless the drywall is crumbling as well (see chapter 14 for repair information).

IN THE BASEMENT

Safety note: Stay out of basements or crawlspaces if there's standing water, wet soil, or substandard electrical wiring! Metal pipes or ductwork could become energized by a short circuit.

Dampness can be mitigated by regularly cleaning gutters and downspouts, as well as grading the soil around the foundation so it slopes away from the house. Musty smells and occasional condensation may be reduced by improving ventilation. Beyond that, fixes get more complex and more expensive.

Wetness is often caused by surface water and may respond to the previous suggestions. Stronger remedies include sump pumps, perimeter drains, and engineered solutions. For example, water seeping through an uphill foundation wall may need to be intercepted and rerouted by drain-pipes or swale drains farther uphill. Such solutions can be expensive.

Cracks range from cosmetic surface lines that you can ignore to larger, deeper fissures caused by water pressure, soil movement, foundation failure, or a combination of those causes.

In general, a serious crack is any gap that runs through the foundation or is at least ¼ in. wide, combined with foundation rotation. (In brief, rotation is caused by horizontal soil pressures that push a foundation wall out of plumb.) Have a structural engineer assess the cause and recommend a solution.

Vertical cracks through a foundation that are wider at the top may be caused by differential settlement. For example, a corner of the house may be sinking because of drainage problems or a second-story addition that's too heavy for the original foundation.

Horizontal cracks through the foundation wall, just below ground level, may be caused by *adfreezing*, in which damp soil freezes to the top of a foundation and lifts it. This condition most often occurs in unheated buildings.

The foundation's bowing-in along horizontal cracks is extremely serious; it's caused by soil movement and strong hydrostatic pressure.

Given the magnitude of the problem, the engineered solution may be very expensive.

Older foundations of unreinforced concrete or brick may be adequate beneath single-story houses on flat lots, but long term, you should plan to replace them. Unreinforced foundations are often poor quality (crumbling) and may have cracks that go all the way through the concrete.

Wood structures are most often damaged by sustained moisture below grade, insects, settling of the foundation, or unwise sawcuts into supporting members during earlier remodels.

If you see signs of water damage or rotted siding, use your pocketknife awl to probe the perimeter of the mudsill—and the studs atop it—for rot or insect damage. To prevent recurrence of rot, you'll need to replace damaged sections with

Knob-and-Tube WIRING

If you see individual wiring secured to ceramic insulators in the basement or attic, that's knob-and-tube wiring. Although outdated, it's generally acceptable if it's in good condition and used only for lighting. To be safe, check with your local code officials or a licensed electrician—and then with your prospective insurance company.

Knob-and-tube wiring is outdated but often serviceable.

ELECTRICAL Safety NEAR WATER

Kitchen receptacles within 4 ft. of a sink and all bathroom receptacles must be protected by GFCIs. They're essential protection against electrical shock. Note, however, that local codes have the final say on what's acceptable in your community.

Use a voltage tester to check receptacles. The inductance tester shown often can detect electrical current nearby—such as through a plastic cover or a lamp socket—without actually touching a conductor.

treated lumber, improve drainage, slope soil away from the house, and so on.

If you encounter sagging girders or joists, the posts and pads supporting them may have failed. In this case, upsize the concrete pads beneath the posts and/or replace the posts. In some cases, you can shorten joist spans by adding girders and posts beneath.

Wooden posts rotting at the bottoms suggest that moisture is wicking up from the ground through the concrete pad—if there is one. Replace the posts, putting a metal or plastic moisture barrier between the bottoms of posts and the concrete pads supporting them.

Joists and girders may have been seriously weakened when they were cut improperly to accommodate ducts or pipes. See "Maximum Sizes for Holes and Notches" on p. 373 for tips on how much and where you can safely cut and drill structural members.

Mechanical Systems

Mechanical systems include electrical, plumbing, and HVAC. Your comfort and safety depend on up-to-date and adequately sized *mechanicals*, as they are sometimes called.

ELECTRICAL SYSTEM

Only a licensed electrician should assess the capacity and condition of your electrical service. In particular, do not remove the covers of service panels. Before examining receptacles, switches, and other devices, always turn off the electricity and make sure it's off by using a voltage tester before handling any electrical device.

If you see scorch marks, rust stains, or condensation on the service panel or damp conditions around it, that service is unsafe. Dampness is particularly unsafe, and many electricians will refuse to work on a panel until surrounding dampness is remedied.

In older homes, electrical service is often undersize. If your house has only two cables running from the utility pole, it has only 120-volt service. After purchasing the house, have your power company upgrade to three-wire, 240-volt service.

A 100-amp, circuit-breaker service panel is considered minimal today. If an older home has a fuse panel, it will typically have a capacity of only 60 amps. It should be upgraded.

Knob-and-tube wiring does not necessarily have to be replaced, but it does not include a ground wire and it has its limitations.

Any electrical cable with cracked or frayed sheathing should be replaced. Deteriorated cable is usually visible as it approaches the service panel and as it runs along joists in attics and unfinished basements. Visible wire splices or cable that sags from joists is unsafe and substandard. Don't handle such wiring. Just note its condition.

Using a voltage tester, you can safely check whether receptacles are operable. If the cover plates of any receptacles are warm, if tester lights flicker, or if there's an odd smell, there may be aluminum circuit wiring in the walls, which tends to overheat and cause fires if incorrectly installed. Have your electrician check for this, too.

To prevent electrical shocks in high-moisture areas, all bathroom receptacles, kitchen receptacles within 4 ft. of a sink, outdoor outlets, and some garage outlets must be protected by ground-fault circuit interrupters (GFCIs). All new 15-amp and 20-amp circuits in bedrooms must have arc-fault circuit interrupter (AFCI) protection.

PLUMBING

Questions suggested under "Kitchen and Bathrooms" earlier in this chapter should be addressed here, too—particularly if you noted water damage around tubs or toilets. By the way, if the house has only a crawlspace, replacing the plumbing will take longer and be more costly than if the house has a full basement.

Drainage, waste, and vent (DWV) pipes should be replaced if they're rusted, corroded, or leaking. Waste pipes past their prime often show powdery green or white deposits along their horizontal runs, where wastes accumulate. Also, if joists around a closet bend (see p. 368) are discolored, probe for rot. If rotted, they'll need to be replaced.

Supply pipes. If water pressure is poor and plumbing is old, it's likely that the pipes are galvanized iron. With a typical life span of about 25 years, the fittings rust out first.

Copper pipe will last indefinitely unless the water is acidic, in which case you'll see blue-green deposits on fixtures and pinhole leaks in the pipe. But if copper pipes aren't too far gone, an acid-neutralizing filter on supply lines may cure the problem.

Copper and galvanized pipe joined together will corrode because of a process called galvanic action. To join these metals, a dielectric union should be installed between them.

A water test by the health department should be part of the purchase agreement; this is especially important if the house has its own well.

Water heaters. Water heaters more than 12 years old probably should be replaced. A manufacturer's plate on the heater will tell its age and capacity. As a rule of thumb, a 40-gal. gas-fired water heater is about right for a family of four. Electric water heaters should have a capacity of 50 gal. because they take longer to recover.

Septic TIPS

If the house has a septic tank, ask when it was last emptied. Most tanks are sized according to the number of users and should normally be emptied every few years. Also inquire how the owner determines the exact location of the tank's clean-out lid, which will usually be buried under more than 1 ft. of soil.

Then walk the area around the tank. If the ground is damp and smelly, the most recent servicing wasn't soon enough. Besides tardy servicing, this could indicate that the tank or the drainage field may be undersize, clogged, or incorrectly installed. A new septic system is a significant expense.

If a cast-iron waste pipe is this rusty, it needs replacing.

PRO TIP

If cleanout traps show fresh wrench marks, suspect recent clogging; if traps are badly scarred, they have been opened many times. This may mean nothing more than children dropping things down the sink or it may indicate an inadequately sized pipe that needs replacing (see "Minimum Drain, Trap, and Vent Sizes" on p. 367).

Corroded galvanized-steel pipe atop a water heater tells you it's time to replace both the heater and the supply pipes.

Tankless water heaters create hot water on demand, so they lack a tank; p. 385 has tips on selecting them. Any gas-fueled water heater—whether tank or tankless—must have an approved vent to exhaust combustion gases. Electric water heaters do not require a vent.

There should be a temperature- and pressure-relief (TPR) valve on or near the top of the water heater. Without TPR protection, a water heater can explode and level the house. If the TPR valve drips, replace it. If you own the home, make sure you have a TPR valve.

Gas lines that smell and corroded gas pipes are unsafe. If you spot either of these problems, call the gas utility immediately (most provide a free inspection). Gas lines are typically black iron pipe with threaded fittings or copper joined by flared fittings. Gas lines should never include PVC plastic pipe, sweated (soldered) copper joints, or compression fittings such as those used for water supply.

HEATING, VENTILATION, AND AIR-CONDITIONING

Heating and cooling systems are varied and complex, so make your house purchase contingent on a professional inspection by an HVAC contractor. In your walk-through, look for the following:

A TPR valve can prevent water-heater explosions caused by excess pressure. The discharge pipe running from the valve should end 4 in. above the floor.

▶ If the bottom of the heating unit is rusted out or if it's 15 to 20 years old, it probably should be replaced. It's certainly not efficient and probably not safe.

▶ Soot around heat registers or exhaust smells in living areas means that the furnace is dirty and poorly maintained or that the furnace heat exchanger is cracked, allowing exhaust gases to escape. If an HVAC specialist can see flame through the heat exchanger, it's definitely time to replace the unit. It may be a fire safety and health hazard.

▶ If your house has forced hot-air heat, your family could develop respiratory problems if the furnace has one of the older, reservoir-type humidifiers, which are notorious for breeding harmful organisms in the always-wet drum. An HVAC specialist can suggest alternatives.

▶ If certain rooms are always cold, an HVAC specialist may be able to balance heat distribution or add registers. That failing, you may need to upsize the furnace or boiler.

▶ If ducts, pipes, or the central heating unit are wrapped with white or gray paper tape, your older heating system may be insulated with asbestos. Do not disturb it—an HVAC specialist can assess its condition and recommend an asbestos-abatement expert.

▶ Air-conditioning (AC) systems that run constantly but don't keep the house cool may need the coolant to be recharged. An AC system that cycles too rapidly and makes the house too cold may simply be too large for the house. Both problems just need adjustments by an HVAC contractor.

FIREPLACE OR WOODSTOVE

Loose bricks in the fireplace firebox or smoke stains between the wall and the fireplace mantel (or surround) could allow flames or superheated gases to ignite wood framing around the fireplace. A mason or chimney specialist can usually make necessary repairs.

If there are gaps between the hearth (firebox floor) and the hearth extension, stray coals could fall into the gaps and start a fire, so repoint gaps with mortar (if there's brick) or grout (tile floors).

Local building codes usually specify minimum distances woodstoves and stovepipes must be from flammable surfaces.

Environmental Hazards

In recent years, building scientists have increased our understanding of how hazardous materials, both naturally occurring and manmade, can impact our health and that of our environment. The hazardous materials that affect our homes include asbestos, lead paint, mold, radon, and a host of volatile organic compounds (VOCs) such as formaldehyde. Many of these materials are no longer produced, but their side effects linger on, hidden in walls, buried in backyards (leaking oil tanks), or invisibly present in the air we breathe.

ASBESTOS

Asbestos is a family of six long-fibered, naturally occurring silicate minerals. From the 19th century on, it was widely added to building materials because of its ability to retard fire, absorb sound, and resist electrical and chemical damage. Its fibers contained a deadly side effect, however: Inhaling them over a prolonged period can cause serious illnesses such as mesothelioma, lung cancer, and asbestosis (a restrictive lung disease similar to miner's lung).

If the asbestos-containing material is intact, generally it isn't a health hazard. However, when it is damaged, disturbed, or begins to deteriorate, its fibers can become friable (easily crumbled) and airborne, which is a hazard.

Heating system components. Because of its fire-retarding and insulating qualities, asbestos can be found on most fuel-oil or gas-fired heating systems installed before 1979. It takes on many forms.

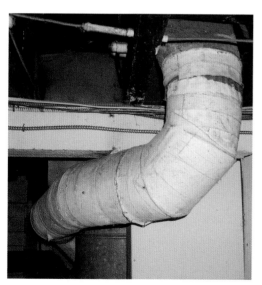

Asbestos is a common wrap on older heating system ducts.

Boilers and hot-water piping in older homes were frequently covered with white-gray asbestos that was applied as fabric strips and/or parged (troweled on) like plaster. Straight sections of hot-water piping were often wrapped with thin corrugated sheets of asbestos, visible when viewed from a pipe end, with asbestos plaster applied to seal joints, elbows, and transitional pieces.

Asbestos *inside* a heating system is a definite health hazard. Damaged asbestos on the outside of ducts may be drawn into the air-supply system, but asbestos inside ducts or the fan/blower housing is a potential hazard even if it is intact. Moving air will, in time, lift and detach asbestos fibers. If you see a flexible white-gray fabric transition between a blower fan box and the main warm-air duct, it may be asbestos, put there primarily to isolate fan noise. Asbestos was also used inside heat supply ducts at registers. When it's time to upgrade an old heating system, a qualified asbestos-abatement specialist should remove the asbestos casing before replacing the boiler or furnace.

Electrical panel enclosures. Old electrical panel boxes (typically, with 60-amp main fuses) often contain asbestos, installed because it is fire resistant and impervious to electrical activity. The asbestos is usually in sheet form, so its edges become more worn and friable each time the panel box door is opened and closed. Such boxes should be removed.

Ceilings. Textured acoustic ceilings ("popcorn ceilings") prompted the first ban of asbestos products in the United States. To reduce shrinking and cracking, asbestos fibers were mixed into a plaster-like slurry that was sprayed onto ceilings. Trouble was, the mixture didn't always adhere, either because of inadequate prep work, an inexact blend, or springy ceiling joists overhead. Before long, those asbestos-impregnated ceilings started to crack, crumble, and create respiratory problems for the people below.

Floor tiles. Vinyl asbestos flooring was very popular in the 1950s, both as resilient sheet flooring and as tiles, typically 9 in. sq. The asbestos is embedded in the vinyl, so it is not generally considered to be hazardous unless the flooring is sanded, drilled, or sawn. Asbestos was also an additive in resilient flooring adhesive, which, 70 years later, is a gray-white substance with a texture similar to dried taffy. If you intend to remove the flooring or the adhesive beneath, contact an asbestos-abatement contractor.

Vinyl asbestos (VA) tile flooring was very popular in the 1950s. The asbestos is encapsulated in the vinyl and is not generally considered to be hazardous unless it is sanded, drilled, or cut.

LEAD

Lead is a neurological toxin, particularly damaging to children six and younger, who seem drawn to lead because it's slightly sweet. Breathing it, eating it, or drinking water with a high level of lead can cause mental retardation in children. In people of all ages, lead can cause headaches, anemia, lethargy, kidney damage, high blood pressure, and other ailments.

Lead paint. In residences, old paint is the principal source of airborne and ingested lead contamination. Lead-based paint adheres to almost any surface and weathers well, so it's not surprising that it can be found in 90% of houses built before 1940. However, as lead's health hazards became known in the 1950s, paint manufacturers began to phase it out, and lead was banned altogether in 1978 by the U.S. government. Lead-paint safety, testing, and abatement is discussed at length on pp. 548–551.

Lead pipes. The deleterious effects of lead have been known for a long time, so finding lead water-supply pipes during an inspection is increasingly uncommon. When lead supply pipes are found at all, they are usually as short lengths near the water main's entry into a house or beneath an old, little-used utility tub in the basement. Lead pipe is typically a dull gray, which when scraped with a penknife or a screwdriver blade gleams silvery.

Homeowners concerned about the presence of lead can draw samples and have them analyzed by a laboratory. EPA lead limits in drinking water are 15 micrograms/deciliter. For more, visit the EPA site at water.epa.gov/ground-water-and-drinking-water.

MOLD

Mold is a diverse family of fungi that causes natural materials to degrade. Mold spores are microscopic and found pretty much everywhere.

Mold needs three things to grow: water, a temperature range between 40°F and 100°F, and organic matter such as lumber or paper. Mold particularly thrives on the paper and adhesive in drywall, and because drywall is so widely used in houses, there are a lot of places for mold to grow when conditions are right.

Old wood frequently contain lead paint, which can become an airborne hazard if the paint is sanded or becomes loose.

Mold often occurs below kitchen and bathroom sinks. If these areas are covered with wood or plaster, they can probably be washed and repainted with a mold-inhibiting paint. But if there is moldy drywall, it should be removed, the area behind washed and dried, and new mold-resistant gypsum board installed.

The easiest way to thwart the growth of mold is to exhaust excess moisture, sending it outdoors. Signs of excessive moisture include condensation on windows, moldy bathrooms or closets, soggy attic insulation, and exterior paint peeling off in large patches. If you see these signs and your nose gets stuffy, your eyes itch or water, or you start coughing when you enter a room, suspect mold. For more, see "Controlling Moisture and Mold" on pp. 435–437.

Vents of this type usually indicate the presence of a buried fuel–oil tank. The tank should be tested for leaks and the ground around it for contamination.

Planning Your Renovation

Planning your renovation can be a lot of fun, especially if you have a creative side. On paper, you can live imperially—poplars lining the drive, marble tile in the bath, teak cladding on the deck. When you tire of that, use an eraser to replace the tiles and replant the trees. If your earthly paradise is of a greener sort, summon up passive solar, thermal mass, and net-zero energy use. Let your right brain romp. When it's time to price out everything, you'll return to earth. So during these early stages, dream big.

A Home for the Long Run

Your house should fit you. As one owner-builder put it, "For me, building is about expressing who you are. A place full of friends and family but also rest and reflection. A place where you can be comfortable with yourself." So it's helpful to begin planning by getting in touch with who you are, which isn't always easy. Shortly, we'll offer some techniques for doing that.

Equally important is creating a home that can accommodate life's changes. In the last decade, there has been a profound (and welcome) shift from how we view our homes—seeing them less as appreciating assets and more as shelters that sustain us. So people are staying put, moving less, and choosing improvements that will make their homes more comfortable, less expensive to maintain and operate, and more flexible now and in the years to come. Instead of grandiose remodels, today's homeowners are planning carefully, spending wisely, and, in many cases, embracing more modest projects that will add lasting value.

To illustrate this return to careful planning and enduring values, we'll conclude with five renovations, all of them modest, including a play-

If you first create a scale drawing, you can use tracing-paper overlays to make quick but accurate sketches.

Creating Lasting Value in Your Renovation

When you plan your renovation, think long term. Consider changes that will make your home safe, comfortable, healthful, functional, cost-effective to operate and maintain, and pleasurable to live in year after year. With this long view, you can add improvements as time and money are available. Moreover, living in a house for a while is the best way to learn its strengths and shortcomings.

It takes time to realize the value of whatever you put into a house. If you live in a house for less than, say, 10 years, you're probably not likely to see much return on the money you spend on major renovations. And keep in mind that each time you sell a house, you'll pay realtors' commissions and transfer taxes, which will also cut into any increase in value you may see.

During the boom years, one could make money by doing little more than refinishing floors and applying a fresh coat of paint. But timing the market is everything—and impossible to predict. Houses also can be illiquid (hard to sell) when interest rates rise and the economy slows. So the best advice we can give is to choose a house that will make a comfortable home and to see renovation projects as investments in the quality of your life.

Home is a place to feed body and soul.

ful kitchen and bath remodel in a Craftsman bungalow; a striking, modern in-law unit created under a hillside garage; and a second-story bump-up that preserved the scale and charm of the original home.

Clarifying Your Goals

Few homeowners are good at drawing or thinking spatially. To get the ideas flowing, one architect asks clients to write up a scenario for a happy day in a perfect house. He says it's surprising how quickly the writing helps people move beyond physical trappings to describing the experiences that make them happy at home. For some, it's waking up slowly while reading in bed or having breakfast on the patio, whereas others tell of puttering in the garden or hosting candlelight dinners for friends.

KEEP A RENOVATION NOTEBOOK

Much as you'd create a shopping list, jot down house-related thoughts as they occur to you and file them in a renovation notebook. A notebook is also a convenient place to stash ideas clipped from magazines and newspapers, along with photos you may have shot. If you have kids, encourage their contributions, too. At some point, consolidate the notebook ideas and begin creating a wish list of the features you'd like in your renovated home. This list will come in handy when you begin weighing design options.

Architects call items on the wish list *program requirements* and consider them an essential first step for planning because they establish written criteria for comparing proposed improvements. The list should contain both objective, tangible requirements (such as the number of bedrooms and baths) and subjective, intangible requirements (such as how the house should eventually feel). If you're now living in the house you'll renovate, you'll undoubtedly have strong opinions about what works and what doesn't, and hence which inconveniences you're willing to tolerate and what you're not. These questions will help you get started.

Comfort. Is the house welcoming? Are there enough bedrooms? Storage? In addition to bedrooms and public rooms, do you have a room of your own? How's the traffic flow? Must you walk through any bedroom to get to another? Is the house warm enough or are some rooms cold and drafty? Does each room get natural light? Are windows placed to take advantage of prevailing winds? Can you shut out street noise? Do you feel safe? Can you see who's on the porch without opening the door? Is the house easy to keep clean?

Rating YOUR ROOMS

As you work up your wish list (program requirements), rate how well each room works. Are bedrooms away, or at least screened, from a noisy street? Is the nursery or a small child's bedroom near a bathroom? Is the home office inside of or detached from the main house? Does each room receive sunlight at optimal times? Are the rooms big enough for your furniture? Also note conflicts within or between rooms because they often generate useful design changes. Your room rating sheet might look something like the one shown here.

Room	Jake's bedroom (he's 15)
Where	Second floor, southeast corner
Size	9 ft. 2 in. x 10 ft.
Also used for	Homework, phone booth
Sunlight	Most of day, but he keeps the shades down (for computer)
Privacy	Door shut most of the time
Noise level	Headphones help, but he's a night owl; can hear him through wall
Nearby rooms	Master bedroom, full bath, MJ's bedroom
Storage/closets	Okay, but clothes on floor, mostly
Traffic issues	Bathroom jam in a.m., last one gets cold shower
Gut reaction	Growing pains, needs his space
Possible solutions	Move Jake to the room off the kitchen? (Northeast corner)
Pluses	Next to kitchen, near laundry; linoleum floor indestructible; more privacy for him, more sleep for us; lots of shelves
Minuses	Not much sun, but he won't care; only half-bath near kitchen now. Bust through pantry wall to add shower?
Definites	Need bigger water heater

Cooking and dining. Does cooking help you unwind? Do you entertain often? Is there enough counter space? Are the sink and appliances close enough to prep areas? Are counters the right height? Can you reach all the shelves without straining? Is there enough storage space? Can you easily transport food to and from dining areas? Can people hang out while you cook? While cooking, do you like to talk on the phone or watch TV? If you recycle cans and bottles, do you have a place to put them? (Pages 390 and 391 have recommended minimum cabinet and counter dimensions and common kitchen configurations.)

Bathrooms. Are there traffic jams outside the bathroom(s) during rush hours? Are you a tubber or a shower person? When everyone showers in the morning, does the last person have enough hot water? Is there a convenient place to store towels and sundries in or near each bathroom? In the tub, can you relax and soak in peace? Is the tub big enough for two? When guests come to stay, do you apologize for the bathroom they use? Is there a place to wash the dog?

Being social. When you entertain, is it formal or informal? Small parties with friends or 30-chair club meetings? Is there room to accommodate those activities? Is there a place to put guests that won't interrupt your routine? Can you get away from people when you need to? That is, when the kids have friends over, do they drive you crazy? (This may have nothing to do with the house.)

Family business. If you have small children, are surfaces easy to clean? Can you quickly stash toys? Are some cabinets childproof? Is there an enclosed, outdoor, safe play area? Are there nearby nooks where children can read or do homework while you're cooking? Do your kids have enough privacy? Will the rooms meet their needs in five or 10 years? When the kids move out, will your empty nest be too big?

Working at home. If you bring work home or simply work at home, do you have a dedicated space for it? Are there enough electrical outlets? Adequate lighting? Can you shut a door, making your workspace safe from pets and toddlers? Are

Creating Floor Plans

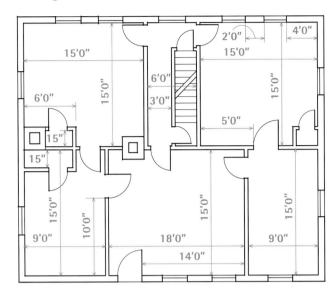

Using a 25-ft. measuring tape, record the dimensions of each room. Graph paper makes the task easier.

layouts or soundproofing such that you can work late without disturbing others?

Outer spaces. Is there a deck or patio for entertaining? A place to cook outdoors that's not too distant from the kitchen or dining area? Is there an outbuilding for lawn equipment and tools? Are you sheltered from the weather while searching for house keys? Is the yard large or sunny enough for a garden? Could the garage accommodate a shop or an in-law unit someday? Need a fence for privacy from the neighbors?

Green dreams. If energy conservation and a healthy environment are priorities for you, include related articles in your scrapbook. There's a plethora of energy-saving products to incorporate into your renovation, as well as less toxic construction materials such as low-volatile organic compound (VOC) adhesives (see p. 87) and formaldehyde-free plywood (see p. 76). If you're serious about researching this topic, subscribe to www.greenbuildingadvisor.com and get a copy of *Green from the Ground Up* (The Taunton Press, 2008), a great guide to environmentally sensitive homebuilding.

Documenting What's There

As you're gathering information about what elements to include in a renovation, take a few hours to draw what's there now. Specifically, create simple but accurate floor plans, noting the location of major appliances, kitchen and bath fixtures, and house systems such as heating and plumbing. Create a simple site map, too. Even if

Bearing and Nonbearing Walls

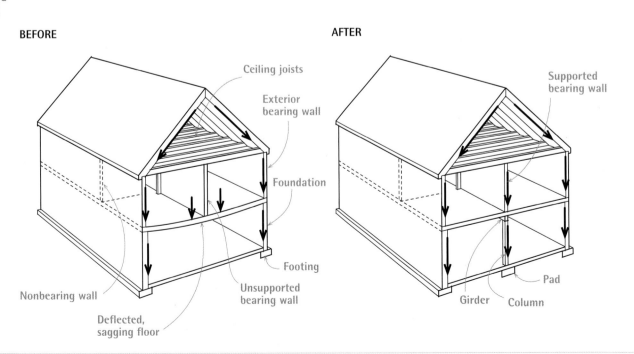

BEFORE

AFTER

Ceiling joists

Exterior bearing wall

Foundation

Footing

Nonbearing wall

Unsupported bearing wall

Deflected, sagging floor

Supported bearing wall

Pad

Girder

Column

you plan to hire building professionals to do everything, it's important to be engaged. Sketching what's there helps you demystify the process, learn the jargon, inform your opinions, and make better decisions.

DRAWING ROOMS

Start by drawing a basic plan of each floor. Using a 25-ft. retractable tape measure, record the overall dimensions of each room, noting the position of doors, windows, closets, fireplaces—anything that affects space. To be most accurate, be consistent in how you measure. Always measure, say, from the insides of window and door jambs to faithfully record the widths of wall openings.

Record these measurements on graph paper. Graph paper is handy because it helps you draw square corners and maintain scale without needing fancy drafting equipment. As to scale, most people find that ¼ in. = 1 ft. is large enough to be detailed yet compact enough to fit on a standard 8½-in. by 11-in. sheet. The other nice thing about having accurate floor plans is that you can use them as templates for quick, accurate sketches. Just place a piece of tracing paper over the floor plan and off you go. Explore as many alternative layouts as you like. Or use a separate tracing tissue for each house system, as described below.

MAPPING STRUCTURE AND SYSTEMS

There are many reasons to map structural elements and mechanical and plumbing systems:

▶ To avoid weakening the structure or disturbing large assemblies that would be expensive, disruptive, or unnecessary to change

▶ To learn where it would be easiest to tie into plumbing pipes when adding, say, a bathroom

▶ To understand which parts of the structure will be easiest to modify and which should be left alone and—once you start exploring designs—which solutions will be the most practical and cost-effective

Using the floor plans you drew earlier as templates, use tracing paper to create a map for, say, major structural members such as girders and bearing walls, another for HVAC system elements, as well as a plumbing map showing fixtures and, at least, drainage-waste-vent (DWV) pipes. While documenting each room, also note water stains, tired windows, outdated fixtures, ungrounded outlets, sagging floors, and other things that need fixing. (Chapter 1 can help you understand what you're seeing.)

Structural elements. Most single-family, wood-framed houses are structurally pretty simple. The

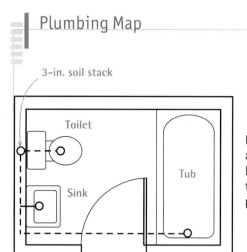

Plumbing Map

3-in. soil stack

Toilet

Sink

Tub

Plumbing fixtures are often grouped around a 3-in. or 4-in. soil stack. Because of their size, the soil stack and the main drain it feeds are the most problematic to relocate.

roof ridge, exterior and interior bearing walls, and girders (if any) run parallel to the long axis of the house, whereas joists and rafters run perpendicular to that axis. As you can see in the drawing on the facing page, interior bearing walls and girders typically support joist grids somewhere near the middle of the house, in effect shortening the distances they must span.

When structural elements are adequately sized and aligned, they transfer loads down to the foundation.

That's the way things are supposed to work.

When houses have been framed in a nonstandard manner, added on to, or remodeled by amateurs, there's no telling what you'll find. To explore your house's structure, start in the basement, where joists and girders are most often visible, or in an unfinished attic, where rafters and floor joists are frequently exposed. In finished living spaces, finish wood floors are typically installed perpendicular to joists, so look at flooring-nail patterns.

In row houses and other narrow structures, interior walls are often nonbearing because joists run the full width of the dwelling. There, interior walls running parallel to joists—or even perpendicular to them—were probably not bearing walls originally. But, here again, they may have become bearing walls if joists were undersize to begin with or if their effective spans were reduced by "remuddlers" cutting into them. Where interior partitions have become bearing walls, floors may slope toward them.

If your plans included removing a nonbearing wall to open up a space, the task may be straightforward. If the wall is load-bearing, removing it is an expensive alteration involving structural engineers to make sure those loads are correctly transferred somewhere else. If that bearing wall is also a *shear wall*, required in earthquake or high-wind

PRO TIP

Digitally photograph the area to be renovated before you start, and store the images on your laptop or iPad. Those photos can be an important source of information as the project progresses—especially when trying to match trim or buy fixtures—and a lot of fun to look back on when it's done.

Using a printout of a Google satellite map of your street, you can quickly create a rough site plan of your property. Site plans help city planners see the scope of your renovation.

easy to route: They're much smaller than DWV pipes and don't need to be pitched because water supplies are under pressure. Being flexible, PEX supply tubing is easy to run almost anywhere.

HVAC systems tend to be big and bulky, especially forced hot-air (FHA) systems, so their presence also will affect your design options. If the system is relatively new—say, installed within the last 20 years—leaving ducts in place and installing a new high-efficiency furnace may be the most cost-effective and least disruptive way to upgrade an FHA system. Modern hydronic (hot-water) heating systems are typically fed by ¾-in. pipe, so they can usually be relocated more easily. Whatever the HVAC system, sizing and configuring it is best left to an HVAC specialist, ideally as part of a whole-house energy assessment. See chapter 14 for a more detailed discussion.

When mapping an FHA system, note the locations of registers and the furnace. Then intuit the locations of ductwork, using dotted lines to suggest duct runs. In many cases, ducts or hydronic heat pipes will be visible where they emerge in the basement—or wherever the furnace is located. For upper-story heat outlets, delivery ducts and pipes generally travel straight up between studs.

Electrical wiring rarely affects design phases unless you intend to move a service panel. Small and flexible, electrical cable is easily routed through walls, floors, and ceilings. Where you don't want to cut into existing surfaces or can't—as with masonry floors—run rigid conduit or track wiring along the surface. Mapping electrical systems and identifying individual circuits is discussed at length in chapter 11.

CREATING A SITE MAP

As part of the permitting process, most towns require a site map—even if your renovation will take place wholly within the existing footprint of the house. Fortunately, accurate site maps are now easy to make. Simply enter your street address into Google Maps, zoom in to choose the closest view, click the "satellite" button, and print out the aerial view of your house and lot. Trees may obscure some of the details, so to simplify things, put tracing paper over the satellite photo, completely outline all structures on your property, and use broken lines to indicate fences or property lines.

If you are planning to build an addition, sketch it onto the site map, too. Be as accurate as you can, but don't agonize. This simple site map will be close enough to discuss your ideas with a city planner (to see how feasible your project is), an architect, or a builder. If you proceed with the renovation, later on you may need to have survey

regions to resist lateral forces, count on extensive plan reviews by the building department.

Plumbing fixtures are often grouped around a 3-in. or 4-in. DWV stack located in a wall near the toilet. Grouping fixtures shortens the distances large pipes must run and thus minimizes the impact on nearby framing. In many houses, large stacks are concealed within *wet walls* framed out with 2x6s to provide enough room. If you can locate new fixtures near existing stacks, you're halfway home.

Layouts that locate new bathrooms far from existing stacks can be problematic because all drains must slope downward at least ¼ in. per 1 ft. to carry off wastes. Thus, a 3-in. toilet drain would need roughly 8 in. of height for a 12-ft. horizontal run: 3½ in. for the exterior diameter of pipe plus 3 in. for slope plus 1 in. of clearance. In other words, you'd need to run drainpipes *between joists* at least 10 in. deep. If existing joists don't provide the height needed to run 3-in. or 4-in. drains, you can add a stack, build up a platform over an existing floor, or frame out false ceilings below—all expensive solutions. You'd do well to consider another design. Likewise, think long and hard about any floor plan that calls for running big drains *across* joist arrays—a nightmare scenario of cutting and drilling, leading to weakened joists.

To get an idea where pipes are running, draw fixtures on your plumbing map, and note where main drains emerge in the basement and where stacks protrude from the roof (connect the dots). Rigid water-supply pipes, on the other hand, are

lines checked and more extensive drawings made, but at this stage the site map can be simple.

Getting Help with Your Renovation

After you've spent time working up a renovation wish list and documenting what's there, talk to building professionals such as architects and general contractors (GCs) to get a reality check.

Building and remodeling projects can get complicated and expensive. Whether you run the whole show, from planning to construction, or you turn the entire thing over to a professional depends on how much time, expertise, and money you have. If you don't have a ton of money, you might want to get involved in as many places as you can. But if you can afford it, it might make sense to hire an architect and let him or her find a good general contractor. Assembling the right team will make the rest of the project go smoothly. Maybe you can do it. Maybe you can't.

Whenever possible, use your personal networks to find building professionals to work with. Perhaps a friend renovated recently and had a great experience. Or maybe you're keen on a contractor who did another project for you. In that case, ask the GC for the names of an architect whom she or he enjoys working with. Or maybe you know an architect but no contractors. In that case, the architect may help you find the rest of the team.

As far as assembling a team to help you get the job done, here's the bottom line: Make sure they are people who can see eye to eye with you and who will work well together. Often, the best teams are ones that have worked together before. The worst thing you can do is bring in a GC long after the plans are all complete, you're in a hurry to get started, and you only have a vague understanding of what kind of person he is and what quality of work he does. You can hope it will work out, but that could be a recipe for disaster.

ARCHITECTS AND DESIGNERS

An architect can refine your design ideas and draw up the floor plans and elevations needed for the *submittal set*—documents and plans submitted to the building department for a permit. But an architect can do much more. He or she will devise solutions that maximize natural light and usable space—always a challenge—and suggest materials and appliances that are stylish, functional, and energy-conserving. An architect also can assemble a team of specialists, from energy specialists to structural engineers.

Confusion Is Costly: A Primer for Owner–Builders

There are many ways you could save money by doing jobs yourself: obtaining permits, ordering materials, hiring and scheduling subcontractors, demolishing walls, hauling rubble, and completing finish work. But that isn't for everyone. And because homeowner skills and experience vary, there are no easy rules for deciding what to attempt yourself. Yet there are times when even scarce money is well spent for a skilled professional, and there are times when regulations require it.

Hire a pro whenever these situations apply:

▶ You're confused and don't know how or where to start a renovation task. This is especially true if you have trouble envisioning spatial solutions—seeing things in three dimensions.

▶ You lack the technical skills to do a job—in which case, learning by working with a pro makes sense.

▶ You're rushed for time and can probably earn more elsewhere (to pay for the work) than you could save by doing it yourself.

▶ Tasks require special or hard-to-find tools.

▶ The job is inherently dangerous. For example, an amateur should not install an electric service panel.

▶ Building codes, bank agreements, insurance policies, or other legally binding documents require that work be done by a licensed professional.

Finally, poorly organized projects and confusing drawings can idle workers, costing you big time. If you're not well organized, patient, and willing to field phone calls at all hours, hire a GC who is. Likewise, if you're not construction savvy and a capable draftsperson, hire a pro to generate final working drawings. It's far cheaper to resolve construction issues on paper, especially if a project is complicated.

As important, a perceptive architect can steer your project through the trickier parts of the approval process. Getting a project approved and built is a hardball game. If your first submittal set isn't accurate, complete, and professional looking, your subsequent efforts will be viewed with skepticism and greater scrutiny. To quote a seasoned builder, "Your documents must not only explain clearly what you want to do, but explain it in the way that planners want to hear it. Few builders are equipped to generate their own set of presentation plans for the site, even if they've got some talent in drawing. It's a lot of work and takes a lot of time."

This succinct observation came from a general contractor: "The main reason to hire an architect is to make sure you're building the right project."

GENERAL CONTRACTORS

To save money, homeowners sometimes try to act as their own general contractor. At first glance, obtaining ("pulling") permits, ordering materials,

When Do You Need a Permit?

Local codes always have the last say about what's permitted, so speak with a planner in city hall or visit your town's building department website to learn the rules.

That noted, these observations hold true for building permits in most areas:

▶ Permits are required for all new construction and most renovations involving structural, mechanical (HVAC), electrical, or plumbing system changes. Permits require inspections at specified stages and/or completion.

▶ A separate permit is usually required for each system being altered.

▶ The person doing the work must pull the permit(s), whether a contractor or a homeowner.

▶ The permit must be posted on the job site where an inspector can see it.

Permits required. As a rule of thumb, renovations that alter the structure or extend an electrical, plumbing, or HVAC system require a permit. If the work involves the creation or enlargement of an opening in an exterior wall—such as adding a door or window—you need a permit. Likewise, if an interior renovation involves cutting into finish surfaces or working inside a wall, a permit is usually required. To cite a few specifics, you need a permit when adding a room, adding or removing walls, finishing any room (basement, garage, attic) to create additional living spaces, adding bathroom or kitchen fixtures, or adding electrical outlets and fixtures.

No permits needed. On the other hand, you generally do *not* need a permit when repairing or replacing an existing fixture or outlet, when the work is largely cosmetic, or when an installation doesn't require cutting into finish surfaces. For example, no permit is required to replace a broken electrical receptacle; replace a toilet or bathroom sink; retile a shower stall; replace an existing garbage disposal or dishwasher; install or replace portable appliances such as washing machines or refrigerators; or replace a doorbell.

Important for safety. Installing or replacing a water heater always requires a permit because a unit's temperature- and pressure-relief (TPR) valve must be installed correctly to prevent an explosion. All work on gas appliances, lines, or vents requires a permit. Installing or altering any vented heating or cooling system needs a permit—including furnaces, woodstoves, gas ranges, range hoods, bath fans, dryer exhausts, and so on. You may also need a permit to repair or replace GFCI-protected receptacles.

PRO TIP

When sketching floor plans, remember to include room for chairs to pull out from tables, appliance doors to open, and furniture drawers to pull out.

▮▮▮▮

---

hiring subcontractors, and calling for inspections doesn't seem so difficult. Well, it is. For one thing, how quickly a sub returns your calls often depends on how long he's known you, and as a homeowner, you probably don't know many subs. And if, for whatever reason, you can't keep the subs, materials, and inspections flowing in the proper sequence, your project can seize up in a big hurry.

GCs have mastered the complicated art of maneuvering and managing—or at least the ones who stay in business do. If your project is simple, maybe you can manage it. But if your renovation is at all complicated or if you're not well organized, patient, persistent, and have plenty of time to do what has to be done, hire a general contractor who is. Contractors with some architectural training may offer "design-build" services under one roof—or in one person.

There are many ways to find a good GC. Start with the recommendations of friends who've remodeled recently. Architects also will know who's good. Once you have two or three names, call each, ask for an hour of his or her time to look at your ideas, and get an idea of what it might cost. (First meetings are usually free; after that, be prepared to pay . . . it's only fair.) Most builders won't want to ballpark a price based on a schematic, but they will give you a price range per sq. ft. based upon construction costs in your area. In any event, if you find a contractor you like, ask for references and a list of jobs she or he has done in the last five years. It's important that the references and jobs be this recent because construction crews change often.

DESIGNER-BUILDERS

Add one more to the mix of people who can help you renovate: the designer-builder. Time was, before the skills of homebuilding and design split apart, they were often embodied in one person, a master builder. That tradition is returning, in part because of tight times but also because clients like having one person who can answer all their questions—whether aesthetic, technical, or budgetary—and accept all responsibility if there's a bump in the road. Designer-builder firms also argue that the transition from design to build is smoother when handled under one roof. Interestingly, two of the case studies at the end of this chapter were executed by design-build outfits and one featured "an architect with a ton of construction experience and a builder who could draw."

DEALING WITH CITY HALL

Because we've invoked city planning and building departments several times, this seems like a good place for an overview of submitting plans and getting permits. Generally, getting something built is a two-step process that begins at the planning department and ends at the building department. In rural areas and small towns, the two departments may be combined, but usually they're separate: different issues, different bailiwicks, and different staff.

Planners are, first of all, the keepers of zoning regulations: how high a house may be or how big; how close it may be to neighboring houses; how much of a lot a house and addition may cover; and what public processes you must go through to make changes to your existing property. In other words, the planning department cares most about how a building looks and how its use affects the community.

The building department focuses on the construction and functioning of the house, including health and safety issues. That list could include how much steel is in the foundation; how walls support the roof; what kind of insulation is in the walls; plumbing, electrical, and HVAC details; or the location of fire sprinklers. The building department makes sure that a renovation project follows code and, ultimately, is safe to live in. Thus building inspectors visit projects periodically to make sure that what's on the plans is what gets built.

For a stripped-down chronology of this process, see the chart at right.

Developing a Renovation Plan, and Getting It Drawn

At this point, you've tried to clarify the scope of your renovation, are getting to know the lingo of building, have a sense of how permitting works, and have spoken to an architect and a contractor or two. You have a fair idea of what's possible, what renovation tasks you can reasonably do, who could help design and build the project, and what it should roughly cost.

PICTURING YOUR RENOVATION

One of the best ways to envision change is to draw it. Start by prioritizing the items on your renovation wish list. Then tape a sheet of tracing paper over the to-scale floor plans you created earlier and start doodling. Start with *bubble diagrams* (as shown on p. 32) to indicate how space will be allotted to different uses and how traffic will flow between work areas or rooms. Do your drawings reflect your wish-list priorities? Bubble diagrams may be as far as you get— but they will be useful to an architect or designer trying to understand your needs.

If you enjoy the process, keep going. It's just tissue paper. Explore alternative layouts and add detail. If you're unsure about the size of an appliance or a piece of furniture, measure again. Be sure to include enough space to open and pass through doors, pull chairs away from a table, open a fridge, or remove food from the oven. Pages 391 and 414 show the minimum clearances needed in kitchens and bathrooms. Have fun, but be realistic: Don't try to fit too much into a small space. Analyze the trade-offs between one floor plan and another. In general, the more problems a design solves, the better.

Most of us live in modest homes with relatively simple floor plans: entry, living room, dining

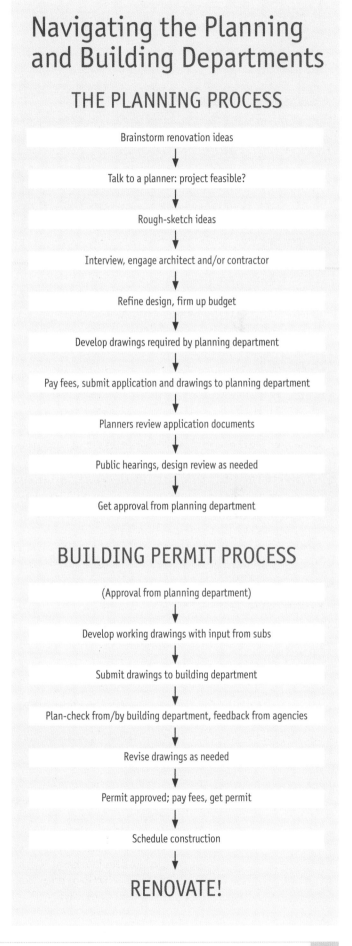

Navigating the Planning and Building Departments

THE PLANNING PROCESS

Brainstorm renovation ideas
↓
Talk to a planner: project feasible?
↓
Rough-sketch ideas
↓
Interview, engage architect and/or contractor
↓
Refine design, firm up budget
↓
Develop drawings required by planning department
↓
Pay fees, submit application and drawings to planning department
↓
Planners review application documents
↓
Public hearings, design review as needed
↓
Get approval from planning department

BUILDING PERMIT PROCESS

(Approval from planning department)
↓
Develop working drawings with input from subs
↓
Submit drawings to building department
↓
Plan-check from/by building department, feedback from agencies
↓
Revise drawings as needed
↓
Permit approved; pay fees, get permit
↓
Schedule construction
↓

RENOVATE!

Evolving Floor Plans

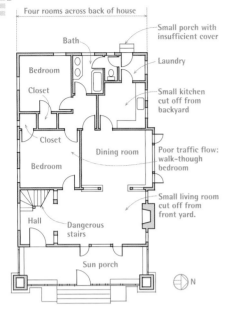

Four rooms across back of house

Bath

Bedroom

Closet

Closet

Bedroom

Dining room

Hall

Dangerous stairs

Sun porch

Small porch with insufficient cover

Laundry

Small kitchen cut off from backyard

Poor traffic flow: walk-though bedroom

Small living room cut off from front yard.

N

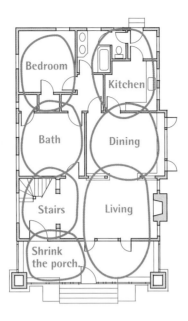

Bedroom

Kitchen

Bath

Dining

Stairs

Living

Shrink the porch.

Porch

Bedroom

Closet   Hall

Closet

Safe stairs

Entry

Porch

Enlarged covered porch

Large kitchen open to backyard

Relocated bath

Dining room

Hall for improved circulation

Enlarged living room open to front yard

FLOOR PLAN: BEFORE
The original floor plan presented a pleasant face to the street, but the back of the house was a hodge-podge of doors, dead spaces, and tiny rooms.

BUBBLE DIAGRAM
Bubble diagrams like this allow you to consider layout alternatives quickly.

FLOOR PLAN: AFTER
After weighing a number of floor plans, the architect settled on the design that solved the most problems.

MAKING THE MOST OF Underused SPACES

Every home has room for improvement. As you allocate space to the activities you enjoy, keep in mind the odd nooks and crannies of your home. The areas under stairs or sloping roofs, for example, may not be tall enough to stand up in, but they may be perfect for built-in closets and cabinets, a bathtub, or a bed alcove. Window seats and built-in breakfast nooks are also handy places to store little-used items. Most kitchen cabinets also can work harder: Adding drawer glides to shelves will enable you to reach items all the way in the back.

More ambitiously, if an empty nest has a few unused rooms, they might be transformed into an in-law suite for an elderly parent, a private apartment for an adult child living at home, or a rental unit. In fact, second units are a hot topic these days, especially to accommodate multigenerational families.

EAST ELEVATION

A painterly elevation. Hand-rendered drawings help give clients a feel for the more subjective aspects of a design, such as textures, shadowing, and how elements relate to each other. This elevation and the following three drawings, created by architect Russ Hamlet, show the second-story addition chronicled on pp. 45–47.

See your design in 3-D. Google's SketchUp software can be downloaded for free, or you can purchase a more robust version. When homeowners couldn't see how a design would look spatially, Russ Hamlet used SketchUp to show the drawing at left in 3-D.

room, kitchen, with a couple of bedrooms and a bathroom or two off to the side. Some layouts will be fairly obvious: dining room next to kitchen, bathroom between bedrooms. And if the house isn't big enough, there are probably only a few directions that an expansion could go—bump out a kitchen wall, add a room or two off the back of the house, bump up, and so on.

DRAWINGS: FROM PRELIMINARY DESIGNS TO CONSTRUCTION SETS

How complicated and complete drawings must be depends on what they are trying to convey and who their intended audience is. The more complicated your renovation, the more drawings you'll need.

Communicating designs to clients. Some people can instantly see drawings in 3-D, without the funny glasses, while others could labor over blueprints for decades and still see only two-dimensional lines on paper. So a good architect will probe to make sure a client can see how a renovated space will look and function. For some clients, the simplest sketch will suffice; other folks will need more inventive approaches.

Designers and architects who love illustrating frequently give elevations a more painterly feel to suggest the rhythm and texture of architectural elements, as Russ Hamlet did in his rendering above. In addition to well-established programs such as Autodesk AutoCAD, architects also can use free software such as Google SketchUp to help homeowners see layouts in 3-D, as shown in

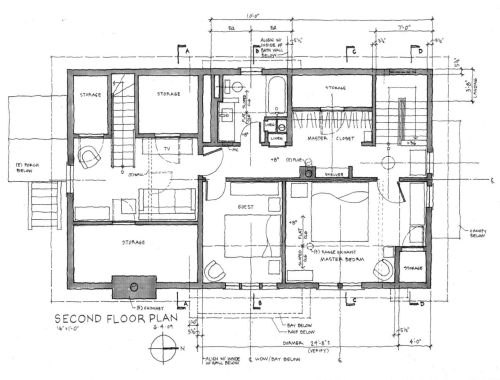

SECOND FLOOR PLAN

When you apply for a permit, the planning department will want to see a set of design development drawings, which must be drawn to scale but need not show construction details.

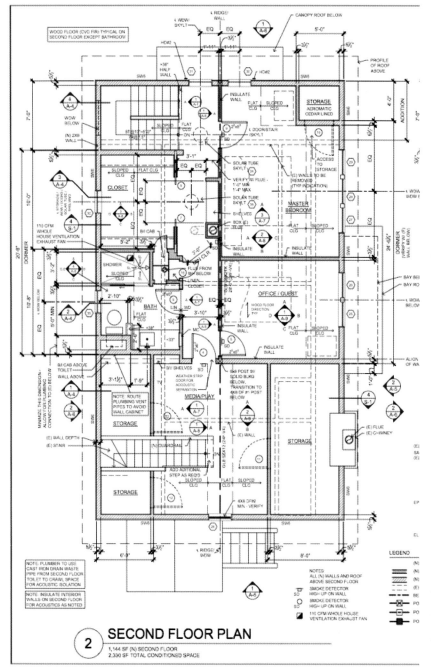

SECOND FLOOR PLAN

2

1,144 SF (N) SECOND FLOOR
2,330 SF TOTAL CONDITIONED SPACE

Working drawings must be extremely detailed because they are the final document that the building department approves and from which the contractor builds.

the top right drawing on p. 33. Learning that a client really doesn't understand (or like) a design until the work is done is a costly and disheartening realization. So homeowners must learn to speak up early and often during the design phase.

Submittal set for the planning department. Homeowners who are thinking of managing the renovation process: Heads up! If you meet with a city planner to discuss your project, the planner probably will give you a summary of town zoning requirements, as well as a "submittal checklist" of documents you must gather should you decide to submit a formal application for a building permit. (The checklist may also be on the town website.)

That submittal set must include a set of *design development drawings*—elevations, a site plan, and a brief written project description. Design development drawings must contain accurate overall dimensions, and their elevations should have windows and doors correctly sized and located. In other words, they are a set of technical drawings that accurately show what the renovation is going to look like but not how it's going to be built.

After you submit your application and the submittal set, the staff of the planning department will analyze your plans. Within a set time period, they will tell you if your plans are complete or, if not, what you must add. The approval process varies from town to town. Typically, someone in the planning department issues an administrative approval, or if a public hearing is required, zoning commissioners issue a ruling. Once a project design is approved, you must submit a set of *working drawings* to the building department.

Working drawings, also called a construction set. The building department has its own checklist of things you must submit, including a set of working drawings. Because this is the final set of drawings from which the contractor will build, the architect and contractor often confer at length before submitting them.

This stage is all about details. If there's a structural component to your renovation—for example, an augmented foundation or a seismic retrofit—the drawing sets will now be reviewed by a structural engineer. Sets will also be sent to electrical, plumbing, and HVAC subcontractors so each can make suggestions and prepare their bids. There may also be soil engineers involved and, routinely these days, an energy consultant.

In a small project, the architect assembles most of this information, and the appropriate subcontractor often works out unresolved details on-site. At this point, there is a lot of back and forth between the subs, the architect, and the general contractor. The structural engineer, for example, may suggest foundation details, but the architect or general contractor might challenge them because of their cost and propose alternatives.

Once your team has heard back from all the subs and specialists and incorporated their suggestions, it will review once more the building department's checklist. That done, four sets of working drawings and supporting documents are printed and the whole package is submitted to the building department. On average, it will take the building department about a month to check the plans, although three months is not uncommon. Once the review process is complete and everything has been approved, your plans are stamped and you get a building permit. In smaller communities, the approval process is typically less formal and doesn't take as long.

Making the Most of Your Renovation

To restate an essential point: How much help you need depends on how experienced you are, how much time you can give to the project, and how ambitious your renovation is. In this section, you'll find suggestions on how homeowners—especially, owner-builders—can most productively work with architects, designers, and contractors. Yes, we're talking about money but also about creating lasting value in a renovation. In the case histories that follow, you'll see such value because, in each, the homeowners were fully engaged in the process, from planning through completion.

AFFORDING AN ARCHITECT

Many people on a tight budget think they just can't afford an architect. Consequently, only 2% to 3% of new homes are architect-designed. But hiring an architect needn't be an all-or-nothing proposition. Increasingly, architects are willing to act as consultants for homeowners who want to manage more of a project themselves. And many architects will reduce their fees if, say, they don't have to provide as many drawings in the construction set—a workable solution if the builder has a decent design sense and can resolve details on-site. Whereas a full-service architect's fees (including site visits) might run 10% to 15% of the total budget, clients willing to do a lot more of the work themselves can often get those fees down to somewhere around 5%.

Following are paraphrased remarks from Duo Dickinson, a Madison, Conn., architect and the author of *Staying Put: Remodel Your House to Get the Home You Want* (The Taunton Press, 2011).

STAYING IN Touch IN THE DIGITAL AGE

Renovations are always unpredictable. But a contractor with a cell phone can share a discovery (say, a rotted sill) with a client or an architect almost immediately and get feedback approval for a change order quickly. This keeps the job flowing and avoids costly delays. One GC notes, "The only way to keep a job affordable is to make it efficient." Alternatively, if homeowners are puzzled by a construction detail, they can photograph it and send it to the architect to make sure the builder is on the right track—without needing a site visit.

In a somewhat related vein, you and your team can also use free software such as Dropbox to gather project info in a single place, where anyone can access it. Evernote, also free, is another place to store job-related photos, site links, and the like.

Here's his advice on how to save money while working with an architect:

▶ *Educate yourself.* Learn everything you can about design and construction before engaging an architect.

▶ *Know your budget.* A good exercise at any time but essential to defining the scope of a renovation.

▶ *Be specific.* The better you know your priorities and spatial requirements, the quicker a design can evolve.

▶ *Get your builder involved early.* If you or your architect is working with pie-in-the-sky budget numbers, your contractor will bring you back to earth. GCs can also spot details that would be problematic to build.

▶ *Don't contract for services you don't need.* But first you must learn what those services are (see the first point).

▶ *Communicate digitally.* Sending emails or digital photos to your architect is much more time- and cost-effective than face-to-face meetings. Plus, you'll have a record of what's been discussed.

KEEPING A CONTRACTOR ON TRACK

Dave Carley, a contractor on Bainbridge Island, Wash., for two decades, did the roof-raising remodel on pp. 45–47. The architect and homeowners of that project raved about Carley, so here's his answer to our question, "How do you get the most out of a contractor?"

"By being a good customer," he said with a laugh. "What frustrates contractors the most is indecisiveness, an inability to move forward, and disorganization by the homeowner. What contractors want to do is move in, do good work, get paid, and go home. . . . We understand that there will be change orders and that every job must be somewhat fluid. But what kills project momentum and morale more than anything is having to stop. Or go backwards in some cases. When you have to take apart your own work or remodel the remodel, it's really tough.

"My favorite customers are the ones who come to meetings prepared and are good decision makers. It allows me to do my job better. If they have everything chosen, I can do better pricing, better ordering, better scheduling. And, of course, better building.

"These days, most people don't go through the full process with an architect. They get a [minimal] set of plans drawn to get them through the city and get a permit and then kind of feel their way through the project. Some people are good at making decisions and others really struggle. I understand their need to contain costs, but good architects prepare the clients, set up realistic expectations, and do their homework so their plans are accurate. Which allows me to move forward with confidence."

TALKING MONEY

Because we briefly addressed money while discussing architects' fees, here are three questions that clients ask contractors all the time. The answers are composites from many builders.

Why shouldn't I just go with the lowest bid?
Reputation is a far better indicator of the kind of work you'll get and the kind of experience you'll have. Initial bids are just baby steps when viewed in the context of an intimate relationship—this is your home, after all—that will go on for months and at times be very stressful. In many cases, competing contractors are fishing from the same pond: buying materials from the same suppliers, paying the same hourly rate, and using competitively priced subs. Low bidders sometimes hope to recover profit on change orders. So check referrals carefully. What was the homeowner's experience? Was the work on time and on budget? How were change orders handled? This information is far more important than an initial bid.

Can I do any of the work to save money?
Maybe. If you're hiring a contractor, the best thing you can do to help control costs is make important decisions in a timely manner so the crew can keep busy. If you do help, it's still important that you stay out of the crew's way so you don't interrupt their work rhythm. You may be able to do some tasks, such as tearing out drywall, pulling nails from demolished lumber, or sweeping up at day's end. But be aware that even during demolition, a good crew will be looking for house conditions that need fixing—such as water damage or insect infestation—so allow the crew to concentrate and observe.

Which is better, a fixed-bid or a cost-plus contract? Until a builder tears off finished surfaces and has a close look behind the walls, there's simply no way to know what he'll find—and hence, what a renovation will cost. Consequently, even fixed bids will contain contingencies (typically 15%) to deal with the unknown. Builders with solid reputations and enough work on their plate often resist fixed-bid situations, however, because they feel that such agreements are unfair, set up unrealistic expectations, generate excessive paperwork, and make for stressful relationships all around. Even careful budgets tend to be a bit fluid anyhow. For example, if demo comes in a little under budget, you have a cushion if the drywaller goes over. In any event, reputable contractors use transparent accounting, regular client meetings, and *progress billings* based on percentage of completion.

Creating Lasting Value: Five Case Histories

Although they vary in scope, the following case histories all attempt to conserve materials and maximize space. Each evolved within a relatively small footprint, made the most of natural light, and created multifunctional areas to optimize space and functionality. Each also employed a number of green-building principles to conserve resources.

CASE HISTORY ONE

This 1925 Craftsman bungalow is modest yet commodious, with an oversize, welcoming porch. Built from kits, such cottages provided affordable housing for working-class families.

Before renovation, the kitchen had few cabinets, almost no counter space, a terrible layout, and was impossible to keep clean. A full-size refrigerator and stove didn't help the cramped space.

To the east, two small rooms further chopped up the area. A breakfast nook, at right, had been turned into a pantry because there was so little room for storage. The table had to be moved into the kitchen. The architect started by removing the partitions to open up the space.

BRINGING A CRAFTSMAN HOUSE INTO A NEW CENTURY

The two-bedroom bungalow "instantly felt like home" to Sandra and David. It had beautiful old windows, wavy glass, unpainted woodwork and built-in china cabinets in the dining room, and the house had been well cared for. Small by today's standards—roughly 1,150 sq. ft.—it had comfortably housed a couple and their three kids when it was new. Its close quarters would be something of a challenge because one of the bedrooms would become David's home office, but the couple welcomed the opportunity to get rid of stuff they didn't need, recycle on a regular basis, and live simply.

The kitchen, however, was a horror: cramped, dingy, and badly out of date. The bathroom was also dark and a bit funky, but they could live with it. The couple lived in the house before starting their renovation, but when they did, they had an ace in the hole. David's cousin, Fran, was an architect who loved a challenge.

Program requirements: "A kitchen you can use without bumping into stuff! More storage, more counters, more room. It's impossible to keep clean, and the old linoleum is shot. The bathroom is dark; it would be nice for both of us to have our own space in it. We rarely use the tub."

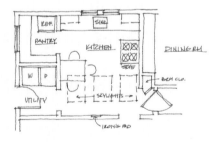

The kitchen, before. Architect Fran Halperin's "before" drawing shows why the kitchen was unworkable: Roughly one-third of the floor space was chopped into two small rooms.

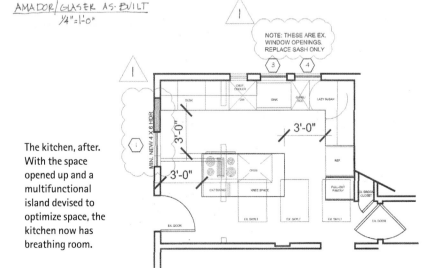

The kitchen, after. With the space opened up and a multifunctional island devised to optimize space, the kitchen now has breathing room.

With natural light from three directions, the renovated kitchen is bright, airy, and easy to navigate. An island with a colorful concrete top conserves space by combining functions: It serves as a cooking and prep area and, with room for three tall chairs, an intimate eating area.

A cheerful corner creates a temporary office for Sandra, a therapist. Lower cabinet drawers are large enough for files, while drawer-pull shapes echo accent tiles on the wall.

Kitchen solutions: "Take out the walls and open it up" to create one large kitchen space. Create a large island whose concrete countertop includes a cooktop, prep areas, and places to eat and hang out. Replace the full-size fridge (17 cu. ft.) with a European (9 cu. ft.) model. Upgrade the sink, and replace the glass in the over-sink window with obscure glass to increase privacy. Replace the old linoleum with Marmoleum. Add a desk in the kitchen for Sandra next to the east window.

Bathroom solutions: Replace the old, encased tub with a glassed-in shower stall (see the photo at right). Instead of installing a second window on the south wall, install a light tube over the shower at a fraction of the cost—and greater privacy. Tile floors with radiant heating. In the small space next to the shower, add a stand-up beauty area just for Sandra—David can have the medicine cabinet over the sink for his stuff.

Green touches: The California cooler in the north wall keeps produce cool naturally, so the refrigerator can be smaller: Shop often, eat fresh. Marmoleum flooring. The light tube is free lighting.

Parting thoughts [Sandra]: "It's such a gift to be able to work with an architect. . . . What I remember most from the renovation was the enormous joy of seeing what you thought it was going to look like turn out even better. And to have the space transformed was like magic upon magic. . . . The island is our favorite place for meals. David and I often sit side by side and hold hands as we eat; it's very intimate."

Project size: 230 sq. ft., kitchen and bath
Architect: Fran Halperin, San Rafael, Calif., www.halperinandchrist.com

Interior design: Sharon Low, San Francisco, Calif., creativerooster@sbcglobal.net
Construction: Eric Christ, Noah Garber, www.halperinandchrist.com
Completed: 2009

A glassed-in shower stall with a light tube upgraded a dark bathroom. To the right of the shower is a stand-up makeup counter with plenty of drawers underneath. Elegant details make the small space feel special: A pomegranate-seed motif in the tiles repeats in the drawer pulls and in a custom tile design on the radiant-heated floor.

A playful cutout provides privacy for a kitty loo. Marmoleum flooring's wide range of colors allowed the designer to duplicate the pattern of a parquet border in the adjacent dining room. Craftsman houses often repeat architectural patterns in floors, cabinets, windows, and doors.

Speck, the Sumners' first love and their inspiration to live compactly on land.

The in-law, to the right of the entry door, coexists nicely with the main house, sharing exterior colors, common siding, a roof overhang, and a brick courtyard. Set in sand, the bricks create a permeable surface that rain can drain through, a welcome detail in a climate with wet winters.

A TALE OF TWO SAILORS

Long before small was beautiful and houses were not so big, there were sailors. More than almost anybody, sailors know about making the most of small spaces because boats, like fish, must be streamlined. Nothing delights sailors more than showing off seats that fold into beds or postage-stamp galleys that can turn out a 10-course meal. That attitude led two sailors to design and build an in-law unit that does everything but float.

When Beth and Gary Sumner aren't sailing, they live in Eugene, Ore., where they own an architectural stained-glass studio. Gary bought a single-story house there in 1978. At first, they housed the studio in the bump-out that had been a single-car garage, but eventually rented a warehouse. Then, one day, they realized that their kids were grown and gone and that two bedrooms and the 360-sq.-ft. studio were sitting idle.

So the Sumners decided to convert the studio into an in-law for themselves and rent out the three-bedroom main house. "Gary and I kid that we're self-unemployed," Beth laughed, "because everything we have is tied up in the business. Renting the main house made sense because it saves money and frees us to travel. Plus, we were curious to see if we could get along in such a small space."

Before sketching a layout, Beth and Gary drew up a short list of requirements:

▶ *No clutter.* The Sumners applied this rule whenever they were uncertain about a design choice. There are no floor or table lamps, for example, and only two movable chairs in the unit—one rocker and a computer chair. They used built-ins whenever possible.

▶ *Nice details.* "Something you'd want to come home to," as Beth put it. The Sumners knew a number of high-end woodworkers, so the windows and doors are clear-grain fir and the custom cabinets are rift-cut white oak. Not surprisingly, there's also a lot of stained glass.

▶ *Open and bright.* This requirement led to two important design choices: a cathedral ceiling whose peak is almost 12 ft. high and skylights that open.

▶ *Full-size kitchen appliances and a roomy shower.* Shipboard cooking and showering are cramped affairs so this is one instance where comfort trumped compact. The Sumners cooked a lot so they wanted a stove and refrigerator up to the task, as well as enough

Everything has its place. A built-in desk sits next to oak doors hiding the Murphy bed. At left, a bookcase covers the soundproofed wall between the house and the in-law.

room for both of them in the kitchen. And there's no nicer antidote to salt-encrusted skin or sandy hair than a blasting hot shower for as long as you like. (They installed a 50-gal. water heater.)

Design squeeze: Beth and Gary's project had a number of constraints, the first of which was the small lot size. The house and studio were built in 1936, before zoning regulations, so the structures were quite close to property lines on the east and west sides of the lot and their combined square footage far exceeded the allowable lot coverage. To retain the *grandfather status* of the studio, the Sumners had to retain the original roof (see the photo below). The roof framing, however, had to be strengthened.

The north end of the in-law was attached to the house, sharing a 6-ft. section of common wall. That end also contained a door that had to stay because it provided access to a covered driveway to the street. The local building code required two exterior doors, so they located the second one where the buildings intersect; the generous overhang of the house would protect the door from rain.

Beth and Gary first considered putting the kitchen in the north end of the in-law, but two things argued against that. First, they wanted to be good neighbors and minimize noise along the section of common wall. Second, on the south end of the in-law was a small shed (2½ ft. by 9 ft.) that would be a great place to park a water heater.

So they decided to locate the kitchen and bathroom on the south end, sharing a *wet wall*. It was a good choice. In addition to housing a water heater, the shed had enough room to create recessed bays for the refrigerator and a stacked washer-dryer in the bathroom. Both the bathroom and the kitchen are full size, which delighted the Sumners.

Bolstering an Old Roof

The good old days weren't all good. Old roofs were often framed with 2x4 rafters, which invariably sagged. If, like the Summers, you must leave an old roof in place, consider their solution: Bolster those 2x4 rafters with 2x6s underneath—creating, in effect, 2x10s.

Those 2x10 rafters have many virtues. They stiffen the roof plane so it doesn't sag. Deeper rafters can accommodate deeper insulation and, as important, provide room for ventilation channels (under the insulation) to vent hot air and excessive moisture. Lastly, 2x10s are strong enough to mount skylights to and deep enough to flare out the sides of the light well and thus distribute light to a larger area.

The original roof couldn't be disturbed, so workers supported it with steel girders while replacing the old concrete floor slab and then building walls up to the roof.

One other constraint led to a creative solution. The windows along the west wall look into the Sumners' courtyard, so its view is sunny and private. On the east, however, the in-law is hard against the property line. Windows on that side, even high up, would have encroached on their neighbor's privacy. And more windows would eat up precious wall space, which the Sumners desperately needed for closets and the Murphy bed. But without windows on the east, there would be no cross-ventilation. What to do?

The solution arrived in a blaze of light: operable skylights in the eastern side of the roof. When open, they'd allow cross-ventilation. Located over the Murphy bed, the skylights make the space feel like a big airy bedroom. On clear nights, the Sumners can look at the stars and plan their next getaway.

Project size: 360 sq. ft. (in-law addition)

Design and construction: Gary & Beth Sumner

Completed: 2008

Thanks to a skylight directly above the Murphy bed, the star shows are spectacular. With operable skylights, there's also plenty of cross–ventilation.

Beautiful wood warms the kitchen and eating areas: clear–grain fir windows, rift–cut white oak cabinets with ebony pulls, a solid teak tabletop, and teak banding to finish countertop edges. Under the bench seats there's room for storage.

In-law Floor Plan

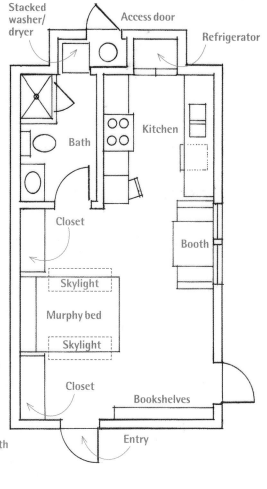

Stacked washer/dryer

Access door

Refrigerator

Bath

Kitchen

Closet

Booth

Skylight

Murphy bed

Skylight

Closet

Bookshelves

Entry

Site Plan

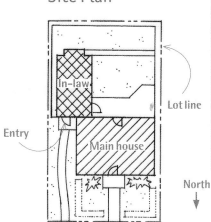

In-law

Entry

Lot line

Main house

North

Constructed by Scandinavian immigrants around 1830, this dovetailed log house north of Ottawa was one of the few survivors of Canada's frontier era.

Workers remove siding before disassembling the log structure. Installed about 20 years after the house was built, the siding protected the logs from the harsh climate.

The log house, overlooking Lake Champlain, echoes the region's historic Adirondack style. Though its poplar-bark shingles look rustic, they are high-tech wonders that are virtually maintenance-free.

DOVETAILING TWO WORLDS

After four decades of restoring historic houses, Gregory and Carolynn Schipa of Weather Hill had a loyal following across the country. Still, they were a bit shocked by a long-time client's sole requirement for a family cottage overlooking Lake Champlain: "Let's do whatever it takes to make it wonderful."

That they could do. For years, the Schipas had wanted to resurrect another dovetailed log building (they had done one in the past), such as those hewn by Scandinavian settlers to Canada in the early 1800s. Dovetailed joinery, in which board ends shaped like trapezoidal tails interlock at corners, is usually a feature of fine furniture. But because such joints rarely pull apart, they

Thanks to the massive logs (8 in. x 14 in. and larger) that comprise the structure, its interior space could be left open to create one "great room" for the family to enjoy.

were also an enduring way to join massive white pine logs culled from primeval forests. All it took was a keen eye, broad axes honed chisel-sharp, and the grit to homestead where winters heaped snow house-high and the mercury plunged to 40 below.

Problem was, few houses from that era survived. "Most of them just rotted away," Gregory explained. "Water seeped into the dovetails or migrated up through unmortared foundations. Fortunately, the log structure we found was in pretty good shape, despite the harsh climate. About 20 years after the house was built, someone covered the logs with clapboards, which cut the wind whistling through the walls and, more important to us, protected the logs. The builders were also smart enough to raise the foundation well above grade,

so that only one or two of the sills needed to be replaced.

"We bought the building from an old guy near Ottawa, who saved historic dovetailed houses wherever he could find them. He didn't restore them but he took them apart, marked each joint so it could be refit exactly, fixed anything that was rotten or bad, and then sold the disassembled logs. After months of e-mail exchanges to figure out what he had (and in what condition), we finally struck a deal and he agreed to oversee the delivery. Once the new foundation and the deck were ready, he showed up with a flatbed stacked with these gigantic 'Lincoln logs.' Our construction crew are pros, so we took it from there.

"This wasn't a purist restoration; we were using historic components to create a modern

"The stair elements were so solid we couldn't *not* use them, though we added a simple rustic balustrade to dress them up a bit," says Greg Schipa.

Although most of the building's exterior is covered with shingles, the dovetailed log walls were left exposed where they were protected by wide porch roofs. The porches, built of rustic cedar logs, look like they have always been a part of the house.

rustic house. What we were after, mainly, was an inspired 'great room' for family gatherings, so our design path was straightforward. We strived to make no further cuts in the logs, to reuse the holes and proportions of the original fenestration. The old front door, for example, became the dramatic bay window on the southern side. We reused doors, much of the flooring, and the stairs—though we added a simple rustic balustrade to dress them up a bit.

"To cut air infiltration and improve energy conservation, we placed a closed-cell foam strip between each 8-in. by 8-in. log. Basically, the log crushes it and fills in the space. Using spray foam, we filled any cracks that were still visible. Then we covered the exterior with Tyvek and 2 in. or 3 in. of Thermax rigid foam before applying rustic poplar-bark shingle siding. Inside, a professional *chinker* filled remaining gaps with foam before applying an elastomeric plaster that can accommodate seasonal expansion and contraction without compromising the air seal.

"We were lucky to have clients who could appreciate such a restoration. Given the skill and materials it took to construct this house—virgin pine forests and the times and circumstances through which it survived—it's rare indeed. That world no longer exists."

Project size: 1,000 sq. ft. (great room), 1,000 sq. ft. (porches)

Design and construction: Carolynn & Gregory Schipa, The Weather Hill Company, Warren, Vt., www.weatherhillcompany.com

Completed: 2010

The west face of the 1928 farmhouse, before the renovation. The porch on the south face, which faces the street, had been the front door of the house when it was new.

The east face of the farmhouse, before the renovation. This face would receive the most radical makeover, including a bump-out for the dining room. The enclosed north porch, at right, would rise to two stories.

RAISING THE ROOF

"There was no huge problem with the house," Robin explained. "And it wasn't like a renovation was going to change our lives or who we are. The house was just showing its age, and our oldest son started sleeping on the couch because he didn't want to share a bedroom with his brother anymore."

"In fact, we loved the way the house looked and fit into the neighborhood," added Mike, "and we didn't want that to change."

To keep that down-home feeling, Robin and Mike turned to two Bainbridge Island, Wash., neighbors, architect Russ Hamlet and contractor Dave Carley, when it came time to renovate. The 1928 farmhouse had the usual old-house ailments: The windows leaked, the furnace rattled and slurped fuel, there was no storage, the dining room was "an afterthought," and there was only one bathroom. How folks live and entertain had changed, too. These days, everyone hangs out in the kitchen, so the front porch had become the back of the house. Getting to the (unheated) attic bedroom meant walking all the way through the house to get to the stairs.

Program requirements: Increase living space without destroying the cozy look and feel of the house. Enlarge the dining room. Add a bathroom upstairs and a guest room somewhere. The family will move out of the house but live on the property, so keep the kitchen and bathroom available. Upgrade the insulation and heating system.

Design solutions: "The old house leaked energy like a sieve, so something needed to be done with the lid of the house," mused Hamlet. "And the upstairs was just not functional." So when a struc-

Reverse CONSTRUCTION

Renovation is the art of whatever works. Sometimes that means reversing the normal order of construction—or deconstruction. Although the roof was to be demolished, the contractor decided to first strip, rebuild, and insulate the exterior walls below, one wall at a time. Here's why:

▶ Washington gets a lot of rain, so he wanted to keep the house covered as long as possible.

▶ The attic framing was seriously undersize, with 2x6 rafters spaced 24 in. on center and 2x6 attic floor joists spanning 15 ft.—way too long for 2x6s bearing live loads. The contractor was concerned that the framing might *rack* unless the exterior walls were bolstered beforehand.

▶ What's more, the walls, built of 2x4 studs 24 in. on center, had never been sheathed and so had virtually no *shear strength*. After removing the asbestos siding, the crew built out the 2x4s to create 2x6 walls, framed new rough openings for replacement windows, added hold-downs to anchor the walls to the foundation, insulated the walls, and covered the 2x6s with 1/2-in. CDX plywood sheathing. *Those* walls could support a second story.

▶ Strengthening the walls helped minimize interior-finish cracking once the attic demo began.

▶ Because so much of the work was done on the outside of the house, the family (living in a converted school bus/RV on the property) could continue using the house kitchen and bath with minimal disturbance.

The west face, gutted.

The Realities of Recycling

Job-site recycling is a great goal, but it's not without its problems.

First and perhaps most important, you'll need to get buy-in from your contractor. Recycling materials means handling them at least twice, which means increased labor costs. If recycling is important to you, be willing to assume the extra costs of doing it. Then work with the GC to figure out how to organize workflow and where to store recycled materials from the start. To do it right, someone should also remove nails before stacking the old lumber.

Second, you'll need room to store recycled materials until there is a large enough load to recycle. Be advised, the volume of dismantled materials can be prodigious.

Third, unless your site is roomy, stored debris can impinge access for workers and suppliers.

Fourth, you may need a couple of trash receptacles to keep recycled materials separate. Know, too, that a fair amount of stuff can't be recycled, such as tar paper and plaster lath.

Fifth, kids seem to find nail-infested piles of lumber irresistible places to play. Can you secure your site?

Despite all that, Mike Derzon was a determined and successful recycler. Evenings often found him pulling nails and stacking lumber so that he wouldn't be in the way during workdays. The house sits on an acre-plus, so there was plenty of storage room. The boys mostly stayed off the piles. The recycling facility was closer than the dump. And because the facility used some debris as fuel for its generators, recycling fees were considerably less than dump fees would have been. All factors considered, recycling was a tiny bit more than straight demolition would have been—and Mike and Robin had done the right thing.

tural engineer approved the old foundation for a second story, up they went. The plan also included adding a cantilevered bump-out to enlarge the dining room, stripping the exterior and reframing the walls, and insulating from the outside.

Structural solutions: The attic joists were undersize for live loads, so the floors were springy and undulated. Cut off the old 2x6 rim joists, install new 2x10 rims, and hang 2x10 joists off them to support the loads of the new second floor. (Leave the original 2x6 joists in place because the finish ceilings on the first floor are screwed to them.) Cantilever the dining room bump-out to increase its size without expanding the foundation.

Green touches: Recycle old lumber to keep it out of the landfill. Reuse all interior trim and preserve most of the interior surfaces. Install salvaged stairs. Build out the walls to make them 2x6s. Blow in R-22 fiberglass insulation. Upgrade all windows. Install a ground-source heat pump for maximum efficiency in a mild climate.

Parting thoughts: "We had heard all the horror stories about remodeling, how our marriage would never survive and all that," said Mike, laughing. "But it was a great experience. We chose friends to be our architect and contractor, and Dave and Russ have a lot of integrity. Plus

A cantilevered bump-out roughly 3$\frac{1}{2}$ ft. by 12 ft. created a dining room that comfortably seats more than a dozen guests. The fir flooring of the bump-out was rescued from the attic demo.

With the addition of full shed dormers on both sides of the roof, the attic was transformed into a light-filled, spacious suite with handsome recycled materials.

they're very decisive. Neither Robin nor I have a lot of spatial planning sense, so, basically, we turned the process over to them. And it worked out beautifully."

Project size: 700 sq. ft. (second-floor addition)

Architect: Russell Hamlet, Bainbridge Island, Wash., www.studiohamlet.com

Construction: Carley Construction, Bainbridge Island, Wash., www.carleyconstruction.com

Completed: 2010

The renovated west face with a full second story off the north end, at right. Because the kitchen is in that end of the house, it has become the main entrance, making it a logical place to add a new set of stairs to the second floor.

The south end of the house looks little changed from the street. Roof profiles gradually step up from the gable end to the shed dormer, minimizing the visual impact of a major remodel.

The new stairs in the north end contained some old elements: The posts, rails, balusters, and foot trim were salvaged from an old home near Seattle.

The 1957 Eichler, before.

A MID-CENTURY MODERN MAKEOVER

Taya and Stephen Shoup bought the house, first of all, for its location—its great schools, beckoning hills, and proximity to Pacific coast hiking trails. The house, designed by visionary builder Joseph Eichler, was also special. "We had grown to love Eichlers: their clean modern lines, single-level living, easy rapport with the landscape, and an indoor-outdoor flow that makes them feel much larger than they actually are."

Stephen, a green designer-builder, appreciated the house's simplicity: "Things were booming after the war, so architects looked for methods and materials that enabled them to build houses quickly and more affordably. Eichlers are all post-and-beam construction. You put a slab down, reinforcing it where you will put your posts. Simple beams. Simple flat roofs. Apart from window-stops to secure large pieces of plate glass, virtually no trim. For its day, the design was pretty avant-garde and, by historical standards, relatively fast to build."

After 50 or 60 years, though, even the best-built houses get a little tired. Yet when the couple decided to update their 1957 Eichler, they encountered a design shortfall that their far-seeing builder possibly never considered.

"The downside of the design," Stephen explained, "is that because it's a slab on grade, there's no crawlspace. And because the underside of the roof sheathing (2x6 tongue-and-groove) is also the finished ceiling, there is no attic. So if you want to move things—say, a kitchen or bathroom—you must either bust into concrete or pierce the roof membrane to run pipes and wires."

Neither is a great option. By developing a master plan, however, you can avoid the disruption and expense of breaking through a stubborn slab or a vulnerable roof more than once.

"Like any homeowners with a small child, a time crunch, and a tight budget, we sometimes did what was expedient rather than what made most sense long-term. But first assessing all major systems and then setting realistic goals was the only sane way to proceed."

The Shoups identified these immediate, big-ticket renovation goals:

▶ *Put in a master bathroom.* The original floor plan had only a small bath off the hall.

▶ *Upgrade the sewer lateral, which runs from the house to the sewer main.* This is a common municipal requirement when a property is purchased or a major remodel is in the works.

▶ *Route utilities, repair and insulate the roof.* (Recoating the roof is shown on pp. 452–453.)

▶ *Replace the house's single-pane windows with IGUs—insulated, tempered, double-pane glass.* The old windows were energy-losers and a safety concern to a family whose young child would soon be careening around the house on her tricycle.

Kitchen wall framing, viewed from the living room. Running wiring and plumbing through one wall reduced the number of holes through the roof membrane.

The kitchen, framed and Sheetrocked, as seen from the dining room.

Those concerns addressed, even a major kitchen remodel required only slight changes to the original floor plan. Removing a small section of wall on the west end of the house improved traffic flow from the kitchen and introduced more light into the living room. Another small but significant innovation was adding an opaque glass backsplash behind the sink to allow light from the kitchen to illuminate part of the living room. Lastly, they added a small bump-out for an audio-visual cabinet next to the fireplace—easily accomplished thanks to the deep overhangs along the west wall. In other words, the roof was already there, so they were able to slip in a couple of short walls and a section of floor, and be done with it.

Taya, a landscape architect, had these thoughts, "We love living in the house. It's low-slung so a lot of sun gets into the backyard. There are raised vegetable beds, a large lawn, and open spaces for an active child. Thanks to the sliding doors, it's easy to wander in and out, whether you're Chloe the dog, Mavis the cat, or our daughter doing laps on her skateboard."

Project size: 1,500 sq. ft. (whole house renovation)

Design and construction: Stephen Shoup, building Lab, Oakland, Calif., www.buildinglab.com

Landscape design: Taya Shoup
Completed: 2013

The finished kitchen seen from the dining room. The walk-through, at left, was added to improve flow between the rooms; the opaque panel behind the sink admits more light into the living room while screening kitchen activities.

Because the house's low profile allows a lot of sun into the backyard, there's plenty of light and room for plants and children to flourish.

The south face of the house, facing the street, is a subtle balance of openness and privacy. The large window at left was added to let more light into the dining room and kitchen.

Floor Plan, after Renovation

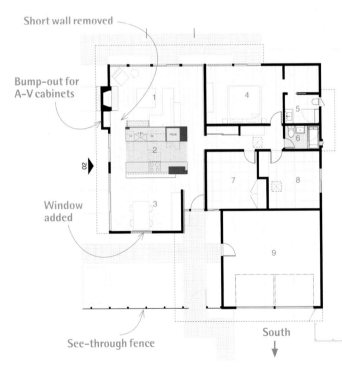

Short wall removed

Bump-out for A-V cabinets

Window added

See-through fence

South

While there is a strong sense of community, individual houses have privacy at the back of the house. The north wall that is all glass lets the family engage more with the landscape.

1 = Living
2 = Kitchen
3 = Dining
4 = Master bedroom
5 = Master bath
6 = Bath
7 = Office
8 = Bedroom
9 = Garage

Tools

3

The tools in this chapter are a subjective collection. Yours should be, too. Choose tools that are right for the scope of your renovation, your experience, your storage space, your budget, and your physical strength. Because tools become an extension of your hand, shop for tools that fit your hand well, have a comfortable grip and controls that are easily reached, and are balanced to minimize muscle strain. More specialized tools are presented in pertinent chapters.

Tool Safety

Few things will slow a job down more dramatically than an injury, to say nothing of the pain and expense involved. Don't be afraid of tools, but respect their power and heed their dangers. The following suggestions come from professional

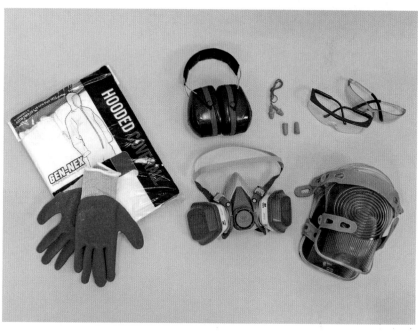

Safety equipment. Clockwise from lower left: work gloves, Tyvek hooded coveralls, over-ear hearing protection, ear plugs, safety glasses, gel knee pads, 3M 6000-series respirator mask.

Many power tools promote worker safety by attaching directly to vacuum systems, such as this Festool orbital sander. Further, many state regulations now require that hazardous dust be captured as it is generated, such as when you're dry drilling, cutting masonry, sanding lead-paint surfaces, and so on.

Lightweight gloves are much cheaper by the case. Rubberized cloth gloves, for example, are about $1.50 per pair in bulk and will last three to five days under normal heavy use.

builders and tool manufacturer owner's manuals, so take them seriously. Always comply with safety tips provided in tool-operating manuals. Some may pertain to features or hazards unique to particular makes and models.

ESSENTIAL SAFETY RULES

Keep the work area clean. This is especially important during demolition.

▶ **Clean up as you go.** You'll accomplish more if you can move freely.

▶ **Pull nails from removed boards at once.** If you're lugging lumber across a work site or descending a ladder in a hurry, you don't want to step on a nail that's sticking up.

▶ **Clean up when you complete each work session.** This seldom takes long and usually means fewer misplaced tools. Plus it lets you get right to work the following morning.

▶ **Set up temporary walls** (see p. 208); they're a great way to isolate the mess and, as important, to provide visual separation and privacy for both workers and family members living in a house being renovated.

Wear safety gear. Wearing safety equipment isn't fun: Tyvek suits are hot, and respirator masks make your safety glasses fog up. It takes some discipline to keep wearing the gear. But protecting your vision or hearing—and avoiding worse injuries—is well worth the discomfort. Building professionals are *required* to wear safety gear, so it's common sense that DIYers should, too. Having safety gear on hand is half the battle, so stock up before you start work.

Buy comfortable gear. If safety gear isn't comfortable, you'll remove it before long. And that's the point at which it won't protect you. Spend a little extra for gear that fits so well that you'll want to keep wearing it.

🚫 **Disconnect electricity.** Be sure to cut off the electricity to the areas you'll disturb. Then use a voltage tester to double-check that current is off in affected outlets. Also avoid cutting or puncturing wires hidden behind wall surfaces.

Plan the job and pace yourself. The job will go more safely and smoothly if you assemble your tools and materials beforehand. Take regular breaks, and you'll stay sharp longer.

Ladder Safety

Don't scrimp on ladders. For greatest safety and durability, buy a type IA, which is a construction-grade ladder rated for 300 lb. Avoid household-grade ladders, which can be unsafe.

Although fiberglass ladders are more expensive than those made from wood or aluminum, most pros prefer them because they are sturdy and nonconductive. Fiberglass ladders are lighter than wood but heavier than aluminum. Wood stepladders are OK for indoor use, but wood ladders used outside can deteriorate quickly. Aluminum ladders are a reasonable compromise in price and weight but are the most electrically conductive of the three. Note: All ladder materials can conduct electricity if they're dirty or wet.

Safe working lengths of ladders are always less than their nominal lengths. When using an 8-ft. stepladder, for example, go no higher than 6 ft., and never stand on the top step—which usually has a label stating, "This is not a step." Likewise, a 32-ft. extension ladder is only 26 ft. to 28 ft. long when extended. Ladder sections overlap about one-quarter, and a ladder leaning against a building should be set away from the wall about one-quarter of the ladder's extended length.

Ladders must be solidly footed to be safe, especially extension ladders. After setting up the ladder so that its sides are as plumb as possible, stand on the bottom rungs to seat the feet. Adjustable leg levelers, as shown, are available for leveling ladders on slopes. If you're at all unsure about the ladder's footing, stake its bottom to prevent it from creeping.

Finally, as you work, always keep your hips within the ladder's sides.

Don't work when sick. Take the day off when you're excessively tired, preoccupied, or taking any substance that impairs your judgment.

Operate tools safely. Follow the instructions in the owner's manual provided by the manufacturer. Never remove safety devices. Avoid electrical tools whose wires are frayed, cut, or exposed. Never force tools—saws can kick back and high-torque drills can knock you off a ladder.

Don't work alone. When you're on a steep roof or a tall ladder, make sure someone is close by. He or she needn't be working with you but should be within earshot if you need help. If you need to work alone, have a cell phone handy.

Keep work areas well lit. Don't work where the light is poor. If you disconnect the power to a work area, run an extension cord and droplight to it.

Isolate the danger. Keep kids away from work sites. If you store equipment at home, lock up power tools, dangerous solvents, and the like.

Tools to Own

Consider buying most of the safety equipment listed here.

SAFETY EQUIPMENT

Hearing protection will prevent permanent ear damage and reduce fatigue while using power tools. There are a number of styles, from over-ear protection (by far the best) to disposable foam plugs, which are easily stowed in your tool bag. Properly fitted, hearing protection should reduce noise 15 decibels (db.) to 30 db. Look for models that meet American National Standards Institute (ANSI) S3.19 specifications.

Eye protection is a must when you're using power tools or striking nails or chisels with a hammer. Safety glasses or goggles that meet ANSI/ ISEA Z87.1-2010 specs are strong enough to stop a chunk of metal, masonry, or wood without shattering the lens. (Most lenses are polycarbonate plastic.) Get eyewear that won't fog up; you can also get combination safety glasses/ sunglasses with UV 400 protection.

A hard hat won't protect you if you don't wear it, so find one that fits well. There are basically two types: Type I protects the top of your head, whereas type II (ANSI Z89.1-2003) offers some additional protection if a blow to the head is somewhat off-center. Both are invaluable when you're handling objects overhead or someone is working above you.

Work gloves are essential when handling caustic, sharp, or splintery materials. Our favorites are lightweight rubberized cloth gloves, which protect your hands while framing or doing finish work, yet are sufficiently tactile to pick up a coin. Gloves will also keep your dirty fingerprints off finish surfaces. Home centers also sell boxes of latex-free nitrile plastic gloves that are inexpensive, reasonably durable, and quite flexible.

A respirator mask with changeable cartridges can prevent inhalation of toxic fumes or dust. The 3M mask shown on p. 51, for example, could protect you when working around toxic materials such as asbestos or lead paint, or when painting with oil-based paint. For most construction dust, inexpensive disposable N-rated respirators are fine; where there may be volatile hydrocarbons, use R- or P-rated masks. To be sure you're using the right mask, consult www.cdc.gov/niosh or product-safety data sheets.

Important: Always read the directions to find out how long cartridges will remain effective before they must be changed. Also, masks should fit tight, so when they get old and stiff, replace them with a new mask whose fittings are supple and will fit well to your face.

Knee pads come in a wide range of styles and costs—from inexpensive foam or rubber to expensive leather or gel pads covered with fabric or hard plastic. Match the knee pad to the task; you can't work all day on your knees without a good pair. Our favorites are gel knee pads, which are soft and extremely comfortable. One veteran floor installer likened gel pads to "a new mattress for your knees."

A headlamp (battery-powered) is invaluable in dark spaces where you need both hands free.

A safety harness should be attached to solid framing when you are working on roofs with a 6/12 pitch or steeper, over open framing, or on any other high, unstable workplace. (Page 92 shows a typical harness.)

A first-aid kit should be secured to a prominent place at the work site so you can find it quickly when you need it. Construction crews should have professional-grade kits that can stabilize major injuries, and everyone should take a first-aid course at some point. Must-haves for any kit: an eyewash cup, fine-point tweezers to pull splinters, chemical cold packs, and, of course, adhesive bandages, gauze pads, tape, and antibiotic ointment.

TALKING Productivity

Having Internet access on the job is a tremendous productivity tool. Notes Chip Harley: "I use a smartphone and Google docs for everything: planning, ordering materials, coordinating suppliers and subs, noting client change orders, scheduling inspections—you name it.

"It used to be that getting accurate information about, say, tub hardware meant sending someone to a plumbing supplier, waiting around while a clerk thumbed through a catalog, printing a spec sheet, and hoping that the item was in stock. Now I can get the same info in two minutes—dimensions, prices, availability—and it's all accurate and up to date.

"I upgrade lists for all my jobs every two weeks to stay current—or to get a jump on what's ahead. If my crew is starting a foundation pour, for example, I am already thinking ahead to framing, siding, calling inspections, ordering fixtures, and so on. So having all those lists with me wherever I go is invaluable."

remediation of lead-based paint in houses built before 1978, when the sale of lead-based paints was banned.

Tarps. Buy at least one 9-ft. by 12-ft. tarp and one 45-in. by 12-ft. runner. Get good-quality, heavy canvas duck. Paint will soak through cheap fabric tarps, and sheet plastic is too slippery to work on.

MEASURING AND LAYOUT TOOLS

You may not need all of the following tools, but it's good to know what each can do.

A framing square with stair gauges is a basic layout tool for plumb and level cuts. It also enables you to set the rise and run for stairs as well as to make repetitive layouts such as for rafter ends.

Mason's string has many uses, whether to support a line level or to temporarily tie things together.

An adjustable square is a smaller version of a framing square and is somewhat less versatile.

A stud finder enables you to locate studs you need to find or want to avoid. Stud finders range from simple magnets that detect screws or nails holding surface materials to framing to electronic multiscanners that can detect wood or metal studs, plumbing pipes, and live electrical cables.

A small combination square fits easily into a tool belt and enables quick and accurate 45° and 90° cuts on small pieces. It also doubles as a depth gauge for getting notches to a certain

PRO TIP

Tarps won't protect finished wood floors from dropped tools and the like. Protect floors by putting down 1/8-in. Masonite or a heavyweight paper such as Ram Board, which comes on wide rolls (use duct tape over the seams).

CONTROLLING THE MESS

Cleanup tools. You'll need a household broom, a push broom, a dustpan, a heavy-duty rubber garbage can, a flat shovel for scooping debris, and a large-capacity (12-gal. to 16-gal.) wet/dry shop vacuum. You'll find Dumpster tips at the end of this chapter.

HEPA vacuums are now essential on any renovation job; their very fine filters can capture the tiny particles that cause respiratory problems. They are also essential to the containment and

Measuring and layout tools:
1. Framing square with stair gauges
2. Mason's string
3. Adjustable square
4. Stud finder
5. Combination square
6. Adjustable bevel gauge
7. Try square
8. Chalkline box
9. Folding rule with sliding insert
10. Tape measure
11. Compass
12. Swanson Speed Square

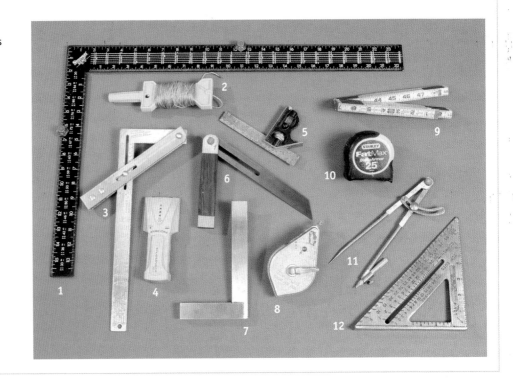

Stud finders. From left: rare-earth magnet, Hanson magnetic stud finder, Zircon StudSensor, Bosch digital multidetector. The two at left use magnets to detect screws or nails in stud edges; the two electronic devices at right use capacitance sensors to detect wood or metal studs—and sometimes pipes and wires—beneath the surface.

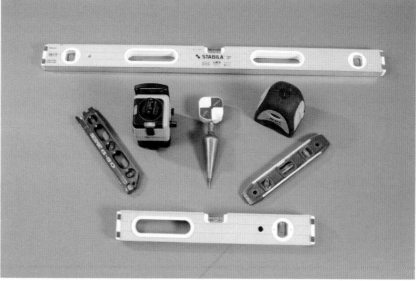

Levels and plumbing devices. Top: 4-ft. spirit level; center, from left: torpedo level with laser, rotating laser, plumb bob with Gammon Reel, five-beam self-leveling laser, and standard torpedo level; bottom: 2-ft. spirit level.

depth, setting door stops to jamb edges, and establishing reveals.

An adjustable bevel gauge copies odd angles and transfers them to workpieces.

An angle-divider allows you to measure, transfer, and bisect angles; essential for cutting accurate miter joints when surfaces do not meet at right angles.

A try square is a precise tool that's more of a shop or bench tool, handy for making sure that a tablesaw blade is perfectly square to the table.

A chalkline contains line and powdered chalk. It's used to snap straight layout lines on sheet goods and make layout lines for framing. The line itself can double as a stringline. And, in a pinch, the box and line can also serve as a plumb bob.

A folding rule with sliding insert is great for accurate, inside measurements such as inner

cabinet or window widths. Because the folding rule is rigid, it will hold the dimension you set. The sliding brass insert doubles as a depth gauge.

Tape measures are a must. A 16-ft. tape will do for most jobs, but the wider tape of a 25-ft. model can span 7 ft. or more without collapsing, allowing you to take an approximate reading across a span opening.

A compass draws circles or doubles as a scribe so you can fit flooring or sheet materials to the curved or irregular profile of a wall, cabinet, or baseboard.

A Swanson Speed Square enables quick 45° and 90° angle layouts. And with a little practice, you can set rafter pitches. This indestructible tool fits snugly in any tool pouch and is easily one of the most popular tools since the 1970s.

PLUMB AND LEVEL

Self-leveling lasers and laser levels have become so affordable that they have largely replaced the older tools below. Lasers give you a consistent level line around a room from which you can work.

A plumb bob takes patience to use, but it's a compact, accurate tool. The plumbed string also is a useful reference line you can measure out from. The Gammon Reel shown in the photo above right automatically reels in the string so it can't tangle up in a tool pouch.

A 4-ft. spirit level is a good, all-purpose level, long enough to level accurately across joists and check for level and plumb of door and window casings.

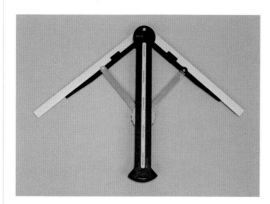

Angle-divider

Circular saws. From left: sidewinder, beam saw, and worm-drive.

Jigsaw and reciprocating saw.

A 2-ft. spirit level enables you to level windowsills, door head jambs, and other tight spaces where a 4-ft. level won't fit.

A standard torpedo level is small enough to fit in a tool pouch. Although it's not as accurate as a longer level, it's good for a quick reference in tight spaces. Magnetic versions can stick to cast-iron pipe and metal conduit.

A torpedo level with a laser has the same limitations as any torpedo level, but its laser allows plumbers to set the tool on a pipe that's pitched correctly and extend the pitch to other pipe sections.

POWER SAWS

Circular saws are generally characterized as either *worm-drives* or *sidewinders* and are often called Skilsaws, the name of a popular brand.

Professional builders, especially on the West Coast, favor worm-drive circular saws for cutting framing lumber. They tend to have bigger motors and more torque. They also spin slower, bind less, and run quieter than sidewinders. If you're right-handed, worm-drive saws make it easier to see the line you're cutting.

Sidewinders are generally lighter and more compact, so they're easier to handle and a good choice for the occasional builder. But because the blade is turning at 90° to the motor shaft, it is more likely to bind if your cut wanders off the line. For that reason, a rip fence is a nice accessory.

Beam saws are called "sidewinders on steroids" because they can accommodate 10-in. blades that cut smoothly through 4x lumber in one pass. It's not a must-buy item, but, wow, what a tool!

DEMOLITION Blades

Bimetal demo blades for recip saws are designed to cut almost everything they meet. They are not, however, indestructible, so here are a few tips on choosing and using them. The average demo blade has 6 to 8 teeth per inch (tpi) and is roughly 0.035 in. thick. Thicker, heavier blades will last longer because they wander, bend, and break less than thin ones.

As best you can, match teeth to the task: 3-tpi blades chomp through wood, but their big teeth will be useless if you hit a nail. Conversely, 12- or 14-tpi blades may soon clog and overheat. (Overheated blades lose their temper.) Hence, 6- to 8-tpi blades will be most versatile. Saw speed and action are crucial, too: A reciprocating saw with orbital action clears debris best from sawkerfs and blade teeth for faster cuts and cooler blades. If a blade isn't cutting, replace it. Never force any saw.

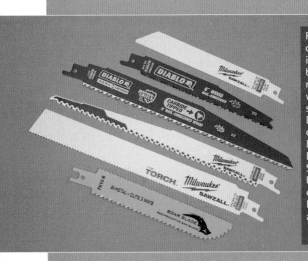

Reciprocating saw demo blades can cut through nails embedded in lumber without destroying the blade. From top: 6-in. Sawzall metal blade will handle nails well, but its fine teeth will be slow going through wood; 6-in. Diablo demo blade can handle embedded nails but won't last as long as its carbide-tipped 9-in. Diablo demo cousin; two 9-in. Sawzall blades can handle nails, but the Torch's finer teeth won't cut quite as aggressively; Boar Blade has different size teeth above and below.

For circular-saw blades in general, the more teeth, the smoother the cut. If you buy just one, make it carbide tipped; it will stay sharp far longer and give cleaner cuts than other types. There are specialty blades for almost anything you'd want to cut: tile, concrete, metal, and wet or pressure-treated lumber (these blades have a Teflon coating). Remodelers' blades cut through wood and the occasional nail without being damaged.

A reciprocating saw, also called a Sawzall after a popular make, is *the* indispensable demolition saw. A marvel in tight spaces, it can remove old pipes, cut through studs or joists, or, with a bimetal demolition blade, cut through nails and studs in one pass. (Blades break, so get extras.) A "recip" saw also is useful in new construction to notch studs for pipes, cut plywood nailed over rough openings for windows and doors, and perform many other tasks.

Features to look for include orbital action, which clears sawdust out of a blade kerf, speeds cutting, and, by reducing heat buildup, extends blade life; variable-speed control for cutting different materials; quick blade-changing mechanism; good sightline so you can see cuts; and reduced vibration.

Jigsaws, sometimes called sabersaws, are useful for notches, curving cuts, and odd-shaped holes. Typical uses include cutting out holes for sinks in countertops and holes in cabinet backs so pipes or ducts can pass through. The blades are thin and prone to snap, so buy extras.

OSCILLATING MULTITOOLS

Oscillating multitools, cordless or corded, have become an essential renovation tool because they are unequalled for cutting materials in place. Often called Fein tools because that company has dominated the niche for decades, multitools have blades that vibrate rather than spin. Thanks to precise (3.2°) oscillations per minute (OPM), they can make fine-kerf, controlled cuts where it would be hard or impossible for most cutting tools to fit. Use multitools to remove a small amount of wood from the bottom of a door jamb; make fine, multiple cuts to fit a threshold to complex trim; or remove a section of finished flooring.

These tools do more than just cut. Diverse attachments allow them to grind, sand, and scrape. They can remove grout without damaging surrounding tiles, scrape out stubborn glazing compound without harming delicate window muntins, and sand into a corner. Cutting blades

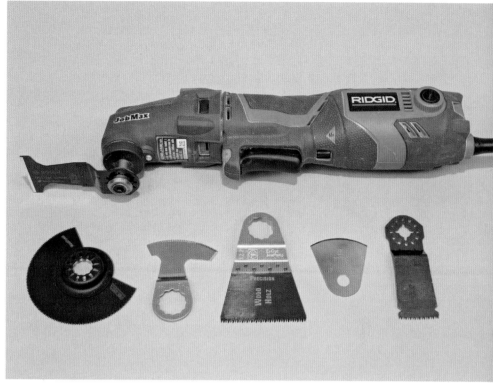

Oscillating multitools and accessories are perfect for making fine cuts in materials that have already been installed. Accessories: on tool, carbide multimaterial plunge cut; bottom, from left: wood/drywall segment blade, boot blade for removing grout, precision E–cut blade (double row of Japanese profile teeth), scraping blade, and plunge cutter for wood.

come in a range of materials, including (from more durable to less): bimetal (BIM), high-speed steel (HSS), and high-carbon steel (HCS). Multitools have variable-speed controls; Fein's MultiMaster 250Q runs 12,000 to 21,000 OPM. Operate cutting blades at highest speeds, scraping blades at medium-high, and sanding attachments at low to medium.

There are a few downsides, most notably (and painfully) the cost of replacement blades, which tend to be proprietary and thus not compatible with other makers' arbor patterns. Some Fein blades can cost $50 a pop, so choose and use them carefully. There's a bit of a learning curve for these tools, and each specialty blade cuts a little differently. Specifically, when starting cuts, blades tend to skitter around until you get the hang of it. Practice on scrap or in an out-of-the-way place. Don't try to make the whole cut at once. Start in a corner of the area to be cut and roll/ease the blade into place. When using multitools, hearing and eye protection are a must, and respirators are highly recommended.

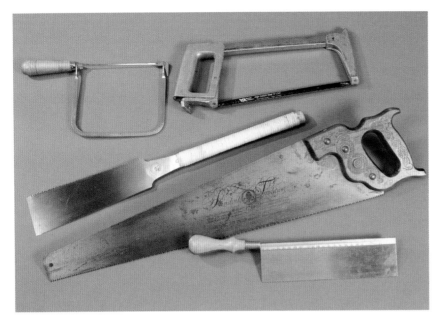

Handsaws. Clockwise, from lower right: dovetail saw, crosscut saw, Japanese saw, coping saw, and hacksaw.

HANDSAWS

The following inexpensive handsaws are handy to have.

Coping saws cut curves into any thin stock, although their primary use is coping trim so intersecting pieces fit snugly. They take both metal- and wood-cutting blades.

A hacksaw is most often used to cut metal, especially bolts or nails. Sawblades will last longer if you use the full length of the blade.

A Japanese saw cuts on the pull stroke. Its thin, flexible blade is perfect for cutting flush shims and other thin stock. Most are two-sided, with rip and crosscut teeth on opposite edges of the blade.

A handsaw is still worth having in your toolbox, preferably a 10-point crosscut saw. Even if you depend primarily on a circular saw, a handsaw is handy for finishing cuts that don't go all the way through a rafter or joist.

A dovetail saw makes clean crosscuts in small molding, doorstops, and casing beads.

A keyhole saw can cut holes in drywall for electrical boxes, without predrilling.

ROUTERS

Full-size routers are probably too expensive for casual remodelers, but trim routers and rotary tools are versatile and reasonably priced. Safety goggles are a must with any router.

Laminate trimmers also are called trim routers. In addition to trimming laminate edges, these lightweight routers are great for mortising door hinges and strike plates.

Plunge routers can be lowered to precise depths in the middle of a workpiece, making them ideal for wood joinery, edge shaping, mortising door hinges, and so on.

Dremel variable-speed rotary tools can dislodge tired tile grout and remove stubborn paint from beaded or ornate woodwork. There are hundreds of specialized accessories for this tool.

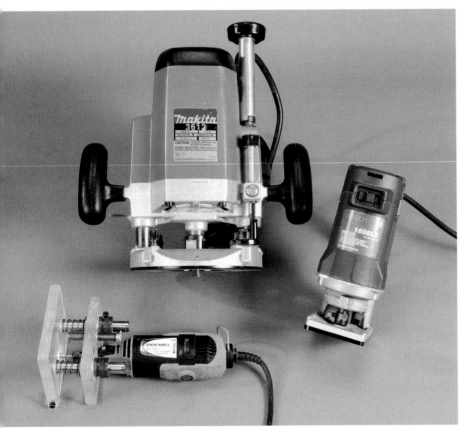

Routers. Clockwise, from top: large plunge router, laminate trimmer, and Dremel rotary tool with plunge base.

HAND CUTTING AND SHAPING TOOLS

Power tools can do a lot, but you often need a hand tool to finish the job.

Chisels clean up the corners of hinge mortises where a router can't reach and quickly notch plates so washers sit flush. Be sure to sheathe cutting edges so they stay sharp and don't cut you when you reach for them.

Mallets can strike chisels without damaging their handles.

Utility knives are indispensable. Quick-blade-change knives dispense fresh blades so you don't need to unscrew the knife's body. Don't use knives with cheap, snap-off blades (often sold at stationery or office-supply stores) to cut construction materials; they can break unexpectedly and injure you.

Rat-tail files smooth and enlarge holes and create an oval slot after two holes are drilled close to each other.

Flat files take burrs off newly cut bolts and the like, so you can start a washer.

Four-in-one rasps contain two flat and two curved rasps in one wood-shaping tool.

Block planes shave off tiny amounts of wood from door edges, casings, and other thin stock, allowing tight, final fits of materials.

Bullnose planes can fine-shave wood edges in tight places. The blade is the same width as the sole of the plane. You can remove the bullnose front piece, allowing you to plane into a corner.

Bahco by Snap-On carbide scrapers are not intended to shape wood, but their blades are so sharp that you can. Use them to remove dried putty or excess Bondo or to clean up the spurs of medium-density fiberboard (MDF) that screws sometimes kick up.

PRO TIP

Claw hammers are designed to pound nails—not other hammers, wrecking bars, or chisels. And claw hammers are generally too weak for extensive nail pulling or prying apart lumber. For those jobs, use a cat's paw or a wrecking bar.

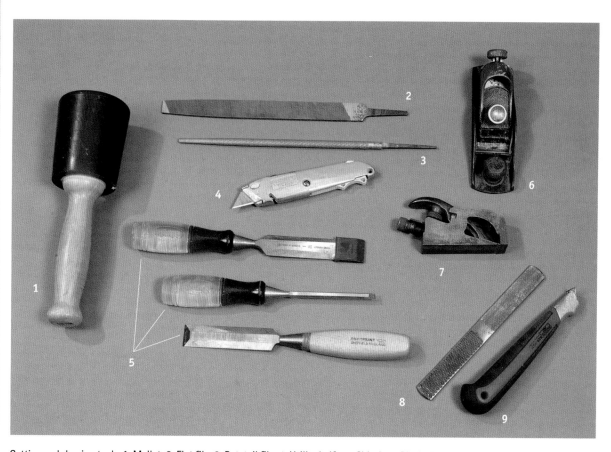

Cutting and shaping tools: 1. Mallet; 2. Flat file; 3. Rat-tail file; 4. Utility knife; 5. Chisels; 6. Block plane; 7. Bullnose plane; 8. 4-in-1 rasp; 9. Carbide scraper

Cordless tools are particularly well suited to work in confined spaces where tool cords can get hung up. If there's no cord to accidentally cut through or extension cords to come apart, there's also less risk of shock. Cordless tools are often the only practical choice when you need to turn off the power or if you're working where outlets are scarce—or not yet energized.

Things to consider

Buy a tool that fits your project's scope, your skill level, and your budget. Research tools thoroughly online before you buy: you'll spend wisely and, more to the point, you'll learn about features you may not know existed. Pro tool reviews are especially helpful in this respect. Then hit the home centers and handle the tools. How long can you hold the tool aloft? Is the grip comfortable? Are there LED lights to help you see what you're drilling or cutting? Is there a spring-loaded chuck to quickly change bits? Does the tool feel well balanced?

▶ **VOLTAGE** is a rough measure of how much power a tool can deliver; most professional-grade cordless tools are 18v to 20v systems, capable of roughing in wiring, boring through studs, drilling into concrete, or handling an extensive rehab.

▶ **AMP HOURS (Ah)** indicate how much battery run-time the tool has: 2.0 Ah means that the battery can deliver 2 amps of current for 1 hr. Pros favor 4.0 Ah (or greater) battery packs. Cordless tools use lithium-ion (Li-ion) batteries, but battery platforms are not interchangeable—you can't swap, say, a DeWalt battery into a Makita drill. For this reason, most people stay with one brand as they add tools.

▶ **BRUSHLESS MOTORS.** These days, better-quality cordless tools have brushless motors that are more efficient (less friction), have longer run times, are generally maintenance-free, and are typically more compact. They are definitely worth the money.

▶ **COMBO KITS.** All cordless tool makers offer kits with some combination of a drill/driver, circular saw or reciprocating saw, two batteries, and a charger. Good-quality kits tend to offer considerable savings above buying tools individually, and you can swap batteries between the tools in the kit. Research tool kits carefully, however, because the quality of individual tools within a kit can vary greatly.

Milwaukee 18v cordless power tools. Clockwise from bottom: Circular saw, hammer drill, Hackzall reciprocating saw, HoleHawg right-angle drill, and impact driver.

Impact DRIVERS

Impact drivers have taken the construction world by storm. Screw guns are fine on tasks where there's little resistance—say, when hanging drywall or driving self-tapping screws into metal studs. But where there's greater resistance, an impact driver with a star bit can knock in screws at lower RPMs, so bits are less likely to slip.

Instead of the gears and clutch of drill-drivers, impact drivers have pairs of hammers and anvils that convert the continuous torque of a motor into intermittent (pulsed) rotational force. More powerful and less prone to kickback than drills, impact drivers are well-suited to jobs with tight spaces and limited accessibility, as when driving lag screws into ledger boards or sinking long structural screws (pp. 81–83) into posts or beams.

Professional builders favor 18v or 20v impact drivers. Better-quality models have brushless motors (more compact, longer run times), spring-loaded hex chucks that enable you to install and release bits quickly, LED lights to illuminate work, multiple speed/torque settings, high-capacity battery packs, and quick-recovery chargers. (Some chargers have USB ports so you can also charge your cell phone.)

Impact drivers are beefy enough to drive bolts into masonry and compact enough to fit into tight spaces.

Corded drills. From top: ³/₈-in. pistol-grip drill and ¹/₂-in. right-angle drill.

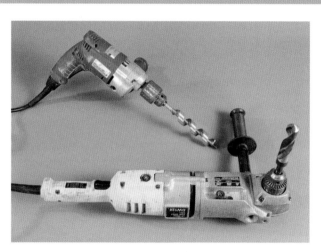

DRILLS

A ³/₈-in. cordless drill is a must. These days, builders use screws to install almost everything from exterior trim to decks and drywall. For most renovators, a 12v or 14.4v drill is optimal; there are models with more voltage and bigger chucks, but 12v or 14.4v has a good weight-to-power ratio. Get a reversible, variable-speed model with a keyless chuck, adjustable clutch, and an extra battery.

A ³/₈-in. corded, pistol-grip drill has the sustained run time that cordless drills lack, more power, and a side handle to help you control its torque. It drills 1-in. or 1¹/₂-in. holes easily, but use a ¹/₂-in. right-angle drill if you're roughing in plumbing lines.

A ³/₈-in. close-quarter cordless drill is best for tight spaces like cabinet interiors. Its right-angle configuration extends your reach when you are hanging upper cabinets.

DRILL BITS

A standard drill nest contains ¹/₁₆-in. to ¹/₄-in. twist drill bits. From there, you're on your own. The following specialized bits are quite useful.

Vix bits have spring-loaded drives that accurately center holes for hinges, striker plates, and window pulls.

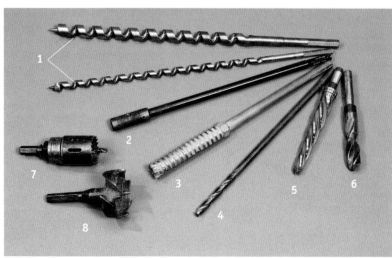

Drill bits: 1. Self-feeding auger bits; 2. Drill bit extension; 3. Rebar-cutting bit; 4. Long twist drill bit; 5. ¹/₂-in. by ³/₄-in. reamer; 6. ³/₄-in. twist drill bit; 7. Hole saw; 8. Plumber's bit (wide self-feeding bit)

Drilling and screwing accessories:
1. Allen wrenches
2. Magnetic bit holders
3. Extension bit holder
4. Flexible bit holder
5. Stubby screwdriver
6. 4-in-1 screwdriver
7. Ratchet-handle bit driver

DRILLING AND SCREWING ACCESSORIES

Magnetic bit holders magnetize drill bits so screws don't fall off. Some types also have a collet that keeps the screw centered as you drive it in.

Extension bit holders enable you to drive screws where drill-drivers won't fit.

Flexible bit holders drive screws at angles drill/drivers can't reach.

Drill bit extensions enable you to drill deeper with the bits you've got.

A ratchet-handle bit driver can turn Phillips- or hex-head screws in tight spaces. They're also great for turning the leveling devices on refrigerator legs.

A stubby screwdriver has a reversible bit: one side Phillips head, the other slotted.

A 4-in-1 screwdriver is the screwdriver to own if you have only one.

Allen wrenches tighten Allen screws on a lot of tools, including drill bit extensions.

CLAMPS

Quick-release bar clamps are a second set of hands on the job site. Use them to hold work to a bench, temporarily join two boards, align stair balusters, or mock up rafter pairs. Their rubber jaws won't mar surfaces on fine work.

Standard bar clamps slide jaws to approximate position and use a threaded handle to draw materials tightly together. They're a little slower than quick-release clamps, but they apply more force.

Spring clamps are the quickest to operate for relatively thin materials that don't require an especially tight grip.

Hand screws apply even pressure to a relatively broad area. Excellent for glue-ups, they hold work well and won't damage wood. Open and close them by using two hands, almost like pedaling a bicycle.

C-clamps apply a strong force and are especially suitable when the workpiece absolutely mustn't move. Insert scrap wood between the jaws and workpiece to protect it from damage.

Clamps. Clockwise, from lower left: spring clamp, bar clamp, quick-release bar clamp, C-clamp, and hand screw.

Self-feeding auger bits drill through posts for bolts and through wall plates for hold-downs. A 12-in. by ¼-in. auger doubles as an exploratory bit.

Hole saws drill large-diameter holes in finish materials such as doors and countertops. The pilot bit in the middle emerges on the backside first, so you can retract the bit and center it to finish drilling from the other side. This lets you avoid splintering wood in a "bust-through."

A plumber's bit is a wide, self-feeding auger good for rough-in framing work.

Reamers are tapered bits that enlarge an existing hole in metal or wood.

Large twist drill bits, also called aircraft bits, are best suited for drilling metal.

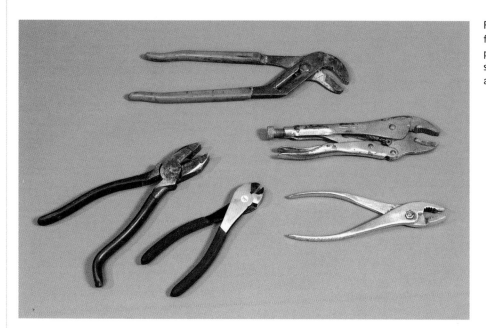

Pliers and cutters. Clockwise, from top: Channellock slip-joint pliers, Vise-Grip locking pliers, slip-joint pliers, side cutters, and lineman's pliers.

PLIERS AND CUTTERS

Slip-joint pliers are old reliables. Our grandfathers used pliers of the same design.

Channellock slip-joint pliers have long, offset handles and jaws that open wide for the slip-nut under the kitchen sink . . . or wherever.

Vise-Grip pliers have an adjustable tension mechanism that lets you lock the tool's jaws around the workpiece, such as a stripped screw. They can double as a temporary clamp, but don't overtighten.

Side cutters are designed to cut wire or small nail shanks. But they're also great nail pullers if you don't squeeze too hard.

Lineman's pliers are an electrician's mainstay, great for twisting and cutting wire.

Aviation snips, also known as tin snips, cut sheet metal. Use them for flashing and ductwork.

HAMMERS

Choose a hammer with a grip and weight that feel right for you. Bigger heads and longer handles can deliver greater impacts when nailing and so require fewer swings to drive nails. But they also require greater effort to use and may cause tendonitis.

Twenty-six-ounce framing hammers are as big as anyone needs. Titanium framing hammers are in vogue these days because they transfer less shock to your arm, although more to your wallet.

Twenty-ounce framing hammers are light enough to double as trim hammers. But, truth is, pros prefer trim guns (pneumatic nailers) for fin-

ish work because they free up one hand to steady the work and don't ding the trim as hammers do.

Sixteen-ounce finish hammers are fine for a small amount of trim.

Hand sledges are handy for knocking shoring or partitions a few inches over and for breaking loose stubborn foundation forms.

Hammer tackers are a quick way to staple building paper, insulation, and sheet plastic.

Hammers. Left: hand sledge; top to bottom: 26-oz. framing hammer, 20-oz. framing hammer, 16-oz. finish hammer, and small hand sledge.

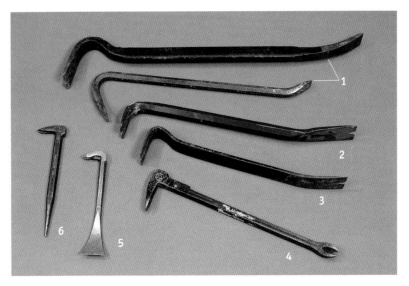

Wrecking and pry bars:
1. Crook-neck wrecking bars
2. L-bar
3. Flat bar
4. Cat's paw
5. 8-in. pry bar/nail puller
6. Cat's paw with punch point

WRECKING AND PRY BARS

🚫 Disconnect plumbing pipes and electrical cables in areas about to be demolished—and check with a voltage tester to be sure the power is off. Be methodical and work slowly.

Wrecking bars have differing lengths and end configurations. Most are crowbars with crooked ends for better leverage. The longer the bar, the better the leverage.

L-bars are wrecking bars with a flat, L-shaped end instead of a crook. Drive the L into lumber that's nailed together and twist the tool to pry the pieces apart.

A flat bar (also called Wonder Bar or handy bar) is the best tool to ease off delicate trim without damaging it. Pry the trim up gradually along its length.

A cat's paw is the tool of choice if you're pulling a lot of nails out of framing. It bites into wood pretty deeply, so don't use this tool on trim.

A small cat's paw with a punch point is small enough to remove finish nails, and its pointed end doubles as a nail punch.

An 8-in. pry bar/nail puller lifts trim gently and pulls finish nails.

MISCELLANEOUS TOOLS

Sawhorses support work at a comfortable height. The metal-leg variety, which nail to lengths of 2x4, are sturdy and easy to collapse and store. The Black & Decker Workmate has an integral clamp in its benchtop; it also folds flat for compact storage and transport.

Electrical and plumbing tools are covered in other chapters. But don't be without a voltage tester to make sure the power is off, slot and Phillips-head screwdrivers with insulated handles, and needle-nose and lineman's pliers. And every toolbox should have a pipe cutter, large and small adjustable open-end wrenches (commonly called Crescent wrenches), slip-nut pliers, and a pair of pipe wrenches.

Earth tools include a round-point shovel, pickax, hatchet or ax, and wheelbarrow.

Tools to Rent

Most contractors own the tools listed in this section, but occasional users should probably rent them. The decision depends on how often or how long you may need the tools and how passionate you are about collecting them.

SAFETY, SCAFFOLDING, AND JACKS

Inside or out, scaffolding gives you secure footing and peace of mind. Instead of hanging precariously from a ladder, you can concentrate on what you're doing. That said, anyone who's not comfortable working at heights shouldn't. As one contractor put it, "If it feels unsafe on a roof, it probably is."

Pipe-frame scaffolding. Have the rental company set up and tear down exterior scaffolding. It takes experience to set scaffolding safely, especially on uneven ground, and units must be attached to the building.

Pipe scaffolding typically consists of two rectangular end frames and diagonal braces secured with wing nuts or self-locking cleats. Once the first stage is assembled, the installer adjusts the self-leveling feet until the platform is level.

To raise successive stages, the installer stacks end frames over coupling sleeves and locks the pieces in place with uplift and cotter pins. Additional lock arms may join the bracing. Platforms should be planked their entire width with 2x lumber or metal planks provided by the rental company. Guardrails are a must on all scaffolding. If your platform is 10 ft. or higher, most safety codes require midrails and toe boards as well.

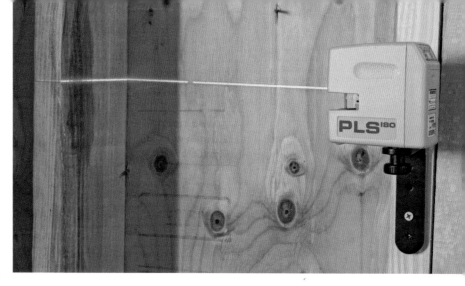

Self-leveling lasers give you a consistently level line around a room from which you can work, essential for setting electrical boxes, installing cabinets, and the like.

Interior scaffolding has rollers that let you move it around a room. Before you mount a platform, always lock the roller locks, and dismount before unlocking the locks for any reason. It's unwise to move an unlocked scaffold while someone is atop it.

Ladder jacks. Ladder jacks offer an inexpensive, quickly adjustable setup, which can be safe if both ladders are well footed. Many jack brackets pivot so that scaffolding planks can rest under or over the ladder. Consult the operating instructions supplied with your ladder jacks. In general, avoid platform heights higher than 8 ft.

Pump jacks. Pump jacks work fine when new, but after a few seasons of rain and rust, they often bind, which produces eye-popping free falls or blind rage when you're 15 ft. in the air and the jacks refuse to go up or down. Consequently, most rental pump jacks have been hammered silly for their failings. Be wary of rental pump jacks.

LASER TOOLS

Laser levels have become more common as their prices have dropped so buy one if your budget allows. You can set a self-leveling laser almost anywhere, and it will project a level reference line all around a room or to a distance of 50 ft. to 100 ft. easily. Laser levels are invaluable for setting electrical outlets or kitchen cabinets at the same height, finding how much a floor is sloping, or checking drainage slope by shooting from the curb to a house foundation.

Laser plumbs (see p. 314) enable electricians to lay out ceiling light-can locations on the floor and then shoot the locations up to the ceiling. It's much easier than repeatedly climbing a stepladder and plumbing down.

Laser tape measures, also known as laser distance meters, bounce a sonic or laser beam between surfaces to get a quick reading of distance. They're not accurate enough for cabinet work, but they're great for drywall estimators, siding installers, and (so I hear) golfers who want to know exactly how far it is to the green.

SLIDING COMPOUND-MITER SAW

A 10-in. sliding compound-miter saw is the ultimate tool for a wide range of finish work. The extended crosscut length, combined with adjustable angle and bevel settings, allows complicated cuts in large materials such as 6-in. by 6-in. deck posts, 10-in.-wide siding boards, and large crown molding.

POWER PLANERS

Moderately priced and incredibly useful, a power planer can plane down studs to create a flat plane for drywall, trim a little off an exterior door, and quickly cut a slot so the nailing flange of an electrical box is flush to the edge of a stud (see the photo on p. 211).

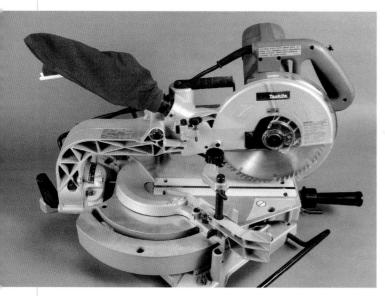

Sliding compound-miter saw.

A pneumatic pin nailer used for finish trim, with finish nails of varying length.

Nailer TRIGGERS

Pneumatic nailers have several types of triggers. The safest is a restrictive trigger, which you must squeeze and release to shoot a nail. A second type, a bounce-fire trigger, shoots a nail each time you depress the gun's nosepiece. Bounce-fire triggers are usually favored for sheathing, which requires a lot of 8d nails (2½ in. long) spaced relatively close to one another. Until you become accustomed to nailers, restrictive triggers are far safer.

RIGHT-ANGLE DRILL

Renting a ½-in. right-angle drill will let you avoid burning out your own drill while roughing in plumbing and electrical runs. The right-angle drill is a godsend in the tight spaces between framing members, and the drill's long handle gives you more leverage to control the torque of this slow-drilling, powerful machine.

When drilling through framing, self-feeding, double-spiral bits clear wood well, but use a hole-cutting bit when bigger holes are required. Whatever bit you use, wear goggles and watch for nails. The better right-angle drills will have a clutch that disengages if the bit meets a certain level of resistance.

PNEUMATIC NAILERS

It may take 20,000 to 30,000 nails just to sheathe an average house. Add to that the nails needed for framing, roofing, and shingling, and you can begin to imagine the number of hammer strokes required. Pneumatic nailers, commonly called nail guns, save a lot of effort.

Control of the workpiece is the other big advantage. When using a hammer and nail, you need one hand for each. A pneumatic nailer delivers the nails, giving you a free hand to hold a stud or top plate in place. The nail goes in quickly without requiring hammer blows that can cause the workpiece to drift out of position.

And pneumatic nailers won't slip and ding expensive trim. Consequently, among professionals, pneumatic finish nailers have all but replaced hand nailing door and window casings.

There are framing nailers, finish nailers, brad nailers, and pin nailers. They are powered by air hoses connected to a compressor and calibrated by a pressure adjustment on the nailer. Staff at rental companies can explain the adjustments as well as safety features and how to use the tools safely.

Hard hats, safety glasses, and hearing protection are musts.

POWDER-ACTUATED TOOLS

Potentially very dangerous, powder-actuated tools are useful for shooting nails into concrete, as when framing an interior wall on a concrete slab or securing pressure-treated lumber to a foundation wall. These connections aren't considered structural. Engineers specify bolts instead for all structural connections to concrete.

A reputable rental company will demonstrate the tool's safe use, describe (and rent) safety equipment, answer your questions, and supply appropriate cartridges and drive pins. Some local codes prohibit renting powder-actuated tools to nonprofessionals.

When using this tool, wear safety glasses, hearing protection, and a hard hat.

HAMMER DRILLS

Hammer drills are a category of tools that combine hammer and rotary functions to drill holes into concrete. The terms hammer drill, impact hammer, and rotary hammer are often used interchangeably, but I'll use hammer drill to denote smaller, less powerful tools and rotary hammers to denote larger, more powerful ones.

Hammer drills (½ in.) typically offer two settings: rotation only and hammering with rotation. They're adequate for drilling small holes in concrete, for anchoring door thresholds to slabs, and for predrilling pilot holes for masonry screws.

A rotary hammer drill (1½ in.) is the tool of choice if you need to drill dozens of ¾-in. holes for standard ⅝-in. anchor bolts. On the hammer setting, the tool punches as it turns, somewhat like a jackhammer. Get a model with padded handles as well as vibration reduction.

Wear safety glasses, hearing protection, heavy gloves, and a hard hat.

CONCRETE BREAKER AND COMPRESSOR

Whenever you need to replace defective concrete, change the configuration of foundations, or remove concrete so you can lay underground drainage pipes, rent a concrete breaker (see the photo on p. 273) and special high-volume compressor.

SOIL TAMPER

Use a gasoline-powered soil tamper tool before you pour a concrete slab, lay a brick walk, or set a stone patio.

DUMPSTERS

Although you can rent Dumpsters by the day or week, carefully plan (and stick to) demolition schedules so your Dumpsters leave the job site promptly. Other people's debris has a way of filling your Dumpster when it sits too long. Don't order one until you're well into tearout and have accumulated a half-week's worth of debris.

If you're demolishing masonry, rent a "low boy," which is a small unit (10 cu. yd.) specially built for the great weight of concrete, brick, and the like. For other jobs, rent the largest size available, usually 20 cu. yd. You'll also pay for the dumping fee the company must pay to your municipality.

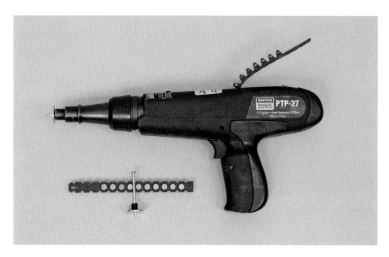

A powder-actuated tool with a drive pin and two strips of cartridges.

Bosch rotary hammer.

4 Building Materials

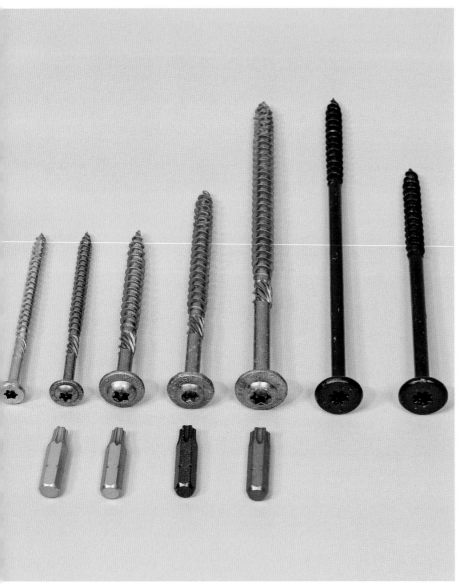

Structural screws are big these days. Typically driven in with an impact driver, they have thicker shanks, higher structural values, greater resistance to withdrawal, and, unlike nails, can be backed out to reposition lumber. Packages of screws come with color-coded star bits.

This chapter offers an overview of the materials needed to frame and sheathe a house. The list includes standard lumber, engineered lumber, sheet materials such as plywood and particleboard, and fasteners, including nails, screws, and construction adhesives. Finish materials used in cabinets are discussed in chapter 13, and finish trim choices can be found in chapter 17.

There's never been a wider choice of building materials or more accessible information on using them, whether you need to size ceiling joists or find an environmentally safe adhesive. Just type your requirements into a website calculator or ask your lumber supplier for a recommendation. Framing techniques are discussed at length in chapter 8.

Standard Lumber

Wood is a superb building material. It is strong, economical, and easily worked. In tree or lumber form, wood can withstand great loads, yet it's resilient enough to regain its shape when loads are removed. *Standard lumber* is lumber sawn from logs in the traditional manner; *engineered lumber* is often an amalgam of peeled, shredded, or reassembled wood pieces and strong adhesives.

LUMBER GRADES

After lumber has been milled, each piece is visually graded according to established performance standards and then stamped. This grade stamp is important because building inspectors won't approve structures built with unstamped lumber. Otherwise, they'd have no way of knowing what loads the wood can support.

In brief, grading is based on the presence of warping, knots, holes, decay, or other imperfections that could weaken the lumber and reduce its load-bearing capacity. Generally, dimension lumber grades are based on strength, appearance, or both. The more imperfections the wood has, the lower the grade.

Grade stamps indicate lumber grade, tree species, moisture content when the lumber was surfaced, sawmill, and regional agency certifying the grading standards. Lumber that is stress-rated by machine will have additional information.

Structural framing lumber grades run from Select Structural (the best looking and strongest) through No. 1, No. 2, and No. 3. An architect might specify Select Structural 4x8s, for example, when beams will be exposed in a living room. In most grading systems, No. 1 and No. 2 are equally strong, although No. 1 has fewer cosmetic flaws. Thus, if appearance is not a factor, you can order "No. 2 and better" without sacrificing strength. No. 3 is the weakest and least expensive grade in the structural category; you won't save much by using it because you'll have to order larger dimensions to carry the same loads as No. 1 and No. 2 grades.

Light framing lumber, which is used for plates, sills, and blocking, has lower strength requirements than structural framing members. Light framing members are 4 in. (thick or wide) or less. Grades are Construction (the best), Standard, and Utility. Contractors often order "Standard and better."

Stud lumber is graded Stud or Economy Stud. In general, avoid Economy grade lumber of any kind. Although it's OK for temporary use, its inferior quality makes it unreliable in any sustained load-bearing situation.

LUMBER SPECIES

Species are denoted by abbreviations such as PP (ponderosa pine), DF (Douglas fir), and HEM (hemlock). Often, manufacturers will group species with similar properties. S-P-F (spruce, pine, fir) is by far the most common Canadian grouping, and HEM-FIR (hemlock, fir) is common throughout the United States. Because lumber is heavy and expensive to ship, lumberyard sources tend to be from nearby mills. Since your lumberyard is likely to carry only a mixed stock of sizes and grades, your choice may be limited to what's on hand.

MOISTURE CONTENT

A mature, living tree can pump a ton of water into the atmosphere each day. So it's not surpris-

ing that when trees arrive at a sawmill, they still contain a lot of moisture. Lumber that's too wet is heavy, difficult to work, and dimensionally unstable—it will shrink excessively and perhaps warp as it dries. So after sawmills rough-cut logs, they air-dry or kiln-dry (KD) the newly milled wood before planing it into finish lumber.

The degree of dryness is expressed as moisture content (MC). According to the Western Wood Products Association, "Moisture content is a weight relationship . . . for example, if a piece of wood has a moisture content of 21%, it weighs 21% more than if all the moisture were removed."

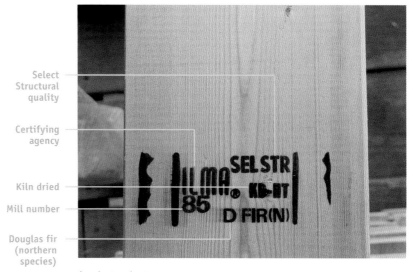

Select Structural quality

Certifying agency

Kiln dried

Mill number

Douglas fir (northern species)

Lumber grade stamps.

Yard TALK

Here's some standard lumberyard lingo:

▶ **BOARDS** are less than 2 in. thick and are used as trim, sheathing, subflooring, battens, doorstops, shelves, and so on.

▶ **LUMBER** (dimension lumber) is 2 in. to 4 in. thick and is used for framing: studs, posts, joists, beams, headers, rafters, stair carriages, and so on.

▶ **FACTORY OR SHOP LUMBER** is wood milled into window casings, trim, and other elements.

▶ **TIMBER** is at least 5 in. thick in its smallest dimension.

▶ **STOCK** applies to any building material in its unworked form, as it comes from the lumberyard or mill.

▶ A **STICK** is jargon for a piece of lumber, such as a 2x4: "If that stick is too warped, go get another."

PRO TIP

Different species have different densities, elasticity, and load-bearing capacities, so their lumber grades are not interchangeable. For example, a No. 1 HEM-FIR 2x10 may not have the same load/span capacity as a No. 1 southern pine 2x10. That's why structural engineers routinely specify both the grade and species of lumber (or species group) needed to satisfy requirements in building codes.

Once it has been dried, lumber is graded according to its moisture content:

S-GRN has MC greater than 19%.

S-DRY/KD/KD-HT has MC less than 19%.

MC-15/KD-15 has MC not greater than 15%.

Kiln-dried lumber tends to cost more because kilns use energy; consequently, most builders use air-dried lumber for new construction. After a structure is framed and sheathed, the wood continues to dry anyway. Thus, S-DRY or MC-15 grades are fine for renovation framing, although if you're framing interior walls inside a tight building envelope, spending a bit extra for KD-15 might be a smart move. The lumber in old houses is often very dry. In any event, avoid S-GRN: Shrinking lumber can wreak havoc with finish surfaces.

LUMBER SIZES

Lumber's final size depends on milling processes. In smaller mills, lumber is often sawn, stickered, and allowed to air-dry for four to six months. If it's not milled further, it's called *rough-cut* lumber. Depending on the accuracy of the sawyer, the size may vary slightly, but the *nominal size* of, say, a rough-cut 2x4 is usually a full 2 in. by 4 in.

However, most lumber is rough-cut and then *surfaced* (run through a planer to achieve uniform thickness) before being kiln-dried. At each stage, the lumber size decreases. When you order a 2x4 (nominal size), you receive a piece

Softwoods AREN'T ALWAYS

Most construction lumber is called *softwood*, which is the lumber industry's term for wood from conifers, needle-leaved evergreens such as pine, fir, spruce, and hemlock. For the most part, these softwoods are softer and less dense than most hardwoods, which come from broad-leaved deciduous trees such as maple, oak, and walnut. That said, some softwoods, namely southern yellow pine, are much harder than some hardwoods, such as basswood.

measuring 1½ in. by 3½ in. Still, you pay for the nominal size.

Another way to size wood, especially hardwood and Select finished woods, is by ¼-in. increments: 2/4, 3/4, 4/4, 5/4 , 6/4, and so on. The nominal actual difference is present here, too: For example, a nominal 5/4 stair tread is actually 1 in. thick.

PRESSURE-TREATED LUMBER

Lumber may also be marked as pressure treated. Such wood, after treatment, may be left exposed to weather, used near the foundation, or otherwise subjected to moisture, insects, or extremes of climate. If the wood will remain in contact with the soil, be sure that it is also rated for Ground Contact. Note, however, that *any* wood sitting on the ground, even redwood, will rot eventually.

At one time, roughly 90% of all pressure-treated lumber was treated with chromated copper arsenate (CCA). But the U.S. Environmental Protection Agency (EPA) determined that CCA leaches arsenic into the soil. As a result, industry leaders agreed to stop using CCA treatment for most residential applications by the end of 2003. More benign types of pressure-treated lumber, such as alkaline copper quat (0.40 ACQ) and copper boron azole (CBA) are available. Both biocides are arsenic-free.

Whatever lumber treatment you consider, consult its product data sheets for the relative safety of the chemicals used and whatever care you should take when handling, storing, and cutting it. In fact, it's smart to capture and safely dispose of the sawdust.

Nominal and Actual Sizes of Softwood

NOMINAL	ACTUAL (in.)
1×2	¾×1½
1×4	¾×3½
1×6	¾×5½
1×8	¾×7¼
1×10	¾×9¼
1×12	¾×11¼
2×4	1½×3½
2×6	1½×5½
2×8	1½×7½
2×10	1½×9½
2×12	1½×11½

ORDERING LUMBER AND CALCULATING BOARD FEET

The price of long, thin pieces of wood, such as molding or furring strips, is based on their length, or *lineal feet* (lin. ft.). Sheet materials such as plywood and composite board are sold by the *square foot*, which is length × width; sheet thickness affects price, but it is not computed directly. Roofing and siding materials are often sold in *squares* of 100 sq. ft. Most yard lumber is sold in *board feet* (bd. ft.), according to this formula:

$$\frac{\text{width (in.)} \times \text{thickness (in.)} \times \text{length (ft.)}}{12}$$

In the two examples below, each board contains 1 bd. ft.:

$$[EQ]\frac{12 \text{ in.} \times 1 \text{ in.} \times 1 \text{ ft.}}{12} = 1 \text{ bd. ft.}$$

$$[EQ]\frac{6 \text{ in.} \times 2 \text{ in.} \times 1 \text{ ft.}}{12} = 1 \text{ bd. ft.}$$

In each of the following two examples, the dimensions given yield 2 bd. ft.:

$$[EQ]\frac{12 \text{ in.} \times 1 \text{ in.} \times 2 \text{ ft.}}{12} = 2 \text{ bd. ft.}$$

$$[EQ]\frac{2 \text{ in.} \times 4 \text{ in.} \times 3 \text{ ft.}}{12} = 2 \text{ bd. ft.}$$

When calculating the total board feet of several pieces of lumber, multiply the numerator (top part) of the fraction by the total number of pieces needed. Thus, here's how to calculate the board feet of 10 pieces of 2-in. by 6-in. by 12-ft. lumber:

$$[EQ]\frac{2 \text{ in.} \times 6 \text{ in.} \times 12 \text{ ft.} \times 10}{12} = 120 \text{ bd. ft.}$$

Salvage Lumber

Salvage materials have striking advantages and disadvantages—the major plus being low cost and the major minus being prep time. Salvaging molding, flooring, and other materials from your own home is a good way to match existing materials, but be picky in selecting materials from other sources.

Reuse centers are popular these days. For example the ReStore resale outlets by Habitat for Humanity offer tax deductions to donors and recycled building materials at a fraction of the cost of new. However, make sure all materials are structurally sound. Use a pocketknife or an awl to test lumber for rot or insects.

Salvaged framing materials may not be worth the effort if they are in small quantities or if, after removing them, you find that they will be too short. For example, by the time you pry 2x4 studs free from plates, remove nails, and cut off split ends, the studs may be only 7 ft. long.

Some materials just aren't worth removing. Siding and other exterior trim is rarely salvageable because it's usually old and weather-beaten. Barn board, in vogue years ago, is hardly charming when it is half-rotted, warped, and crawling with carpenter ants. If there is any danger of destroying a piece of salvage by removing it, leave it alone. Parts of many beautiful old places that were restorable have been ruined by people who didn't know what they were doing.

Moreover, if you have any qualms about the structural strength of a building, stay out of it. Dismantling a building is a special skill, and inexperienced people who undertake the task can get hurt. Perhaps the best advice for would-be users of salvage materials is to buy it from a salvage yard. In this case, somebody has already done the dirty and dangerous work of removal.

Engineered Lumber

Like any natural product, standard lumber is quirky. It has knots, holes, and splits, and it twists, cups, and shrinks. As mature old-growth timber was replaced by smaller, inferior trees,

Seven Tips for Deconstruction

If you are determined to do on-site salvage, here are a few suggestions:

▶ Get a tetanus shot, and wear a long-sleeved shirt, heavy pants, thick-soled shoes, safety glasses, disposable respirator, sturdy work gloves, and hard hat.

▶ Always cut electrical power to areas being demolished or dismantled. Then use a voltage tester in outlets, fixtures, and switches to make sure there's no current flowing through them.

▶ Don't hurry. Look at construction joints closely and disassemble pieces slowly.

▶ As you free each piece, remove nails immediately. Remember, footing on construction sites is chancy at best, and you don't want to land in a bed of nails when descending from a ladder in a hurry.

▶ If the piece is complex, such as a fireplace mantel, photograph it and then label each element before removing it.

▶ Before cutting into framing lumber, scrutinize it for nails. The best tool for cutting through both lumber and nails is a demolition blade (see p. 56) in a reciprocating saw. Safety glasses are a must.

▶ Most salvaged wood is old and dry, so if you plan to reuse it, get it under cover immediately. It can absorb water like a sponge and rot before you know it.

lumber quality became less reliable—much to the dismay of builders.

In response, the lumber industry combined wood fiber and strong glues to create engineered lumber (EL), including I-joists, engineered beams, plywood, and particleboard. EL spans greater distances and carries heavier loads than standard lumber of comparable dimensions. In addition, EL won't shrink and remains straight, stable, strong, and—above all—predictable.

Engineered *beams*, on the other hand, have two main drawbacks: EL beams of any size are quite heavy and so dense that they must often be predrilled. Moreover, they cost considerably more than sawn lumber. But the comparison is somewhat like comparing apples and oranges because you'd be hard-pressed to find sawn lumber 24 ft. long, whereas ordering engineered lumber up to, say, 40 ft. is pretty common. In fact, the expansive, open spaces that modern homeowners crave probably wouldn't be possible using only traditional lumber.

Alternatives to Solid-Wood Joists

I-JOIST

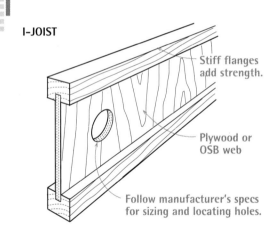

Stiff flanges add strength.

Plywood or OSB web

Follow manufacturer's specs for sizing and locating holes.

OPEN-WEB FLOOR TRUSS

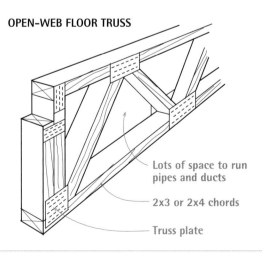

Lots of space to run pipes and ducts

2x3 or 2x4 chords

Truss plate

TRUSSES

Until recently, the most common truss was the prefabricated *roof truss,* which serves as the roof's support structure. Trusses' short, weblike members are fastened together with steel truss plates. Trusses are lightweight, inexpensive, quick to install, and strong relative to the distances they span. They eliminate the need for deep-dimensioned traditional roof rafters and complex job-site cutting.

▶ *Advantages:* When designed for tract housing, trusses can be prefabricated for almost any roof contour, trucked to the job site, and erected in a few days. In addition, you can route ducts, pipes, and wiring through openings in the webbing. But roof trusses are less useful in custom remodels: The more complex an existing roof's framing is, the more difficult a roof truss would be to configure and install in tight spaces.

▶ *Disadvantages:* Roof trusses leave little living space or storage space in the attic. Adding kneewalls along the sides will gain some height, but design options will be limited. Factory-built roof trusses should never be modified, unless an engineer approves the changes; otherwise, unbalanced loads could cause the trusses—and the roof—to fail.

Floor trusses, on the other hand, are often open webs spaced 24 in. on center. Although they can span roughly the same distance as I-joists of comparable depth, it's much easier to run ducts, vents, wiring, and plumbing through open-web trusses.

I-JOISTS

I-joists are commonly called TrusJoists, or TJs, after a popular brand (a subsidiary of Weyerhaeuser). Typically, I-joists are plywood or oriented strand board (OSB) webs bolstered by stiff lumber flanges top and bottom, which add strength and prevent lateral bending.

Although I-joists look flimsy, they are stronger than solid-lumber joists of comparable dimensions. Whereas solid joists are spaced 16 in. on center, I-joists can be laid out on 19½-in. or 24-in. centers. They also are lightweight, straight, and stable. Floors and ceilings constructed with I-joists stay flat because there's virtually no shrinkage; hence, there are almost no drywall cracks, nail pops, or floor squeaks.

Installing I-joists is not much different from installing 2x lumber, but blocking between I-joists is critical. (They must be perfectly perpendicular to bear loads.) You can drill larger holes in I-joist webs than you can in solid lumber, but religiously follow manufacturer guidance on hole size and placement. And *never* cut into the flanges.

Manufacturers continue to develop more economical I-joist components. Webs may be plywood, particleboard, or laminated veneer lumber (LVL). Flanges have been fabricated from LVL, OSB, or—back to the future!—solid lumber (2x3s or 2x4s) finger-jointed and glued together. I-joists with wider flanges are less likely to flop and fall over during installation. Plus they offer more surface area to glue and nail subflooring to.

ENGINEERED BEAMS

The most daunting part of using engineered beams may be the wide selection. Fortunately, lumberyard staff can usually explain the merits of each type and help you determine correct size.

Glulams, or glue-laminated timbers, are the granddaddy of engineered beams. They've been used in Europe since the early 1900s. In North America, they're fabricated from relatively short pieces of dimension lumber (often Douglas fir or southern pine), which are overlapped or finger-jointed, glued, and pressure clamped. Glulams come in stock widths of 3⅛ in. to 6¾ in., but you can obtain them in almost any size or shape, including curves and arches, as well as pressure treated.

Glulams are expensive, but their stability and strength make them suitable for high loads in clear spans as great as 60 ft. Obviously, you'd need a crane to move such a behemoth.

LVL (Microllam) is fashioned from thin layers of wood veneer glued together—much like plywood, except the wood grain in all LVL layers runs parallel. It's stronger than sawn lumber or laminated strand lumber of comparable size, although it's roughly twice the cost of sawn lumber.

LVL is usually milled as planks 1¾ in. wide, so it's typically used as rim joists, cantilever joists, or in-floor headers and beams. It's a good choice for medium-span beams up to 16 ft., and because individual beams are easy to handle, a small crew can join LVL planks on site to create a built-up girder. LVL is available in other widths, from 3½ in. to 5½ in. Depths range up to 20 in.

A drawback of LVL is that it can't be pressure treated and shouldn't be used on exteriors. If it gets wet, it will cup. For this reason, keep it covered until you're ready to use it.

PSL (parallel strand lumber, Parallam) is created from wood fiber strands 2 ft. to 8 ft. long, running parallel to each other, and glued together under tremendous pressure. PSL is the strongest and most expensive of any structural composite lumber; on many projects, PSL is the material of choice.

Standard PSL sizes are 7 in. to 11 in. wide, up to 20 in. deep, and they can be fabricated to vir-

This engineered beam is a 4-in. by 14-in. Parallam girder secured with a Simpson CCQ column cap.

Heavy STEEL FRAMING

More and more renovations rely on a handful of steel columns and a steel beam or two to support things. With a good steel fabricator, the costs are actually reasonable for what you get. The trick is ordering the right size steel and coordinating installation dates correctly into your construction sequence. When the steel shows up, it's best to pop it into place at once.

tually any length—66 ft. is not uncommon. Because they're stronger than glulams, PSLs are built without camber (a curve built in to anticipate deflection under load), so they're easier to align during installation.

PSL beams can be pressure treated and thus can be used outside.

LSL (laminated strand lumber, TimberStrand) is fabricated from 12-in. wood strands from fast-growing (but weaker) trees, such as aspen and poplar, and then glued together in a random manner. Consequently, LSL carries less load than the beams noted previously, and it costs less. Still, it is stronger than sawn lumber, although more expensive.

Raising a large beam is inherently dangerous, so rent a pair of Genie Lifts to raise it safely. On p. 271, workers use Genie Lifts to raise an I-beam assembly that weighs more than 600 lb. Most rental yards that stock lifts will show you how to operate them correctly.

LSL is available in 1¾-in. to 3½-in. widths and in depths up to 18 in., but it's most often used as short lengths in undemanding locations, such as door or window headers, wall plates, studs, and rim joists.

LSL headers are stable, so they'll probably reduce nail pops and drywall cracks around doors and windows. But for small openings of 10 ft. or less and average loads, sawn-lumber headers are usually more cost-effective.

Framing with Steel

The use of steel framing in residential renovation is increasing, but it's still relatively rare, limited to applications where the greater weight or size of wood framing would be a problem—such as furring down a ceiling (see p. 215), building soffits, or framing out chases in which to run pipes, ducts, or electrical wiring. Metal studs may also be specified in residential situations where fire is a concern.

LIGHT STEEL FRAMING

Light steel framing consists primarily of C-shaped metal studs set into U-shaped top and bottom plates, joined with self-drilling pan-head screws. Fast and relatively inexpensive to install, light steel framing (20 gauge to 25 gauge) is most often used to create non-load-bearing interior partitions in commercial work. With drywall attached, metal studs become rigid, so, in effect, drywall panels become structural agents.

Advocates argue that more residential contractors would use light steel framing if they were familiar with it. In fact, light steel framing is less expensive than lumber; it can be assembled with common tools, such as aviation snips, screw guns, and locking pliers; and it's far lighter and easier to lug than dimension lumber. To attach drywall, use type S drywall screws instead of the type W screws specified for wood. One big plus: Because metal-stud walls are assembled with screws, they can be disassembled easily and recycled completely.

If you want to hide a masonry wall, light steel framing is ideal. Masonry walls are often irregular, but if you use 1⅝-in. metal framing to create a wall within a wall, you'll have a flat surface to drywall that's stable and doesn't eat up much space.

That said, light steel is quirky. You must align prepunched holes for plumbing and wiring before cutting studs, and, for that reason, you must measure and cut metal studs from the same end. If you forget that rule, your studs become scrap. Further, if you want to shim and attach door jambs and casings properly, you need to reinforce steel-framed door openings with wood. And finally, safety glasses, hearing protection, and sturdy work gloves are essential when working with light steel framing—edges can be razor sharp.

FLITCH PLATES

Flitch plates are steel plates sandwiched between dimension lumber and through–bolted to increase span and load-carrying capacity. Flitch plates are most often used in renovation where existing beams or joists are undersize. (You insert plates after jacking sagging beams.)

Ideally, a structural engineer should size the flitch plate assembly, including the size and placement of bolts. Steel plates are typically ⅜ in. to ½ in. thick; the carriage bolts, ½ in. to ⅝ in. in diameter. Stagger bolts, top to bottom, 16 in. apart, keeping them back at least 2 in. from beam edges. Put four bolts at each beam end. To ease installation, drill bolt holes ¹⁄₁₆ in. larger than the bolt diameters.

Flitch plates run the length of the wood members. The wood sandwich keeps the steel plate on edge and prevents lateral buckling. Bolt holes should be predrilled or punched—never cut with an acetylene torch. That's because loads are transferred partly through the friction between the steel and wood faces, and the raised debris around acetylene torch holes would reduce steel-to-wood contact.

STEEL I-BEAMS

Although lately eclipsed by engineered-wood beams, steel I-beams, for the same given depth, are stronger. Consequently, steel I-beams may be the best choice if you need to hide a beam in a relatively shallow floor system—say, among 2x6s or 2x8s—or if clearance is an issue.

Wide-flange I-beams are the steel beams most commonly used in houses, where they typically range from 4⅛ in. to 10 in. deep and 4 in. to 10 in. wide. Standard lengths are 20 ft. and 40 ft., although some suppliers stock intermediate sizes. Weight depends on the length of the beam and the thickness of the steel. That is, a 20-ft. 8x4 I-beam that's 0.245 in. thick weighs roughly 300 lb., whereas a 20-ft. 8x8H I-beam with a web that's 0.458 in. thick weighs 800 lb. If you order a nonstandard size, expect to pay a premium.

Before selecting steel I-beams, consult with a structural engineer. For installation, use an experienced contractor. Access to the site greatly affects installation costs, especially if there's a crane involved.

Sorting Out Panel Names

Structural panels:

▶ **Plywood** is a sandwich of thin veneers sliced from logs, with veneers stacked perpendicularly to one another (cross-grain) in alternating layers and glued. Each layer is a ply. Alternating wood grain direction adds stiffness, dimensional stability, and strength.

▶ **OSB** (oriented strand board) is made from logs shredded into long strands. The strands are oriented in the same direction, mixed with resins, and pressed into thin sheets. As with plywood, strands in alternating layers run perpendicularly.

Nonstructural panels:

▶ **Particleboard** (also known as chipboard) is fabricated from mill waste, mixed with resins, and hot pressed. Because of its stability and uniform consistency, particleboard is an excellent core material for veneered cabinets, laminated counter-tops, and bookcases.

▶ **MDF** (medium-density fiberboard) is a mixture of fine, randomly oriented wood fibers and resins, hot pressed for a smooth surface. It is used as interior trim and cabinetry stock.

▶ **Hardboard** (such as Masonite) is a high-density fiberboard created by steaming wood chips and then hot pressing them into sheets. The hard, smooth surface is well suited for underlayment, interior trim, and paneling. Hardboard used as exterior siding has been plagued by warping, delamination, and other moisture-related problems.

YET MORE Panel STAMPS

▶ **T&G:** tongue and groove
▶ **G2S:** good two sides
▶ **G1S:** good one side
▶ **PRP 000:** performance-rated panel (number follows)
▶ **SEL TF:** select tight face
▶ **SELECT:** uniform surface, acceptable for underlayment

Structural and Nonstructural Panels

Plywood and OSB are the structural panels most often specified to sheathe wood framing and increase its shear strength. For example, a 20-ft. wall sheathed with ⁷⁄₁₆-in. plywood can withstand more than a ton of lateral force pushing against the top of the wall.

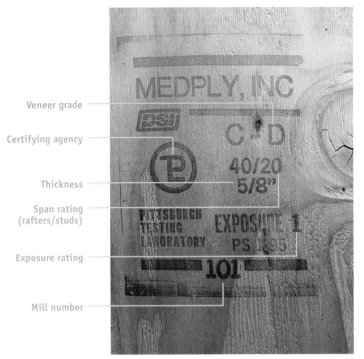

Veneer grade

Certifying agency

Thickness

Span rating
(rafters/studs)

Exposure rating

Mill number

Plywood grade stamps.

PLYWOOD

Structural plywood is made by laminating soft-wood plies. Each panel is stamped to indicate veneer grade, species group or span rating, thickness, exposure durability, mill number, and certifying agency. The Engineered Wood Association, which oversees about two-thirds of structural panels in North America, also has stamps that specify a panel's intended use, such as "rated sheathing," or installation details, such as "sized for spacing," to remind carpenters to leave ⅛-in. expansion gaps between panels.

Veneer grades. Veneer grades range from A to D, with letters appearing in pairs to indicate the front and back veneers of the panel. "A/B Exterior," for example, has a grade A front veneer, a grade B back veneer, and grade C inner plies. When you buy CDX (C/D exterior grade), it's advisable to place the grade C side toward the weather—or up, if used as subflooring.

Most roof and wall sheathing and subflooring is CDX. If a panel is also stamped PTS, its imperfections have been plugged and touch sanded. Lower veneer grades have more plugs and bigger knots.

Grade D is the lowest grade of interior plywood panels; it should not be exposed to weather.

Species grade or span rating. Plywood's strength may be indicated by two marks. One is a species group number (1–5). Group 1 is the strongest and often contains Douglas fir or southern yellow pine.

The second mark, a span rating, is more common. The two-digit rating looks like a fraction, but it's not. Rather, a rating of 24/16 indicates that a panel can sheathe rafters spaced 24 in. on center and studs spaced 16 in. on center. Span ratings assume that the panel's long axis runs perpendicular to framing members.

Another common stamp is Struc I, which stands for Structural I sheathing, a five-ply CDX that's tested and guaranteed for a given shear value. If an engineer specifies Struc I, it must be used. Plywood used for structural sheathing must have a minimum of five plies. Avoid three-ply, ½-in. CDX: Although it is widely available and cheaper than five-ply, it's vastly inferior.

Thickness and length. Panels rated for Struc I wall sheathing, roof sheathing, and subflooring range from ⅜ in. to ²³⁄₃₂ in. thick. Although 4x8 panels are the most common, 4x9 or 4x10 sheets enable you to run panels vertically from mudsills to the rim joists atop the first floor, thereby reducing the shear-wall blocking you might need behind panel edges and greatly improving the shear strength of the wall. (*Shear walls* are specially engineered walls that brace a building

Reduced-Formaldehyde AND FORMALDEHYDE-FREE PANELS

Composite panels such as plywood, OSB, and MDF have long been bonded with urea-formaldehyde (UF) resins, which outgas (give off) noxious gases. Volatile organic compounds (VOCs) are a problem for homeowners with chemical sensitivities, especially as houses become more airtight. Also, formaldehyde is classified as a probable human carcinogen by the EPA, which, while alarming to homeowners, could prove fatal to installers.

In response, manufacturers investigated less-toxic resins. To date, they have been able to greatly replace UF resins in structural panels such as CDX plywood, primarily by switching to *phenol* formaldehyde resins (PFs), which are less toxic than UFs. Typically, these panels will be stamped NAUF (no added urea-formaldehyde). It is hoped that manufacturers will replace all formaldehyde resins in structural panels someday, but presently there are no alternatives with the same bonding strength and resistance to the elements.

Fortunately, there are formaldehyde-free options for the sheet materials used inside a house—as molding and paneling, as well as cabinet frames, doors, and shelves. These low-VOC materials greatly help the quality of indoor air. Medite II, Medex, and SierraPine are three of the better-known brands of MDF; visit the Internet or consult your local lumberyard for more choices.

against lateral seismic and wind forces.) Although the square-foot prices of 4x9 and 4x10 panels are higher than that of 4x8s, the larger panels speed up the job.

Exposure durability. How much weather and moisture a wood-based panel can take is largely a function of the glues used. *Exterior-grade* panels can be exposed repeatedly to moisture or used in damp climates because their plies are bonded with waterproof adhesives. *Exposure 1* is suitable if there's limited exposure to moisture—say, if construction gets delayed and the house doesn't get closed in for three to six months. *Exposure 2* panels are suitable for protected applications and brief construction delays. *Interior-grade* panels will deteriorate if they get wet; use them only in dry, protected applications.

OSB PANELS

OSB and plywood have almost exactly the same strength, stiffness, and span ratings. Both are fabricated in layers and weigh roughly the same. Both can sheathe roofs, walls, and floors. Their installation is almost identical, down to the blocking behind subfloor edges and the need for H-clips between the unsupported edges of roof sheathing. Exposure ratings and grade stamps are also similar.

In some respects, OSB is superior to plywood. It rarely delaminates, it holds screws and nails better, and it has roughly twice the shear values. (That's why I-joists have OSB webs.) So given OSB's lower cost (10% to 15% cheaper, on average), it's not surprising that OSB grabs an increasing market share every year.

But OSB has one persistent and irreversible shortcoming: Its edges swell when they get wet and appear as raised lines (ghost lines) through roofing. To mitigate this swelling, OSB makers seal the panel edges. However, when builders saw panels, the new (unsealed) edges swell when wet. Buildings under construction get rained on, so edge swelling is a real problem. Swollen edges can also raise hell in OSB subflooring or underlayment if it absorbs moisture, as commonly occurs over unfinished basements and uncovered crawlspaces. Thus, many tile and resilient-flooring manufacturers insist on plywood underlayment.

Given the huge market for OSB, however, count on solutions before long. At this writing, J.M. Huber AdvanTech, Louisiana-Pacific TopNotch, and Weyerhaeuser Structurwood are all tongue-and-groove–edged OSB panels purported to lie flat, install fast, and have minimal "edge swell." Stay tuned.

NAILING STRUCTURAL PANELS

Your local building code will have the final say on sizing structural panels. To accommodate heavy loads, choose a panel rated for a higher span. A span rating of 32/16 indicates that the panel is strong enough to sheathe rafters spaced 32 in. on center and studs 16 in. on center and, therefore, can support far greater live loads than a 24/16-rated panel, even though a 32/16 panel is only 1/32 in. thicker.

Nailing schedules for different uses of plywood are the same: Nail every 6 in. around the perimeter, not closer than 3/8 in. to the edge; elsewhere, nail every 12 in. For subflooring, annular ring or spiral nails hold best; use hot-dipped galvanized nails for all exterior purposes. An 8d nail is sufficient for 1/2-in. to 3/4-in. plywood. For structural shear walls, follow the engineer's specifications for nailing. Shear walls often require tighter nailing around the edges of the panel—and, sometimes, thicker nails.

If a panel stamp says "sized for spacing," leave an expansion gap of 1/8 in. between sheets—or whatever the stamp specifies. (Tongue-and-groove panels may not need gaps.) For greatest strength, run the long axes of panels perpendicular to structural members and stagger butt ends. In the intervals between joists, support plywood edges with solid blocking; on roofs, place *blocking clips* (also called ply-clips or H-clips) beneath the panel edges unsupported by solid wood.

Pneumatic nailers are widely used to nail down plywood, and they save a lot of time. But one thing a nailer won't do is "suck up" a piece of plywood to framing. This is worth noting because almost all plywood is warped to some degree. So after you nail down plywood with a pneumatic nailer, go back over the surface and give each nail an additional rap with a framing hammer. The hammerhead, being larger than the striker of the nailer, will help drive the plywood down as well.

As important, don't drive a nail too deeply. If a pneumatic nailer's pressure is set too high, the nail may be driven through the face ply, diminishing the shear value and holding capacity of the nail. Set the nailer's pneumatic pressure a little lower than would be needed to drive the nail flush. Then finish each nail with a hammer blow.

HARDWOOD PLYWOOD

Hardwood plywood is not intended to be structural, but because your renovation may need some, here's a brief overview. As with softwood plywood, there's a great variety, classified by species, face plies (appearance), core material (medium-density fiberboard [MDF], LVL, particleboard), and glues.

Bamboo plywood has beautifully figured faces and edges and is a great favorite of many cabinetmakers. A durable, sustainably produced hardwood, bamboo is as green as green gets.

Varieties such as birch, Russian birch, and marine plywood have many thin, dense laminations that can be quite handsome when finished and left exposed as shelves. These types are also quite strong compared to, say, CDX. Consequently, many builders now use 1-in. hardwood plywood as a substrate beneath stone countertops, especially where generous stone overhangs might split if inadequately supported.

Specify the grade of both faces, and check the stock carefully for damage—especially edges. Hardwood plywood is expensive, so sheets are often used right up to the edges. Here's a list of hardwood grades:

▶ *Specialty.* You can special-order closely matched flitches (veneer surfaces that can be laid together in sequence) that allow repetitions of face grain for visual effect.

▶ *Premium (A).* Grain patterns and colors are matched precisely.

▶ *Good (No. 1).* Colors of matched veneers on a face do not vary greatly, but patterns are less closely matched than premium grade.

▶ *Sound (No. 2).* Although colors and patterns are not matched, there are no open flaws.

▶ *Utility (No. 3).* These may have small flaws, tight knotholes, discoloring, and splits that can be filled but no rot.

▶ *Backing (No. 4).* Defects are allowed as long as they don't weaken the sheet or prevent its use; the backing side may be from a different tree species than that of the exposed face.

Metal Connectors

If wood is the universal building stock, metal is the universal connector. Nails of many types, bolts, and screws are discussed in this section—notably, structural screws, whose use has increased dramatically in renovation. Specialty plates that reinforce structural members also are described. Later in this chapter is a review of construction adhesives, which, some say, are destined to supplant metal connectors.

Nail Names

▶ **Common:** The workhorse of construction. Basic uncoated nail; flat head.

▶ **Spike:** A common nail "on steroids," 40d or 60d. Rarely used these days.

▶ **Box:** Same length and head size as a common nail but with a thinner shank.

▶ **Sinker:** Shank about the same size as a box nail; flat head, countersunk. Usually cement coated.

▶ **Cooler:** A sinker with a bigger head.

▶ **Finish:** Same length as a box nail but with a thinner shank. Brad head (not much wider than shank).

▶ **Casing:** Similar to a finish nail, but shank is thicker and head is slightly larger; countersunk.

▶ **Duplex:** Double-headed nail for temporary nailing to depth of first head, which holds wood down. Protruding top head is easily gripped for removal.

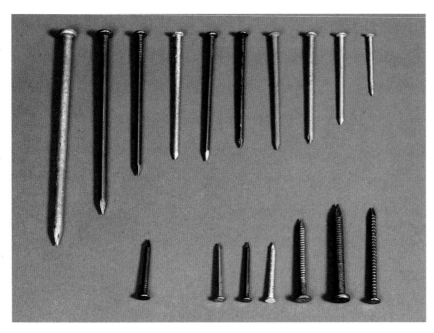

Nail types. Top row, from left: 60d galvanized spike, 40d common, 20d common, 16d galvanized, 16d common, 12d vinyl-coated sinker, 12d galvanized box, 10d galvanized common, 8d galvanized box, and 4d galvanized siding. Bottom row, from left: concrete nail and six joist-hanger and metal-connector nails (also called Teco nails). Longer nails may be required when sheathing covers framing.

Nail Sizes

NAIL LENGTH (in.)	PENNY SIZE (d)
1	2
1½	4
2	6
2½	8
3	10
3¼	12
3½	16
4	20
4½	30
5	40
6	60

NAILS

As they're driven in, nail points wedge apart wood fibers. The ensuing pressure of the fibers on the nail shank creates friction, which holds the joint together. Nails also transmit shear loads between the building elements they join. Where nails join major structural elements, such as rafters and wall plates, the loads can be tremendous; where nails attach finish elements, such as trim, loads are usually negligible.

There are hundreds of different nails, which vary in length, head size, shank shape, point, composition, and purpose.

Length. Length is reckoned in penny sizes, abbreviated as *d*. The larger the nail is, the greater the penny rating. Nails 20d or longer are called spikes.

Heads. The shape of a nail's head depends on whether that nail will be exposed or concealed and what type of material it's designed to hold down. Small heads, such as those on casing, finish, and some kinds of flooring nails, can easily be sunk below the wood surface. Large heads, such as those used to secure roofing paper or asphalt shingles, are needed to resist pull-through.

Shanks. Nail shanks are typically straight, and patterned shanks usually have greater holding strength than smooth ones. For example, spiral flooring nails (with screw shanks) resist popping, as do ring-shank nails. (By the way, it takes more force to drive spiral nails.) Spiral and ring-shank nails are well suited to decks and siding because changes in wood moisture can reduce the friction between wood fibers and straight shanks.

Points. Nails usually have a tapered four-sided point, but there are a few variations. For example, blunt-point nails are less likely to split wood than pointed nails because the blunt points crush the wood fibers in their path rather than wedging them apart. You can fashion your own blunt points by hammering down a nail point. However, the blunt point reduces the withdrawal friction on the nail shank.

Composition. Most nails are fashioned from medium-grade steel (often called mild steel). Nail composition may vary, according to the following situations:

▶ *Material nailed into.* Masonry nails are case-hardened. That's also true of the special nails supplied with joist hangers and other metal connectors. Do not use regular nails to attach metal connectors.

▶ *Presence of other metals.* Some metals corrode in contact with others because of galvanic action (see "Galvanic Action" on p. 95). Try to match nail composition to the

Recommended Nailing Schedule*

APPLICATION	FASTENER
Joist to sill or girder (toenail)	3-8d
1×6 subfloor or less to each joist (face-nail)	2-8d (or two 1¾-in. staples)
Wider than 1×6 subfloor to each joist (face-nail)	3-8d
2-in. subfloor to joist or girder (blind- and face-nail)	2-16d
Sole plate to joist or blocking (face-nail)	16d at 16 in. o.c.
Top plate to stud (end-nail)	2-16d
Stud to sole plate (toenail)	4-8d
Sole plate to joists or blocking	3-16d at 16 in. o.c.
Doubled studs (face-nail)	10d at 24 in. o.c.
Doubled top plates (face-nail)	16d at 16 in. o.c.
Doubled top plates, lap spliced (face-nail)	8-16d
Continuous header, two pieces	16d at 16 in. o.c., along each edge
Rim joist to top plate (toenail)	8d at 6 in. o.c.
Ceiling joists to plate (toenail)	3-8d
Continuous header to stud (toenail)	4-8d
Ceiling joists, laps over partitions (face-nail)	3-16d
Ceiling joists to parallel rafters (face-nail)	3-16d
Collar tie to rafter (face-nail)	3-10d
Rafter to plate (toenail)	3-8d
Built-up corner studs	16d at 24 in. o.c.
Built-up girders and beams	20d at 32 in. o.c., along each edge

* From the 2016 California Building Code Table 2304.10.1.

Pilot HOLES

Whenever you need to avoid bending nails in dense wood, such as southern pine, or when you're worried about splitting a board by screwing or nailing too close to the end, drill a pilot hole. There's no absolute rule to sizing pilot holes, but 50% to 75% of the nail shank diameter is usually about right.

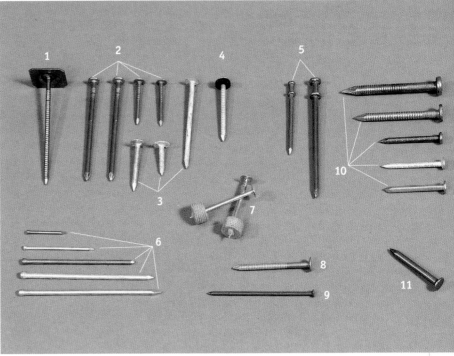

Specialty nails: 1. Simplex nail for roofing underlayment and thin foam insulation; 2. Copper flashing nails; 3. Galvanized roofing nails; 4. Gasketed nail for metal roofing and some skylight flashing; 5. Duplex nails; 6. Finish nails (the middle one is vinyl coated); 7. Furring nails for stucco wire; 8. Ring-shank nail; 9. Stainless-steel nail color matched to wood siding; 10. Joist-hanger nails (Teco nails); 11. Case-hardened masonry nail

metals present. The choice of nails includes aluminum, stainless steel, brass, copper, monel metal, and galvanized (zinc coated).

▶ *Exposure to weather and corrosion.* These days, professional builders use stainless-steel nails almost exclusively for outside attachment. They are more expensive than galvanized nails, but galvanized heads chip when hammered and so must be sealed with primer. You can get coils of stainless-steel nails to use in nail guns, and they come in a wide range of colors to match most siding stains. If you use redwood or cedar siding, colored, small-head stainless-steel nails are perfect. You don't have to set (sink) these nails to diminish their visual impact, as you would shiny nails. Just nail colored nail heads flush and they'll all but disappear.

▶ *Holding power.* Nails that are rosin coated, cement coated, or hot-dipped galvanized hold better than uncoated nails. Vinyl-coated nails both lubricate the nail shaft as you drive it in (friction melts the polymer coating) and act as an adhesive once the nail's in place.

Sizing nails. Please see the nailing chart on the preceding page. Most nailing schedules give you choices, depending upon the nail's length—

and thickness. That is, nails with thicker shanks can be spaced farther apart. In general, framing and sheathing nails should penetrate at least 1½ in. into the framing member. Thus, increase the nail length to accommodate the thickness of the material being attached, whether plywood, gypsum board, and so forth.

The workhorse of framing is the 16d common, although 12d or 10d nails are good bets if you need to toenail one member to another. Use 10d or 12d nails to laminate lumber, say, as top plates, double joists, and headers.

When nailing near the edge or the end of a piece, avoid splits by using the right size nail, staggering nails, not nailing too close to the edge, blunting nail heads (p. 79), and drilling pilot holes. Box nails, which have smaller-diameter shanks than common nails, are less likely to split framing during toenailing.

Pneumatic framing nails. When you've got a lot of nailing to do—say, to sheathe an addition— rent a nail gun (pneumatic nailer). Gun nails are typically a bit shorter and thinner than the traditional box and common nail sizes. Although some framing nailers can shoot a true common-nail thickness of 0.162 in., pneumatic nail shanks are typically 0.131 in. (Professional builders should follow the engineer's nail-sizing and spacing requirements on structural plans.)

Pneumatic NAILER MISCELLANY

▶ When using pneumatic nailers, set AIR PRESSURE so that nail heads stop just shy of a panel's face ply. Then use a framing hammer to drive each nail flush.

▶ Many pneumatic framing nails are VINYL coated to make them hold well.

▶ Pneumatic nail heads are frequently COLORED to denote the size you're loading.

▶ As a rule, avoid using "CLIPPED HEAD nails," which are not permitted for shear-nailing plywood.

Typical pneumatic nail sizes for a few common tasks:

▶ Nailing ⅜-in. or ½-in. shear plywood: 2¼ in. x 0.148 in. (shank diameter)

▶ Nailing ⅝-in. or ¾-in. plywood: 2½ in. x 0.148 in.

▶ Through-nailing 2x studs and plates: 3¼ in. x 0.131 in. or 3½ in. x 0.131 in.

▶ Toenailing 2x lumber: 3¼ in. x 0.131 in. or 3½ in. x 0.131 in.

SCREWS

Screws have revolutionized building. Thanks to a flood of specialized screws, builders can now detach, reposition, and reattach most building materials. This is especially important in renovation. When scribing cabinets or setting door-casing to walls that aren't plumb, for example, adjusting and repositioning materials is a fact of life. Likewise, when raising and attaching large beams and posts, it is often difficult to place things perfectly on the first try. With screws, you can put the lumber roughly where you want it, back out screws to fine-tune its position, and then screw it securely in exactly the right spot.

Screws also excel when you're working in spaces too tight to swing a hammer, thread a nut onto a long bolt, or gain access to both sides of a beam. But by using a right-angle drill or an impact driver, you can almost always find enough room to drive a structural screw.

Heads. The widespread use of impact drivers (p. 61) and hex-head screws has made slotted screws almost obsolete. When torque is applied, an old-fashioned screwdriver blade can slip out of a slotted screw head, whereas screw heads with centered patterns surround the point of the driver tip, holding it in place. Phillips-head screws were among the first types to "capture" the driver tip; today, there are many more, including square-drive, hex-head, Torx-head, and star-head screws.

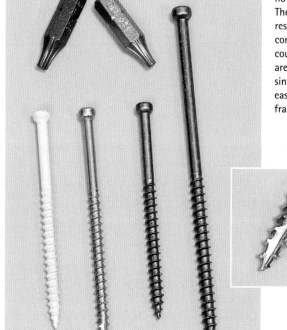

This handful of exterior trim-head finish screws shows how sophisticated and specialized screws have become. The white and brass-colored screws are climate-coated to resist weathering, whereas stainless steel is impervious to corrosion. Though all are intended for exterior use, they could also be used for interior work. Because their heads are small, just drive them flush and leave them; no need to sink or prime screw heads. The 4-in. screw at right could easily attach 2-in. fascia trim through sheathing to the framing behind.

Even screw tips have been engineered. The stainless-steel screw tip has a self-drilling slot that cuts out a little bit of wood as it goes in, so predrilling isn't needed and screws enter easily.

Today's screws are designed for specific uses. Trim-head screws have small heads like casing nails so they can be sunk easily. Drawer-front screws have integral washers so they won't pull though. Deck-head screws are designed to minimize "mushrooming" of material around the screw hole. Some structural screws have washer heads with beveled undersides so the screws will self-center in predrilled hinges or connector plates.

Threads. Screw threads are engineered for the materials they join. Traditionally, screws for joining softwoods are made of relatively soft metal, with threads that are steeply pitched and relatively wide in relation to the screw shaft. Screw threads for hardwoods tend to be low-pitched and finer. (The steeper the thread pitch, the more torque is needed to drive the screw.) When attaching dense particleboard or MDF, one should predrill and then use Confirmat-style screws, which have thick shanks and wide, low-pitched threads.

Many screws—especially structural screws—essentially drill their own pilot holes. Some screw types have points with serrated threads that look like tiny saw points, which cut through wood fibers as they advance. Screws with such self-tapping features require less torque to drive, so they go in faster and easier.

There are even screws that cut into concrete. Granted, you need a hammer drill to predrill pilot holes, but once the pilot is drilled, an impact driver will drive the screw home. The threads grab the concrete and hold fast; the trick is not overtightening and breaking screws.

Coatings and colors. Coatings matter most on screws used outdoors or in high-humidity areas. Although galvanized screws resist rusting and are relatively inexpensive, they won't last much more than 8 to 10 years on a deck—fewer if used near saltwater. One manufacturer, GRK Fasteners, promises "25 years in most applications" for its Climatek coated screws. Makers of epoxy-, polymer-, and ceramic-clad screws offer varying life spans.

Screws are available in a range of colors. The white trim-head screw in the photo on p. 81 would be unobtrusive against white siding or light gray material such as Boral siding. There are also brown trim-head screws, which will all but disappear when screwed into cedar siding; red trim-heads will do the same in redwood.

STRUCTURAL SCREWS

"Structural screws" include a broad range of screw types, but we'll focus on three:

▶ **Framing screws,** load-rated as alternatives to framing nails;
▶ **SDS and SD screws,** which attach hardware connectors to framing;
▶ **Concrete anchors,** primarily used to attach wood to concrete.

All structural screws are engineered and rated for the loads they must withstand, such as *withdrawal loading* (screws pulling out), *lateral load parallel to grain* (such as shear stress during seismic activity), *lateral load perpendicular to grain*,

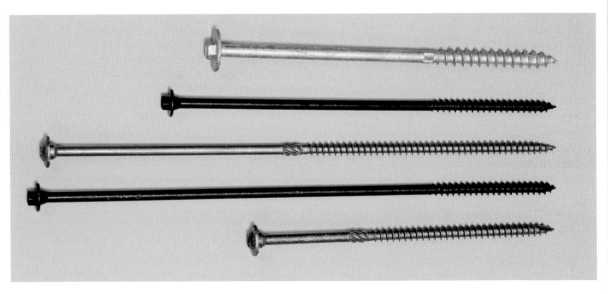

Climate-coated structural screws were designed to replace framing nails. In addition to having thicker shanks and higher structural values than nails, structural screws are less likely to pull out and can be backed out to reposition timbers exactly. The integral washered heads give the screws a larger bearing area.

and *multi-ply connections* (forces that tend to pull apart spliced members). Industry catalogs such as Simpson Strong-Tie's "fastening systems" spell out product load ratings in great detail, but local building authorities have the final say on what fastening systems are code-compliant.

Framing screws such as those shown on the facing page were designed and load-rated to replace framing nails such as 8d, 10d, 12d, and larger. A 6-in. structural screw, for example, is roughly the same length as a 20d spike. If you're screwing a 6-in. structural screw into, say, a header, it will be long enough to sink 2 in. or 3 in. into a stud.

Longer structural screws (12-in. and 14-in. lengths are readily available) might be used to hold timbers together, such as screwing beams into the end grain of posts in trellis construction. It's difficult to get a good end-grain connection (down through the top of a beam) that will resist withdrawal, but the deep threads of structural screws have a very positive grab. Such lengths are substantial enough to grab through the sides of headers, through beams into posts, or to toe-screw, say, a 6x12 beam to a 6x6 post. The large, washered heads of framing screws maximize the bearing area and prevent driving the screw too deeply into the wood.

In most situations, you do not need to predrill structural screws. In framing lumber, redwood, cedar, and other exterior finish woods, a ½-in. drill or an 18v impact driver should drive in screws cleanly. When driving a screw longer than 10 in., however, predrilling a pilot hole may strain your drill less and speed assembly. A correctly sized pilot hole should be the same diameter as the (unthreaded) shank of the screw. Once seated, the screw's wide, sharp threads and huge shear strength combine to create a joint that won't be pulled apart—although, of course, the screw can be backed out to reposition timbers.

SDS and SD screws are hot-dipped galvanized screws designed to be used with Simpson Strong-Tie connectors (pp. 84–86), a system of hardware designed to reinforce wood-to-wood and wood-to-masonry connections. Specifically, SD screws are designed to replace *nails* in certain connectors such as joist hangers, though Teco nails are still widely used because they are less expensive. And ¼-in.-diameter SDS screws replace *bolts* formerly used to attach heavier Simpson hardware such as post caps, post bases, hold-downs, and heavy straps.

Here again, SDS and SD structural screws were developed because they are quicker and easier to install and best suited to applications in

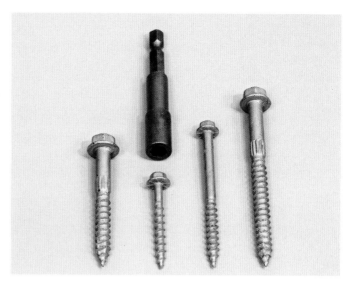

SDS and SD hex-head structural screws, whose shanks are sized to fit in the holes of various Simpson hardware. At top, a hex driver. The two outer screws are ¼-in.-diameter SDS screws; the two inner screws are SD screws, typically used instead of Teco nails to attach joist hangers.

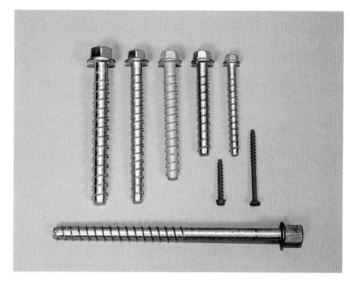

An assortment of concrete anchors, which are specialized screws that attach wood to concrete. The larger ones are Titen bolts, which can be used when anchor bolts are not required, such as when attaching the sills of non-load-bearing interior walls to concrete. The huge bolt at the bottom of the photo is a ³/₄-in. specialty hanger bolt, which has a threaded hole in its oversize nut.

tight or inaccessible spaces. The problem with bolting assemblies was that there frequently wasn't enough room for the drill, bit, and bolt once the framing was installed, so builders had to anticipate the process and predrill holes in lumber before it was framed out. Failing that, they often had to put the drill in the next stud bay out and use a drill extension to drill through one stud before they could reach the bay in which the

PRO TIP

Backing out a screw and repositioning a timber is usually straightforward unless the new hole is so close to the old one that the screw tip "seeks" it. Thus, when repositioning, you might want to move the screw over an inch or two so the new hole is well away from the old one.

hardware was to be bolted. It was often an exercise in frustration.

Instead, with a hex-head bit in a right-angle drill, you simply fit the bit over the hex head of an SDS or SD screw and the screw is snug in 15 seconds.

Concrete anchors (AKA masonry anchors) join wood to concrete. All concrete anchors require pilot holes—and very precise ones at that. Usually when you buy a box of masonry anchors, it will include a bit correctly sized for the pilot hole. (Usually, the shank size is the bit size, so the threads cut through the concrete along the sides of the pilot holes.) For the pilot hole, use a rotary hammer, but use an impact driver to drive in the masonry anchors.

If you run the drill bit in and out a few times, it will clean out most of the concrete dust and debris in the hole. It's important that the pilot hole is deep enough so the anchor can go all the way in. If the hole isn't deep enough, the anchor may get stuck part of the way in and be a monster to get out.

Titen bolts, the larger screws shown in the bottom photo on p. 83, anchor wood or metal to concrete. They are used instead of epoxied-in or mechanical anchor bolts. Because you drill a pilot hole beforehand, you can use almost any tool with enough torque to drive these anchors

in—the resistance of the threads cutting into the sides of the pilot holes is relatively slight. Use an impact driver or a ½-in. right-angle drill to drive them in.

The huge bolt at the bottom of that photo on p. 83 is a specialty hanger bolt, which has a threaded hole in its oversize nut, into which you could attach an all-thread rod. A bolt this thick (¾ in. diameter) might also be called for where there are significant lateral forces—say, to keep a mudsill from sliding off a foundation during seismic activity.

And this brings us back to screw coatings. When attaching pressure-treated lumber such as a mudsill, it's best to use hot-dipped galvanized screws and connectors, which will resist being corroded by the chemicals in the lumber.

Bolts

Bolts are used to join major structural members, although as noted previously, structural screws are supplanting them. In general, machine bolts and carriage bolts have nontapering, threaded, thick shanks.

Some bolts are more than 1 in. in diameter and longer than 2 ft. Allthread (threaded) rod comes in lengths up to 12 ft. and can be used with nuts and washers at each end. Carriage bolts have a brief section of square shank just

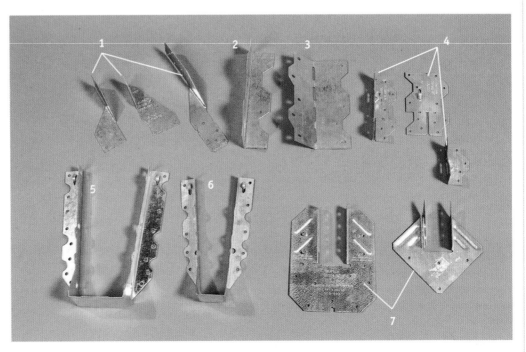

Wood construction connectors (contractors generally refer to them by their Simpson catalog numbers): 1. H2.5, H4, and H8 hurricane ties; 2. L90 reinforcing angle; 3. LS70 skewable angle (bend one time only); 4. A35, A35F, and A34 framing angles; 5. U410 face-mount hanger for 4x10 beam (or double 2x10s); 6. LU28 face-mount hanger for 2x8 joist; 7. H10 and H1 hurricane ties

below the head. Lag screws (also called lag bolts) have a hex head, but the lower half of the shank tapers like a wood screw.

WOOD CONSTRUCTION CONNECTORS

Wood construction connectors are commonly called Simpson Strong-Ties after the company that popularized them. For a complete overview of available connectors from Simpson, go to www.strongtie.com. Professionals swear by these ingenious connectors for three main reasons:

▶ They offer wood-to-wood connections superior to most traditional construction methods. For example, unlike toenailing, metal connectors are unlikely to split lumber ends or loosen under stress. These galvanized steel connectors are strong and durable.

▶ They greatly strengthen joints against earthquakes, high winds, and other racking forces. They can tie rafters to walls, walls to floor platforms, and the substructure to its foundation.

▶ Most can be attached to existing framing, a great boon to renovators, and in many cases steel connectors are the only cost-effective way to bolster the existing structure and tie additions to the original structure.

Joist hangers are indispensable in renovation when you want to add joists but can't end-nail, either because access is limited or because you're using engineered lumber, which is too thin in cross section to end-nail successfully. (Sawn-lumber joists and I-joists require different hangers.) There are joist hangers for single joists, double joists, 4x10 beams, joists intersecting at a 45° angle, and so on. You also have the choice of face-mount or top-flange hangers. Top flanges are popular because they effortlessly align the tops of I-joists with the top of a header or beam.

Strap ties come in myriad shapes—tees, right angles, twists—but all help keep joints from pulling apart. Install flat strap ties where wall plates are discontinuous or where rafter pairs meet at the ridge. Strap ties also keep floor platforms from separating, much as shear walling does. *Hurricane ties*, or twist straps, have a 90° twist to join rafters to top plates, thereby fighting the tendency of roofs to lift during a strong crosswind. T-straps and L-straps are face-nailed to members joining in a right angle.

Framing angles are used extensively to reinforce wood connections at 90° intersections. For seismic strengthening and on shear walls, framing angles help prevent floor framing from slipping off walls and supports during an earthquake.

Simpson Top-Flange Hanger

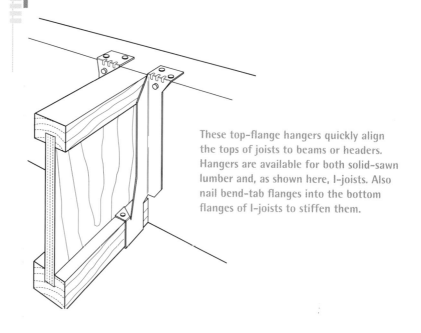

These top-flange hangers quickly align the tops of joists to beams or headers. Hangers are available for both solid-sawn lumber and, as shown here, I-joists. Also nail bend-tab flanges into the bottom flanges of I-joists to stiffen them.

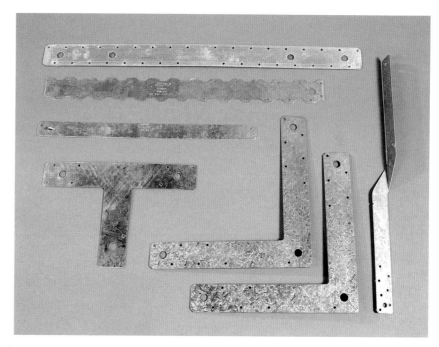

Straps, ties, and angles. Clockwise, from lower left: T-strap (post-to-beam connector), SST22 light-gauge strap, ST6224 (24-in.) strap, MST24 heavy-gauge strap, twist strap, and heavy-gauge L-straps.

Hold-downs are massive steel brackets that anchor framing to foundations and, when used with long threaded rods, join framing on different floors. When retrofitting to a foundation, use epoxy to attach hold-down bolts to concrete, as shown in the right photo on p. 278.

Caps and bases: 1. A CBSQ44 post base anchors a 4x4 to concrete; 2. BC6 post cap and base; 3. AC6 post-to-beam connector, which can be attached after the members are in place; 4. BC460 half-base; 5. BC46 post cap and base

Made of 16-gauge steel, nail plates protect plumbing pipes and electrical wires from being punctured by nails.

Clips vary by function. *H-clips* are an alternative to solid blocking when installing roof sheathing. They also act as $\frac{1}{16}$-in. spacers so roof sheathing can expand. *Drywall clips* allow you to eliminate some blocking in corners, but it's best to use these clips sparingly; solid blocking is much stronger. *Deck clips* are nailed to deck joists, and then 2x4 decking is driven onto the sharpened point of the clip. You can lay down decking without having to face-nail it; the clip also acts as a spacer so that water can clear.

Post bases and caps provide strong connections while eliminating the need to toenail posts, which tends to split them. Post bases are typically set in concrete, with posts then bolted or nailed to the base. Bases also reduce post rot, for their raised *standoffs* elevate the post and double as a moisture shield. Post caps resemble a pair of U-brackets set at right angles to each other: One U, upside down, straddles the top of the post, while the other, right side up, receives the beam on which the joist will sit.

Miscellaneous metal connectors are often needed. Where you absolutely must notch stud edges to accommodate plumbing, a metal *shoe plate* reinforces the stud and protects the pipe from errant nails. Speaking of protection, this discussion would be incomplete without mentioning *nail plates*, which protect wires and pipes from stray nails when finish walls go up.

Adhesives, Caulks, and Sealants

Common adhesives, caulks, and sealants look the same, but their formulations are complex and carefully formulated for specific materials and expected conditions. They all generally come in 10-oz. cartridges that fit into an applicator (caulking gun). You simply cut the cartridge nozzle to the desired diameter and squeeze the long pistol-grip trigger to lay down beads of the stuff.

CONSTRUCTION ADHESIVES

Construction adhesives bond to a variety of materials, including standard lumber, treated lumber, plywood and OSB panels, drywall, wall paneling, rigid insulation, concrete and masonry, tile, metal, and glass.

Construction adhesives are a boon to builders. Instead of nailing sheathing every 6 in. around panel edges and every 10 in. in the field, builders using adhesives need nail only every 12 in. Being flexible, adhesives fill surface irregularities and double as sealants. Structurally, panels bonded with adhesive are stiffer and capable of bearing greater loads than panels that are only nailed. Floor sheathing and stair treads bonded with construction adhesives are far less likely to flex, pop nails, or squeak. Drywall ceiling panels bonded with adhesives do a better job of deadening sound and cutting air infiltration.

A number of factors should determine your choice of adhesives: the materials being joined, strength, durability, flexibility, shrinkage, job-site conditions (especially temperature and humidity), workability, curing time, ease of cleanup, odor, and—increasingly important—an adhesive's toxicity and green profile. Accordingly, there is now a wide selection of solvent-free, low-VOC adhesives that are nontoxic. To help you make your choice, many manufacturers now offer interactive, online product selectors.

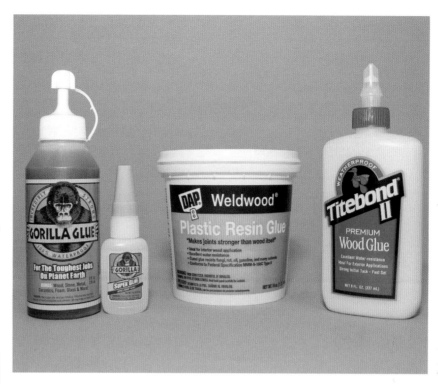

Four popular adhesives, followed by their chemical names. From left: Gorilla Glue (polyurethane), Gorilla Super Glue (cyanoacrylate), Weldwood plastic resin (urea formaldehyde), and Titebond II wood glue (polyvinyl acetate).

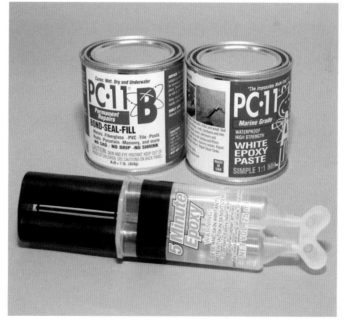

Epoxy resins are typically mixed from two components, whether as a paste mixture used to fill sizable areas or as a dual-cylinder injector where the ingredients meet at the point.

Sticky NAMES

▶ CAULKING was the first on the market, so that's the default word for materials that fill gaps and keep out weather.

▶ SEALANTS generally refer to high-performance compounds such as silicones and polyurethanes.

▶ ADHESIVES stick things together; their fancy name is quite a mouthful: elastomer-based extrudable mastics. ELASTOMER indicates that the product will stay flexible and rubbery, and MASTICS is a general term denoting any pasty or gooey adhesive. That noted, adhesives also seal and sealants adhere, although that's not their main mission.

PRO TIP

Choosing the right adhesive is tough. These compounds are cutting-edge chemistry. Formulations of familiar brands change often—and often without warning. So here are four commonsense tips: (1) Use manufacturers' online product selectors to narrow your choice of adhesives. (2) Get friendly with the resident adhesive expert at the lumberyard, who may have gotten a lot of feedback from the field. (3) Test a tube before you buy a case. (4) "But it's cheap" is always a fool's bargain.

As a rule of thumb, both solvent-free and solvent-based adhesives have the same holding power once they've cured. However, it should be noted that solvent-free adhesives, which cure by evaporation, can be affected by rain and job-site temperatures outside a range of 40°F to 100°F. Also, solvent-free adhesives may set poorly between two nonporous surfaces and won't bond to some metals. On the other hand, solvent-based adhesives, which cure via a chemical reaction, can bond to most surfaces and are little influenced by temperature or the presence of water. Chronic exposure to airborne VOCs, however, can cause serious health problems.

Many new waterproofing materials are self-adhering (have adhesive backing), so if you are using a waterproofing membrane, make sure that any sealants or caulks that you intend to use with that membrane are compatible (visit manufacturers' websites for information). The chemical agent in a caulk could react badly and compromise a seal. You're probably pretty safe using brands if they have the same basic chemical composition (e.g., contain a polyurethane or acrylic base). A lot of companies make their own lines of sealants and caulks, so choosing their system will probably guarantee compatibility—and that warranties will be honored.

A POTPOURRI OF ADHESIVES

Here's a primer on common adhesives.

Polyurethanes are a great all-purpose waterproof adhesive, capable of bonding wood, stone, metal, ceramics, and so on. Strong, versatile, and easy to use, polyurethanes such as Gorilla Glue and PL Turbo are favorites with builders and woodworkers. Because a chemical reaction cures these glues, they are not affected by moist or oily surfaces; they can even join pieces that have been finished. Before polyurethane glues dry, they can be removed by a solvent such as acetone or mineral spirits.

Polyvinyl acetates (PVA) are a broad category of glues that range from water-based brands such as Elmer's white glue to Titebond II and III, yellow glues whose emulsions combine water and polymers. PVA glues are further divided into three types: Type I is waterproof, Type II is water-resistant, and Type III is not water-resistant. Beyond that, all PVA-glued joints need to be clamped until they cure, and, once dry, they all have about the same holding strength.

Cyanoacrylates (CA) are one-part or two-part glues that set in 10 seconds to 10 minutes and include the popular brands Krazy Glue and Super Glue. Because they set strong and quick, CAs are prized by carpenters working one-handed, say, to hold a miter joint in place until the glue sets. More expensive and much faster setting than polyurethane or PVA glues, CAs are not well suited to joining large surfaces.

Epoxy resins are famous for their strength. Typically mixed from two components, epoxies can bond to materials on which almost nothing else will—that is, when the surface areas to be bonded are small or when dampness is extreme. Epoxy products are especially important in foundation repairs and seismic strengthening (for more information, see chapter 10). Unlike most other adhesives, epoxies also maintain structural strength across a gap.

Styrene-butadiene is a good all-purpose exterior and interior glue for joining materials of low porosity, such as tile and masonry.

Hot-melt glues are applied using an electric glue gun and are excellent for tacking surfaces quickly—they're the glue of choice for building templates out of thin plywood strips (see chapter 13).

SEALANTS AND CAULKS

As noted in "Sticky Names" on p. 87, sealants and caulks do pretty much the same thing: fill gaps, keep moisture at bay, and reduce air infiltration. Sealants tend to last longer, perform better, and cost more. Here's a look at the strengths and weaknesses of three major types of caulk.

Although caulks aren't quite as diverse as construction adhesives, they do have varying formulations and properties. For specifics, go online or visit your lumberyard.

Silicones are arguably the most durable and most water-resistant of any caulk. They are especially suitable for window glazing and slick bathroom and kitchen surfaces.

▶ *Advantages:* **Silicones are incredibly tenacious on nonporous materials such as glass, glazed ceramic tiles, and metal. There's little shrinkage, and they can be applied at –40°F. Silicone sealants specified for metal flue pipe function at 500°F. And silicones have the best long-term flexibility, UV resistance, and weatherability. Also, molds won't grow on them.**

▶ *Disadvantages:* **Silicones are messy to work with; wear rubber gloves to protect your skin. Once silicones have cured, it's almost impossible to remove them. Plus, they are bond breakers—that is, because nothing will** stick to an area they've tainted, think twice about trying them on wood, concrete, or other porous surfaces they don't adhere well to. Avoid inhaling acetoxy silicones, and don't use them on metal because they'll corrode it. Pure silicones can't be painted, although siliconized acrylics can.

Polyurethanes are versatile multipurpose caulks but are not as tenacious as silicones.

▶ *Advantages:* Because they attach equally well to wood, masonry, and metal, they're good for caulking joints where dissimilar materials meet. Polyurethanes won't corrode metal. They're easy to work, although polys get pretty stiff as temperatures approach freezing. Shrinkage is negligible. They're great for skylight flashing and metal roofs. And they can be painted. Also, they're easier to work than silicones, even though they're solvent based.

▶ *Disadvantages:* Polyurethanes have poor UV resistance, but additives or painting can improve that dramatically. Although they are a good all-purpose caulk, they don't have the durability or shelf life of silicones.

Latex acrylics are a good balance of performance, price, and workability.

▶ *Advantages:* Latex acrylics are water based, hence nontoxic, largely odor free, and easy to apply (you can shape caulk joints with your finger). They clean up with soap and water. They adhere well to a range of materials, have good UV resistance, and can be painted. Durable once cured, they are best used in protected areas in temperate climates.

▶ *Disadvantages:* Expect significant shrinkage (up to 30%) and long curing times. Although good as bedding caulk under door or window casing, they're iffy as exterior caulk or shower and tub caulk. Properties vary widely from product to product. Although some manufacturers tout spectacular performance specs, check out online contractor chat groups for real-life performance ratings.

5 Roofs

A roof is a building's most important layer of defense against water, wind, and sun. Properly constructed and maintained, a roof deflects rain and snowmelt, and routes them away from other house surfaces. Historically, roof materials have included straw, clay tile, wood, and slate. Although many of these materials are still used, most roofs installed today are asphalt-based composites.

If roofs consisted simply of two sloping planes, covering them would be easy. But today's roofs have protruding vent pipes, chimneys, skylights, dormers, and the like—all potential water dams and channels that need to be *flashed* to guide water around them. Then, as runoff approaches the lower reaches of the roof, it must be directed away from the building by means of overhangs, drip-edges, and—finally—gutters and downspouts.

This chapter assumes the foundation and framing are stable. Because structural shifting or settling can cause roofing materials to separate and leak, you should fix structural problems before repairing or replacing a roof.

Roof ventilation is another important part of house health and comfort. Venting excessively moist air can forestall condensation, mold, rot, and, in colder climates, ice dams. Heat buildup under the roof can also be mitigated by adequate ventilation. The topic is so important that it is discussed here and in chapters 7 and 14.

Roof Safety and Equipment

Among the building trades, roofing is considered the most dangerous—not because it's inherently hazardous but because it takes place high above the ground. The steeper the roof pitch is, the greater the risk. If heights make you uneasy or if you're not particularly agile, hire a licensed contractor.

Although these wood shingles seem randomly placed, the installer is taking great pains to offset the shingle joints between courses and maintain a minimum exposure of 5 in. so the roof will be durable as well as distinctive.

A gauge stop on the base of a pneumatic nailer can be used to set the correct shingle exposure. When the gauge stop is snug to the butt of a shingle, the next shingle above—placed against the nose of the tool—will be positioned correctly.

COMMONSENSE SAFETY

▶ Stay off the roof unless you have a compelling reason to be on it. Besides being hazardous to you, walking on a roof can damage roofing materials.

▶ If you must work on a roof, make sure there's a second person within earshot in case you fall or need help.

▶ Don't venture up when the roof is wet or near freezing or extremely warm. When wet, most roofing materials are slippery. Cold asphalt shingles are brittle; warm asphalt can stretch and tear. Always wear shoes with soft, nonslip soles.

▶ Position ladder feet securely away from the building about one-quarter of the ladder's extended length. Never lean sideways from a ladder. If you can't reach something while keeping your hips within ladder sides, move the ladder.

▶ When installing a roof, use scaffolding with a safety rail. The most dangerous part of a roofing job—apart from tearing off shingles and underlayment—is applying the first few courses along the eaves.

▶ When walking on a roof, try to "walk on nails." In other words, try to walk directly over the rafters, where the sheathing is nailed. The roof will be less springy over rafters, and you'll be less likely to break through rotten sheathing.

▶ Follow the roofing manufacturer's installation instructions, which often provide time- and money-saving tips. Moreover, if roofing has been correctly installed, manufacturers are more likely to honor their warranties.

EQUIPMENT

Unless you are installing roofing systems that need to be "torched" (heat-sealed with a propane torch), you won't need a lot of specialized equipment or tools. Most of the items discussed here are safety related.

Footwear should be sneakers or other soft-soled shoes that grip well on a roof. Old-time roofers prefer boots with thick soles that are less likely to be punctured by stray nails, but they are inflexible and don't grip as well.

Scaffolding can make applying the first few courses along the eaves far safer. After the lower courses are installed, the scaffolding serves mainly as a staging area for materials and tools.

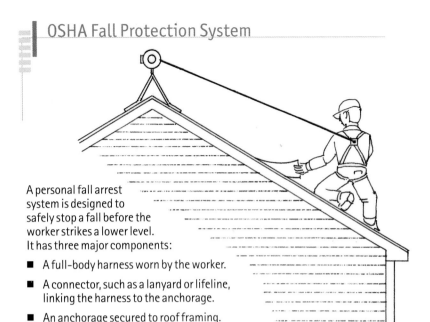

OSHA Fall Protection System

A personal fall arrest system is designed to safely stop a fall before the worker strikes a lower level. It has three major components:

- A full-body harness worn by the worker.
- A connector, such as a lanyard or lifeline, linking the harness to the anchorage.
- An anchorage secured to roof framing.

PRO TIP

As you prepare to install roofing components, imagine rainwater running down the roof. To prevent water running underneath underlayment, flashing, or shingles, roofing materials upslope must always overlay those below.

Roofing jacks enable you to work safely on roofs with a 6-in-12 pitch or steeper. (Several different roof pitches are shown in "Gutter Lip and Roof Pitch" on p. 190.) Jacks also are indispensable platforms for storing materials. As you work up a roof, install additional pairs of jacks whenever you must stretch to nail the next course. Many roofers leave jacks installed until the job is almost completed.

Full-body harnesses, also known as personal fall-arrest systems (PFAS), are required for most roofing work. Though state OSHA regulations typically cite a "trigger height" of 20 ft. for roofing work, a height as little as 7½ ft. above ground may trigger regulations for "work on unprotected platforms, scaffolds, or edges of structures." For the most recent regulations, go to: www.dir.ca.gov/dosh/dosh_publications/Fall-Protection-in-Construction-fs.pdf.

Knee pads, sturdy ones with integral plastic cups, will spare you a lot of pain. In addition, get a foam kneeling pad such as those gardeners use; it will allow you to sit or kneel while shingling and will protect shingles from abrasion.

A pneumatic nailer, which you can rent, speeds the job along. Have the rental company recommend nails and explain the nailer's correct use.

Shingling hatchets in the hands of a pro can fasten shingles almost as fast as a pneumatic nailer.

Miscellaneous tools include a utility knife with hooked blades, a straightedge or framing square, a hammer, a chalkline, a tape measure, caulk guns, work gloves, safety glasses, and hearing protection.

Preparing to Install a New Roof

Daily temperatures on a roof or in an unventilated attic can swing from 50°F to 150°F, thereby causing tremendous expansion and contraction of roof materials. Improving ventilation under the roof, as shown in "Roof Venting" on p. 102, can prolong shingle life somewhat, but the key to a long-lasting roof is the quality of the materials and attention to details, especially flashing around chimneys, skylights, and vent pipes.

In the long haul, shingles with a 30-year warranty are a smarter buy than 20-year shingles because they last significantly longer, even though they cost only a little more. Typically, 75% of a roofing job's cost is labor.

WHEN IT'S TIME TO TEAR OFF

Short term, you can save money by installing a new roof over an old one if local codes allow. However, new roofing applied over old (see p. 109) rarely lasts as long as roofing installed on a stripped and properly prepared substrate.

You *must* tear off existing roofing under these conditions:

The roof already has two roofing layers. Two is the limit for most local codes because it's virtually impossible to install a third layer that will lie flat. Even if you could, three layers would be a nightmare to flash and nail correctly. Underlying shingle layers are a springy substrate to nail through, and old wood shingles often split and migrate. Besides, if the bottom layer is wood shingles over *skip-sheathing* (1-in. boards with

Roof Longevity

MATERIAL	LONGEVITY (years)
Slate and tile*	80
Wood shingles or shakes	20–25
Metal	40–50
Asphalt shingles	20–50 (depends on warranty
Three-ply built-up roof	15–20
Four-ply built-up roof	20–25

Underlayment quality also determines how many leak-free years you can expect. With 15-lb. building paper underneath, a tile roof might start leaking in 15 years; modified bitumen under tile could help create an 80-year roof.

spaces between), only half of the new roofing nails would be likely to hit sheathing. Additional layers would be poorly attached and therefore wouldn't last. Also, multiple layers of shingles put a tremendous strain on framing and sheathing.

Sheathing and rafters show extensive water damage. When you can't determine exactly what's been causing leaks, it's time to strip. The previous roofers may have installed flashing incorrectly or not at all. Or roofers may have left tired old flashing in place. Whatever the cause, if the remedy is stripping back extensive sections of roofing to replace faulty flashing, reroofing may be the most cost-effective cure.

Rafters and sheathing are undersize. If rafters are too skimpy, the roof will sag, especially along the ridge. If the sheathing is too thin, the roof will sag between rafters and look wavy. The remedy may be stripping the roof and nailing ½-in. plywood over old sheathing or bolstering undersize rafters with new lumber, but let a seasoned professional make the call.

Shingles are prematurely worn, curling, or missing. If a roof is relatively new and these symptoms are widespread, suspect product defects, inadequate ventilation, faulty installation, or a combination of those factors. A layer of new shingles won't lie flat over curling ones. So if shingles are curling—even if there is only a single layer of roofing—tear them off.

Adjacent roof sections must be replaced. This is a judgment call. When a house has additions that were roofed at different times, they probably will need reroofing at different times. Likewise, south-facing roof sections age 20% to 30% faster than north-facing ones. If you see signs of leaks, strip the whole roof, install flashing, and reroof.

STRIPPING AN OLD ROOF

Stripping a roof is one of the nastiest, dirtiest, most dangerous jobs in renovation. If you can afford it, hire an insured contractor for this. Most roofing contractors know of tear-off crews that will obtain permits, rip off the old roof, and cart away the debris, or you might want to subcontract the job through a roofer. Professional stripping takes at most a couple of days. It's money well spent.

If you must strip the roof yourself, remove all roof gutters and then minimize the mess by buying a heavy 6-mil plastic tarp to catch shingles and old roofing nails. So you won't be picking shingle shards and nails from the lawn for years to come, lay tarps from the house to the Dumpster as well. To protect plants around the house, place sawhorses or 2x4 frames over them

and cover with cloth drop cloths. *Caution:* Don't cover plants with plastic or they'll bake. Finally, lean plywood in front of windows so falling objects don't break them. When the job is done, rent a *magnetic roller* (also called a magnetic nail broom) and roll the lawn to locate roofing nails—before your lawn mower does it for you.

And don't forget the inside of the attic. Spread plastic tarps over attic floors, especially if there's insulation between the joists. During tearoff, an immense amount of debris and fine dust falls into an attic. Unless you catch it in plastic and remove it, you could breathe it or smell asphalt-shingle residue for years.

Other than that, stripping is mostly grunt work. Most strippers use a specially designed tear-off shovel, starting at the top and working down, scooping shingles as they go. Tear-off shovel blades have a serrated edge that slides under nail heads and a fulcrum underneath that pops nails up. Be sure to tear off all old building paper (felt or rosin paper), too.

SHEATHING

Once you've stripped off roofing, survey the sheathing for damage and protruding nails. As you pound down nails, be sure to place your feet directly over rafters. Probe suspect sheathing and replace any that's soft. Cut bad sections back to the nearest rafter centers. For this, wear safety glasses and use a circular saw with a carbide-tipped, nail-cutting blade because the blade will hit a lot of nails. Replacement pieces of sheathing should be the same thickness as the original.

If the old roof was wood-shingled, it probably had skip-sheathing, which consists of 1x4s spaced 5 in. on center. Skip-sheathing allows air to circulate under wood shingles. If the boards are in good shape, you can nail on new wood shingles after stripping old ones. But many contractors prefer to sheath over the 1x4s with ½-in. exterior-grade plywood (for rafters spaced 16 in. on center) or ⅝-in. plywood (for rafters 24 in. on center). This stiffens the roof and makes it safer to work on, but plywood virtually eliminates airflow under shingles. Consequently, some builders install a synthetic mesh, Cedar Breather, over plywood to increase circulation, before nailing on wood shingles.

Run plywood lengths perpendicular to rafters, centering plywood edges over rafter centers. Nail every 6 in. with 8d galvanized nails. Between rafters, use H-clips to support panel joints and create ¹⁄₁₆-in. expansion gaps. Sweep the roof well, and hammer down nail pop-ups.

A stripping shovel is a flat–nose shovel whose blade is serrated to peel off shingles and grab nail shanks. The fulcrum on the back increases leverage as you pry up old nails and roofing.

PRO TIP

One of the hazards of stripping a roof is its mind-numbing tedium. Besides staying alert, earlier-mentioned safety considerations apply here as well: scaffolding with a safety rail, proper footwear, and another person within earshot. If the roof is steep, wear a body harness secured over the ridgeline (see the drawing on the facing page).

UNDERLAYMENT

Once limited to building paper, underlayment now includes self-adhering rubberized sheets that replace metal flashing in some cases.

Weather-resistant underlayment. Traditionally, underlayment has been 36-in.-wide, 15-lb. or 30-lb. felt paper used as a *weather-resistant layer*. It keeps sheathing dry until shingles are installed, serves as a backup layer when water gets under shingles or flashing, and prolongs shingle life by separating sheathing and shingles. (Without underlayment, shingle asphalt can leech into wood sheathing, or resins in sheathing can degrade the shingle.) Heavy-duty, 30-lb. felt paper is often specified in high-wear, high-water areas such as eaves and valleys. A standard roll of 3-ft.-wide, 15-lb. felt paper covers roughly 400 sq. ft.; the same-size roll of 3-ft.-wide, 30-lb. paper covers only 200 sq. ft. because it's roughly twice as thick.

Because unreinforced lighter grades of building paper (15 lb.) tear easily and wrinkle when wet, some types are now reinforced with fiberglass. Moreover, all asphalt-impregnated building papers dry out and become less water-resistant when exposed to sunlight, so the sooner they're

Some WSU membranes may be degraded by petroleum-based roofing cements. If you need to caulk near WSU, urethane caulks are probably the best choice, but check product specs to be sure.

A hammer tacker allows you to staple building paper quickly so the paper won't slide down the roof as you roll it out. Later, secure the paper with tabbed roofing nails.

covered by shingles, the better. Building paper was never intended to be an exterior membrane.

When installing building paper on a sloped roof, have a helper and work from the bottom up. As you roll the paper out, it will tend to slide down the roof, so be sure to unroll it straight across the roof. The first course of paper should overlap a metal drip-edge nailed along the eaves. Align the paper's lower edge to the lower metal edge and unroll the paper, stapling as you go. Staples are only temporary fasteners to keep the paper from bunching or sliding. Paper should be nailed down with tabbed roofing nails (also called Simplex nails) shown on p. 80. Along roof edges and where the ends of the building paper overlap, space nails every 6 in.; don't nail within 1 in. of the edge. In the field, place nails in a zig-zag pattern, spacing them 12 in. to 15 in. apart.

For steep-slope roofs (4-in-12 or steeper), overlap horizontal courses of building paper 2 in. Overlap ends of seams (end laps) at least 6 in. To prevent water backup on low-slope roofs (less than 4-in-12), building codes often specify two plies of 36-in.-wide underlayment, with horizontal seams overlapped at least 19 in. and (vertical) end seams overlapped 12 in. Check your local code to be sure, for it may also specify self-sticking waterproof shingle underlayment along the eaves.

If you're installing wood shakes or slate, use shake liner, 18-in.-wide rolls of 30-lb. building paper alternated between roofing courses.

Waterproof shingle underlayment (WSU). WSU is a heavy peel-and-stick bituminous membrane that protects areas most likely to leak because of concentrated water flows in valleys, ice dams at eaves, or high winds at eaves and rake edges. Many building codes also specify WSU in lieu of building paper where asphalt shingles are

Eaves, Rake, and Underlayment Details

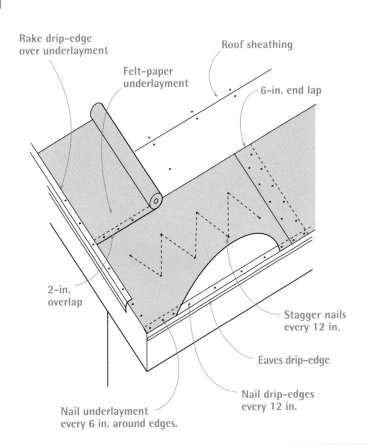

Rake drip-edge over underlayment

Felt-paper underlayment

Roof sheathing

6-in. end lap

2-in. overlap

Stagger nails every 12 in.

Eaves drip-edge

Nail drip-edges every 12 in.

Nail underlayment every 6 in. around edges.

Temporarily staple one side of the WSU to the sheathing, fold it in half lengthwise, and peel off the release sheet for that half. Roll out the untacked half, smoothing it from the valley outward. Pull the stapled half free and repeat.

installed on low-slope roofs. In addition to being self-adhering, WSU also self-seals around nails, making it a truly waterproof membrane.

WSU comes in 9-in. to 36-in. widths. Brand names include CertainTeed Winterguard, W.R. Grace Ice & Water Shield, and GAF Weather Watch.

It's vastly easier to install WSU along valleys than it is to use roll roofing or aluminum flashing. And with WSU, there's no need to trowel on wide swaths of roofing cement between 30-lb. building-paper layers to protect eaves from ice dams. Snap a chalkline to position the WSU. Then unroll and cut the membrane to length, align it to the chalkline, peel off its release-sheet backing, and press the material down. Most manufacturers recommend rolling it once it's down to make sure it's well adhered.

WSU is more easily installed with two people, but if you're working alone, fold the WSU in half lengthwise, and temporarily staple one edge of the membrane to a chalked guideline. Peel off the release-sheet backing from the unstapled half of the WSU, and flop the adhesive side of the membrane over onto the sheathing. Finally, yank the stapled edge free, peel off the backing from the second half, and stick it to the sheathing. Because WSU is self-adhering, you needn't nail it; you need to use staples only to keep the sheet from sliding around before sticking it down.

Galvanic ACTION

A number of metals, if paired, will corrode one another in a process called galvanic action. To be safe, use nails or clips that are the same metal as the flashing you install. Because water is an electrolyte, any moisture present will increase corrosion. The following metals make up a group known as the electrolytic sequence.

1. ALUMINUM
2. ZINC
3. STEEL
4. IRON
5. NICKEL
6. TIN
7. LEAD
8. COPPER
9. STAINLESS STEEL

When you pair up materials, the metal with the lower number will corrode faster. If you must pair two different metals, you can retard galvanic action by insulating between the metals with a layer of 30-lb. building paper.

Roof Flashing

Because underlayment directs water away from sheathing, it's technically flashing, too. More often, however, *roof flashing* refers to sheet metal that protects building seams or edges from water penetration or diverts water around pipes, chimneys, dormers, and other penetrations. Metal flashing is widely used because it's durable and relatively easy to cut and shape. Always replace old flashing when installing a new roof.

MATERIALS

Various sheet materials are suitable for flashing. Unformed, they come in sheets 10 ft. long or in rolls of varying lengths, widths, and gauges. Copper is the longest lasting and most expensive. Lead is the most malleable but also the most vulnerable to tears and punctures. Galvanized steel ranks second in longevity, but it's so rigid that you should buy it preshaped or rent a metal-bending brake to use on site.

Lightweight aluminum is commonly shaped on site, and it's a good compromise of cost and durability. There's also painted steel flashing, in case you don't like the glare of bare metal.

When installing flashing, use the fewest nails possible and avoid nailing in the center of a flashing channel, where water runs. If possible, position nails so the heads will be covered by roofing. Where you must leave nail heads exposed—for example, when installing skylight flashing or wall cap flashing—put urethane caulk under the nail heads before driving them down or use gasketed roofing nails.

DRIP-EDGE FLASHING

Drip-edge diverts water away from roof edges so it won't be drawn by capillary action back up under shingles or sheathing. Drip-edge also covers and protects sheathing edges from gutter splashback and ice dams along the eaves and gives rake edges a clean, finished look. The crimped edge of drip-edge flashing also resists bending and supports overhanging shingles.

Drip-edge is sold in varying widths and comes in an L shape or a lopsided T shape. Install drip-edge along the eaves first, nailing it directly to sheathing, using big-head roofing nails. Space nails every 12 in. or so. Underlayment along the eaves will overlap the top of the drip-edge. Along the rake edges of a roof, install underlayment

Drip-edge flashing allows water to drip free from roof edges. Here, copper drip-edge is being retrofitted under an existing roof before being fastened with 3d copper flashing nails. An uninstalled piece is shown lying atop the shingles.

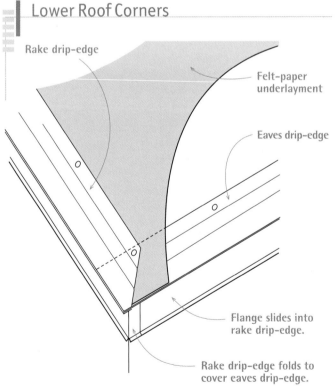

Lower Roof Corners

Rake drip-edge

Felt-paper underlayment

Eaves drip-edge

Flange slides into rake drip-edge.

Rake drip-edge folds to cover eaves drip-edge.

Underlayment runs over the eaves drip-edge, and rake drip-edge runs over the underlayment. Thus, at lower corners the rake drip-edge will overlay the eaves drip-edge.

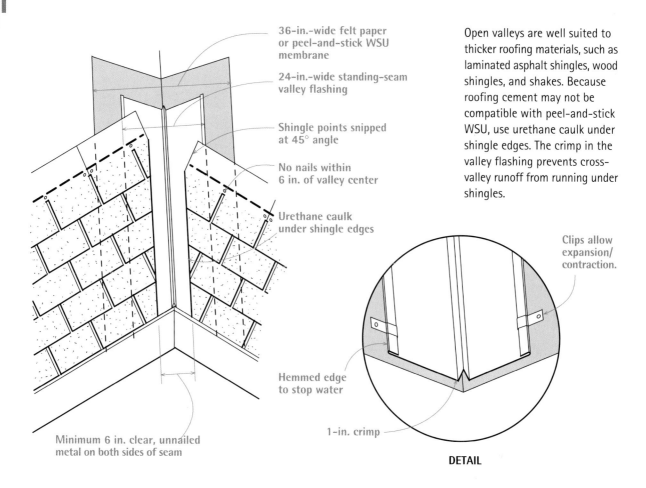

36-in.-wide felt paper or peel-and-stick WSU membrane

24-in.-wide standing-seam valley flashing

Shingle points snipped at 45° angle

No nails within 6 in. of valley center

Urethane caulk under shingle edges

Minimum 6 in. clear, unnailed metal on both sides of seam

Open valleys are well suited to thicker roofing materials, such as laminated asphalt shingles, wood shingles, and shakes. Because roofing cement may not be compatible with peel-and-stick WSU, use urethane caulk under shingle edges. The crimp in the valley flashing prevents cross-valley runoff from running under shingles.

Clips allow expansion/contraction.

Hemmed edge to stop water

1-in. crimp

DETAIL

before applying drip-edge. In corners, where rake edges meet eaves, run the rake drip-edge over the eaves drip-edge. Slitting the vertical leg of the rake drip-edge makes it easier for you to bend it over the leg of the eaves drip-edge.

In general, drip-edge flashing 6 in. wide or wider is better than narrower flashing because it enables you to nail well back from the edge of the flashing—always desirable.

VALLEY FLASHING

There are basically two types of valleys: *open,* where the valley flashing is exposed, and *closed,* where flashing is covered by shingles. Each has advantages. Open valleys clear water well, are easy to install, and work especially well beneath wood shingles, shakes, and laminated asphalt shingles, which are thicker and harder to bend than standard three-tab shingles. In woven valleys, shingles from both roof planes meet in the valley in alternating overlaps and are slower to install but offer double-shingle protection and are favored for low-slope roofs. And there are

variations, such as the *closed-cut valley* as shown in "Closed Valleys" on p. 98.

Prepare valleys by sweeping away debris, hammering *all* sheathing nails flush, and then lining the valley with underlayment. Traditionally, this lining was 30-lb. building paper, but peel-and-stick WSU, although more expensive, is simpler to install and generally more durable. Install 36-in.-wide WSU for valleys, centering a single piece down the length of the valley and overlapping the drip-edge flashing at the bottom.

If you line the valley with 30-lb. building paper, run a continuous piece of 36-in.-wide paper down the valley or overlap pieces by at least 6 in. Use tabbed roofing nails to secure the paper, keeping nails 6 in. away from the center of the valley. The 15-lb. building paper underlayment used elsewhere on the roof will overlap the outer edges of this heavier "valley paper."

Install metal valley flashing that's 18 in. to 24 in. wide on most slopes so that each side of the valley is 9 in. to 12 in. wide. If the roof pitch is steep

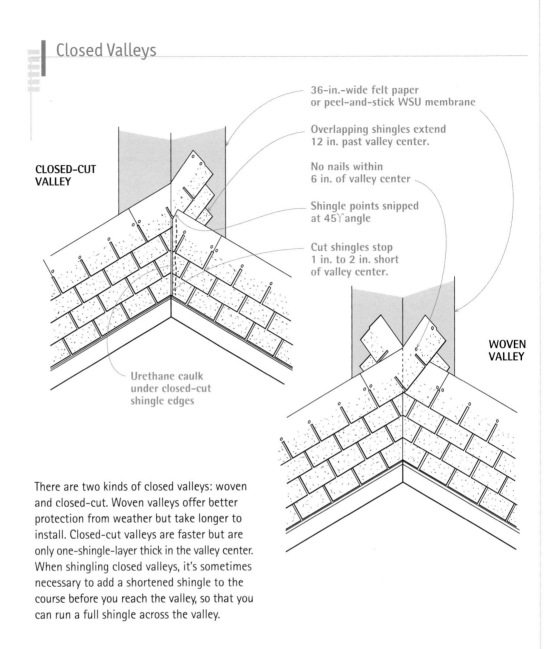

CLOSED-CUT VALLEY

36-in.-wide felt paper or peel-and-stick WSU membrane

Overlapping shingles extend 12 in. past valley center.

No nails within 6 in. of valley center

Shingle points snipped at 45˚ angle

Cut shingles stop 1 in. to 2 in. short of valley center.

WOVEN VALLEY

Urethane caulk under closed-cut shingle edges

There are two kinds of closed valleys: woven and closed-cut. Woven valleys offer better protection from weather but take longer to install. Closed-cut valleys are faster but are only one-shingle-layer thick in the valley center. When shingling closed valleys, it's sometimes necessary to add a shortened shingle to the course before you reach the valley, so that you can run a full shingle across the valley.

or if you live in an area that gets a lot of rain, have a metal shop fabricate valley flashing with an inverted V-crimp down the middle. This crimp helps prevent heavy runoff from one side running across the valley and up under shingles on the other side. Overlap lower sections of flashing 8 in. to 9 in. The heavier the metal, the more durable the valley: 26 gauge is standard for prefabricated pieces, but the heavier 24 gauge is better.

Avoid driving nails through metal valley flashing. Instead, place nail shanks snugly against the edge of the flashing and drive nails until the heads touch the flashing. Don't dent the metal. Space nails every 12 in. to 16 in. along both edges, or use clips that interlock seams along the edge. Not nailing through the metal allows it to expand and contract freely and leaves no nail holes for water to penetrate. To prevent corrosive

galvanic action, use nails or clips of the same metal as the flashing.

VENT-PIPE FLASHING

Vent-pipe flashing (also called jack flashing) is usually an integral unit with a neoprene collar atop a metal base flange. Some pros prefer all-metal units because UV rays won't degrade them and their taller collars are less likely to leak on low-slope roofs.

Neoprene combos are easier to install. In both cases, shingle up to the base of the vent pipe and slide the unit down the pipe. Nail the top edge of the base, then overlap it with shingles above. Neoprene collars slide easily over the pipes, but metal collars must be snipped and spread to receive the pipe before being caulked with ure-thane to prevent leaks. For either base, don't nail the lower (exposed) edge; instead, apply a bead of urethane caulk beneath the flange to seal it to the shingles beneath.

CHIMNEY FLASHING

Chimneys must be *counterflashed*. The upper pieces of counterflashing are usually tucked into chimney mortar joints and made to overhang various pieces of *base flashing*, which are nailed to the roof deck. Counterflashing and base pieces overlap but aren't physically joined, so they can move independently yet still repel water. (This independence is necessary because houses and chimneys settle at different rates, causing single-piece flashing to tear and leak.)

Replace counterflashing and base flashing when reroofing. To avoid damaging the chimney, use the gentlest possible method to remove counterflashing. If mortar is weak and crumbling, you may be able to pull the flashing out by hand; in that case, repoint the mortar after replacing counterflashing. If the mortar is sound and the counterflashing is firmly lodged, try using a cold chisel or a carbide-tipped bit in a pneumatic air chisel to cut out the flashing and as little of the mortar as possible.

Base flashing should be removed because you'll strip the roofing and building paper at the same time. Base flashing has several components: a continuous sheet-metal *apron* across the chimney's downslope face, L-shaped *step-flashing*

Vent-pipe flashing overlaps shingles below and is overlapped by shingles above. To keep water from entering the snipped metal collar around the pipe, seal the joint with urethane caulk.

1. As you roof along a chimney, alternate shingles and L-shaped pieces of step-flashing. Counterflashing will cover the tops of the step-flashing. For extra protection, run a bead of urethane between the step-flashing and the chimney. Press the flashing into the urethane to achieve a positive seal.

3. Counterflashing is held in place by a folded lip jammed into the mortar joint. For good measure, the mason hammered masonry nails into the mortar and then used a cold chisel to tap them in deeper.

running up both sides, and (when the chimney sits below the roof ridge) a *cricket* (or saddle) running across the upslope face. A cricket is sloped like a tent roof to deflect water around the chimney. Use a claw hammer, a flat bar, or a cat's paw to pry up old base flashing from the roof sheathing. Then hammer down any nails you can't pull.

Reattach base flashings first. As shingles butt against the chimney's downslope face, place the apron over them. The apron's bottom flange should overlap shingles at least 4 in.; its upper flange should run at least 12 in. up the face of the chimney. Prefabricated aprons usually have "ears" that wrap around chimney corners and are nailed to the sheathing. As shingles ascend both sides of the chimney, they overlap the bottoms of L-shaped pieces of step-flashing.

Keep nails as far back from the flashing crease as possible. Use a single nail to nail down

2. This grinder is poised to grind out an old chimney mortar joint. Once the abrasive wheel hits mortar, you'll see nothing but grit. Wear safety glasses or goggles.

4. Finally, run a bead of urethane caulk such as Vulkem 116 to fill the joint and seal out water. Once the caulk has set a bit, you can tool it with your thumb.

PRO TIP

Before inserting chimney counterflashing, blow or brush out any debris from the mortar joints. If your air-compressor hose can reach the roof, use it. Otherwise, insert a piece of plastic tube in the joint and just blow. Wear safety glasses— and when the tube is in place, don't inhale.

each piece of step-flashing and the shingle covering it. When shingle courses along both sides of the chimney reach the back (upslope) face of the chimney, the lower flanges of the cricket overlap them.

If a self-supporting cricket is fabricated from heavy 20-gauge galvanized steel, predrill the nail holes in the cricket's lower flange. Nail it down with gasketed roofing nails spaced every 6 in., down 2 in. from the top edge. Then cover the top edges of the cricket flange with a strip of peel-and-stick bituminous membrane, and overlap that with shingles. Finally, caulk the top edges of apron, cricket, and step-flashings with urethane caulk to seal them to the chimney.

Chimney Flashing

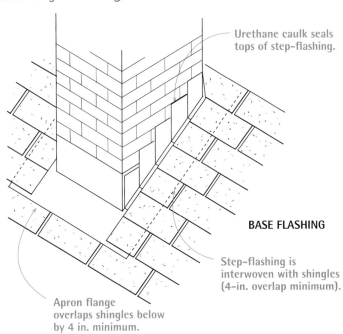

Urethane caulk seals tops of step-flashing.

BASE FLASHING

Step-flashing is interwoven with shingles (4-in. overlap minimum).

Apron flange overlaps shingles below by 4 in. minimum.

COUNTERFLASHING

Cricket flange extends up under shingles by 6 in. minimum.

Tops of apron, cricket, and step-flashing are covered by counterflashing.

The apron flashing along the downslope of the chimney wraps around the corners of the chimney and is itself overlapped by step-flashing coming down each side. The cricket on the upslope side is also complex; it wraps corners and overlaps step-flashing below. Have a sheet-metal shop solder all seams so they'll be watertight.

AN Alien ON THE ROOF

You often see TV antennas strapped to chimneys. Terrible idea! Whipped by winds, an antenna stresses the mortar joints and causes leaks. A chimney is designed to be a freestanding unit that safely carries hot gases out of the house. Don't ask more of it.

The transition step-flashing that turns the corner on a dormer sidewall is complex and inclined to leak, so have it prefabricated by a sheet-metal shop. Seal the top of the step-flashing with 9-in.-wide peel-and-stick membrane, then overlap that with housewrap before covering both with siding.

There are several ways to install counter-flashing. Counterflashing should overlap the base flashing by 4 in. Traditionally, a mason used a *tuck-pointing chisel* to remove chimney mortar to a depth of 1½ in. and then inserted a folded lip of counterflashing into the mortar joint. The joint was then packed with strips of lead to hold the flashing in place, followed by fresh mortar applied with a *striking tool* (also known as a slick). This method works well, but you need to be careful not to damage the surrounding bricks. Caulk with urethane caulk once the mortar has set.

Alternatively, you can use an abrasive wheel in a cordless grinder to cut narrow slots in the mortar joints; then insert counterflashing with its lip folded back so sharply that it resembles the barb of a fishing hook. This barbed lip friction-fits tightly into the slot, so mortar is unnecessary. Instead, fill the slot with urethane caulk, which adheres well and seals out water.

FLASHING ADJOINING STRUCTURES

Leaks are common where a shed roof abuts a wall, if the joint isn't flashed. At the very least, you'll need to remove enough siding nails so you can slide the upper leg of L-shaped flashing up at least 4 in. under both underlayment and siding—8 in. in snow country. Because the bottom of the siding must clear roof shingles by at least 1 in., however, wise builders will strip all siding above the roof and reinstall it so courses are evenly spaced above the required 1-in. clearance. (Cutting 1 in. off the bottom course would be faster but won't look as good.)

The lower leg of the L-shaped flashing goes over the uppermost course of shingles on the roof. Because that leg will be exposed, caulk under it and use gasketed roofing nails to secure it, placing the nails at least 1 in. above its lower edge.

Where a gable-end addition abuts a vertical sidewall or a sloping roof meets a dormer wall, install step-flashing. Fold in half 10-in. by 10-in. pieces of sheet metal to create L-shaped step flashing that goes 5 in. up the wall and 5 in. out onto the roof. Alternate pieces of flashing and shingles. Again, you may need to pull siding nails or remove courses of siding to fit the upper legs of the flashing up under both underlayment and siding. Place nails as far as possible from the flashing folds. Use two nails to attach each piece of step-flashing: one nail (into the sidewall) 1 in. from the top edge of the upper leg and the other nail through the bottom leg and the shingle overlapping it. Apply urethane caulk under any flashing legs or shingle edges that don't lie flat.

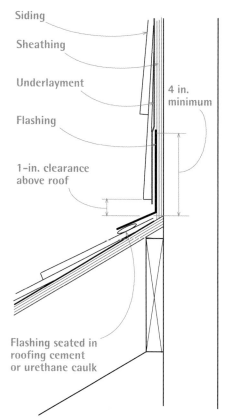

Flashing a Shed Roof That Abuts a House Wall

Siding

Sheathing

Underlayment

Flashing

4 in. minimum

1-in. clearance above roof

Flashing seated in roofing cement or urethane caulk

SKYLIGHT AND RIDGE FLASHING

Skylight flashing is shown in chapter 6. As a general rule, follow the instructions in the flashing kit provided by the skylight manufacturer.

Traditionally, ridges were flashed with a continuous strip of 12-in.-wide, 30-lb. building paper folded lengthwise, which straddled the ridge and overlapped the top courses of shingles. The building paper was then covered with a shingle saddle or overlapped ridgeboards. Metal flashing was sometimes used instead of building paper.

These days, ridges are often covered with ridge vents that allow hot air to escape, as shown on pp. 102–103.

Asphalt Shingles

Until recently, most asphalt shingles were three-tab shingles with two slots dividing the exposed part of the shingle into thirds. Today, laminated shingles (also called architectural and dimensional shingles) are the de facto standard. Consisting of two bonded layers, laminated shingles are thicker, more wind-resistant, and somewhat easier to install because they have a random

Venting a Roof

Roof vents allow air to flow beneath the roof deck, thus moderating attic temperatures, extending the life of roofing materials, reducing ambient moisture in the house—a big source of mold—and preventing ice dams from forming along the eaves in cold climates.

Typically, in a house with passive ventilation, air enters through soffit vents, flows up through an unfinished attic or over insulation between rafters, and exits through either gable-end vents or a ridge vent. Passive ventilation is desirable because it consumes no energy, but intake and exit vents must be balanced and large enough to allow adequate flow, and channels must be continuous to the exit vent. Minimum venting requirements are summarized below.

It can be difficult to retrofit existing roofs to add ventilation. Some houses lack soffits or have eave trim too shallow to add intake vents; blown-in insulation may fill the eaves and block intake air. In those cases, adding *rooftop intakes* may be a viable way to draw in cool air. Rooftop intakes are typically installed a few feet up from the eave; they include the *eyebrow ventilator* shown on the facing page and slotted *intake vents* such as that shown below. Both admit air, but the slotted vent will admit a greater volume and more evenly distribute flow of intake air—provided that there are no impediments between rafters to block the rising air. Slotted intake vents are, essentially, a one-sided ridge vent. They require several courses of shingles to be removed, a 1-in. slot cut through the sheathing, and the intake vent to be integrated into the shingle field so that it overlaps the course below and is overlapped by the course above.

Adding intake vents, however, may not be a viable solution if rafters are not deep enough to accommodate both insulation (R-30 requires 8 in. of insulation) and 2 in. of clear space above the insulation to allow airflow. Solid wood blocking between rafters also will stop airflow, creating hot spots of stagnant air between rafter bays. At some point, it may be more cost-effective to create an unvented roof by closing all vents to the exterior, spraying its underside with foam insulation, and turning the attic into conditioned space. Doing so may increase the amount of living space you have, provided the attic is framed to support live loads and is accessible. Unvented roofs and conditioned attics are discussed further in chapter 14 beginning on p. 430.

MVR = Minimum Venting Requirements

Chapter 14 offers more information on ventilation. In brief, you need a minimum of 1 sq. ft. of ventilation per 300 sq. ft. of attic space. Half the vent area will be intake vents, and half ridge or gable-end exit vents. For example, if attic floor space totals 2,500 sq. ft., total vent surfaces should be 2,500 ÷ 300, or 8.33 sq. ft. Ridge vents would therefore be half that, or 4.16 sq. ft. That area corresponds roughly to 33 lin. ft. of ridge vents, based on net free vent area (NFVA) charts.

Roof Venting

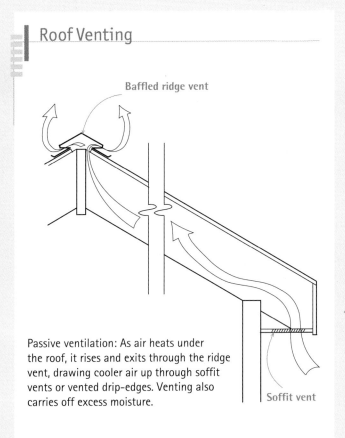

Passive ventilation: As air heats under the roof, it rises and exits through the ridge vent, drawing cooler air up through soffit vents or vented drip-edges. Venting also carries off excess moisture.

Intake vents take many forms, including slotted vents that resemble half of a ridge vent. Intake vents are typically installed a few feet up from, and parallel to, the eaves.

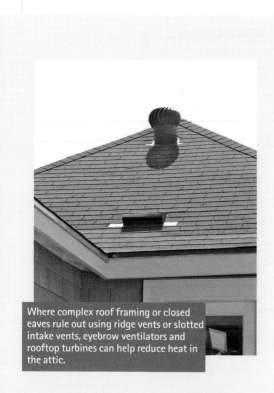

Where complex roof framing or closed eaves rule out using ridge vents or slotted intake vents, eyebrow ventilators and rooftop turbines can help reduce heat in the attic.

Have a boom truck deliver roofing supplies directly to the roof. Stack materials along the ridge so the rest of the roof will be clear to work on.

Ridge vents allow hot air—and excess moisture—to exit the building. Here, a perforated plastic ridge vent gets capped with shingles. For more on ridge vents, see pp. 110–111.

pattern, with no shingle cutouts to line up. When installed, they look distantly like wood shakes.

MATERIALS: ROUGH NUMBERS

Shingle dimensions vary by maker. *Three-tab shingles* are typically 12 in. by 36 in. Laminated, *architectural* shingle dimensions are often metric, roughly 13 in. by 40 in. Most shingles are installed with a 5-in. exposure, although shingles with metric dimensions may specify a $5\frac{5}{8}$-in. exposure. Always follow the manufacturer's exposure recommendations, no matter what type of shingle it is.

Calculating shingles needed. Begin by measuring the roof accurately, making a sketch to scale on graph paper as you go. Note valleys, ridges, chimneys, skylights, plumbing vents, and other elements that require flashing, waterproof membranes, or special attention. From that sketch, a building supplier can develop a final materials list for shingles, nails, underlayment, flashing, vents, and so on.

Asphalt shingles come three to five bundles to the *square* (100 sq. ft.), depending on shingle dimensions. To calculate the number of squares you'll need to shingle the *field,* add up the square footage of roof surface and divide by 100. In addition, you'll need materials to reinforce shingles along eaves and rake edges—either by installing a double layer of shingles along the roof perimeter or by applying a heavy *starter strip* before shingling.

PRO TIP

When reroofing an old house, make sure rafters can support the weight of new shingles. Some thick "luxury shingles" weigh 480 lb. per square versus 250 lb. per square for 30-year architectural shingles or 200 lb. for standard three-tab singles.

If you're installing laminated shingles, use three-tab shingles as an underlayer along the eaves and rakes. For this purpose, three-tab shingles are far cheaper than laminated shingles and will lie flatter. If you're installing woven valleys, you'll interweave roughly one bundle of shingles per 16 lin. ft. of valley. Finally, add two extra bundles for waste, ridge and hip caps, and future repairs.

Shingle colors often vary from one production lot to another. To avoid having a new roof with a patched-together look, specify that all bundles come from the same lot when you order. When your order arrives, check the lot numbers on the bundles and open a few bundles from different lots. If lot numbers don't match and the color variation is noticeable, call the supplier. If the color varies only slightly, you might mix lots every other shingle during installation. Finally, have shingles delivered directly to the roof. Many suppliers will place bundles on the roof by means of truck booms or conveyor belts.

Roofing nail quantities vary according to method: hand nailing, power nailing, or some combination of the two. Typically, use four nails per shingle, or about 2 lb. of nails per square if you're hand nailing. However, high-wind areas require six nails per shingle, or 3 lb. per square. Roofing nails come in 5-lb. and 50-lb. quantities. Boxes of pneumatic nails typically contain 120 nails per coil and 60 coils per box. At four

nails per shingle, you'll need about 3⅓ coils (400 nails) to attach a square of shingles.

Use corrosion-resistant roofing nails at least 1¼ in. long for new roofs; use 1½-in. nails if you're roofing over a previous layer. Ideally, nails should sink three-quarters into sheathing or stop just short of penetrating all the way through for ½-in.- to ⅝-in.-thick sheathing. If the roof has an exposed roof overhang (you can see the underside of the sheathing), use ¾-in. nails (along the overhang) for a new roof and 1-in. nails for roofovers.

When ordering, don't forget tabbed roofing nails for underlayment and metal-compatible nails for attaching flashing or valley clips.

SHINGLE LAYOUT

We'll assume that the roof has been stripped of old shingles, that failed sheathing has been replaced, and that the roof is safe to walk on.

Reconnoiter the roof. Use a tape measure to see whether the roof is square, the ridge is parallel to eaves, the rake edges are parallel, and whether—overall—the width of the roof requires shifting shingle courses left or right. *To determine square,* measure diagonally from both ends of the ridge down to the opposite eave corner; if the readings are roughly equal, chances are the roof is square. *If the ridge is parallel to eaves* within ½ in., run shingle courses right up to the ridge. But if ridge-to-eaves readings differ by ¾ in. or more, you'll

Shingling Terms

These definitions will help you make sense of roofing terms.

▶ **Course:** A horizontal row of shingles.

▶ **Butt edge:** The bottom edge of a shingle.

▶ **Exposure:** Typically, the bottom 5 in. of the shingle, left exposed to weather. Shingles with metric dimensions are usually exposed 5⅝ in.

▶ **Cutouts:** Slots cut into the exposed part of a three-tab shingle to add visual interest and allow heat expansion.

▶ **Offset:** The distance that shingle slots or ends are staggered from course to course.

▶ **Self-seal strip:** The adhesive on the shingle face, which, when heated by the sun, fuses to shingles above and prevents uplift.

▶ **Fastener line:** On shingles with a 5-in. exposure, a line roughly 5⅝ in. up from the butt edge. Nails along this line will be covered by the shingles above. (If shingles don't have such lines marked, nail just below the self-seal strip.)

▶ **Control lines:** Chalklines snapped onto underlayment to help align courses and cutout lines.

▶ **Underlayment:** A water-resistant sheet material—usually building paper—that covers the roof sheathing.

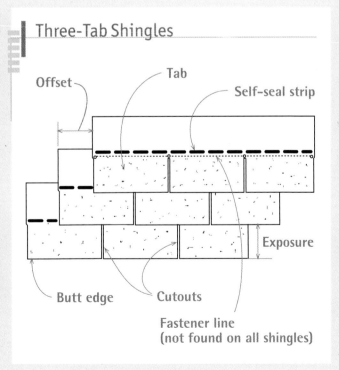

Three-Tab Shingles

Offset · Tab · Self-seal strip

Exposure

Butt edge · Cutouts

Fastener line (not found on all shingles)

need to compensate by adjusting shingle exposures as you approach the ridge.

The last two measurements, for parallel rakes and roof width, are of most concern if you're installing three-tab shingles. Because three-tab shingle patterns align vertically, avoid cutting shingle tabs less than 2 in. wide along either rake edge—tabs that short look terrible. It's far better to shift the shingle layout (and thus the vertical control lines) slightly right or left so the shingle piece is larger. For additional strategies for installing three-tab shingles, see Mike Guertin's fine book, *Roofing with Asphalt Shingles* (The Taunton Press, 2002).

Establishing control lines. After installing drip-edge along the eaves, many pros measure only once, to establish a horizontal *control line* parallel to the eaves, to which they nail the shingle *starter course.* After the starter course is down, they use only the exposure gauge of their pneumatic nailer or shingle hatchet to position successive courses.

To make sure courses stay evenly spaced and straight, however, snap horizontal chalklines at regular intervals onto the underlayment (building paper). Snapping chalklines every second or third course won't take much time and will ensure professional-looking results.

If you're installing three-tab shingles, it's also wise to snap a pair of parallel, vertical control lines 6 in. apart to line up the slots of alternating shingle courses, as shown in "Shingle Layout" on p. 107. With a 6-in. offset, the slots of every other course line up, creating a strong visual pattern. If slots don't align, the installation will look sloppy. Laminated shingles have no slots to align, so you won't need vertical control lines.

INSTALLING SHINGLES

After attaching drip-edge to the eaves and rolling out underlayment over the drip-edge, install the starter course along the eaves. You'll cover the starter course with the first course of shingles. Running starter courses along the rakes isn't imperative, but it's smart because starters stiffen the overhanging shingle edges and create a cleaner sightline from below.

The starter course. First, determine how much the starter course will overhang the drip-edge: ¼ in. to ¾ in. overhang is typical, but some roofers allow as much as 1 in. if eave or rake boards are bowed.

Along the eave, extend your tape measure past the drip-edge the amount of the overhang. If that overhang is 1 in., make crayon marks on the underlayment at 7 in. and at 12 in. Do this at both ends of the roof, and snap chalklines

After installing drip-edge along the eaves, double the first course of shingles or, as shown, install a starter strip. The lower edge of the starter strip overhangs the drip-edge by ¼ in. to ¾ in.—or even 1 in. if bowing eaves or rake boards require it.

Instead of doubling shingles along the roof rakes, install a starter strip. It's stiff enough to resist wind uplift. And, seen from below, it presents a much cleaner, straighter line than individual shingles.

PRO TIP

Many roofers prefer hand nailing the first half-dozen shingle courses rather than pneumatic nailing. That's because the hose of a pneumatic nailer makes footing more treacherous along the eaves, where there's little time to catch yourself before rolling over the edge.

Score shingles on the back-side, using a utility knife. If you try to cut through the granules on the front, the blade will go awry and soon dull. When you score along a straightedge and snap along the cut, you'll get a straight, clean edge.

through both sets of marks. The 7-in. line indicates the top of the starter course; the 12-in. line indicates the top of the first course of shingles.

Starter courses can be three-tab shingles with the bottom 5 in. cut off or a *starter strip* that comes on rolls in various widths. Starter strips have the advantage of stiffening shingles above and, viewed from below, providing a clean, unbroken line. Still, trimming three-tab shingles is cheaper, so here's a quick look at that method.

Using Pneumatic Nailers

Because pneumatic nailers can easily blow nails through shingles, some codes specify hand nailing. And it's safer to hand-nail the first five or six courses along the eaves, where stepping on a pneumatic hose could roll you right off the roof. It is advisable to wear safety glasses when using nailers.

Those concerns aside, pneumatic nailers are great tools if used correctly. Here's how:

▶ Don't bounce-fire a nailer until you're skilled with it. (To bounce-fire, you hold the trigger down and press the nailer's nose to the roof to fire the nail.) Shingles must be nailed within a small zone—below the sealer strip but above the cutouts, if any—and it's hard to hit that zone if the nailer is bouncing around. Instead, position the nailer nose where you want it, then pull the trigger.

▶ Trigger-fire the first nail of every shingle. Do this to keep shingles from slipping, even if you're skilled with pneumatic nailers. Once the first nail is in, you can bounce-fire the remaining ones.

▶ Hold the nailer perpendicular to the roof so nails go in straight, and keep an eye on nail depth as the day wears on. Nail heads should be flush to the shingle; if they're underdriven or overdriven, adjust the nailer pressure.

▶ Follow a standard nailing schedule of four nails per shingle or six nails for high-wind areas. Trimmed-down shingles must have at least two nails. Place the first and last nails in from the edges at least 1 in. All nails must be covered by the shingle above.

Nailing schedule: four nails per shingle is standard; use six nails for high-wind areas. Trimmed-down shingles must have at least two nails. Place the first and last nails in from the edges at least 1 in. All nails must be covered by the shingle above.

Traditionally, the starter course was just a full shingle turned upside down so its tabs faced up, but that placed the shingle's self-seal strip too high to do much good. It's far better to measure down 7 in. from the top of the shingle, trim off the bottom 5 in., and snap off the shingle tabs.

Align the top edge of the starter shingles to the 7-in. control line and nail them down, four nails per shingle. If you're installing a starter strip, align its top edge to the 7-in. control line. Next, install starter courses over rake drip-edges, using the same overhang you used for eaves. Rake starter strips overlap eave starter courses when they meet at lower corners. Along both ends of the roof, measure up from the 12-in. control line and snap horizontal chalklines for the shingle courses to come. As noted earlier, a horizontal chalkline every two or three shingle courses is plenty. Once that's done, you're ready to install the first course of shingles.

One final note: Building-paper underlayment tears pretty easily if it gets too much traffic, so only roll out one strip of building paper at a time. Install shingle courses until you are close to the upper edge of the felt paper, then roll out the next strip so that its lower edge overlaps the one below.

Shingling the field. Install the first course of shingles over the starter course. If you're right-handed, start at the left side of the roof and work right; otherwise, you'll be reaching across yourself continually. Left-handers, of course, should start right and work left.

There are many ways to lay out and install shingles. If you're installing laminated shingles, a pyramid pattern is best. With this method, you precut a series of progressively shorter shingles, based on some multiple of the offset dimension. Because each successive course is, say, 5 in. or 6 in. shorter, the stepped pattern looks like a pyramid. Typically, pyramids start along a roof edge, with the first shingle in each course flush to the rake starter strip.

Once the pyramid is established, the job goes quickly. Just place a full shingle against each step in the pyramid and keep going. Because the offset is established by those first shingles, you can install full shingles until you reach the other end of the roof. But most roofers prefer to work up and out, maintaining the diagonal. If there are color variations among bundles, they'll be less noticeable if the shingles are dispersed diagonally.

The frequency of a pyramid pattern's repeating itself depends on how random you want shingles to look. Traditionally, patterns repeat every fourth course, that is, every fourth course begins with a full shingle. Whatever pattern you choose,

Roofing Jacks

Roofing jacks provide safe platforms on slopes. Typically, install a pair of jacks for every six to eight courses. To attach the jacks, level the pair by aligning the jacks to horizontal chalklines, and drive two 10d galvanized common nails through each jack into rafters. (Jacks nailed only to sheathing are unsafe.) Place nails above the fastener line on the shingle because the nails will stay in place after jacks are removed.

If you space each pair of jacks 8 ft. apart horizontally, they can be spanned by 10-ft.-long planks that allow a 1-ft. overhang on each end. However, install the two shingle courses above the jacks before you add the plank; otherwise, the plank will prevent you from nailing those courses. To prevent a plank from sliding out of a jack, nail through the hole in the front of the jack arm into the plank.

Plank-and-jack removal is a two-person job, especially in windy weather. Above all, play it safe. First, move the plank to a secure location. To remove a jack, sharply hammer its bottom upward, thereby driving the slotted jack holes off the 10d nails. Then, while being careful not to disturb the overlapping shingles, slide the jack out. It's a good idea to drive jack nail heads flush, but it's not imperative. To drive them flush, slide a flat bar under the overlapping shingle, placing it atop the nail head, then strike the bar's handle with a hammer. This may require several blows.

trimmed pyramid shingles should be at least 8 in. wide; otherwise, they'll look flimsy.

Keeping things aligned. As you work up the roof, align the tops of shingles to horizontal chalklines. Chalklines wear off quickly, so don't snap them too far in advance; snapping chalklines each time you roll out a new course of building paper is about right. However, to get the measuring done all at once, you can measure up from that original 12-in. line and use a builder's crayon to mark off exposure intervals along the rake edges on both ends of the roof, and then snap chalklines through those marks later.

Alternatively, if you snap chalklines only every second, third, or fourth course, use the gauge on the underside of your pneumatic nailer or on your shingle hatchet to set exposures for intervening courses. If your shingling field is interrupted by dormers, always measure down to that original 12-in. line to reestablish exposure lines above the obstruction.

Finally, if the ridge is out of parallel with the eaves by more than ¾ in., stop shingling 3 ft. shy of the ridge and start adjusting exposures so that the final shingle course will be virtually parallel with the ridge. For example, if there's a discrepancy of 1½ in., then at 3 ft. below the ridge, you'll need to reduce exposures on the narrow end of the roof by ¼ in. in each of six courses.

Shingle Layout

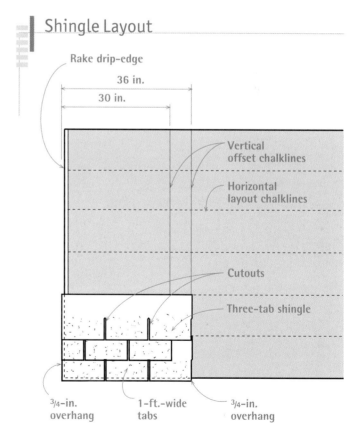

Aligning three-tab shingle cutouts can be as easy as snapping two vertical chalklines 6 in. apart. Because individual shingle tabs are 1 ft. wide, a 6-in. offset will line up shingle cutouts every other course.

Using a Pyramid Pattern Layout

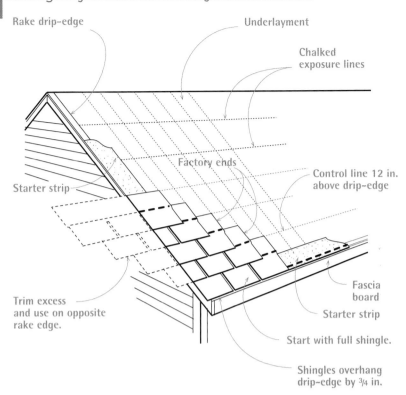

Rake drip-edge
Underlayment
Chalked exposure lines
Factory ends
Control line 12 in. above drip-edge
Starter strip
Trim excess and use on opposite rake edge.
Fascia board
Starter strip
Start with full shingle.
Shingles overhang drip-edge by ¾ in.

A pyramid pattern is best for installing laminated shingles that don't have distinct cutouts. As you build the pyramid along the rake edge, trim excess from the rake end so a factory end always leads off each course. Align shingles to horizontal chalklines you snap, not to lines printed on building paper.

Valleys. Both open valleys (in which metal valley flashing is exposed) and closed valleys should be lined, as described in "Underlayment" on pp. 94–95. Closed valleys are more weather-tight but slower to install, so they've become less popular. Open valleys are faster to install and better suited to laminated shingles, which are too bulky and stiff to interweave in a closed valley.

Once you've installed valley flashing, snap chalklines along both sides to show where to trim overlying shingles. Locate chalklines at least 3 in. back from the center of the valley; oncoming shingles cover valley flashing at least 6 in. When nailing shingles, keep the nails back at least 6 in. from the valley centerline—in other words, 3 in. back from the shingle trim line—so nails can be covered by shingles above. To seal shingle ends to the metal flashing, run a bead of roofing cement under the leading edge of each shingle, and put dabs of cement *between* shingles. Ideally, you should not nail through the metal at all, but that could leave an inordinately wide area of shingles unnailed. Besides, self-adhering waterproofing membranes beneath the metal flashing will self-seal around the nail shanks.

As a shingle from each course crosses a chalk-line, use a utility knife to notch the shingle top and bottom. Then flip the shingle over and, using a straightedge, score the back of the shingle from notch to notch. Or, to speed installation, run shingles into the valley, and then when the roof

---

PRO TIP

As shingles run diagonally into an open valley, their leading edges often end in sharp points, under which water can run. To prevent that, use a utility knife with a hooked blade to remove about 1 in. of the point, cutting perpendicular to the bottom edge of the shingle.

1. Carefully align the first shingle (which is uncut) to the edges of both eaves and rake starter strips.

2. Shorten the second shingle in the pyramid by the offset dimension. Use the gauge stop on the bottom of the pneumatic nailer to establish the correct exposure between courses.

3. At this point, you could run courses all the way across the roof, but most roofers prefer to work up and out, maintaining the diagonal. Although this veteran roofer didn't need to snap horizontal chalklines across the roof to keep courses straight, novices should.

This classic wood shingle roof has an open valley. Shingle caps cover the roof hip.

section is complete, snap a chalkline along their ends to indicate a cutline. To avoid cutting the metal flashing underneath, put a piece of scrap metal beneath shingle ends as you cut, using a hooked blade in a utility knife or snips.

Finally, codes in wet or snowy regions may require that valleys grow wider at the bottom. In that case, move the bottom of each chalkline away from the valley center at a rate of ⅛ in. per ft.

Closed-cut valleys are faster to install than woven ones because you don't need to weave shingles from two converging roof planes at the same time.

New Roofing over Old

Although placing new roofing over old may be allowed by some codes, "roofovers" tend to be inferior. As noted earlier, you must strip a roof if there are two or more layers or if existing shingles are curling.

Before beginning, replace or flash over old flashing as follows:
▶ Use aviation snips to cut away flashing around plumbing vents.
▶ Use a cold chisel to remove chimney counterflashing.
▶ Along the eaves, install a new drip-edge over the butt ends of existing shingles.
▶ Along the rakes, install a new J-channel drip-edge, which wraps around the old rake flashing and starter courses.
▶ Leave the old valley flashing in place, and install new over it.

To level out an existing asphalt-shingle roof, rip down (reduce the height of) two courses of shingles, as shown in the drawing. Because asphalt shingles are routinely exposed 5 in., rip down the first strip (the starter course) so that it is 5 in. high. Then lay it over the original (old) first course so that the strip is flush against the butts of the original second course above. Next, rip down a second strip (second course) 10 in. high, and put it flush against the butts of the original third course. You now have a flat surface along the eaves. The third course of new shingles—and all subsequent shingles—need not be cut down, just butted to an original course above and nailed down. Use 1½-in. roofing nails for roofovers—or whatever length is necessary to reach the sheathing.

New Shingles over Old

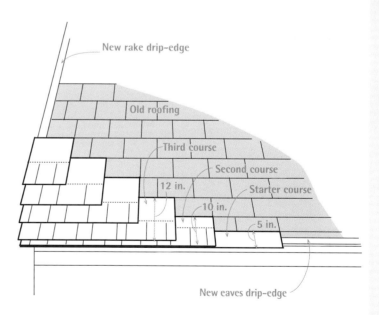

When installing a new roof over an old one, avoid a wavy-looking job by reducing the height of the first two courses to establish a flat surface.

A Dissent on the RIDGE VENT

Says a West Coast contractor: "Although a ridge vent is a perfectly valid detail for much of the U.S., it may not be accepted in high-wind or earthquake zones. There, building codes and exacting engineers may want the top of the roof sheathing to be nailed to the ridge board to complete the roof diaphragm and stiffen the structure. In fact, engineers may require that the sheathing be nailed to ridge beams with a shear-nailing pattern. Thus, if there's a vaulted ceiling, we almost always opt to do a spray-foam, nonvented roof cavity. If there's an attic, we would probably try to use eyebrow vents rather than ridge vents."

FINISHING THE RIDGE

Where composition shingles converge at the ridge, they should be capped to keep out weather elements and create a neat, finished joint.

Traditionally, roofers created *cap shingles* from leftover three-tab shingles, by extending the cutouts across the width of the shingle. Each standard shingle yielded three cap shingles, which the roofer gently folded to straddle the ridge and cover the tops of oncoming shingles. Two roofing nails per cap shingle are sufficient, with the nail heads covered by successive shingles—until the last cap shingle, which gets four nails. Two important details: Cap shingles must be placed so that prevailing winds blow *over* shingle overlaps, rather than into them, which could lift and loosen them. And roofing nails must be long enough to penetrate the several shingle layers at the ridge and sink into the sheathing.

These days, there are more options for capping the ridge, notably preformed ridge and hip shingles. These specialty shingles come prefolded and so are unlikely to split along the fold—a real problem with laminated or luxury shingles, which are thicker and stiffer than three-tab varieties. Most composition shingle makers offer preformed ridge shingles to match the shingles you use in the field. There also are plastic and metal ridge caps, which, although clearly a different material, can usually approximate the color of field shingles.

Lastly, many modern roofs have vents along the ridge, which allow air to rise and exit the house. In tandem with soffit vents, ridge vents encourage a flow of air beneath the roof deck, helping to keep the attic cooler in summer and lessening the chance for ice dams in winter. To install a ridge vent, cut sheathing back at least 1 in. on either side of the ridge board as shown in the bottom drawing on the facing page. Run underlayment and shingles to the edge of the sheathing. Then nail the ridge vent over the opening, straddling the shingles on both sides. In most cases, the ridge vent is then covered by cap shingles—with care taken not to block vent openings.

ASPHALT SHINGLE REPAIR

Most roofs are repaired in response to leaks caused by a missing shingle or, more often, worn-out or missing flashing. Or, in some cases, it's necessary to disturb shingles to install a roof vent or a vent pipe.

When removing a shingle or a course of shingles, disturb surrounding shingles as little as possible. First, break the adhesive seal between courses by sliding a mason's trowel or a shingle ripper under the shingles and gently slicing through the adhesive strips (see photo 1 on p. 162). It's best to do this when shingles are cool and the adhesive is somewhat brittle and easier to break. If you attempt this when the roof is hot, you're more likely to tear the shingles. Actually working with the shingles—lifting them to remove nails or slide in new shingles—is best done when the shingle is warm and flexible.

To remove a damaged shingle, raise the shingles above and tear the old one out. If it doesn't tear easily, use a utility knife to cut it out. Remove the nails that held the damaged shingle by inserting a shingle ripper against the nail shafts and prying up. Keep in mind that those nails are actually going through two shingle courses—the one you're trying to remove and the top of the course below. If you extricate the shingle and the nails don't come up with a reasonable amount of trying, just knock the nails down with your hammer and a flat bar.

Fill old nail holes with roofing cement, and slide the new shingle into place. Gently lift the course above, and position nails so they'll be overlapped by that course. Once you've inserted a new shingle, use the flat bar to help drive them down, placing the flat bar atop the nail head and striking the bar with a hammer.

Wood Shingles and Shakes

There's romance in wood shingles and shakes. Despite wood's tendency to cup, split, rot, grow moss, and catch fire, people still love it. Insurers, however, don't because it can catch fire and often refuse to insure homes with wood roofs in fire-prone areas. Keep in mind, too, that wood shingles take longer to install than asphalt shingles—and shakes even longer—so labor will cost more. Regarding durability, Eastern white cedar and Western red cedar are about the same—though each region asserts that its shingles last longer. Old-timers opine that split shakes last longer than sawn shakes.

PREPARING THE ROOF

Be sure to read the earlier sections on sheathing, underlayment, and flashing. And review the methods of asphalt-shingle installation, for they have some things in common with wood shingling.

If the old roof was once covered with wood shingles, they were likely nailed to skip-sheathing, which consists of widely spaced 1-in. boards that allow air to circulate under the shingles and dry them. These days, most roofers cover skip-sheathing with plywood because it stiffens the roof and is safer to walk on. But nailing shingles directly to plywood or building paper impedes air

Wood Shingle BASICS

No matter how wide the wood shingle, use only two nails—placed ³/₄ in. from the edge and 1¹/₂ in. above the exposure line. To allow for expansion, leave a ¹/₄-in. gap between shingles, unless they are wet, in which case you can place them snugly against each other. Offset joints between successive courses at least 1¹/₂ in. Shingle joints that line up must be separated by two courses. In other words, shingle joints can line up every fourth course but not sooner. Finally, nail heads should touch but not crush shingle surfaces; any deeper and they may split the wood.

circulation and may lead to cupping (shingles' undersides will dry much more slowly than the tops), rotting, and shortened shingle life.

One answer to this dilemma is a layer of ¹/₄-in.-thick synthetic mesh between the building paper and the shingles. Cedar Breather is one brand, which comes in 39-in.-wide rolls. Roll the mesh out over 30-lb. building paper, tack or staple it down, and you're ready to shingle. The mesh retains enough loft to allow air to circulate freely under the shingles so they can dry fully. To attach shingles over the mesh, you'll need longer nails: 6d shingle nails should do, but check the product's literature to be sure.

Flash a wood shingle roof as you would an asphalt roof, including WSU along the eaves, rakes, and valleys and metal drip-edge along the eaves and rakes.

ESTIMATING MATERIALS

Use only No. 1 (blue-label) shingles on roofs because they're free of sapwood and knots. Lesser grades are fine for siding but may leak on a roof. Shingles come 16 in., 18 in., and 24 in. long, with recommended exposures of 5 in., 5¹/₂ in., and 7¹/₂ in., respectively, on roofs with a 4-in-12 slope or steeper. Ultimately, slope determines exposure and thus the number of bundles per square (100 sq. ft.).

In general, four bundles will cover a square. To calculate the number of squares you'll need, calculate the square footage of the roof and divide by 100. Because shingles are doubled along eaves and rakes, add an extra bundle for each 60 lin. ft. of eaves or rake. For valleys, add an extra bundle for each 25 lin. ft. For ridges and roof hips, buy preassembled ridge caps, sold in

Wood Shingle Details

If the old wood shingle roof was attached to skip-sheathing, cover it with plywood for safety and stability. A layer of synthetic mesh allows the underside of the wood shingles to dry out.

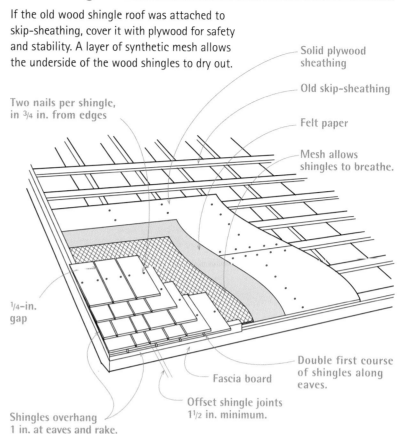

Two nails per shingle, in ³/₄ in. from edges

Solid plywood sheathing

Old skip-sheathing

Felt paper

Mesh allows shingles to breathe.

¹/₄-in. gap

Double first course of shingles along eaves.

Fascia board

Offset shingle joints 1¹/₂ in. minimum.

Shingles overhang 1 in. at eaves and rake.

Ventilating a Wood Shingle Ridge

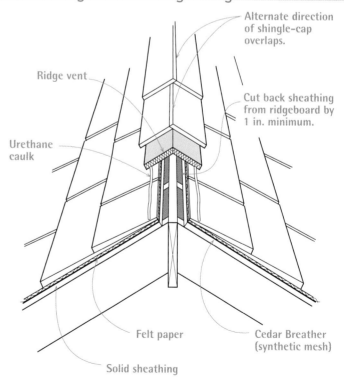

Alternate direction of shingle-cap overlaps.

Ridge vent

Cut back sheathing from ridgeboard by 1 in. minimum.

Urethane caulk

Felt paper

Cedar Breather (synthetic mesh)

Solid sheathing

Use a pneumatic nailer to attach preassembled ridge caps over the ridge vents because hand-nailing the caps could split them.

After installing drip-edge flashing, double the first course of wood shingles along the eaves. Along eaves and rakes, the drip-edge overhangs sheathing or trim boards by 1/2 in. to 3/4 in., and shingles extend at least 1 in. beyond the drip-edge.

bundles that cover 16 lin. ft. One Canadian supplier, WoodRoof (www.woodroof.com), has an especially informative website about hip and ridge caps, precut "fancy butt" shingles, specialty tools, and more.

You'll need 2 lb. of 4d or 5d galvanized shingle nails per square of shingles. For shingle caps along ridges and hips, use 6d shingle nails to accommodate the greater thickness of materials. Have your supplier deliver the materials in a lift-bed truck so that you can unload the shingles directly onto the roof.

INSTALLING WOOD SHINGLES

Note the offset requirements in "Wood Shingle Basics" (p. 111), which are essential for a long-lasting roof.

Double the first courses of shingles along the eaves, extending them beyond the drip-edge by 1 in. Double shingles along rakes, overhanging rake trim by 1 in. In addition, install metal drip-edge along eaves and rakes. When nailing the bottom course of doubled shingles along the eaves, nail them about 1 in. up from the butt edges; if possible, sink the nails into the edge of the fascia board. Those two nails will be covered by the top course of doubled shingles. Nail that course with two nails placed 1½ in. above the exposure line.

You could snap chalklines to indicate exposures for successive courses, but chalk can be unsightly and slow to fade on wood shingles. Instead, get a straight board as wide as the shingle exposure (5 in., for example), place its bottom edge flush to the bottom of the last shingle course, and position the next course of shingles by simply placing them atop the board. Move loose shingles around until all their joints are correctly offset to the courses below, then nail them down. A single shingle nail at either end of the "nailing board" will tack it in place.

When you must stretch to install courses above, install roof jacks as described on p. 107. Because wood shingles have random widths (unlike the fixed lengths of asphalt shingles), several people can work on a course at the same time. As you get within 6 ft. of the top, measure up to the ridge. If the ridge is not parallel to the eaves, there may be a discrepancy of several inches between measurements at one end of the roof and the other. If so, start adjusting exposures so that the top course of shingles will be more or less parallel to the ridge. If there's a discrepancy of 2 in., for example, you should reduce exposures on the narrow end of the roof by ¼ in. per course.

Open valleys. Refer to previous sections in this chapter on underlayment and valley flashing before you start. For most wood-shingle roofs, an open valley is the way to go. The exposed metal flashing clears water effectively and it's not likely to clog with debris. In addition, shingles running alongside an open valley require less fitting and cutting than those in closed valleys.

You can shingle out from a valley or into a valley from roof planes on either side. In either case, start by snapping parallel chalklines along both sides of the metal valley flashing; each line should be at least 3 in. from the center fold of the flashing. As each course approaches the valley, don't nail the last four or five shingles immediately; just arrange them so that wide shingles end in the valley. Where shingles cross a chalkline, use a utility knife to notch the shingle at those points.

Keep nails as far as possible from the valley center: 5 in. away is minimum. A bead of urethane caulk under the leading shingles should keep the edges from lifting.

Closed valleys. Run shingles into the valley until they meet oncoming shingles from the other side. At that juncture, rough-cut each shingle about ¼ in. *proud* to establish the correct angle. Then use a block plane to back-bevel the leading edge. By cutting and planing, you create a compound angle so that the shingles fit tightly; this can be done only on the roof, shingle by shingle—a slow job. For best weathering, alternate miters right and left from course to course. Build up several courses in the valley, and shingle out to the rest of the field.

Ridge treatments. Be sure to read the comments on venting a ridge on pp. 102–103. In brief, cut back sheathing at least 1 in. on either side of the ridge board. Then run underlayment and shingles to the edge of the sheathing before nailing a ridge vent over the opening.

The ridge vent can be finished in one of two ways:

▶ *Shingle caps.* Use a pneumatic nailer to attach preassembled shingle caps over the vent. Because the mesh underlayment and the ridge vent are compressible, they would move if you tried to hand-nail them. Shoot 2-in. to 2½-in. galvanized shingle nails through the shingle caps; nails should penetrate the roof sheathing at least ½ in.

▶ *Ridge boards.* These butt to each other; for a weathertight fit, they should be mitered. To establish the miter angle, lap two pieces of scrap at the peak and, using an adjustable bevel, transfer this angle to your tablesaw. Test-cut several pieces of scrap until the fit is tight, then rip down the ridge boards on the tablesaw. Because ridge boards should be as long as possible, get help nailing them down. If it takes several boards to achieve the length of the ridge, bevel end joints 60° and caulk each with urethane caulk. Using 8d galvanized ring-shank roofing nails, nail the ridge boards to the rafters, using two nails per rafter. Then go back and draw the beveled joint together by nailing it with 6d galvanized box nails spaced every 12 in. As you work, push down on the ridge boards to force them together. To avoid splitting boards, predrill them or use a pneumatic nailer to shoot nails through the roofing layers into the rafters.

SHINGLE REPAIRS

To remove wood shingles, use scrap blocks to elevate the butt ends of the course above. Work the blade of a chisel into the butt end of the defective shingle, and with twists of your wrist, split the shingle into slivers. Before fitting in a new shingle, remove the nails that held the old one. Slide a hacksaw blade or, better, a shingle ripper (also known as a slate hook) up under the course above and cut through the nail shanks as far down as possible. If you use a hacksaw blade, wear a heavy glove to protect your hand.

Wood shingles should have a ¼-in. gap on both sides, so size the replacement shingle ½ in. narrower than the width of the opening. Tap in the replacement with a wood block. If the replacement shingle won't slide in all the way, pull it out and whittle down its tapered end, using a utility knife. It's best to have nail heads covered by the course above, but if that's not possible, place a dab of urethane caulk beneath each nail head before hammering it down. Use two 4d galvanized shingle nails per shingle, each set in ¾ in. from the edge.

A Medley of Roofing Types

Although this section covers modest repairs a novice can make, most of the roofs discussed here should be installed by a specialist. You'll also find suggestions for determining the quality of an installation as well as a few tips.

FLAT ROOFS

Actually, no roof should be completely flat, or it won't shed water. But "flat roof" is a convenient term for a class of multimembrane systems. At one time, built-up roofs (BURs) once represented half of all flat roof coverings. BURs consisted of alternative layers of heavy building paper and hot tar. Today, modified bitumen (MB) is king, with cap membranes torched on to fuse them to fiberglass-reinforced interplies or base coats. For that reason, future roofs are likely to employ hot-air welding, cold-press adhesives, and roll membranes with self-sticking edges. Increasingly, professionally installed sheet-material systems such as EPDM (synthetic rubber), TPO (thermoplastic olefin), and PVC systems are gaining market share.

Causes of flat-roof failure. Whatever the materials, flat roofs are vulnerable because water pools on them, people walk on them, and the sun degrades them unless they're properly maintained. Here are the primary causes of membrane damage:

LAYING A FLAT ROOF

Once the granular membrane is down, its overlapping edges are often lifted and torched again to ensure sound adhesion and a waterproof seam.

In the final phase of a modified bitumen roof, an installer torch-welds a granular surface membrane to an interply sheet or directly to a base sheet. The granular surface is somewhat more expensive at installation, but it is cost-effective in the long run because it doesn't need periodic recoating.

Roofers refer to the molten material being forced out by the pressure of the trowel as wet seams—the mark of a successful installation.

The intersection of flat and sloping roof sections is worth extra attention. Run MB membranes at least 10 in. (vertical height) up the sloping section. Then overlap those membranes with the underlayment materials and the composition roofing material so water can run down unimpeded.

▶ Water trapped between layers because the roofing was installed too soon after rain or when the deck was moist with dew. Trapped water expands, resulting in a blister in the membrane. In time, the blister is likely to split.

▶ Inadequate flashing around pipes, skylights, and adjoining walls.

▶ Drying out and cracking from UV rays— usually after the reflective gravel covering has been disturbed.

▶ Foot traffic or roof decks placed directly on the roof membrane. Roof decks should be supported by floating posts fastened to rafters through the sheathing and flashed correctly.

Repairing roof blisters. If there are no leaks below and the blister is intact, stay away from it. Don't step on it, cut it, or nail through it. However, if it has split, press it to see what comes out. If the roof is dry, only air will escape; if the roof is wet, water will emerge. In the latter case, let the inside of the blister dry by holding the side open with wood shims; if you're in a hurry, use a hair dryer. Once the blister has dried inside, patch it.

Professionals repair split blisters with a three-course patch, which requires no nails:

▶ Trowel on a ¼-in.-thick layer of an elastomeric mastic, such as Henry 208 Wet Patch, carefully working it into both sides of the split. Extend the mastic at least 2 in. beyond the split in all directions.

▶ Cut a piece of "yellow jacket" (yellow fiberglass roofer's webbing) slightly longer than the split and press it into the mastic; this reinforces the patch.

▶ Apply another ¼-in. layer of mastic over the webbing, feathering its edges so it can shed water.

A three-course patch also is effective on failed flashing, where dissimilar materials meet, and for other leak-prone areas.

TILE AND SLATE ROOFS

Tile and slate are relatively brittle, expensive, and easily damaged if you don't know what you're doing. And they're slippery when wet. In most cases, hire a pro to make repairs.

Tile and slate are so durable that they often outlast the underlayment and fasteners. So when repairing or replacing these roofs, prolong the life of the installation by using heavy underlayment

(30-lb. building paper or a self-adhering bituminous membrane), copper attachers, and copper flashing. Although these materials are expensive, if the roof lasts upward of 60 years without leaking, it's money well spent.

Roof tiles weigh about three times as much as asphalt shingles. So if you're thinking of installing a new type of tile or tiling a roof for the first time, have a structural engineer evaluate your situation. Your roof framing may need bolstering to support the additional load.

Tiles are available as two-piece *mission-barrel* tiles; as one-piece *low-profile concrete* tiles; and as *flat-shake tiles*, which mimic wood shakes. Tiles overlap to direct rain into tile channels. Traditionally, tiles were set without nails on wood battens or skip-sheathing. But today they are commonly nailed to plywood sheathing, especially if the roof pitch is 5-in-12 or steeper. Two-piece mission tiles are nailed in two manners: *Trough tiles* are nailed directly to sheathing. *Cap tiles*, which sit atop trough tiles, are attached to intervening copper wires nailed to the roof.

Slate roofs are most often damaged by tree limbs. You can usually make repairs safely if the damage is along an edge and hence reachable without requiring that you walk on the roof. If you note a number of missing slates and rust stains on the roof, the nails may be rusting through, creating an extremely dangerous situation. Call a slate specialist at once.

To repair incidental damage, buy a slate hook. Work the head of the tool up under the damaged piece until the tool's hooked head catches a nail shank. Then strike the handle of the tool with a hammer to cut through the nail shank. When both shanks are severed, slide out the damaged slate, being careful not to disturb adjacent pieces.

Once the damaged slate is on the ground, transfer its dimensions onto a replacement slate. Then rent a tile-cutting saw to cut the new piece, wearing goggles. Or ask a tile dealer to cut the piece for you. Ideally, the new piece should be the same size as the one it's replacing, but old nail

shanks may prevent sliding a full-length replacement piece into place.

That done, align the bottom edge of the replacement slate to others in the course. Predrill two nail holes, each one 1 in. from the edge and about 1 in. below the course above; you'll probably be drilling through two layers of slate, so take it slow. (Use a cordless drill with a carbide-tip bit.) Size the drill bit just slightly larger than the thickness of the nail shank. Because this pair of nails won't be covered by the slate above, caulk all holes with urethane before inserting nails. Drive the nails down just snug and no more so you don't split the slate.

Individual mission-barrel tiles are irregular, so it's smart to snap horizontal chalklines to align tile courses and vertical lines to line up trough tiles, which are nailed directly to the roof deck. Cap tiles, which lie atop trough tiles, are secured to the roof deck with copper wire. To prevent galvanic corrosion, use nails and wires of the same material.

If properly installed over a durable membrane—such as the fiberglass-reinforced underlayment shown here—a tile roof can last 80 years.

When replacing sections of slate or slatelike materials, alternate slate courses with strips of 30-lb. building paper. Century-old sheathing boards can become pretty hard, so if copper nails bend or deform, use stainless-steel roofing nails instead.

Tile-Roof UNDERLAYMENT

A rubberized asphalt underlayment reinforced with fiberglass, layfastSBS is getting a lot of buzz among professionals. Specified for tile roofs, it's installed in two layers (double-papered) with 36-in.-wide sheets overlapped 19 in. Tiles often gouge building paper underlayment during installation, but not this stuff, which is also specified for shake, shingle, and metal roofs.

6

Doors, Windows, and Skylights

We ask doors, windows, and skylights to do a lot. They must be durable yet movable, let in light yet keep out rain, admit guests but deny drafts. Mediators of outside and in, they largely decide how comfortable, healthful, and energy-efficient a home is. Choosing and installing these units can be complicated, and you'll find related information in several other chapters: flashing in chapters 5 and 7; openings in exterior walls, repairing rot, and structural elements in chapter 8; installing casing in chapter 17; and energy conservation and air quality in chapter 14.

We'll take this chapter in order: doors first, then windows, and finally skylights. Proceeding from simple to more complex tasks, we'll start with installing prehung interior doors and then move on to exterior doors, which require additional weatherproofing steps. But first a few words about selecting and sizing doors.

Ordering Doors: An Overview

Door *frames* consist of several pieces: two side pieces, or *side jambs,* and a *head jamb* (or *frame head),* running across the top; exterior doors also have a *sill* spanning the bottom. (The sill may also have a *threshold,* but more about that later.) Jambs are further distinguished by the hardware they bear: The jamb on which the door is hung is the *hinge jamb,* whereas the jamb that receives

Getting ready for liftoff: The rough opening, wrapped in flashing tape, awaits a prehung triple-casement window. After removing the protective plastic and shipping blocks, the crew will test-fit the unit in the opening.

the latch is the *latch jamb* (also called *strike jamb* or *lock jamb*).

On a common frame-and-panel door, the thicker vertical elements are called *stiles;* hence, *hinge stile* and *latch* (or *strike*) *stile.* Horizontal elements are called *rails.* Glass panes in French doors are called *lights*, and the thin wood strips between lights are called *muntins.*

Consider the following factors when ordering doors:

Interior versus exterior. Exterior doors are generally thicker (1¾ in. versus 1⅜ in.), more expensive, more weather-resistant, and more secure than interior doors. Exterior doors may have water- or UV-resistant finishes and often are insulated and weatherstripped. Don't use interior doors outside—they won't last.

Prehung. Prehung (preframed) doors come fitted to a frame, with hinges mortised into a jamb. Ordering prehung doors can save huge amounts of time. However, if doorways are already framed, specify unframed doors (see "Hanging a Door to an Existing Frame" on p. 140).

Knockdown prehung doors arrive with the frame head cut to the correct width and all other parts milled with correct clearances around the door, but the parts are not assembled. This allows you to trim the jambs down to the right length for your flooring and threshold heights. Suppliers will cut exterior sills to fit if you ask them to, but many contractors prefer to buy sills separately and fit them on site.

Width. Door widths increase in 2-in. increments. When door dimensions are stated as a pair of numbers, width always comes first—for example, 2 ft. 8 in. by 6 ft. 8 in. (this is sometimes abbreviated as 2868).

Standard *interior* doors are 2 ft. 6 in. and 2 ft. 8 in. wide. For doors leading to busy hallways, or if you need extra room for a wheelchair or walker, architects often specify 2 ft. 10 in. or 3 ft. 0 in. You can also special-order interior doors 3 ft. 6 in. wide. Narrow doors (2 ft. 0 in. to 2 ft. 4 in.) are available for half-baths and closets; even narrower ones (1 ft. 4 in. to 1 ft. 10 in.) are for linen closets and such.

Standard *exterior* doors are 3 ft. 0 in. wide, although side doors are sometimes 2 ft. 8 in. or 2 ft. 10 in. wide. You can special-order extra-wide 42-in. exterior doors, but their greater weight requires larger hinges, and, of course, extra-wide doors need a greater area of free space when they swing open.

Height. Standard door height is 6 ft. 8 in., for both interior and exterior doors on newer houses. Older houses (1940s and earlier) sometimes had

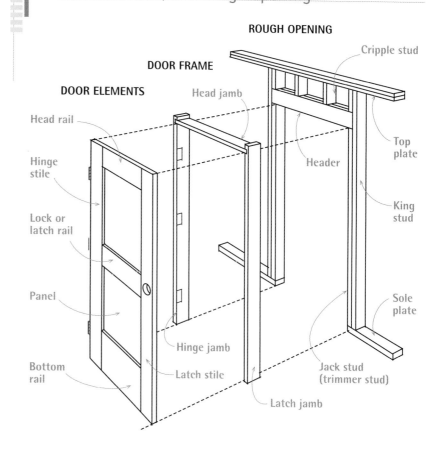

Door, Door Frame, and Rough Opening

ROUGH OPENING

DOOR FRAME

DOOR ELEMENTS

Cripple stud
Head jamb
Head rail
Hinge stile
Top plate
Header
Lock or latch rail
King stud
Panel
Sole plate
Bottom rail
Hinge jamb
Latch stile
Latch jamb
Jack stud (trimmer stud)

doors 7 ft. high, so that size is still widely available. Of late, 8-ft.-high French doors are in vogue because they allow light to penetrate far into living spaces. Of course, you can special-order a door of virtually any size if you're willing to pay for it. Salvage yards are excellent sources of odd-size doors.

Jambs. Wall thickness determines the width of door jambs if you order prehung units. Interior door jambs are commonly 4⁹⁄₁₆ in. wide, which can span a 2x4 stud wall (actual width, 3½ in.) with ½-in. drywall on both sides—leaving ¹⁄₁₆ in. to spare. Typically, interior door jambs are built from ¾-in.-thick stock (nominal 1 in.).

For 2x6 walls (actual width, 5½ in.), specify 6⁹⁄₁₆-in.-wide jambs, which can accommodate two wall-sandwiching layers of ½-in. drywall or—on an exterior wall—½-in. drywall and ½-in. plywood sheathing. Jambs for prehung exterior doors are usually fashioned from 1½-in.-thick stock, rabbeted with an integral doorstop to receive the door when closed.

Standard-width jambs won't work if your old house has full-dimension lumber and plaster walls or if you're covering walls with ⅝-in. drywall. Your choices then become (1) *jamb extensions* to increase the width of standard jambs,

Interior Wall Cross Section

½-in. drywall

2x4
(actually 3½ in. wide)

Doorstop

Shim

Jamb casing

Frame jamb

A standard frame jamb, which is 4⁹/₁₆ in. wide, can span a stud wall sandwiched between ½-in. drywall panels.

(2) *custom-milled jambs*, including ⅛ in. extra to accommodate wavy walls or twisted lumber, and (3) *split jambs*, which are interlocking half-jambs that can be adjusted to the width of a wall. (An integral doorstop covers the gap between sections.)

Swing. Door swing indicates which side you want the hinges on. Imagine facing the door as it swings open *toward you*: If the doorknob will hit your right hand first, it's a *right-handed* door; if your left hand grabs the knob, then it's a *left-handed* door. Another big distinction—usually ignored—is whether a door swings in or out. The vast majority of residential exterior doors are *inswings*, but *outswing* doors are better at sealing out drafts and water, as explained on p. 138. To avoid confusion when ordering doors, include a small architectural icon (sketch) to indicate which way each door swings. Some hardware is also right- or left-handed, so clarify what you want by sending the door manufacturer a photo.

Type and style. *Hinged single doors* are by far the most common type, but they need room to operate. If space is tight, consider *sliding doors*, *pocket doors*, or *bifolds*.

Try to match existing doors in the house or those on houses of a similar architectural style. In general, frame-and-panel doors tend to go well with older houses and flush doors have a more contemporary look.

If your exterior door has glass panels, they should be double glazed at the least; triple glaz-

Installation, in a Nutshell

MOST doors, windows, and skylights are preassembled in factories and delivered *prehung*, which makes installation much easier. Basically, you screw or nail the unit into a rough opening (RO). If the opening is in an exterior wall, weatherproof it first—wrap it with building paper, apply rigid or flexible flashing, and install cap flashing over the top. Rough openings are typically ½ in. to 1 in. wider and taller than the outside dimensions of the door or window frame that you are installing.

Rough openings are rarely square or perfectly sized, so you need to insert pairs of shims (thin, tapered pieces of wood) between the square frame and the often out-of-square opening. Shimming takes patience. But if you install shims carefully, doors, windows, and skylights will operate freely, without binding. Once a prehung window or door is installed in its RO, stuff loose insulation or spray low-expanding foam insulation between the door or window frame and the framing to seal air drafts, then cover those gaps with casing.

As you level and plumb windows and doors, use pairs of tapered shims to hold units in their rough openings. Installers often use 8d finish nails to tack jambs in place, but this photo shows a carpenter using trim-head screws, which are easier to remove when adjusting shims and less likely to bend than finish nails. Their small heads are easy to sink, fill, or cover with a stop.

ing is more energy efficient but costs more. Double glazing and a storm door may be a better choice. Finally, prefinish exterior doors with a UV- and water-resistant finish; at the very least, prime or seal all sides and edges.

Hardware. Although hinges for prehung doors are installed at the factory, doorknobs, locksets, locks, and the like are usually ordered separately. However, if you tell the door manufacturer what kind of hardware you will be using and specify where you want the bores or mortises to go, the factory will machine the door to your specs. Then, when the doors are on-site, attach the mortised lock, bored lockset, deadbolt, and so on, to the locations you specified.

As indicated in "Sizing Hinges" below, hollow-core or light solid-wood interior doors up to 1⅜ in. thick can be supported by two 3½-in. by 3½-in. (opened size) hinges. Wider, heavier interior doors and all 1¾-in. solid doors require a minimum of three 4-in. by 4-in. hinges. Extra-heavy or thicker exterior doors may need even bigger hinges with ball bearings or grease fittings.

As shown on pp. 134–135, *exterior locksets* are most often cylinder locks, inserted in a hole drilled in the face of the door, or mortise locks, housed in a rectangular slot cut into the latch edge of the door. *Interior locksets* are almost always some kind of cylinder lock: *passage locks* on doors that don't need to be locked and *privacy locks* on those that do.

Installing an Interior Door

Prehung doors come preassembled in a frame, with the door hung on hinges and held shut (for shipping) by a single screw through the latch jamb into the edge of the door. Or the door may be secured with a removable plastic plug through a predrilled hole where the lock will go.

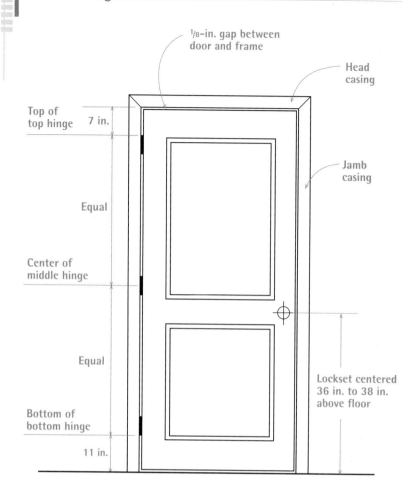

1/8-in. gap between door and frame

Head casing

Jamb casing

Top of top hinge — 7 in.

Equal

Center of middle hinge

Equal

Bottom of bottom hinge

11 in.

Lockset centered 36 in. to 38 in. above floor

Sizing Hinges

OPENED HINGE SIZE (in.)	DOOR THICKNESS (in.)	DOOR WIDTH (in.)
3½	1⅛–1⅜	Up to 32
4	1⅛–1⅜	32–37
4½	1⅜–1⅞	Up to 32
5	1⅜–1⅞	32–37
5, extra-heavy	1⅞ and up	37–43
6, extra-heavy	1⅞ and up	43 and up

Reading the Situation

We can't say it too often: There's no one right way to hang an interior door. But by carefully examining what's going on around an opening—*reading the situation*—you can usually figure out a solution. Especially when remodeling old houses, you'll often find that a wall is out of plumb in several different ways. Your goal is to have the closed door line up to the edge of the jamb, but invariably some part of the jamb sticks out beyond—or into—the wall plane.

The best way to solve such a problem—if you know in advance—is to oversize the jamb stock a little. That way, the jamb will be wide enough to straddle the drywall (or plaster) on both sides of the wall and any gaps in between. In the real world, though, we usually notice that a wall is out of plumb while we're trying to hang a pre-framed door in it—rarely in advance. In that case, we will try to "split the difference" so that the jamb is in a little here and out a little there. If one side of the wall is more important (more visible), we make sure the jamb edge is *proud* (above the wall surface) so the casing will lie flat and look good. (We don't care how the casing looks where it won't be seen—say, inside a closet.)

Don't beat yourself up. Situations that baffle you would probably baffle a pro, too.

PRO TIP

For aesthetic reasons, the top of a new door should be the same height as that of nearby doors and windows. Try to match that height exactly because the eye notices even slight discrepancies.

When your doors arrive at the work site, store them in a dry place out of harm's way, with their packaging undisturbed until you're ready to install them. Because millwork can absorb moisture, store the door(s) in conditioned space if possible.

There are as many ways to install a prehung door as there are carpenters. In general, opinions fall into one of two camps: those who favor leaving the door on the hinges while installing the frame in the rough opening (method 1), and those who favor pulling the hinges and removing the door, installing the frame, and then replacing the door in the frame (method 2).

There's no one right way. Method 1 is somewhat quicker, but it requires a second set of hands and constant communication between you and your helper on the other side of the door. Leaving the door in the frame also reduces chances that you'll damage the door. However, if you're working alone, method 2 is probably the way to go. The frame will be a lot lighter and less unwieldy if you take the door out first. And by being able to see both sides of a jamb as you shim and adjust it, you'll be more likely to get it right the first time.

The photo sequence on pp. 121–123 is a variation of method 1. Learned by a carpenter who earned his stripes building tract homes in 100°F summers, it favors speed and economy of movement.

CHECKING AND PREPPING THE OPENING

It's helpful to assess the rough opening (RO) closely and develop a plan of attack. If you know exactly what you need to do, you can reduce the number of times you put the frame in and out of the opening, which eats up time.

1. Measure the height and width of the rough opening. ROs are typically 2 in. wider than the door, which allows ¾ in. for the thickness of each jamb and about ¼ in. to shim on each side. RO height is usually 82 in. from the subfloor, which

Prep Steps

Start by surveying the rough opening. Check the dimensions of the rough opening, the thickness of walls from finish surface to finish surface, whether trimmer studs are plumb, and make sure the floor is level.

. . . and use a square to mark the bottom of the high-side jamb. Cut off the amount that side is high, reinsert the frame in the opening, and check the head jamb for level again.

For best support, side jambs should rest on the subfloor or finish floor. Some carpenters start their installation by standing the frame in the opening to check the head jamb for level. If it is not level, note by how much, remove the frame . . .

Use a square to see if jamb edges are flush to finished walls. If the jamb edges are flush or slightly proud (projecting beyond drywall), casing corner joints will meet. However, if the jamb is shy (shallower than the drywall), mitered joints will gap.

leaves ¾ in. for the thickness of a finish floor, approximately ¼ in. of clearance under the door, a standard 80-in. door, the ¾-in. thickness of the head jamb, and room to shim.

2. Measure the thickness of the wall, from finish surface to finish surface. Interior 2x4 walls covered with ½-in. drywall are 4½ in. thick, so standard jamb stock is 4⁹⁄₁₆ in. wide, providing an extra ¹⁄₁₆ in. to accommodate wall irregularities when you install the door casing. If your building plans call for ⅝-in. drywall, which would produce a wall 4¾ in. thick, specify jamb stock 4⅞ in. thick when ordering a prehung door.

3. Check the opening and remove anything that could interfere with installing the door. Occasionally, installers will run drywall past the trimmers, so cut it back so it's flush with the edge of the opening. Likewise, if a sole plate protrudes into the opening, use a reciprocating saw to cut it flush.

4. Check that the subfloor or floor is level and the trimmer studs are plumb. If the floor slopes, you will shorten the jamb on the high side so the head jamb will be more or less level. Whatever installation method you use, the crucial step is plumbing the hinge jamb of the frame. So, as you assess the RO, pay special attention to the trimmer stud on the hinge side. If it's not plumb, note which way it leans.

SETTING THE FRAME, AND TACKING THE HINGE JAMB

Again, the photo sequence depicts method 1, in which the door stays in the frame as it's installed.

1. Stand the door frame in the RO. Starting with the hinge jamb, use a Speed Square to *margin* the frame—that is, to center the door frame within the thickness of the wall. If the jamb stock was correctly sized, the jamb edges will be more or less flush to finish surfaces on both sides.

2. As one person holds the frame in place against the hinge jamb, the other checks the height of the unit and the RO. Ideally, there should be ½ in. of clearance between the bottom of the door and the finish floor. If the floor is level, remove the frame from the opening and, as needed, use a circular saw to trim an equal amount from both jamb legs. Replace the frame in the opening, and use a level to check the head jamb for level.

3. *If the floor slopes,* you'll need to remove more from the high-side jamb. Remove the frame from the opening, use a circular saw to trim the bottom of the jamb, replace the frame in the RO, and check the head jamb for level. Here again,

WHERE Dissimilar FLOORING MEETS

If a doorway is a juncture between different finish floors—say, wood and tile—a square cut across the bottom of the jambs won't be possible because the floors' thicknesses will vary. In that case, let the jambs rest on the subfloor and notch the flooring around the jambs. Cover the joint with some kind of transition strip, typically a prefinished oak threshold trimmed underneath along one side to accommodate the thicker flooring. Run a bead of silicone under both edges of the threshold so it will stay put.

allow ½ in. of clearance between the bottom of the door and the finish floor.

4. If the finish floors are not yet installed, stand the side jambs on scrap blocks of flooring. When it's time to install the flooring, just pull out the blocks and slide the finish flooring under the jambs. Otherwise, the flooring contractor would have to notch around the jamb profile, which is time-consuming. Alternately, you could let the jambs rest on the subfloor and, later, the installer could use a flush-cutting saw to trim the jamb bottoms so he could slide the flooring under.

5. Having trimmed the jambs as needed, once again margin the hinge jamb in the RO, and tack it to the trimmer stud, using three 8d (2½-in.) finish nails, one nail below each of the three hinges. Now you're ready to plumb the hinge jamb. (As the door is mounted and closed at this point, open the door to nail off the jamb.)

SECURING THE FRAME, AND FOLLOWING THE REVEAL

1. Holding a 6-ft. level to the hinge jamb, drive shims between the door jamb and the RO until the jamb is plumb. Because finish nail shanks are small, one nail won't offer a lot of resistance as you drive a pair of shims behind it. In other words, the nail will hold the jamb in place but can be pulled out slightly. To avoid bowing the jamb, place shims directly behind the hinges if possible.

2. Once the hinge jamb is plumbed, tack it with trim-head screws, using two screws per shimming point. Especially if walls are out of plumb, being able to back out a screw and reset shims is a godsend. In addition, remove the mid-

PRO TIP

In general, use pairs of shims with their tapers pointing in opposite directions to create an even gap between a door jamb and a stud. Wood shingles make great shims because they taper. Occasionally, when a stud is slightly twisted and so creates a tapered gap, you'll use just one tapered shim to fill that space.

A. Plumbing the hinge jamb is the critical part of most door installations, so get a sense of how much shimming you'll need to do to make that side plumb.

Trimming Doors

When the door swings open, it should not "pattern" your carpet or abrade the finish floor. If it does, trim the bottom rail of the door. To register the height of the carpet on the base of the door, slide a flat builder's pencil across the carpet. The pencil, being flat, won't sink into the carpet as much as a round pencil would. Add ⅛ in. of clearance to that rough line and score the final cutoff line onto the door, using a utility knife drawn along a straightedge. Scoring the door is important because it prevents wood grain from splitting and veneer from lifting and splintering. Run the circular saw a whisker below that scored line and you'll get a nice, clean cut.

For best results, use a circular saw with a sharp Mastercut blade, which has a close configuration of at least four fine teeth and a raker to clear chips. Use a straightedge clamped to the door to guide the blade. Clean the saw sole (base plate) well: Degum it with turpentine (or paint thinner) and steel wool, then rub it with a metal-polishing cloth or paraffin to help it glide across the wood. Smooth the cut, and ease the edge with 220-grit sandpaper, sanding with the grain.

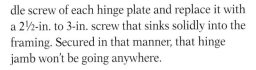

B. If you've got help, leaving a prehung door in its frame is generally a faster way to install it.

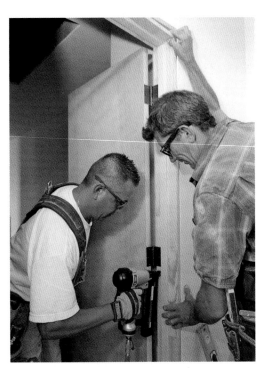

C. In this method, margin (center) the hinge jamb in the wall, then tack it to the trimmer stud with just three finish nails—one beneath each hinge. Or use three trim-head screws, which can be backed out easily if you need to adjust the jamb's fit or to reset the shims.

dle screw of each hinge plate and replace it with a 2½-in. to 3-in. screw that sinks solidly into the framing. Secured in that manner, that hinge jamb won't be going anywhere.

3. Between the door and the hinge jamb, you'll notice a slight *reveal* (gap), typically about ⅛ in., created at the factory so that the door would fit closely but without binding. If you maintain that ⅛-in. reveal between the door and the frame as you shim the head and latch jambs, the door will operate freely and look good. This method is called *following the reveal,* and it works well because your eye notices the gap first and foremost. So you can relax about abstractions such as plumb and level. Set your level aside if you

D. Then, holding your level against the hinge jamb, gently drive pairs of shims between the hinge jamb and the stud to bring the jamb into plumb. Because the finish nails don't offer much resistance, the jamb will move into position easily yet remain tacked to the RO.

E. Once the hinge jamb is plumbed and shimmed, margin the head jamb and tack it with a single nail or trim-head screw near the top of the latch jamb. This holds it loosely in position.

F. With your helper on the other side of the door, shut the door and "read the reveal"—the gap between the door and the jamb edge. Working from both sides, drive pairs of shims until the reveal along the head jamb is even (usually ¹⁄₈ in.). Nail shim points.

like. (If you are using method 2, once you've tacked the hinge jamb and the head jamb, you can set the door back on its hinges.)

4. Because you've already margined the hinge jamb, that corner of the head jamb should be centered as well. So, using a Speed Square, margin the latch corner of the head jamb, then use two 8d finish nails to tack it to the header. Shut the door so you can read the reveal between the top of the door and the head jamb; add shims and adjust the head jamb so there's an even ¹⁄₈-in. reveal along its length.

5. Next, follow the reveal to the latch jamb. Margin the latch jamb, tack it in place, and add three pairs of shims—one at each corner and one behind the latch strike plate. This step goes best with a worker on either side of the closed door—one reading the reveal and advising whether to push shims in or out at each point. When the reveal is even along the latch jamb, nail it off with two 8d nails through each set of shims. If the jamb stock is straight, three pairs of shims should suffice. However, if the jamb bows between shimming points or doesn't seem solidly anchored, add two more pairs of shims, each pair spaced equidistantly between the others.

G. Margin the latch jamb, tack it to the stud, shut the door, and continue reading the reveal between the door and the latch jamb, adding shim pairs until the reveal is even. Keep tweaking.

H. Once the reveal is even on all sides, nail the frame securely, using a pair of finish nails at each shimming point. To add holding power, some installers also replace each middle hinge screw with a 2¹⁄₂-in. screw that sinks into the framing.

I. Trim shims flush and you're ready to add casing. Here, an oscillating multitool does the trimming.

If you must reduce the width of a door, use a power plane on the hinge stile of the door. Planing down the lock stile is not advisable because locks have specific setbacks from door edges and you'd need to move face bores, too. So, it's far easier to plane hinge stiles.

After planing down the hinge stile, use a small router (laminate trimmer) to mortise the hinge gains deeper. Clamp scrap to the door edge to provide a wider base for the router. You could use a chisel, but a router will get the depth exactly right. Use a chisel to square up hinge-gain corners.

6. Check the reveal around the door one last time, make sure all shims are snug, and then trim the shims flush, using a utility knife, a Japanese saw, or an oscillating multitool. That done, you're ready to install handles and locksets (see pp. 133–136) and door casings (p. 130 and chapter 17).

Installing an Exterior Door

Installing a prehung exterior door is much the same as installing an interior door, so consult the preceding section if any step below is insufficiently explained. Exterior door units are inherently more complex to install, however, because of the need to make them weathertight. If you discover rot in the existing door frame or the surrounding framing, attend to that before installing a new unit.

PREPPING THE OPENING

Most of the prep work involves leveling and weatherproofing the bottom of the opening. That's where water is most likely to enter and damage flooring, subflooring, and framing. When installing an exterior door, cover the work area with tarps to contain the mess and heavy cardboard to protect finish floors.

1. Before ordering the new door unit, measure the height and width of the RO and the thickness of the wall. Check the trimmer studs for plumb, the header and sill for level, and corners for square. Generally, the RO will be 2 in. wider than the door—or ½ in. to 1 in. taller and wider than the framed unit. (When the prehung unit arrives, check all these measurements again—and the dimensions of the unit as well.)

2. Determine the height of the finish floor. In most cases, prehung doors come with a combina-

tion sill/threshold already attached—usually screwed to the bottom of the jamb legs. In new construction, this combo sill/threshold sits directly on the plywood subfloor, although in some installations the doorsill sits directly on top of the floor joists with subflooring and finish flooring butting up against it.

3. Install a sill pan, even if the doorway is protected by an overhang. Given the popularity of rubberized self-adhesive flashings these days, most builders fashion a sill pan from membrane such as the Moistop Corner Shield. Because most of these products are flexible and self-sealing, they can be easily fitted into corners or rolled to create water dams. If your region gets driving rains, a prefabricated metal or plastic sill pan is another option. Whatever the pan material, however, fold its ends and back edge up—as shown in the drawing below—so the pan will confine any water that gets under the sill. All sill pans should be caulked well inside corners and along seams. If you use a bituminous membrane, make sure the sealant you use is compatible with the membrane. Butyl-based sealants are usually a safe bet, but check manufacturer's specs for both products.

Sill Pan for Exterior Door

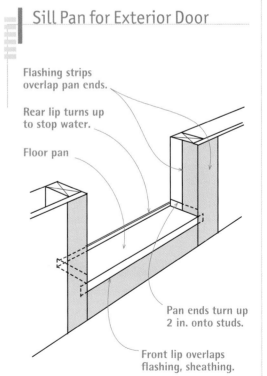

Flashing strips overlap pan ends.

Rear lip turns up to stop water.

Floor pan

Pan ends turn up 2 in. onto studs.

Front lip overlaps flashing, sheathing.

Whether fabricated from sheet metal or bituminous membrane, floor pans can protect doorway openings from getting soaked by standing water.

If the subfloor isn't level—common in older homes—one option is shimming beneath a doorsill and then filling between shims with a cementitious mix. Cedar shims are OK over a dry concrete subfloor, but if the doorway is damp or unprotected, use plastic shims.

4. Before installing the door, weatherproof the sides of the RO by folding building paper or housewrap into the opening (see p. 131), making sure that it overlaps the upturned edges of the sill pan. Alternatively, if you haven't yet installed a water-resistive barrier, you can attach flexible, self-adhesive flashing tape to the sides of the RO that will overlap the tops of the sill pan. Flashing tape comes in widths up to 12 in. and can be easily shaped and smoothed around the corners of the RO. Finally, to direct water away from framing, the outer lip of the sill pan must overlap any exterior flashing beneath the sill. To keep water out, *above always overlaps below.*

DRY-FITTING THE DOOR

Once you've prepped the opening, place the door unit in it to see what needs doing. This is a dry run, so leave the door in the frame. You'll need a helper for this sequence.

1. Test-fit the door frame in the opening. Center it right to left in the RO; there should be ¼ in. to ½ in. of clearance around the frame so it can be shimmed. Then, margin the frame—center it within the thickness of the wall. Jamb edges inside and out should be flush to wall surfaces, or ¹⁄₁₆ in. proud to accommodate framing quirks.

2. Is the door frame square? If it is, there will be an even reveal (gap) between the door and the jambs all the way around the door. If not, the frame is skewed. To square it, remove the frame from the opening, and tilt the frame to one side until acute corners (less than 90°) become square and the reveal is even all around. Factory-made

Leveling a Doorsill

In standard construction, the doorsill sits on the subfloor and the finish floor butts to the sill, but in many older houses, the sill sits directly on top of the joists and the subfloor and finish floor butt to the inside edge of the sill. Now and then, you even see the sill notched into the joists.

There are several ways to level an exterior doorsill. After you've installed a sill pan, place a spirit level across the bottom of the rough opening and insert pressure-treated wood shims under the sill until it reads level. Tack shims to plywood sub-flooring to keep them from drifting, and then install the prehung door unit. If the subfloor is concrete, use construction adhesive to spot-tack the shims. Or you can place the door frame into an out-of-level opening, use a flat bar to raise the low end of the frame sill, and insert shims under the sill until it's level. To prevent flexing between shims, space shims every 12 in. beneath aluminum/combination or oak sills, or every 6 in. to 8 in. beneath aluminum sliding-door sills.

The problem with any of these solutions is that many sills—especially hollow aluminum ones—are so thin that they may flex between shim points. If there's ¼ in. to ½ in. of space under the sill, use a ¼-in. tuck-pointing trowel (see the bottom photo on p. 252) to push *dry-pack mortar* into the spaces between the shims, compacting the mortar as best you can. (Dry-pack mortar has a minimal amount of water—just enough to activate the cement. If you squeeze a handful, it will keep its shape but there will be no water glistening on its surface or on your hand.) Or fill under the shimmed-up sill with nonshrinking mortar, such as an epoxy mortar/grout.

The only way to correct an out-of-level opening without shims is to rebuild it, which is rarely simple. For example, you can remove the subfloor and power-plane the joists until they are level, then install pressure-treated plywood. But if nearby floors are out of level, you may need to level them next.

exterior door frames are made from heavy stock, milled to exact specifications, and reinforced with sturdy sills, so they rarely arrive out of square. Place the unit back in the opening and check that the reveal is even.

3. Assuming the reveal is even, remove the unit from the RO. If the bottom of the opening is not level, attend to that now, as described in "Leveling a Doorsill" above. Then caulk the bottom of the opening with a triple bead of caulk.

INSTALLING AN EXTERIOR FRAME

There is no one right way to install a prehung exterior door. Some carpenters prefer to install door frames with doors in them. Others prefer to remove the door, plumb and attach the hinge jamb, rehang the door, then attach the other jambs. Although the photo sequence shows the door removed, both methods are described here.

Entry Door Essentials

Start by selecting an entry door whose architectural style fits your house. Then refine your choices by considering climate, the orientation of the door toward prevailing winds or strongest sun (south and west), and to what extent the entry is protected by, say, a porch roof or eave. As a rule of thumb, wood door makers recommend that an overhang stick out from the house an amount equal to at least half the distance from the doorsill to the underside of the eave. If the depth of the overhang is less than that, a fiberglass or steel door may be a more durable choice for that location.

Materials

Aesthetics aside, picking an entry door is a trade-off of durability, maintenance, price, and energy efficiency. Long term, a cheap, leaky door will cost you plenty.

Wood. Quality wood doors feature panel-and-frame construction, with mortise-and-tenon joints at the corners for greater durability. Traditionally, panels have floated free within the frame to allow for expansion and contraction; today's energy-efficient models also add elastic sealants around the perimeter of the panel to cut air leaks and moisture penetration. To minimize warping, many panels have engineered-wood cores with $1/16$-in.-thick veneers, whereas their frames have solid *stave cores* of the same hardwood species—mahogany is a favorite because it wears well. Wood doors can be clear-finished, stained, or painted. Our advice: Get a *factory finish* with an exterior-grade, oil-based varnish or polyurethane with UV-protection.

Energy profile: On average, wood doors have an R-2 value, although more efficient ones achieve R-5. Interestingly, doors with *low-e glass* have higher R-values than wood doors without glazing.

Fiberglass. Spurred by energy efficiency and surface details that are almost indistinguishable from wood, fiberglass doors have been gaining market share for the last decade. Typically, fiberglass skins are applied to a wood-composite, laminated veneer lumber (LVL) or steel frame to lend rigidity; the core of the door is then filled with insulating foam. Fiberglass doors are dimensionally stable and surprisingly sturdy. Better-quality models look very much like wood because they were created from molds taken from actual wood doors—including period moldings and wood grain. Once painted, they'd stump an expert, and several manufacturers offer unfinished fiberglass doors that can be stained.

Energy efficiency: Among the best doors, expect R-8 or higher. As fiberglass expands and contracts at the same rate as glass, glazing seals are especially stable in such doors.

Steel. Steel entry doors are economical, durable, energy-efficient, and easy to maintain. The knock is that they don't look like wood, and so they look out of place on older homes. But that's not a negative on a newer house with sleek lines. Most residential steel doors are made from 24-gauge steel skins over an engineered-wood or steel frame; internal cavities are filled with an expanding foam, typically polyurethane. Where houses are located in a fire zone, codes will specify the

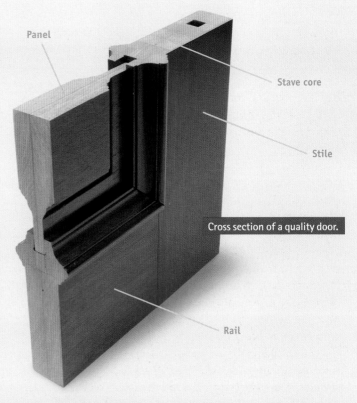

Panel

Stave core

Stile

Cross section of a quality door.

Rail

core insulation, the thickness of the steel skin, and so on. Steel doors need repainting every three to five years. If the steel skin begins to rust, it can go fast.

Energy profile: Not quite as good as fiberglass but close. Some models hit R-8.

Weatherstripping and Thresholds

For best results, buy an entry door system with integral weatherstripping, in which a *kerf* (slot) is milled into the door frame. Integral systems have close tolerances, tight fits, and nylon-jacketed foam stripping that can be compressed repeatedly without deforming or losing its resilience.

There are many sill-threshold combinations to choose from. First, examine the existing sill because, especially in older houses, there's a lot of variation in how sills were detailed. (See "Leveling a Doorsill" on p. 125.) If your present sill is nonstandard, take photos of it and share them with the lumberyard person who's selling you a replacement door. Ideally, you can find a sill/threshold combination that will not only be weathertight but also relatively easy to install. On outswing doors, a *bumper threshold* (p. 137) has a built-up curb; the door closes against a weatherstripped section to seal against air infiltration and moisture.

Sweep gaskets seal the bottom of the door, typically with a neoprene gasket in a metal channel that screws to the bottom door rail. The sweep also may be mated to a gasketed channel in a threshold. Better-quality sweeps are adjustable and have replaceable gaskets, as they become abraded in time. In general, avoid bristle-type sweeps because they don't block air or water as successfully and aren't durable.

Entry Door Hardware

There's a plethora of door hardware, but apart from finishes, styles, and other aesthetic considerations, hardware choices can be boiled down to a few essential differences between hinges and lock assemblies.

Hinges. Any prehung door you buy from a reputable supplier should have hinges matched to the weight and width of the door. Modestly priced exterior doors usually come with standard 4-in. by 4-in. *butt hinges.* However, if your entry door is oversize or wood (which tends to be heavier than fiberglass or steel doors), consider paying a bit more for 4½-in. or 5-in. *ball-bearing hinges.* You can also get *adjustable hinges,* which are mortised into the edge of the door and can accommodate vertical and horizontal adjustments—rather like the "German hinges" (see p. 401) used in good-quality cabinets. Adjustable hinges seem like a good idea for houses subject to seasonal shifting, such as those built on adobe soil.

Lock assemblies. Exterior locksets are most often *cylinder locks,* which require a 2⅛-in. hole drilled through the face of the door, or a *mortise lock,* which is housed in a rectangular mortise cut into the latch edge of the door. For a modest bump in price, you can add a *dead bolt* and a *reinforced strike plate* (see p. 135), which can't be dislodged by a kick—the preferred method of entry by breaking-and-entering men. If you are ordering good-quality, factory-built doors, multi-point locking systems (which secure two sides of the entry door to the door frame) are often a standard option. If you purchase a lockset separately, give its specifications to the factory so it can machine the door specifically for that lockset.

Before installing a prehung exterior inswing door, remove the door and test-fit the frame in the opening. The 2x4 cleats nailed to the sheathing act as depth gauges, so jamb edges will be flush to sheathing.

Shim and secure the hinge jamb first, checking for plumb often. Here, installers place shims slightly above and below the hinges so the hinge screws sticking through the jamb won't snag on the shims.

Ordering EXTERIOR DOORS

When ordering an exterior door, measure the rough opening carefully—noting also if the opening is racked or corners are out of square. Discuss those measurements with the exterior-door factory salesperson, because all door sills and thresholds will add to the height of the door assembly. You want to make sure that the door will fit in the opening you have, whether rough opening or existing doorway. Ask the salesperson for details of construction for the door unit, including jambs, sill, door shoe, and so on. Older houses often have low head heights, so the factory or door shop may need to cut down its door stock to fit. So, again, you must be concise about the size of the opening you have.

Method 1: The door stays hung

1. Pry off the shipping blocks, remove the plastic plugs from the lock bore, and place the unit into the RO. (At this point, assume the doorsill is level.) Approximately center the unit left to right in the opening, then margin the hinge jamb while a helper keeps the unit from tipping. Near the top of the hinge jamb, drive in a 10d finish nail through the face of the jamb to tack it to the trimmer. (You may have to open the door to do this.) Then hold a 6-ft. level to the edge of the jamb and then to the face of the jamb to see if the hinge jamb is plumb.

2. If the sill is level and the frame has remained square, you should get a plumb reading. Also, sight along the frame to make sure the jamb stock is straight. If the hinge jamb is plumb, insert pairs of shims between the hinge jamb and the trimmer stud—one pair of shims behind each of the three hinges and two more pairs spaced equidistant between the hinge shims. Because you and your helper will have a door between you, you'll need to communicate constantly as you insert shim pairs from both sides and continually check for plumb.

3. If, on the other hand, the hinge jamb is not plumb, the frame may be racked slightly. Again, assuming that the sill is level, insert a flat bar between the frame and the trimmer near the top of the frame, and pry the frame out on one side until the hinge jamb is plumb. Then tack the hinge jamb as described. If the hinge jamb is bowed, that could give you an off-plumb reading, too. Adjust shims and use nails to adjust the bowed section.

4. Once you've plumbed and secured the hinge jamb, *read the reveal*—make sure there is an even clearance between the door and the jambs—to position the head and the latch jambs. Use nails to draw the frame toward the trimmer or adjust the shims so the reveal is uniform—typically about ⅛ in. As you work the jambs, also use the edge of your Speed Square to make sure frame edges are margined in the wall.

One final aside: *Precased* door units can be shimmed only from the inside. In that case, insert a shim, fat end first, until it butts against the back of the casing, then slide additional shims, thin edge first, until shims are tight. Nail or screw below each stack of shims, then close the door and see if it seats evenly against the frame.

Method 2: Remove the door

1. After prepping the opening, pull the hinge pins and remove the door from the frame. (As is the case in the directions above, assume the doorsill is level.)

2. Center the frame left to right in the opening, and margin the frame. If you're working solo and want to make sure the frame stays margined, tack temporary *cleats* to the outside of the frame and to the exterior sheathing, as shown in the top photo on p. 127. Tack it lightly though, because you'll probably need to reposition the frame as you plumb its sides.

3. Start with the hinge jamb. Having margined the frame, drive a 10d finish nail in the middle of the hinge jamb about 6 in. down from the top; leave the nail head sticking out. Using a 6-ft. level to check for plumb, shim the jamb behind each hinge: top, bottom, and then middle hinge. If you nail just below the shims, they'll be easier to adjust. Use two 10d finish nails for each shimming point and, again, leave the nail heads sticking up. As you hold your level against the jamb hinge, note whether the jamb stock is straight. If it bows, you'll need to nail or adjust shims to pull the jamb into line. In all, use five pairs of shims (and nails) to secure the jamb hinge.

4. Using a framing square to ensure that it's roughly square to the hinge jamb, tack the head jamb to the header. The head jamb's position is approximate at this point because you'll use the door to fit things more exactly. With the aid of a helper, set the exterior door back on its hinges.

PRO TIP

Hand-nail the finish nails used to tack a frame to the RO and leave nail heads sticking up. If you use a nail gun for this operation, you'll likely drive the nails all the way in—making subsequent adjustments difficult.

After plumbing and securing the hinge jamb, rehang the door so you can use it as a gauge to align the head jamb and the latch jamb correctly. This method is particularly helpful if the door is slightly twisted or warped because you can align the latch jamb to the door.

As you adjust latch and head jambs, make sure there's an even gap (typically ⅛ in.) between the door and the jambs all around.

Shim the head jamb after plumbing the latch jamb. If the frame is uncased, you can insert shims from both sides and easily slide them in and out. The white line inside the jambs is kerf-in weatherstripping.

Close the door and note how it fits the frame. Without weatherstripping, there should be an even reveal, about ⅛ in. wide, around the door.

5. If the door hits the edge of the latch jamb, that jamb may be bowed or the shims behind the hinge jamb may be too thick. Adjusting an uncased exterior door frame is similar to "working" an interior door frame. Shim the head jamb and then the latch jamb—fine-tuning the jambs so the ⅛-in. reveal between the door and the frame is constant. Operate the door to make sure it opens and shuts without binding. The latch jamb should have four or five pairs of shims, and the head jamb should have at least three. Then fit insulation between the frame and the RO, install casing, and flash the unit, as described in the following section.

SEALING, CASING, AND FLASHING THE FRAME

If any edges or faces of the frame are still unfinished, prime or paint them before installing exterior casing.

1. If the cavity between the jambs and the framing is accessible only from the outside, fill it with insulation before installing the exterior casing. I favor packing the cavity with loose fiberglass or recycled cotton insulation. Should you need to adjust the frame at some future time, you will be able to do so easily. If you fill the cavity with spray foam—even low-expanding foam—you are, in effect, gluing everything together and creating a huge mess for the next person to work on the frame. Avoid high-expansion foam at all costs: It's so powerful that it can easily bow jambs into the opening and bind doors and windows.

Shimming Exterior Doors

Shim exterior door frames at five points along each side jamb. Along the hinge jamb, shim behind each of the three hinges—or as close as possible if hinge-screw points stick out of the back side—and add two more sets of shims spaced equidistant between hinge shims. Space shims along the latch jamb at roughly the same intervals, but don't shim directly behind strike plates or dead bolts. Shim the head jamb midway and at both corners.

Exterior door frames are often installed with pairs of 10d galvanized finish nails that go through or slightly below each pair of shims. Using two nails at each interval keeps the frame from twisting. Other builders favor pairs of 2½-in. stainless-steel trim-head or plated flathead screws because they grip better and can be removed easily. In addition, many builders remove the middle screw of the top hinge, shim behind it, and replace the original screws with 2½-in. or 3-in. screws that sink deep into framing.

Here are some fine points to consider:

▶ It doesn't matter whether you screw or nail below shims or through them, as long as the shims are snug. (Shimming below hinges allows them to be adjusted later—which you can't do easily if you nail through them.)

▶ Even if trimmer studs are plumb, shim between the door frame and the rough opening anyhow. That is, don't nail jambs directly to the framing: A shimmed frame will be easier to modify or replace later.

▶ Always shim the head jamb or it may bow into the opening or jump when you nail casing or drywall to it.

▶ To cut shims flush to finish surfaces, score the shims with a utility knife. Then snap off the waste. You can also use a Japanese saw or an oscillating multitool.

2. Before attaching casing, air-seal the gaps around the frame. There are two primary ways to do this. The first is to apply self-adhesive flashing tape so that it straddles the gap between the jambs and the housewrap. The second way is to run beads of siliconized acrylic caulk around the jamb edges and the RO, then press 6-in.-wide fiber-reinforced paper flashing strips into the caulk, as shown in the center photo on p. 130. These strips aren't self-adhesive, so staple edges that overlap the sheathing. Whether you use tape or strips, align their edges back ¼ in. from the inside edge of the frame so they won't be visible. (Strips are favored by builders who want a thinner material between frame jambs and casing, or those who have not yet installed housewrap and want to slide flashing under the housewrap at the head and sides.)

3. If the unit's doorsill has horns that extend beyond the side jambs, trim each horn to match similar details elsewhere on the house. Typically, horns line up with the outer edge of side jambs or protrude ¼ in. beyond them. Because the sill is pitched, you'll need to cut the bottoms of the

PRO TIP

Prime and paint all six sides of exterior doors—especially the top and bottom edges—before putting on the hardware and weatherstripping. Protect unfinished wood with at least one coat of primer and two coats of good-quality oil-based paint. Also, carefully prime lock holes, leaf gains, and all edges. Finally, caulk panels after priming and before painting so there's no place for water to penetrate.

Secure the frame, trim the shims flush, remove the temporary cleats, and apply a bead of caulk to the jamb edges. Keep the caulk back at least 1/4 in. from jamb faces.

Apply flashing strips to the edge of each side jamb so that it beds in the caulk. However, before applying exterior casing, apply a second bead of caulk over the flashing paper. Double-caulking virtually eliminates air and water infiltration. Staple the other edge of the strip to the sheathing.

As you install casing, drive a finish nail through corner miter joints to keep them from separating. You can also glue the joint, but this joint will be kept in place by the stucco that surrounds it. Note the 1/4-in. reveal between the casing and the edge of the jamb.

Install cap flashing atop the head casing, nailing its upper flange as high as possible. This metal flashing will be overlapped by a self-adhesive flashing tape and stucco. Cap flashing should overhang the casing slightly along the front and at the ends of the casing so water drips free.

Door-Casing Reveal

Finished wall

Side jamb

Gap between door frame and RO

Casing

Head jamb

1/4-in. reveal

Doorstop

When installing casing, set it back 1/4 in. from the jamb edges. This setback, called a *reveal*, tricks the eye: Even if jambs or casings are not straight, their joints look straight.

casing at the same angle; use an adjustable bevel gauge to transfer the angle to the casing.

4. Attach the casing. Set the inside edges of jamb casing back 1/4 in. from the inner edges of the jambs to create a 1/4-in. reveal. Then nail up the head casing. If casing corners are mitered, nail through the joint to draw it tight, as shown in the bottom photo above. *Note:* A flashing strip or piece of flashing tape will be applied over the upper leg of the cap flashing to direct water away from the sheathing. (*Above overlaps below.*)

5. Once you have installed the head casing, attach the head flashing (cap flashing) that was supplied by the door manufacturer. If your prehung door didn't come with head flashing, any sheet-metal shop can fabricate a piece of galvanized head flashing with the proper offset for the thickness of the head casing so water drips beyond it. Many lumberyards also carry a variety of preformed flashing, and brickworks carry specialized head casing.

6. Caulk the head casing/sheathing joint, press the cap flashing down onto the casing, then use large-head nails to nail the top flange of the flashing to the sheathing. Apply self-adhesive flashing tape or strip flashing over the top leg of the head flashing, and run housewrap and siding down over that.

Installing French Doors

Installing double doors requires more plumb and level readings, shim adjustments, and—above all—more patience than hanging a single door, but the procedure is much the same. So we'll zip through the steps covered earlier to get to the adjustment most often required when installing double doors: figuring out why the doors don't meet perfectly in the middle and what to do about it.

INSTALLATION: A QUICK SUMMARY

▶ Measure the RO to make sure it's large enough to accommodate the prehung unit.

▶ Using a 6-ft. level, see if the bottom of the RO is level and the sides are plumb. Leveling the bottom is critical: French doors are wide and heavy and must rest solidly on a level opening. "Leveling a Doorsill" on p. 125 covers the process. But if the sill/threshold is made out of aluminum, it won't have much inherent strength. To prevent flex, you'll first need to create a level pad of shims spaced every 6 in. to 8 in. and then set the sill on top of them. Or you might rip down a beveled piece of treated wood to attain a level surface. Whatever works.

▶ Check the wall faces on both sides of the RO for plumb—so you'll know what to adjust if the wall isn't plumb. The frame *must* stay square for the doors to operate correctly. You may need to adjust the casing to reconcile a plumbed frame to a not-plumb wall. Fortunately, caulk will fill gaps between casing and jambs.

▶ The doors *must* come together in plane; otherwise, they won't seal or lock right. In other words, doors meeting in the middle is more important than having casing hit jambs perfectly. You may have to split the difference—push the corners of the jamb in or out slightly to get the doors to meet correctly. That done, center the frame in the RO, margin the frame in the wall, and tack the frame in the opening using one trim-head screw at the top of each jamb.

▶ Double doors have two hinge jambs, of course, so it doesn't really matter which one you plumb first. An old-timer we know likes to shim the top hinge on each side first, to

Because French doors are wide and heavy, it's crucial that the subfloor of the opening be level. Insufficiently supported, French-door thresholds can flex, become misaligned, and leak.

Just above grade, this wide opening needed some extra attention. A self-adhesive bituminous waterproofing membrane seals the sheathing/foundation joint (just visible, lower left); this flexible membrane, in turn, is covered by galvanized sheet metal. The bottom of the rough opening also is wrapped with foil-faced peel-and-stick flashing, which extends 4 in. up the studs at each end.

Tilt French doors into place. Tack the cleats to the upper corners of the frame to keep the unit from falling through the opening and to ensure that the jambs will be flush with the sheathing.

When jambs are bowed, a plumb bob gives more accurate readings than a spirit level. Hang the bob from a nail near the top of the frame. When the jamb is plumb, readings to the string will be equal along its length.

When there's too much of a gap where French doors meet in the middle, use a flat bar to ease the jambs toward the middle so you can slip shims behind them. Tweak and reshim the doors until they seat correctly and close evenly in the center.

If doors are not protected by an overhang, apply flexible, self-adhesive flashing tape over the top flange of the head casing after the doors are set and the casing is installed. This flashing overlaps the fiber-reinforced flashing strips on each side. To facilitate painting, the manufacturer premasked door lights with plastic film.

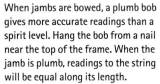

PRO TIP

Before drilling, chiseling, or routing a prehung door, insert wedges beneath it so it can't move. Alternately, you can pull the hinges, remove the door from the frame, and support it in a pair of homemade door bucks, also called door-holding jigs.

establish the correct ⅛-in. gap between the doors, high up. Then he plumbs and secures one side completely, then the other, keeping an eye on the gap as he goes. He adjusts shims continually on both sides. When he's done, there are five pairs of shims (and five pairs of 10d finish nails) on each jamb and five shim-nail pairs across the head jamb. Lastly, he removes a middle screw from the top hinge on each jamb and replaces it with a 3-in. screw that sinks deeply into the framing. Hanging doors is a little like having a religion—it doesn't matter *what* you believe, really, just that you believe in something.

FINE-TUNING FRENCH DOORS

Now it's time for fine-tuning the installation or, as the pros call it, *working the frame*. Here are a few problems you might run into and what to do about them:

▶ **Doors hitting in the center.** Ease off the shims behind the hinge jambs. If the problem is a bowed jamb, use additional nails or, as needed, 2½-in. screws to pull it back toward the outside of the RO.

▶ **Doors too far apart in the center.** Shim out from the RO more. If that doesn't work or if you must shim so much that the center of the jamb

bows, the head jamb was probably milled too long and should be cut down. The gap between the double doors should be ⅛ in. or whatever the manufacturer suggests.

▶ **Wider gap between doors near the top.** Shim out behind a top hinge on one side.

▶ **Uneven door heights, binding in a corner.** One corner of the frame is lower than the other. Try driving a shim under the jamb on the low side. But, given the importance of solid shimming under the length of the sill, you may want to remove the frame and reset all the shims. A less likely explanation: One jamb is too long. In which case, use an oscillating multitool (see p. 57) to cut the jamb in place, then pull the nails tacking the jamb to the RO so you can reposition the jamb.

▶ **Doors aren't in plane.** The frame, doors, or RO may be twisted. Use a level to see which element is out of plumb, then sight along the straight edge of the level to see if any surface is bowed. If the RO is plumb, the jambs or the door may be warped. Contact your vendor to see about getting a replacement. Doors *must* meet in plane.

▶ **The door won't stay shut, or one hinge binds while the others work fine.** See if the jamb is twisted. You may need to reset the shims until the jamb is square to the door. Otherwise, the hinge may be irregular. To correct it, use an adjustable wrench or locking pliers to bend the knuckles on one of the hinge leaves. Bend the leaf on the door, though, because you'll probably split the jamb if you try to bend a leaf attached to it.

Installing Hardware

Specifics vary, but most locksets come with paper templates that locate the center of the holes drilled in the *face* of the door (face bores) for handle-spindles or cylinders, and holes drilled into the *edge* of the door (edge bores) for latch assemblies. A second paper template locates holes drilled in the *latch jamb* of the door frame. Although the directions given in this section are typical, always follow the directions supplied by your lock maker. *Note:* Measure the door thickness before buying locksets or key cylinders. Some mechanisms are adjustable, whereas others fit only specific door thicknesses.

MORTISE LOCKSETS

Mortise locksets house latch bolts and dead bolts in a single casing, so the door stile must be solid wood to receive this type of lock. If possible, order the door machined specifically for the lockset. If you intend to make the holes yourself, a door-boring jig will be a big help.

1. Using the template, mark the outline of the lock case on the edge of the door. Then mark a line in the exact center of the edge. Along this line, use a 7/8-in. spade bit to drill holes to the depth of the lock case. Overlap holes slightly.

2. Use a chisel to square up the edges of the lock-case mortise. As you chisel, test-fit the lock case periodically to avoid chiseling away any more wood than necessary. When the lock case fits all the way into the hole, trace the outline of the main latch plate onto the edge of the door. Use a router to mortise the latch plate. If the door edge is beveled, adjust the tilt of the main latch plate to match the bevel beforehand.

3. Remove the lock case. Using the template, mark knob/spindle and key/cylinder holes on the face of the door stile. Use a hole saw to cut the cylinder hole and a Forstner bit or spade bit to cut the smaller spindle hole, holding drill bits perpendicular to the stile. Drill the holes until the point of the bit just starts through the other side. To prevent splintering of the stile face, back the drill out and finish drilling from the other side.

4. Reinsert the lock case, and screw it to the edge of the door. Then insert the spindles, slide the escutcheons over the spindles, attach the handles or knobs to the spindles, and see if they turn freely. Once they do, screw on all the trim

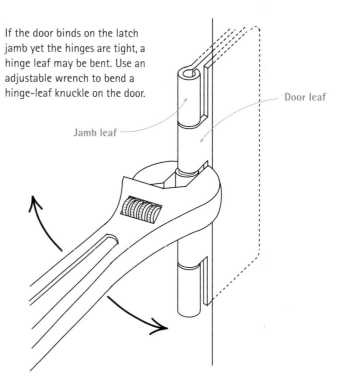

Bending a Hinge Knuckle

If the door binds on the latch jamb yet the hinges are tight, a hinge leaf may be bent. Use an adjustable wrench to bend a hinge-leaf knuckle on the door.

Door leaf

Jamb leaf

PRO TIP

Because manufacturers often create a general hardware template for several different door styles, the template provided may be inaccurate. You may not need all the holes indicated, or you may need to reposition the template to accommodate a door edge bevel. So examine the door hardware and think things through before you mark or drill the door.

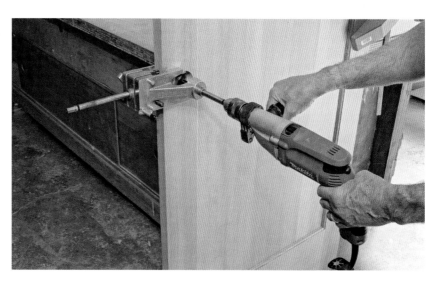

A door-boring jig will keep boring devices perpendicular to the door edge or face. This is crucial when boring the edge because there will be no more than 7/16 in. of wood on each side of the mortise casing. Moreover, a lockset set askew may not work correctly.

Mortise Lockset

Thumb turn

Main latch plate

Dead bolt

Lock case

Knob spindle

Knob

Key cylinder

Trim plate

Finished latch plate

Latch bolt

Thumb lever

Thumb-lever shaft

Handle

Door edge

Mortise locksets combine security and convenience, because you can use a single key to operate both a latch bolt and a dead bolt.

A. Use the paper template supplied with your lockset to center face bores on the door stile and edge bores on the door edge. The template gives the exact setback and hole sizes. Use an awl to mark the hole centers. Prehung doors often come with lock cases prebored.

B. Drill the face bores, which are positioned with a paper template. Use a hole saw for the larger, key/cylinder hole and a Forstner bit for the spindle or thumb-lever hole. The small, round level taped to the top of the drill helps the installer drill perpendicular to the door face.

C. Once you've mortised the lock case into the edge of the door, use a router and a template to mortise the latch plate into the edge.

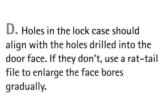

D. Holes in the lock case should align with the holes drilled into the door face. If they don't, use a rat-tail file to enlarge the face bores gradually.

E. Use a chisel to square up the rounded corners of the latch-plate mortise.

F. Strike plates are closely matched to the locksets they're supplied with. Typically, the larger opening receives the dead bolt. For added strength and security, use 3-in. screws that will reach framing.

hardware. Typically, door handles or thumb levers are 34 in. to 38 in. high.

5. You also will find a strike-plate template. Depending upon the depth of the strike-plate assembly, use a router or a combination of drill and chisel to mortise the plate into the jamb. To more accurately position the leading edge of the strike plate, rub pencil lead on the latch edge; when the latch is released against the jamb, it will leave a pencil mark. For greater security, buy a unit with a *strike-plate reinforcer* and 3-in. mounting screws.

CYLINDER LOCKSETS

Cylinder locksets (also called tubular or key-in-knob locks) are popular because they're inexpensive and easy to install. Better models have a spring-loaded dead latch that prevents the bolt from being retracted by slipping a plastic credit card between the door edge and the frame. But no cylinder lock is secure because all can be snapped off with a pry bar or a swift kick. To be safe, install a dead bolt, too.

1. Using the template supplied by the manufacturer, mark the centers of holes to be drilled into the face of the door (face bore) and the edge (edge bore). Use a 2⅛-in. hole saw to drill the face bore. But after the tip of the hole-saw bit emerges on the other side, prevent splitting by backing the bit out and finishing the hole by drilling from the other face.

2. Use a ⅞-in. spade bit to drill the edge bore, keeping the bit perpendicular to the edge. Insert the latch/bolt assembly into the hole, and use a utility knife to trace around the latch plate. Rout the inscribed area so that the plate is flush to the edge of the door.

3. Screw down the latch plate and insert the lock mechanism through the latch assembly. Try the handle; it should turn freely. Next, position the strike plate on the jamb. To locate the strike plate exactly, rub a pencil on the end of the latch bolt, shut the door, and release the bolt against the jamb.

4. Using a ⅞-in. spade bit, drill a latch hole ½ in. deep into the jamb. Center the strike plate over the hole, and trace around it with a utility knife. Use a router to mortise the strike plate. When the door is shut, the latch bolt should descend into the strike-plate hole; the small spring-loaded plunger next to the latch bolt should not. The plunger should be stopped short by a lip on the strike plate.

5. For greater security, install a unit with a strike-plate reinforcer and 3-in. mounting screws.

Reinforced Strike-Plate Assembly

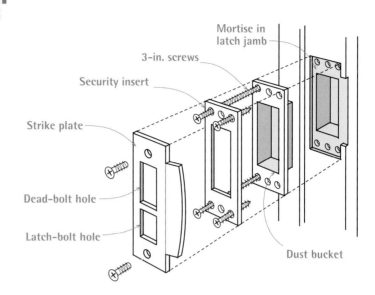

Because 3-in. screws anchor this assembly to framing behind the door frame, this strike plate can't be dislodged by a kick.

Cylinder Lockset

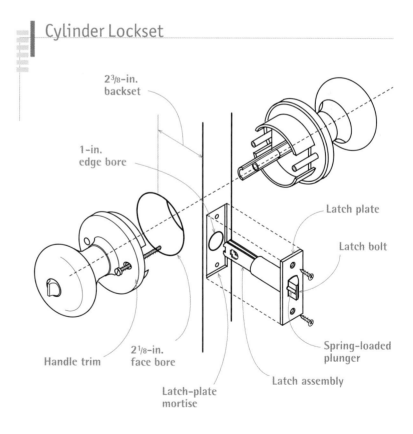

Cylinder locks are relatively inexpensive and easy to install. Many interior doors come with the large face bore predrilled.

To install the reinforcer, you'll need to drill through the jamb into the framing behind; the extra-long screws will grab the framing.

Weatherstripping Door Frames

Air leaks can account for 20% to 30% of the total heat loss of an insulated house. If your budget is tight, sealing seams (see chapter 14) and holes in the building envelope, packing insulation into the gaps around windows and doors, and installing

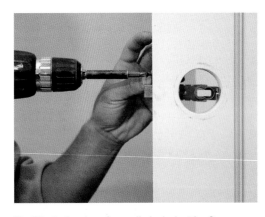

The 2¹⁄₈-in. face bore for a cylinder lockset is often predrilled at the mill. You may need only to screw the latch plate to the edge of the door . . .

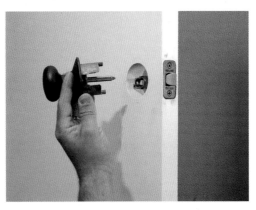

. . . then insert the lock body into the spindle hole of the latch assembly, and screw the two handles together. Follow the instructions supplied with your lockset.

weatherstripping should be your first priority. The single most crucial piece of weatherstripping is a tight-fitting door threshold.

WEATHERSTRIPPING JAMBS

Today, there are three main types of weather-stripping: tubular, metal-leaf, and kerf-in. Most are easy to install and require few special tools. Prehung exterior doors typically come with weatherstripping attached, which can be easily removed before installing the unit so you can better see the ⅛-in. reveal between the door and the jambs. Door shoe gaskets (which seal the bottom of a door) are removed for the same reason.

Tubular is the easiest to retrofit to old doors and the least expensive type of permanent weather-stripping. The reinforced part of the strips is usually metal, with slots for screws; slots allow you to adjust the stripping so it fits tight to windows or doors. To install tubular weatherstripping, shut the door and press the strip's flexible seal against the door, then screw the reinforced part to the jamb. Don't buy tubular stripping that nails up or has round holes (not slots) because it can't be adjusted.

Metal-leaf, commonly called a V-bronze or metal tension strip, is a thin metal strip folded lengthwise and nailed with brads to door jambs. When the door shuts, it compresses the metal, stopping drafts. Metal tension strips are durable and, because they fit between the door or window and the frame, are hidden when the door is shut. When installing it, place the leaves flush to the doorstop, spacing brads every 3 in. Install the head piece first, then the sides. To keep the leaves from snagging where they meet in the corners, snip them back at a slight angle (5°). In time, the leaves flatten, but they can be raised by running a flathead screwdriver down the center of the fold.

Kerf-in features flexible stripping (silicone, vinyl, or nylon) that slides into a kerf (slot) between the jamb and the doorstop. Kerf-in is the pre-ferred weatherstripping for prehung exterior doors because it seals tightly and can be easily replaced if the stripping gets compressed over time. To cut kerfs into existing frames, use a *kerfing tool,* which looks like a laminate trimmer on an angled base. (There's an interesting kerf-in sequence on p. 157.) Silicone stripping is a good choice for retrofits because it compresses so small that old doors shut easily without doorstop adjustments. At head-jamb corners, cut stripping at 45° angles so it lies flat.

Door Weatherstripping

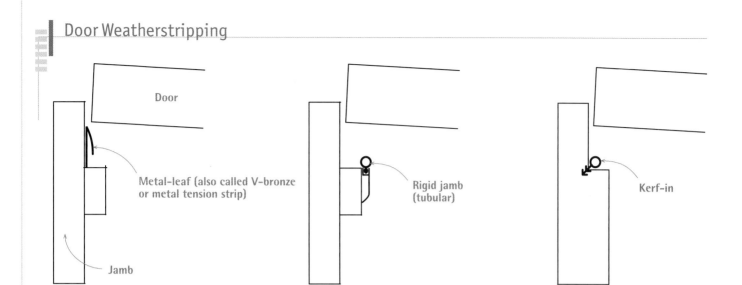

Door

Metal-leaf (also called V-bronze or metal tension strip)

Jamb

Rigid jamb (tubular)

Kerf-in

Weatherstripping Thresholds

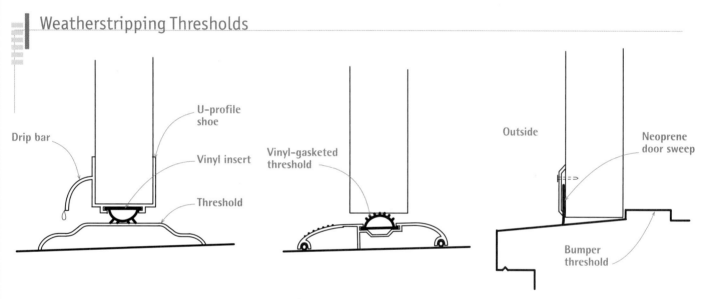

Drip bar

U-profile shoe

Vinyl insert

Threshold

Vinyl-gasketed threshold

Outside

Neoprene door sweep

Bumper threshold

SEALING UNDER THE DOOR

Weatherstripping at the bottom of a door must create a tight seal and be durable. Elements that attach to door bottoms are called *shoes* or *sweeps*; those that attach to doorsills are *thresholds*. Although primarily intended to reduce drafts, they also must resist water seeping or blowing in under the door. If your region gets a lot of precipitation, a roof or porch overhang should be your first line of defense—properly sized, the overhang should extend out from the building half the distance from the doorsill to the underside of the eave. You'd also be well served to retrofit the *water-return threshold* described on p. 138.

Attaching sweeps and shoes. If you're fortunate enough to have an outswing door, install a compressible gasket to the integral stop in the threshold and a flap neoprene sweep to the out-side of the door. On an outswing door, cut the sweep to length and hold it against the bottom edge of the door so the sweep just touches the threshold when the door is shut. Flap sweeps are not as durable as kerf-in sweeps, but they're inexpensive, easy to install, and reasonably durable.

If you have an inswing door, install a door shoe or a gasketed threshold. To do so, measure the distance, at several points, between the bottom of the door and the top of the threshold. Typically, you'll need about ½ in. of space between the door and the threshold to accommodate the thickness of the shoe and its vinyl insert—check the manufacturer's specs to be sure. If there's not enough clearance, plane or cut down the door.

After measuring, pull the door's hinge pins and lift the door out. Then, using a hacksaw, cut the shoe to length—typically, the shoe stops short of the doorstops on either side, so it will be 1 in.

The vast majority of exterior doors on residences are *inswing doors*, which swing into living spaces. In a perfect world, they would all be *outswing doors*, which are inherently better at blocking air and water because, when they are closed, outswing doors butt against a stop in the threshold that is similar to the stops in jambs. Add a compression gasket to that threshold stop and water will have to defy the laws of gravity to get around it. But our imperfect world includes burglars, who can easily pop the hinge pins of an outswing door to gain entry—unless you install nonremovable hinges. Food for thought.

cardboard or "door skins" (thin wood veneer), create a template, transfer those profiles to the underside of the threshold stock, and use an oscillating multitool with a metal-cutting blade to make cuts. Dry-fit the threshold to ensure a good fit, then remove it, caulk the sill, and screw the unit down. Install and adjust the threshold's gasket, rehang the door, and test the fit.

Retrofitting a water-return threshold. This is a bit more work. A three-piece threshold can be installed over an existing threshold. Consisting of a sill cover, a drain pan, and a threshold with weep holes, the assembly is sloped to send water back outside, hence the name water return.

The retrofit requires accurate measurements and careful cutting so all three pieces fit snugly to the existing door frame. To start, measure the widest point of the doorway opening and rough-cut the sill cover to that length. Next, hold one end of the sill cover square to the base of a jamb, and scribe the jamb's profile onto the metal sill. Use an oscillating multitool with a metal-cutting blade to cut the first end before scribing and cutting the other end. The sill cover must slope about 1/8 in. toward the outside so it will shed water. Hold a torpedo level on the sill cover while you temporarily shim the back edge up.

Measure and cut the threshold (not the drain pan) next. Usually, the threshold lines up to the outside edge of the doorstop, so when the door shuts, it fits snugly to the threshold as well. As you did with the sill cover, scribe the profile of the jambs onto both ends and cut them out. With the threshold resting on the sill cover, lightly scribe the outside edge of the threshold onto the sill cover, using an awl or a pocketknife. The outer edge of the drain pan should sit just shy of this scribed line. Remove the threshold and scribe the profiles of the jambs onto the drain

shorter than the width of the door. Shoes have screw slots that let you adjust their height; attach the shoe with one screw at each end. Rehang the door, lower the shoe until its seal makes solid contact with the threshold, tighten the screws at that height, and then open and shut the door. The door should drag slightly as you operate it, and you shouldn't see daylight under the shut door. Finally, insert and tighten the rest of the screws in their slots.

Installing a gasketed threshold is similar. If there is an existing (ungasketed) threshold, remove it and measure from the underside of door to the top of the sill. Remove the door so you have complete access to both ends of the doorway. The trickiest parts of fitting a threshold are the complex trim profiles at either end—doorstops, jamb casing, and so on. Using heavy

Door shoes are usually sized for the 1 3/4-in. thickness of the exterior door. The shoe is cut back about 1/2 in. from both sides of the door so the shoe's drip cap will clear the thicker part of the jamb as the door closes.

By turning the screws of this adjustable threshold, you can raise and lower the oak strip to get a good seal to the door bottom, thus stopping drafts and water.

Water-Return Threshold

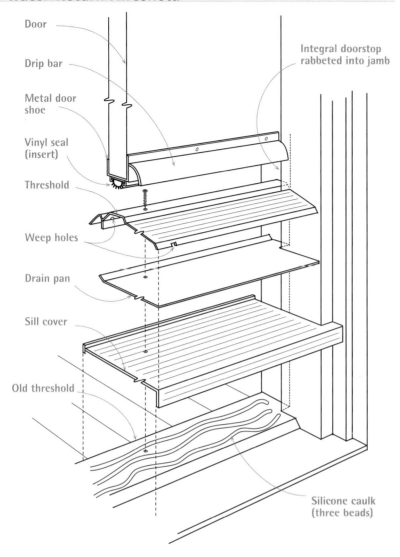

- Door
- Drip bar
- Metal door shoe
- Vinyl seal (insert)
- Threshold
- Weep holes
- Drain pan
- Sill cover
- Old threshold

Integral doorstop rabbeted into jamb

Silicone caulk (three beads)

This three-piece threshold fits over an existing threshold and can withstand just about anything short of Noah's Flood. For success, carefully fit and caulk the unit's pieces.

PRO TIP

If hinges are loose because the screw holes are stripped, fill the holes with epoxy putty. Allow that to harden before drilling new pilot holes and replacing the screws. Alternately, you can enlarge the hole by drilling it with a ¼-in. bit and drive in a tapered dowel of the same thickness to fill the hole. Glue the dowel, and let the glue dry before drilling new pilot holes.

pan; use aviation snips or an oscillating multitool to cut the pan—it's very thin.

With all three threshold pieces in place (and sloping toward the outside), measure up ½ in. onto each jamb and make pencil marks. The ½ in. indicates the thickness of the door shoe you'll attach to the bottom of the door. Remove the threshold and the drain pan, shut the door, and spread a scriber to the distance from the top of the sill cover to the mark(s) on the jambs. Then scribe the bottom of the door to indicate the amount that you'll need to trim off. Cut down the door, vacuum the existing threshold thoroughly, apply three beads of silicone caulking across the opening, and press the sill cover into it. Then do the same with the sill drain pan. Also caulk the jamb-and-pan joints. Finally, screw down the threshold and install the shoe on the bottom of the door, as described in the preceding section.

Keeping the Weather Away: A Primer on Flashing

Midway in our look at exterior doors and windows seems like a good place to discuss *flashing*, specialized materials that help seal out water and—in conjunction with insulation, sealants, and exterior casing—cut drafts. This short section shows some of the materials used as flashing and how they are applied around the perimeter of a rough opening to make exterior doors and windows weathertight.

(continued on p. 142)

Hanging a Door to an Existing Frame

Hinge Setbacks

Hanging a new or recycled door to an existing frame is a common renovation task but not always feasible. At the very least, the hinge jamb must be straight. If planing down or screwing down won't straighten a bowed or twisted section of hinge jamb, you may need to remove the old frame (and its casing) and replace it with a prehung unit. If possible, try to leave the frame and its casing in place—especially exterior casing. Retrofitting weatherproofing elements such as head flashing is tedious and tough to do right.

Creating a Template

The most reliable way to fit a door to an out-of-square frame is to create a template. (Read "The Beauty of Templates" on p. 404 for tips.) Use 4-in.-wide, 1/8-in.-thick strips of plywood; staple the plywood strips to the edges of the doorstops the door will seat against. Where strips cross, join them with fast-setting hot-melt glue. In addition, run horizontal strips across the top and bottom of the door opening from jamb to jamb, and diagonally cross-brace the strips so they will retain the

outline of the frame opening after you pull the staples.

As you transfer the outline of this template to the face of the door, subtract 1/8 in. from the top and sides of the template to create a 1/8-in. clearance between the door and the jambs; subtract 1/2 in. from the bottom edge of the template for clearance above the finish floor or threshold. If the floor isn't level, you may need slightly more clearance. Place a spirit level across the opening to see how much out of level the floor is.

A spring-loaded, self-centering Vix bit is the best way to center pilot holes for hinge screws. Otherwise, screws that drift off-center can cause hinges to twist and misalign with the other hinges.

Trimming the Door

Trim the door rails of the door, then the stiles. To prevent splintering, first score along cutlines with a utility knife (see the photo on p. 122). Then use a 7 1/4-in. circular saw with a sharp, 40-tooth carbide blade to make the cut. For best results, cut 1/16 in. beyond the cutline. Then use a belt sander with 80-grit sandpaper to trim the edge exactly to the cutline. Some carpenters trim door edges with a handplane or power plane instead. If you use a power plane, go slowly to remove the wood gradually, making several shallow passes.

Cutting the Hinge Gains

Cutting hinge gains (recesses) is best done with a router and a template, as shown on p. 134, but a hammer and chisel will do fine if you're hanging just one door. If the door frame already has hinge gains, transfer their locations to the door by marking them right onto the template. Alternately, you can use shims to wedge the door snugly against the hinge jamb, then use a utility knife to mark the hinge positions on both the jamb and the door.

Typically, the top of the top hinge is 7 in. from the top of the door; the bottom of the bottom hinge is 9 in. to 11 in. from the bottom of the door (see "Positioning Door Hardware" on p. 119). If there's a third hinge, it's equidistant between the other two.

Pull the hinge pins so you can work with one hinge leaf at a time. Set the hinge leaf slightly back from the edge of the door, as shown in "Hinge Setbacks" at left. The setback from the hinge to the doorstop should be slightly greater than that from the hinge to the edge of the door, so there is room for several coats of paint. Use a combination square to mark setback lines on doors and jambs. Finally, when mortising hinges, it's best to set the router slightly shallow and then use a chisel to pare away the last little bit of wood so the hinge leaf is just flush. Setting hinges too deep can cause the door to bind.

Hanging the Door

Hinge screws can be diverted easily by wood grain, resulting in crooked screws and misaligned hinges. The pros avoid this problem by using a Vix bit, a spring-loaded bit that centers pilot holes for hinge screws perfectly. With hinge leaves perfectly set on both the door edge and hinge jamb, lift the door, align the leave pairs, and slide in hinge pins, starting with the top hinge. After inserting the pins, eyeball the hinges as you open and shut the door. If the gap between the door and the hinge jamb is excessive, remove the door and set hinges a little deeper.

Hinge Setbacks

Set hinge back 1/4 in. from door face.

Hinge jamb

Doorstop

1/16-in. clearance

Door

Set hinge back 5/16 in. from stop.

Leave a slight gap between the door and the doorstop, so the door won't bind.

Quick Door Fixes*

SYMPTOM	CAUSE OR SIGNIFICANCE	WHAT TO DO

Hinged doors, general

Door binds against top of latch jamb or scrapes floor	Loose hinges allowing door to sag into opening	Rescrew hinge to jamb, replacing inner screws with ones long enough to reach studs, if needed
Hinge-screw holes in jambs are stripped	Larger-diameter screws won't fit holes in hinges	Use longer screws or fill holes with wood plug glued in, let dry, and rescrew
Door binds along latch jamb but hinges are tight	Hinge may be bent	Use adjustable wrench to bend hinge-leaf knuckles on door (p. 133)
Door binds on latch jamb but hinges are tight; big gap seen along hinge jamb	Hinges not mortised deep enough into door or frame	Remove hinges, chisel hinge gains (recesses) deeper, and reattach hinges
Door binds along hinge jamb	Hinge leaves set too deep	Remove hinges, place cardboard shims under hinges, and reattach
Door binds because door frame is racked (out of square)	Seasonal shifting or foundation has settled	Scribe and trim door to fit skewed opening or replace old frame with squared, prehung door unit
Door shuts but won't latch	Strike plate is misaligned	Raise or lower strike plate

Pocket doors

Door slides roughly	Built-up dirt or floor wax on floor	Vacuum track thoroughly
Door slides roughly; floor abraded under door	Top track sagging or mechanism needs adjustment	Remove trim to expose top-hung mechanism; adjust to raise door
Door does not slide at all; hard to operate	Door has fallen off track	If bottom track, lift door back onto it; if top track, remove trim and set tracking wheels up onto overhead track
Door drags, balking at certain points; wheels squeal	Wheels not turning freely or are rusty; track bent or broken	Remove trim, swing door out, and oil or replace wheels; use flashlight to examine track inside pocket
Door face abraded; door difficult to operate	Door off track or stud has bowed into pocket	Lift door onto track; if problem persists, remove finish wall on one side—may need to plane down or replace stud

Exterior doors

Drafts around door	Door not fitting tightly to frame	Install weatherstripping or new threshold
Water damage to wood doorsill, finish floors, and subfloor	Water collecting around doorsill area, soaking wood	Replace damaged materials; install overhang outside and water-return threshold beneath door
Water stains on interior walls, especially around top of door	Absent or poorly installed cap flashing on exterior	Remove siding above top of door frame and retrofit head flashing
Heavy condensation on metal sliding door; floor is water damaged	Metal frames conduct cold; moisture condenses on them	Replace with clad door unit with better insulating properties
Door frame not square; casing tilts; large diagonal cracks at corners of doors or windows	Possible foundation settlement (p. 262)	Have structural engineer check foundation

* For additional quick-diagnostic charts such as this one, see House Check: Finding and Fixing Common House Problems by Michael Litchfield with Roger Robinson (The Taunton Press, 2003).

Pocket door detail #1: Top-mounted mechanisms need oiling and ingenuity to keep rolling. Someone retrofitted a new machine bolt and sawed a slot in its head so the bolt could be easily turned to raise or lower the door.

Pocket door detail #2: The mortise in this pocket door's bottom rail will house an adjustable wheel mechanism. The hole bored in the face allows access to a screw that raises or lowers the wheels and thus the door.

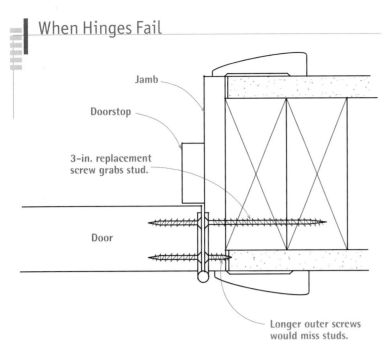

Jamb

Doorstop

3-in. replacement
screw grabs stud.

Door

Longer outer screws
would miss studs.

Replacing a short hinge screw with one long enough to grab
the framing should be your first "cure" when hinge leaves
pull free from door jambs.

The order in which you install weatherproofing materials depends on several variables: Is exterior sheathing exposed or covered by building paper (or housewrap) and siding? Are window units to be installed *precased* (with exterior casing already attached) or *uncased*? I will get into particulars, but the overriding idea behind installing flashing correctly is *above always overlaps below*. Falling water is always directed away from living spaces, away from framing and sheathing and window frames. Remember that rule of thumb and you can't go wrong.

Imagine a rough opening awaiting a door or window. If the sheathing is exposed, apply housewrap to the wall and run it into the opening, stapling it to the sides of the studs, and top of the rough sill. However, leave the housewrap over the top of the rough opening unstapled because it will overlap the window's cap (head) flashing later. Alternately, you can first flash the perimeter of the RO with *flashing strips* or *self-adhesive flashing tape* (p. 169 explains the difference).

FLASHING FOR A WINDOW

First, flash the sill of the rough opening. Windows generally have sloping sills to shed water, but water still finds its way underneath. In rainy areas, builders often install rigid sill pans, but these days you can easily shape a watertight pan from flexible, self-adhesive flashing tape. Create a lip toward the back (inside) of the sill by dou-

bling the tape's back edge, and run the tape at least 4 in. onto the sides of the RO. Apply extra flashing tape as needed to make corners watertight. If you use (nonadhesive) flashing strips instead, you can fold the strip into the bottom of the RO as shown in the bottom right photo on the facing page, caulk the edge, and set the windowsill atop it. If there is a *drip kerf* cut into the underside of the windowsill, you can also caulk and tuck the top of the flashing strip into the kerf. This provides an unbreachable seal, which apron casing or siding will cover as well.

Now you're ready to install a prehung window unit into the rough opening. Succeeding pages describe installations in great detail, so let's focus here on flashing. You've shimmed and secured the window. Now you need to start sealing the gaps between the window frame and the framing.

To seal side jambs, you can use self-adhesive flashing tape as shown in the sequence on pp. 154–155. Or you can use the strip-and-caulking method shown in the photo sequence on pp. 143–144. Apply siliconized caulk to each jamb edge, embed a flashing strip in the caulk, and then run another bead of caulk atop the strip before installing the casing. This caulk-and-spline sandwich stops infiltration effectively. Finally, caulk the edge of the head jamb before installing the head casing, but do not insert a flashing strip between the head jamb and the casing—that could direct water behind the head casing.

Apply the exterior casing to the window frame, then install rigid cap flashing over the head casing. Careful builders will first caulk the upper leg of the cap flashing to prevent water from running behind it. (This caulk is especially important with stucco siding, which is water permeable and often collects water between the stucco and the building paper.) Lastly, run building paper or housewrap over the top of the cap flashing before installing the siding.

Should side flashing go over or under the building paper covering the sheathing? It doesn't really matter because the side flashing is extra protection. But under the windowsill, leave the lower edge of the flashing strips unstapled (or self-adhesive flashing tape unattached) so it can overlay the building paper or housewrap.

Choosing Windows

Choosing the right windows can be daunting, so let's break the decisions into several manageable categories: (1) window styles, (2) frame materials, (3) making sense of energy-efficiency labels, and (4) choosing the right replacement window. This sequence leads us to installing three types of

WOOD-CASED WINDOW

WINDOW WITH INTEGRAL NAILING FLANGE

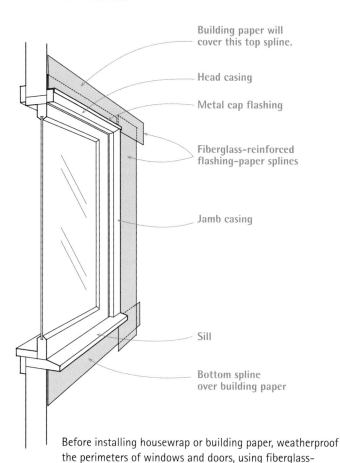

Building paper will cover this top spline.

Head casing

Metal cap flashing

Fiberglass-reinforced flashing-paper splines

Jamb casing

Sill

Bottom spline over building paper

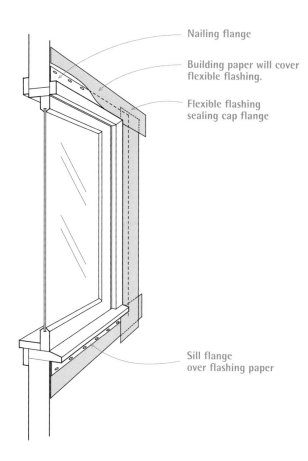

Nailing flange

Building paper will cover flexible flashing.

Flexible flashing sealing cap flange

Sill flange over flashing paper

Before installing housewrap or building paper, weatherproof the perimeters of windows and doors, using fiberglass-reinforced flashing paper splines (strips). In addition, cap wood-cased units with metal cap flashing; head casing on units with nailing flanges should be overlapped with a flashing paper spline or sealed with flexible flashing.

FLASHING AN UNCASED WINDOW

1. Dry-fit the window first to make sure it fits in the RO and that there's enough space to level and shim the unit. The temporary diagonal cleats at the upper corners keep the jambs flush to the sheathing.

2. Set the window aside, apply flashing to the sill of the RO, and run a bead of siliconized acrylic latex caulk along the sill edge. You'll later caulk the underside of the windowsill, too, to cut air infiltration.

3. Reinsert the window into the RO, shim it to plumb and level, check to be sure it operates freely, then screw or nail the jambs to the framing. That done, cut shims flush inside and out.

4. Next, carefully caulk jamb edges and overlap those edges with flashing strips.

5. Before installing exterior casing, run another bead of caulking atop the flashing strips.

6. Rigid cap (head) flashing diverts water that might otherwise dam up behind the head casing and allows it to drip free.

7. Cover the top leg of the cap flashing with flexible, self-adhesive flashing tape or, as shown, a fiber-reinforced flashing strip. The siding—in this case, stucco—will overlay the flashing strip.

replacement windows and then to a photo essay of refurbishing an old double-hung window.

WINDOW STYLES

In general, choose new or replacement windows that match the style already on the house, especially on the facade that faces the street. Likewise, choose frame and sash materials that have the same approximate thickness. Keep in mind when you make your selection that not all frame materials can be painted.

Double-hung windows are the traditional choice, with two sashes that slide up and down; if the top sash is fixed, it's a *single-hung* window.

Pros: Double-hung windows offer the widest choice of sash patterns, from single pane to tiny four-over-four arrays separated by delicate *muntins.* Snap-in muntins enable affordable double-glazing and easy cleaning, but they won't fool an experienced eye.

Cons: Sliding sashes are tough to weatherproof well, and old sashes rattle and leak air; if sash ropes break, windows cease to function reliably. Meeting rails between the sashes have two

Window Styles

Double Hung

Casement

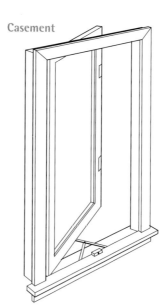

Tilt-and-Turn

Fixed

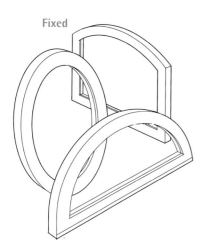

Horizontal Slider

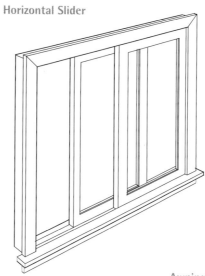

Hopper

Awning

Aluminum Wood Clad Vinyl Fiberglass

faces exposed to outside air, which hastens heat loss, convection, and condensation. And only half the window area offers ventilation.

Casement windows are side hinged and swing outward, usually operated by a crank. There are also push-out casement windows, which use friction to hold their position.

Pros: A single large pane of glass maximizes solar gain, a tight compression seal minimizes air leaks, and the whole area ventilates when open. It's an excellent choice for egress.

Cons: A casement opening onto a deck or walkway is hazardous to those walking by.

Awning windows are hinged at the top, and the bottom swings outward.

Pros: A single large pane of glass maximizes solar gain, and you can leave the window open for ventilation even when it's raining—the window acts as an awning.

Cons: A sash that swings onto a deck or walkway is hazardous, and the screen is on the inside.

Hopper windows, which tilt from the bottom, are like upside-down awning windows.

Pros: It has the same energy profile and compression seal as casement and awning windows.

Cons: It's hazardous if it swings down at head height, and it's a poor choice for egress.

Tilt-and-turn windows are hybrids that swing like a casement or bottom-tilt like a hopper.

Pros: This type of window is easy to clean, is a good choice for an egress, has a tight compression seal, and has a good energy profile.

Cons: Two moving parts mean two things that can break or jam. Roller shades and curtains can interfere with the window's operation.

Horizontal slider windows are like miniature patio doors; sashes slide in tracks.

Pros: This is a great choice for egress, and because there are no muntins, it has good solar gain.

Cons: Sliding sashes are tough to weatherstrip, crud collecting in tracks can impede operation, and only half of a window's area offers ventilation.

Fixed windows don't open.

Pros: A fixed window can't be beat for being airtight, and it's less expensive than an operable window of the same design. With acoustic glazing, it's the best choice to reduce outside noise.

Cons: It offers no ventilation and no egress. You can't wash the outside from the inside.

FRAME MATERIALS

In the last few decades, there have been so many improvements in insulated glass that the R-value of the glazing generally exceeds that of the frames. In other words, windows with a larger percentage of glass (and a smaller percentage of frame)

Window Words

Many of the terms used to describe doors are also used for windows. Window frames consist of *jambs*—side jambs and a head jamb—and a sloped *sill*. Window frames also have *stops* to guide sash movement and provide a seal. Window *sashes*, like doors, have horizontal *rails* and vertical *stiles*. And *casing* is applied, inside and out, to limit air infiltration and impart a trim, finished look to the perimeter of windows. Lastly, many styles have terms that describe particulars of construction, such as the *muntins* and *meeting rails* of double-hung windows.

If you need doors and windows that exactly match the architectural details of existing house elements, a custom shop will probably have greater flexibility to match that style. That said, factory-built windows are generally tighter, better insulated, come with better locks, and are rigorously tested to meet state and federal energy-conservation standards. Custom windows and doors may be good quality, but they probably won't have NFRC labels such as that shown on p. 149 and may cost more than an equivalent factory window. Finally, custom-built windows rarely come with cladding, so you'll be buying windows that you'll have to paint.

Double-Hung Window Elements

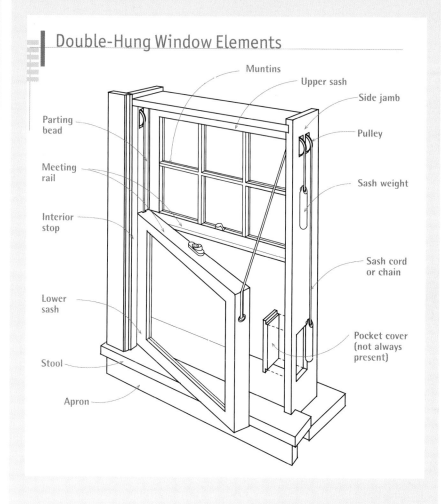

Muntins
Upper sash
Side jamb
Parting bead
Pulley
Meeting rail
Sash weight
Interior stop
Sash cord or chain
Lower sash
Pocket cover (not always present)
Stool
Apron

are better energy performers. In response, window makers are hustling to make window frames less conductive (of heat and cold), more airtight, and more durable. To do so, they have developed frames with a wide range of core materials, cladding, finishes, and in some cases insulation.

The chart on p. 148 will give you an overview of how different frame materials perform, but to be honest, the best way to select a window is to choose a frame style that fits your house, calculate the energy performance your window must deliver (hence, the type of glazing, etc.), and then talk to builders and building-supply staffers about which manufacturers they favor. Try out windows as well, opening and shutting display models to see how tightly they fit, how smoothly they operate, and how sturdy they feel. Look especially closely at frame corners, glazing seals, and weatherstripping gaskets, for that's where windows fail first. And when scrutinizing window cross sections, try to envision cold air encountering that frame: Is there a *thermal break*

Window Frame Materials

FRAME MATERIAL	DURABILITY	INSULATING LEVEL	MAINTENANCE	COST	COLORS	COMMENTS
Aluminum	Good	Poor	Low	$	Limited	Most conductive of cold, but OK in mild climates
Wood	Fair	Good	High	$$ to $$$	Paintable	Needs painting every 10 years, cleaning
Clad	Good	Best	Low	$$$$	Custom colors cost extra	As attractive as wood but easy to maintain; choice of claddings
Vinyl	Fair	Fair to good	Fair	$	Limited	Degrades if overheated; otherwise, it's durable
Fiberglass*	Good	Good	Low	$$$	Paintable	Least conductive, can also be insulated with foam

*Fiberglass's rate of expansion and contraction is virtually the same as that of glass, which helps glazing seals last longer.

that will slow or prevent cold air from chilling the frame and causing conductive heat loss?

Lastly, "you get what you pay for" is especially true when buying windows. Even frames fabricated from less expensive materials, such as vinyl, offer options such as chambers filled with foam insulation that raise performance—and cost.

READ THE LABEL: UNDERSTANDING WINDOW EFFICIENCY

R-value, the ability of a material to resist heat flow through it, became a household word in the 1970s, when soaring energy costs led to a boom in home insulation. The higher the R-value, the better, we learned. Given the push to make houses ever more energy-efficient, the next frontier was windows because single-pane glass is a terrible insulator. Shortly, there was a blizzard of energy-saving technologies applied to windows, doors, and skylights—and a flurry of new terms to describe them.

Chief among this thicket of terms is the *U-factor*, an aggregate measure of how well non-solar heat flows through a window's glazing and frame. Simply put, U-factor is the inverse of R-value (1 divided by the R-value), so *the lower the U-factor, the better*. The U-factor is a rough measure of conductive heat loss and gain. But the more elaborate window science became, the more crowded labels grew—as you can see in the drawing on the facing page. This information is all good, but to figure out which window features

make economic sense for your region, have a look at www.efficientwindows.org.

Well, that will get you started. And, of course, there's a wealth of professionals, from architects to energy consultants, who can help you balance the competing claims of cost, performance, and a plethora of features that there's no room for here—including impact-resistant glazing that's required in storm-prone areas, acoustic windows whose ability to reduce noise is rated with a sound transmission coefficient (STC), and even "self-cleaning" coatings that claim to slow the buildup of dirt on glass.

CHOOSING A REPLACEMENT WINDOW

If you need to replace windows, whether because they are drafty and waste energy or because they show signs of deterioration, you have three basic choices.

1. Replacement sashes. If the frame and sill are in decent shape, you can install jamb liners along the inside of the frame and insert new sashes into the liners. In general, this option is the most economical way to make windows tighter and more energy-efficient. This is such a popular option that many window makers offer replacement-sash kits that fit easily into old frames. Detractors of this option tend to dislike plastic jamb liners and doubt that they provide much of an air seal. That aside, it's tough to knock the value for the money.

2. Window insert. This option features a rigid frame and sash assembly that sits inside an existing frame and generally seats against an interior or exterior window stop. The existing windowsill stays put. Inserts offer a better air seal and greater energy efficiency than a replacement sash, yet the space between the insert frame and existing jambs must be caulked or insulated in some manner. Inserts cost slightly more than a new-construction window but less to install because you leave the casing in place. The downside of inserts is that a frame within a frame reduces the amount of light that can enter and cuts the view.

3. New-construction window. When there's water damage and rot present, you must remove the existing window entirely, including its sill. A new-construction unit thus sits inside the rough opening and has a full-dimension sloping sill. To install this unit "by the book," you'll need to replace old head and pan flashings, and use self-adhesive flashing tape and compatible sealants to make the window weathertight—and thus replace the exterior casing. But real life doesn't always go by the book, as you'll see on pp. 154–155. Sometimes disturbing the existing casing is both inordinately expensive and disruptive and not warranted when a less-disruptive method will be sufficiently effective.

INSTALLING REPLACEMENT SASHES

The Marvin Tilt Pac shown in the photo sequence on pp. 151–152 is but one of many sash replacement and liner kits. For best results, consult the manufacturer's instructions that come with your unit. Each window maker also will have specific instructions for measuring the rough opening. In the sequence shown, the installer measured width and height at the frame's centerpoint—the point at which the upper and lower sashes meet.

The installation sequence shown is pretty straightforward, but a few things are worth emphasizing: Be sure to remove any impediment

PRO TIP

Retrofitting cap flashing takes dexterity. Cut back the siding nails 2 in. to 4 in. above the opening, and pry up the siding so you can slide the top leg of the cap flashing under the building paper and siding, while simultaneously holding up the window unit. You can nail the top leg of the flashing through the siding, but that's not imperative. Wedged into place, the cap flashing won't go anywhere.

Energy-Efficient Ratings

Reading the NFRC label on a window is an education in itself. NFRC, for starters, is the National Fenestration Rating Council, a nonprofit organization that independently rates the energy efficiency of windows, doors, and skylights. Here's what the label is telling you.

Double glazing: Today's standard insulated glass unit, consisting of two glass panes separated by spacers, with a sealed, airtight space between them. Triple glazing is often specified for cold climates. *Glazing,* by the way, is the fancy name for glass panes, the space between them, their airtight seals, and the putty or caulk that seats them in the frame.

Gas filling: Clear, inert, nontoxic gas that insulates between panes. Argon and krypton, for example, are less conductive than air and hence better insulators.

U-factor: The sum of a window's insulating values, drawn from all its parts (glazing, frame, sashes). Measuring nonsolar heat flow, the smaller a U-factor number, the better.

Visible transmittance (VT): The amount of visible light that enters. Values range from 0 to 1; higher is better. Wide sashes and frames block light, lowering VT numbers.

Condensation resistance (CR): Higher ratings (from 0 to 100) predict less condensation.

Low-emissivity (low-e) coatings: Thin, nearly invisible window coatings that selectively reflect heat back into a room (to conserve energy) or block sunlight to reduce solar gain. Low-e coatings also filter damaging UV rays.

Solar heat gain coefficient (SHGC): The percentage of solar heat that passes through the glazing. Higher SHGC numbers indicate greater passive solar gain—desirable in cold regions. Conversely, lower SHGC numbers mean lower air-conditioning bills in hot climates.

Air leakage (AL): The movement of air through a window system, measured in cu. ft./sq. ft. Lower is better; <0.03 is optimal.

NFRC
National Fenestration Rating Council®
CERTIFIED

World's Best Window Co.
Millennium 2000+
Vinyl-Clad Wood Frame
Double Glazing • Argon Fill • Low E
Product Type: **Vertical Slider**

ENERGY PERFORMANCE RATINGS

U-Factor (U.S./I-P)	Solar Heat Gain Coefficient
0.35	**0.32**

ADDITIONAL PERFORMANCE RATINGS

Visible Transmittance	Air Leakage (U.S./I-P)
0.51	**0.2**

Condensation Resistance
51

Manufacturer stipulates that these ratings conform to applicable NFRC procedures for determining whole product performance. NFRC ratings are determined for a fixed set of environmental conditions and a specific product size. Consult manufacturer's literature for other product performance information.
www.nfrc.org

Sizing Windows

Window catalogs have several ways of denoting window dimensions, including a *callout size,* which usually indicates the size of rough opening required, or a *unit size,* which denotes the outer dimensions (width and height) of the window frame. Window dimensions are stated as pairs of numbers; the first number is always width.

So, a window with a callout size of 4030 denotes an RO of 4 ft. 0 in. by 3 ft. in. 0 in. Typically, the unit size of a window (frame) is ½ in. less in height and width than the callout size, but always check the manufacturer's specs to be sure. Window sashes are usually 1³⁄₈ in. thick unless otherwise noted.

When ordering window units, measure the width of the ROs in three places from top to bottom and the height in three places from side to side. Window manufacturers usually prefer the smallest reading in each direction, but, again, follow the manufacturer's ordering instructions to the letter. Also measure the thickness of the walls, from interior finish surfaces to exterior sheathing; you may need jamb extensions as well.

Measure openings and order windows well in advance. If your window units are in odd sizes or otherwise unusual, they may require a special order, which takes longer.

Measure Windows in Three Directions

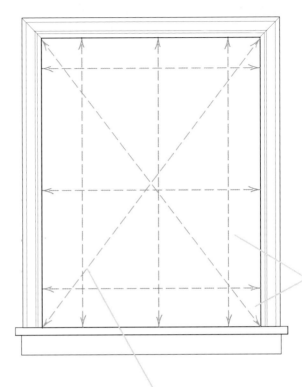

Measure horizontally and vertically, and use the smallest measurement in each direction so replacement jambs will be sure to fit the opening.

Measure diagonally to see if the old window is square.

Existing Trim: Let It Be

If you need to upgrade windows, try to choose an option that avoids disturbing casing. House exteriors consist of interconnected elements that must remain intact to be weathertight. Flashing over head trim is especially difficult to retrofit because, ideally, its top leg goes under building paper and siding and is caulked as well. You can slide a recip saw blade under the siding to cut nail shanks and slide a piece of metal head flashing up underneath. But it's rarely that simple. Disturbing a window frame and sill usually means replacing interior trim.

Here, the lower roof leaked because the corner of the window was too close and a lazy roofer didn't bother to install step-flashing (see p. 100). Also, the window lacked cap flashing. As you can see, retrofitting flashing is no simple matter.

with the frame that would prevent the jamb liners from lying flat. Jamb liners are backed with compressible foam, which will bridge minor irregularities. But to ensure that sashes slide freely and the unit is as airtight as possible, all jambs must be straight, plumb (sides) or level (head), and the frame must be square. The jamb and head channels can accommodate minor discrepancies but not large ones. In the sequence shown, the installer packs cotton insulation around the frame; loose fiberglass also would work. Low-expanding foam is another option, but one I don't recommend.

Another reason for the Tilt Pac's popularity: After installation, you can pivot sashes to wash the outside of the window.

INSTALLING WINDOW INSERTS

Window inserts can be installed from inside or outside. If you favor an exterior installation, you will first need to cut the window's exterior blind-stops flush with the casing so the new insert frame can be fitted into the existing frame. Insert frames can be installed with the sashes left in or removed. Although the photo sequence on pp. 153–154 shows them removed, in this description, assume the sashes remain in the frame.

As shown in the photos, remove sashes and parting stops, as well as anything sticking out of the jambs, such as weatherstripping, stray nails,

1. Remove the interior stops from the side jambs, disconnect sash weights or other balancing devices, and remove the bottom sash. Avoid damaging the stops so they can be reused.

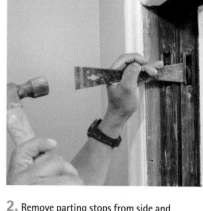

2. Remove parting stops from side and head jambs, lift out the top sash, then remove any remaining hardware such as pulley wheels or metal weatherstripping.

3. After surveying the opening for rot and checking to be sure jambs are straight and plumb, insulate between the frame and the rough opening. Here, cotton insulation is being stuffed into old pulley weight cavities.

4. The new jamb liners, which are 3¹/₂ in. wide, will seat against the exterior stop. To attain that width, the installer here discovers that he must trim the edge of the sill trim about ¹/₄ in.

5. Jamb liners are held in place by a series of metal installation clips placed approximately 4 in. below the head jamb, 4 in. above the sill, and spaced evenly in between.

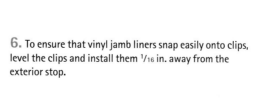

6. To ensure that vinyl jamb liners snap easily onto clips, level the clips and install them ¹/₁₆ in. away from the exterior stop.

7. Sashes seat in the jamb liners via a locking terminal assembly. Before installing the sashes of a double-hung unit, you must first use a screwdriver to pull the four assemblies into position.

8. Each sash has a "pin" on both sides.

9. To install each sash, position a sash pin above a locking terminal, then raise the other side.

10. When both pins are aligned, slowly tilt up the sash, then press down until you feel the pins snap into the locking terminals. Then swing the top of the sash up until it seats in the liners.

11. Repeat this procedure with the bottom sash. After checking to see that sashes slide freely and seat well at the top and bottom, reinstall the interior stops.

and the like. Old jambs should be as flat as possible. Inspect the opening for rot, and repair any you see, then once again ascertain that jambs are plumb and the frame is square. If they are not, you'll get an idea of where you'll need to shim. Don't overshim, as that could bow the frame.

Before installing the insert frame, bend the sill fin on the bottom of the unit so that its angle matches that of the existing sill. *Note:* Depending on whether you install the insert from outside or inside the house, its frame will seat against the *interior sash stop* or the *exterior sash stop* (also called a *blind stop*). Center the unit in the opening, plumb it, take diagonal readings to be sure the insert frame is square, and shim the corners

of the frame. Using screws provided with the kit, tack the top of the frame in place—there will be a predrilled hole nearby. Raise the bottom sash and lower the top one a few inches and check the frame for square again.

Using a tape measure, take multiple readings from jamb to jamb to ensure that the jambs are equidistant and the sashes slide without binding. The key adjustment device in the Marvin insert system shown here are two jamb *jack screws* midway up each jamb. Drill pilot holes for both jack screws, and screw them into the old frame behind until they seat. You can turn each jamb jack screw clockwise to move jambs away from a sash, or counterclockwise to move them toward a

Measuring FOR INSERTS

Old window frames are frequently out of square, even if their jambs are parallel. So when you measure the frame's height and width, note which way it leans as well, so you can order the biggest rectangle that will fit into that opening. Makers of replacement windows manufacture units in ¼-in. or ½-in. increments, so when you measure the inside of a window frame, be accurate. Take three vertical measurements—from the high point of the sill to the head jamb—and three horizontal measurements, and select the shortest of each. Also measure the frame from diagonally opposite corners to determine if it's square. The two readings will be equal if it is square.

By the way, measure from jamb faces, not from parting stops; the stops will be removed before you install the new jamb liners or frame.

1. Carefully remove interior stops so you can reuse them, then pull out the lower sash. If sash cords are attached, detach them and lift the sash out of the frame.

2. Older windows frequently have metal weatherstripping tacked to jambs, that needs to be removed—as do parting stops between upper and lower sashes. A flat pry bar is the best tool for the job; wear work gloves. Remove the upper sash.

3. If there are access panels for sash weights, open them, lift out the weights, and pack the cavity with loose insulation. In general, don't use spray foam here: Even low-expanding foams tend to fix elements in place, making it difficult to adjust the width of a frame later.

4. After checking the old frame for rot, dry-fit the insert frame into the cleaned-up opening. Many installers prefer to leave the sashes in the insert, but this installer took them out so he could easily eyeball the perimeter of the insert frame inside and out to see how it fit.

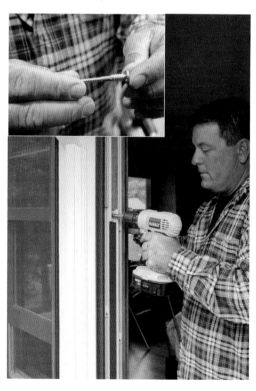

5. Center the unit in the opening, plumb it, take diagonal readings to be sure the frame is square, and shim the corners as needed. To move jamb frames in and out, this insert unit relies on a pair of adjustable jack screws.

6. These sashes have pivot pins on each side (near the bottom rail), which seat into a clutch assembly in the frame jambs.

7. Once sash pivot pins are engaged in jamb mechanisms, swing the upper rail of the sash up into place, where a pair of spring-loaded tilt latches will also seat in jamb channels. By retracting the latches, you can later tilt out sashes for washing. After checking to be sure sashes operate freely, replace interior stops.

PRO TIP

Always replace the parting stops of double-hung windows. Old stops are invariably gunked up and not worth stripping. Besides, they usually break when you remove them.

sash. Work the frame until everything is to your liking. Trim shims flush and caulk the perimeter of the insert frame, using backer rod as needed to fill larger gaps. If you removed interior stops, reinstall them now.

INSTALLING NEW-CONSTRUCTION WINDOWS

When rot is present in a window frame or the framing material around a rough opening, you'll need to remove the old unit, cut out the rotted section, and replace it with new wood or, if it's limited, use two-part epoxy putties (see p. 175) to rebuild the decayed part. As important, you will need to figure out how water got into that area and correct that as you prepare the opening and install a new window.

Just how you install this unit, however, will depend in part on what's there. To install a new window by the book, as shown on pp. 143–144, you should replace old head and pan flashings, and, in particular, slide the top leg of the head flashing up under exterior siding. That will require sliding a reciprocating saw blade under the siding to cut nail shanks. Or you could tear off and then reinstall the siding above the opening to ensure the building paper and siding correctly overlap the head flashing. It's quite a job. Moreover, if it's stucco and you don't want an obvious patch, this repair may mean restuccoing the whole face.

But if there's one thing that renovation teaches you, it's that there's more than one way to solve a problem, even if the solution is not in the book.

1. This old multiple-casement unit had rot, so a new-construction window replacement was in order. When the crew removed casing inside and out, they discovered only small finishing nails holding the frame to the rough opening. So this one was easy to lift out.

2. After scraping and vacuuming the exposed rough opening, install a rigid sill pan or use flexible, self-adhesive flashing tape to construct one. To create a raised lip that will dam any water that gets in, double the flashing lip along the inside edge of the sill.

3. Apply self-adhesive flashing tape around the perimeter of the opening as the first step toward creating an air seal around the new window. (Note the siding to the right and top of the opening: It was not painted because the exterior casing was applied *over* the siding—an odd but not uncommon way of doing things in the late 19th century.)

4. To compensate for a crown in the RO sill, installers flipped the unit and used a router to remove a thin strip of material from the bottom of the windowsill. Note the simple but effective use of a straight board (at right) to create a level routing surface on a sloping sill.

5. Whenever you cut into a primed surface, seal the exposed wood before installing it.

6. The frame, trimmed and resealed, is being set back into the flashed opening. Center the unit in the opening, check for plumb and square, tack upper corners to hold the frame temporarily, and you're ready to start shimming in earnest.

7. Shimming goes best with someone on both sides of the opening. There's always a lot to do, whether prying or pushing the frame into position, adjusting shims, or sinking screws. Here, the frame was secured with 3-in. stainless-steel flathead screws countersunk and filled.

8. Once the frame is shimmed and secure, the crew applies a second layer of flexible, self-adhesive flashing to seal any gaps between the jamb edges and the siding. Tape edges are trimmed back 1/2 in. from jamb edges so the tape will be covered by casing.

9. After applying a bead of compatible sealant around the perimeter of the flashing tape, the crew installs the exterior casing over the siding, as was done throughout this Craftsman-era house. The head casing installed here was, in turn, covered by a piece of metal cap flashing.

10. After the exterior casing was complete, all seams were sealed as well. Nice trick: If you hold a finger against the nozzle while pulling the caulking gun along the seam, you will simultaneously compress and smooth the sealant.

In the replacement sequence shown, for example, a house built in late Victorian times has its exterior casing nailed *over the siding.* How do you *correctly* flash a head casing in that instance? Casing and flashing a replacement window in the modern way is not an option because that orphan window would be glaringly obvious on a period house. Fortunately, the crew was experienced. By using modern materials such as flexible, self-adhesive flashing tape and a lot of ingenuity, they fashioned a weathertight solution. Sometimes, you've just got to do what works.

Refurbishing Old Double-Hung Windows

Windows built 50 to 100 years ago often were constructed from fine-grained, rot-resistant fir, cypress, or redwood, woods that are no longer available. For that reason alone, it makes sense to refurbish rather than replace them when they get tired and don't work so well.

To undertake this task, here are a few common-sense suggestions to supplement the techniques shown in the photo sequence on these two pages.

▶ **Dress for the job.** Sturdy gloves with rubberized palms are essential to avoid cuts and to provide a secure grip on heavy window sashes.

Safety glasses with wraparound lenses will protect your eyes from glass shards, irritating dust, and dangerous projectiles should a power tool strike a hidden nail or screw. Because older windows often contain lead-based paint, a HEPA-rated respirator and a HEPA vacuum to capture dust are must-haves.

If children live in the house, it's especially important to follow the EPA Lead Renovation, Repair, and Painting Rule (RRP Rule). Old windows are a major source of flaking lead-based paint; lead poisoning can devastate a child's neurological development.

▶ **Be methodical.** It might make sense, say, to remove and repair all sashes at the same time but only if you've carefully noted the location of all trim pieces and sashes beforehand. Also, if there are people in residence, move cutting and sanding operations outside whenever possible, lay tarps to protect finish floors, and vacuum periodically to minimize dust indoors.

▶ **Finally, three tips.** First, for operable sashes, there should be a $^3/_{16}$-in. space between the sash and jamb on each side—$^3/_8$ in. total—so there's room for weatherstripping as well as movement. Second, refurbish jambs, too. Pull any old nails or screws sticking out, sand jambs smooth, prime all bare wood, and then, when the paint is dry, rub paraffin—a candle stub will do—along the jambs to allow windows to move easily. Do the same with the sides of refurbished sashes. Lastly, coating jamb channels with a high-gloss marine enamel works almost as well. If you're looking for a wide choice of airtight seals for old windows, visit www.conservationtechnology.com.

PRO TIP

If double-hung sashes require too much effort to operate after you've reattached sash weights, the weights may be too small.

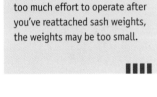

1. Refurbishing starts with detective work. Note what's working and what's not and figure out why. After removing the parting stops and the bottom sash of a double-hung window, move the upper sash to see why it's sticking. Here, the sash had swollen and was too wide for the frame.

2. If there's not an access panel in the jambs, remove the casing carefully to get at sash weights. In general, pulling finish nails *through* the casing causes less damage than pulling nails from the front.

3. The pros weigh sashes to ascertain how much weight is needed to offset them. Ideally, weights should weigh about the same as the sash they counterbalance so that the sash will stay at the height you set it.

4. Sash cords should be long enough to knot at both ends—one knot around the sash weight, the other inserted into the side of the sash—yet short enough to keep the weight from touching the bottom of the cavity. Weights that hit bottom can jam.

5. If sashes are swollen or out of square, they can be trimmed accurately by using a Festool system, which features a guide rail that clamps to the sash and a circular saw whose base is keyed to rail channels.

6. Remove loose or built-up finishes from the sash before priming and painting it. If there's cracked glass or tired glazing compound, attend to it now. This scraper's debris never becomes airborne because the tool is attached to a HEPA vacuum.

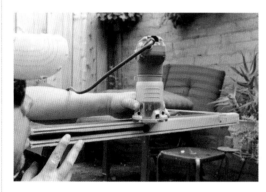

7. To add weatherstripping to movable sashes, use a router to cut two slots along each sash stile; set back slots ¼ in. to ⅜ in. from stile edges.

8. Most weatherstripping friction-fits into slots. This type has a "barbed" neoprene side that resists pulling out, whereas its other side is a feathery nylon brush that slides smoothly.

9. Prime all bare wood surfaces before returning sashes to their frames. If you have time, paint them now, too.

10. Before reattaching sash cords, lubricate pulley wheels so they won't squeal.

11. As you replace each sash, make sure it glides smoothly and weights rise and fall freely.

12. Once sashes and weights are moving well, close up the wall cavities. If that means renailing interior casing, first pull old nails through the casing. Fill nail holes with color-matched wood putty, and use a tiny paintbrush to apply finish just to puttied spots.

GLAZING

Wear safety glasses and gloves when removing old glass and putty (glazing compound). Try not to damage the sash when removing putty. Although a glazier's chisel will remove most putty, try to soften tougher stuff with a hair dryer first. If that doesn't do it, lay the sash flat and pour a small amount of rubbing alcohol on the putty and let it sit overnight. (Do *not* combine heat and alcohol.) After removing old putty, glazier's points, and damaged glass, sand the frame lightly, using 180-grit sandpaper. Prime and paint bare wood before proceeding.

To cut glass, pull a glass cutter along a straightedge as shown in photo 1 below. After scribing the glass, gently rap the ball end of the cutter along the underside of the cut until a clear line develops. Then, holding the cutline directly over a table edge, snap the waste portion free. If the waste piece is too small to grasp, use glass pliers. Cut the glass ⅛ in. smaller than the length and width of the frame opening.

Using a caulking gun, apply a bead of acrylic latex with silicone around the perimeter of the frame. Place the glass in the frame, and press down so it seats evenly in the sealant. Next, use a putty knife to push glazier's points into the muntins to secure the glass; space points every 6 in. around the perimeter, using at least two points per side. Glass in metal-frame windows is typically held in place by metal spring clips, which can be reused.

When the glass is secure, apply putty. For the best combination of skin protection and dexterity, wear nitrile disposable gloves for this operation. Scoop out a generous palmful of putty, and knead it until it is soft and pliable. Roll it up into a fat snake, then use your thumb to press the putty around the perimeter of the pane. You'll trim excess putty, so use a lot to ensure a good seal. The pros apply putty and trim it in one sweep, but the rest of us should do it in two passes. Once a side has been puttied, rest one point of the putty knife on the glass, and holding the blade at

1. After removing old glass, cleaning up the frame, and priming it, cut replacement panes ⅛ in. smaller than the dimensions of the opening. Holding a glass cutter perpendicular to the pane, pull it evenly along a straightedge. Safety glasses and gloves, please!

2. Rap the ball end of the cutter along the underside of the cut until a clear line develops. Then, holding the cutline directly over a table edge, snap the waste portion free.

3. Apply a bead of acrylic latex with silicone around the perimeter of the frame lip on which the glass will rest. Press the glass into the sealant so it beds evenly.

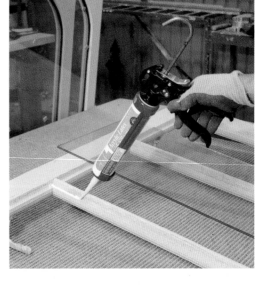

4. Drive glazier points around the perimeter of the opening, using at least two points per side. Here, a glazing gun makes the task easier.

5. When the glass is secure, apply putty generously to each side, pressing it into the frame. Pull the blade through the putty to cut away excess, holding it at a 45° angle to the glass.

a 45° angle to the glass, pull the knife steadily to cut a furrow through the putty.

Touch up corners after removing trimmed putty. If putty doesn't stick, either you're holding the knife handle too high or there is dust on the glass. If the problem is the drag of the knife blade, a bit of saliva on the blade will lubricate it. Allow the putty to cure for a week before painting, then overpaint it slightly (onto the glass) to seal the putty joint.

Installing a Skylight

By letting in light, skylights can transform a room, making a bathroom seem larger, a kitchen warm and cheery, or a bedroom a place to watch the stars. And because of marked improvements in flashing and quality control, skylights can be installed without fear of leaks from the outside or of excessive heat loss from within.

Where you put the skylight is partly aesthetic and partly structural. In a kitchen, installing the skylight in an east-facing roof will help it catch the morning sun. If placed in a hallway, even a small unit provides enough light to let you do without electric light during the day. In a bathroom, privacy is the main issue. Avoid skylights in south-facing roofs unless you're designing for passive solar gain; you'll roast in summer. Similarly, skylights in west-facing roofs may create too much heat in the South and Southwest United States. Avoid skylights that straddle a ridge because they are difficult to flash and likely to leak.

Skylights dramatically change rooms. Because drywall joints and nail holes will be illuminated, too, apply joint compound generously and sand carefully. Heat buildup in the light well can cause compound to shrink, so you may need to apply an extra coat.

SELECTING A SKYLIGHT

Consider a number of things when buying a skylight, not the least of which is the manufacturer. Although sizes have become pretty much standard, quality varies greatly—so ask local contractors or lumber suppliers which brands they prefer. Here are a few aspects to investigate:

▶ Do you want *ventilating* (openable) or *fixed* (closed) skylights? Skylights that open can vent excess heat. And *turn-tilt* models pivot so they're easy to clean.

▶ Is it energy efficient? Most units come with doubled thermal-pane glass, but units should also have a *thermal break* to minimize loss by conduction. A metal frame that's continuous (inside to out) will wick out a lot of indoor heat. A better bet is sealed wood.

▶ Does it have step-flashing along its sides (desirable for shingle roofing) or strip flashing? Strip flashing must be caulked to keep water out—and thus is less reliable. If the unit is wide (more than 32 in.), does the maker supply a cricket to deflect water around the skylight head or will you need to have one fabricated?

▶ Does the unit have tempered glass? All skylights should have tempered or safety glass because they are overhead. Double-glazed units typically have tempered glass on the top and safety glass on the bottom.

▶ If the unit will be installed beyond reach, how easy is it to open and close? Remote-controlled units with motors can be programmed to open at a given temperature and shut when an electronic sensor detects rain, but they are more expensive and more temperamental than manually operated skylights.

▶ Can you get units with screens, blinds, shades, or polarizing tints? *Low-e coatings* selectively admit light while reflecting heat.

▶ Velux now offers a solar-powered operable skylight with a remote control, which doesn't need house power to operate.

SIZING SKYLIGHTS

Folks often order skylights larger than they need. Keep in mind that even the smallest unit brightens a room greatly. Moreover, much of the light gain comes from reflections off the sides of the light well (or light shaft), which is why wells are usually painted white. You can increase the amount of light markedly by flaring out the sides of the well. If you need more light than one narrow skylight will yield, consider ganging several, side by side, in adjacent rafter bays. Smaller skylights are easier to frame, and the fewer rafters you disturb, the better.

Most skylights are sized incrementally to fit between rafters spaced 16 in. or 24 in. on center, so they routinely come in 24-in., 32-in., and 48-in. widths. Skylights 32 in. wide are the most popular size because you need to cut only one rafter to accommodate the unit. Velux, for example, has several models whose inside curb dimensions are 30½ in. wide—the same distance between rafters spaced 16-in. on center, if one rafter in between is removed. Aligning the inner faces of skylight curbs and rafters also makes installing drywall much easier. Attached to the roof sheathing with L-shaped mounting brackets, the sides of Velux units sit right over rafters.

FRAMING A SKYLIGHT OPENING

This section provides general guidance related to framing procedures shown in the photos. Usually, it doesn't matter whether you cut ceiling joists or rafters first, as long as they're adequately supported. Some pros prefer to frame the light well completely before opening up the roof. Others install the skylight first and work down from there.

Insulation. Remove the ceiling insulation. Then disconnect and cap any wires and pipes that will need to be rerouted around the opening. (Use a voltage tester to be sure the power is off.) Where possible, work from a stepladder rather than sit-

> **PRO TIP**
>
> The area under the roof is a messy place to work, especially if the ceiling's insulated. Use sheet plastic to isolate the area below the light well and a dustpan and trash bags to store the insulation for reuse. Wear gloves, a dust mask, and goggles. To protect your head from roofing nails sticking through the sheathing, wear a hard hat. (You should already have gotten a tetanus shot.)

Skylight Positioning

This roof is framed for a deck-mounted skylight (see pp. 164-165). For a curb-mounted skylight, the illustration would include a 2x4 or 2x6 on edge, perpendicular to the roofing, above the doubled headers reinforcing the rafters.

Temporary braces support cut-through rafter.

Doubled headers reinforce cut rafters, cut joists.

Remove roofing, sheathing.

Bevel top edge of these doubled headers.

Plumb lower wall.

Temporary brace

Rafter

Strongback

Upper wall at 90° angle to roof plane

Ceiling joist

Doubled headers

Trim drywall flush to doubled headers.

Strongbacks keep cut-through joists from sagging.

Before you cut anything, take time to think through and lay out the rough openings and light-well walls. Also add strongbacks and braces.

ting on ceiling joists; that way, you'll be less likely to crack the finish ceiling. The job will go faster if one worker on a ladder measures carefully and calls out measurements for headers, trimmers, light-well studs, and the like to a second worker on the floor, who does the cutting.

Before cutting ceiling joists, support them with a *strongback*, which is a piece of dimension lumber nailed to the tops of ceiling joists to keep them from sagging, as shown in "Skylight Positioning" on the facing page. The strongback should run perpendicular to the joist grid, be placed within 1 ft. of the cut joist ends, and rest on uncut joists beyond the opening to distribute the load. To tie the strongback to the joists, you can use steel hurricane ties (see the photo on p. 84).

Joists. Cut out the ceiling joists after attaching the strongbacks. Use a framing square to make sure the cuts are square. Cut the ceiling joists so their ends will be set back 3 in. from the final opening in the ceiling. This setback ensures that the doubled headers (3 in. wide) at both ends will be flush with the edges of the finish ceiling. Use three 16d common nails to nail the first board of each doubled header to the ends of the joists you have cut; nail the second header of each pair to the face of the first. Once the headers are in place, nail into their ends, through the trimmer joists. Use steel double-joist hangers for greater strength. Then double the trimmer joists along each side, making sure the nails don't protrude into the opening.

Roof opening. Cutting out the roof opening is relatively easy if the roof pitch is low (5/12 or less). But if it's steep, use roof jacks to create a safe working platform. Once on the roof, snap chalklines through the four 16d nails you drove through earlier, as described in "Skylight: Sloping Roof, Flat Ceiling" on p. 163. Then use a hooked blade in a utility knife to score the asphalt shingles along the chalklines. Scoring allows you to remove the shingles within the RO without disturbing those around the perimeter.

Once you've removed the shingles within the RO, use a circular saw with a demolition blade set to the depth of the roof sheathing. If you're skilled, you can use a reciprocating saw with a bimetal blade to cut through the sheathing, as the pro is doing in the top center photo on p. 162, but don't cut through a rafter while you're standing on it! Whatever tool you use, wear eye protection because you're likely to hit nails. After you've cut around the perimeter of the RO, use a claw hammer or a flat bar to pry out the sheathing.

Next, slide a shingle ripper under the shingles around the RO, gently breaking the self-sealing

Skylight Framing

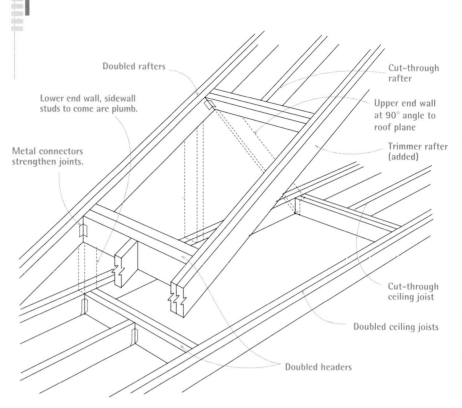

Doubled rafters

Lower end wall, sidewall studs to come are plumb.

Metal connectors strengthen joints.

Cut-through rafter

Upper end wall at 90° angle to roof plane

Trimmer rafter (added)

Cut-through ceiling joist

Doubled ceiling joists

Doubled headers

In addition to doubling headers at the ends of rough openings, double up rafters and joists running along the sides of openings. Framing out a light well will be somewhat easier if sidewalls and the lower end wall of the light well are plumb.

shingle spots, and pull out any nails within 1 ft. of the opening. This nail-free perimeter will enable you to properly flash the skylight curb with building paper or self-adhering bituminous flashing. Finally, if you're installing a skylight with L-shaped mounting brackets, cut back the shingles an additional 1½ in. to 2 in. around the RO so the brackets can sit on a uniformly flat surface and be screwed easily into sheathing.

Framing. Frame the roof opening and the light well. Before cutting through the rafters, install temporary braces to the ceiling joists below; those rafter braces stay in place until the rafters are headered off. If the upper wall of the light well will be perpendicular to the rafters, use a framing square to mark a square cut back 3 in. from the edge of the opening (a doubled header is 3 in. wide).

The lower end of the light well is a bit trickier to frame because it intersects the roof plane at an obtuse angle (greater than 90°). Hold a 4-ft. level plumb against the lower edge of the roof RO, and mark that angle onto the face of the trimmer rafters on each side of the opening. Bevel-cut the

1. After snapping chalklines to outline the rough opening, use a utility knife with a hooked blade to cut the shingles. Then use a shingle ripper (left) or a flat bar (right) to pry up the shingle nails and remove the shingles within the cutout area. Knee pads are a must for most roof work.

2. Unless you are highly skilled with a reciprocating saw, like this pro, use a circular saw to cut through the roof sheathing. To avoid gumming up a blade, remove the building paper before cutting. Be sure to wear eye protection.

3. Standing outside the cutout area, pry up the sheathing and pass it through the hole to your helper inside. By the way, many old-timers don't like sneakers because nails can pierce the thin soles; on the other hand, sneakers can improve your footing. Your call.

4. After removing the sheathing, the installer used a hooked blade to cut back the shingles more precisely so the Velux skylight's mounting brackets could sit on the flat plane of the roof sheathing, rather than on an uneven shingle surface.

top edges of the doubled headers at that same angle. Once you've headered off the top and bottom of the RO, double up the trimmer rafters along the sides of the opening, using as long a board as possible. (Space is tight in an attic.)

Finally, install studs running from the RO in the roof to the RO in the ceiling below. If you install the four corner studs first, you can run taut strings between them to align the intervening studs. If the light-well sidewalls are plumb, rather than flared out, you will have saved yourself a lot of work.

PRO TIP

Cut the length of the light well a little long. It's difficult to know exactly where the grooved lower edge of the skylight curb will meet the finish surface on the plumbed lower wall, so you're better off cutting the opening long and shimming it up as needed, using thin pieces of plywood. Ultimately, ½-in. drywall edges should fit perfectly into curb grooves.

Here, all four light-well walls will flare out toward the bottom (none plumb). Using his level as a straight-edge, the installer marks the rafter cuts; the level runs from the edge of the roof opening to the doubled headers in the ceiling opening. After cutting the rafters, he'll attach a doubled header behind the cutline.

Different skylight, same installer. Here, he frames out the light well before cutting a hole in the ceiling because it was raining. First, he headered off the ceiling joists, then the rafters around the roof opening, before angle-cutting the studs between the two openings. (He cut the four corner studs first.)

The same light well after the roof has been opened: Note the doubled headers around all sides of the opening. Because all walls flare, this is a complicated piece of framing.

Skylight: Sloping Roof, Flat Ceiling

Positioning a sloping skylight above a flat ceiling is an inexact science. Situating the skylight between rafters is easy enough, but because light wells typically flare out, sizing and positioning the ceiling opening can be tricky if you've never done it before. Here are a few tips to demystify and simplify the process.

▶ Don't overdo. As with hot sauce, so too with skylights: A little goes a long way. Letting light into a space dramatically transforms it, so err on the conservative side when sizing them. Skylights placed on north- and east-facing roofs will yield the most even light, whereas openings on south- and west-facing roofs can get quite harsh and hot, making shades must-haves on those sides. When adding skylights, give a thought to how those ceiling openings will look at night, too—in too many cases, like a large black hole. Some designers add tiny LED lights inside a light well to give it a softer, romantic light.

▶ Size them small. The most common sizes are 14½ in. or 30½ in. wide—which fit in the spaces between two or three rafters set 16 in. on center, respectively. (The fewer rafters you cut into, the better.) Let's say that both skylights are roughly twice as long as they are wide: The smaller skylight (14 in. by 28 in.) will have an area of 2.7 sq. ft., whereas the larger one will have 12.5 sq. ft. Things get interesting, though, when you add a flared light well from the skylight to a ceiling below. Measure a flared lampshade's width top and bottom and you'll see how flaring radically increases surface area. So the bottom of a flared light well can easily be double or triple the area of a skylight.

▶ Get it on paper. Using graph paper, make a roughly to-scale drawing (side view) that shows your roof (with the correct slope), a skylight in it, and the distance between the head of that skylight to the ceiling plane. Because the angle of the light-well flare is undecided at this point, use the drawing on p. 160 as a guide. From this sketch, extrapolate how large the opening in the ceiling will be.

▶ Get it on the ceiling. Using pieces of painter's tape, mark the four corners of that imaginary light well onto your ceiling. The four sides of the light-well rectangle should be parallel to walls in the room. Move the tape until you like the size and shape of the opening-to-be.

Then push a nail up through each corner of the rectangle into the space above.

▶ Go into the attic or space above to find the four corner nails and determine if there are wires, pipes, and ducts that would be a problem to relocate. Avoid positioning skylights where roof planes converge, by all means. If the space over the ceiling is inaccessible, turn off the electricity and use a cordless recip saw to punch a hole big enough for your head, so you can take a better look before enlarging the hole further.

▶ If you can position skylight openings to avoid cutting rafters or ceiling joists, do so. If you must cut more than one rafter, have an engineer review your plans. Otherwise, double up headers and trimmers around the rough opening to redistribute loads, and use steel connectors to ensure solid connections.

▶ Refine your design. Do you like the light-well shape? (You might want to make a foam-core model.) Remember, you can flare the sides of the light well, too—there's no law that says they must be vertical—but that will simplify construction. Compound-angle cuts on all studs can be a monster to do correctly.

▶ When you have a better sense of the light well's location, enlarge the hole in the ceiling, but don't cut it to its final size yet. To mark the corners of the opening in the roof, drive 16d nails up through the sheathing. *Note:* Thus far you've cut into finish surfaces only, not into framing.

▶ Parting thoughts: Operable skylights can be a great source of natural ventilation, especially in attic bedrooms and the like. Skylights with center pivots can be turned around for easy cleaning, whereas units high in the ceilings can be opened and closed via tiny motors activated by wireless controls. There are lots of accessories and options, so visit www.velux.com or some other makers' sites to see what's available. And don't forget to look into glazing options that provide somewhat greater energy efficiency.

INSTALLING AND FLASHING SKYLIGHTS

Follow the manufacturer's instructions when installing your skylight so the warranty will be honored. On the following pages, a Velux deck-mounted skylight is being installed.

Screw the mounting brackets to the sides of the skylight curb; most brackets are adjustable. Then, with one worker on the roof and one in the attic, pass the skylight out through the opening. As the worker outside raises or lowers the brackets until the top and bottom of the unit are level, the worker inside centers the unit in the opening.

That done, the outside installer screws the bottom legs of the brackets to the sheathing.

Wrapping the curb with building paper or self-adhering membrane underlayment gives you an extra layer of protection before installing the unit's apron, side flashing, and head flashing. Apply 12-in.-wide strips of underlayment, one on each side, folding each so that it runs up onto the curb about 2 in. Apply the bottom piece first, which overlaps the shingles below, then place the side pieces and, finally, the top piece. If possible, slide the top piece of self-adhering membrane up under the building paper so you adhere it directly to the sheathing. Slit the folded strips

A cordless framing nailer is great for framing skylights, especially for toenailing 16d nails when there's little room to swing a hammer. And you won't need to drag a 100-ft.-long compressor hose up into the attic.

where they overlap the curb corners so the strips lie flat, and apply a dab of roofing cement to adhere the slit pieces.

Install the apron flashing first. It runs along the lower edge of the skylight and overlaps the shingles below it. Holding the apron snugly against the curb, attach it to the curb—not the roof—using a single screw (or nail) on each side. Although it's usually not necessary to caulk under the apron's lower flange, follow the manufacturer's advice. Ideally, the apron will line up with a course of shingles, but in renovation work that's not always possible.

Install step-flashing along the sides, weaving the L-shaped flashing between shingle courses. Ideally, shingle courses should stop ¼ in. shy of the curb so that water can run freely along the sides. The vertical leg of each piece of step-flashing should extend up high enough so it will be protected from rain by curb caps; the horizontal leg should extend under the shingles at least 4 in. As with all flashing, avoid nailing in the chan-

nel where water will run. Rather, nail overlapping pieces of step-flashing to the curb only and high enough so the nails will also be protected by curb caps. If necessary, trim the last (uppermost) pieces of step-flashing so they don't extend beyond the curb.

Install the head flashing by slipping its upper flange under the next full course of shingles above and pressing the head flashing snugly to the curb. Attach the head flashing to the curb with one screw or nail at each corner. Then install a strip of building paper—or a second strip of self-adhering membrane—over the head flashing's flange before shingling over it. *Note:* Stop the flashing strip and shingles 2½ in. to 4 in. above the skylight curb so there is a clear expanse of metal to clear leaves and accelerate runoff.

Installing cladding (curb caps) is the last step. These pieces cover and seal the tops of apron and side (step-) flashing and, on some models, the top of the head flashing as well. Follow your skylight's installation guide religiously. In some fixed

Cladding for a Deck-Mounted Skylight

Cladding covers and seals the top legs of apron and step-flashing and—on some operable models—it covers the head flashing as well. Read installation instructions closely. Numbers on the cladding indicate the installation sequence.

models, the head flashing is the last piece to go on, covering the upper ends of the side cladding pieces. Details vary, but caps snap or screw on. Better-quality units have self-sticking foam gaskets that are applied to the top of the curb before the caps are installed, which reduces air and dust infiltration.

Finish off with a couple more tasks. To reflect light, light wells should be covered with drywall and painted a light color. Insulate around the light wells to reduce heat loss and condensation. Because condensation is common around skylights, cover the light well with water-resistant (WR) drywall, often called greenboard for its color. Stiffen the well's outside corners with metal corner beads.

Velux units have integral curbs and proprietary flashing kits. This one is "directional"—that is, with a top and bottom. Having slipped head flashing under shingles above, the installers will slip the unit's head underneath. Small L-shaped brackets along the curb will mount to the sheathing.

Head flashing has two pieces, a base and a cap. Slip the base piece under the course of shingles above as shown, then hold the base snug to the curb and attach it with a single screw or nail at each corner.

Once the unit's mounting brackets are screwed to the sheathing, wrap the curb with building paper or self-adhering membrane as described in the text. Here, the apron's bottom flange overlays the shingles below.

Working from the bottom up, apply L-shaped step-flashing along the side. Each overlaps the one below and is nailed with a single nail to the side of the curb. The uppermost pieces of step-flashing slide up under the head flashing. When these individual pieces of step-flashing are in place, a continuous cap piece will cover each side.

7 Exteriors

It still takes a skilled eye to install siding, but pneumatic nailers have largely replaced hammers and drill drivers. It is advisable to wear safety glasses when using this tool.

The exterior wall of a house is a multi-layered membrane that weatherproofs a house in much the same way a roof does. In addition to protecting underlying elements from damage by sun and wind, the exterior intercepts and directs water away from sheathing and framing.

The visible exterior layer consists of siding and trim. Beneath the siding, ideally, is a *water-resistive barrier*, typically building paper or plastic housewrap. In addition, flashings seal transitions from one material to another or direct water around potential dams, such as window and door headers, vent hoods, and outdoor outlets. Finally, various sealants or caulks fill gaps, bond materials together, or cut air infiltration. Of course, windows and doors are exterior elements, too, but they are discussed in greater detail in chapter 6.

Although gutters may be considered part of the roof, they are discussed at the end of this chapter because gutters are attached to the exterior and help protect it. Gutters also direct water away from foundations, thus reducing moisture in basements and crawlspaces and forestalling mold.

Water-Resistive Barrier and Window Flashing Details

As you install the WRB and flashing, always do so in a manner that will divert water away from the sheathing and framing behind it. Because water flows "downhill," materials above should overlap those below.

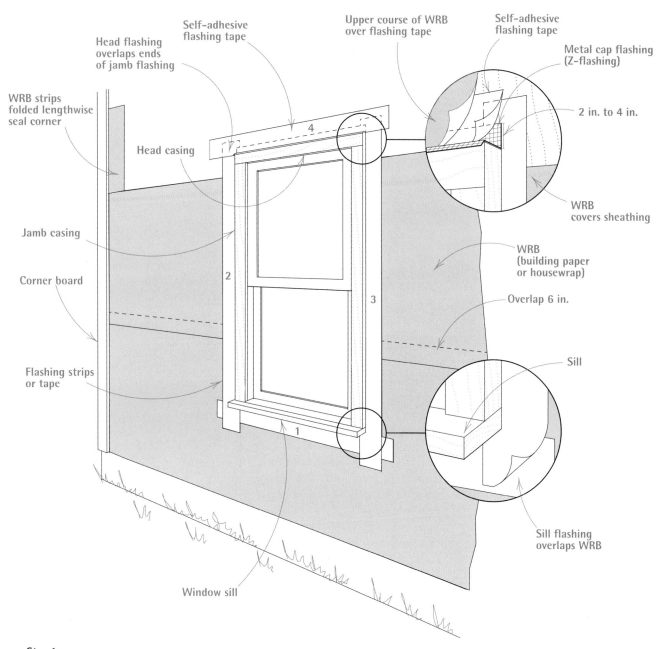

Head flashing overlaps ends of jamb flashing

Self-adhesive flashing tape

Upper course of WRB over flashing tape

Self-adhesive flashing tape

Metal cap flashing (Z-flashing)

Head casing

2 in. to 4 in.

WRB strips folded lengthwise seal corner

WRB covers sheathing

Jamb casing

WRB (building paper or housewrap)

Corner board

Overlap 6 in.

Flashing strips or tape

Sill

Window sill

Sill flashing overlaps WRB

Step 1
Fold the top edge of the sill flashing over the bottom of the rough opening. Split each end of the sill flashing: fold the upper "ears" up onto the sides of the opening, and extend the lower ears out about 8 in. beyond the opening, onto the sheathing. Leave the lower edge of the sill flashing unattached for now.

Steps 2 and 3
Fold jamb flashing around both sides of the rough opening and over the ends of the sill flashing. Lap jamb flashing over the ends of the sill flashing but leave the bottom edges unattached so they can overlap the WRB (building paper or housewrap). Once the WRB is installed, overlap and attach the bottom edges of the sill flashing and the jamb flashing to the WRB.

Step 4
Install rigid cap head (cap) flashing first so that it weather-seals the gap between the top of the window frame and the sheathing. The ends of the head flashing overlap jamb flashing on both sides. Next, install self-adhesive flashing tape over the top of the head flashing to further ensure the seal. Last, install the WRB layer over the head flashing and the flashing tape.

When installing traditional membranes such as felt paper or red rosin paper, overlap horizontal courses by 6 in. The overlap for housewrap is similar, but you must also tape seams with a compatible tape.

PRO TIP

Plastic housewrap is slippery, so don't lean unsecured ladders against it. Secure the ladder or, better yet, use scaffolding if you're installing housewrap and siding to a second story. Always install housewrap with its printed side facing out: It's engineered to allow migration of water vapor in one direction only.

As with other building systems, maintenance is crucial. Each autumn after the leaves have fallen and each spring, clean gutters and downspouts; if needed, do this more often to keep them flowing freely. Every year, survey and recaulk building seams as needed. Paint or stain wood siding periodically.

Water-Resistive Barriers

No matter how well siding or trim is installed, sooner or later water will work its way behind it. Typically, this happens when storms drive rain into building seams or gaps around doors or windows. For this reason, builders cover exterior sheathing with a water-resistive barrier (WRB). In older houses, the most common WRB is 15-lb. asphalt-impregnated building paper—also called *felt paper*.

In newer houses, plastic housewraps such as Tyvek or R-Wrap are commonly used. However, the category of breathable membranes has experienced explosive growth, so research it carefully. Some of the more popular products include Henry Blueskin self-adhered, water-resistive, vapor-permeable air barriers; VaproShield breathable membrane systems for roofs and walls (has UV inhibitors); and Vycor Plus self-adhered flashing for windows, doors, and other exterior details. Though specs vary, this new class of products is easily integrated with housewraps and other water-resistive barriers and tends to be vapor-permeable. Thus it gives you options when detailing tricky assemblages such as rain screens,

where intentionally gapped boards admit sunlight (and UV rays) behind the siding.

Both building paper and the newer housewraps are permeable enough to allow excessive moisture behind the siding to escape, yet sufficiently water-resistive to protect sheathing from wind-driven rain. Bottom line: It doesn't matter whether you use building paper or housewrap as long as it's correctly installed and conforms to local building codes.

In new construction, water-resistive barriers are often installed before windows and doors have been inserted into rough openings. Often, builders just cover the whole expanse of a wall with housewrap—including rough openings. Then, using a utility knife, they slit the housewrap within each opening (like a giant X), fold it into the sides of the opening, and staple it. Precased windows and doors are then installed and their perimeters weatherproofed with self-adhering flashing tape or flashing strips. As shown in the drawing on p. 167, the outermost WSB layer *overlays* cap flashing and fits *under* sill flashing. Cap flashing redirects water that might otherwise dam up behind door or window head casing, leading to stains and mold on interior surfaces, peeling paint and rot. All flashing is important, but flashing the head of a window or door is *the* critical detail to get right.

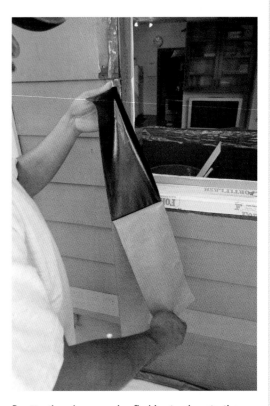

Be attentive when removing flashing tape's protective paper backing because its self-adhesive side will stick aggressively to whatever it comes in contact with.

Of course, renovation is rarely as straightforward as new construction. If you're replacing the siding around an opening and installing new WRB, then sliding the top leg of cap flashing underneath is easy. But what if the siding is in great shape and slipping cap flashing underneath would require removing siding nails and, in the process, destroying the siding? Or what if long-ago builders installed the exterior trim *over* the siding, as they often did on Victorian houses? There, the cure of installing cap flashing by the book might be worse than the disease. Consequently, builders trying to keep a job on time and on budget may bend the rules. Occasionally, they will even surface-mount cap flashing, as shown on p. 155. If a large overhang protects a window, this solution works just fine.

INSTALLING BUILDING PAPER

Installation techniques for housewrap and building paper are much the same. In this chapter, both materials are often referred to generically as *building paper*.

Before starting, survey the sheathing and knock down protruding nail heads and wood slivers that could puncture the paper.

Installation is a two-person job, especially if you're working with 9-ft.-wide rolls of Tyvek. As one person unrolls the material, the other aligns and staples it. Start at the bottom of the wall, overhanging the bottom edge of the sheathing by about 1 in. Position the building paper, tack an upper corner in place, roll out about 8 ft., and raise or lower the roll until the edge of the building paper is roughly parallel to the bottom edge of the sheathing. When it's in position, staple the paper every 16 in. along the edges and every 24 in. in the field; use ¼-in. or ⁵⁄₁₆-in. staples. You can also use tabbed nails or FasCap plastic caps and staples to attach building paper, although they are more expensive and slower to install than staples. Thereafter, roll out 3 ft. to 4 ft. at a time, stapling as you go. Trim the 1-in. overhang later.

As you roll out the building paper, make sure it lies flat. Puckered paper can elevate the siding or trim and compromise its weathertightness. Typically, the person holding the roll maintains a slight tension on the building paper to prevent puckers. Also, take care not to tear the membrane while handling and stapling it—more of a problem with paper than with housewrap, which stretches.

Overlapping and taping seams. Overlap vertical end seams by 4 in. to 6 in.; overlap horizontal seams 6 in. to 8 in. As you work up the wall, upper courses should always overlay those below so any water that gets behind the siding is directed down and out. Building paper also should overlay cap flashing at doors and windows.

Seal building-paper seams, tears, and punctures with seam tape; smooth down the tape to make sure it adheres well. If you're installing housewrap such as Tyvek, use a seam tape recommended by the manufacturer. Last, install siding as soon as possible after installing the building paper because wind can lift and stretch it, especially housewraps.

Corners. Take special pains with corners. Although a 4-in. overlap is often adequate for corners, you'll be safer to overlap outside corners by 1 ft. In addition, many builders reinforce corners with flashing tape to keep water out of gaps between corner boards and siding. To prevent the building paper from bunching in inside corners, use a straight board to press the paper into the corner before stapling it.

FLASHING TAPES AND STRIPS

As building science improved its understanding of air and moisture infiltration, many codes specified additional measures to seal building joints and exterior openings such as windows, doors, vents, outside lights, and electrical boxes. Rigid cap flashing over windows and doors had been used for a long time, of course, to direct water away from these potential dams. But a new category of *flexible* flashings developed whose primary intent is to *seal* gaps where air and moisture might get in.

To weatherproof windowsills, fold flashing paper strips into the rough opening, staple the paper down, and apply caulk. When the window is installed, the sill will compress the caulk, creating a seal. Preformed corner pieces, such as Moistop Corner Shield, also are available. Tuck building paper or housewrap under sill flashing.

Every Person's Guide to Flashing Lingo

There are hundreds of flashing products to choose from, and reading the labels isn't always enlightening. Here's a sampler of terms from flashing labels, and what they're really saying:

Self-adhesive, self-adhering, self-adhered: Flashing, membrane, or tape with sticky stuff on the back; also known as flashing *tape*.

Flashing *strips*: Nonsticky-backed, precut flashing pieces.

Mechanically attach flashing solution: Staple it.

Flexible flashing: Not metal.

Rigid flashing: Metal or, less often, plastic.

Elastomeric: Rubber based or rubberlike.

Water-resistant, weather-resistant, moisture-resistant: Our lawyers won't let us say *waterproof*.

Through-wall penetrations: Holes in walls, which need to be flashed.

Compatible with most sealants and caulks: Read the fine print.

To retrofit French doors, it was necessary to cut back stucco roughly 1 ft. around the new opening so the unit could be correctly flashed. Gaps along both sides were covered with flashing strips stapled to the sheathing and caulked to the jamb edges; the gap above the frame head was flashed with foil flashing tape, which is self-adhesive. These flashings will, in turn, be covered by fortified building paper tucked under the stucco. Three coats of stucco will then be applied to lath nailed over the building paper.

PRO TIP

Trim on older buildings is rarely level or parallel. Thus new trim may look better if it's installed slightly out of level so that it aligns with what's already there. For example, when stretching a chalkline to indicate the bottom of the water table, start level and then raise or lower the line until it looks right in relation to nearby windowsills and the like. Once the chalkline looks more or less parallel to existing trim, snap it on the building paper and extend it to corner boards.

Flexible flashing takes predominantly two forms: Precut *flashing strips*, reinforced with a polymer to resist tearing, are stapled to rough openings or sheathing, and are often used in tandem with a sealant or caulk. The second group, self-adhering *flashing tapes*, are typically butyl-based or modified bitumen; some are foil-faced as well. Because flashing tapes adhere and seal aggressively, they usually aren't used with caulks or sealants. Self-adhesive flashing tapes seal so well that they are by far the favorite with builders.

Both flashing tapes and flashing strips are available in rolls 4 in., 6 in., 9 in., and 12 in. wide and are used for essentially the same tasks. Their differences are subtle but telling. Because flashing strips are thinner than flashing tapes, for example, strips are often specified when sealing the perimeter of an uncased window (see pp. 143–144), where the flashing is sandwiched between jamb edges and casing. On the other hand, the greater thickness (20 mil to 45 mil) and adhesion of butyl-based flashing tapes enable them to seal around nail shanks, thus preventing leaks. Installing (and retrofitting) flashing around door and window openings is discussed at length in chapter 6.

Exterior Trim

However ornate or complex it may be, exterior trim's basic function is to cover critical building seams, keeping weather out and reducing air infiltration.

GENERAL PREPARATION

Exterior trim can be applied in many different ways, depending on the design of the house and the type of siding. For wood shingles, clapboards, or cement-board siding, use trim boards that are thicker than the siding. For flat shiplap and board-and-batten sidings, apply trim boards over the siding. In general, try to use the same materials and installation methods that were used on the house originally.

Solid-wood exterior trim should be a rot-resistant species such as redwood, cedar, or hard pine and sufficiently dry to avoid shrinkage, cupping, and checking. For those reasons, avoid sugar pine, knotty pine, hemlock, fir, and the like. If you'll be painting the trim, you may find it cost-effective to use *finger-jointed* trim stock fabricated from shorter lengths of high-grade wood. It's widely available and can be durable if you keep it sealed with paint. For best results, specify vertical-grain heartwood because it resists decay, holds paint well, and is the most stable dimensionally. *Caution:* If this trim absorbs moisture, its finger joints may separate, so don't use it next to stucco.

BACK-PRIMING AND PAINTING

Apply primer to all faces and edges of exterior wood (and engineered wood) siding and trim, including the back faces. Back-priming is critically important because wood will cup (edges warping up) when the sun dries out the exposed

Eaves Trim

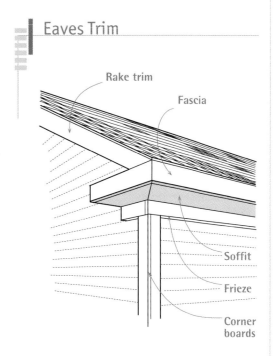

Fascia, soffit, and frieze boards are collectively called the eaves trim.

Engineered Trim: A Primer

Just as engineered lumber revolutionized structural materials, there's now engineered trim rated for exteriors. Many of these types are perfectly straight, flat, and factory primed on all six sides. If you're accustomed to working with wood, engineered trim will take some getting used to, and the jury is still out on its long-term stability and durability, especially that of hardboard. Here's an overview of types:

Laminated veneer lumber (LVL) is made from thin wood veneers glued so the grain runs in the same direction, faced on one side with a medium-density overlay (MDO) of resin-impregnated paper. LVL trim is available in 1-in. and 1¼-in. (5/4) thicknesses and in standard widths from 4 in. to 12 in., in 2-in. increments. Standard lengths are 16 ft. and 20 ft. It's easy to cut, drill, nail, and handle, and its MDO face paints wonderfully. However, LVL is expensive. Clear Lam is one well-known brand.

Hardboard, fiberboard, or wood-fiber composite is fashioned from wood chips that are steamed, pressed, and glued. There are many brands in this category, including MiraTEC, ChoiceTrim, and ProTrim, and features vary greatly. Composites tend to be available in 1-in. and 1¼-in. thicknesses, in standard widths of 4 in. to 12 in., and in 16-ft. standard lengths. Some are primed on all sides, some not; some have MDO-like paper facing, some not—so check with your suppliers. Prime all sides and the cut edges, drive corrosion-resistant nails just flush to the trim surface (nails driven deeper may need caulk to keep out moisture), and gap butt-joints ⅛ in. to allow for expansion. Caulk and paint the trim ASAP.

Boral TruExterior siding and trim is a composite that contains 70% fly ash. It can be cut, shaped, and trimmed like wood and is largely unaffected by water: It won't swell or warp. Typically, it is not available as molded trim but as flat stock: 1-in. boards, 5/4, and 2-in. (e.g., 2x6 or 2x8). In strong southern exposures, Boral's stability makes it a superior choice to wood. It's also a particularly green product because it recycles fly ash waste from cement factories.

Fiber-cement trim is wood pulp mixed with portland cement and sand, and it's virtually indestructible. Few people would mistake it for wood, but it's strong, lightweight, stable, and resistant to rot, insects, and fire. HardieTrim is available in thicknesses from ⁷/₁₆ in. to 1 in.; in widths of 4 in., 6 in., 8 in., and 12 in.; and in lengths of 10 ft. and 12 ft. See p. 183 for working with fiber-cement siding and trim.

Cellular PVC plastics are a rapidly growing category of exterior trim, shaped molding, and sheet materials that can be worked using regular woodworking tools. For optimal results, use sharp, carbide-tipped sawblades and drill bits when working either type. The top-selling cellular PVC trim, Azek, is a free-form type, as are Fypon, Kleer, and CertainTeed. In some respects, PVC trim is easier to work than wood. You can screw or nail close to the edges and ends of stock and not worry about splitting; unlike wood, PVC has no grain. There are, however, several critical differences when working with PVC: (1) It tends to expand and contract along its length, so when joining long boards, bevel-cut their ends so they overlap. (2) Join trim with PVC cement, which joins them chemically rather than mechanically, as is the case with wood joints. (3) PVC doesn't absorb moisture, so paint takes longer to dry than with wood.

front face if the back retains moisture. The greater the moisture differential between front and back faces, the more likely the cupping.

While cutting trim or siding, keep a can of primer and a cheap brush nearby to seal the ends after every cut; unprimed end grain can absorb a lot of moisture. (It's especially easy to forget to prime cut edges when you're using preprimed trim.) Ideally, apply at least two topcoats of acrylic latex paint after priming to seal trim and siding. If you want stained or clear-finished trim or siding, use cedar or heart redwood.

ATTACHING TRIM

For best attachment, secure exterior trim to framing. Where you have only sheathing to attach trim to, drive nails at an angle or use stainless-steel trim-head screws.

Prime all faces and edges of exterior trim and siding, including the back. Back-priming is especially important because moisture trapped between back faces and sheathing can lead to paint or sealer failure, cupping, or—in extreme cases—rot. After cutting trim or siding, be sure to prime the cut edges as well.

Choosing fasteners. Pick a nail meant for exteriors. If you'll be using a transparent finish, making nail heads visible, stainless-steel nails are the premier choice; though expensive, they won't rust. Aluminum nails won't stain but are somewhat brittle and more likely to bend. Galvanized nails are the most popular because they're economical, stain minimally, and grip well. Many nail types (including stainless steel) also come in colors matched to different wood types—cedar, redwood, and so on. Ring-shank nails hold best.

For stained exteriors, some contractors prefer galvanized finish or casing nails because their heads are smaller and less visible. Box nails are a good compromise. Their larger heads hold better than finish nails, yet their shanks are smaller than those of common nails, making box nails less likely to split wood. There are also "splitless" siding nails that come with preblunted points to minimize splits. (The blunt point smashes through wood fibers, rather than wedging them apart.)

Where you want maximum grip, use stainless-steel trim-head screws instead of nails.

Pneumatic nailers. Most pros use pneumatic nailers to attach exterior trim. Using a finish nailer with galvanized nails allows you to tack up trim exactly where you want it. Anyone who has spent time trying to simultaneously hold and nail

a 16-ft. corner board in place while balancing on a ladder will appreciate this tool. Nailers also drive nails quickly and accurately, reducing splits and eliminating errant hammer blows that mar trim. After setting the trim with finish nails, you can always go back and hand-nail with headed nails to secure the trim further. Or you can use headed siding nails in the nail gun.

Nailing schedules. To face-nail nominal 1-in. trim (actual thickness, ¾ in.), use 8d box nails spaced every 16 in. Nail both edges of the trim board to prevent cupping, placing nails no closer than ½ in. to the edge. If the trim goes over siding, say, at corners, use 8d to 10d box nails. To create a tight joint between corner boards, use 6d nails spaced every 12 in. and drive them in at a slight angle. If you'll be painting the trim, also caulk this joint or glue it using an exterior urethane glue, such as Gorilla Glue or Titebond III (which is less expensive and won't stain your hands).

About nail heads. Taking the time to line up nail heads makes the job look neater. For example, when nailing up jamb casing, use a combination square to align nail pairs. If you're putting up a long piece of trim that runs perpendicular to studs, snap chalklines onto the building paper beforehand so that you'll know stud positions for nailing. If the trim will be painted, take the time to set the nail heads slightly below the surface, using a flathead punch. Then use exterior wood filler to fill the holes. If you don't set the heads slightly, they may later protrude as the wood shrinks, compromising the paint membrane and admitting water. On larger jobs, carpenters are usually expected to set nail heads. Painters fill and paint them.

EAVES TRIM

Because eaves trim is often complex and can impact framing, roofing, ventilation, and the house's aesthetic integrity, draw a cross section of it as early as possible.

There is no single correct way to construct the eaves, but the boxed eaves on the facing page are a good place to start. First, a fascia board that overhangs a soffit by ⅜ in. to ½ in. enables you to hide rafter irregularities—rafters are rarely perfectly straight or cut equally long. Second, that overhang accommodates a rabbeted fascia–soffit joint, which protects the outer soffit edge, even if the wood shrinks slightly. Third, if you rabbet the back edge of a frieze board or build it out using blocks, the frieze will conceal the top edge of the siding. A frieze also creates an inconspicuous space to install an eave vent.

Ventilation channels at eaves allow air to flow up under the roof and exit at ridge or gable-end vents. This airflow is beneficial because it lowers attic temperatures and helps remove excess moisture from the house, thus mitigating mold, ice dams, and a host of other problems. To keep insects out, soffits need screening. In a wide soffit, there's plenty of room for screened vents in the middle. In a narrower soffit, you may need to leave a ¾-in. space at the front of the soffit or at the back hidden behind a built-out frieze board.

If the house has exposed rafter tails rather than soffits, you have fewer ventilation options that will look good. Consider leaving the eaves sealed and adding rooftop intake vents, as described on pp. 102–103.

WATER TABLE

A water table is horizontal trim running around the base of a building below the siding. Not all houses have it. Depending mostly on regional preferences, the water table takes several forms. In the West, it typically looks like windowsill ears (the beveled parts that stick out) and is often

Water-table trim often finishes off the bottom of a wall and provides a level base for the first course of siding. To forestall rot, cap the water table with metal or vinyl flashing before installing the siding.

Boxed Eave: Detail 1

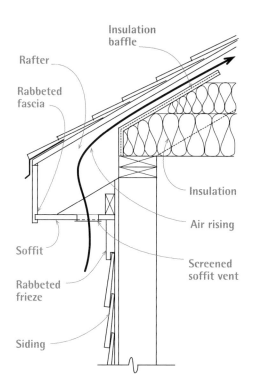

Rafter
Rabbeted fascia
Insulation baffle
Insulation
Air rising
Soffit
Screened soffit vent
Rabbeted frieze
Siding

A strip of continuous screen in the soffit allows air to circulate into the attic. The rabbeted frieze conceals and protects the top of the siding.

Boxed Eave: Detail 2

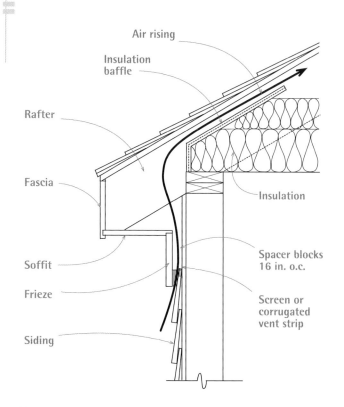

Air rising
Insulation baffle
Rafter
Insulation
Fascia
Soffit
Spacer blocks 16 in. o.c.
Frieze
Screen or corrugated vent strip
Siding

You can create ventilation channels behind the frieze by spacing it out ¾ in., using blocks spaced every 16 in. o.c. This also conceals the top of the siding.

used to separate different types of siding materials, such as shiplap siding from wood shingles above. Typically, 1½ in. by 1½ in., this type of water table runs continuously around the building and is mitered at the corners. It often has a rabbeted heel, which fits over the top of the wood siding below, and a beveled top, which is overlain with shingles or clapboards.

In the East, water tables are also called *splashboards*; they usually are 1-in. boards 8 in. to 12 in. wide and may be capped to shed water. Splashboards are most common in wet regions, where roof runoff often splashes back along the base of a house. (Some primal carpenter may have reasoned it would be easier to replace a single rotted board than to disturb several courses of siding, or that a thicker board would simply last longer.)

Whatever the shape of the water table, flash the top with a metal drip-edge that extends at least ½ in. beyond the face of the board. The section where corner boards sit atop the water table is especially prone to rot. Prime and paint the boards thoroughly.

CORNER BOARDS

Corner boards are usually 1-in. boards butted together. Siding is then butted against them, making an attractive and weathertight corner.

Not all buildings have horizontal trim below the bottom of the siding. For example, the first (bottom) course of shingles is often doubled and overhangs the sheathing slightly. In that case, run corner boards 2 in. to 3 in. below the bottom edge of the sheathing; then, after you nail up the first course of shingles, trim the corner-board ends level to the shingles' butts. If the house has a water table or splashboard, measure from its top edge up to the underside of the soffit to determine the length of the corner boards.

If you're installing shingles, whose overlapping courses have a higher profile than clapboards, use 5/4 corner boards, which are a full 1 in. thick. To give the illusion that corner boards are the same width at each side, rip down the overlapped board by the thickness of the stock. And for a crisp, straight corner, preassemble corner boards before installing them, as shown in the photo below.

Occasionally, corner boards are nailed over siding. This can be a problem because nails driven through the trim may split the siding underneath. Besides, corner boards can't seal well if

Corners

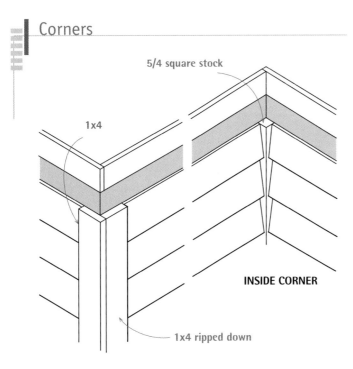

Butt siding to the corner boards to avoid complex miter cuts. In outside corners, rip down the overlapped board by the thickness of the stock, and both boards will look equally wide.

Preassemble corner boards, soffit and fascia boards, and other exterior trim on the ground whenever possible. The joined pieces will be tight and square, even if the framing and sheathing behind them isn't.

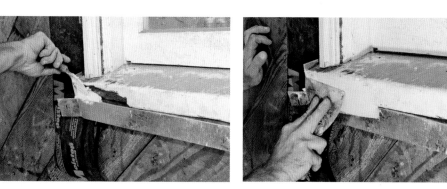

After wire-brushing away loose material, inject liquid consolidant into the wood until saturated. After allowing the consolidant to set, apply the pastelike filler to build up the damaged area.

Use a taping knife to compress and smooth the filler. The galvanized metal tacked to the sill keeps the filler in place until it dries and doubles as a screed strip to which you can level the filler.

PRO TIP

Epoxy filler can be applied with a putty knife. But the filler is easier to shape to match existing contours by hand—hands protected by latex gloves, that is. Restorationist Tom O'Brien suggests donning three or four pairs of disposable latex gloves at the start of the job and peeling them off as they become gunked up.

nailed over an irregular surface. It's an awkward way to resolve a corner, but this method was commonly used on Victorian homes with flat, shiplap siding. Careful nailing and liberal doses of caulk will help ensure a weathertight seal. To minimize splits, predrill the board nails.

The corner boards described thus far cover outside corners. Inside corners aren't as exposed to weather, so wide trim boards aren't necessary. Instead, nail 1-in. by 1-in. strips (or 1¼-in. by 1¼-in. strips) to the inside corners, and butt the siding to them. That's much faster than cutting compound miters in the clapboards or interweaving shingles.

REPAIRING EXTERIOR TRIM

Although it may be tempting to rip out exterior trim that's badly weathered or rotten, repair is often a better option if the original trim would be difficult to remove or replacement trim is too expensive. Before deciding, survey the extent of the rot and address the root cause. Otherwise, you're treating only the symptom.

Replacing rotted sections is a reasonable option when the bottom of an otherwise sound trim board has rotted away. Flat and square trim is easier to replace and match than molded trim. Rotted bottoms of corner boards and splash-boards are usually easy to cut free and replace, whereas punky doorsills or windowsills are probably best replaced or repaired in place using epoxy, as described in the next section.

Replacing the bottom end of a rotted board is straightforward. Draw a line across the face of the board, 6 in. above the bad section. After setting your circular-saw blade to the thickness of the board, use a Speed Square to guide the saw shoe, making a 90° cut. Wear safety glasses, and use a demo blade because there may be hidden nails. The replacement piece should be the same thickness, width, and—preferably—species as the original trim. To join the new section to the old, use a biscuit joiner (see p. 514) to cut a biscuit slot in both board ends. Dry-fit everything, prime all surfaces with epoxy primer, and allow the primer to dry well. Then epoxy the pieces together. Hold the boards in place with a piece of scrap screwed to both. Give the epoxy a day to cure (or whatever the manufacturer suggests), and you're ready to sand and paint. Alternately, you could use polyurethane glue or Titebond III as long as both pieces of wood being joined are sound.

In-place epoxy repairs are appropriate when the rotted area is relatively small (epoxy is expensive) and the trim would be difficult or costly to replace. Rotted windowsills or sashes, for example, are tough to remove. Epoxy applications vary considerably, so visit manufacturers' websites (try Abatron, ConServ Epoxy, and Advanced Repair Technology, for example) for specifics or get recommendations at your local home center.

Use a chisel to dislodge loose, crumbling wood. Suck up debris with a shop vacuum. Allow the wood to dry thoroughly before proceeding. Although it's often desirable to cut back to solid wood, soft, punky wood can be reinforced by impregnating it with liquid two-part epoxy. Typically, you'd drill a series of small-diameter holes into the wood and then inject liquid epoxy into them until the wood is clearly saturated. After the epoxy cures, the impregnated wood becomes as hard as a rock.

For the best bond, apply the puttylike filler while the consolidant is still tacky. Avoid getting epoxy on your skin, and by all means wear a respirator when applying or sanding it. After the filler dries and you've sanded it to its final shape, prime and paint it. Although otherwise tough, some epoxies are degraded by UV-rays, and whatever original wood remains still needs protection from the elements.

PRO TIP

Some pros align shingles by tacking up a straight 1x2 for each course and resting shingle butts on it. This allows nailing to go a bit faster, but you must still snap a chalkline or measure up from the bottom course periodically to level the 1x2. Most pros prefer to snap a chalkline and then barely cover that line with the shingle butts so the chalk doesn't show.

Wood siding is pleasant to work with and requires few specialized tools. Although power nailing has largely replaced hand nailing, there's still plenty of handcrafting and fitting to do, such as the shingle shaving shown here.

Siding

This section shows how to install four of the more common types of siding: wood shingles, wood clapboards, fiber-cement siding, and stucco. Vinyl and aluminum siding also are cost-effective alternatives; they're durable and virtually maintenance-free if correctly installed. Consequently, vinyl siding in particular continues to gain ground when homeowners faced with replacing wood siding take a close look at vinyl's benefits. However, vinyl siding is most efficiently handled by specializing contractors and isn't addressed here.

The information that follows assumes that building paper or housewrap covers the wall sheathing and that windows, doors, and exterior trim are already installed.

Aligning Siding Courses

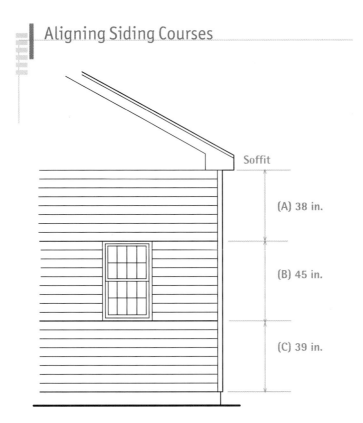

Soffit

(A) 38 in.

(B) 45 in.

(C) 39 in.

For best visual effect, align siding courses with the top and bottom of windows and doors. This may require three separate calculations for exposures, as represented by A, B, and C and explained in the text. Whether installing clapboards or shingles, it's customary to mark course lines lightly on corner boards and then snap chalklines between them.

LAYING OUT WOOD SHINGLES AND CLAPBOARDS

For clapboards and shingles, the two most important parts of the layout are establishing level for the bottom course and varying the exposure of subsequent courses so that they align with door or window trim, if possible. If a leveled water table has already been installed, skip the next section.

Establishing level. If there's no water table (see p. 173), the bottom of the first course of siding typically overhangs the sheathing by ¾ in. to 1 in. Because the bottom edge of sheathing is often not level, use a laser level or a water level to establish a level base line. (Laser levels are relatively inexpensive and offer quick layout over long distances without the need for a helper.) Lay out the front of the building first. Pencil in an appropriate siding overhang on one corner board. Then transfer that mark to all the other corner boards. When you're done, use a combination square to draw light lines through your pencil marks, across the faces of the corner boards. These lines indicate cut-off lines for the corner boards and the bottom of the first course of siding. To align that first course of siding, stretch a taut line through the marks and "eyeball" the bottom of the siding to that line, or simply align the siding to the laser line. If you use a taut line, place clapboards or shingle butts slightly above, but not touching, the line so it won't be distorted.

Varying subsequent courses. By aligning siding courses to window and door trim, you can minimize funky-looking notch cuts at door and window corners. (When installing wide-board siding, however, notch cuts are sometimes unavoidable.) Achieve these alignments by increasing or decreasing the exposure of individual courses. Of course, there are physical restrictions. For example, clapboards must overlap at least 1 in. But as long as exposure adjustments are no more than ¼ in. between courses, they'll look evenly spaced.

The following steps refer to the illustration at left:

1. Measure the full height of the wall, from the cutoff at the base of a corner board to the underside of the soffit or frieze trim. Let's say the height is 10 ft. 2 in. (for calculation purposes, 122 in.). Because shingle exposures are customarily 5 in., that wall will have roughly 24 courses.

2. The wall has three windows and a door. Fortunately, their head casings happen to align 84 in. above the base line. This creates three separate areas for which shingle exposures need to be adjusted, as shown in the drawing: (A) from

Cross-grain cutting is a snap with shingles. Score them with a utility knife...

... and snap them sharply over your knee. Use a cordless jigsaw for complex cuts around windowsills and exterior light fixtures.

the top of head casings to the soffit, 38 in.; (B) from the top of the window head casing to the bottom of sills, 45 in.; and (C) from the bottom of windowsills to the base line, 39 in. Total: 122 in.

3. Calculating exposure adjustments is easy. Round off each measurement to the nearest increment of 5 in. Then increase or decrease the shingle exposure accordingly. Thus area A yields a 4¾-in. exposure (38 ÷ 8 = 4¾); area B is exactly 5 in. (45 ÷ 9); and area C is roughly 4⅞ in.

The easiest way to keep track of these adjustments is by penciling them onto a *story pole*, a long, straight board (a 1x2 is fine) whose length equals the distance between the top of the water table (if any) to the underside of the soffit or frieze. First, mark the tops and bottoms of window and door casings onto the story pole, then the adjusted course heights between.

As you work around the house, align the bottom of the story pole to each corner board and transfer marks from the pole to each board. If the house has windows set at varying heights, story-pole marks will better align with casing on some walls than on others; give precedence to the house's most prominent facade. Where courses just won't line up with casing joints, notch the siding around them.

INSTALLING WOOD SHINGLES

Before you start shingling, make sure that windows and doors are correctly flashed, and that sheathing is covered with building paper, and that exterior trim is installed. A plastic mesh underlayment such as Cedar Breather, which allows air to flow behind shingles, is installed over building paper.

Materials. No. 2 grade shingles are fine for walls because they receive less weathering than roofs. (Roof shingles should be No. 1 grade.) For a standard 5-in. shingle exposure on walls, figure four

bundles per square (100 sq. ft.). Always inspect the visible shingles on a bundle to make sure they're uniformly thick (⅜ in.) at butt ends, of varying widths (on average, 6 in. to 12 in.), knot-free, and reasonably straight grained. Installing shingles requires a lot of trimming, so you don't want to be fighting knots and wavy grain. Shingle butts should also be cut cleanly—not ragged. Feel

Wood Shingle Details

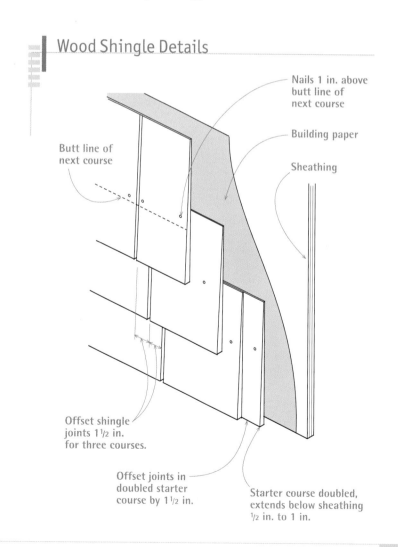

Nails 1 in. above butt line of next course

Building paper

Sheathing

Butt line of next course

Offset shingle joints 1½ in. for three courses.

Offset joints in doubled starter course by 1½ in.

Starter course doubled, extends below sheathing ½ in. to 1 in.

free to send back bundles that look inferior or contain mostly narrow shingles. Standard-milled shingles usually taper slightly: Their butt edges are not perfectly square to sides. Thus, as you nail up each course, you must approximate the butt line. If you want a crisper, more exacting look, ask your lumberyard for R&R (resquared and rebutted) shingles, whose sides are sawn parallel and whose butts are squared off. R&R shingles cost more but, being squared, are easier to install.

To save a little money, you might want to use No. 3 shingles for the bottom layer of doubled starter courses. In this case, order one bundle per 50 lin. ft. of wall. Typically, the starter course of shingles along the bottom is doubled, with vertical joints between the two shingle layers offset by at least 1½ in.

When buying shingles, lift a few bundles to see how dry they are. Relatively wet shingles are fine, as long as they're good quality, but they'll shrink. In fact, most shingles shrink. Although how-to books are fond of telling you to leave a ¼-in. gap between shingles during installation, many shinglers don't bother; unless the shingles are bone-dry, installers assume that all shingles will shrink some.

Use two 1¼-in. galvanized nails or staples per shingle, whatever the shingle's width. Because nails must be covered by the course above, place

The Art of Weaving Corners

To weave shingles tightly on corners, nail them up and shave them in place. As you place each shingle, use a torpedo level to level the butt, overlapping the adjacent wall by about 1 in., and nail the shingle down. As before, use stainless-steel, brass, or bronze nails wherever the nails will be exposed; ring-shanked nails hold best.

Use a utility knife to rough-trim the excess. Then continue with a block plane until the shingle edge is almost flush to the shingle it overlaps on the adjacent wall. To prevent splits and to draw overlapping

shingles tight, drill a pilot hole in ¼ in. from the edge and up 2 in. from the butt of the shingle. A push drill (also called a Yankee drill) is perfect for the task. Alternate the edge that overlaps from course to course.

Because there's not enough room to use a block plane on the bottom and top course of corners, use a cordless jigsaw to precut shingle edges or butts. And because the top course of shingles will be quite short (about 4 in. long) and susceptible to splits, caulk the back sides with a durable urethane caulk, in addition to nailing.

Alternate shingles overlap at the corners. Nail each shingle up, then use a utility knife to trim all but ¼ in. of excess before finishing up with a block plane. Stop planing when the edge is barely proud of (above) the oncoming shingle. On the top course, where a block plane won't fit, use a jigsaw to precut the shingle to the correct angle.

Draw the shingle corners together with a single nail driven through the overlapping shingle. To prevent splits, predrill with a push drill.

Because the nails joining corners or finishing the top course of shingles will be exposed, use either stainless-steel, brass, or bronze nails at those junctures.

shingle nails in ¾ in. from the edge and 1 in. above the exposure line. Where nails will be visible—say, on interwoven corners or the top course below a windowsill—use siliconized bronze ring-shank nails or stainless-steel nails. Again, 1¼-in. nails are fine, unless you're also nailing through a gypsum layer to reach the sheathing on a fire-rated wall.

Installation. If you've got a water table, set your first course of shingles atop it, even if it's not level. That way you eliminate unsightly gaps along the trim, and it's easy enough to level the next course of shingles. If there are corner boards, snap chalklines between them to mark shingle courses, and off you go.

However, if there are no corner boards, weave shingles at the building corners, alternating shingle edges every other course. This requires more skill and patience than just butting shingles to trim boards but produces corners that are both handsome and weathertight. Weave the corners first, then nail up the shingles in between. Because the starter course overhangs the bottom of the sheathing ½ in. to 1 in., measure down that amount at each corner, using a laser level to establish level. After establishing the correct exposure, as described above, shingle up each corner. As you work up the wall, snap a chalkline from corner to corner to line up shingle butts.

As you did on the first course, offset the vertical shingle joints at least 1½ in. between successive courses. If you have a partner, you'll find it easier if each of you works from a corner toward the middle. Only the last shingle will need to be fitted.

When you weave the corners first, leveling the shingle courses in between is largely a matter of snapping chalklines between shingle butts at either end of the wall. If you snap the chalkline slightly high, as shown, the shingle butts will cover the chalk.

Replacing SHINGLES

If you split a shingle while installing it (or if you need to remove shingles to install an exhaust vent for a fan), break out the shards, hammer down the nail heads, and replace the shingle. To remove a few damaged shingles on an otherwise intact wall, use a shingle ripper (also called a slate hook). Slide its hooked head up under the surrounding shingles until you can feel it hook around a nail shank. Then hammer down on the tool's handle until the hook cuts through the shank. To avoid damaging the replacement shingles as you drive them into place, hold a scrap block under the shingle butt to cushion the hammer blows.

Fit shingles closely to window and door casings. The top of the shingle course under a windowsill should butt squarely to the sill. Because this course needs to be shortened and will be susceptible to splits, caulk the back sides. It's also wise to caulk the shortened top course of shingles under the eaves. Ideally, the tops of those shingles will also be protected by a rabbeted-out or built-up frieze, as shown on p. 173, or by an apron trim covering the tops of the shingles.

If you need to angle-cut shingle butts for use along gable-end walls and dormers or need to angle-cut shingle tops to fit under rake trim, use an adjustable bevel to capture the roof angle and transfer it to shingles. Such angle cuts are best made all at once, on the ground, using a tablesaw or miter saw. To notch shingles around windowsill ears and the like, use a cordless jigsaw. Finally, leave a ½-in. to 1-in. gap beneath dormer-wall shingles and adjacent roofing; otherwise, shingles resting directly on roofing can wick moisture and rot (see p. 101).

INSTALLING CLAPBOARDS

The following discussion assumes that you've read this chapter's earlier sections on layout and that you've installed door and window flashing, building paper, and exterior trim. It also assumes that the building has corner boards that you can

PRO TIP

Clapboards have a planed, smooth front side and a rougher back side. Using 100-grit sandpaper, lightly sand the smooth side to help finish coats of paint adhere, even if the clapboard is preprimed.

Rain-Screen Walls

In humid regions where housewrap and back-primed siding are not enough to pre-vent rotted siding, some builders have retrofitted rain-screen walls to remedy paint failure and soaked sheathing. Basically, rain-screen walls employ furring strips to space clapboards out from the building paper, allowing air to circulate freely behind the siding and dry it out. The reasoning is sound, and field reports are encouraging.

As sensible as this solution is, it's not for every renovation. Rain screens require careful detailing and a skillful crew. For example, if furring strips raise the siding roughly 3/8 in., existing trim needs to be oversize already (5/4 stock) or built up to compensate for the increased thickness of siding layers. Another option: Home Slicker or Cedar Breather, a thin layer of nylon mesh, keeps siding off the building paper.

Rain-screen walls allow air to circulate behind the siding and encourage drying. Here, thin wood furring strips raise the siding above the building paper; each strip is centered over a stud.

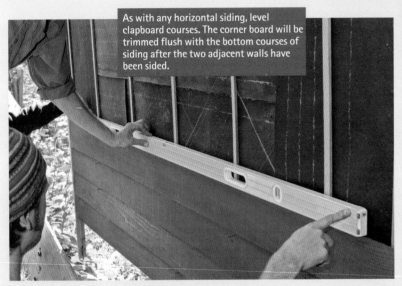

As with any horizontal siding, level clapboard courses. The corner board will be trimmed flush with the bottom courses of siding after the two adjacent walls have been sided.

Note the rain-screen corner stop. These clapboards were too thin (7/16 in., butt end) to miter, and the builder didn't want corner boards. So a clever carpenter fashioned this corner stop using his tablesaw. The stop legs are the same thickness as the furring strips affixed to the wall studs. To avoid splits, the corner stop was predrilled and attached with stainless-steel screws.

A layer of nylon mesh over the building paper allows better air circulation under wood shingles and siding. The mesh and building paper overlay the cap flashing.

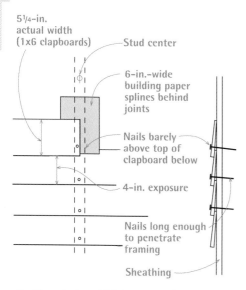

5¼-in.
actual width
(1x6 clapboards)

Stud center

6-in.-wide
building paper
splines behind
joints

Nails barely
above top of
clapboard below

4-in. exposure

Nails long enough
to penetrate
framing

Sheathing

Position clapboard joints over stud centers. For the most weathertight joints, bevel-cut ends. Note: For clarity in this drawing, building paper between clapboards and sheathing isn't shown.

Shiplap and other horizontal lapped sidings are less likely to cup if nailed to stud centers—indicated here by the vertical red chalkline. Install strips of building paper or waterproofing membrane behind joints to keep water out.

The Science of Nailing Clapboards

Even though nailing up clapboards isn't rocket science, four carpenters will give you five opinions on how to do it. Here's what you need to know:

▶ Nail roughly 1¼ in. from the bottom so nails don't pierce the upper edge of the clapboard underneath, especially if you're installing 1x8 or wider clapboards. Wide clapboards are more likely to split if they are inadvertently nailed at top and bottom.

▶ Nail clapboards to stud centers. For guidance, snap vertical chalklines on the building paper over the stud centers; offset the clapboard butt joints by at least 32 in. Lined-up nails look better, especially if you're using a clear finish.

▶ If you'll be painting the clapboards, take note. Carpenters set nails; painters fill and paint them. Setting nails makes painters cranky, and if there are 1,000 to set, painters will miss some.

butt the clapboards to. Otherwise, the clapboard corners will require compound-miter joints, which require a fair amount of work and skill.

Materials. Clapboards are a beveled siding milled from redwood, red cedar, or spruce; for best results, use Grade A or better. Preprimed finger-jointed clapboards, made from shorter lengths of high-quality wood, are a cost-effective alternative. Clapboards come in varying widths and thicknesses, but all are nominally 1-in.-thick boards that have been planed down. Thus, a 1x6 is actually ⅜ in. thick (at the butt) by 5¼ in. wide; a 1x8 is actually ½ in. by 7½ in., and so on. Traditional clapboards come in varying lengths, whereas finger-jointed products are manufactured in 16-ft. lengths; all are sold by the lineal foot.

To estimate the amount you need, calculate the square footage of your walls, less window and door openings. Then consult the table on p. 182, which assumes ½ in. of overlap for 1x4 clapboards and 1-in. to 1¼-in. overlap for all other sizes. It also factors in 5% waste. Order preprimed (or prestained) clapboards. Prepriming seals out moisture, saves tons of time, and keeps the job moving. You will need a small amount of primer on hand to touch up newly cut ends.

Nails. Buy 5d stainless-steel, ring-shank siding nails, whether you're painting the clapboard or not. Stainless-steel nails cost four to five times as much as galvanized nails, but it's worth the peace of mind. Galvanized nails are fine 99% of the time, but if the coating breaks off, the nail will rust. Moreover, the tannins in cedar and redwood can chemically react with galvanization, which can cause staining. Same with galvanized staples. For every 1,000 lin. ft. of siding, buy 5 lb. of 5d nails.

PRO TIP

If painted clapboards outside kitchens and bathrooms chronically peel, excess interior moisture may be migrating through the siding. Add an exhaust fan, and drive plastic shims under clapboards in the affected area so moisture can exit. It's worth a try.

Factory-Finished Siding

If you're thinking of installing cedar or fiber-cement siding, consider having it factory finished. You can choose virtually any color, have paint applied uniformly to all sides of the siding in a dust-free setting, then have it delivered ready to install. You'll only have to make occasional touch-ups and seal cut ends, cutouts, and such. The price for the whole labor-saving operation is often less than it would cost you to pay someone to paint it, and finishes routinely come with a 25-year warranty against paint failure.

A lumberyard or paint manufacturer should be able to recommend a factory finisher. This recommender can help specify the siding materials and the particulars of the paint job, including price, and will be a valuable ally in the unlikely event that you need to make a claim on the warranty. Regarding specs, wood siding usually gets an alkyd oil primer to prevent bleed-through, and fiber-cement siding gets primed with latex. For all siding types, there are acrylic-latex finish coats.

In addition to its durability, factory-finished siding is predictable. The cost of a job is based solely on the amount of siding being painted and the number of coats. You won't pay more if your house has one story or three because there's no painters' scaffolding required. And the caveats are relatively minor. To minimize retouching, installers must take great care when handling the siding. New cuts must be painted immediately after cutting so moisture doesn't enter the dry siding. And color-matched siding nails are slightly more expensive. But pay attention to the details, and you may not need to paint again for a quarter century.

Homemade Clapboard Gauge

Scribe this line.

To fit clapboards closely to jamb casing, make a gauge from 1-in. stock. Slip the gauge over the clapboard, slide it next to the casing, and mark the casing edge onto the clapboard.

Installation. It's worth repeating: Standard clapboard exposure is 4 in. for 1x6 clapboards (actual width, 5¼ in.), but you may want to vary that exposure by ¼ in. or less between courses to help align the clapboards with the window and door casings.

The first course of clapboards typically sits atop a water table. First, flash the top of the water table with metal drip-edge to prevent rot. To establish the correct pitch for that first course, rip a 1¼-in.-wide beveled starter strip from the top of a clapboard. (Save the 4-in.-wide bottom waste rip for the top of a wall.) Tack the strip atop the water table, and you're ready to nail up the first course. The water table may not be level, but that's OK; it's better to avoid a noticeable gap between a level first course and an off-level water table. In that case, take pains to level the second and all successive courses.

Start at one corner board and work all the way across the wall, nailing clapboards to each stud center they cross. Use only one nail at each stud crossing, locating nails just above the top of the clapboard below, as shown on p. 181. (If you nail through the clapboard below, you'll split its thin, tapered edge.)

All butt joints should be square and centered over a stud so that the ends of both boards can be securely nailed. Remember to stagger joints by at least 32 in. To further weatherproof the joints, back them with strips of building paper; the paper overlaps the top of the clapboard beneath by ½ in.

For clean, square cuts, rent or buy a 10-in. sliding miter saw with a 40-tooth or 60-tooth carbide-tipped blade. And be prepared to recut joints. When fitting the second board of a butt joint, leave it a little long until you're satisfied with the joint. If it isn't perfectly square on the first try, you'll have excess to trim.

When butting clapboards to corner boards and jamb casings, use the homemade gauge shown at left. Using the gauge to hold the clap-

Clapboard Needed to cover 100 Sq. Ft.

CLAPBOARD SIZE	LINEAL FEET
1x4	440
1x6	280
1x8	200
1x10	160

board tight to the trim, scribe the cut-off line with a utility knife. Never fit clapboards so tightly to the casing that you need to force them into place. If you cause window jambs to bow in, sashes might bind. Where top courses abut the underside of eave or rake trim, rabbet or build out the trim to receive the top edges of the clapboards, as illustrated on p. 173. Caulk all building joints well with latex acrylic or urethane caulk before nailing up the top course of clapboards.

Fiber-Cement Siding

One of the first engineered building materials, fiber-cement (FC) siding has been around since the 1940s. It fell out of favor after one of its components, asbestos, was found to be a carcinogen. However, new fiber-cement products are asbestos-free, consisting of cellulose fiber (wood pulp), sand, and portland cement. Available in a wide range of sizes and textures, fiber-cement products now include trim boards, clapboard siding, and panels manufactured to simulate wood-shingle courses. All can be factory-finish painted in the color of your choice.

Managing the silica dust created by cutting fiber cement continues to be *the* workplace issue, but better respirators, eye protection, and power saws with integral dust catchers have made FC much safer to work with. Fiber-cement siding paints like a dream, is impervious to insects and rot, and won't burn. Once installed, it is stable, durable, and virtually maintenance-free. Moreover, it's about one-third the cost of red cedar. Given the declining quality and availability of lumber—and the energy required to ship huge blocks of cedar to China to be milled into siding—one can argue that fiber cement is a viable green alternative to wood siding.

Fiber-cement clapboards come in planks 5¼ in. to 12 in. wide and from 5⁄16 in. to 5⁄8 in. thick. Because FC is roughly three times the weight of wood, 12-ft. planks are an optimal length to work with.

WORKING WITH FC CLAPBOARDS

Unlike wood clapboards—which are lightweight, long-fibered, and springy—fiber-cement siding is heavy, short-fibered, and brittle. So although its layout is essentially the same as that described for wood clapboards, handling fiber-cement planks is quite different.

Whenever possible, carry fiber-cement boards on edge—with the width of the board perpendicular to the ground. Boards will flex less and crack less if carried that way. For the same reason, always have help when handling or installing long boards—it's not a one-man job. Especially

Because fiber-cement siding and trim stock are thin and brittle, carry planks on edge to avoid cracking them. It's also advisable to have two workers handling planks, especially when cutouts further weaken them.

after it's been notched or cut out, a fiber-cement clapboard is likely to snap if handled incorrectly.

Most fiber-cement clapboards don't taper and are rather thin—typically 5⁄16 in. thick. That they don't taper is something of an advantage, however, because it creates a triangle of open space behind each course's overlap through which air can circulate and keep things dry. Consequently, it's not essential that you install FC boards over a rain screen. Fiber-cement clapboards typically overlap the board below by 1¼ in., so nail each clapboard about 1 in. below its top edge so nail heads will be covered. *Note:* Use only one nail per stud crossing, preferably 1¾-in. large-head nails, which look somewhat like roofing nails.

Another important detail: Leave ⅛-in. gaps between the butt ends of boards to allow for expansion and contraction. Behind each butt joint, affix a piece of flashing tape to the building paper, then caulk the joint with a siliconized acrylic caulk, which is paintable and typically comes with a 45-year rating.

SPECIALTY TOOLS

In theory, you can cut fiber-cement planks with hand tools and all but eliminate harmful dust. But if you're siding a whole house, a few specialized power tools will speed the job while greatly reducing the dust.

A circular saw with a dust-collection container will ideally have a housing over the blade, with a dust port at the rear that connects to a large canister vacuum. Makita has a 7¼-in. corded fiber-cement saw that is among the priciest available

PRO TIP

Keep fiber-cement siding covered. It's cement based and will absorb water, even when preprimed. Moreover, if you install it wet, the outside will dry faster than the back, which could result in paint bubbling later on.

When you get fiber-cement siding delivered to a job site, ask the supplier to unload the truck by hand rather than rolling the load off, which could crack the clapboards. If your shipment is too large for hand-unloading, ask the supplier to put a layer of framing lumber—2x8s or 2x12s—under the banded bundles to protect the siding as it rolls off the truck.

Minimizing airborne silica dust is the key to working safely with fiber cement. Specially designed to contain such dust, this circular saw features a plastic cover with a back port that connects directly to a canister vacuum. Always wear a respirator and safety glasses to further reduce exposure.

Use electric shears—also called nibblers—for cutouts in fiber cement. Cordless shears are especially helpful when you're working up on scaffolding, angle-cutting boards to fit under rake trim, and you don't want to climb down to use a circular saw.

Once you've correctly set the air pressure of a pneumatic coil nailer, you'll be less likely to crack brittle edges. When using a pneumatic nailer, it's not necessary to predrill siding nail holes.

but also one of the most effective in collecting dust. Of course, you should still wear a respirator while using it. FC circular-saw blades typically have four to eight polycrystalline diamond teeth, which are far more durable than carbide-tipped blades.

Electric or pneumatic shears are another must-have tool for cutting curves or inside corners in fiber-cement siding. Because shears (also called

nibblers) don't create much dust, some installers argue that you can do without a circular saw. But I beg to differ: A nibbler can cut across a plank, but a circular saw is faster and yields a cleaner edge. For those not siding a whole house, the Malco TurboShear attachment converts most drill/drivers into FC shears for about half the cost of dedicated shears.

A pneumatic coil nailer designed for fiber cement is a big-ticket item, but it's worth the money on a whole-house job. This nailer is also a great asset when working with fiber cement: Once you've established the correct air pressure, you're unlikely to crack this brittle siding. Moreover, although predrilling is recommended when you're hand-nailing FC siding, that's not necessary when pneumatic-nailing if the air pressure is correct and you place nails far enough back from the edge. Pneumatic nailers also have exposure gauges that enable you to quickly position siding courses before nailing them up, much as roofers set shingle courses (see p. 91).

Miscellaneous tools include carbide-tooth hole saws for dryer-vent or electrical-box cutouts and a carbide scoring knife or hand snips for incidental cuts when you don't feel like going back down the ladder to get your shears. You should also have an N95 NIOSH-certified respirator whenever you're using power tools with FC siding.

Stucco Repairs

Stuccoing a whole house requires skills that take years to learn, but stucco repairs are well within the ken of a diligent novice. If you spend a few hours watching a stucco job in your neighborhood, you'll pick up useful pointers.

A BASIC DESCRIPTION

Stucco is a cementitious mix applied in several layers to a wire-lath base over wood-frame construction or to a masonry surface such as brick, block, or structural tile. Like plaster, stucco is usually applied in three coats: (1) a base (or scratch) coat approximately ⅜ in. thick, scored horizontally to help the next coat adhere; (2) a brown coat about ⅜ in. thick; and (3) a finish coat (called a dash coat by old-timers) ⅛ in. thick. For repair work and masonry-substrate work, two-coat stucco is common.

The mix. The mix always contains portland cement and sand, but it varies according to the amount of lime, pigment, bonders, and other agents. See "Stucco Mixes" at right, for standard mixes. The consistency of a mix is easy to recognize but hard to describe. When you cut it with a shovel or a trowel, it should be stiff enough to retain the cut mark yet loose enough so it slumps into a loose patty when dropped from a height of 1 ft. It should never be runny.

Building paper. Stucco is not waterproof. In fact, unpainted stucco will absorb moisture and wick it to the building paper or sheathing underneath. Always assume that moisture will be present under stucco, and apply your building paper accordingly.

Basically, you want to cover the underlying sheathing with two layers of building paper before attaching the metal lath. Two layers of Grade D building paper will satisfy most codes, but you're better off with two layers of a fiberglass-reinforced paper such as Super Jumbo Tex 60 Minute. Although 60-minute paper costs more, it's far more durable. Typically, the stucco sticks to the first layer of paper, exposing it to repeated soakings until it largely disintegrates; the second layer is really the only water-resistant one, so you want it to be as durable as possible.

Stucco Details

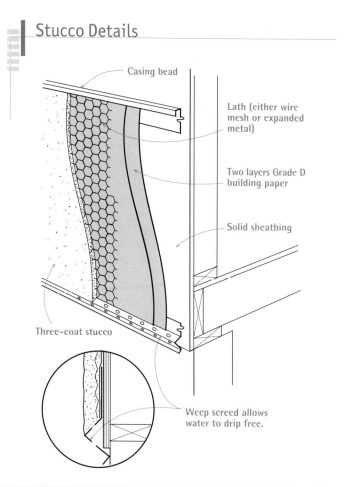

Casing bead

Lath (either wire mesh or expanded metal)

Two layers Grade D building paper

Solid sheathing

Three-coat stucco

Weep screed allows water to drip free.

Stucco Mixes

PORTLAND CEMENT	MASONRY CEMENT	LIME	SAND
1 part	—	¼ to 1 part	3¼ to 4 parts
1 part	1 part	—	3¼ to 4 parts

The perfect stucco mix is stiff enough to retain a trowel mark yet loose enough to slump into a loose patty when dropped from a height of 1 ft. This mason is using his trowel to load stucco mix onto his hawk, which he'll carry to the wall. With a trowel, he'll apply the mix in sweeping motions.

Take care not to tear the existing paper around the edges of a patch. Tuck the new paper under the old at the top of the patch, overlapping old paper at the sides and bottom of the patch. If the old paper is not intact or the shape of the patch precludes an easy fit, use pieces of polymer-reinforced flashing strips as "shingles," slipping them up and under the existing stucco and paper and over the new. Caulk new paper to old at the edges to help keep water out.

Lath. Metal lath reinforces stucco so it's less likely to crack, and it also mechanically ties the stucco to the building. Lath is a general term; it encompasses wire mesh or stucco netting (which looks like chicken wire) and expanded metal lath (heavy, wavy-textured sheets). When nailing up wire mesh, use galvanized furring nails with a furring "button" that goes under the mesh. When you drive these nails in, you pinch the wire mesh between the nail head and the button, creating a space behind the mesh into which the scratch coat oozes, hardens, and keys. Don't use aluminum nails because cement corrodes them. Use about 20 nails or staples per square yard of lath, spacing nails at least every 6 in. Overlap mesh at least 2 in. on vertical joints, and extend it around corners at least 6 in.

Expanded metal lath is a thicker, stronger lath used in situations requiring greater strength—for example, to cover soffits, where you're fighting gravity while applying stucco. That is, expanded metal lath won't sag. It typically comes in 2-ft. by 8-ft. sheets, is somewhat more work to install, and costs more. Expanded metal lath is stapled or nailed up; there's no need for furring nails because it's self-furring.

Base coat. Here's how to apply the base (scratch) coat:

1. Cover the sheathing with building paper and attach the lath.

2. Establish screed strips, which are guides for the stucco's final surface thickness. Screeds can be existing window edges, corner boards, or strips manufactured for this purpose.

3. Mix and trowel on a thick first coat, pressing it to the lath.

4. When the mud has set somewhat, screed it (meaning get it to a relatively uniform thickness) using screed strips as thickness guides.

5. Even out the surface further with a wood- or rubber-surfaced float.

6. Press your fingertips lightly against the surface; when it is dry enough that your fingers no longer sink in, steel-trowel the surface. Steel troweling compacts the material, setting it well in the lath and driving out air pockets.

7. Scratch the surface horizontally.

Brown coat. Installing the brown (second) coat requires the most skill, care, and time because this stage flattens the surface and builds up the stucco to within ⅛ in. of its final thickness.

Applying the brown (second) coat takes a lot of skill. The mason starts with an irregular scratch coat and builds it up until the surface is flat and within ⅛ in. of its final thickness.

After applying the rough, brown coat with a hand trowel, the mason steel-trowels the surface to make the thickness uniform and the surface relatively flat. After the mortar sets, he'll rough up the surface slightly with a wood- or rubber-faced float.

Diagnosing STUCCO PROBLEMS

Here are some common symptoms of stucco problems and their probable causes.

▶ **DIAGONAL CRACKS** from the corners of door or window trim indicate a shifting foundation; call a structural engineer.

▶ **CRUMBLING STUCCO** along the base of a wall suggests standing water and probable deterioration of the sheathing. Cut back the damaged area and install a weep screed so water can exit.

▶ **LARGE PATCHES FALLING OFF** suggest faulty flashing, rotted sheathing, or lath that's insufficiently attached. *Note:* If you see a shiny surface beneath a section of stucco that fell off, the installer likely waited too long to apply that coat, mistakenly letting the bonder dry. In this case, roughen the surface with a chisel before applying new bonder and a stucco patch.

To apply the brown coat, trowel on the stucco, screed it to a relatively uniform thickness, float the surface further, and steel-trowel to improve the uniformity. Then roughen the surface slightly with a wood or rubber float. As the stucco sets up, you will be able to work it more vigorously to achieve an even, sanded texture that will allow the finish texture coat to grab and bond. Do not leave the brown coat with a smooth, hard-troweled surface. Otherwise, the finish coat won't stick well.

Finish coat. The finish coat is about ⅛ in. thick and textured to match the rest of the building. When attempting to match an existing texture, you may need to experiment. If at first you don't achieve a good match, scrape off the mud and try again until you find a technique that works. Textures are discussed at the end of this stuccoing section.

Stucco trim. If doors or windows in stucco walls are cased with wood trim, flash head casings with flashing tape. Metal windows in stucco usually have no casing to dam up water and need no head flashing; metal windows usually have an integral nailing flange that serves as flashing after being caulked. If you need to cut back stucco siding to repair rot or install a new window, install flashing tape over the head casing, and install polymer-reinforced flashing strips along the sides and under the sill.

HELPFUL MATERIALS

The following materials are particularly useful for repair work and are available from any masonry supplier.

▶ *Weep screed* is a metal strip nailed to the base of exterior walls, providing a straight edge to which you can screed stucco. Because it is perforated, it allows moisture to "weep" or migrate free from the masonry surface, thus allowing it to dry thoroughly after a rain. Weep screeds are an easy way to make the bottom edge of stucco look crisp and clean. And because the weight of the stucco flattens the screed down against the top of a foundation, the screed provides a positive seal against termites and other pests. Weep screed is also a good solution for the frequently rotted intersection of stucco walls and porch floors. Weep screed isn't difficult to retrofit, but you'll need to cut away the base of walls 6 in. to 9 in. high to flash the upper edge of the screed strip properly.

▶ *Wire corners* are preformed corners (also called corner aids) that can be fastened loosely over the wire lath with 6d galvanized nails. Set the corner to the finished edge, taking care to keep the line straight and either plumb or level.

▶ *Latex bonders* resemble white wood glue. They are either painted into areas to be patched or mixed into batches of stucco and troweled onto walls. To ensure that the mixture is uniform, stir the bonder into water before mixing the liquid with the dry ingredients. To help a patch adhere well, brush bonder full strength all around the edges of the hole or crack you'll fill with new stucco. Merely applying patch stucco without bonder creates a "cold joint," which is more likely to fracture.

▶ *Prepackaged stucco mix* is helpful because it eliminates worry about correct proportions among sand, cement, and plasticizing agents. However, you need to add bonder to the mix.

▶ *Elastomeric finish coats* for stucco are acrylic-based and so, unlike paints, they are more flexible, less likely to crack, and their porosity allows moisture to escape. They come in a bucket, premixed—typically with sand added to match stucco texture—and are available in a wide range of rich colors. Acrylic finish coats are somewhat more expensive than a regular cementitious coat, but because they come precolored, you won't have to hire a painter after the finish coat of stucco has been

PRO TIP

Two ways to reduce stucco shrinkage cracks: (1) Add nylon fiber to the scratch coat. (2) After the top stucco coat has cured at least three weeks, paint it with an elastomeric paint, which flexes as materials expand and contract.

Cut back stucco to replace and correctly flash windows. Here, the roof leaked because the corner of the window was too close to the roof, where a lazy roofer just stopped the step-flashing. The replacement window won't be as tall, so its sill will be at least 4 in. above the roof.

applied—so you can actually save some money.

▶ *Color-coat pigment* (also known as LaHabra color or permanent color topcoat), is a cementitious finish coat available in a limited range of colors. However, precolored topcoats usually aren't of much use to renovators because their limited colors aren't likely to match older colors on a house.

▶ *Masonry paint and primer,* which is alkali-resistant, can be used on any new masonry surface. You should still wait at least two to three weeks for the stucco to "cool off" before painting it. (Follow the manufacturer's recommended wait times.) Use two coats of primer and two coats of finish paint.

THE REPAIR

Before repairing damaged areas, first diagnose why the stucco failed. Then determine the extent of the damage by pressing your palms firmly on both sides of the hole or crack. Springy areas should be removed. Continue pressing until you feel stucco that's solidly attached. When removing damaged areas, be deliberate and avoid disturbing surrounding intact stucco. Avoid damaging existing lath so you can attach new lath to it. Also avoid ripping the old building paper if possible. *Safety note:* Whether removing old stucco or mixing new, wear eye protection, heavy leather gloves, and at least a paper dust mask.

You can use a hammer and a cold chisel to remove a small section of damaged stucco. But for larger jobs, rent an *electric chipping hammer* with a chisel bit. *Important:* The bit should just fracture the stucco, not cut through it. Ideally, the underlying wire mesh and building paper will remain undamaged.

Let's say you're removing stucco to expose a rotted mudsill. Using a chipping hammer, fracture the stucco surface in two roughly parallel lines. On the first pass, delineate the top of the stucco to be removed. Then make a second pass 6 in. lower. Basically, you'll eventually demolish all the stucco below the top cutline and restucco it after you replace the mudsill. But if you carefully remove the top 6 in. of the damaged stucco, you'll preserve the wire mesh in that section, giving you something to tie the new mesh and stucco to.

Cut through the wire mesh exposed by the second pass, insert a pry bar under the stucco, and pry up to detach the stucco from the sheathing. Because the first pass of the chipping hammer separates damaged stucco from intact stucco,

prying up this 6-in. corridor of stucco will not disturb the solid areas above it.

Now the strenuous work begins. Using a mason's hammer or a beat-up framing hammer, carefully pulverize the corridor of pried-up stucco. Stucco is hard stuff, so whacking it in place with a hammer is more likely to drive it into the wood sheathing than to pulverize it. However, if you pry up the stucco slightly, you can slide a hand sledge under it to serve as an anvil. Then, between a hammer and a hard place, the stucco will shatter nicely.

After you've removed all old chunks of stucco, you'll have a 6-in.-wide section of unencumbered wire mesh and, ideally, a layer of largely intact building paper under that. Once you've replaced the rotted framing (and sheathing), insert new paper, tie new mesh to the old—just twist the wire ends together—and nail both to the sheathing. Now you're ready to apply the new scratch coat.

Used correctly, an electric chipping hammer fractures stucco without destroying the underlying wire mesh and waterproofing membrane.

To remove stubborn chunks of old stucco from the wire mesh, pry up the mesh enough to slide the head of a hand sledge underneath so it can serve as an anvil. Hammered stucco will then pulverize. Wear eye protection and a paper respirator mask.

If only the finish coat is cracked, wire-brush and wet the brown coat, apply fresh bonding liquid, and trowel in a new finish coat. But if the cracks are deeper, the techniques for repair are much the same as those for patches, except that you need to undercut the cracks. That is, use a cold chisel to widen the bottom of each crack. This helps key in (hold) the new stucco. Chip away no more than you must for a good mechanical attachment. Then brush the prepared crack well with bonding liquid.

TEXTURING AND FINISHING

To disguise new stucco patches, it's often necessary to match the texture of the surrounding wall. Before texturing the finish coat, steel-trowel it smooth and let it set about a half hour—although the waiting time depends on temperature and humidity. Cooler and more humid conditions delay drying. Stucco allowed to cure slowly is far stronger than fast-cured stucco. So, after applying the finish coat, use a hose set on a fine spray to keep the stucco damp for 3 days. Here are descriptions of the four common textures:

▶ *Stippled.* For this effect, you'll need rubber gloves, an open-cell sponge float like those used to spread grout in tiling, a 5-gal. bucket, and lots of clean water. After dampening the sponge float, press it into the partially set finish coat and quickly lift the float straight back from the wall. As you lift the float, it will lift a bit of the stucco material and create a stippled texture looking somewhere between pebbly and pointy. Repeat this process over the entire surface of the patch, feathering it onto surrounding (old) areas as well to blend the patch in. Rinse the float often. Otherwise, its cells will pack with stucco, and the float won't raise the desired little points when you lift it. Equally important, the sponge should be damp and not wet. If you want a grosser texture than the float provides, use a large open-cell natural sponge. If you notice that the new finish is pointier than the old surrounding stucco, that's probably because the old finish has been softened by many layers of paint. To improve the match, knock the new texture down a little by lightly skimming it with a steel trowel.

▶ *Swirled.* Screed (level) the patch's finish coat to the surrounding areas, and feather it in. After you've made the patch fairly flat, comb it gently with a wet, stiff-bristled brush. For best results, use a light touch and rinse often; otherwise, you'll drag globs of stucco

After completing repairs, texture the wall to blend the patch to the surrounding stucco. Here, plastering cement, LaHabra color, and water were mixed repeatedly, poured into a hopper, then sprayed onto the wall to create a stippled effect. (This look is also known as "spray dash.")

out of the hole. By varying the pressure on the brush, you can change the texture.

▶ *Spanish stucco or skip troweled.* This texture looks rather like flocks of amoebas or clouds. To achieve this look, screed off the patch so it's just 1/16 in. below the level of surrounding areas. Then, using a steel trowel, scoop small amounts of stucco off a mortarboard and, with a flick of the wrist, throw flecks at the wall. Skim the flecks with a swimming pool trowel because its rounded edges are less likely to gouge the stucco as you flatten the flecks slightly. As the trowel spreads the material, it skips over the sand particles, leaving flattened patches of texture and gaps.

▶ *Sand float* is achieved by rubbing the final coat with a rubber float. This creates an even, sandy texture that looks rather like suede—even from 20 ft. away. It's a very popular texture.

Rain Gutters

Gutters direct water away from the house, preventing water from collecting next to the foundation and thereby possibly undermining it. The two most common gutter profiles are half-round and K-style, in which the gutter has a squared-off back and an ogee front. For appearance's sake, try to match the profile of new gutters to old.

To clear water adequately, gutters must be sized properly and cleared of leaves and debris twice a year—in spring and in fall. Your lumber-yard probably has elaborate gutter-sizing charts based on regional rainfall, roof square footage, and pitch. But you might remedy chronically overflowing downspouts simply by upsizing the gutters from a standard 5-in. width to a 6-in. model and installing larger downspouts. To keep roof runoff from running behind gutters and rotting fascia, extend the roof drip-edge flashing so it overhangs the gutter.

If gutters are spaced too far from the roof edge or slope away from the drip-edge, you can place an L-shaped piece of metal flashing over the back of the gutter and tuck its upper edge up under the drip-edge. (Notch that flashing so it fits around the roof hangers.) Some K-style gutters also come with an integral flange that runs under the roofing and serves the same function as a drip-edge.

Begin any gutter installation by checking the slope of roof edges and trim and then measuring the areas to receive gutters. Typically, gutters extend beyond the roof section 1/2 in. on each end. As you reconnoiter, consider where downspouts will be least obtrusive.

MATERIALS

Metal gutters can be fabricated on site by a gutter specialist with a mobile machine. Or you can assemble them from 10-ft. or 20-ft. prefab lengths, using pop rivets and exterior caulking. Plastic gutter sections need cementing. Gutter runs that are longer than 40 ft. need an expansion joint to keep them from buckling. Here are the materials most commonly used.

Aluminum. By far the most popular, aluminum resists corrosion, is easily worked, is reasonably priced, and is durable—though it will dent. It comes prepainted in a range of colors. Standard thickness is 0.028 in., but spending a little more for 0.032 in. is prudent, especially if heavy snows and ice dams are common in your area.

Galvanized steel. Stronger and harder to dent than aluminum, galvanized steel rusts if you don't keep it painted. There are prepainted varieties, generally in the same colors as painted metal roofs. The minimum thickness is 26 gauge.

Copper. Handsome when new, copper acquires a beautiful green patina as it weathers. It's malleable, durable, and about five times as expensive as aluminum. This gutter is usually formed from 16-oz. sheet copper. Copper resists salt air but may be corroded by cedar-shingle runoff.

Plastic. Plastic comes in 8-ft. and 10-ft. lengths, with matched fittings. You can join sections either with liquid cement or with neoprene gaskets. It's virtually maintenance-free and durable, if the plastic contains a UV inhibitor. Its expansion joints can accommodate a wide range of movement.

Gutter Lip and Roof Pitch

FLAT

1 in.

5/12

3/4 in.

Place gutters below the roofline, so they won't be torn off by sliding snow.

7/12

1/2 in.

12/12

1/4 in.

Most metal gutter stock is relatively lightweight, so it can be cut, drilled, and attached easily with a modest assortment of tools. When using a hacksaw on gutters, cut from the back and bottom sides forward, so you'll be cutting the shaped (thicker) edges last. Aluminum and copper are soft and easy to work; galvanized steel is more challenging.

Gutter Hangers

Gutter sections need to be supported by hangers at least every 32 in.; closer if there's a heavy snow or ice load. The many variations can be grouped into two general types: *roof mounted*, which employ a strap nailed to roof sheathing, and *fascia mounted*, which screw or nail directly to fascia boards or rafter tails. Whatever type hanger you use, gutters are less likely to pull free if you nail or screw the hangers to framing behind the fascia or roof sheathing. Here are profiles of four common hanger types:

▶ **Spike-and-ferrule hangers** nail directly into rafter ends or through fascia boards. Although this is a simple system, its detractors point out that 7-in. spikes leave large holes, encourage rot, and—in the end—don't hold well.

▶ **Roof-mounted strap hangers** support gutters well and are an alternative to end-nailing rafters—in fact, they're the only option when there's no fascia. If you're reroofing, nail them to the roof sheathing and apply shingles over them. Or, if the rafter tails are exposed, nail the straps atop the rafters, and install flashing over the straps to forestall rot.

▶ **Hidden hangers** are favored for hanging K-style aluminum gutters. They can be inserted into the gutters on the ground and, thanks to integral screws, attached to the fascia one-handed. But because they clip inside the gutter channels —rather than supporting them underneath—these hangers are best used with heavy, 0.032-in. gutter stock, which is stiffer and less likely to flex or sag than the lighter stock.

▶ **Bracket hangers** are usually screwed to fascia boards. They range from plain 4-in. brackets that snap over the back gutter lips to cast-bronze brackets ornamented with mythical sea creatures. Brackets simplify installation because you can mount them beforehand—snap a chalkline to align them—and then set gutters into them.

Prefab Gutter Pieces

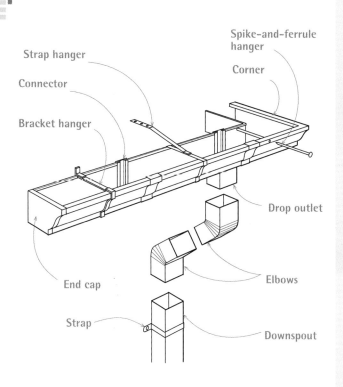

Prefabricated gutter pieces facilitate assembly.
Use the hanger type most appropriate to your eaves detailing.

Hidden Gutter Hanger

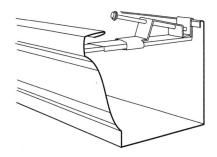

Hidden hangers are commonly used with K-type gutters. They can be prepositioned in the gutter and quickly screwed or nailed to the fascia. Because hangers are not visible, gutter lines are clean.

SLOPING AND PLACING GUTTERS

Ideally, gutters should slope down toward downspouts 1 in. per 16 ft., but this is not always possible. For starters, this may not look good: Next to a level fascia board, the steeper the slope of the gutter, the more it looks out of whack. As long as there is a slight pitch—say, ½ in. in 20 ft.—with no low spots en route to the downspout, a gutter will drain. If a house settles so that its roof edge or trim—and its attached gutters—slope away from downspouts, either install new gutters with downspouts properly located or reattach gutters so they slope toward existing downspouts.

Place the front lip of the gutters below the roof plane, low enough so the sliding snow won't tear them off yet high enough so the rain runoff won't overshoot them. The distance below the projected roof plane varies with pitch: For a gently sloped 5-in-12 pitch, place the front lip of the gutter ¾ in. below the projected roof plane; for a steeper 7-in-12 pitch, ½ in. below the plane; and for a 12-in-12 pitch, ¼ in. below the plane. The front lip of a gutter should always be about 1 in. lower than the back. That way, if the gutter overflows, water will spill over the front lip rather than soaking the fascia and siding behind.

INSTALLING GUTTERS

Measure the length of the roof the gutter will service, and check the fascia (if any) and roof edge for level. Try to place the downspouts in an inconspicuous place, away from foot traffic. In positioning downspouts, the biggest challenge is usually on the uphill side of the house, where downspouts often require underground drains to carry water beyond the outside house corners.

Next, determine where you want seams, which should also be placed inconspicuously. Because gutter stock comes in 10-ft. or 20-ft. lengths, it might look better to join a 15-ft. length and a 10-ft. length to achieve a gutter 25 ft. long, rather than tacking a 5-ft. length onto the end of a 20-ft. length, if that joint would be near the front door.

Although gutter sections are light enough for one person to carry, the job is safer and more predictable with two workers, especially if it's windy. Snap a light chalkline along the fascia to indicate the level of the hanger brackets, and install a bracket at either end of the roof. Alternatively, you can snap a chalkline to indicate the back lip of the gutter. If your gutter hangers fit into the gutters, position them before you carry gutter sections aloft; also, preassemble the end caps, downspout takeoffs, and so on.

Once you've secured either end of the gutter and checked its position, add hangers every 32 in. (every other rafter); in snow country, install a hanger every 16 in. The hangers you choose will determine exactly how you secure gutters. Most modular gutter systems can be cut to length with a hacksaw and joined with pop rivets or self-tapping sheet-metal screws. A disadvantage of screws: Their points protrude, snagging leaves and causing blockages. An advantage: Screws can be removed to disconnect the sections.

PRO TIP

Although it's desirable to secure the gutter hangers to the framing, you might not be able to see the rafter-center marks if you're holding the gutter section over the fascia. Use a builder's crayon to mark the framing centers along the edge of the roofing in advance.

Use a hole saw in a cordless drill to create openings for downspout outlets. In a pinch, you can also start holes by hammering an old chisel into the metal, then use aviation snips to complete the cutout.

A pop riveter is indispensable for joining gutter sections. Unlike screws, pop rivets don't intrude into the gutter or downspouts and won't snag leaves and cause clogs. Predrill pop-rivet holes.

Don't try this at home on a windy day. Pros know how to raise and attach long sections safely, but the rest of us will do well to have a helper or two.

After securing gutters, use screws to attach the downspouts to the outlets. Screws allow you to disconnect this joint later. Have the downspout brackets handy so you can immediately secure the downspouts to the siding.

Consequently, use rivets to join gutter sections, downspout outlets, and miter strips. And use the shortest screws feasible to join the downspouts to the gutter outlets. To avoid galvanic corrosion, use screws that are the same material as the gutter; otherwise, use stainless-steel screws. Because elbows slow water and tend to clog, use as few as possible. Apply gutter caulk freely to seal the joints, rivet holes, and the like. And where you see holes left by earlier gutter hangers, fill them with exterior wood filler or color-matched acrylic latex caulk.

GUTTER REPAIRS

If gutters are rusty but otherwise intact, use a wire brush to remove rust. Then rinse well and allow the gutters to dry. Paint gutters with an elastomeric roof coating such as GacoFlex acrylic latex, which can handle the expansion and contraction of metal gutters.

You may be able to get a few more years from metal gutters beginning to rust through by patching them with a compatible-metal patch. First, vigorously wire-brush the rusted area until you uncover solid metal, wipe the area clean with a rag damped with paint thinner, and prime with metal primer. After the primer dries, spread epoxy around the hole and press the patch into it. Or simply wire-brush the rusted area clean and apply a piece of foil-faced flashing tape, which is often used to flash skylights, plumbing pipes, and other roofing elements. Flexible flashing tape is easily shaped to the contour of a gutter.

Wooden gutters should be inspected every year for deterioration and repainted every two to three years. They must be thoroughly dry before painting; otherwise, paint will seal in moisture and promote rot. So it's best to paint gutters after a dry spell. Let morning dew evaporate.

Begin work by sanding the wood well and wiping away grit with a rag dampened with paint thinner. Next, apply a water-repellent preservative, prime, and apply two finish coats of paint. If you find rot, your problem is compounded if the gutter also doubles as exterior trim and abuts sheathing or framing. Short of replacing such integral gutters, you may be able to prolong their life by lining them with flashing tape.

Unless you're an experienced sheet-metal worker, buy preformed corners (such as this one) as well as preformed miter strips, and so on. After pop-riveting such connections, caulk them liberally, including the rivet holes.

8 Structural Carpentry

This chapter is mostly about wood, the king of building materials. Built amid virgin forests, the first wood houses were fashioned from massive ax-hewn timbers that took half a neighborhood to raise. Because iron was scarce, those great post-and-beam frames were joined without nails. Instead, they were fitted tightly and then fastened with whittled wooden pegs. The technology was crude, but the houses survived, in large part because of the mass and strength of wood.

Early in the 19th century came plentiful iron nails and circular-sawn lumber of uniform, if smaller, dimensions. Although such lighter components needed to be spaced closer than rough-hewn timbers, their reduced weight made it possible for three or four people to raise a wall. Balloon framing was the earliest of milled lumber houses, with long studs running the full height of the wall, from foundation to eave, and is rarely used today. Since the beginning of the 20th century, platform framing (also called western framing) has been the most widely used method. Here, each story is capped with a floor platform. Because the studs of a platform-framed house run only one story, they are shorter and easier to handle.

If there's room, assembling a wall on a flat surface and walking it upright is the way to go. This crew nailed restraining blocks to the outside of this second-story platform beforehand, so the sole plate couldn't slide off the deck.

Understanding Structure

A house must withstand a variety of loads (forces): the *dead load* of the building materials, the *live loads* of the people in the house and their possessions, and the *shear loads* from earthquakes, soil movement, wind, and the like, which exert racking (twisting) forces on a building. There are other, finer distinctions, including *point loads*, where concentrated weights dictate that the structure be beefed up, and *spread loads*, in which a roof's weight, say, pushes outward with enough force to spread walls unless counteracted.

Loads are transferred downward by *framing members*, primarily by exterior walls sitting atop a perimeter foundation and by interior *bearing walls*, often supported by a secondary foundation consisting of a girder, posts, and pads. Generally, a *girder* runs the length of the house and supports floor joists running perpendicular to it. *Nonbearing walls*, as their name denotes, are not intended to bear anything but their own weight. *Headers* (or lintels) are bearing beams that carry loads across openings in walls. A *partition* is any interior dividing wall, bearing or not.

Before you decide to demolish old walls or frame up new ones, determine what is a bearing wall and what is not. This will influence how you frame up, for example, the size of headers, whether you need *shoring*, whether you need additional support below the walls being removed, and whether you should disturb the structure at all. Get as much information as you can before you commit to a plan because there are always surprises once you start. If you plan to remove walls, be sure to hire a structural engineer to review your plans.

Framing Walls

Framing walls is arguably the most common carpentry task in renovation, and it employs a variety of layout and assembly techniques. For a deeper look at carpentry, consult Rob Thallon's *Graphic Guide to Frame Construction*, Mike Guertin and Rick Arnold's *Precision Framing*, and Larry Haun's *The Very Efficient Carpenter*, all published by The Taunton Press. Also highly recommended: Joseph Lstiburek's "The Future of Framing Is Here" in *Fine Homebuilding* issue #174.

Basically, wood-frame walls are an array of vertical studs nailed to horizontal top plates and sole (bottom) plates. Depending on whether walls are bearing or nonbearing, plates may be doubled. Although 2x4 walls are by far the most commonly framed, 2x6 walls allow thicker insulation and make routing pipes easier.

Exploring Your Options

To assess the framing hidden behind finish surfaces, go where it's exposed: the basement and the attic. Joists often run in the same direction from floor to floor.

Generally, a *girder* (also called a carrying timber or beam) runs the length of the house, with joists perpendicular to it. Some houses with crawlspaces may have framed *cripple walls* (short walls from the top of a foundation to the bottom of the first-floor joists) instead of a girder. Main bearing walls often run directly above the girder, but any wall that runs parallel to and within 5 ft. of a girder or cripple wall is probably bearing weight and should be treated accordingly.

Bearing walls down the middle of the house also are likely to be supporting pairs of joists for the floors above. That is, most joists are not continuous from exterior wall to exterior wall—they end over bearing walls and are nailed to companion joists coming from the opposite direction. This allows the builder to use smaller lumber—2x6s rather than 2x8s, for example—because they cover a shorter span. If you cut into a bearing wall without adding a header, the joists above will sag.

Large openings in obvious bearing walls are often spanned by a large beam or a header that supports the joists above. These beams, in turn, are supported at each end by posts within the wall that carry the load down to the foundation. These *point loads* must be supported at all times. Similarly, large openings in floors (stairwells, for example) should be framed by doubled headers and studs or posts that can bear concentrated loads.

Finally, it may be wise to leave a wall where it is if pipes, electrical cables, and heating ducts run through it. Look for them as they emerge in unfinished basements or attics. Electrical wiring is easy enough to remove and reroute—but disconnect the power first! However, finding a new home for a 3-in. or 4-in. soil stack or a 4-in. by 12-in. heating duct may be more trouble than it's worth.

HANDLING LUMBER

Here are a few lumber handling tips that will save labor and make the job go quicker.

Minimize moves. Lumber is heavy. Tell your lumberyard to load the delivery truck so that the lumber you'll use first—say, floor joists—will be on the top of the load. Clear a level place close to the work site where the truck can unload. Many suppliers have boom trucks that can unload lumber stacks directly onto a work deck.

Sort your lumber. Lumber today is often bowed, so eyeball each piece for straightness—and sort it into like piles. Save the straight stock for kitchen and bathroom walls—especially those that will get cabinets or be tiled—and for corners, top wall plates, and jack studs along rough openings. Pieces with slight bows (¼ in. in 8 ft.) can be used as studs and joists, but draw an arrow on the face of the lumber to indicate which way the lumber bows so you remember to place slightly bowed joists crown (bow) up, so they'll be less likely to sag under weight. Set aside stock that bows more than ¼ in. in 8 ft., which you'll cut shorter and use as headers, cripple studs, and blocking. Return corkscrew (twisted) studs to the lumberyard for credit.

Be methodical. Snap chalklines onto floors to mark wall plates. Cut top and bottom plates, then mark stud locations and ROs for doors and windows onto the plates. Make a cutting list: By cutting same-length lumber all at once you'll save a ton of time. If possible, cut lumber or plywood right atop the stack that the lumberyard truck delivered.

NAILING IT

On bigger projects, pneumatic nailers do most of the work, but it's worth knowing how to use a hammer correctly. Then you'll create fewer bent nails, splits, and dings (dented wood when the hammer misses the nail), and perhaps avoid a smashed thumb, tendonitis, and joint pain.

The perfect swing. If you're driving large nails such as 16d commons, start the nail with a tap. Then, with a relaxed but firm grip on the end of the hammer handle, raise the hammerhead high and swing smoothly from your shoulder. If you're assembling a stud wall, spread the pieces out on a deck, put one foot on the lumber to keep it in place, bend forward slightly, and let the falling hammerhead's weight do some of the work. Just before striking the nail, snap your wrist slightly to accelerate the swing.

However, if you're driving smaller nails (6d or 8d), you won't need as much force to sink them. So choke up on the handle and swing from the elbow. Choking up is particularly appropriate if you're driving finish nails because you'll need less force and be less likely to miss the nail and mar the casing. It's also helpful to choke up when there's not enough room to swing a hammer freely.

To nail in a hard-to-reach spot, start the nail by holding it against the side of the hammer and smacking it into the wood. Once the nail is started, you can finish hammering with one hand.

A blunted nail point is less likely to split wood because it will crush the wood fiber in its path rather than wedge it apart, as a sharp nail point does.

A wood block under your hammerhead makes nail pulling easier.

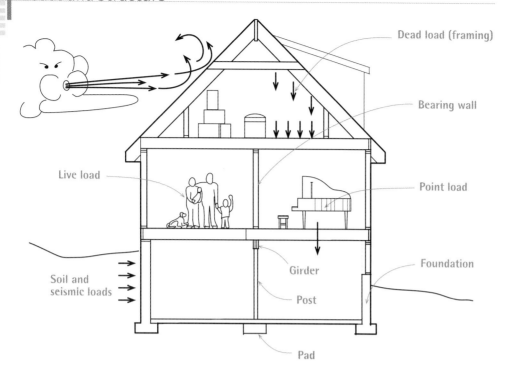

Dead load (framing)

Bearing wall

Live load

Point load

Girder

Foundation

Soil and
seismic loads

Post

Pad

NAILING Plywood SHEATHING

To nail plywood panels ½ in. thick or less, most codes specify 6d common nails spaced every 6 in. along the panel edges and every 12 in. in the field. Panels thicker than ½ in. require 8d common nails spaced in the same pattern. (Structural shear walls are typically nailed with 10d nails 4 in. to 6 in. on center along the edges and 12 in. on center in the field, but an engineer should do an exact calculation.)

Don't overdrive nails. Ideally, nail heads should depress but not crush the face ply of the plywood. Panel strength isn't affected if nail heads are overdriven by 1/16 in. or less, but if more than 20% of the nail heads are 1/8 in. or deeper, APA—The Engineered Wood Association recommends adding one extra nail for each two overdriven ones. Pneumatic-nailer pressure that's too high is the most common cause of overdriven nails. It's far better to set the nailer pressure so the heads are flush, and then use a single hammer blow to sink each nail just a little deeper.

Assembling a wall on the ground is no guarantee that lumber edges will line up. Here, a builder uses his hammer to raise the top plate flush to the header before nailing. It is advisable to wear safety glasses when using a pneumatic nailer.

The right nail. The nailing schedule given in "Recommended Nailing Schedule" on p. 198 suggests the size and number of nails you need for various framing tasks, but local building codes will have the final say. When joining two pieces of framing lumber, nails should be long enough to penetrate the second piece of wood, without sticking out the other side. Properly sized nails also are less likely to split board ends. As the table indicates, use two 16d nails to end-nail a stud through a plate, and four 8d nails, which are shorter and skinnier than 16d nails, to toenail stud ends to sole plates. To reduce splitting, hammer nail points to blunt them before driving them in.

Removing nails. Everybody bends nails now and then, especially when nailing at an odd angle or in a tight space or nailing into a hard wood like southern pine. To remove a bent nail, slip a block under a claw hammer head to increase the leverage as you pull out the nail; if the nail head is buried too deep to grasp with a claw hammer, use a cat's paw, which has pointed claws, to dig it out.

LAYING OUT WALLS

Wall layout varies, depending on whether you're erecting walls in open space (say, framing an addition) or within existing space (adding a partition). In both cases, use house plans to position walls. Snap chalklines onto subfloors (or floors) to indicate the location of sole plates. If you're building a partition, next measure from existing

Use a Speed Square to mark stud locations on a pressure-treated mudsill (sole plate) and a top plate. An X usually indicates regular studs 16 in. on center; a J, jack studs; a K, king studs; and a C, cripple studs. The sole plate has been predrilled so it will fit over the anchor bolts when the wall is assembled and lifted into place.

framing to determine the height and length of the new wall, as described in "Reinforcing and Repairing the Structure" on p. 211. Finally, mark the locations of ROs, wall backers (blocking you nail drywall corners to), and studs onto plates.

The easiest way to frame a wall is to construct it on a flat surface and tilt it up into place. Once the wall is lifted, align the sole plate to a chalkline on the floor. This construction method is arguably stronger because you can end-nail the studs to plates rather than toenailing them. End-nailing uses larger nails (16d) driven into the center of each stud, which will better resist lateral forces on a wall. Toenailing better resists uplift, but the four smaller nails (8d) angled through the corners of stud ends are more inclined to split the wood. These distinctions may be splitting hairs, however, because both methods are time-proven. In any case, in renovation, there's often not enough room to tilt up walls—which will be addressed later in this chapter.

Mark rough openings. Place the sole plate, face up, next to a chalkline, then place a top plate next to it so that edges butt together and the ends align. Use a square to mark the top plate and the sole plate at the same time. (If the top plate is doubled, there's no need to mark the upper top plate.) Using a tape measure, mark the ROs for doors and windows. Rough openings are so named because they are roughly 1 in. taller and wider than prehung doors or windows (so frame jambs can be shimmed snug), and thus 2 in. taller and wider than unframed units.

As you mark the width of the RO on the plates, keep in mind that there will be a king stud (full length) and a shortened jack stud (also called a trimmer stud) to support the header on each side of the opening. After marking the ROs, mark the corner backers (also called wall backers)—extra blocking for drywall where partitions intersect with the wall you're framing. "Corner-Stud Layouts" on p. 200 shows several backer configurations.

Mark studs on the plates. After marking the first stud, which is flush to the end of the plates, mark the subsequent studs ¾ in. back from the red 16-in.-interval highlights on your measuring tape. (In other words, mark stud edges at 15¼ in., 31¼ in., 47¼ in., and so on.) By marking stud edges ¾ in. back, you ensure that stud centers will coincide with the edges of drywall or sheathing panels, which are usually some multiple of 16 in., for example, 48 in. by 96 in.

Mark studs every 16 in. on center on plates—through the ROs as well—so that drywall or sheathing panels running above or below openings can be nailed to cripple studs at regular intervals. At window openings, you'll mark cripple studs on both the top and the bottom plates. But on door openings, you'll mark cripple studs on the top plates only. *Note:* For door openings, where 16-in. on-center studs occur within 2 in. of a king stud, omitting the 16-in. on-center stud will not weaken the structure.

HEADERS

Every opening in a bearing wall must have a header over it, and it's common practice to put headers in all openings in exterior walls, whether bearing or not. Headers for 2x4 walls are usually constructed by sandwiching a piece of ½-in. CDX plywood between two pieces of 2x lumber; as 2x lumber is actually 1½ in. thick, the header package is thus 3½ in. thick—the same as the width of a nominal 2x4 on edge. Headers must be able to carry a cumulative load and transfer it downward without flexing or pulling away from the sides of the opening. Thus headers must be sized according to the loads they carry and the distances they span.

That noted, many builders use this rule of thumb when sizing headers for single-story buildings with 2x4 walls and a 30-lb. live load on the

Stud-Wall Elements

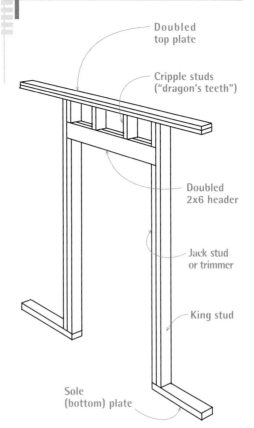

If you use doubled 2x6s for your header in a rough opening in standard 8-ft. wall framing, you need cripple studs to support the doubled top plate. If you use a 4x12 instead, it will support the top plates. Although code may allow a single top plate for nonbearing walls, a single plate offers little to nail drywall to if the ceiling is finished before the wall.

PRO TIP

When partitions run perpendicular to the joists, center the studs over the joists whenever possible. Aligning studs and joists creates straight, open channels from floor to floor, so plumbers and electricians can easily drill through plates and run wires and pipes. Partitions that ran parallel to and directly over the joists would be a big problem if you needed to run wires or pipes. Instead, consider moving the partition 1 in. to 2 in. to avoid the joist.

Standard FRAMING

For standard 8-ft. wall framing, cut studs 92½ in. long—or you can buy them as "precuts." One bottom plate and two top plates will be roughly 4½ in. thick, creating a wall height of 97 in. This height accommodates one drywall panel (½ in. to ⅝ in. thick) on the ceiling and two 4-ft.-wide drywall panels run horizontally on the wall. If you install a 4x12 header directly under the top plates of a standard wall, the rough opening height will be 82½ in. This is just right for a 6-ft. 8-in. door, and door head heights will match that of the windows.

Stud Layouts

STUD-AND-PLATE LAYOUT

Mark stud edges on plates so that stud centers will be spaced 16 in. o.c.

Distances (o.c.)

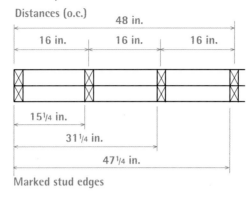

Marked stud edges

CORNER-STUD LAYOUTS

Corners require at least three studs to provide adequate backing for finish materials. In the first example, the middle stud need not be continuous, so you can use pieces.

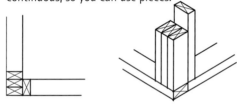

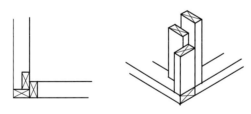

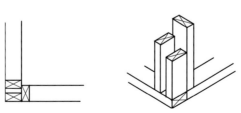

roof: The nominal depth of a 4x header in inches equals the span in feet of the opening.

Oversizing headers. Span tables establish minimum requirements. In the field, however, many builders often oversize headers—using 4x12s, say, to span all exterior openings—because it's quicker than building header "sandwiches" from two pieces of lumber and plywood between. Plus, you don't need to cut *dragon's teeth*, the short cripple studs between a header and the top plate. Quicker it may be, but it's an extravagant use of wood and it increases *thermal bridging* (see p. 204). That noted, oversize headers have some advantages:

▶ Same-size headers ensure that the tops of most exterior openings will be at the same height, which is aesthetically pleasing. (Again, you could achieve the same door and window height by using headers of different depths and cripple studs, but they'd take longer to assemble.)

▶ The additional cost of using an oversize beam is offset by the peace of mind it brings. That is, there won't be any cracks in finish surfaces caused by undersize beams.

▶ Even in nonbearing walls, the header is the weakest point structurally. Each time you shut a door, compressing the air in the room, the wall flexes a little. The more solid wood you've got to nail to, the stronger the connection. (Code requires at least five 16d nails through-nailed into each end of a header.)

Insulating headers. Wood is wonderful to build with but is a mediocre insulator. Consequently, many builders now use exterior foam sheathing and, whenever possible, try to slip some insulation into header assemblies. In a header for a 2x4 wall, this is easily done by replacing the ½-in. plywood with ½ in. of rigid foam insulation, typically expanded polystyrene (EPS) or extruded polystyrene (XPS).

Douglas Fir Header Spans

HEADER SIZE (in.)	SPAN (ft.)
4×4	4
4×6	6
4×8	8
4×10	10
4×12	12

Headers in 2x6 walls offer an opportunity to add insulation, but the perfect solution has not yet appeared. One can build a sandwich of three pieces of 2x lumber and two intervening ½-in. pieces of rigid foam to attain the 5½-in. thickness of a 2x6 on edge—and an R-5 value for the header. But that squanders wood. Alternatively, one could create a header sandwich of two pieces of 2x lumber with a foam center 2½ in. thick. That would give you a whopping R-12 for the header, but that header may be structurally suspect because the 2x lumber pieces on the outside of the sandwich aren't really tied together and so offer little resistance to lateral or racking forces. See what local building codes—and local builders—recommend.

ASSEMBLING THE WALL

After marking the top and bottom plates and cutting full-length studs, start assembling the wall. Place the plates on edge, roughly a wall height apart. Then insert the studs on edge between them. Again, use straight studs at wall ends and cabinet locations; elsewhere, place slightly bowed studs crown (bow) up so that stud ends will rest on the deck when it's time to nail them to the wall plates.

Nailing studs. Position studs to the squared marks along the plates. Then end-nail studs through the sole plate, using two 16d common nails at each end. Space nails ½ in. to 1 in. from the edge of the plate. If you stand on the stud as you nail it, it will stay put. As you nail, be sure that stud and plate edges are flush, or the resultant wall plane won't be flat. Once you've nailed all studs to the sole plate, nail them to the top plate. *Important:* If the sole plate will sit on concrete, it should be pressure-treated lumber. Moreover, all nails set into redwood or pressure-treated plates must be galvanized so the nails won't be corroded by chemicals in the wood.

Framing the rough opening. After cutting the header, end-nail the king studs (through the plates) on both sides of the RO. If you're installing a full-height header such as a 4x12, insert the header between the king studs, and nail down through the top plate into the header to draw it tight to the top plate. Then nail through the king studs into the ends of the header, using at least five nails per end. Next, determine the length of jack (trimmer) studs by measuring from the underside of the header to the top of the sole plate. Cut jack studs slightly long, tap them into place, and face-nail them to the king studs, making sure their edges are flush.

When nailing the top plate to the studs, lift the studs as needed so their edges align with the top plate. Flush studs and plates allow drywall finishing to go smoothly. This exterior wall will have a doubled top plate; the second plate is nailed on once the wall is up, tying this wall to another.

Because a 3x4 mudsill is too thick to end-nail through, toenail studs instead. Here, pneumatic nailers really shine: They nail so quickly that studs won't drift off stud marks, as they frequently do when you're hand-nailing them.

After using five 16d nails to end-nail the header through the king stud, face-nail the trimmer stud to the king.

Half-Cutting the Sole Plate

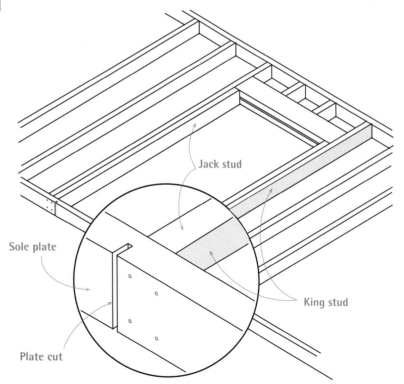

Jack stud

Sole plate

King stud

Plate cut

A stud wall with a continuous sole plate is less likely to flop around as you raise it, but it's difficult to cut through a plate once it's resting on a subfloor. The solution? Cut halfway through the sole plate before you nail it to studs, and finish the cuts after the wall is up.

Raising Walls

Two workers can raise an unsheathed stud wall 8 ft. to 10 ft. long. But if it's much longer than that or if it's sheathed, assemble a larger crew or use wall-lifting jacks to raise it.

Raising walls safely takes prep work: Clear the deck of tools, scrap lumber, and other things you might trip over. Nail the top of a diagonal brace 1 ft. below the top plate, using a single 16d nail so that the brace can pivot as you raise the wall, and nail a 2x block into floor framing so you can quickly nail the bottom of the diagonal brace once the wall is plumb. If you are raising an exterior wall, nail 2x stops to the outside of the platform so the bottom plate can't slide off the deck during the operation.

If you are raising a partition within an existing structure, expose the ceiling joists or end-wall studs you'll nail the partition to. If joists run parallel to the new partition, add blocking between the joists beforehand, as shown in "Partition Parallel with Joists" on p. 212 and "Nailing Off Sole Plates" on p. 214.

Here are three tips for raising walls:

▶ As shown below, several workers straddling the top plate can drive their hammer claws into the top plate, lift in unison, and slide 2x blocks beneath the top plate. This will enable them to get a good grip before actually lifting.

▶ Before you lift a wall, toenail the sole plate to the deck—from the outside of the building. This will keep the wall from inadvertently sliding off the deck when you lift it. The toenails will get squished under the plate, but that's no big deal.

▶ Screwing 2x4 blocks to the outside of the deck to create a curb of sorts can also keep a wall from sliding off the deck when raised. Use structural screws.

After leveling and toenailing the lower plate of a windowsill, end-nail cripple studs under the sill. After all cripple studs are in, face-nail the upper plate of the sill. Doubled sill plates are common in the western United States; elsewhere, they're usually reserved for wider windows only.

If the header requires cripple studs between it and the top plate, install king studs, then jack studs, then the header. Holding the header tight to the top of the jack studs, nail through the king studs into header ends. If the header is laminated from pieces of 2x lumber, each piece should get two or three nails per end. Then cut and toenail the cripple studs that run between the top of the header and the top plate. If you're framing a rough opening for a door, you're done.

However, if you're framing a rough opening for a window, your final steps will be leveling and toenailing the sill (also called a saddle) to jack studs and then nailing cripple studs between the underside of the sill and the sole plate. Again, space cripples according to the 16-in. on-center markings along the plates.

Leaving the sole uncut. You're now ready to tilt up the assembled wall. Note, however, that sole plates haven't yet been cut and removed within door ROs, and for good reason: It's far easier to raise a wall if its sole plate is continuous. Cut only halfway through the sole plate while it's flat on the deck, as shown in "Half-Cutting the Sole Plate" on p. 201. Finish the cut once the wall is up and nailed down.

After constructing a wall on the ground and walking it up, the crew must now lift it up 3 ft. and align the mudsill holes to the bolts in the foundation wall. Have a lot of workers on hand, and lift with your knees, not your back.

Several workers hold the wall plumb while one worker adds diagonal braces on both ends. After plumbing and bracing both ends and tightening sill bolts, they'll run a string between ends to ensure the wall is straight. The top plate needs to be straight so the rafters will line up.

PLUMBING AND SECURING THE WALL

Once the wall is up, nail the bottom of the brace so the wall will stay upright as you fine-tune its position. Use a sledgehammer to tap the sole plate until it aligns with your chalkline on the floor. As you adjust, continuously check for plumb, using a 6-ft. level. If you unnail the brace to plumb the wall, have workers support the wall until you've renailed it.

Once the bottom plate lines up with the chalkline, drive two or three 20d nails through the plate into the joists or blocking below so the wall can't drift. Methods for securing the top of the wall vary. If you're framing an addition and have wide-open space, typically two walls intersecting at right angles are raised, plumbed, and braced, and then tied together by overlapping top plates.

But if you're raising a partition in an existing room, you'll usually nail the top plate to ceiling joists. Invariably, space is tight indoors, and you'll often need to gently sledgehammer the partition into place, alternating blows between top and sole plates until the wall is plumb. Alternatively, you can gain room to maneuver by first nailing the upper 2x4 of a doubled top plate to the exposed ceiling joists—use two 16d nails per joist—before raising the wall. Tilt up the wall, slide it beneath the upper top plate, plumb the wall, and then face-nail the top plates together using two 16d nails per stud bay. Finally, finish nailing the sole plates, driving two 16d nails into the joists or blocking below. In the corners, use 10d or 12d nails to toenail the corner studs to blocking or existing studs; use 16d nails if you can face-nail them. Adding blocking to existing framing is discussed on pp. 211–212.

ALTERNATIVE FRAMING METHODS

In renovation, it's not always possible to assemble a wall on the deck and tilt it up. There may not be enough room, shoring may be in the way, or sloping floors may frustrate attempts to cut studs accurately in advance. In those cases, it may be easier to build the wall in place, piece by piece.

Building a wall in place. Start by positioning the plates and tack-nailing them to joists (or blocking) above or below. Although it's most common to snap a chalkline on the floor and plumb up to the top plate, it doesn't really matter which plate you attach first, unless there's a compelling structural or design reason. If you're erecting a bearing wall, for example, center its sole plate over the appropriate girders or bearing walls below. But if you're trying to align a non-bearing partition with a rafter above, set the top plate first and plumb down to establish the sole plate. If possible, face-nail the plates with two 16d nails at each joist crossing.

Mark stud intervals onto the plates, and then—especially if floors or ceilings slope—measure the stud lengths individually. Cut the studs slightly long (say, 1/16 in.) so that they fit snugly. Toenail each end of the studs with three 10d nails or four 8d nails, angling them roughly 60° from horizontal. Use a spirit level to level the headers. Use three 16d nails to end-nail a header through the king studs on either side. Then face-nail jack studs to the kings, staggering 10d or 12d nails every 16 in.

Framing beneath slopes. Framing beneath stair stringers or rafters isn't difficult if you measure carefully and use an adjustable bevel to transfer the angle of the slope. Mark off 16-in. intervals along a 2x4 sole plate, nail it to the floor, and then plumb up to the underside of the rafter or stringer to mark the top plate. Cut the top plate to length, and nail it to the underside of the sloping rafter or stair stringer. Then use a laser or a plumbed board to transfer the 16-in. intervals marked on the sole plate up to the top plate.

To establish the angle at which you'll cut the top of the studs, plumb a piece of 2x stock in front of the top plate and use an adjustable bevel to duplicate the angle at which they intersect. (Set your circular saw to the angle of that bevel.) Once you've cut two adjacent studs, the difference in their lengths—represented by the X in "Sizing Gable-End Studs" on p. 206—will be constant for all successive pairs. Toenail the studs with four 8d nails on each end.

Establishing kneewalls. Kneewalls are short partitions, about knee-high, which isolate the largely unusable space where the rafters approach the top plates of the exterior walls. (This space can be great for storage and built-in closets, however.) Kneewalls usually run parallel to the roof ridge and consist of a single top plate and sole plate, with studs spaced 16 in. on center. Position the sole plate and plumb up to the underside of the rafters to mark the location of the top plate. Using an adjustable bevel, copy the angle at which the rafters intersect, using a plumbed spirit level or board.

Nail the top plate with two 16d nails per rafter and the sole plate with two 16d nails per joist.

A Smarter Way to Frame

Renovation carpentry will always be a mixed bag of standard methods and whatever works in nonstandard situations. For the most part, the methods discussed in this chapter are pretty conventional.

And pretty wasteful, building scientist Joseph Lstiburek would add. Not only are three-stud corners, 16-in. o.c. framing, and oversize headers not needed structurally, but they're also wasting energy, he maintains. Compared with fiberglass, cellulose, or foam sheathing, wood is a pretty lousy insulator. So anywhere there's wood—when there could be insulation—you're creating *thermal bridges* that waste energy.

A case in point is three-stud corners. They reduce the amount of batt insulation you can fit into a corner, and cold air moves easily in the odd spaces and cavities three-stud framing creates.

Three-Stud Corners Waste Wood and Energy

1/2-in. plywood or OSB, R-1

Wood escorts cold into the house.

Cold air can circulate freely, making the insulation ineffective.

Cold spots can condense water vapor.

1-in. foam sheathing, R-15

Full-width wall cavity doesn't compress insulation

Drywall clips

Two studs, Lstiburek points out, are adequate structurally and, combined with R-5 foam sheathing and full-width cavities full of insulation, they will eliminate cold spots and condensation. Moreover, using superfluous studs to back drywall is not a smart choice when *drywall clips* (p. 465) would allow panels to float slightly and crack less.

Spacing rafters, 2x6 studs, and joists 24 in. o.c. and lining up framing members so loads are transmitted straight down to the foundation are a few things Lstiburek recommends to save wood and make it easy to route plumbing, wiring, and the like. His observations build upon optimum-value

engineering studies done by the NAHB Research Council in the 1970s. Lstiburek's article "The Future of Framing Is Here" (*Fine Homebuilding* #174), in which these illustrations originally appeared, has more on the subject, as does his website, www.buildingscience.com.

Incorporating these framing suggestions into an old-house renovation could be tricky, but they're certainly worth considering in an addition of any size.

A Better Way to Frame

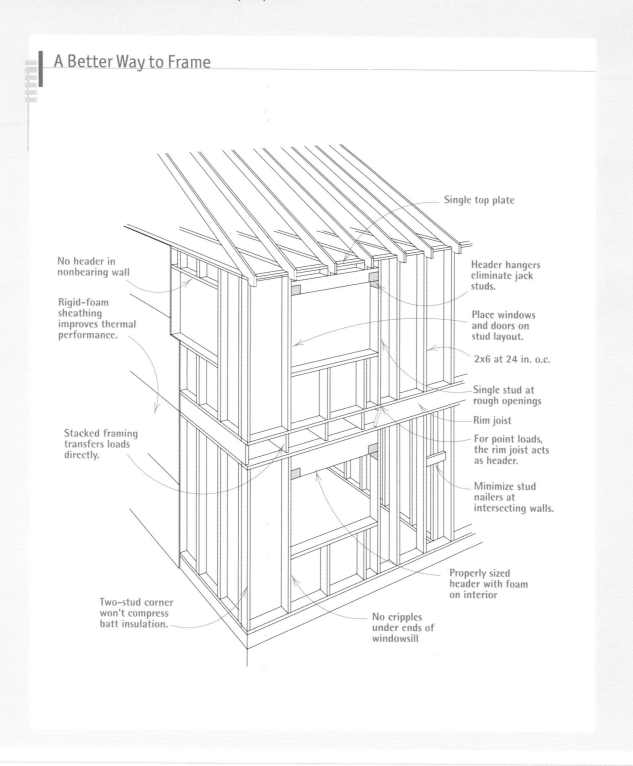

Single top plate

No header in nonbearing wall

Rigid-foam sheathing improves thermal performance.

Header hangers eliminate jack studs.

Place windows and doors on stud layout.

2x6 at 24 in. o.c.

Single stud at rough openings

Rim joist

Stacked framing transfers loads directly.

For point loads, the rim joist acts as header.

Minimize stud nailers at intersecting walls.

Properly sized header with foam on interior

Two-stud corner won't compress batt insulation.

No cripples under ends of windowsill

Sizing Gable-End Studs

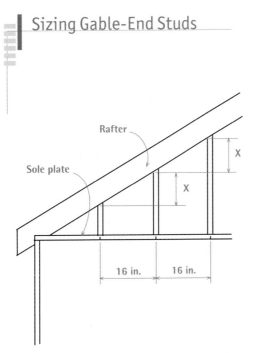

After marking off 16-in. stud centers onto the sole plate, plumb up and transfer the stud marks to the rafter or top plate. Once you've cut two consecutive studs, you'll know the difference in length between adjacent studs, indicated by "X."

Kneewalls

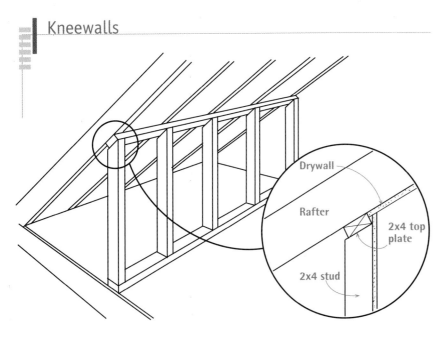

When nailing a kneewall top plate to the underside of a rafter, align the leading edge of the top plate with the inner edge of studs. Drywall will cover the small triangular voids where the plate meets the rafter.

Demolition is incredibly dusty and disruptive, so seal off one room where you can store your stuff.

Cut across the faces of the studs in the angle established by the bevel gauge. To attach the studs to the plates, toenail each end with 8d nails.

Demolition

Before demolishing an old wall, determine whether it's a bearing wall. If so, erect shoring if needed, and have a plan for rerouting electrical cables, pipes, or heating ducts in existing walls. Above all, shut off the electrical power before cutting into finish surfaces—and use a voltage tester to be sure the power is off.

GETTING READY FOR DEMOLITION

Living in a house that's being torn apart and renovated isn't fun, and it can be murder on marriages. But you can minimize the stress caused by disruption, noise, confusion, and dirt. If you don't need to be around the house during the demolition, don't; house-sit or camp out at a relative's if you can't afford to rent another place. But if you must live where you renovate, create a clean zone—usually a bedroom—in which you do no work at all.

Isolate the zone by covering the doorway with sheet plastic held up by duct tape or by installing temporary plastic walls with zippered doorways such as the ZipWall System. Situate your Shangri-La upstairs if you can because dust settles downward. If you are beneath a room being renovated, particularly one in bad repair, tape plastic to the ceiling. In the clean zone, store clothes, stereo equipment, art—anything that could get ruined by the omnipresent dirt of tearout. At the end of the day, go there and relax.

LAST-MINUTE PRELIMINARIES

Take care of these items before you start:

▶ Notify gas, water, and electric companies if you haven't done so already. Utility representatives can tell you what temporary hookups are safe and who must do them.

▶ If you have newly purchased the building and if an inspection found signs of insect infestation, have a pest-control professional treat the condition before you move in.

▶ Make sure your general contractor is properly covered by insurance. Or, if you're acting as your own GC, check with your insurance company to make sure you and any helpers are covered.

▶ If you don't have a cell phone, have a telephone installed so you can call for help in an emergency.

▶ Have a first-aid kit and fire extinguisher handy. Has everyone had a tetanus shot?

▶ On older houses, assess the job site for asbestos and lead paint and remediate as needed. See pp. 21–22 on environmental hazards and p. 548 for RRP lead abatement.

USEFUL TEAROUT TOOLS AND EQUIPMENT

Many of the tools mentioned here are shown and discussed at greater length in chapter 3.

Safety. Be sure you have a voltage tester, hard hat, goggles or safety glasses that wrap around the sides of your face, a respirator mask with HEPA filters, sturdy work gloves, a droplight, and shoes with thick soles. Make sure your tetanus shot is up to date, your first-aid kit is complete, and your cell phone is charged.

Dismantling. Your kit should include wrecking bars of various sizes (see p. 64), a flat utility bar, a cat's paw, a sledgehammer, and heavy scrapers. Reciprocating saws with demolition blades (see p. 56) are the workhorses of renovation. Demo blades typically have pointed ends that enable plunge-cutting (gradually lowering a blade into a surface) and bimetal construction that can cut through nails without shattering.

Work aids. Rent scaffolding if you're doing a lot of tearout over your head. To protect finish floors, you'll need a good supply of sheet plastic, painter's tape, duct tape, and Ram Board (thin,

durable cardboard) or ⅛-in. Masonite hardboard. A HEPA vacuum's fine filters will capture dust, rather than recirculate it as ordinary shop vacs do. Disposable hooded coveralls are a good way to leave dust at the job site. Get heavy rubber trash cans, wheelbarrows, push brooms, dustpans, square-nose shovels for scooping debris, and heavy canvas tarps to drape over plants or manage debris outside. Rent a Dumpster for big jobs.

ELECTRICAL SAFETY

Before you cut into finish surfaces, *always shut off electrical, water, and gas service* to that area. After disconnecting electrical power and *using a voltage tester to make sure the power is off,* start slowly and proceed carefully. If electricity-related procedures described in the following sections are unclear, read chapter 11 and especially "Using a Voltage Tester" on p. 294 before continuing. Particularly useful are *inductance testers* (see p. 294), penlike voltage detectors that can detect electrical current through a wire's insulation. In other words, inductance testers do not need to touch a bare conductor to detect electrical current.

Circuits. Identify circuit breakers or fuses controlling electricity to the construction areas. This will require one person at the panel to flip breaker switches or unscrew fuses while another person watches a light fixture or a voltage tester inserted in a receptacle to see if the light goes out. If you use cell phones to communicate, you won't need to scream instructions between floors.

Receptacles and switches. Voltage testers allow you to see if a receptacle is energized. Always test a tester first on an outlet you know is hot to be sure the tester is working correctly. Usually it's sufficient to insert the tester probe into a receptacle to get an accurate reading, but occasionally devices fail or wires become detached. To be absolutely certain an outlet is not live, remove the cover plate and—being careful not to touch bare metal—apply tester probes to wire ends. To do so safely, follow the procedures shown on pp. 294 and 327.

Junction boxes. As you break through drywall or plaster surfaces, you may find junction boxes. To get at the wires within, remove the junction box cover. The wires inside will either be spliced together with wire nuts or wrapped in tape. Using pliers with insulated handles, carefully pull wire groups out of the box and remove wire nuts or tape to expose wire ends. (If you are at all uneasy about handling wires, turn off all the electricity in the house and remove the wire nuts before proceeding.) Touch your voltage tester to

Visually Isolating the Work Area

Chip Harley, *Renovation*'s co-author and a contractor for more than three decades, offers these thoughts about isolating the mess of a renovation when the homeowners are living in the house:

"When a job is going to run a week or longer, we isolate the work area by erecting temporary walls out of 2x2s, 2 ft. on center, and covering them with ⅛-in. door skin over a layer of plastic. The walls are strong but lightweight; the whole assembly is only 50 to 60 lb. If you cut the 2x2 studs a bit long, you can pressure-fit them so there's little damage to finish surfaces. Just a couple of screws will keep the walls from walking.

"On a really short job (a day or two) in which we make only a couple of cuts, we may use an expandable pole-and-plastic system. To be fair, they're fast to put up, but they're not durable. If the poles get bumped you need to reset them, and it seems like you spend a lot of time taping and retaping the zipper entry or repairing holes where somebody's tool belt snagged when he entered.

"The other big advantage of our temp walls over plastic is that they give both the homeowners and the workers a safety barrier and a visual barrier—and hence a modicum of privacy. The longer a project runs, the more important that sense of separation becomes. It's stressful having strangers in your house all day. And for a worker, privacy means being able to take the time you need to think through a tricky detail without having a client looking over your shoulder wondering, 'Why has he been scratching his head for the last five minutes?'

"Thinking is some of the hardest work there is, and the best carpenters stop and think all day long."

CONTAINING THE MESS

Managing the mess is crucial to a successful renovation and, from a contractor's point of view, tangible proof to your clients that you're trying to keep them safe and comfortable. Torn-out plaster and drywall are nasty to handle and create noxious dust. The dust gets everywhere, and the volume of debris is prodigious. So especially if part of the house is going to be occupied, isolating the work area and containing the dust is job one. Visually isolating the work area is also a smart move, as explained at left.

Here are some other tips:

▶ Isolate the demolition area by taping clear plastic over door and window openings. Clear plastic lets in light. Extendable ZipWall poles allow you to raise a plastic wall quickly, and they have integral zippers in their systems. In front of the zipper door, hang a second sheet of plastic to create an airlock of sorts. Inside the work area, put a fan in the window farthest from the entry, blowing out to move dust away from the entry.

▶ Finish floors are particularly vulnerable to grit that isn't swept up and to stray nails.

the black and white wires simultaneously, then to each wire group and the metal box.

Hidden wires. If you unexpectedly discover cables running through a wall you are demolishing, stop and turn off all power in the house. Then snip the cable in two with a pair of insulated wire cutters—never do this when the power is on. After testing both ends of the snipped cable with a voltage tester to make sure that they are not energized, separate the black and white wires, and wrap individual wire groups with electrical tape or cap them with wire nuts. While the power is off, also pull any staples holding the cable to the studs so that you can remove the studs later without damaging the cable.

With the cable severed, you may proceed with the demolition. If the cable is to be discarded, have a licensed electrician disconnect it from the entrance panel. If the cable is to be reconnected, reroute it after the structural work is complete and house all new connections in a junction box. Per code, all junction boxes must be accessible—that is, not buried in a wall. Again, for your safety, have an electrician do this.

A zipped plastic entry is a good first step toward isolating the mess of a demolition. Use painter's tape to attach the plastic to woodwork, but don't leave the tape on past its day-rating or it will become gummy and lift off the finish.

Clean the floor thoroughly, then cover it with RamBoard (a thin but durable cardboard). Cover the entire floor, from baseboard to baseboard, overlapping RamBoard sheets an inch or two, and use duct tape to tape sheets to each other.

▶ Be deliberate when you demolish. Have a battle plan. Know what tools and helpers you'll need, where you'll start tearout, and which route is best for removing debris. As much as possible, remove materials in manageable sizes so you won't strain lifting it or endanger others as it's being carried out.

▶ Clean up as you tear out. Don't allow trash to accumulate underfoot. Lumber or plaster lath with nails sticking out are particularly hazardous. If you're working at any height—say, standing on a stepladder—know there's a safe area to land if you need to come down in a hurry.

▶ When demolishing outside or carrying rubbish out to the Dumpster, drape heavy tarps to protect your plants and to avoid a lot of raking later on. After discarding large pieces of debris, two people can lift the tarp and shake the remnants directly into the trash.

▶ Organize debris. Maximize Dumpster loads by putting in dense materials such as plaster, concrete, and soil first. Place lighter, bulkier materials on top. Check with local landfills: Many will sort and separate recyclable materials so you don't need to spend time sorting debris beforehand.

TEAROUT

Before you construct anything in renovation, it's usually necessary to tear out part of what's there and beef up what remains. No sooner have you torn out plaster than you're nailing up blocking for partitions to come. This natural flow from demolition to construction is a little frustrating for how-to writers who like to pigeonhole everything, but it's a fact of life if you're renovating. Frequently, you're doing both at the same time.

Conserve when you can. If you are gutting only part of a room, avoid damaging adjacent areas that are sound. When replacing a window, adding an opening, removing loose plaster, or exposing framing, try to isolate the area to be renovated. For example, if you're adding a medicine cabinet, set your circular-saw blade to the depth of the old plaster and lath, and cut back those materials to the nearest stud centers on both sides so the plaster edges can be renailed before you patch the opening.

On the other hand, if at least half of a room is to be demolished or existing plaster walls are cracked and damaged, it may be better to gut the room. With wires, pipes, and old studs exposed, new framing and drywall will go in faster. It's tough and time-consuming to patch extensively.

REMOVING DOORS AND TRIM

Doors, hardware, and wood trim make a house distinctive. If you want to reuse casing and molding, carefully remove and store them until you are done with tearout and rough framing. If you're working with a crew, mark these items "Salvage" or "Save" so they don't get tossed. Most of the time, it's easy to pop hinge pins and lift doors out of the frame. But if hinge leaves are encrusted with paint, use an old screwdriver or a chisel good for little else to chip away paint from the screw heads. Or apply paint stripper.

To remove wood trim without damaging it, first run a putty knife between the wall and the trim to break the paint seal. A utility knife isn't as good because its sharp blade can easily slice trim. Then gently tap a small flat bar (also called a painter's pry bar) behind the trim, as shown in the photo below. Pry up along the entire length of the trim, raising it little by little. Be patient. Photograph woodwork assemblies beforehand. As you remove trim pieces, use a permanent marker to number the back of each so you can reinstall the trim correctly.

PRO TIP

Before storing salvaged trim, remove finish nails by pulling them through the back of the wood to avoid splintering the face. Likewise, if it's difficult to pry off trim without breaking it (rusty nails are often the culprit), use a fine nail set to drive the finish nails all the way through the wood. Holes created by a nail set are small and easy to fill.

Old casing is likely to be dry and brittle, so be patient when removing it. Score along the edge to break the paint seal, then use a pair of flat bars to pry it up, gradually raising it along its entire length.

REMOVING PLASTER AND DRYWALL

Whether you're cutting a large hole for a skylight or gutting the whole ceiling, try to minimize the mess. By using a reciprocating saw with a demo blade to cut out 2-ft. by 2-ft. sections of drywall or plaster, you'll create a lot less dust and have a compact load to carry to the trash.

That's not always possible. If the old plaster falls off the lath as you try to cut out sections, go ahead and break it out. Use a crowbar to pull the lath and plaster down together or use a hand sledge or a framing hammer to dislodge plaster from the attic space above (place planks across the attic joists) or from the back side of the wall, if it's exposed. After separating and bundling the lath, shovel the plaster into buckets.

Ceilings. Hard hats, please! If ceiling joists are exposed in the attic, first take out the insulation from the area where you'll be working. If it's loose insulation, use a dustpan to shovel it into garbage bags.

If you work on the ceiling from below, use movable scaffolding for ceilings 10 ft. or higher. Otherwise, stand on 2-in.-thick planks straddling sawhorses or stepladders. If the plaster is solidly adhered to the lath, use a reciprocating saw to "outline" sections. Then rock them from side to side until the nails holding them to joists work free. Hand the removed section to a helper on the floor, then proceed to the next section. If plaster is sound but sagging in a few spots, you may be able to reattach it with washered screws (see p. 477) or cover it with ¼-in. drywall, as described in chapter 15.

Walls. Walls are easier to gut than ceilings because debris won't rain down on you. Start at the top of each wall and work down, periodically carting out debris before it restricts your movements. Again, use a reciprocating saw to cut out 2-ft.-sq. sections if possible; otherwise, break them out. Tile walls must be broken out. If you'll be putting up new drywall or plaster, this is the time to pull old nails. Likewise, remove any old wires and pipes.

Before removing bearing walls, first shore up the joists or other loads they support. But if you're removing a nonbearing partition, you can do so after stripping plaster or drywall. Cut through the middle of each stud using a reciprocating saw; its thin blade is less likely to bind than a circular saw's blade. With studs cut, pull them away from their plates. To remove plates, pry them up with a wrecking bar. Use a metal-cutting blade in a reciprocating saw to cut through any remaining nail shanks.

REMOVING WOOD FLOORING

If old wood strips are solidly attached, floors are generally left in place to be refinished later or floored over. However, it's sometimes necessary

When demolishing, be as deliberate as possible. Here, a worker carefully breaks through old plaster so he can remove plaster and lath in manageable chunks.

A skillful integration of old and new framing: After stripping interior surfaces and removing a wall to enlarge the room, carpenters installed a new 4x4 window king post and a third top plate to raise the old wall's framing. Note: Old and new window headers line up.

to pull up a few boards so you can install joists or blocking, run wires, or patch-repair floor sections elsewhere. Partitions installed over finish flooring make it difficult to pry out floorboards.

If you'll be reinstalling the floorboards, try to pry them up in an inconspicuous spot, such as along the base of an existing wall. Remove the baseboard and try to insert a flat bar under the leading edge of a floorboard. You may need to destroy the first row of boards to get them out if they're face-nailed or, at the very least, break off the tongue on tongue-and-groove flooring. Successive courses will likely be toenailed through the tongue.

If you're gutting wall surfaces, the space between studs is a good place to fit the curved head of a wrecking bar to pry up a first row of floorboards.

Reinforcing and Repairing the Structure

This section focuses on upgrading nonbearing structural elements: adding blocking, leveling ceilings, straightening stud walls, bolstering joists, and treating rotten or insect-damaged wood.

ADDING BLOCKING

In renovation, it's sometimes necessary to add blocking (short pieces of wood) to bolster existing joists or studs, to provide a nailing surface for new framing, or to provide backing for drywall or plaster lath to come. If you are gutting finish surfaces in a bathroom, that's a great time to add blocking for tub grab bars, diverter valves, towel bars, toilet paper holders, and the like.

Attaching top plates. To attach the top plate of a new partition, first cut back finish surfaces to expose ceiling joists. Snap two parallel chalklines to indicate the width of the top plate. If joists run perpendicular to the partition, cut out a 4-in.-wide slot to receive the top plate. Remove plaster or drywall sections, relocate insulation (if any), and pull nails sticking out of the joists. Use a utility knife to clean up ragged edges before nailing up the top plate, using two 16d nails at each point the plate crosses a joist.

If joists run parallel to the partition, cut back finish surfaces to joist centers on either side of the proposed plate so you can add blocking. Snap chalklines to indicate joist centers, and cut along those lines. (Set a plaster-cutting circular-saw blade to the thickness of the ceiling drywall or plaster.) Install blocking that's the same depth as the joists, spaced 24 in. on center. Cut blocking square for a tight fit, and make sure that its lower

edges are flush to the underside of the joists. If there's access, end-nail each block with three 12d nails through adjacent joists. If you toenail them, use four 8d nails on each end. A pneumatic palm nailer is ideal for driving nails in such tight spaces.

Finally, add backing for the ceiling patch to come and reattach plaster or drywall edges along joist centers, as needed. Metal drywall clips (see p. 465) are a good alternative to blocking. Nail them to the top edge of the top plate.

Blocking for walls. To effectively nail off a new wall where it abuts an existing one, first cut into the existing wall to expose the framing. Start with a small exploratory hole to determine exactly where the studs are. Then cut back finish surfaces to the nearest stud center on either side. Even if your new wall runs directly to a stud in place, add blocking for metal drywall clips to reattach drywall patches.

If, as is more likely the case, there are no studs in the spot where you need a nail-off, add them, as shown in the drawing "Where Walls Meet" on p. 213. These nailers will be stronger if you preassemble them and then sledge them into place. Face-nail them together with 16d nails staggered every 16 in. Full-length nailers should be toenailed with three 10d or four 8d common nails top and bottom. Or prenail metal L-angles to tie nailers to plates.

Blocking for sole plates. Nail partition sole plates to the framing below, not merely to flooring or subflooring. If the partition runs perpendicular to the joists, use two 16d nails at each point the sole plate crosses a joist (see "Nailing Off Sole Plates" on p. 214).

When studs bow into the room, use a power planer to plane down the high spots. Make the first pass over the high point of the bow, then make several successively longer passes to feather out the surface.

Here, new joists are sistered to both sides of existing joists with a 3-ft. overlap. The red chalkline down the center of the joists indicates the center of a new girder to come.

However, if the wall is parallel to the joist grid, try to locate it over an existing joist. If that's not possible, add blocking between the joists so there's something solid to nail the sole plate to. If the partition is nonbearing, use blocking the same depth as the joists, spaced on edge every 24 in. on center. Cut the blocking square so that it fits snugly, flush to the underside of the sub-flooring. Use two or three 16d common nails to end-nail blocking through the joists. Blunt the nail points to prevent splits.

Note: Bearing walls should be supported by two full-length joists directly under the sole plate. Add blocking to adjacent joists to keep the new joists from rotating, and attach both ends with a double-joist hanger. Because doubled joists are, in effect, a girder, they may also need post support beneath; see "Beam Span Comparison" on p. 270, which offers sizes and spans. But because local codes have the final say, consult a structural engineer in your area.

Attaching Top Plates

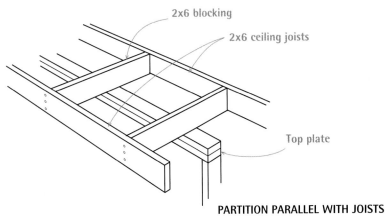

Add backing to nail off finish surfaces.

Cut back ceiling.

Top plate

Finish surface

Joists

PARTITION PERPENDICULAR TO JOISTS

2x6 blocking

2x6 ceiling joists

Top plate

PARTITION PARALLEL WITH JOISTS

Where a partition runs perpendicular to joists, nail its top plate at each joist crossing. Remember to add backing for drywall or plaster lath, to which you can screw finish materials when you patch the ceiling. Where a partition runs parallel to joists, add 2x blocking to nail top plates to.

STRAIGHTENING STUDS

Before installing drywall on recently gutted or newly erected stud walls, scrutinize them to make sure they're flat. Stud variations of ⅛ in. (from flat) are generally acceptable, unless they're in bathroom or kitchen walls—where studs should be within 1/16 in. of flat. There, plumbed cabinets will make high and low spots glaringly obvious. Granted, you can scribe cabinet backs to fit wavy walls, but it's easier to straighten studs while they're still exposed.

Eyeball walls for obvious discrepancies. Then stretch strings across the studs at several heights. If the studs aren't flush to the top or sole plates, hammer them flush and screw on steel reinforcing angles (see the photo on p. 85) to attach studs to the new position; more toenailing might split them. Next, stretch a string, chest high, across the wall to find high (protruding) and low (receding) spots. Mark them with a pencil. Finally, use a 6-ft. or 8-ft. level or straightedge to assess individual studs for bowing. Scribble symbols directly on stud edges, indicating high spots to be planed down (where studs bow toward you) and low spots to be built up (where studs bow away from you). Use special cardboard *furring strips* to build up the low spots.

Plane down high spots. Before power planing the high spots, use a magnet to scan old studs for nails. Nails will destroy planer blades, so if nails are too rusty or deep to pull, use a metal-cutting blade in a reciprocating saw to shave down the stud edges—or drive the nails deeper with a nail set. If studs are nail free, plane down the high spots in several passes, starting at the middle of

Where Walls Meet

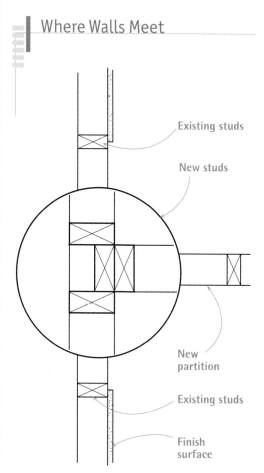

Where a new partition abuts an existing wall, cut back finish surfaces to the centers of the nearest stud on either side, and add studs to nail the partition to.

the high spot and gradually tapering out. Because knots are hard, they'll take more passes. Use your straightedge to check your progress, and use taut strings to check the whole wall again after building up or planing down the studs. *Caution:* Wear eye protection when using a power planer.

Common sense and "feel" are a big part of this process: If all the studs in the wall have a slight bow inward or outward, the wall won't be flat, but drywall covering it may look flat. In that case, leave the studs as they are.

BOLSTERING JOISTS

Widespread sagging or excessive springiness in a floor is probably caused by joists that are too small for the span or by post, pad, and foundation failure, as covered in chapter 10. Isolated joist failure is usually caused by insect or water damage, alterations to the joist, or a point load from a heavy piece of furniture or bathtub. If there's infestation or rot, correct that condition first.

Sistered joists. The most common way to reinforce a weakened joist is to nail a new one to it—a "sister" of the same dimension. The new sister needn't be the exact length of the original but should be long enough to be supported on both ends by the perimeter foundation or a girder. For this reason, short sections "scabbed on" don't work and are usually prohibited by local building codes.

To insert the new joist, remove blocking or bridging between the affected joists and bend over or snip off flooring nails protruding from the underside of the floor. Then eyeball the new sister joist and note its crown: If its arc is excessive, power plane it down so that you don't bow up the floor as you drive the joist into place. Beveling the leading top edge of the joist will also make sledging into place easier. Once the new joist is in position, use bar clamps to draw it tight to the old joist. Face-nail the two together, staggering 16d nails every 12 in. If there's no room to swing a hammer, use a pneumatic palm nailer to drive the nails most of the way.

Angled-end joist. Where joists will rest on a foundation mudsill at one end and hang from a girder at the other (rather than sitting atop it), cut away the top corner of the end that will rest on the mudsill so it will fit between the mudsill and subfloor. Cut the other end square to butt to the girder. Place the angle-cut end of the joist on edge over the mudsill, then lift the squared end and slide it toward the girder until it butts against it. Angle-cut joists must be a few inches shorter than the original joist you're sistering to.

You may need a plumbed adjustable column or a screw jack to raise the joist until it's flush to the underside of the subflooring. Once it is flush, use a double-joist hanger to join the new joist and its sister to the girder. Face-nail the two joists, staggering 16d nails every 12 in. Remove jacks and replace blocking between joists. Should the double-joist hanger overlap a hanger already there, predrill the metal so that you can nail through both hangers with case-hardened hanger nails. To learn more about jacking safely, see chapter 10.

Flitch plates. Steel flitch plates are sometimes used to reinforce undersize beams or joists. Because they are typically 3/8 in. to 1/2 in. thick and must be predrilled, they're not well suited to casual installation by nonspecialists. There's more on flitch plates on p. 75.

METAL STUDS

Lightweight steel studs are still relatively rare in home construction, but they are appropriate for certain tasks. They're dead straight—one big

Use 1/16-in.-thick cardboard furring strips to build up low spots on bowed studs. Strips typically come in 2-in. by 100-ft. rolls or bundles of 50 precut strips 36 in. or 45 in. long. Cut strips with a utility knife, and attach them to the studs with hammer tackers, using 3/8-in. staples. (Fortifiber is one manufacturer.)

A palm nailer is ideal in tight spaces where you don't have room to swing a hammer.

Nailing Off Sole Plates

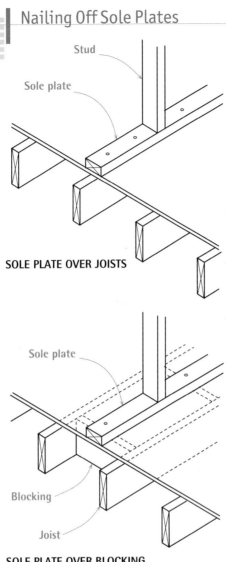

SOLE PLATE OVER JOISTS

SOLE PLATE OVER BLOCKING

When a partition runs parallel to the joists, try to position it over a joist so you'll have something to nail the sole plate to. If you can't reposition the wall, add blocking between the joists.

advantage over much of the wood framing that's available today. If you work with them, eye and hearing protection are a must—as are rubberized cloth gloves to protect your hands.

Metal studs are sometimes specified in residential situations where fire is a concern, say, in common walls between living spaces and a garage. They are also good for framing kitchen lighting soffits or for forming *chases*, essentially long, shallow boxes in which to run mechanicals such as pipes, ducts, or wiring. Lightweight steel studs are also ideal for furring out a foundation wall to attach a finish surface to—especially

1⅝-in.-wide studs, which won't take up much room in space-starved basement remodels.

Metal stud walls are also a snap to take out because they're assembled with self-tapping sheet-metal screws. That's why they're used so often on commercial or retail spaces. A green delight, they can also be recycled completely.

Leveling ceilings. As shown on the facing page, metal studs really shine when you need to level a ceiling whose slope would be especially noticeable against, say, cabinets with crown molding. Because metal studs are so light, they go up quickly, yet they are sturdy enough to support a new drywall ceiling. This procedure is called "furring down a ceiling," and using metal studs is much easier than ripping tapered wooden strips to create a leveled ceiling, which is *hugely* time-consuming and often doesn't look very good.

1. Start by tearing out the existing drywall or plaster ceiling to expose the ceiling joists. (If ceiling joists are undersize or rotted, fix those conditions first.)

2. Using a laser level, establish a level line around the room, ¼ in. below the lowest point of the ceiling joists.

3. At this height, drive nails into the corners of the room. Stretch string perpendicular to the ceiling joists at both ends of the room and roughly every 3 ft. or 4 ft. in between.

4. With the aid of a helper, lift the steel studs over the leveled strings, and lower each stud until its bottom edge is ⅟₁₆ in. above the strings. (The ⅟₁₆-in. gap is necessary to avoid moving the alignment strings.)

5. Once each steel stud is correctly positioned, use 1-in. self-drilling screws to attach it to a joist face; space screws every 16 in. along the length of the stud. Place screws back at least ½ in. from the lower edge of the joist. Use aviation snips or a metal-cutting blade in a reciprocating saw to cut lightweight steed studs.

TREATING INSECT INFESTATIONS

If you see signs of an infestation, hire a pest-control professional to assess and remedy it. Pesticides are often toxic, and anyone unfamiliar with insect habits may not destroy all their nesting sites or may apply pesticides inappropriately or unnecessarily. Moreover, local codes may require a professional.

It's often difficult to tell whether an infestation is active. For example, if a subterranean termite infestation is inactive, a prophylactic treatment may suffice. But if the infestation is active, the remedy may require eliminating the condi-

This wall sheathing shows evidence of fungi and insect damage. After failed window flashing allowed water behind a stucco exterior, wood-destroying fungi grew quickly. Then subterranean termites tunneled, consuming the moist, fungus-damaged wood. Cross-grain checking just above center is a typical sign of fungus damage. A beetle larva showed up, too.

Leveling a Ceiling with Steel Studs

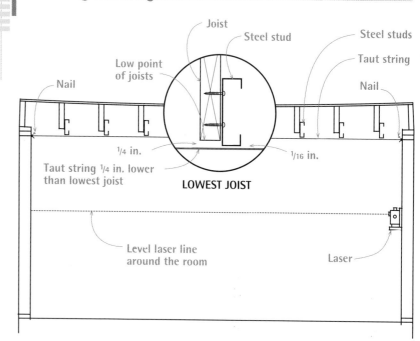

Steel studs let you establish a level plane of nailers for eventual drywall. As detailed in the text, measure up from a level laser line, drive a nail at each room corner, and stretch a string perpendicular to the joists at each end of the room.

tions that led to the infestation (such as excessive moisture and earth–wood contact) and an aggressive chemical treatment. Treatment usually consists of applying a chemical barrier on the ground that repels the termites or a "treated zone" whose chemical doesn't repel them initially but later kills or severely disrupts them.

Termites, the most famous of insect pests, include drywood, subterranean, and Formosan types. Because subterranean termites need access to moisture in the soil, they build distinctive dirt tubes up along the surface of foundations. When they eat into the wood, they usually proceed with the grain. Termites swarm in spring or fall. Discourage the return of subterranean termites by lowering soil levels around foundations, footings, and the like.

Drywood termites hollow out chambers separated by thin tunnels and often travel cross-grain; they eject fecal pellets through holes, with the pellets forming pyramid-shaped piles. Fumigation is effective for drywood termites, but it's ineffective for treating subterranean termites because their colonies are located in the ground and fumigation gas does not penetrate the soil.

Formosan termites, whose colonies may exceed 1 million individuals, are wreaking havoc along the Gulf Coast of the United States; they live in the ground or in buildings and build huge, hard nests.

Carpenter ants are red or black, ¼ in. to ½ in. long. Sometimes confused with termites, these ants have narrow waists and, when winged,

wings of different sizes. Although they do tunnel in wet or rotting wood, they do not eat it as food and are therefore less destructive than termites. To locate their nests, look for borings rather like coarse sawdust. Professionals will often drill into nests and spray them with an insecticide safe enough for inside use; dusting with boric acid is another common treatment.

Powder-post beetle holes look like tiny BB-gun holes; their borings resemble coarse flour. Because these insects favor the sapwood, evidence of borings may be only superficial until you prod with a pocketknife. Still, holes are not a sure sign of an active infestation. One approach is to remove the damaged wood, sweep up borings, paint the area, and monitor it for a year. If holes reappear, it's an active infestation: A professional will need to fumigate the wood or apply a pesticide.

DEALING WITH ROT

If wood is spongy, contains wispy fibers that look like cotton, or disintegrates without reference to the grain, it is infested by a rot fungus. The white strands are spores.

Sistering Joists

SISTER JOIST SITS ATOP SILL AND HANGS OFF GIRDER

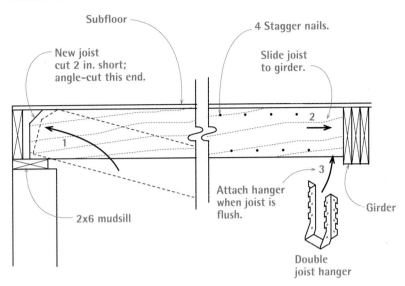

To a sister a new joist when one end will rest on a mudsill and the other will hang from a girder, angle-cut the mudsill end and slide it into place. Then raise the other end until it's flush with the girder before attaching with a hanger. Join sistered joists with 10d nails staggered 8 in. o.c.

SISTER JOIST SITS ATOP SILL AND GIRDER

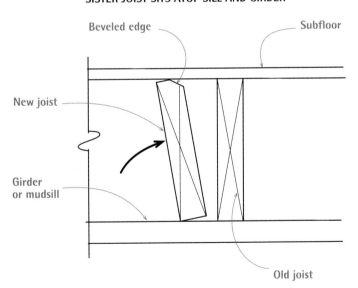

To "sister" a new joist to one that sits atop a mudsill on one end and a girder on the other, bevel the leading top edge so it can be "sledgehammered" into place.

If the rot is limited to a few spots, suspect leakage and look for other clues above. Water may be entering from inadequate roof flashing, blocked gutters, or—quite common in older houses—windows or doors that lack flashing altogether, in which case there will probably be fungus damage to walls and ceilings inside. Also, look for structural damage beneath tubs, sinks, and water heaters; if you find any, make sure the damage hasn't progressed into subflooring and floor framing.

Rotted wood, unsightly as it is, poses no threat to healthy wood nearby, as long as you control the moisture that encouraged the fungus in the first place. Cut out deteriorated sections, but don't bother to remove discolored wood that is solid; leave it in place, especially if flooring is

Water, Insects, Rot, and Mold

Whether structural damage is caused by insects or rot fungi, excess water is usually at the heart of the problem. Before treating the specific agent causing the deterioration, reduce excess water by maintaining gutters, improving drainage, grading the soil away from the building, eliminating wood–soil contact, and improving ventilation.

The fungi that rot wood reproduce by airborne spores, so they're virtually everywhere. But they can't establish colonies on wood with a moisture content (MC) less than 28%, and they go dormant if the MC drops below 20% or the air temperature drops below 40°F. Household molds, also caused by fungi, thrive in a similar moisture and temperature range. So if moisture-meter readings in your basement or crawl-space are too high, reducing excess moisture may solve both wood rot and mold problems.

How you reduce moisture, however, is something experts can't agree on (see pp. 284–285 and chapter 14 for more on mitigating moisture and mold). Many building codes recommend covering dirt floors in crawlspaces or basements with sheet plastic and installing screened vents to circulate air and disperse moisture: 1 sq. ft. of vent per 100 sq. ft. of floor space is the standard formula. Whereas another group of builders argues that it makes more sense—especially in the humid South—to close vents, insulate crawl-space or basement walls, seal air leaks, and install a dehumidifier. That, they argue, will stop mold from colonizing and migrating to living spaces. Best bet: See which approach builders in your region favor.

nailed to it. You can reinforce existing joists by sistering on new ones, but before doing so, allow the old joists to dry thoroughly.

The two-part epoxy repair discussed on p. 175, though interesting, is primarily a cosmetic fix. Most epoxy fills are inappropriate for structural elements. Some companies, such as ConServ Epoxy, make products specifically for structural repairs.

Stair Repairs

Stairs are complicated to build, and problems can be tricky to diagnose. For example, it may be possible to repair squeaky stairs with glue and a few screws. But if squeaking is widespread, stairs tilt to one side, and there's a gap along a stairwell wall, the diagonal supports beneath the staircase may be failing. In that case, you'll probably need to expose those supports to find out what's going on.

SQUEAKY STEPS

If the underside of your staircase is covered and you have only a few squeaks, try fixing them without tearing out finish materials. Do *not* use finish nails to nail down squeaky treads: The nails may split the nosing, and they'll almost certainly pull loose in time. Instead, drill pilot holes for trim-head screws to pull stair joints together. If stairs will be painted or carpeted, first try caulking the riser-tread joint with a construction adhesive such as PL 400.

If that doesn't stop the squeaking, expose the underside of the staircase by cutting into the drywall or plaster covering it. If there are blocks glued along a riser-tread joint, it's likely the glue has failed. After you've identified which step is squeaking, try to squeeze some glue (such as Titebond aliphatic resin) or construction adhesive into the joint. Then go upstairs, drill pilot holes, and use trim-head screws to pull tread and riser together.

Or you can choose to live with the squeaks. It is usually very difficult to repair stairs without breeching finish surfaces, pulling things apart, refinishing the treads, and so on. If you've got kids, a squeaky tread will let you know when they are sneaking in after curfew.

REPLACING BALUSTERS

Broken balusters often can be doweled, glued, and filled. But if you're disappointed with the repair, see if your lumberyard can order a replacement in the same pattern. Stair parts have been mass-produced for a century or more, so there are catalogs full of stock balusters. Custom

As shown below, stringers and carriages support steps. Stringers serve as the diagonal support frames on each side. And carriages carry by means of a sawtooth pattern cut into them. Stringers and carriages are often fastened together. Another option is a housed stringer, in which a stringer has routed grooves that receive and support tread and riser ends.

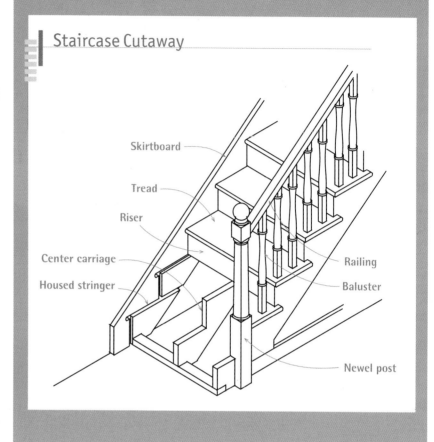

Staircase Cutaway

Skirtboard

Tread

Riser

Center carriage

Housed stringer

Railing

Baluster

Newel post

mills can create new balusters to match old ones, but the process is labor intensive and costly. If you have many damaged or missing balusters and can't find stock replacements, consider replacing all of them with another pattern.

If balusters are intact but shaky, remove and reglue them; simply nailing them won't do much good. To get at balusters, pry free the *return molding* at the end of the tread by inserting a putty knife or small chisel into the nosing seam. Once you start prying, you'll see small finish nails holding the nosing in place; remove these nails. Then gently tap out the bottom of the baluster, which is usually joined to the tread by a dovetail joint or a dowel. The top of the baluster is held in place by *fillet strips* that fit tightly between baluster tops and are toenailed with tiny finish nails to the *plow* (routed channel) in the underside of the railing. Pry out fillets first.

The doweled end of this square-cut baluster fits precisely into a hole predrilled into a stair tread.

The rough cap atop this baluster section of a custom-built staircase will be covered by a three-piece railing assembly. The short fillet strips between the baluster tops can be removed should balusters need replacing later.

This three-piece railing assembly mockup has two skirt (side) pieces that cover the tops of the balusters and a cap. This unusual assembly will create a massive, magisterial look.

Anatomy of a Hollow Newel Post

Cap

Star nut

Railing

Washered nut

Threaded rod

Plug

Hollow newel post

Unlike modern newel posts, older ones are often hollow and attach to railings in various ways. One common way is a star nut centered in the end of a railing, which is accessed by removing a plug on the underside. The bottom of the post may be screwed to a stringer or held fast by an adjustable rod-and-plate assembly running down the middle of the post.

To fit the baluster back in place tightly, lightly coat both ends with carpenter's glue and replace the tenoned or doweled end first. Replace fillets to space the tops of the balusters evenly. Replace the return molding and wipe off the excess glue. To prevent marring, use a rubber mallet or cushion hammer blows with scrap wood.

TIGHTENING NEWEL POSTS

If many of the balusters are loose, check the railing and the newel post: They may not be firmly attached. Or if the upper end of the railing dead ends into a wall on the floor above, the railing may be anchored with a bracket beneath. Make sure this bracket is tight.

If the newel post is shaky, try shimming underneath its base or screwing the post down with predrilled 3-in. Torx screws. How you do that will depend on how exposed the base of the post is. Ideally, drill down at an angle, through the bottom of the post, through the flooring, into the framing below. If this doesn't suffice, see if the internal hardware needs tightening. Newel posts often are hollow, with a long, threaded rod inside, as shown in "Anatomy of a Hollow Newel Post" at left. You may be able to tighten the upper end of this rod, concealed by the post cap, by turning a nut against a restraining plate. Because you may have difficulty finding the cap joint under many years of polish and grime, loosen the cap by rapping the side of it with a rubber mallet. The bottom end of the threaded rod often emerges on the underside of the subflooring—if it's exposed, have a look.

On occasion, newel posts also are connected to another plate-and-rod assembly on the inside of the nearest stair carriage. About the only way to get at that assembly (if it exists at all) is to pull up the first tread. Where the railing meets the newel, the railing is held tight by wood joinery or by a double-ended hanger bolt accessible through a plug on the underside of the railing.

REPLACING STAIR TREADS

Treads crack because they aren't supported adequately or they weren't made from good stock. To replace them, you'll need to pry or cut them out. Prying is preferable but rarely possible, especially because the treads are usually rabbeted to risers or housed in stringers.

To cut a tread out, start by removing the balusters from the open end of the tread. If it is not possible to pry out the tread in one piece, drill holes across the middle of the tread and, driving a chisel parallel to the wood grain, split out the old tread. But try to remove intact the end of the tread in which the balusters fit so you can reuse it as a template for the replacement tread. Clean up any old glue or wood fragments. After fabri-

With skirt pieces glued and clamped to both sides of the rough cap, the finish railing cap is test-fitted.

cating the new tread and testing its fit, apply glue to its edges and to the tops of the carriages on which it will sit. To each carriage, screw down the tread with two or three trim-head screws, predrilled to prevent splitting and counterbored for a plug to hide the screw head. Reinsert and glue balusters and nosing.

SAGGING STAIRS

If the staircase has several of the ailments described in preceding sections, it also may have structural troubles underneath. Investigate further. If the stairs tilt to one side, the carriage on the low side is having difficulty; that is, nails or screws holding it to the wall may be pulling out, the wood may be rotting or splitting, or the carriage may be pulling free from the stringer. Sagging on the open side of a stairway is common, for there's no wall to bolt the carriage to. If there are large cracks or gaps at the top and bottom of the stairs, you're seeing symptoms of a falling carriage.

To learn more, remove the finish surface from the underside of the staircase. Because you don't know what you'll find, go slow. Use a utility knife to cut drywall, or a diamond blade in an angle grinder for plaster. Cut no deeper than the thickness of the finish material—you don't want to cut into stair carriages. Wear safety glasses and a respirator mask, cutting out 2-ft.-square sections to keep things manageable.

You can probably save decorative plaster molding along the staircase by cutting parallel to it—about 1 in. from its edge, thus isolating the section of lath nailed to the underside of the outer carriage. Leaving a 1-in. strip will also make it easier to disguise the seam when you reattach the ornamental border after repairing the stairs.

With the underside of the stairs exposed, you should be able to see exactly what the problem is. If the carriages have pulled loose from adjacent walls, you'll see a definite gap. Replace wood that is rotted or badly cracked, especially wood cracked across the grain. If the wood sags or is otherwise distorted, bolster it with additional lumber; it also may need to be reattached. All of these repairs are big ones. To do them right, you'll need complete access to the substructure, from one end of the carriages to the other.

Starting at the top, remove all nosing, balusters, treads, and risers. You could theoretically bolster undersize carriages without removing all the treads, risers, and balusters, but it's better to remove them. Otherwise, misaligned or distorted carriages will be held askew by all the

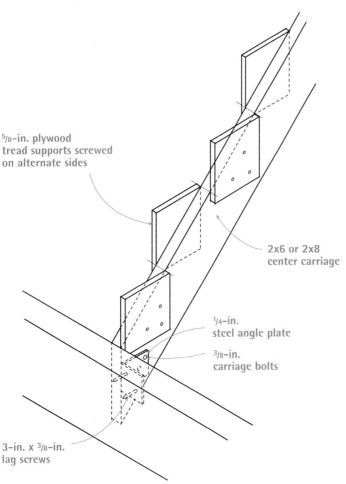

⅝-in. plywood
tread supports screwed
on alternate sides

2x6 or 2x8
center carriage

¼-in.
steel angle plate

⅜-in.
carriage bolts

3-in. x ⅜-in.
lag screws

If the center carriage isn't notched, it may not support treads well. In this case, screw plywood tread supports to alternate sides. Also, if the center carriage was only nailed to the face of a header originally, it may have slipped down. If so, jack it up and reattach with a ¼-in. steel angle plate.

pieces nailed to them. So remove the treads and risers, and jack up the distorted carriages to realign them. (You may want to stretch string lines as an alignment aid.) Number all parts as you remove them, grouping pieces according to the step number.

REALIGNING AND REINFORCING CARRIAGES

If a stringer or carriage has bowed outward, use a 2x4 jammed against a near wall to push the carriage back into place. Alternately, use an adjustable screw column horizontally to push the carriage back, but nail the column's top plates so it can't fall. To keep the bowed element in place after the pressure is released, add blocking to a nearby solid framing element.

Where a carriage has separated from its stringer, clamp the pieces together, and then add two ¼-in. lag screws, staggering the pairs of screws every 18 in. along the length of the boards. If a carriage has pulled free from a stud wall, reattach it with washered lag screws. Where a carriage is attached to a masonry wall, drill through the carriage into the masonry, using a carbide-tipped masonry bit. After drilling the carriage, use a Titen concrete screw to attach the failed carriage to the masonry wall. To forestall rot, slip a piece of 30-lb. building paper behind the carriage before bolting it down.

Occasionally, stringers or carriages come loose at the top and bottom. In a well-built staircase, the upper ends of carriages are nailed to the inside of the header above; the lower ends of those carriages sit on, and are nailed to, the doubled joists of the rough opening below.

However, sometimes the lower ends of center carriages are mistakenly nailed to the inside of a RO header; in time, the nails pull free and the carriages slip down. Jack up the fallen carriages using a plumbed, adjustable column securely footed on the floor or atop a 4x8 beam on edge. To create a flat jacking surface for the top of the column, screw a triangular piece (with the same slope as the stairs) to the underside of the carriage. Should old nails resist your effort, cut through them with a metal-cutting blade in a reciprocating saw.

Jack up the center carriage and join its lower end to the header with steel connectors or ¼-in. right-angle mending plates. Secure the plates to the header with ⅜-in. lag screws and through the bottom of the carriage with ¼-in.-dia. carriage bolts. It's not usually necessary to use mending plates on the upper end of the carriage because the lower end is bearing most of the weight.

Inadequate support for the middle of a staircase can lead to split treads or major failures. Where a center carriage is not sawtoothed to receive treads (and in some older houses they aren't), add plywood supports beneath each step. Cut support blocks from scrap plywood ⅝ in. thick. Then glue and screw them to alternate sides of the carriage—one per tread. If the stair sags in the middle and has no center carriage, add one.

You can replace finish surfaces after the carriages are bolstered and reattached and the stairs and balusters are reinstalled. Be sure that the nailing plane on the underside of the carriages is flat, shimming as needed. To reattach plaster lath or drywall, use type W drywall screws (hammering drywall nails can crack surrounding materials).

Shoring

Shoring temporarily supports loads formerly carried by bearing walls while you modify them—say, to add a window or a door opening. Typically, shoring is installed after removing finish surfaces and rerouting pipes and wires but before cutting into a bearing wall. If you're not sure if the wall is bearing or whether it can be safely modified, have a structural engineer inspect the house and review your remodeling plans. This is hard-hat work.

For first- and second-floor walls, two types of shoring are common: screw jacks used with top and bottom plates, and temporary stud walls built from 2x4s. In either case, position shoring back 2 ft. to 3 ft. from the wall you're working on so you'll have room to move tools and materials. There are two ways to approach it:

▶ **If you're using screw jacks,** doubled 2x6 top plates will distribute loads better. Here's how to laminate the top plates in place: Use two or three 16d common nails to nail the upper 2x6 directly to the ceiling joists, then face-nail the second 2x6 to it. Ideally, the top plates should extend one joist beyond the new opening on both sides. Don't overnail; you're just holding up the plates until you get jacks underneath. Plumb down to mark the location of the single 2x6 sole plate. Place jacks every 4 ft., and plumb them. Tack-nail the top of each jack so it can't drift out of plumb. Then raise one jack in tiny increments before moving to the next. Raise ceiling joists no more than 1/8 in.—just enough to take pressure off the bearing wall.

▶ **Building a temporary stud wall** is similar: Tack-nail the top plates, then plumb down to mark the bottom plate. (To keep the bottom plate in place, tack-nail it to the joists underneath.) Cut studs 1/4 in. longer than the distance between the plates because, here, the studs do the lifting. Toenail the studs to the top plate on 16-in. or 24-in. centers. Then use a sledge to rap the bottom of each stud until the stud is plumb. Recheck each stud for plumb as you progress, and monitor them periodically.

Once shoring supports the loads above, remove the studs from the bearing wall as needed to enlarge openings, add headers, and the like. If the bearing wall transfers loads from upper stories down into girders or foundation walls, study the lengthy section on jacking and shoring in chapter 10 beginning on p. 263.

Screw jacks and a doubled 2x6 top plate pick up loads so a window opening can be safely enlarged. The floor shown is concrete. If yours is wood, use a 2x6 plate under the jacks as well.

In the foreground is a 2x4 shoring with a horizontal brace.

Structural Remodeling

Once shoring is in place, it should be safe to remove bearing walls. Check with a structural engineer if you have any doubts. Again, wear safety gear (hard hat, eye protection, work boots with thick soles), and test the electrical outlets to be sure the power is off. If you need to do any jacking, read chapter 10.

FRAMING A DOOR OR WINDOW OPENING

After cutting back interior surfaces to expose the framing in the exterior wall, outline the RO by snapping chalklines across the edges of studs. If you can incorporate existing studs into the new opening—an old stud might become the king stud of the new opening, as shown in the right photo on p. 210—you can save time and materials.

To remove old studs within the new opening, use a sledge to rap the top plate upward, thus creating a small gap above the studs (and old header, if any). That should create enough space to slip in a metal-cutting reciprocating-saw blade and cut through nails holding the studs to the top plate. Although sheathing or siding may be nailed to the studs, they should still pull out easily.

Start framing the new opening by toenailing the king stud on both sides, using three 10d nails or four 8d nails top and bottom. Laminate the header package or cut it from 4x stock. (The procedures described here employ terms illustrated in "Stud-Wall Elements" on p. 199.) Precut the jack studs and face-nail one to a king stud; lean the second near the other side of the opening. Place one end of the header atop the jack stud in place. Then slide the second jack under the free end of the header. Raise the header by tapping the second jack into place. Or, as an alternative, you can use a screw jack to hold the header flush to the underside of the top plate. Check for level, then measure and cut both jack studs to length.

If there are cripple studs over the header, nail up one jack stud, use a level to establish the height of the jack stud on the second side, and nail it up. Install the header, then cut the cripple studs to length and install them. If you're framing a window opening, there will also be cripple studs under the sill. So level and install the sill next, using four 8d nails on both sides for toenailing the sill ends to the jack studs. End-nail through the sill into the top of the cripple studs; toenail the bottoms of the cripple studs to the sole plate.

PRO TIP

Today, rough openings are usually 82½ in. high, which accommodates a standard 6-ft. 8-in. preframed door. But if your house is nonstandard, try to line up new or enlarged openings in exterior walls to the tops of existing doors and windows. The underside of a new header will usually be 2 in. to 2½ in. above the window or door frame, but check your unit's installation instructions.

IS IT A Bearing WALL?

As noted in "Exploring Your Options" on p. 195, bearing walls and girders usually run parallel to the roof ridge and perpendicular to the joists and rafters they support. And in most two-story houses, joists usually run in the same direction from floor to floor.

Things get tricky, however, when rooms have been added piecemeal and when previous remodelers used nonstandard framing methods. For that, you'll need to explore. Use an electronic stud finder or note which way the heating ducts run (usually between joists) to figure out joist direction. If all else fails, go into a closet, pantry, or another inconspicuous location, and cut a small hole in the ceiling to see which way joists run.

Finally, nonbearing walls sometimes become bearing walls when homeowners place heavy furniture, bookshelves, appliances, or tubs above them. If floors deflect—slope downward—noticeably toward the base of such walls, they're probably bearing.

To mark the rough opening outside, cut through the sheathing and the siding using a reciprocating saw. Or, if you want to strip most of the siding in the affected area first, drill a hole through each corner of the RO. Outside, snap chalklines through the four holes. Remove siding within that opening—plus the width of the new exterior casing around all four sides. Nail the sheathing to the edges of the new frame. Finally, run a reciprocating saw along the chalklines to cut sheathing flush to the edges of the RO. Now you're ready to flash the opening and install the door or window. Chapter 6 will guide you from there.

REPLACING A BEARING WALL

Bearing-wall replacements should be designed by a structural engineer and executed by a contractor adept at erecting shoring and handling heavy loads in tight spaces. In the two methods presented in the following text, beam-and-post systems replace bearing stud walls. In the first method, the bearing beam is exposed because it supports joists. In the second method, the beam is hidden in the ceiling, and joist hangers attach joists to the beam.

Once you've cut electrical power to the affected area, installed shoring on both sides of the existing bearing wall, and inserted blocking under support post locations, you're ready to remove the bearing wall and replace it with a new beam.

However, if you're installing a hidden beam, your job will be easier if you leave the old wall in place a bit longer to steady the joist ends as you cut through them.

Installing an exposed beam is the easier of the two methods. Because ceiling joists sit atop an exposed beam, it's not necessary to cut the joists—as it is when installing a hidden beam. After removing the bearing wall, snap chalklines on the ceiling to indicate the width of the new beam—say, 4½ in. wide for a beam laminated from three 2x10s or 2x12s. Cut out the finish surfaces within this 4½-in.-wide slot so the joists can sit directly on the beam. Chances are, the slot won't need to be much wider than the width of the top plate of the wall just removed.

Because the beam extends into end walls, notch the beam ends so they will fit under the end-wall top plates, which may also support joists. Notching ensures that the top of the beam, the top plates, and the bottom of the ceiling joists will be the same height. If end walls have doubled top plates, the notch will be 3 in. to 4 in. deep. Before notching the beam, eyeball it for crown and place it crown up. Before raising the beam, be sure to have blocking under each post to ensure a continuous load path down to the foundation.

A laminated 2x12 beam can weigh 250 lb., so rent a Genie Lift (p. 74) to raise it safely. Once the top of the beam is in place, flush to the underside of the joists above, temporarily support it with plumbed screw jacks or 2x4s cut ¾ in. long and wedged beneath the beam—have workers tack-nail and monitor the 2x4s so they can't kick out. (Put 2x plates beneath the jacks or the wedged 2x4s to avoid damaging finish flooring.)

Measure from the underside of the new exposed beam to the floor or subfloor. Then cut 4x4 posts slightly longer than the height of the opening, and use a sledgehammer to tap them into place. (Ideally, cut posts the *exact* length, but a little long is preferable to a little short.) Plumb the posts, and install metal connectors such as Simpson Strong-Tie A-23 anchors to secure the post ends to the top and sole plates. Add studs to both sides of each post, as shown in "Supporting an Exposed Beam" at right, to "capture" it and keep it from moving; nail these studs to the plates and to the 4x4s as well.

Installing a hidden beam (see photo at right) takes more work than installing an exposed beam but yields a smooth ceiling. To summarize, after erecting stud-wall shoring on both sides of the bearing wall to be replaced, cut all the ceiling joists to create a slot for the hidden beam, assemble the beam on the ground, and then lift it into place. Here, joists will hang from the sides of the

Supporting an Exposed Beam

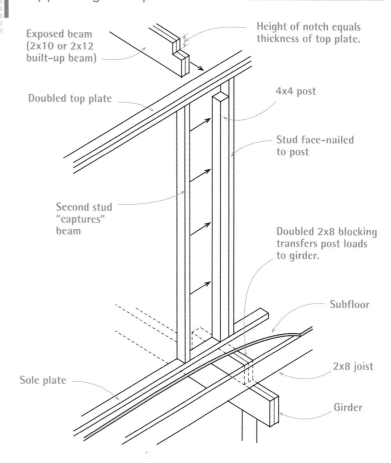

After notching and then raising an exposed beam with jacks, cut 4x4 posts to support it at both ends. There must also be blocking under each post to transfer post loads to the girder and other foundation elements.

As a chain fall and a nylon sling support one end of the beam, master builder John Michael Davis and his helper lower the other end into the tight space between newly cut joist ends. Both ends of the beam will sit on a doubled-top-plate-and-post assembly that transfers loads down to foundation components.

Hidden Beam

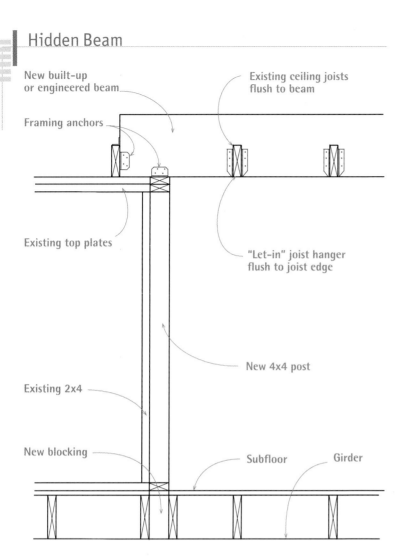

New built-up or engineered beam

Framing anchors

Existing ceiling joists flush to beam

Existing top plates

"Let-in" joist hanger flush to joist edge

New 4x4 post

Existing 2x4

New blocking

Subfloor   Girder

A hidden beam allows you to remove a bearing wall and still have a smooth expanse of ceiling. A continuous load path—from the beam, through the posts, to the foundation—is crucial to a successful installation.

1. Make the attic workspace safe and comfortable. Place 3/4-in. plywood walkways on both sides of the area where you'll insert the beam. (Scrap plywood 2 ft. wide is ideal.) Tack-nail walkways so they can't drift. Clamp work lights to rafters and add a fan to keep air moving. Even in winter, it can be pretty hot in an attic. If roofing nails protrude from the underside of roof sheathing, wear a hard hat.

If there's enough room in the attic, assemble the beam in place and lower it down into the slot you'll create by cutting back joist ends. However, if you must assemble the beam on the floor below—or if you're raising an engineered beam—use a nylon web sling and a chain fall (see the photo on p. 223) bolted through the rafters to raise the beam up through the cutout in the ceiling. Angle-brace rafters to keep them from deflecting under the load, and don't attach a chain fall to rafters that are cracked or already sagging.

2. Snap a chalkline to mark the beam location onto the top edges of joists. Snap a first line to mark the centerline of the beam. Then measure out half the beam width plus 1/8 in. on both sides, and snap chalklines to indicate cutlines on the joists. Using a square, extend these lines down the face of each joist. Use vivid chalk so the marks will be visible.

(see the photo on p. 223)

PRO TIP

Before attaching joist hangers, make a single pass of a power plane across the underside of each joist end where it abuts the new beam. The planed area, roughly the width and thickness of a joist hanger "stirrup," ensures that the joist hangers will be flush to the bottom of the joists and the beam and that the patched ceiling will be evenly flat.

beam rather than sitting atop it, so the hidden beam will rest on top of end-wall top plates.

Install 4x4 posts between the top and the sole plates—and blocking under the posts—before raising the beam. Snap chalklines on the ceiling to indicate the width of the beam plus 4 in. extra on each side so you can slide joist hangers in later. Cut out drywall or plaster within that slot to open up the ceiling and expose joists. After installing shoring, as explained in earlier sections, go into the attic. You could use the same method to replace a bearing wall on the first floor of a two-story house, but that would involve tearing up the finish floor above to expose the floor joists and making an unholy mess. If you're gutting the house anyhow, that might not be a big problem.

Checking THE LOAD PATH

Because loads will be concentrated on support posts beneath each end of the new beam, those posts must be supported continuously all the way down to girders and to concrete pads, footings, or foundation walls. To make sure there is adequate support below proposed post locations, strip the bearing wall to its studs and insert shoring along both sides before removing corner studs at both ends of the bearing wall. Leave the rest of the wall alone for now.

Using a long auger bit (18 in. by 3/8 in.), drill down through a sole plate at each end, where a post will stand. (In fact, the posts may be hidden in end walls at either end of the bearing wall.) If the bit hits a girder, posts should have adequate support. But if the bit hits air or only a single joist, add solid blocking. That blocking may be a 6x6 atop a girder or a new post and concrete footing—but let a structural engineer decide. Fit the blocking tight to the underside of the subflooring beneath the posts so there can be no deflection when loads are transferred to them.

Because thin reciprocating-saw blades wander, use a circular saw to ensure square cuts across joists. It's hard to see the cutline of a saw you're lowering between two joists, so clamp a framing square to each joist to act as a guide for the saw shoe. Some renovators prefer a small chainsaw for this operation, but hitting a single hidden nail in a joist can snap a chainsaw blade and send it flying. Whatever you use to cut the joists, wear hearing and eye protection, and, ideally, have a similarly protected helper nearby shining a light into the cut area.

3. Get help to raise the beam one end at a time. If your cuts are accurate, you should be able to raise the beam between the severed joists and onto the top plate of one end wall and then the other. But, invariably, the beam will get hung up on something. Here, a chain fall is invaluable because it allows you to raise and lower one end of the beam numerous times without killing your back or exhausting your crew.

When the first end of the beam is up, nail cleats to both sides of the beam so it can't slip back through the opening as you raise the other end. Raise and position the other end of the beam atop the other end wall and directly over the 4x4 support post. Then use a metal connector such as a Simpson BC4 post cap (similar to the BC6 shown in the top photo on p. 86) or an A34 framing angle (see the bottom photo on p. 84) to secure the beam to the top plates.

4. Fine-tune the height of individual joists until their lower edges are flush to the bottom of the beam. This operation is easiest with one worker downstairs using a 2x4 to raise or lower the joist ends as a worker in the attic directs. As each joist is correctly positioned, attach it to the beam using joist hangers and Teco nails. Before attaching joist hangers, however, use a power planer to cut a shallow slot into the underside of each joist to let in the hangers so they're flush to the underside of the joists. If there's not much room to swing a hammer between joists, use a pneumatic palm nailer to drive the nails most of the way. Finally, along the edges of the beam slot, center and end-nail 2x4 backing between the joists for the finish-surface patches to come.

For more details on this complex operation, see John Michael Davis's article "Removing a Bearing Wall" in *Fine Homebuilding* #152.

Heavy Metal

Steel connectors are an important part of renovation carpentry, often joining new and old framing members.

Straps such as Simpson Strong-Tie LSTA strap ties are frequently used where wall plates are cut, at wall intersections, and as ridge ties. Pros often use them to splice new rafter tails to existing rafters (top photo below) to replace sections that rotted at the wall plate. After cutting rafter tails at the correct angle, toenail them to the top plate, and use 12-in. or 18-in. strap ties to tie new tails to the existing rafters. After the sheathing is nailed on, such reinforced rafter tails will stay in line indefinitely.

The 4x14 Parallam beam sitting atop a 2x4 top plate (bottom photo below) is much like the hidden beam discussed in "Installing a Hidden Beam" in the text, starting on p. 223. In the old days, beams of this size would have been merely toenailed to top plates. So the Simpson BC4 post cap now specified by engineers is quite an improvement. It's simple to install and strong enough to resist uplift and lateral movement. Note, too, the 4x4 post directly under the beam—it's part of the load path that goes all the way down to the foundation.

A Post-and-Beam Portfolio

Timber-frame buildings engage us in many ways. They can be as beautiful and enduring as any fine antique. The conciseness of their joinery—whether simple pegged braces or corners as intricate as a Japanese puzzle box—shows a keen intelligence at work. And ax or adz marks inscribed along a post or beam recall the sturdy backs and steady hands of those who hewed the line and transformed virgin forests into homes for future generations.

Yet as enduring as such well-built houses are, they are also records of change. They reflect the skills and building customs that each wave of immigrants brought with them, the boom and bust of business cycles, and the relentless evolution of building materials and techniques.

In the pages that follow, two master joiners discuss the changes, in their own words, in building methods they've observed while restoring timber-frame structures in Vermont and Illinois. We think it's a fascinating story whether you build homes or just live in them because it reflects the social, economic, and technical forces that have shaped this country—and continue to do so. Though it's tempting to see the past through a haze of nostalgia,

Post-and-Beam Repairs:
BEST LEFT TO THE PROS

Timber-framed buildings have fewer structural members than stick-framed ones, so the loads on individual members tend to be greater. Consequently, modifying or moving such elements inexpertly could lead to a catastrophic failure or serious injuries. Understanding load paths and bracing timber structures for repair require specialized training, deep on-site experience, and, frequently, heavy equipment. If you have a post-and-beam that needs structural modification, we strongly suggest that you contact a firm that specializes in dismantling and restoring such structures.

Post-and-Beam House

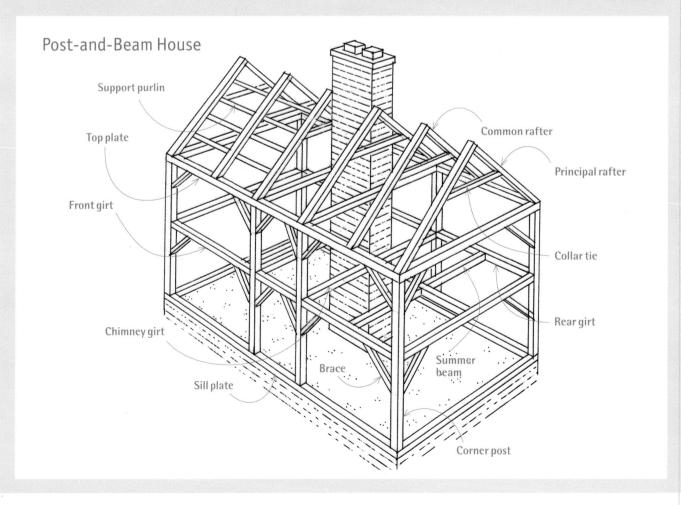

Support purlin

Top plate

Front girt

Chimney girt

Sill plate

Brace

Common rafter

Principal rafter

Collar tie

Rear girt

Summer beam

Corner post

the photos that follow tell quite a different story: of a young country busting at the seams, eager to get to the next big thing, and builders shedding traditional methods for newer, faster ways to build—200 years ago.

Early New England Post-and-Beam Construction

Gregory and Carolynn Schipa are principals of the Weather Hill Company in Warren, Vermont, a nationally known architectural firm specializing in restoring period homes, inns, farms, and the like. In 30 years, they have completed more than 200 projects, several of them post-and-beam structures that they disassembled, moved, reconstructed, and painstakingly restored. (One is shown on pp. 42–44.) Following a youthful apprenticeship working with master craftsmen on historic houses in Nantucket, Greg moved to Vermont and began prowling the woods for houses that deserved saving.

Gregory Schipa
ON FINDING A GEM IN THE ROUGH

"The days when you could buy an abandoned timber-frame house for $500 are long gone, but there are surely old places out there worth saving. You can tell a lot about a house's condition from a quick scan of its exterior. If the ridge is straight, the top plates are likely intact and the tops of the walls are not spreading. If the roof is good, even better, because that suggests that the house has been maintained and maybe lived in recently. The sills, on the other hand, are almost always shot. In Vermont, old foundations were typically dry-laid stone set at grade or on crawlspaces that didn't elevate the wood understructure and sills much above grade. So when we saved houses, we always replaced the sills and put the frames on a brand new deck."

"This is a very early frame in beech—not pine—as the earliest frames were hardwood. A house this primitive in Nantucket might date to 1750, but because Vermont was wild, remote, and less sophisticated architecturally, this one was probably built around 1790. All of its rafters were hand-hewn, and there's no ridge beam because early builders raised one pair of rafters at a time. There's not an ounce of fat on it: Hand-hewn 8x8 floor joists ran the width of the frame, holding in top plates and doubling as collar ties. So the rafters stayed up and the ridge stayed straight for two centuries. Throughout, the joinery was tight: rafters met in full lap joints and were pegged, floor joists were half-lapped to top plates, the rafters sat atop the ends of floor joists—which doubled as cornice blocks. These were simple people trying to get a roof over their heads before winter hit, so they built spare, but with great integrity."

(continued on p. 228)

"This Cape is later than the beech frame discussed earlier because it has pole rafters *trued* (flattened) on one side, which was faster than hewing four sides. But because longer, skinnier rafters tend to sag in the middle, an experienced joiner pegged them to a *purlin* (wind beam), underneath. He then added a diagonal brace to stiffen the purlin, and a second brace to the purlin brace. You could put quite a load on this roof and it wouldn't go anywhere."

"This is a corner joint, lying down. The timber to the right, on which the mallet lies, is an 8x8 top plate. To the left, an 8x14 *gunstock post*. Both are first-growth pine, which is tight-grained and incredibly rugged. If you picture the wall upright, the top plate would rotate 90 degrees to join a top plate from the adjacent wall; the tenon of the post slid into a mortise where the plates intersected and was pegged fast. The older the building, the more amazing and complex the joints. To disassemble such buildings we first had to figure out in what order the joints went together—often a challenge. As long as moisture didn't rot them, such joints held indefinitely: With a big enough crane, you could have lifted the entire frame."

"This is a Greek Revival house built around 1820-1830. Your first hint is that it's got knee walls—so there's more headroom upstairs than in a Cape—and taller walls. The second hint is that all the studs and collar ties are sawn, not hewn. But it was constructed by a builder with considerable integrity: his wind-bracing in the gables, for example, is old-school and extra strong. Only an older builder would have done that; newer builders would have figured out ways to do less. The hand-hewn posts and beams in the corners are also beautifully joined."

"This house is later yet because all the joists are all sawn, as are the posts and bracing. The only hand-hewn element visible is the top plate. In the photo below we see how house construction was evolving toward worse, not better.

"The *girt* (girder) carrying the floor joists and tying the middle of the house together is joined to the post in the foreground solely by a pegged through-tenon. There is a tremendous load on that tenon. Even more telling, there is a huge spread-load on the top plate, atop that post. Yet nothing is holding that top plate in except this post! Typically, such a post would be an 8x8, but it doesn't look much deeper than the diagonal brace, so it may only be a 4x8.

"As you increase the load on that top plate (and post), you break out the tenon—a common failing during this transitional building period from timber frame to sawn stick-framing, which caused many houses to collapse. And when the tenon goes, the ridge sags, the walls spread, and it's all downhill from there. Finally, the short studs coming down from the underside of each brace were once full-length studs that someone sawed off to add a window or to change something downstairs—further weakening the structure."

"To end on a positive note, the Josiah Durkee house (below), built in 1836, survived long enough for me to restore it because its builder was a master of the old school. Sawn lumber was common at that time, but the frame was all hand-hewn. And though that date was well into the Greek Revival architectural period, the house had a lot of details—such as moldings, dentils, and a classic mantel—from the earlier, Federal period. Having taken a lifetime to hone his skills, this earnest joiner was in no hurry to give them up."

(continued on p. 230)

Repairing Rot at the Eaves

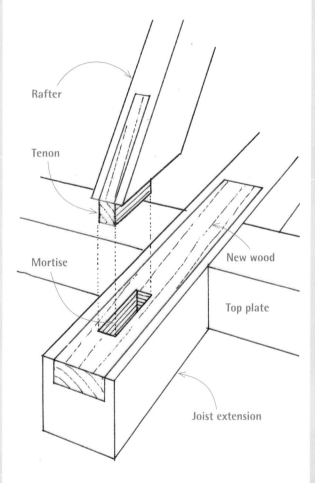

Rafter

Tenon

Mortise

New wood

Top plate

Joist extension

The most damaging leaks are those at the eaves, because they destroy the joint at the rafter base and at the end of the joist or girt, causing a collapse from the natural outward thrust of the rafter weight. The sensitive point at which the 8x8 attic joists lapped the eave plate on Durkee's house had taken some moisture on the south side. I let in new wood on the top of each joist, replacing questionable or punky places and creating new, strong mortises. The rafter tenons were in good shape and could be reused. When they are too far gone, we remove them by cutting right up the middle of the rafter for 3 ft. or 4 ft. We then let in a new 2-in.-thick hardwood plank to the full depth of the rafter, like replacing the contents of a sandwich. The tenon is then recut at the extending end of this piece. (Reprinted courtesy of *Fine Homebuilding* magazine.)

Big Timber in the Heartland

Rick Collins began Trillium Dell Timberworks in the fall of 1996 after graduating from the University of Illinois with a B.S. in Forestry/Wood Science and serving six years in the U.S. Marine Corps Reserve as a combat engineer. One of only 30 registered journeymen in North America, Rick has traveled extensively throughout the U.S. and Europe studying the methods and tooling used by Europeans who settled the Midwest from the 1600s through the 1800s, and working with some of the greatest master builders on both continents. A native of western Illinois, he lives on a small farm with his family in Knox County, just outside of Galesburg, Illinois, the home of poet and Lincoln biographer Carl Sandburg.

Collins has a perspective on builders and building that is refreshingly unsentimental and, historically, quite broad. Quiz him on some seemingly mundane building detail, and you may suddenly find yourself in medieval France or ancient Egypt watching some ancestral builder struggling with a problem that plagues us still. In response to a question about why the timber-frame structures in the photos shown here had rotted out he said, "Rafter plate rot is almost always due to a leak in the roof. With the sill plate, rot typically occurs because a sill is too close to the ground or the builder doesn't put a moisture barrier between the wood sill and the stone foundation—not that they didn't know better but because they didn't want to take the time or pay the money to install a barrier. That's what happened in the photo below.

"At least from medieval times, builders were using sheets of lead between a timber frame and its stone foundation. Historically, American builders cut corners in ways that Europeans didn't. There are several theories why, but I

think part of it was resource-driven. Maybe sheet lead was just too scarce or expensive on the frontier. Or maybe settlers on the move didn't worry about—or didn't even see—buildings as long-term things. There was so much wood around—virgin forests, in many cases—that they saw no need to conserve it.

"Although the wood frame on p. 230 sat directly on a stone foundation, its rot was *not* caused by moisture wicking up from the foundation. There, the rot came from above. The barn in Pennsylvania had clapboard siding and trim nailed directly to the frame. Had it been correctly detailed, the back edges of the corner trim would have been rabbeted so that they overlapped and protected the ends of the clapboards. But that didn't happen. So when the clapboards shrank, creating a $^1/_8$-in. gap, water got in and rotted the frame. It was just negligence. These were sophisticated builders, but someone didn't want to spend the time and money to detail corners correctly.

"The posts of the Illinois barn shown at right were 10x10s, 30 ft. long—white pine from Wisconsin rafted

down the Mississippi and sawn. Problem was, mills typically did not saw boards longer than 24 ft., so there was a seam where the 24-ft. siding boards above met the 8-ft. boards below. That seam was not flashed, so water got in and rotted the girts (horizontal members) immediately below—all the way around the barn—and several posts as well. (The sills were also gone, but we elected not to replace them when we rebuilt.) Then, once the timbers started rotting, mice and rats moved in and tunneled through the softened wood.

"To repair the damaged portion of the corner post, we used a variation of a traditional *scarf joint*—in this case, a

bladed cogged joint. A scarf joint is commonly used for vertical connections because it's quite strong; this joint is basically a half-lap joint with a two-way tenon. Because the cog and blade are perpendicular to each other, they resist bending in two directions. To resist separation, the joint is pegged together. The assembly requires precise joinery, which we prefer to cut in our shop. Because this building was slated to be moved, we took the frame apart, cut out bad pieces back in the shop, cut and fit replacement sections, and then reassembled it all on a new site.

(continued on p. 232)

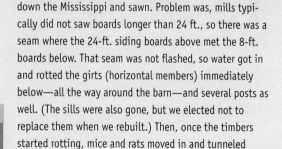

Corner Repair: THREE OPTIONS

Real-world repairs are usually a trade-off between how pervasive the rot is, what loads the section bears, and how expensive repairs are. Before attempting any repair, install shoring to temporarily take loads off the section being repaired.

An in-place repair is usually the simplest. Cut away the rotted section and laminate new boards to the existing timber. To repair an 8x8, one would typically laminate a 2x8 or a 3x8 onto the back. New wood is less likely to kick out if it is under-squinted (angled slightly rather than being installed perfectly on edge) and structurally screwed to the good wood.

If an in-place repair isn't possible, the next option is removing the rotting section and joining new wood to old with a vertical scarf joint, as shown at right.

If a vertical scarf joint isn't possible, completely replace the old post with a new one.

If you are considering an in-place repair, read up on epoxies. If for example, you have a post bottom that must carry some load, a *compressive epoxy* might be appropriate. If, on the other hand, you want to reinforce the back of a doty timber so that nails have something solid to bite into, use a *filler epoxy* such as Abatron (www. abatron.com).

"When a leak in the roof is within 10 ft. or 15 ft. of a rafter plate (top plate), the water often runs down the rafter and pools on the rafter plate. This problem is particularly acute when plate sections meet in a step-lap joint (also called a *skew notch*), because the joint makes a trough in the top of the plate, in which water pools. If plates are white pine, they can rot out in 10 years; if white oak, 25 to 30 years. In the section shown above, a valley rafter ended at a post top and rotted it out as well.

"Once you've fixed the cause of the rot (such as a leaky roof), it is generally OK to leave old rafters in place—as long as their appearance is not an issue. In barns, for example, owners often do not want to spend the extra time and money to remove old rafters. In that case, place solid new rafters against the old ones, *sister* them snug with structural screws, and move on.

"Here, new replacement wood clearly shows the path that water took as it ran down and rotted an octagonal barn in Wisconsin. It all started with undersized rafters, a common failing of 19th-century Midwestern roofs. Because the *purlin plate* couldn't support the load of the sagging rafters, the roof bellied, shunting water down the center of the section, where it rotted the rafter plate. It next soaked the post so thoroughly that it became as hollow as a downspout, dumping water onto the sill. Accelerating the rot at the top were pine shingles that never dried out where the roof bellied.

term is a *Jupiter scarf joint;* the German term is a *lightning-bolt scarf joint.* Here, both new and old sections are white pine, held fast with oak pegs.

"The vertical strap is a Simpson strap (photo below left) that the mason embedded in the new foundation to act as a hold-down for the frame. The horizontal strap with the nut on it is a repair of an original strap, dating from the 1880s, that tied corner elements together. Many Americans think of structural steel as something new, but it's an old wives' tale. Builders have used iron straps and tension rods for millennia—especially in multifaceted buildings or those with clear spans that required joinery. Romans used iron in trussed constructions throughout the empire, and in big buildings all across Europe—in the cathedrals, for example, you'll find iron.

"Where it's essential that horizontal joints not pull apart—as when joining sections of a top plate or a sill—carpenters rely on a scarf joint (shown above). It's been used for centuries. I've seen joints like this in Europe dating to the 13th or 14th century, where there's five or six of these in a row. In English, this is called a *stop-splayed under-squinted wedge scarf;* the translation of the French

"Laminated beams are also quite old. To reinforce a mezzanine in the large timber building below, we employed *laminated wedged beams* that were common in Europe in the 13th and 14th centuries. Joined together, the individual beams have greater depth and greater load-bearing capacity because inclined shear blocks keep them from sagging—as do the compression bolts that hold things together. Bearing blocks support both ends of the beam."

9 Masonry

By deeply tooling mortar joints with a square–edged raking tool, this mason created dramatic patterns of light and shadow.

Modern masonry employs a range of materials, including stone, brick, tile, concrete, and other minerals that become strong and durable when used in combination. The craft of masonry is ancient. The oldest surviving buildings are stone, but stone is heavy and difficult to work with. Brick, on the other hand, is less durable than stone but lighter and easier to use. And clay, the basic component of brick, is found almost everywhere.

Technologically, the switch from stone to brick was a great leap in several respects. First, masons began with a plastic medium (mud and straw) that they shaped into hard and durable building units of uniform size. Second, brickmaking is one of the earliest examples of mass production. Third, basic bricklaying tools, such as trowels, were so skillfully designed that they've changed little in 5,000 years.

Terms, Tools, and Tips

Unless otherwise specified, mixes and methods in this chapter are appropriate for brickwork as well as concrete-block work. But most of this chapter is about brick and poured concrete because concrete-block work is uncommon in renovation.

TERMS

Here's a handful of mason's lingo that's frequently confused:

Portland cement. The basic component of all modern masonry mixtures. When water is added to cement, it reacts chemically with it, giving off heat and causing the mix to harden, thus bonding together materials in contact with the mix. By varying the proportions of the basic ingredients of a concrete mix, the renovator can alter the concrete's setting time, strength, resistance to certain chemicals, and so on. Portland cement is available in 94-lb. bags.

Masonry cement. Also called mortar cement, a mix of portland cement and lime. Exact proportions vary. The lime plasticizes the mix and makes it workable for a longer period. Once dry, the mix also is durable.

Aggregate. Material added to a concrete mix. Fine aggregate is sand. Coarse aggregate is gravel. Concrete aggregate is typically ¾-in. gravel, unless specifications call for pea gravel (⅜-in. stone).

Mortar. Used to lay brick, concrete block, stone, and similar materials. As indicated in "Mortar Types" on p. 237, mortar is a mixture of masonry cement and sand or of portland cement, lime, and sand. It's typically available in 60-lb. bags.

Grout. A mix of portland cement and sand or of masonry cement and sand. Mixed with enough water so it flows easily, grout is used to fill cracks and similar defects. In tiling, grout is the cementitious mixture used to fill joints.

Non-shrink precision grouts (NSPG) are useful for a number of renovation tasks such as patching holes in concrete floors, building up a small area of concrete subfloor, or filling under the base plates of structural steel columns. Quikrete and RapidSet make NSPGs, which set up in approximately 15 minutes, are incredibly strong (7,000 psi), and easy to mix, apply, and set. Most NSPGs can be colored or tinted to make patches less obvious.

Concrete. A mixture of water, portland cement, sand, and gravel. Supported by forms until it hardens, concrete becomes a durable, monolithic mass.

Reinforcement. The steel mesh or rods embedded in masonry materials (or masonry joints) to increase resistance to tensile, shear, and other loads. In concrete, the term usually refers to steel rebar (reinforcement bar), which strengthens

A bricklayer's tool kit (clockwise, from upper left): 4-ft. brass-bound level, tool bag, 6-ft. folding rule, statistical booklet (which covers estimating bricks and blocks, portland cement types, metric conversion, and a glossary of masonry terms), 11-in. steel trowel, 5½-in. pointing trowel, brick hammer, two convex jointers, 4-in. brick set, box of line clips, and yellow stringline.

foundations against excessive lateral pressures exerted by soil or water.

Admixtures. Mixtures added to vary the character of masonry. They can add color, increase plasticity, resist chemical action, extend curing time, and allow work in adverse situations. Admixtures are particularly important when ordering concrete because mixes may contain water reducers, curing retardants, accelerants, air entrainers, and a host of other materials that affect strength, curing times, and workability.

BASIC MASONRY TOOLS

Most of the tools listed in this section are hand tools. Chapter 3 describes impact drills, rotary hammers, and other useful power tools. *Important:* Wear safety *goggles* and a respirator when striking, grinding, or cutting masonry. Errant chunks of masonry can blind you, and masonry dust is not stuff you want to breathe. Cement is also highly caustic, so heavy gloves are essential—preferably rubberized ones.

Trowels are indispensable masonry tools. If you have no other tool, a trowel can cut brick, scoop and throw mortar, tap masonry units into place, and shape mortar joints. A good-quality trowel has a blade welded to the shank. Cheap trowels are merely spot welded. Bricklayer's trowels tend to have blades that are 10 in. to 11 in. long. Pointing trowels, which look the same, have blades roughly 5 in. long; they're used to shape masonry joints. Margin trowels are square-bladed utility trowels used for various tasks.

THE Point OF IT ALL

When you see the term *pointing* in masonry texts, someone is doing something to mortar joints—usually shaping or compressing them so they weather better. *Repointing* or *tuck-pointing* refers to adding (and shaping) new mortar after old, weak mortar has been partially removed from a joint, usually with a tuck-pointing chisel or a tuck-pointer's grinder.

PRO TIP

Online videos are a great way to see masons' *movements* as they work. Whether spreading plaster, throwing mortar, or finishing surfaces, the masters move with an economy and fluidity that may take years to learn.

Jointers (striking irons) compress and shape mortar joints, some of which are shown in "Mortar Joints" on p. 241. The most common are bullhorn jointers, shown in the bottom photo on p. 243, and convex jointers, shown in photo 5 on p. 240. The half-round, concave mortar joint they create sheds water well.

Tuck-pointing trowels are narrow-bladed trowels (usually the width of a mortar joint, ⅜ in.) used to repoint joints after old mortar has been cut back. Because it packs and shapes mortar, this tool is both trowel and jointer and has more aliases than an FBI fugitive: tuck-pointing trowel, jointing tool, repointing trowel, striking slick, slicker jointer, slicker, and slick.

Tuck-pointing chisels partially remove old mortar so joints can be repointed (compacted and shaped) to improve weatherability. Angle grinders and pneumatic chisels also remove mortar.

Mason's hammers score and cut brick with the sharp end and are used to strike hand chisels with the other. The blunt end is also used to tap brick down into mortar.

Brick sets have a cutting edge beveled on one side, so you can cut bricks precisely or dislodge deteriorated brick without damaging surrounding ones.

Brick cutters are rentable levered tools that precisely cut or "shave" brick, as you often must do when fitting firebricks to fill gaps in a firebox.

Line blocks (or pins) secure a long, taut line to align masonry courses. They're less important in renovation masonry, where you're often filling in between existing courses or laying short runs, such as the sides of a chimney.

Mason's levels are indispensable for leveling courses and assessing plumb. Generally 4 ft. to 6 ft. long, better-quality levels have an all-metal casing and replaceable vials. As you work, be sure to wipe wet concrete or mortar off a level before it hardens.

Brick tongs enable you to carry up to 10 bricks comfortably, as if they were in a suitcase.

Concrete tools include floats (used to level concrete), finishing trowels (for smoothing surfaces), and edgers (short tools that contour edges). You'll also need a strike-off board (usually a straight 2x4) for leveling freshly poured pads. Photos of these tools in use appear in chapter 10.

Miscellaneous tools include safety goggles, knee pads, rubber gloves, rubber boots (concrete work), a flat bar, and a homemade mortarboard (a platform that holds mortar near the work) made from scrap plywood. You'll also need sheet plastic to cover sand or cement, a concrete mixer or a mortar pan, a wheelbarrow, square-nose shovels, buckets, a garden hose, and stiff-bristle brushes, among other things.

PREP TIPS

The following tips will help your job go smoothly.

Code. Check local building codes and get necessary permits.

Water. Protect materials from rain. Because water causes cement to set, sacks of portland or mortar cement left on the ground—or on a seemingly dry concrete floor—will harden and become useless. If you store materials outdoors, elevate sacks on a pallet or scrap lumber and cover the pile with sheet plastic, weighting down the edges with rocks.

Although bricks should be wetted before being laid, don't leave them uncovered in a downpour. They will absorb too much water, which can dilute the concrete and weaken the bond. (Concrete blocks, on the other hand, should be laid dry. Don't wet them beforehand.)

Sand and gravel are little affected by water, but if they absorb a lot of water, you'll need to reduce the amount of water you add to a mortar or concrete mix. Damp sand won't ball up when you squeeze a fistful; it contains about 1 qt. of water per cubic foot. Wet sand will ball up and will contain about 2 qt. of water per cubic foot. Dripping-wet sand oozes water when you squeeze it and will contain about 3 qt. of water per cubic foot. Of greater concern is the purity of these aggregates: Unload them onto an old sheet of

Brick tongs let you carry bricks as if they were in a suitcase.

plywood or a heavy (6-mil) plastic tarp to keep them from being contaminated with soil or other organic matter.

Weight. Masonry materials are heavy. To save labor, have materials delivered close to the work site. Likewise, have a mortarboard within 3 ft. of your work area and about waist high so you don't need to bend over to scoop mortar. For this reason, scaffolding is a sensible investment if you will be working higher than shoulder height. Divide materials into loads you can handle without straining, and use ramps and wheelbarrows when possible. As you lift, get close to the object and lift with your knees, not your back.

Game plan. Before mixing mortar, complete preparatory work, such as chiseling out old joints, removing old brick, and brushing dust off receiving surfaces.

Curing. Give masonry time to cure. Because freezing compromises strength, plan your work so the mortar joints or new concrete will set before temperatures drop that low. Admixtures can extend the temperature range in which you can work, but exterior masonry work is easiest when the 24-hour temperature range is 40°F to 80°F. On hot summer days, start early—preferably on a shady side of the house—and follow the shade around as the day progresses. Cover fresh work with burlap sacks, dampened periodically, or with sheet plastic. The longer masonry stays moist, the stronger it cures.

Protecting surfaces. Spread tarps to catch mortar droppings. And if you're working on a chimney, tack plywood over the windows to protect glass from falling bricks or tools.

Cleanup. At the end of the day, clean tools well. Wet them down and use a wire brush as needed to remove hardened materials. Before lunch breaks or at the end of the day, run a few shovelfuls of gravel and a few buckets of water in the concrete mixer to loosen caked materials. Then dump it out, ensuring that the barrel wall and mixer blades are clean.

Working with Brick

Common brick-related repairs include repointing mortar joints, repairing chimney tops, rebuilding chimneys and fireboxes, and cleaning bricks. You may also have to repair or add flashing where the chimney meets the roof, as shown in chapter 5. Less common repairs include filling openings after the removal of doors or windows. If you want to create an opening in a brick wall, leave that to a structural engineer and a licensed and insured mason.

Bricklaying Terms

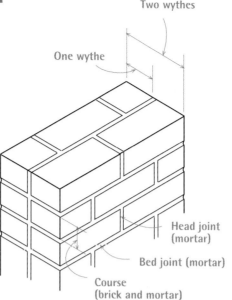

Two wythes

One wythe

Head joint (mortar)

Bed joint (mortar)

Course (brick and mortar)

Mortar Types

Mortar is usually classified according to its strength and weatherability. The table on p. 240 describes the correct proportions of ingredients for each.

▶ **Type M** has the highest compressive strength, at least 2,500 pounds per square inch (psi). This durable mix is recommended for load-bearing walls, masonry below grade, and masonry not reinforced with steel.

▶ **Type S** has a relatively high compressive strength (1,800 psi) and the best tensile strength of any mortar listed here, so it best resists wind and soil movement.

▶ **Type N** offers medium compressive strength (800 psi) and is suitable for all above-grade uses, including those subject to heavy weathering, such as chimney mortar.

▶ **Type O** has a low compressive strength (325 psi) and is limited to non-load-bearing interior uses. However, it is sometimes specified for repointing chimneys with soft, old brick that would be destroyed by stronger mortar (see "The Mortar Mix" on p. 242 for more information).

▶ **Type K** is an extremely low-strength (100 psi) mortar and is not recommended.

Of the mortar types listed here, type N is the most versatile. A simplified version of its proportions is 1 part portland cement, 1 part lime, and 6 parts sand (or 1 part masonry cement and 3 parts sand). Before portland cement became widely used in the 19th century, mortar was usually a mixture of lime and sand (animal hair was often added to reduce cracking). If brickwork 100 years old or older needs repointing, use type O so it won't destroy the brick (roughly, 1 part portland cement, 2 parts lime, and 4 parts fine sand).

TWO WAYS TO CUT BRICK

Using a mason's hammer, score all the way around the brick, then strike the scored lines sharply. Cuts will be more accurate if you place the brick on a bed of sand.

A brick-cutting tool is safer and quieter than using a diamond blade in a power saw or a grinder, throws off almost no dust, and can precisely slice small amounts off a brick. But if you have only a few bricks to cut, score the brick with a grinder and break it by striking the scored line with a mason's hammer or the claw end of your hammer. Safety glasses, please!

To conserve resources and get the best-looking results, respect existing masonry. Match existing bricks and mortar as closely as possible, including the width of mortar joints. When repointing brick, choose a mortar of appropriate strength.

TYPES AND TERMS

Of the many types of brick available, renovation calls mainly for building brick, also called common brick. Building brick is classified according to its weathering grade: SW (severe weathering), MW (moderate weathering), and NW (nonweathering). SW grade should be used where brickwork will be below grade—that is, in contact with the soil and hence subject to freezing in cold climates. Use SW on all floors, whether indoor or outdoor. MW grade is used indoors or on exteriors above grade. NW is used only indoors, though not as flooring.

Standard-size brick is nominally 8 in. by 4 in. by $2\frac{2}{3}$ in., but it is actually $7\frac{5}{8}$ in. by $3\frac{5}{8}$ in. by $2\frac{1}{4}$ in. to accommodate mortar joints $\frac{3}{8}$ in. thick. Thus, three courses of brick (and mortar) will be approximately 8 in. high.

Brick also is named according to how it is positioned, whether it is laid on its face, end, or side. *Stretcher* and *header* are the most common placements, with *rowlock* patterns often being used to finish courses beneath windowsills or to cap the tops of walls where coping isn't used.

In masonry work, the word *bond* has several meanings. Mortar bond denotes the adhesion of brick (or block) to mortar. Structural bond refers to the joining or interlocking of individual units to form a structural whole. If there are two wythes (pronounced w-EYE-ths) of brick (a double wall), the wythes may be bonded structurally by steel ties, by header bricks mortared into both wythes, or by grout poured into the cavity between the two *wythes*. Finally, pattern bond indicates brick placement, as shown in "Bond Patterns" on the facing page.

If you're laying up a typical brick pattern—say, running bond—you will need about $6\frac{3}{4}$ bricks per square foot of wall; figure 7 bricks per square foot to have enough extra for waste. As you handle bricks, inspect each for soundness. All should be free of crumbling and structurally significant cracks. When struck with a trowel, bricks should ring sharp and true.

Brick Names Based on Positioning

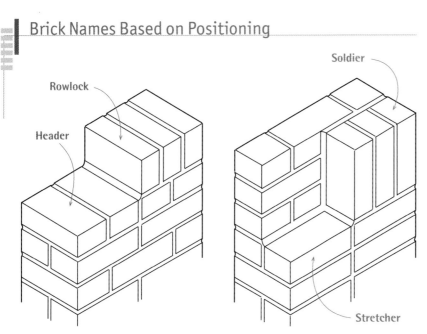

Rowlock

Header

Soldier

Stretcher

BASIC BRICKWORKING TECHNIQUES

You should wet bricks before using them so they won't absorb moisture from the mortar. Hose down the brick pile once a day (more often in hot, dry weather), but don't overdo it. If the bricks become too wet, they will slide around on the mortar bed.

Cutting bricks. Wear eye protection when cutting bricks. If you're cutting across brick faces, rent a brick cutter, a levered tool that cuts easily. Otherwise, cut bricks by hand.

An experienced mason can score and cut bricks with only a trowel, but you'll probably find that a brick set works better. Placing the brick on a bed of sand makes the brick less likely to jump when you strike the brick set. After marking the cutline on the face of the brick, hold the brick set perpendicular to the brick and strike it with a 4-lb. hand sledge. Because the edge of the brick set is beveled on one side, keep the bevel on the waste side of the line. You can also use a mason's hammer as shown in the top photo on the facing page, controlling the cut by using the hammer point to score entirely around the brick. Then rap the scored line sharply to break the brick.

Dry mixing. When mixing mortar, mix the ingredients dry first to ensure a uniform mixture. Once that's done, create a pocket in the middle, and add water gradually. As you add water, be fastidious about turning out the material in the corner of the mixing pan so that there will be no dry spots. Mortar should be moist yet stiff. A batch that's too wet will produce a weak bond. Once the mix is nearly right, its texture will change radically if you add even a small amount of water.

Mortar will remain usable for about two hours, so mix only about two buckets at a time. If the batch seems to be drying out, "temper" it by sprinkling a little water on the batch and turning it over a few times with a trowel. As you seat each course of bricks in mortar, use the trowel to scrape excess mortar from joints gently and throw it back into the pan or onto the mortarboard. Periodically turn that mortar back into the batch so it doesn't dry out. Don't reuse mortar that drops on the ground.

Trowel techniques. Hold a trowel with your thumb on top of the handle—not on the shank or the blade. This position keeps your thumb out of the mortar, while giving you control. Wrap your other fingers around the handle in a relaxed manner.

There are two basic ways to load a trowel with mortar. The first is to make two passes: Imagine that your mortar pan is the face of a clock. With the right-hand edge of the trowel raised slightly, take a pass through the mortar from 6:30 to 12:00. Make the second pass with the left-hand side of the blade tipped up slightly, traveling from 5:30 to 12:00. According to master mason and author Dick Kreh, a trowelful of mortar should resemble "a long church steeple, not a wide wedge of pie."

The second method is to hold the trowel blade at an angle of about 80° to the mortarboard. Separate a portion of mortar from the main pile and, with the underside of the trowel blade, compress the portion slightly, making a long, tapered shape. To lift the mortar from the board, put the trowel (blade face up) next to the mortar, with the blade edge farthest away slightly off the board. With a quick twist of the wrist, scoop up the mortar. This motion is a bit tricky: If the mortar is too wet, it will slide off.

To unload the mortar, twist your wrist 90° as you pull the trowel toward you. This motion spreads, or *strings*, the mortar in a straight line.

Bond Patterns

Running

Common

Roman third

Stretcher

Flemish

Dutch cross–bond

Stacked soldier

Stacked header

Running, common, Roman third, Flemish, and Dutch cross-bond are stronger because their head joints are staggered.

Mortar Types (ASTM C 270-68)*†

TYPE	PORTLAND CEMENT	MASONRY CEMENT	HYDRATED LIME or LIME PUTTY	AGGREGATE‡
M	1	1	—	Not less than 2¼ and not more than three times the sum of the combined volumes of lime and cement used
	1	—	¼	
S	½	1	—	
		—	>¼-2	
N	—	1	—	
	1	—	>½-1¼	
O	—	1	—	
	1	—	>1¼-2½	
K	1	—	>2½-4	

* Adapted from the publications of the American Society for Testing and Materials, as are the compression figures given in the text.

† Parts by volume.

‡ Measured in a damp, loose condition.

It is a quick motion, at once dumping and stringing out the mortar, and it takes practice to master. If you are laying brick, practice throwing mortar along the face of a 2x4, which is about the same width as a wythe of brick. Each brick course gets a bed of mortar as wide as the wythe.

After you've strung out the mortar, furrow it lightly with the point of the trowel to spread the mortar evenly. Trim off the excess mortar that hangs outside the wythe, and begin laying brick.

Laying brick. If the first course is at floor level (rather than midway up a wall), snap a chalkline to establish a baseline. Otherwise, align new bricks to existing courses.

Throw and furrow a bed of mortar long enough to seat two or three bricks. If you're filling in an opening, "butter" the end of the first brick to create a head joint, as shown in photo 2 below. Press the brick into position and trim away excess mortar that squeezes out. Both bed and head joints are ½ in. to ⅝ in. thick until the brick is pressed into place, with a goal of compressing the joint to about ⅜ in. thick.

Use both hands as you work: One hand maneuvers the bricks, while the other works the trowel, scooping and applying mortar and tapping bricks in place with the trowel handle. If you use a stringline to align bricks, get your

FROM THE ARCHIVES: OLD-SCHOOL BRICKLAYING

1. Experienced masons lay up bricks from the corners in, moving string guides up as they complete each course.

2. After using a bricklayer's trowel to throw and furrow a mortar bed, butter one end of the brick to create a head joint.

3. Before the mortar is compressed, it is ½ in. to ⅝ in. thick, as shown. Press the brick into the mortar to create a good bond, compressing the mortar to ⅜ in. thick.

4. After using the end of the trowel handle to tap the brick down, trim off the excess mortar.

5. After the mortar joints have set enough to retain a thumbprint, strike (tool) them to compress the mortar and improve weatherability. Strike the head joints, as shown, before striking the bed joints. (This tool is a convex jointer.)

thumb out of the way of the string just as you put the brick into the mortar bed. As you place a brick next to one already in place, let your hand rest on both bricks; this gives you a quick indication of level.

When you have laid about six bricks in a course, check for level. Leaving the level atop the course, use the edge of the trowel blade to tap high bricks down—tap the bricks, not the level. Tap as near the center of the bricks as the level will allow. If a brick is too low because you have scrimped on mortar, it's best to remove it and reapply the mortar.

Next, plumb the bricks, holding the level lightly against the bricks' edges. Using the handle of the trowel, tap bricks until their edges are plumb. (Hold the level lightly against the brick, but avoid pushing the level against the face of the brick.) Finally, use the trowel handle to tap bricks forward or back so that they align with a mason's line or a level held lightly across the face of the structure.

The last brick in a course is called the closure brick. Butter both ends of that brick liberally and slide it in place. The bed of mortar also should be generous. As you tap the brick into place with the trowel handle, scrape excess mortar off, ensuring a tight fit. If you scrimp on the mortar, you may need to pull the brick out and remortar it, perhaps disturbing bricks nearby.

Striking joints. Striking the mortar joints, also called tooling the joints, compresses and shapes the mortar. Typically, a mason will strike joints every two or three courses before the mortar dries too much. To test the mortar's readiness for striking, press your finger into it. If the indentation stays, it's ready to strike. If the mortar's not ready for striking, wet mortar will cling to your finger and won't stay indented.

Use a jointer to strike joints. First, strike the head joints and then the bed joints. The shape of the joint determines how well it sheds water. As suggested in "Mortar Joints," joints that shed water best include concave (the most common), V-shaped, and weathered joints. Flush joints are only fair at shedding water. Struck, raked, and extruded joints shed poorly because they have shelves on which water collects.

REPOINTING MORTAR JOINTS

Even materials as durable as brick and mortar break down in time, most commonly near the top of a wall or chimney, where masonry is most exposed to the elements. Often, the structure wasn't capped or flashed properly. If the bricks are loose, remove them until you reach bricks that are solidly attached. If joints are weathered

Throwing Mortar

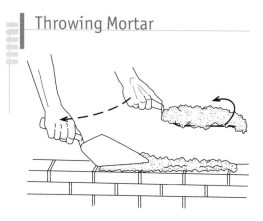

As you turn the trowel to unload the mortar, pull it toward you quickly, thus stringing the mortar in a line.

Mortar Joints

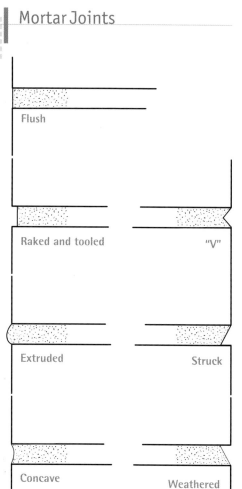

Flush

Raked and tooled

"V"

Extruded

Struck

Concave

Weathered

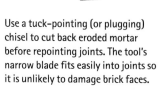

Use a tuck-pointing (or plugging) chisel to cut back eroded mortar before repointing joints. The tool's narrow blade fits easily into joints so it is unlikely to damage brick faces.

but bricks remain firmly attached, repoint (or tuck-point) the joints by partially cutting back the mortar, adding new mortar, and shaping the joints. If the brick is painted, see "Stripping Painted Brick" on the facing page. Finally, if vertical or diagonal cracks run through several courses, there may be underlying structural problems, which must be corrected before repointing. In this case, consult a structural engineer.

Raking old mortar. For best results, rake out (scrape out) mortar joints in an inconspicuous area as a test, starting with the least destructive tool. If the mortar is soft enough, an old screwdriver may be all you need. But if the mortar is as hard as the brick, you'll need to be patient. Cut joints 1 in. deep, and try to cut a square trough (not a V-groove) in the old mortar.

Once you've raked the joints to the correct depth, brush them out well, using a whisk broom or a wallpaperer's brush, which you can also use to wet down the joints before adding fresh mortar. Remove debris with an air hose or a heavy-duty vacuum. Of course, wear safety glasses and a respirator for this work.

The mortar mix. Try to match the old mortar mix when repointing an older brick building. Before portland cement was widely used, mortar joints were usually a resilient mix of hydrated lime and sand, which compressed slightly as the bricks expanded during summer and expanded slightly as the bricks contracted during cool weather. Soft, lime-rich mortars also show autogenous healing, an ability to self-repair hairline cracks caused by seasonal temperature shifts. Mortar joints with portland cement, on the other hand, are relatively hard and inflexible: As old

bricks heat up, they have no room to expand, so they crack and spall (flake).

Although mortar analysis is the best way to match old mixes—historic preservation agencies can suggest mortar analysts—type O mortar (described in "Mortar Types," p. 237) should be a close match for most old mortars. For this, mix 1 part portland cement, 2 parts hydrated lime, and 8 parts fine sand. The mix should be fairly stiff as mortar mixes go, keeping its shape when squeezed into a ball. If the mix becomes too stiff to work, periodically sprinkle on and stir in small amounts of water.

If you're repointing only a section of a structure, however, and don't want it to stick out like a sore thumb, experiment with combinations of mortar dye, cement, sand, and lime, carefully labeling the proportions of each batch and allowing it to dry for a month before committing to a recipe. Masonry-supply houses stock such materials; they also carry new bricks manufactured to look old should you need to replace bricks as well.

Repointing technique. Using a spray bottle or a brush, dampen the newly cleaned-out joints before applying fresh mortar. There are two ways to fill joints with new mortar. If you're repointing a relatively small area, use a bricklayer's trowel as a palette for the mortar and a tuck-pointing trowel to push the mortar into the joint (see the bottom photo on p. 252). Press the mortar firmly so that it will stick.

If you're repointing a large area, use a grout bag to squeeze the mortar into the joint. A grout bag looks like the pastry bag used to dispense fancy icing onto cakes. You force the mortar out by twisting the canvas bag. But you'll need strong hands and forearms to twist the filled 5-lb. to 10-lb. bag. And you may need to thin the mix slightly so it will flow easily through the bag. After using a grout bag, you'll still need to tuck-point the mortar joints.

When the mortar has dried enough to retain the imprint of your thumb, tool the joints. In most cases, use a jointer that creates mortar joints the same shape as the old ones. Point the head joints first, then the bed joints. As you work, use a trowel to clean the mortar from the brick faces, but don't disturb the mortar joints. Then wait two to three hours before using a stiff plastic-bristle brush to remove the mortar still stuck to the brick faces.

CLEANING AND SEALING EXTERIORS

Use the gentlest, least damaging cleaning agents, chemicals, steam cleaner, or water pressure that works. Determine this by testing in an inconspicuous area. If the gentlest method doesn't work, move to the next strongest. If mortar joints are eroded, a pressure washer may make them leak. After cleaning, allow the brickwork to dry for two to three days, then caulk gaps around doors and windows and replace worn flashing. *Safety note:* Whatever cleaning method you choose, wear a face shield, rubber gloves, protective clothing, and a respirator.

Cleaning brick surfaces. Get bids on hiring an authorized cleaning service. Because cleaning solvents can be hazardous and must be disposed of according to Environmental Protection Agency (EPA) and local environmental standards, hiring professionals will spare you those headaches. Moreover, the service will be responsible for achieving the desired results, however long it takes. If you've got a tight budget or an adventurous spirit, consider the DIY options described next; they're listed more or less in order of gentleness. Wear safety goggles and rubber gloves for all procedures, and read operating manuals carefully before using pressure washers, steam cleaners, and the like.

Use a garden hose to soak the surface, and then scrub with a nylon scrub brush. The warmer the water and the longer the soak, the more dirt you'll remove.

If the hose wash isn't sufficient, try a pressure washer on a low setting. Increase the pressure slightly—say, to 300 psi to 400 psi—and you'll remove yet more. *Note:* If you see sand in the runoff water, lower the machine's pressure settings immediately. Otherwise, you may be stripping the mortar joints. Likewise, monitor the inside of the building, especially around windows, for leaks; it's easier to lower the pressure than to replace drywall.

Steam cleaning is especially effective if surfaces are mossy or have ivy "trails" or built-up grime. Although somewhat slower than pressure washing, steam doesn't generate the volume of runoff and won't penetrate as deeply into brick surfaces or cracks.

If you're in an urban area where soot and auto exhaust have soiled the building, try a nonionic detergent with a medium pressure (1,000 psi) next. Nonionic detergents such as Rhodia's Igepal and Dow's Tergitol and Triton won't leave visible residues, as household detergents and trisodium phosphate (TSP) will. Again, scrub with a synthetic-bristle brush, and rinse well.

If these methods don't produce the results you want, proprietary chemical cleaners are the next step. They usually involve a three-step process— wetting surfaces, applying the cleaner and scrubbing it in, and then rinsing—repeated as many times as needed. If you apply the cleaner, follow

WORKING MORTAR INTO JOINTS

Option 1: Let your bricklayer's trowel serve as a palette as you scoop mortar from it with your smaller pointing trowel and press mortar into joints. Two trowels are useful when repointing old joints.

Option 2: Or you can use a grout bag if you've got a lot of joints to fill. But this is hard work: The bag is heavy, and you need to twist hard to force the mortar out a small opening—a bit like wringing water from a stone.

Here, a bullhorn jointer compresses mortar joints.

Stripping PAINTED BRICK

All in all, sandblasting a painted brick wall is much more destructive than using a pressure washer to strip paint. For this reason, we recommend never sandblasting an *exterior* brick wall. Once brick pores have been enlarged or mortar joints have been eroded by sandblasting, an exterior wall will be too fragile to withstand weathering, especially in wet, cold climates where moisture can soak brick, freeze, expand, and eventually destroy it.

Interior walls are another story because they aren't exposed to weathering. Renting a pressure washer (1,800 psi to 2,500 psi) will be easier on brick and is appropriate if there are only one or two coats of paint to remove. If there are five or six coats of paint, however, sandblasting is about the only way to get it off. Be advised, though, that either method will be incredibly messy. Brick dust and paint scraps get everywhere even when you think you've sealed everything. Cleanup seems to go on for months, but the exposed brick is gorgeous.

the manufacturer's instructions to the letter. Instructions will be quite specific about safety garb, dilution rates, dwell times (how long the chemical remains on), washer settings, and temperature ranges (most don't work well below 50°F). Before committing to a cleaning system, visit the manufacturer's website and call its tech-support number.

Sealing exteriors. Are water-repellent or waterproof coatings necessary on exterior masonry walls above grade? Mostly, no. There may be a few 200-year-old buildings in every city whose porous brick would benefit from being coated, but most masonry exteriors won't admit water if rain is directed away from the structure by gutters, downspouts, and other standard drainage details and if the masonry is properly flashed, caulked, and detailed.

Exterior sealants, loosely divided into water-repellent and waterproofing coatings, don't fully seal masonry surfaces, nor would you want them to. A perfect seal could trap water inside the walls. Moreover, masonry walls with water trapped inside and walls that are wicking moisture from the ground will, in time, exude soluble salts in the masonry as powdery white substances called efflorescence.

Water-repellent coatings, which are typically clear, penetrate masonry pores and keep rain from penetrating to a large degree, while allowing water vapor from living areas to escape through the wall. Most water-repellent compounds are water based, formulated from silanes, siloxanes, and silane/siloxane blends. Both premixed and concentrated coatings are available—

typically applied in several coats. Some water-repellent coatings double as graffiti barriers, although they tend to be shiny.

Waterproof coatings come closer to being true sealers because they're usually pigmented or opaque and form a thin elastomeric (flexible) film. Chemically, they run the gamut from water based to bituminous. Bituminous varieties are widely applied below grade on building foundations and, to a lesser extent, to interior basement walls where mild leaks have occurred.

Chimneys

Masonry chimneys are freestanding units that carry exhaust gases out of the house. To prevent superheated gases from escaping, chimneys should be tile lined and free from cracks or gaps, or they should have insulated stainless-steel flues. Annual inspections and maintenance are crucial to chimney health: If you discover mortar or flue tiles that are cracked or missing, the chimney is unsafe. Chimney flashing and roof safety are discussed further in chapter 5.

These days, new and retrofit chimneys are often nonmasonry. There are several reasons for this transition: building code and insurance requirements; a shrinking pool of qualified masons; and the inherent inflexibility and tendency of masonry to crack and compromise safety when structurally stressed. Perhaps most important, a host of safe, cost-effective, and easily installed insulated metal chimneys are now available. That noted, the review here is limited to masonry chimneys.

Openings in Brick Walls

If you want to add a door or window to a brick wall, hire a structural engineer to see if that's feasible. If so, hire an experienced mason to create the opening; this is not a job for a novice. If the house was built in the 1960s or later, the wall will likely be of brick veneer, which can be relatively fragile because the metal ties attaching a brick veneer to wood- or metal-stud walls tend to corrode, especially in humid or coastal areas. In extreme cases, steel studs will rust, and wood studs will rot. When opening veneer walls, masons often get more of a challenge than they bargained for.

Brick homes built before the 1960s are usually two wythes thick (with a cavity in between), very heavy, and very likely to have settled. Undisturbed, these walls may be sound, but openings cut into them must be shored up during construction, adequately supported with steel lintels, and meticulously detailed and flashed. Moreover, creating a wide opening or one too close to a corner may not be structurally feasible, so a structural engineer needs to make the call.

To close off an opening in a brick wall, remove the window or door and its casings, and then pry out and remove the frame. Prepare the opening by toothing it out—that is, by removing half bricks along the sides of the opening and filling in courses with whole bricks to disguise the old opening. The closer you can match the color of the existing bricks and mortar, the better you'll hide the new section. As you lay up bricks, set two 6-in. corrugated metal ties in the mortar every fourth or fifth course, and nail the ties to the wood-frame wall behind. Leave the steel lintel above the opening in place.

CLEANING A CHIMNEY

Chimneys and their flues should be inspected at least once a year and cleaned as needed—ideally, before the heating season. Better chimney-cleaning services will get up on the roof, inspect the chimney top, and in some cases lower a video camera to inspect the flue linings. That video is helpful if the chimney needs relining because homeowners can see the damage for themselves and make an informed decision.

Because chimney cleaning takes serious elbow grease, working atop the roof is often the most effective way to brush clean a chimney. But working on a roof is inherently dangerous to you and your roof shingles, which can be easily abraded, torn, and dislodged, leading to leaks. Because many people put off cleaning a chimney until it's almost heating season, they frequently go aloft when the weather is inclement or the roofs are slick after a rain. For all these reasons, you're probably better off hiring an insured professional who is certified for cleaning and inspecting.

If you are determined to clean the flues yourself, start by turning off the furnace and other appliances (such as water heaters) that vent to the flues and disconnecting their vent pipes. Using duct tape, tape plastic over the thimbles that open into living spaces, to prevent dislodged soot from entering. If you have a fireplace, open its damper to allow dislodged soot to fall into the fire pit. Then firmly tape sheet plastic around the fireplace opening. But before you start, suit up. Dislodged creosote and soot are highly carcinogenic, so wear a respirator with replaceable cartridges, tight-fitting goggles, gloves, and disposable coveralls.

To clean a chimney thoroughly, you'll need special brushes, which scrub flue surfaces without damaging them. Today, many professional sweeps favor polypropylene brushes to clean sooty flues and stiff steel-wire brushes for flues with heavy use and creosote buildup. These brushes come in various sizes to match the most common flue shapes. You can screw them onto a series of 3-ft. to 4-ft. rod sections or to a continuous flexible rod (on a reel) up to 50 ft. long.

After you've brushed the flues well, allow the dust and debris to settle before removing the plastic covering the fireplace and other openings. Shovel up the soot and debris at the bottom of each flue and from the fireplace, then vacuum all areas thoroughly. Don't forget the soot that may be resting in thimbles or on the fireplace smoke shelf (see p. 252).

About Chimney Fires

The root cause of a chimney fire is the imperfectly burned material in wood smoke that condenses and sticks to the inside of a chimney. Creosote is a sticky brown or black substance that may harden to the consistency of glass. Incomplete combustion also produces tar, ammonia, methane, carbon monoxide, toluene, phenol, benzene, and eventually, turpentine, acetone, and methyl alcohol.

Creosote can combust in a flash fire, often producing temperatures in excess of 2,000°F. For homeowners, a chimney fire is a terrifying experience, for it may literally roar for extended periods inside the entire flue, flames shooting skyward from the chimney top as though from an inverted rocket. If there are cracks in mortar or flue tiles—or no flue tiles at all—those superheated gases can "breach the chimney" and set fire to wood framing. At that point, the whole house can go up in flames.

You can prevent chimney fires simply by inspecting and cleaning the chimney regularly. In general, don't burn green (unseasoned) or wet wood. Give a fire enough air to burn completely. Each time you start a fire, open the dampers and air controls until the fire is burning well. Don't burn Christmas trees (whose unburned resins collect as sticky masses inside flues), wrapping paper, or glossy-coated papers because their emissions can corrode stovepipes and attack mortar joints. Above all, never use chimneys whose tiles or mortar joints are cracked or chimneys that have no flue lining.

If you're considering buying a house, have its chimney professionally inspected if you see signs of a chimney fire such as creosote flakes on the roof or the ground; scorched or cracked flue liners or chimney crowns; warped dampers; or charred studs or joists near a chimney. Many local codes require inspections before homeowners fire up new wood-burning appliances.

This flue tile and mortar cap were cracked by a chimney fire in a flue that was overdue for cleaning.

REPLACING A CHIMNEY CROWN

A masonry crown is a beveled flue collar at the top of the chimney, sloping gently to direct water away from flue tiles. When crowns weather and crack, water can drain between flue tiles and brick, seep into mortar joints, and freeze, thereby cracking flue tiles, bricks, and mortar joints. In

A professional chimney sweep needs a variety of brushes. The reel at right contains a 50-ft. flexible rod that can push and pull brushes.

warmer seasons, this water can leak into living spaces, stain walls, and linger as acrid combustion smells. Replacing the crown is easy enough if the chimney is not too tall—and the roof not too high or too steeply sloped. Otherwise, you'll need rooftop scaffolding—which a pro should install—before tearing down the chimney to sound masonry and rebuilding from there.

In warm regions, uncovered flues and crowns are common. However, for winters in cold climates, flues should be capped to prevent the entry of rain and sleet that can damage flues during freeze-thaw cycles.

In most cases, a few hammer blows will dislodge deteriorating mortar crowns. Put the debris into a bucket. Then sweep the top of the chimney clean. If some mortar joints need repointing, attend to that. If many mortar joints are soft and badly eroded, tear down the chimney to the roofline, clean the bricks, install new flashing, and rebuild it. If bricks are cracked or broken, replace them with new SW grade bricks. It's okay to reuse old bricks if they're solid, but if you must replace more than a handful, rebuild the chimney with new bricks. They'll look more uniform and last longer.

Installing a new crown. Before you start, spread sheet plastic around the base of the chimney to catch falling mortar. There are essentially two types of crowns. If your region gets a lot of precipitation, pour an in-place concrete crown, which overhangs the chimney 1 in. to 1½ in. and acts as a drip cap, keeping rain and sleet off bricks near the top. Otherwise, use a trowel to build a sloping mortar crown that runs flush to the chimney faces. A flush crown isn't as durable as an overhanging crown but is much quicker to build.

To construct an overhanging crown, as shown in the drawing at left, measure the outside dimension of the chimney top and build a frame from 2x2s that slides snugly over the chimney top. Shim from below to wedge the 2x2 frame in place so its upper face is flush to the top course of bricks. This 2x2 frame (actual dimension, 1½ in. by 1½ in.) creates a 1½-in. overhang. Next, cut strips of plywood 3½ in. wide and as long as the sides of the frame; using a screw gun or an impact driver, screw these plywood strips to the 2x frame so that the strips stick up 2 in. above the top of the frame. The resultant plywood frame keeps the concrete in place and creates a 2-in.-thick edge.

Wet the bricks with a brush or a spray bottle, and you're ready for the concrete. Use a cement-rich, fairly stiff concrete mix: 1 part portland cement, 2 parts sand, and 2 parts ⅜-in. gravel is about right. To prevent cracks, mix in a handful (¼ cup) of fiberglass fibers, which concrete suppliers carry. As you place the concrete into the plywood frame, use a trowel to force it into the corners and to drive out air pockets.

Important: Whether you build a mortar crown or pour a concrete one, wrap the flue liners with polyethylene bond-breaker tape or closed-cell foam strips. This prevents the mortar and concrete from bonding to the tile liners and thereby provides an expansion joint. Without this gap, heat-expanded flue tiles can crack a new crown in a single heating season.

Overhanging Chimney Crown

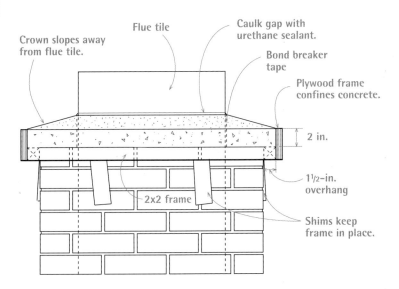

Crown slopes away from flue tile.

Flue tile

Caulk gap with urethane sealant.

Bond breaker tape

Plywood frame confines concrete.

2 in.

1½-in. overhang

2x2 frame

Shims keep frame in place.

A simple 2x2 and plywood frame creates a 1½-in. overhang and a 2-in.-thick edge for this site-built concrete crown. Join the frame with drywall screws, and it will be easy to disassemble.

If you're building up a sloping mortar crown, use a premixed mortar mix. For a slightly more flexible, crack-resistant crown, substitute a liquid latex fortifier for part of the water.

Both types of crowns should be sloped away from the flue liners and troweled to a smooth finish. In addition, both will cure slower and stronger if you cover them with damp burlap or plastic to protect them from rain and sun. Caulk the gap between the flue liners and the crown with a good urethane sealant.

Chimney caps should match the style of the house: Chimneys on colonials and capes are often protected by slabs of bluestone bedded into corner tiers of mortared brick. On newer homes or those with stainless-steel liners, a stainless-steel or copper cap may be more appropriate. If there are multiple flue liners, you may need multiple caps or an overall custom cap.

REBUILDING A BRICK CHIMNEY

Brick chimneys take a lot of abuse. High on the roof there's little protection from searing sun or driving rains, and winter cycles of soaking and freezing will soon degrade both brick and mortar, to say nothing of wind-whipped TV antennae that stress chimney joints (p. 11) or flue tiles so caked with creosote that when they finally ignite, it sounds like a 747 taking off.

Safety, from the ground up. Roof work is risky, so even if you're comfortable with heights, plan the job carefully to limit the time you spend on the roof. For starters, you can assess most of a chimney's failings from the ground using binoculars. Scan the top of the chimney for cracks in its concrete crown, flue tile(s), or mortar joints. Can you see daylight between bricks where mortar is missing? Are any bricks awry or obviously loose along the top course?

Scan down the chimney, noting flaking brick faces or recessed, sandy-looking mortar joints, especially at chimney corners, where the most weathering occurs. At the base of the chimney, is the flashing intact or slathered with roofing cement? Under the roof, dark, tarry-smelling stains on the chimney or nearby framing may signal a cracked or missing flue tile, either of which is a serious risk because superheated flue gases could escape and set the house ablaze.

Working safely on the roof. On the roof, test the soundness of chimney crowns and mortar joints by tapping them with a mason's hammer

or a hand sledge and a brick chisel. Go slowly—it will soon be apparent what needs doing. If the crown is solid, it may just need patching. If, on the other hand, bricks are loose and you can dislodge mortar by scraping it with a chisel, the chimney probably needs to be torn down to the roof line, where flashing often protects mortar joints below from weathering. If flue tiles are cracked or missing, replacing them typically requires a complete tear-down.

Protect roofing with heavy tarps or plywood around the base of the chimney before proceeding with the tear-down. As you disassemble the chimney, you will need a solid platform on which to stand and stack removed materials and, later on when you rebuild the chimney, a place to put your mortar pan, tools, buckets, and, as needed, new bricks and flue tiles. At this point, you should devise a plan for lowering masonry debris and bricks to the ground—and for keeping yourself working safely aloft. Search online to find sources for *roof scaffolding* and *roof hoists,* as well as OSHA-approved *fall-protection devices.*

CHIMNEYS and the Code

Woodstoves, fireplaces, and brick chimneys are a 16th-century technology that is increasingly at odds with modern building codes. Especially in seismically sensitive or fire-prone regions, authorities may specify that replacement chimneys be insulated metal pipe housed in a non-combustible chase rather than masonry. So before you spend a dime, consult your local code.

Using a hand sledge and a brick chisel, strike the mortar joints to dislodge the bricks. Remove bricks till you reach a course that is stable.

To hold flashing in place till the mortar sets, gently force the edge of the lead into a few brick holes. Here, apron flashing runs along the lower face of the chimney.

Many masons set corner bricks and then fill in. Using a trowel handle, tap bricks down to achieve level and compress the mortar. Level along the length and width of each brick.

Get a hard hat for your helper on the ground, where much of the work will be taking place. Because workspace on a roof is limited (and precarious), place masonry rubble and old bricks into a sturdy metal bucket as you remove them from the chimney. Old bricks are heavy (about 4 lb. each), so keep bucketloads light and before lowering each one, warn your helper that it's on the way. It's easier and much safer to mix fresh mortar and to clean old mortar from bricks on the ground, then send materials up when the mason is ready for them.

Speaking of old bricks, you may not know whether you can reuse them until you reach a stable course of brick. In general, using old and new bricks on a chimney won't look good. So if too many old bricks are damaged or their mortar is too tenacious to remove, it's best to tear down the chimney to the roofline and rebuild it with all new bricks. Newer, extruded bricks with holes in them are lighter and bond well because mortar "keys" into the holes; forcing lead flashing into a few holes (use a trowel point) holds it in place till the mortar sets.

Slow and steady as you go. Done right, teardown is slow work. Remove rain caps, damper lids, or other accessories good enough to reuse. Chances are, a damaged crown will be stuck to the top flue tile, in which case you won't be able to save it. But if the mason used a *bond breaker* (p. 246) between the crown and the tile, the flue may be undamaged. Work your way down the chimney, striking the mortar with your chisel to dislodge it without damaging bricks.

When you get to a stable course of bricks, clean the bricks by scrubbing them with a wire brush. You want a flat course of brick, free of old mortar, which would interfere with the application of new mortar and the bonding of courses above.

For basic bricklaying techniques, see pp. 239–241. When rebuilding the chimney, put down a generous amount of mortar so it

As you lay subsequent brick courses, overlap adjacent pieces of cap-flashing along the sides by about 3 in. Check for plumb and level as you go.

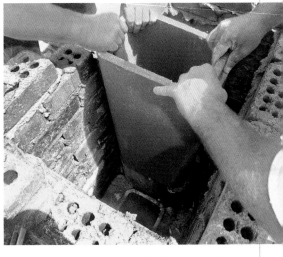

Set new flue tiles in a generous bed of mortar, reaching inside each tile to scrape off excess mortar and smooth joints with your hand. Ragged mortar joints can collect creosote.

will compress down to a 3/8-in. mortar joint. Don't scrimp on mortar: Full-height joints will avoid voids in the front and back faces of courses. Masons often set corner bricks first, level and plumb them, and then lay the bricks between the corners. Check plumb and level often, using the trowel handle to tap bricks down to achieve level and compress the mortar. Level along the length *and* width of the brick.

If you tear down to the roof line, install the apron flashing along the lower chimney face, using the trowel point to form the upper edge of the lead flashing into the outer row of holes in the brick. As you progress, install side pieces of cap-flashing, overlapping adjacent pieces by about 3 in. To ensure adequate draft, extend the chimney at least 3 ft. above the highest point where it exits the roof; it should be at least 2 ft. higher than any part of the roof within 10 ft.

Flue tiles typically stick up 2 in. above the top of the chimney crown, although architectural styles vary. Important: Leaving a void between brick and flue tiles will allow tiles to expand as they heat up, so avoid dropping mortar into the space between those surfaces. Alternatively, you can place a bond breaker between chimney brick and flue tiles to prevent flue cracking. After mortar joints have set a bit, use a 3/8-in. jointer to compress and smooth them. Cover the rebuilt chimney with plastic to allow mortar joints to cure slowly. When curing is done, coat the chimney with a transparent penetrating sealer to protect its surfaces and extend its life.

RELINING A CHIMNEY

While inspecting a chimney, you may find that it has no flue-tile lining or that existing tiles are cracked or broken and too inaccessible to replace. Because superheated gases can escape through gaps, such a chimney is unsafe to use. In this case, your options are:

▶ **Seal up the chimney so it can't be used, and add a new, properly lined chimney elsewhere. Or tear out the old chimney and replace it.**

▶ **Install a poured masonry liner. In this procedure, a heavy-gauge tubular rubber balloon is inflated inside the chimney, and the void is then filled with a cementitious slurry. After the mixture hardens, the tube is deflated and removed. Poured masonry creates a smooth, easily cleaned lining and can stiffen a weak chimney. Poured masonry systems are usually proprietary, however, and must be installed by someone trained in that system. This method is expensive.**

▶ **Install a stainless-steel flue, a sensible choice if you want a solution that's readily available, quick, effective, and about one-third the price of a poured masonry liner. Interchangeable rigid and flexible pipe systems enable installations even in chimneys that aren't straight.**

Installing a stainless-steel liner. Steel flue liners and woodstoves are often installed in tandem, correcting flue problems and smoky fireplaces at the same time.

Start by surveying the chimney's condition, including its dimensions. After steel flue pipe is installed, there should be at minimum 1 in. clearance around it. Thus a 6-in. pipe needs a flue at least 7 in. by 7 in. Note jogs in the chimney that might require elbows or flexible sections. Also note obstructions, such as damper bars, that must be removed before you insert the pipe.

If you're installing a woodstove, too, measure the firebox carefully to be sure the stove will fit and that there's room for the clearances required by local code and mentioned in the stove manufacturer's instructions. You'll also need room to insert the stove, with or without legs attached, and raise it up into its final position. Stoves are heavy—300 lb., on average—so give yourself room to work. Fireboxes often need to be modified to make room for a fireplace insert or stove. Install the stove or fireplace insert before installing the flue liner.

Assemble the flue pipe on the ground, joining pipe sections with four stainless-steel sheet-metal screws per joint so the sections stay together as you lower them down the chimney. Although pop rivets could theoretically be used, they are likely to fail under the stress and the corrosive chemicals present in wood smoke.

Next, insulate the flue pipe, as necessary, with heat-resistant mineral batts and metal joint tape. Heat ratings vary. Temperatures inside flue pipes intermittently reach 2,000°F. Flue pipes are insulated to keep temperatures constant inside and prevent condensation, which also prevents accretion of creosote, which is corrosive. Generally, the first pipe section coming off the woodstove is not insulated because temperatures are so high that there's little danger of condensation. Toward the top of the pipe, stop the insulation just before the pipe clears the chimney—you don't want to expose the insulation to the elements.

Carry the flue-pipe assembly onto the roof and lower it down the chimney. This is a two-person job, especially if it's windy. Once the lower end of the flue pipe nears the woodstove, a coworker below can fit the lower end over the woodstove's outlet.

As two unseen helpers on the other side of this double-sided fireplace steady the Franklin stove, the mason tips it upright. She placed heavy sheet metal over the hearth and slid the stove on its back into the fireplace.

Although installation details vary, a metal top plate centers the pipe within the chimney and is caulked to the chimney top with a high-temperature silicone sealant. The juncture between the pipe and the top plate is then covered with a storm collar, which typically uses a band clamp to draw it tight. That's caulked as well. Finally, cap the top of the flue pipe. The monsoon cap shown below in the bottom right photo maintains a fairly uniform updraft even when winds shift suddenly.

REBUILDING A FIREBOX

If you can see broken firebricks or missing mortar inside your fireplace, it's time to rebuild the firebox. You'll need to decide which bricks to leave and which to replace. But you'll almost certainly need to replace the back wall, which suffers the highest temperatures as well as the most physical abuse.

This job requires a respirator, eye protection, and—at least during demolition—a hard hat. A head sock is also a good idea because you'll be sitting in the dusty firebox during most of the repairs. Finally, you'll need a droplight that can withstand abuse.

Measure. Measure the firebox before you start tearing out old bricks. Note its height, width, depth, and angle at which sidewalls meet the back of the firebox. And if the back wall also tilts forward, take several readings with a spirit level to determine the slope. Finally, note the height and dimensions of the chimney throat, the narrowed opening at the top of the firebox usually covered by a metal damper. Knowing the location and dimensions of the throat is particularly helpful—it tells you the final height of the back wall of the firebox. To keep all of this information straight, you'll do well to take digital photos of the firebox, print them out on typing paper—and write angles, dimensions, and whatnot directly on the prints.

Tear out. Starting with the back wall of the firebox, use a flat bar to gently dislodge loose firebricks—most will fall out—and place them in an empty joint-compound bucket for removal. Use only new firebricks to rebuild the back wall of the firebox, because that's the portion that gets the most intense heat. Remove the damper, and if it's warped, replace it. As you remove firebricks from the back wall, you may find an intermediate wall of rubble brick between the firebox and the

INSTALLING AN INSULATED FLUE PIPE

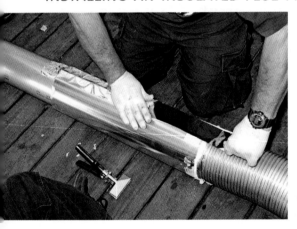

To keep internal temperatures constant and prevent condensation, insulate the lowest 6 ft. to 8 ft. of stainless-steel flue liners with heat-resistant mineral wool batts and metal tape. Note: The flexible flue section that attaches to the woodstove outlet does not need to be insulated.

Don't try this on a windy day. The entire length of flue liner is preassembled and screwed together on the ground, carried aloft, and then lowered into the chimney ...

... while the flexible lower section and an adjustable elbow will enable you to thread the pipe through a slightly offset chimney and still connect to the woodstove outlet.

To increase directional draw and prevent rain blow-in, the big monsoon cap is clamped to the top of the metal flue liner. A steel top plate sealed to the top of the terra-cotta tile centers the steel flue liner in the opening and stabilizes it.

This is a partially dismantled fireplace firebox. To the left, the firebrick sidewall is solid enough to be left in place, although mortar joints need repointing. The back-wall firebricks have already been removed, revealing the back of the chimney. (An intervening wall of rubble bricks also was removed.)

NINE Fixes FOR SMOKING FIREPLACES

▶ Open a window. New houses are often so tightly sealed and insulated that there's not enough fresh air entering to replace the smoke going up the chimney, and smoke exits sluggishly if at all. Alternatively, install an air-intake vent near the hearth.

▶ Use dry wood. Burning wet or green wood creates a steamy, smoky fire whose low heat output doesn't create enough of an updraft and promotes creosote buildup.

▶ Clean chimneys at least once a year. Cleaning also removes obstructions, such as nests. A screened spark arrester will also keep birds out.

▶ Make sure the flue is sized properly. Flues that are too large won't send volatiles upward at a fast enough rate and often allow smoke to drift back into living spaces. Although flue pipes are sized to match woodstove flue outlets (6 in. or 8 in.), sizing fireplace flues is trickier. In general, a fireplace flue's cross section should be one-eighth to one-tenth the area of a fireplace opening.

▶ Reduce air turbulence inside the smoke chamber above the metal damper by giving the corbeled bricks (p. 252) on the front face a smooth parge coat. To do this, brush, vacuum, and wet the corbeled bricks before applying a heat-resistant mortar such as Ahrens Chamber-Tech 2000. (You'll need to remove the damper for access.)

▶ Replace the chimney rain cap. Clogged or poorly designed metal or masonry caps can create air turbulence and prevent a good updraft.

▶ Increase the height of the chimney. A chimney should be a minimum of 3 ft. above the part of the roof it passes through and a minimum of 2 ft. above any other part of the roof within 10 ft.

▶ Rebuild the firebox with Rumford proportions. Count Rumford was a contemporary of Ben Franklin and almost as clever; however, he bet on the British and left the colonies in a hurry—but not before he invented a tall, shallow firebox that doesn't smoke and radiates considerably more heat into the living space than low, deep fireboxes. Search the Internet for companies that sell prefab Rumford-style fireplace components—or build your own.

▶ Install a fireplace insert. Charming as they are, fireplaces are an inefficient way to heat a house. Install an efficient, glass-doored stove and you can watch the flames without getting burned by wasted energy costs.

outer wall of the chimney. And, as likely, the rubble bricks also will be loose, the mortar turned to sand. You can save, clean, and reuse these bricks when you rebuild the rubble wall.

Next, remove loose or damaged firebricks from the sidewalls and floor of the firebox. But, again, if the bricks are intact, it's a judgment call. If repointing the joints is all that's needed, leave the bricks in place. Rap bricks lightly with the end of your trowel handle, however, to make sure bricks are sound. If they're crumbling or cracked, replace them. Interestingly, firebricks on the floor, which are protected by insulating layers of ash, often only need repointing. Once you've removed loose bricks, sweep and vacuum the area well (rent a shop vacuum). Using a spray bottle, spritz all surfaces with clean water until they're damp.

Select bricks and mortar. Firebricks (refractory bricks) are made of fire clay and can withstand temperatures up to 2,000°F. They're bigger and softer than conventional facing bricks and less likely to expand and contract and hence are less likely to crack from heat. Yet, because they are soft, they can be damaged when logs are thrown against them. Firebrick walls need tight joints of 1/16 in. to 1/8 in. thick and thus require exact fits. To achieve this, rent a lever-operated brick cutter.

For firebricks, two kinds of mortar are used. The first type, hydraulic-setting refractory mortar, cures rather than dries. Once its curing time has elapsed—typically 48 to 72 hours—the mortar is impervious to water and is acid resistant. It's the best all-around mortar for setting firebricks and parging the fireplace throat, and it's the only refractory mortar type to use outside or to set clay tile liners.

The second type, air-set or air-drying, is water-soluble and when sold premixed in pails is roughly the consistency of joint compound. Once it has dried, however, it is durable and can withstand high temperatures without degrading. Typically, air-set refractory mortars set more quickly, so if you're new to bricklaying, a hydraulic-setting mortar will be more forgiving. On the other hand, some masons report that because air-set mortars are water-soluble until they dry, you don't need to be as fastidious applying mortar—

Here, the mason is building a rubble-brick wall between the firebox and the back wall of the chimney. Firebrick mortar joints are thin, typically ¹/₁₆ in. to ¹/₈ in. thick, because the firebricks do virtually all of the insulating.

Traditionally, the back walls of fireboxes start tilting forward at the third or fourth course to reflect heat into the room. To create this tilt, make the mortar bed thicker at the back, canting the course of bricks forward. Periodically scrape off excess mortar.

Gently use a mason's hammer to seat firebricks in the mortar. At this point, the back wall's forward tilt is similar to that of the sidewall, shown on p. 251. Angle-cut firebrick to fill any voids between the sidewalls and the back wall. (A rented brick cutter is ideal for this.)

After cutting back deteriorated mortar joints, pack them with fresh mortar. Fill a margin trowel with refractory mortar, as shown. Then use a thin tuck-pointing trowel to scrape mortar from it into the joint. Refractory cement is so sticky that it will cling to the margin trowel's blade even if held vertically.

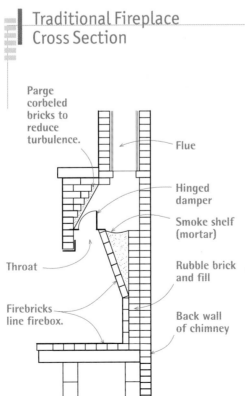

Traditional Fireplace Cross Section

Parge corbeled bricks to reduce turbulence.

Flue

Hinged damper

Smoke shelf (mortar)

Throat

Rubble brick and fill

Firebricks line firebox.

Back wall of chimney

you can wet-sponge the firebox later to clean up errant mortar. (I'd be careful about diluting the new mortar, though.)

Build up. The rest is basic bricklaying technique. String a bed of mortar as wide as the edge of a firebrick across the back of the firebox and press the brick(s) firmly into it, working from one side to the other. Bricks should be damp but not wet. Butter the ends of each brick to create head joints, and when you've laid the first course, check for level. Use the handle of your trowel or mason's hammer to tap down bricks that are high. Typically, you'll need to cut brick pieces on each side of the back wall to "tooth into" the staggered brick joints on the sidewalls, but that step can wait until the back wall is complete. As you lay up each course of firebrick, lay up the rubble brick courses, which needn't be perfect, nor do you need to point their joints.

Unless yours is a tall, shallow Rumford fireplace, firebricks in the back wall should start tilting forward by the third or fourth course. To do that, apply the mortar bed thicker at the back. Build up the firebox and rubble-brick walls until you reach the throat opening. Then fill in any space between the firebox and rubble-brick walls with mortar, creating a smoke shelf. The smoke shelf can be flat or slightly cupped.

Once the back wall is up, fit piece bricks where the back wall meets sidewalls. Clean and repoint the mortar joints as needed. With a margin trowel serving as your mortar palette, use a tuck-pointing trowel to "cut" a small sliver of mortar and pack it into the brick joints. Allow the mortar to dry for a month before building a fire. Make the first few fires small.

Dressing Up a Concrete Wall

If you're bored with the drab band of foundation concrete around the bottom of your house, dress it up with a glued-on brick or stone facade. A number of adhesive materials will work well.

In the project shown here, the mason used SGM Marble Set, intended for marble or heavy tiles, but epoxies would work, too. Whatever adhesive you choose, check the manufacturer's instructions for its suitability for exterior use in your area, especially if you have freezing winters. Use exterior-grade bricks.

How traditional or free-form you make the facade depends on your building's style and your sense of fun. The clinker brick, tile, and stone facade shown completed below right nicely complemented the eclectic style of the Craftsman house. It would probably also look good on the foundation of a rambling brown shingle, a Gothic revival house, or a more whimsical sort of Victorian.

Not relying on mortar joints to support the courses gives you a certain freedom in design, but it's still important that you pack joints with mortar and compress them with a striking tool so they shed water—especially if winter temperatures in your region drop below freezing.

Today's masonry adhesives are so strong that they can adhere heavy materials—such as brick, stone, and tile—directly to concrete. Freed from needing to support much of anything, mortar joints can be as expressive as you like.

Butter the backs of masonry elements with adhesive, in this case, a mortar designed to adhere marble and heavy tile to concrete.

Use short sticks to space bricks, stones, and tiles. This prevents any slippage before the adhesive sets and creates a joint wide enough to pack mortar into. Compress and shape the mortar to make it adhere and keep the weather out.

Masonry needn't always be straight lines. Here, clinker bricks, fieldstones, and tile playfully conceal a drab concrete retaining wall.

Laying Concrete Block

Concrete block is a good choice for retaining walls, basement partitions, and utility structures. Because blocks are large, walls go up quickly. Concrete blocks are manufactured both solid and hollow, but hollow types are by far the most commonly used because they weigh less and are easier to handle and their hollow cores can be filled with insulation to improve a wall's energy profile, or with rebar and grout to improve its lateral strength.

Blocks (also called concrete masonry units, or CMUs) come in numerous shapes, including standard stretcher, half, corner, jamb, half-height, partition, bond beam, and so on. Nominal dimensions of a stretcher block are 16 in. x 8 in. x 8 in.; but actual dimensions are ⅜ in. less, to allow for ⅜-in.-thick mortar joints. Special precast units, such as lintels to span window or door openings, are also available.

Note: Concrete blocks have a top and bottom. The tops of edges and webs (cross pieces) flare out slightly to make them easier to pick and to provide a wider surface on which to spread mortar. The flared center web is easy to grab one-handed, but a standard block's 30-lb. weight will soon tire fingers and forearms. So lift blocks with two hands.

FOOTINGS AND FIRST COURSES

As a rule of thumb, footings for concrete block walls must be at least twice as wide as the width of the block (min. 16-in.-wide footing for 8-in.-wide block) and extend below the frost line. For bearing walls, starter bars (to which vertical rebar tie) should be set at least 6 in. deep in the footings. For bearing or nonbearing walls, use Type S mortar to lay block. Local building codes will specify the type and interval of reinforcement required for concrete block walls, whether steel rebar, bond beams, grout-filled cores, or other reinforcement.

Check the footing for level. Masons still use traditional spirit levels to check individual blocks for plumb and level, but many also use a laser level (often, with a grade rod) to check the height of each course.

If a footing is slightly out of level—say, ¼ in.—you can throw a thick bed of mortar and adjust the first course of block till it is level. If the footing slopes by an inch or more, you can add more mortar and use pieces of block to shim up low sections. Alternatively, if the top course of block must hit a precise elevation—say, to match the height of other walls—you can cut the bottoms of the first course of blocks to make them level.

There are many ways to cut block, but a carborundum or diamond blade in a circular saw (or an angle grinder) is easy to find and reasonably accurate. If you're cutting a lot of block, get a diamond blade. As you cut, it's essential to wear eye protection and proceed slowly. Draw a cutline around all four sides of a block and make a shallow pass—¼ in. to ½ in. deep—along one side before flipping the block and continuing the cut on the next. Once you've scored all four sides, rap a cutline (with a mason's hammer) and the block should split cleanly. If not, lower the blade and make a second pass around the block.

LAYING BLOCK: FIRST DRY, THEN MUD

Start by laying out the blocks dry, without mortar, to see if they fit the space. If not, you must vary the thickness of end-joints slightly or cut the end of a block to make the course fit. When you are satisfied with the layout, snap chalklines onto the footing along both sides of the course. Lift blocks from between the lines and set them on-end nearby. Then dampen the footing so it won't suck moisture out of the mortar.

Between the chalklines, spread a 1-in.-thick bed of mortar. Using the tip of the trowel, furrow the mortar (make parallel V-grooves in it) to distribute it evenly and better bond blocks to the footing. The mortar should be workable but stiff enough to support the weight of the block. To test mortar consistency, scoop up a moderate amount and hold the trowel vertical: The mortar should be sticky enough to stay on the blade.

After setting, leveling, and plumbing the first course of block, apply 1 in. of mortar to it and start the second course, using a half block at the end to establish a running-bond pattern.

Trim excess mortar as you go, returning it to the mortar pan for reuse.

PRO TIP

As you trim excess mortar, throw it back into the pan and reuse it unless it's dirty or dried out. If the mortar in the pan starts to stiffen, periodically mix in a little water to keep it workable.

Using the handle of your trowel to tap blocks down, check for level along the length and across the width of blocks.

Though there is no one right way to set blocks, many masons prefer to set corner blocks and then fill in the course between them. Start with a full block in a corner, aligning it to the chalklines. Using your trowel, tap the block down till the mortar beneath it compresses to ⅜ in. As you position the block, place your level atop it and keep checking for level along the block's length *and* across its width—then check for plumb.

Next set the opposite-corner block and work toward the center of the wall or, alternatively, just build out from the first corner. "Butter" the ends of each successive block, place it snugly against the one before, and tap it level into the mortar bed. Check for plumb and level as you go, also using the tool as a straightedge to align each block to the face of the one(s) next to it. If you fill

Plumb block faces with a spirit level, and use a line level (not shown) to ascertain course height.

After applying a generous bed of mortar for the next course, the mason "butters" the ends of nearby blocks so he can set them quickly.

in from the corners, the last block in the course—the closure block—gets both ends buttered. As you work, use your trowel to trim off excess mortar.

Up you go. To establish a running bond pattern (p. 239), apply a 1-in.-thick mortar bed atop the first course, and start the second course by placing an 8-in. x 8-in. x 8-in. half block in the corner, with its smooth side facing out. The half block ensures that vertical mortar joints will be offset by 8 in. from one course to the next. Set another half block in the opposite corner and after leveling and plumbing both corner blocks, run a taut mason's line between them to align the remaining blocks. Move up the mason's line as you start each new course. When the mortar has set enough to retain a thumbprint, use a jointing tool to smooth the joints—head joints first, then bed joints. Then brush the joints to remove debris from the surface.

As noted earlier, local building codes will specify the materials required to cap and rein-force walls—whether rebar, steel mesh, mortar grout, precast units, or other material—and the intervals at which they must be placed.

A taut mason's line level is a quick, accurate way to make sure block faces are in the same plane. Raise the line as you start each new course.

When mortar is thumbprint hard, use a jointing tool to smooth and compress joints.

To increase the lateral strength of a block wall, insert rebar in block cores and fill them with grout—as specified by local codes.

Knock-out Bond Beam

By removing a block's cross webs, you can create a bond beam course—through which you can run rebar and fill with grout to reinforce the wall.

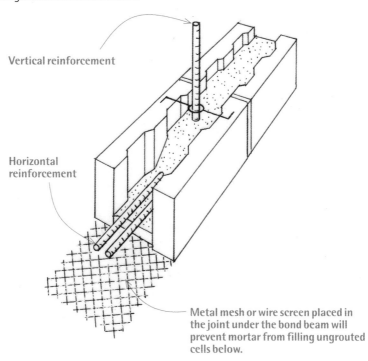

Vertical reinforcement

Horizontal reinforcement

Metal mesh or wire screen placed in the joint under the bond beam will prevent mortar from filling ungrouted cells below.

10 Foundations and Concrete

Foundation issues can be complex. Before starting extensive remedial work, such as replacing failed foundation sections or adding a second story to your house, ask a structural engineer to evaluate your foundation. In addition, bring in a soils engineer if the site slopes steeply or if the foundation shows any of the following distress signs: bowing, widespread cracking, uneven settlement, or chronic wetness. Engineers also can assess potential concerns such as slide zones, soil load-bearing capacity, and seasonal shifting.

An Overview

A foundation is a mediator between the loads of the house and the soil on which it rests. A well-designed foundation keeps a house's wood underpinning above the soil so it doesn't rot or get eaten by insects. And it should be sturdy enough to keep walls plumb and floors level despite wind, water, soil movement, and earthquakes.

FOUNDATION TYPES

Foundations should be appropriate to the site. For example, on sandy well-drained soil,

These concrete forms are half complete, showing oiled inside formboards, a new mudsill nailed up, and the bottom outer formboard in place. Rebar, as shown, will be wire-tied to anchor bolts after they've been inserted into predrilled holes in the mudsill.

unmortared stone foundations can last for centuries. But an unstable clay hillside may dictate an engineered foundation on piers extending down to bedrock.

The tee, or spread, foundation is perhaps the most commonly used type, so named because its cross section looks like an inverted *T*. It's remarkably adaptable. On a flat site in temperate regions, a shallow tee foundation is usually enough to support a house, while creating a crawlspace that allows joist access and ventilation.

Where the ground freezes, foundation footings need to be dug below the frost line, stipulated by local codes. Below the frost line, footings aren't susceptible to the potentially tremendous lifting and sinking forces of freeze–thaw cycles in moist soil. (Thus, most houses in cold climates often have full basements.)

When tee foundations fail, it's often because they're unreinforced or have too small or too shallow a footprint. Unreinforced tee foundations that have failed are best removed and replaced. But reinforced tees that are sound can be *underpinned* by excavating and pouring larger footings underneath a section at a time.

A mat slab is a massive pad of extensively reinforced concrete, poured simultaneously with a slightly thicker perimeter. A mat slab designed for a single-family dwelling is typically 12 in. to 18 in. thick, with a double layer of rebar, so the entire slab is basically a continuous footing that can bear column and wall loads at any point. Because of a mat slab's ability to transfer loads across a broad expanse, it is often used where subsoils are weak. A mat slab should not be confused with a slab on grade, though at first glance they look similar. Slabs on grade are thinner, their pads contain only a single layer of rebar, and typically bear loads only along perimeter footings.

Grade beams with drilled concrete piers are the premier foundation for most situations that don't require basements. These foundations get their name because pier holes are typically drilled to bearing strata. This foundation type is unsurpassed for lateral stability, whether it's a replacement foundation for old work or for new construction. Also, concrete piers have a greater cross section than driven steel piers and hence greater skin friction against the soil, so they're much less likely to migrate. The stability of concrete piers can be further enhanced by concrete-grade beams resting on or slightly below grade, which allows soil movement around the piers, without moving the piers.

The primary disadvantages of drilled concrete piers are cost and access. In new construction, a backhoe equipped with an auger on the power

Foundation Types

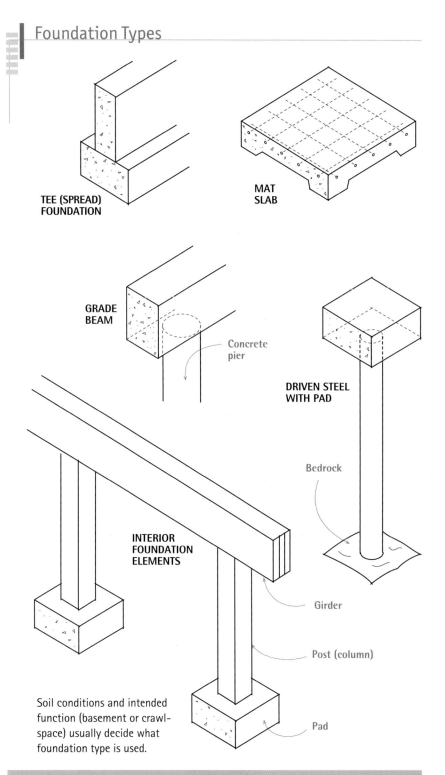

TEE (SPREAD) FOUNDATION

MAT SLAB

GRADE BEAM

Concrete pier

DRIVEN STEEL WITH PAD

Bedrock

INTERIOR FOUNDATION ELEMENTS

Girder

Post (column)

Pad

Soil conditions and intended function (basement or crawl-space) usually decide what foundation type is used.

Pouring VERSUS PLACING CONCRETE

Every trade has its jargon. Concrete snobs, for example, insist on using the phrase *placing concrete*, though concrete coming out of a 4-in. hose looks more like a pour, albeit a sluggish one. Perhaps "placers" want to emphasize that concrete is so weighty that you should place it as close to its final location as possible. Point taken. But to denote the general movement of concrete from truck to forms, *pour* us a tall one.

This pier and grade-beam foundation was built on a sloping site with expansive clay soil, so the engineer specified parallel-grade beams and a more massive grade-beam perimeter. (The piers go down 15 ft.) Integral concrete post piers atop grade beams support 4x6 joists spaced 3 ft. on center.

takeoff requires 10 ft. or 12 ft. of vertical clearance. Alternatively, there are remote-access portable rigs that can drill in tight quarters, even inside existing houses, but they are labor intensive to set up and move, increasing the cost.

Driven steel pilings are used to anchor foundations on steep or unstable soils. Driven to bedrock and capped, steel pilings can support heavy vertical loads. And, as retrofits, they can stabilize a wide range of problem foundations. There are various types of steel pilings, including *helical piers*, which look like giant auger bits and are screwed in with hydraulic motors, and *push piers*, which are hollow and can be pushed in, strengthened with reinforcing bar, and filled with concrete or epoxy.

IMPORTANT ELEMENTS

In many parts of North America, building codes don't require steel reinforcement in concrete foundations, but steel is a cost-effective means of avoiding cracks caused by lateral pressure on foundation walls.

Steel reinforcement and fasteners. *Steel reinforcing bar* (rebar) basically carries and distributes loads within the foundation, transferring the loads from high-pressure areas to lower-pressure areas. It thereby lessens the likelihood of *point failure*, either from point loading above or from lateral soil and water pressures. *Anchor bolts,* or threaded rods, tied to rebar, attach the overlying structure to the foundation. *Steel dowels* are usually short pieces of rebar that pin foundation walls to footings or, epoxied into existing foundations, are used to anchor new sections.

There also are a number of metal connectors —such as Simpson Strong-Ties—that tie joists to girders, keep support posts from drifting, and hold down mudsills, sole plates, and such. Several are shown in chapter 4.

Before the pour, the rebar spine of this foundation wall is still visible. The green bolt holders will position anchor bolts 3 in. from the outside face of the foundation; as the 2x6 mudsill lines up with the outside of the foundation, the anchor bolts will be centered in the mudsill.

THE FOUNDATION Within

Most perimeter foundations have a companion foundation within, consisting of a system of *girders* (beams), *posts* (columns), and *pads* that pick up the loads of joists and interior walls and thus reduce the total load on the perimeter foundation. By adding posts, beams, and pads, you can often stiffen floors, reduce squeaks, avoid excessive point loading, support new partitions, and even avoid replacing a marginally adequate perimeter foundation.

Quality. Concrete quality is critically important, both in its composition and in its placement. Water, sand, and aggregate must be clean and well mixed with the cement. Concrete with compressive strength of 2,500 pounds per square inch (psi) to 3,500 psi is common in residential foundations, yet there are many ways to achieve that strength, including chemical admixtures. Discuss your needs with a concrete supplier who's familiar with soil conditions in your area, and read "Ordering Concrete: Be Specific" on p. 281.

Drainage. On far too many job sites, drainage is an afterthought. It should be top of mind, for when the ground next to foundations becomes super-saturated, clay in the soil will expand and contract, soften, and ultimately succumb to the weight of the building, resulting in cracked foundations, sloping floors, doors that don't close, and porches that lean. Pervasive moisture will also encourage mold, dry rot, and termites in crawlspaces and basements. A good drainage system is the only way to ensure that these spaces stay dry. Ideally, before backfilling basement walls, install French drains (shown at right), waterproof the foundation walls with a below-grade waterproofing membrane, and backfill with loads of clean, crushed drain rock. Give water an easier place to go and it will flow away from the house.

INDOOR SYMPTOMS OF FOUNDATION FAILINGS

Most foundations that fail were poorly designed, poorly constructed, or subjected to changes (especially hydrostatic pressure or soil movement) that exceeded their load-bearing capacities. Exact causes are often elusive.

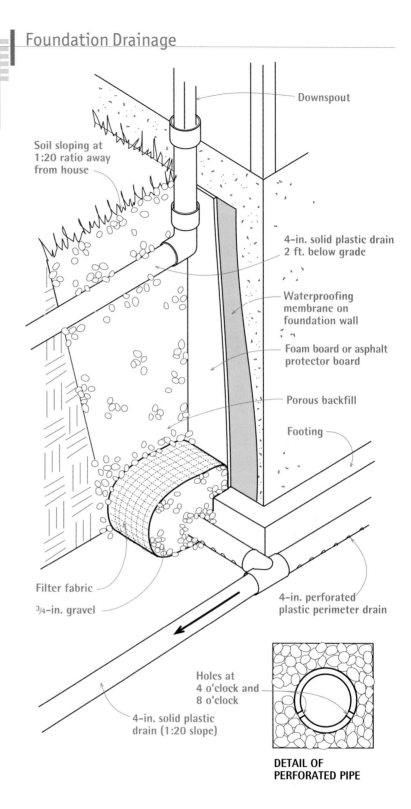

Soil sloping at 1:20 ratio away from house

Downspout

4-in. solid plastic drain 2 ft. below grade

Waterproofing membrane on foundation wall

Foam board or asphalt protector board

Porous backfill

Footing

Filter fabric

³/₄-in. gravel

4-in. perforated plastic perimeter drain

Holes at 4 o'clock and 8 o'clock

4-in. solid plastic drain (1:20 slope)

DETAIL OF PERFORATED PIPE

If you're willing to excavate, you can retrofit a first-rate drainage system such as this.

This dramatic crack through the corner of this building was caused by a downspout and drainage system that was clogged for decades. Instead of runoff being directed away from the house, water collected at the base of the foundation, undermined the footings, and caused a corner to sink.

PRO TIP

If you're not sure that a foundation crack is active (moving), epoxy a small piece of glass to both sides of the crack. If the glass breaks over two months' time, the crack is active. A glass microscope slide is perfect for this test. Or substitute a scrap of window glass.

Localized springiness or low spots in flooring are probably caused by an undersize pad or by a deteriorated or absent post beneath a girder. If you find wet rot or insect damage at the base of the post, correct that situation before doing anything else.

Widespread springiness in floors and joists sagging in midspan are caused by joists that are too small for a span or by a failed or absent girder. If an existing girder seems sound, adding posts or new pads may fix the problem. Otherwise, add a girder to reduce the distance joists span.

Failure of all or part of a perimeter foundation often explains flooring that crowns above a girder, sloping downward toward the outside walls; doors and windows that are difficult to open; and cracking at the corners of openings.

Foundation cracks often signal foundation failure. Cracks may range from short surface cracks to through-the-wall cracks that should be examined by a structural engineer. Here are some common symptoms and remedies:

▶ Narrow vertical or diagonal surface cracks that are roughly parallel are likely caused by foundation settlement or soil movement but are probably not serious. If water runs from cracks after a storm, fill them with an epoxy cement, and then apply a sealant.

▶ Wide cracks in foundations less than 2 ft. tall indicate little or no steel reinforcement, a common failing of older homes in temperate climates.

▶ Large, ½-in. or wider vertical cracks through the foundation that are wider at the top usually mean that one end of the foundation is sinking—typically at a corner with poor drainage or a missing downspout.

▶ Large vertical cracks through the foundation that are wider at the bottom are usually caused by footings that are too small for the load. You may need to replace or reinforce sections that have failed.

▶ Horizontal cracks through a concrete foundation midway up the wall, with the wall bowing in, are most often caused by lateral pressure from water-soaked soil. This condition is common to uphill walls on sloping lots.

▶ Concrete-block walls with horizontal cracks that bulge inward are particularly at risk because block walls are rarely reinforced with steel. If walls bulge more than 1 in. from vertical and there's a chronic water problem, foundation failure may be imminent.

▶ In cold climates, horizontal cracks through the foundation, just below ground level, are usually caused by adfreezing, in which damp soil freezes to the top of the foundation and lifts it. If these cracks are accompanied by buckled basement floors, the foundation's footings may not be below the frost line.

Gaps between the chimney and the house are usually caused by an undersize chimney pad. If the mortar joints are eroded, too, tear the chimney down and replace it.

If your foundation upgrade will require big steel I-beams, hire house movers. They'll have the cranes, beam rollers, cribbing, and expertise to handle I-beams safely.

Jacking and Shoring

Jacking refers to raising or lowering a building so you can repair or replace defective framing or failed foundations or to level a house that has settled excessively. *Shoring* refers to a temporary system of posts, beams, and other structural elements that support building loads. *Temporary* is the crucial word: Shoring supports structural elements between jackings. Once repairs are complete, you need to lower the house and remove the shoring as soon as possible. If repairs are extensive—say, replacing foundation sections—have a structural engineer design the new sections, specify jack size, and specify the posts, beams, and bracing needed to safely jack and shore the building.

Jacking a house is nerve-wracking. It requires a deep understanding of house framing and how structures transfer loads. It also requires superb organizational skills and a lot of specialized equipment. For that reason, foundation contractors routinely subcontract house-raising to house movers with crews who know what they're doing plus have on hand the heavy cribbing blocks, hydraulic jacks, and cranes to lift steel I-beams for bigger jobs. Structural engineers usually will know qualified house movers. (By the way, these specialists are still called *house movers* even when the house stays on the site.)

MATERIALS AND TOOLS

Basically, timbers used for jacking and shoring stages are the same size. For example, once you have jacked the building high enough with 4x4 *jacking posts*, you can plumb and insert 4x4 *shoring posts* next to the jacking posts. Then, with shoring posts solidly in place, you can slowly lower the jacks of the jacking posts and remove each jacking component.

Posts and beams. To support a single-story house, 4x4 posts with 4x8 or 4x10 beams should be adequate. Because harder woods compress less when loaded, shoring should be Douglas fir, oak, or a wood with similar compressive strength. You'll also need footing blocks under each post: typically, two 2-ft.-long 4x12s placed side by side, with a 2-ft.-long 4x6 placed perpendicular atop the 4x12s. Footing blocks of that size are wide enough to accommodate a jacking post and a shoring post side by side while you make the transition from jacking to shoring.

For two-story houses or heavy single-story houses, most professionals specify W6x18 or W8x18 steel I-beams to support the loads. (A heavy single-story house might have a stucco exterior, a plastered interior, and a tile roof.)

Patching Foundation Cracks

Determine the cause of the crack and fix that first; otherwise, the crack may recur. Shallow foundation cracks less than $1/8$ in. wide are usually caused by normal shrinkage and needn't be patched, unless their appearance disturbs you or they let water in. However, you should repair any cracks that go all the way through the foundation: Probe with a thin wire to see if they do.

Of the many crack-repair materials, there are three main types: cement based, epoxy, and polyurethane foams. When working with any of these materials, wear disposable rubber gloves, eye protection, and a respirator mask with changeable filters.

Cement-based materials such as hydraulic cement are mixed with water and troweled into cracks. To ensure a good connection, first use a masonry chisel and hand sledge to enlarge the crack; angle the chisel to undercut the crack, making it wider at the back. Then wire-brush the crack to remove debris. Next, dampen the surfaces, fill the crack with hydraulic cement, and feather out the edges so the repaired area is flat. Work fast because most hydraulic cement sets in 10 to 15 minutes and expands so quickly that it can stop the flowing water of an active leak.

Epoxies range from troweled-on pastes to *injection systems* that pump epoxy deep into cracks. Application details vary, but many injection systems feature *surface ports*, which are plastic nozzles inserted into the crack along its length. You should space ports 8 in. apart before temporarily capping them. Then seal the wall surface with epoxy gel or hydraulic cement, which acts as a dam for the epoxy liquid you'll inject deep into the wall through the ports. Working from the bottom, uncap each surface port, insert the nozzle of the applicator, and inject epoxy until it's visible in the port above. Cap the port just filled, and then move up the wall, port by port.

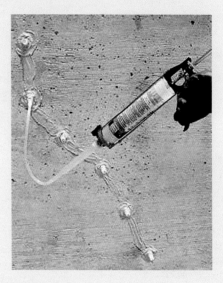

Epoxy is famously strong. The manufacturer of Simpson Crack-Pac claims that its injected epoxy achieves 11,000 psi compressive strength when cured for seven days. (Foundation concrete averages 3,000 psi to 4,500 psi.) Consequently, injected epoxy, which bonds to both sides of the crack, is a true structural repair, not just a crack filler. There are a couple of disadvantages: cost and curing time. Epoxy takes hours to harden, so it can ooze out the back of the crack if there's a void between the soil and the foundation wall—as there often is. If your main concern is water leaks and not structural repairs, polyurethane foam is probably a better choice.

Polyurethane foam is applied in many ways, including the surface-port injection just described for epoxies. Polyurethane sets up in minutes, so it's unlikely to sag or run out the back of the crack. It's largely unaffected by water, so you can inject it into a damp crack. Unlike epoxy or cement-based fillers, polyurethane is elastomeric (meaning it stays flexible), so it's great for filling foundation cracks that expand and contract seasonally. One disadvantage is that it has little compressive strength and hence does not create a structural repair.

PRO TIP

Shoring timbers are heavy, so get help. It takes three strong workers to raise a 16-ft. 4x8. A 4x12 that same length requires four people.

Because steel beams can span greater distances than wood, they require fewer posts underneath, which frees space under the house and improves access for workers. However, if you use steel beams, stick with wood posts: Fir 6x6s are less likely to migrate than steel columns.

Cribbing. Cribbing refers to a framework of usually squared timber (often 6x6s) stacked in alternate layers to create a stable platform for jacking or shoring house loads. In earthquake country, foundation contractors "shear-wall" cribbing higher than 8 ft.—that is, they temporarily nail ½-in. plywood to the cribbing using duplex nails. The precaution is worth the trouble: In 1989, a California house resting on 13-ft.-high shear-walled cribbing remained standing through a 7.1 quake.

Braces and connectors. To keep posts plumb and prevent structural elements from shifting, builders use a variety of braces and connectors such as these:

▶ Diagonal 2x4 braces 3 ft. or 4 ft. long are usually nailed up with double-headed nails for easy removal.

▶ Plywood gussets are acceptable if space is limited.

▶ Metal connectors such as Simpson hurricane ties, post caps, and post-to-beam connectors are widely used because they are strong and quick to install.

Jacks. House-raising screw jacks and hydraulic jacks are by far the most common types. For safety, all jacks must be placed on a stable jacking platform and plumbed.

Screw jacks vary from 12 in. to 20 in. (closed height) and extend another 9 in. to 15 in. Never raise the threaded shaft more than three-quarters its total length because it would be unstable beyond that. Screw jacks are extremely stable: Of all types, they are the least likely to fail or lower unexpectedly under load. But they require a lot of muscle and at least 2 ft. of space around the jack for operation.

Hydraulic jacks are the workhorses of foundation repair and are rated according to the loads they can bear, such as 12 tons. In general, hydraulic jacks are easier to operate than screw jacks, and they fit into tighter spaces. They are lowered by turning a release valve, so they can't be lowered incrementally. Because hydraulics release all at once, many house movers use hydraulics to raise a house and screw jacks to lower it gradually.

Because the head of a hydraulic jack is relatively small, you need to place a 4-in. by 4-in. by ¼-in. steel plate between it and the wood it supports so the head doesn't sink in during jacking. *Safety note:* As the jack is lowered, have a helper keep a hand on the steel plate so it doesn't fall and injure someone. Alternatively, have the plate predrilled so you can screw or nail it to the underside of the beam.

Unsuitable for raising a house. Post jacks employ a screw mechanism but are of flimsier construction than the compact screw jack described in the previous section. They tend to be fashioned from lightweight steel, with slender screws that could easily distort when loaded beyond their capacity. So use post jacks only for low-load, temporary situations.

Jacks. Left: a hydraulic jack, the workhorse of house-raising. Right: a screw jack. Textured heads on both reduce chances of post slippage.

Unsafe! Although post jacks such as this are widely used as temporary shoring, they are not strong enough for house loads. Here, footing blocks are undersize, and the post is badly out of plumb.

JACKING SAFELY

For safe jacking, you need to proceed slowly and observe the following precautions.

Preparatory steps

▶ Survey the building, noting structural failings and their probable causes as well as which walls are load-bearing. Also determine whether joists or beams are deflecting because of heavy furniture, such as a piano; which pipes, ducts, or wires might complicate your repairs; where the gas pipe shutoff is; and so on.

▶ Have a plan. If excavation is necessary, who's going to do it? And where will you put the displaced dirt? (Disturbed dirt has roughly twice the volume of compacted soil.) Will you need to rent equipment, such as a compressor, a jackhammer, or jacks? Where will you store materials? How will rain affect the materials and the work itself? Can a concrete-mixer truck reach your forms or will you need a separate concrete pumper, an auxiliary pump on wheels that pumps (pushes) the concrete from the mixer truck to the pour?

▶ Assemble safety equipment. This is mostly hard-hat work. You'll also need safety glasses that don't fog up, sturdy knee pads, and heavy gloves. For some power equipment, you'll need hearing protectors. Update your tetanus shot. Set up adequate lighting that keeps cords out of your way—and, on a post near a suitable light, mount a first-aid kit. Even though a cell phone is handy if trouble strikes, never work alone. Workers should stay within shouting distance.

▶ Have all necessary shoring materials on hand before you start jacking. Remember, jacks are for lifting, not supporting. Within reason, level the ground where you'll place footing blocks or shoring plates. As soon as a section of the house is raised to the proper level, be sure to set, plumb, and brace the shoring. Hydraulic jacks left to support the structure too long may slowly "leak" and settle or—worse—kick out if bumped or jostled.

Jacking basics

▶ Support jacks adequately. The footing blocks or cribbing beneath the jacks must be thick enough to support concentrated weights without deflection and wide enough to distribute those loads. It's difficult to generalize how big such a support must be; a 4x12 footing block 3 ft. long or two layers of 4x4 cribbing should adequately support a jack beneath the girder of a single-story house. In this case, the soil must also be stable, dry, and level. If the soil isn't level, dig a level pit for the cribbing,

Jacking Components

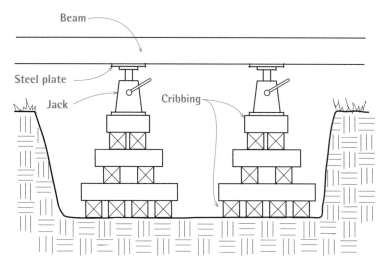

Jacks must be supported on a level, stable platform. Here, cribbing beneath the jacks and steel plates atop them disperse loads to larger surface areas. Without the steel plates to spread the load, jack heads can sink into wood beams.

DON'T DO THIS!

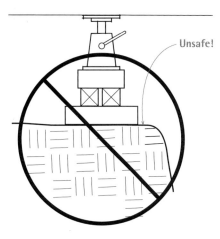

When the jack is loaded, bearing blocks placed unsafely close to the edge of an excavation can cause it to collapse.

FOOTING PIT

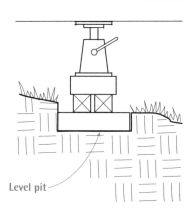

Where you must jack on a slight incline, dig down into the soil to create a level pit. The pit will also surround the bearing blocks, giving them no place to go when loaded.

bottom right drawing on p. 265, so footings or
cribbing can't migrate under pressure.
(Typically, the foundation contractor digs the
pits and prepares the site before the house
mover arrives to install the cribbing.)
Surrounded by the walls of the pit, the bearing
blocks have no place to go.

▶ Keep checking for level and plumb as you
jack. If supports sink into the soil, posts tilt, or
the jack starts "walking" under pressure,
lower the jack, reset the supports, and begin
anew. First thing each day, check jack supports
and shoring for plumb and level. Diagonal
bracing, plywood gussets, and metal
connectors will each help posts stay plumb.
When cross-bracing temporary posts and
beams, use screw guns or pneumatic nailers to
attach braces. Hand-nailing braces could
knock posts out of plumb or cause beams to
rotate or jack heads to migrate.

▶ Raise jacks in small increments—say,
¼ in. per day—to minimize damage to finish
surfaces inside the house. When you're jacking
a structure to be repaired, as when replacing a
mudsill, jack just enough to lift the weight off
the sill to be removed. If many jacks are
involved, raise them simultaneously if
possible, so excessive stress (and damage)
doesn't build up above any one jack.

Steps in jacking and installing shoring.
Setting jacking equipment varies according to the
type of jack, the structural elements to be raised,
and site conditions, such as ceiling height,
access, and soil stability. That noted, the following observations hold true in most cases.

1. Position the jacks and jacking beams as
close as possible to the joists, girders, or stud
walls you're jacking. If you're adding posts under
a sagging girder, support may be directly under
the girder, but more often, it will need to be offset
slightly—say, within 1 ft. to 2 ft. of joist ends—to
give you working room. In other words, place
them close enough to joist ends so they won't
deflect, yet far back enough to let you work.
Again, don't put jacks or shoring where they
could be undermined by unstable soil later.

2. Level and set the footing blocks or cribbing
on compacted soil. Each jack base should be
about 2 ft. by 2 ft. Or, if you're using a single timber block, use a 4x12 at least 3 ft. long or, if the
soil is crumbly, at least 4 ft. long. If you spend a
little extra time leveling the footings, the posts
will be more likely to stay plumb. To support a
single-story house, set posts every 5 ft. or so
beneath an adequately sized beam—typically, a
4x8 or 4x10 set on edge.

as shown on p. 265. (Avoid precast concrete
piers as jacking blocks because their footprints
are too small and the concrete could shatter
when loaded.)

▶ Don't place jacking or shoring platforms
too close to the edge of an excavation. Otherwise, the soil could cave in when the timber is
loaded. The rule of thumb is to move back 1 ft.
for each 1 ft. you dig down. Also, don't put
jacks or shoring where they could be
undermined later. For example, if you need a
needle beam to support joists parallel to the
foundation, excavate on either side of what
will be your new foundation and place jacking
platforms in those holes. In that manner, you
can remove foundation sections without
undercutting the jacking platforms.

▶ Level support beams, and plumb all
posts. The logic of this should be evident:
When loads are transmitted straight down,
there is less danger that jacks or posts will kick
out, injuring someone and leaving shoring
unsupported. Accordingly, cut the ends of the
posts perfectly square, plumb the posts when
you set them, and check them for plumb
periodically as the job progresses.

▶ Where the ground slopes, dig a level
footing pit into the soil, as shown in the

Needle Beams

When replacing a mudsill or sections of a foundation whose joists run perpendicular to the adjacent foundation wall (see p. 268), place a 4x8 or 4x10 carrying beam on edge under the joists within 2 ft. of the foundation. A jack every 6 ft. under the beam should suffice.

When joists run parallel to the foundation wall being replaced, you'll need to run *needle beams* through exterior walls and support each beam with one post underneath the house and a second post outside, roughly 2 ft. beyond the foundation wall. For this, you'll need to remove sections of siding so you can insert a beam every 6 ft. to 8 ft. If the siding is stucco, you'll need to punch large holes through it. To keep the rim (outer) joist from deflecting under the load, nail a second rim joist to it, doubling it before jacking. Also, add solid blocking from those doubled rim joists to the first adjacent joist. Use metal connectors to affix the blocking and 10d nails to face-nail the rim joists.

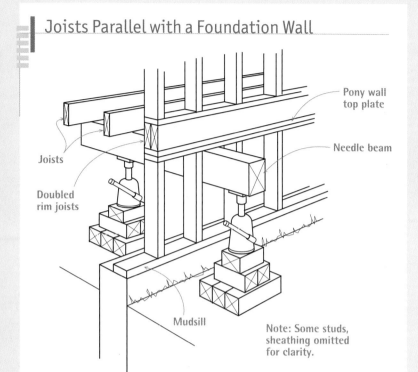

Joists Parallel with a Foundation Wall

Pony wall top plate

Needle beam

Joists

Doubled rim joists

Mudsill

Note: Some studs, sheathing omitted for clarity.

Where joists run parallel to the foundation wall, remove a section of siding and run a temporary needle beam through the wall as shown. To prevent its deflection under pressure, double the rim (outer) joist and run solid blocking to the next joist inward.

This big needle beam needed to fit under the top plate of the pony wall (short wall) on the right. As a result, the top of the beam needed to be built up with a 4x4, which now supports the floor joists. The metal connector atop the post keeps it from drifting out of plumb.

Joists Perpendicular to a Foundation Wall

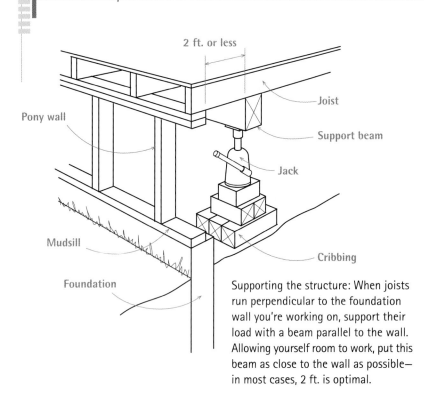

Supporting the structure: When joists run perpendicular to the foundation wall you're working on, support their load with a beam parallel to the wall. Allowing yourself room to work, put this beam as close to the wall as possible—in most cases, 2 ft. is optimal.

To determine the height of the shoring posts, jack the beam to the desired height, level the footing block, and measure between the two. Nail the post cap on before inserting the post. Because this needle beam is simply holding a wall in place—not raising it—it was jacked just snug to a wall plate.

PRO TIP

If a new foundation section isn't terribly long or deep, use an electric demolition hammer with a shovel bit to excavate the trench. This tool is particularly helpful when there's no room to swing a pick. Shovel bits cut a clean edge and dig themselves in, even in heavy clay. Then you simply shovel out the loosened soil.

3. Positioning a jacking beam requires prep work. Ideally, the crew should raise the beam into place and then immediately plumb and set the jacks. But they may need to catch their breath or gather equipment before setting the jacking posts. In that event, cut two 2x4s approximately ½ in. longer than the distance from the underside of the beam to the top of the cribbing plates, and use a sledgehammer to wedge a 2x4 under each end of the beam. *Caution:* This setup is inherently unstable, so workers should monitor both 2x4s continually to make sure they don't kick out.

4. As soon as the beam is in place, cut posts 10 in. shorter than the distance from the underside of the beam to the top of the footing blocks. This 10 in. is roughly the (closed) height of a hydraulic jack plus a little room to move. With a helper, place jacks and posts under both ends of the beam, plumb the posts, and start jacking. Center each jack on its footing blocks so there's plenty of room for the shoring post(s) that will follow shortly. (If you use hydraulic jacks, you can position the shoring posts 3 in. to 4 in. from the jacking posts.) As you jack, try to raise both ends of the beam evenly, using a 4-ft. or 6-ft. spirit level to check for level.

When the beam is at the desired height, measure down to the tops of footing blocks to deter-

mine the height(s) of the shoring posts. (If you jack up an additional ⅛ in., you'll find it easier to slide shoring posts in.)

5. After you're done jacking, install the shoring posts, which are more stable than jacks on posts. To keep the shoring posts in place, nail steel caps to their tops before installing them. Once you've placed the posts under the footing beams, attach the post caps to the beams and add cross bracing or plywood gussets to keep the beam from rotating. Once you've plumbed the shoring posts and braced the beams, lower the jacks slowly until they no longer bear weight, and then remove them. With shoring supporting all necessary bearing members, you're ready to begin repairs.

Note: Some foundation contractors install two shoring posts—one on either side of the jacking post—for greater stability. Once the jack is removed, nail two 3-ft.-long 2x4 diagonal braces between the two shoring posts; for this, use a pneumatic nailer. Hammer blows could dislodge the posts.

6. When your repairs are finished, begin to remove the shoring by reinserting the jacks and then simultaneously raising all the jacks slowly and evenly to take weight off the shoring. Leave the cross bracing in place until those loads are removed. Then, keeping the jacking posts plumb, carefully remove the post-and-beam bracing and lower and remove those elements. Gradually lower the building onto its new pads, posts, and foundation, then remove the jacks.

Minor Repairs and Upgrades

The category *minor repairs* includes anything short of replacing a failed foundation, which is covered in the next section. Repairing surface cracks is explained on p. 263.

REPLACING POSTS AND PADS

If floors slope down to a single point, there's a good chance that a post or pad has failed. If a floor slopes down to an imaginary line running down the middle of the house, there's probably a girder sagging because of multiple post or pad failures. Fortunately, the cures for both conditions are relatively straightforward.

Post repairs. The most common cause of wooden post failure is moisture wicking up through a concrete pad, rotting the bottom of the post. To replace a damaged post, use the techniques just described in "Jacking and Shoring." Place footing blocks as close as possible to the existing pad and

jack just enough to take the load off the post—plus ⅛ in. Remove the rotted post, measure from the underside of the girder to the pad (remembering to subtract the ⅛ in.), and cut a new post— preferably from pressure-treated lumber.

To keep this new post from rotting, cut a 22-ga. sheet aluminum plate and place it under the bottom of the post. A dab of silicone caulking will hold it in place while you plumb and position the post, then lower the jack so the new post bears the load. Or replace the wooden post with a preprimed metal column. However, if basement floors are wet periodically—suggested by sediment lines along the base of the walls—build up or replace the existing pad with a taller one to elevate the base of the post. Add a sump pump, too, as explained on p. 285.

Replacing pads. Replace concrete pads that are tilting or sinking because they are too small for the loads they bear. Likewise, you'll need to pour a new pad if there was no pad originally and an overloaded post punched through the concrete floor. Pads for load-bearing columns should always be separated from floors by isolation joints.

Cutting into a Concrete Floor

To enlarge an existing load-bearing pad or create a new one, you may need to cut through a concrete floor. Depending on the condition and thickness of that floor, the job will range from nasty to horrible. Cutting concrete is noisy, dirty, and dangerous, and the tools are heavy and unpredictable. Wear safety glasses, gloves, hearing protectors, and a respirator mask. Adequate ventilation and lighting are a must.

If the floor was poured before the 1950s, you'll likely find that it is only 3 in. to 4 in. thick and is without steel reinforcement. The floor also may be badly cracked. In this case, you can probably break through it with a pickax, but to minimize floor patching later, rent an electric concrete-cutting saw with a diamond blade to score around the opening. Then finish the cut (the sawblade rarely cuts all the way through) with a hand sledge and a chisel.

Be advised, however, that a concrete-cutting saw cuts dry and thus throws up an extraordinary amount of dust. Therefore, you may need to seal off the basement with plastic barriers and then spend an hour vacuuming afterward. Alternatively, you can rent a gasoline-powered wet-cut saw, which keeps down the dust but fills the basement with exhaust fumes. And, if the concrete floor is a modern slab that is 5 in.

thick and reinforced with rebar, you can spend a day accomplishing very little. Well . . . you get the picture.

Fortunately, for a few hundred bucks you can hire a concrete-cutting subcontractor to cut out a pad opening in about an hour. (Don't forget to allow for the thickness of the formboards when sizing the opening.) The subcontractor also can bore holes needed for drainpipes and such.

A gasoline-powered, wet-cut saw has a diamond blade and is connected to a water supply (via the yellow fitting), which reduces airborne dust when cutting concrete.

JOIST SIZE		TYPICAL SPAN (ft.)
2×6	→	8
2×8	→	10
2×10	→	12
2×12	→	14

Load-bearing pads should be 24 in. by 24 in. by 12 in. deep, reinforced with a single layer of No. 4 (½-in.) rebar arranged in a grid. Pads supporting a greater load (such as a two-story house) should be 30 in. by 30 in. by 18 in. deep, with two layers of No. 4 rebar; in each layer, run three pieces of rebar perpendicular to three other pieces. In either configuration, keep the rebar back 3 in. from the edges of the pad.

ADDING A GIRDER

Adding a girder under a run of joists shortens the distance they span, stiffens a springy floor, and reduces some loading on perimeter foundations. If your floors are springy and joists exceed the following rule-of-thumb lengths, consider adding a girder.

An engineer can size the girder for you. "Beam Span Comparison" below shows maximum spans for built-up girders in two-story houses.

Ideally, the new girder should run beneath the midpoint of the joist span, but if existing ducts or drainpipes obstruct that route, shift the girder location a foot or two. Once you locate the girder,

snap a chalkline to mark its center and plumb down from that to mark positions for pads and posts. Place posts at each end of the girder and approximately every 6 ft. along its length. If you create a girder by laminating several 2xs, keep at least one member of the "beam sandwich" continuous over each post.

Size and reinforce pads as described in the preceding section. After the concrete pad has cured for a week, bring in the girder or laminate it on site from 2-in. stock. Prescribed widths for built-up girders are usually three 2x boards (4½ in. thick when nailed together). For built-up girders and beams, the *Uniform Building Code* recommends the following nailing schedule: 20d nails at 32 in. on center at the top and bottom and two 20d nails staggered at the ends and at each splice.

Whether solid or laminated, if the girder has a crown, install it so it faces up. Installing a new girder is essentially the same as positioning a temporary shoring beam, except that the girder will stay in place. Have helpers to raise the girder and support it until permanent support posts are in place. Properly sized, the pad will have more than enough room for jacks and posts, so place jacking posts as close as possible to the permanent post's location. Raise the girder approximately ⅛ in. higher than its final position to facilitate insertion of the new posts.

STEEL BEAMS

If there's limited headroom or clearance under the house, steel beams provide more strength per equivalent depth than wood beams. If you're

Beam Span Comparison

Typical joist
Header or other support
Beam
Header
y/2
y/2
x/2
x/2
Joist span y
Joist span x
x/2
Beam supports ½ of each joist span, or x/2 + y/2. See table at right.
Header supports ½ of single joist span.

Use this drawing and table for estimating beam sizes and comparing beam types for uniform floor loads of a 40-psf (pounds per square foot) live load and a 15-psf dead load. Have a structural engineer calculate your actual loads.

	JOIST SPAN (x/2 + y/2)			
	8 FT.	10 FT.	12 FT.	14 FT.
BEAM TYPE	BEAM SPAN (ft.)			
(2) 2x8 built-up beam →	6.8	6.1	5.3	4.7
4x8 timber →	7.7	6.9	6.0	5.3
3⅛-in. x 7½-in. glue-laminated beam →	9.7	9.0	8.3	7.7
3½-in. x 7½-in. PSL beam →	9.7	9.0	8.5	8.0
(2) 1¾-in. x 7½-in. LVL (unusual depth) →	10.0	9.3	8.8	8.3
4x8 steel beam (W8 x 13 A36) →	17.4	16.2	15.2	14.1

Replacing a Wooden Girder with an I-Beam

As level building lots continue to disappear, we'll need to make the most of sites once thought unbuildable, such as the remodel in the photos here, which takes place on a steep, south-facing slope above San Francisco. The lot's high side bordered a city street, so a garage right on the street became the portal to descending stairs, an elevated walkway, and a dramatic, modern house set downhill about 50 ft. from the road.

In time, the homeowner realized that the cavernous space under the garage could accommodate an in-law suite, so he hired Stephen Shoup, principal of Building Lab in Emeryville, Calif., to design and build it. The great challenge of the project involved transferring the loads borne by the main girder—which supported a garage floor above—to a steel I-beam set flush to the finished ceiling that would span the 22-ft. width of the suite without intervening support posts.

The weights were considerable. The garage floor system consisted of 2x12 Douglas fir joists spaced 12 in. on center, a 3/4-in. plywood subfloor, and a 4-in.-thick reinforced concrete slab. The clearances were also quite tight when it came time to cut through the garage floor joists: The slot cut was roughly 3/16 in. wider than the width of the I-beam—allowing 1/16 in. on each side of the I-beam and 1/16 in. of "wiggle room." The fit had to be that exacting so joist ends could be hung off the 2x12s bolted to the sides of the I-beam.

Unfinished space under the garage. The 6x12 Douglas fir girder, at top, spanned 22 ft. with the aid of 2x6 beam pockets at both ends and a 6x6 post at midspan. The white sticks about halfway up the wall indicate the height of the subfloor to come.

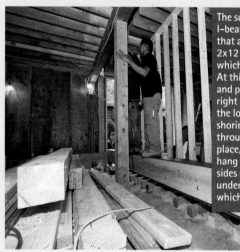

The subfloor in place, with an I-beam waiting to be hoisted. Note that a slot has been cut into the 2x12 ceiling-joist array above, into which the W12x26 I-beam will fit. At this transitional point, the girder and post seen in the photo above right are now carrying roughly half the load—the joists at left—and shoring, at right, supports the cut-through joists. Once the I-beam is in place, the severed joist ends will hang from the 2x12s bolted to both sides of the I-beam. Peeking up from under the I-beam is cribbing on which the shoring sole plate rests.

An I-beam on the way up. At left, shoring; upper right, a rectangular hole into which the beam end will fit once it has been fully raised. At either end, posts will be cut to exact length, set into the wall, and bolted to the legs of the I-beam cap, which is an integrally welded assembly of 1/4-in. steel plate. Those posts will transfer loads down to the perimeter foundation.

One end almost in place. Raising a beam of this length and weight should be attempted only by professionals with specialized equipment. A W12x26 I-beam weighs 26 lb./ft., so its total weight is 572 lb. As the worker at right raises the end of the I-beam, his partner, on the other side of the shoring, uses a ratcheted come-along to pull the beam end toward the opening in the wall.

going to use steel beams, hire an engineer to size them and a specialist to install them: Steel I-beams are expensive and heavy, and they can be problematic to attach to wood framing, without special equipment to drill holes, spot-weld connectors, and so on. For commonly available sizes and some sense of the weights involved, see "Steel I-Beams" on p. 75.

Replacing a Shallow Foundation

If you decide to replace a shallow foundation, begin by checking local building codes for foundation specs appropriate for your area. Before beginning foundation work, be sure to review this chapter's earlier sections on shoring and jacking. Then survey the underside of the house and the area around the foundation for pipes, ducts, and other potential obstructions. If you can reposition jacks or move shoring slightly to

avoid crushing or disconnecting drains, water pipes, and the like, do so.

Remember, jacking timbers and shoring are *temporary* supports. Complete the job and lower the house onto foundation elements as soon as possible. Work within your means, skills, and schedule: If you can't afford a house mover to raise the house and replace the whole foundation, have him do it one wall at a time.

REPLACING MUDSILLS

Mudsills are almost always replaced when foundations are. With the framing exposed, it's easy to install new pressure-treated mudsills that resist rot and insects. At the same time, replace rotted or insect-infested pony-wall studs. (If just a few studs are rotted, cut away the rot and nail a pressure-treated sister stud to each. If the bottom 1 in. to 2 in. of many studs has rotted, you might also install a thicker mudsill to make up for the amount you cut off stud bottoms.) If the siding is in good shape, remove just enough to expose the mudsills and rotten studs; the siding holds the pony-wall studs in place and keeps them from "chattering" while you cut them. You'll also need to punch through the siding to install temporary needle beams, discussed earlier in this chapter.

Once you've jacked up and shored the house framing, lay out the height of the new sill by snapping chalklines across the pony-wall studs. Use a laser level to indicate where the chalk marks should go or, if the old foundation is level, measure up from it. Although the line should be as level as possible, small variations will be accommodated when the concrete is poured up to the bottom of the mudsill.

With the siding removed, use a square to extend the cutoff marks across the face of the studs; a square cut optimizes load bearing from the stud onto the mudsill. Use a reciprocating saw to make the cuts. If the first stud chatters as you attempt to cut through it, tack furring strips to all the studs, just above the cut-line to bolster successive studs. Then remove the old mudsill and rotted stud sections. Chances are the old mudsill will not be bolted to the foundation.

The replacement sill should be pressure-treated Douglas fir or yellow pine to resist insects and moisture. It should be end-nailed upward into the solid remnants of each stud, using two 16d galvanized common nails. Use a pneumatic nailer to nail up the new mudsill; it does the job quickly. However, predrill anchor bolt holes into the new mudsills before nailing them to stud ends. Anchor bolts will secure house framing to the foundation after you pour it.

Girder (Beam) Supports

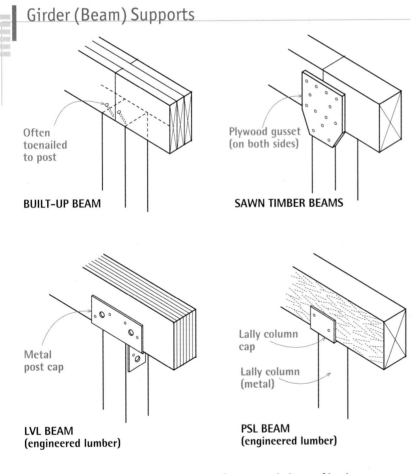

BUILT-UP BEAM

Often toenailed to post

SAWN TIMBER BEAMS

Plywood gusset (on both sides)

LVL BEAM
(engineered lumber)

Metal post cap

PSL BEAM
(engineered lumber)

Lally column cap

Lally column (metal)

When it's necessary to construct a girder from several pieces of lumber, support each girder joint by placing a post or column beneath it. Many building codes also require metal connectors or plywood gussets at such joints to join posts to beams.

REMOVING OLD FOUNDATIONS

To tear out an old foundation, you've got several options. All require safety glasses, hearing protection, heavy gloves, a mask, patience, and a strong back. Before acquiring heavy and expensive equipment, try to break out a section of the old foundation using a 9-lb. sledgehammer and a 6-ft. pointed steel bar. Old concrete without rebar is often cracked and soft. Once you've removed a small section, the rest may come out easily.

If the concrete is too thick and hard, rent a towable air compressor and jackhammer. A 90-lb. jackhammer will break almost anything, but it's a beast to maneuver; a 60-lb. hammer is light enough to lift onto a foundation wall and almost always strong enough to break a wall apart. A 60-lb. electric jackhammer is less powerful than a compressor-driven one, but it may have enough muscle to get the job done.

Or you can rent a gas-powered saw with a 10-in. concrete-cutting blade that cuts 4 in. to 5 in. deep, letting you cut the concrete into manageable chunks. A third option is roto-hammering a line of ⅝-in. holes across the foundation and then splitting along that line with a large mason's chisel and a hand sledge.

Should you encounter rebar, you'll either need an acetylene torch to cut through it or a metal abrasive wheel in a circular saw or grinder. Cutting rebar is hard work. Once the old foundation is removed, excavate the footing as necessary and square up the trench.

Leveling a House

In theory, you can level a house using individual hydraulic jacks. And if the house has only one or two low spots, you may succeed. However, the framing of a house will usually have sagged and settled, increasing the likelihood that jacking one area will raise an adjacent area too much. Heavily loaded points in multistory houses may resist being raised at all, and when they finally do move, it's often sudden, loud, and frightening.

Leveling a house is far more likely to succeed if done by a house mover with a *unified hydraulic jacking system*, in which jacks are interconnected, via hoses, to a central console that monitors the load on each jack. Instead of 12 workers trying to turn 12 jacks at exactly the same time, a single operator at the console can ensure that the jacks rise at the same—or variable—rates to the desired height.

The desired height is determined beforehand by the foundation contractor, house mover, structural engineer, and—on occasion—the architect. Most often, the house mover works from a master reference point outside, against which house corners are read to determine whether they need raising or lowering. (For example, corner 1 might be listed as +¾ in., corner 3 as –½ in., and so on.) Once the corners are leveled, the framing in the middle of the house is fine-tuned.

Even when professionals level a structure, there's invariably damage to the finish surfaces inside, such as cracked plaster or popped drywall seams, door latches that no longer meet strike plates, trim that's askew, and windows and doors that bind as they're opened and closed. Consider all this before you jack. Raising only the most out-of-level areas may be more cost-effective than leveling floors perfectly. Moreover, gently sloping floors may add character to an older house.

This old mudsill rotted out because the foundation was too close to the ground. After using a laser level to transfer the height of the new foundation, this builder snapped a chalkline across the pony-wall studs to indicate the height of the new sill.

Once the needle beams are in place to pick up the loads that were previously carried by the pony wall, you can cut the pony-wall studs and remove the sill.

A 60-lb. jackhammer is powerful enough to bust concrete yet light enough to lift onto the foundation. Jackhammering is bone-rattling work, so have workers take turns at it.

Concrete Forms for a Shallow Foundation

Joist

Blocking

Rim joist

Top plate
of pony wall

Plumbed
2x4 form-hangers

Sheathing

1½-in.-thick
spacer

Brace

Anchor bolt

Mudsill

2x4 brace

3-in. clearance for
concrete pump hose

Termite
shield

1½-in.
formboards

Rebar

Stake

Minimum 8 in.
above grade

Wedges

Form ties

Perforated
steel stake

Dobie blocks keeping
rebar 3 in. above soil

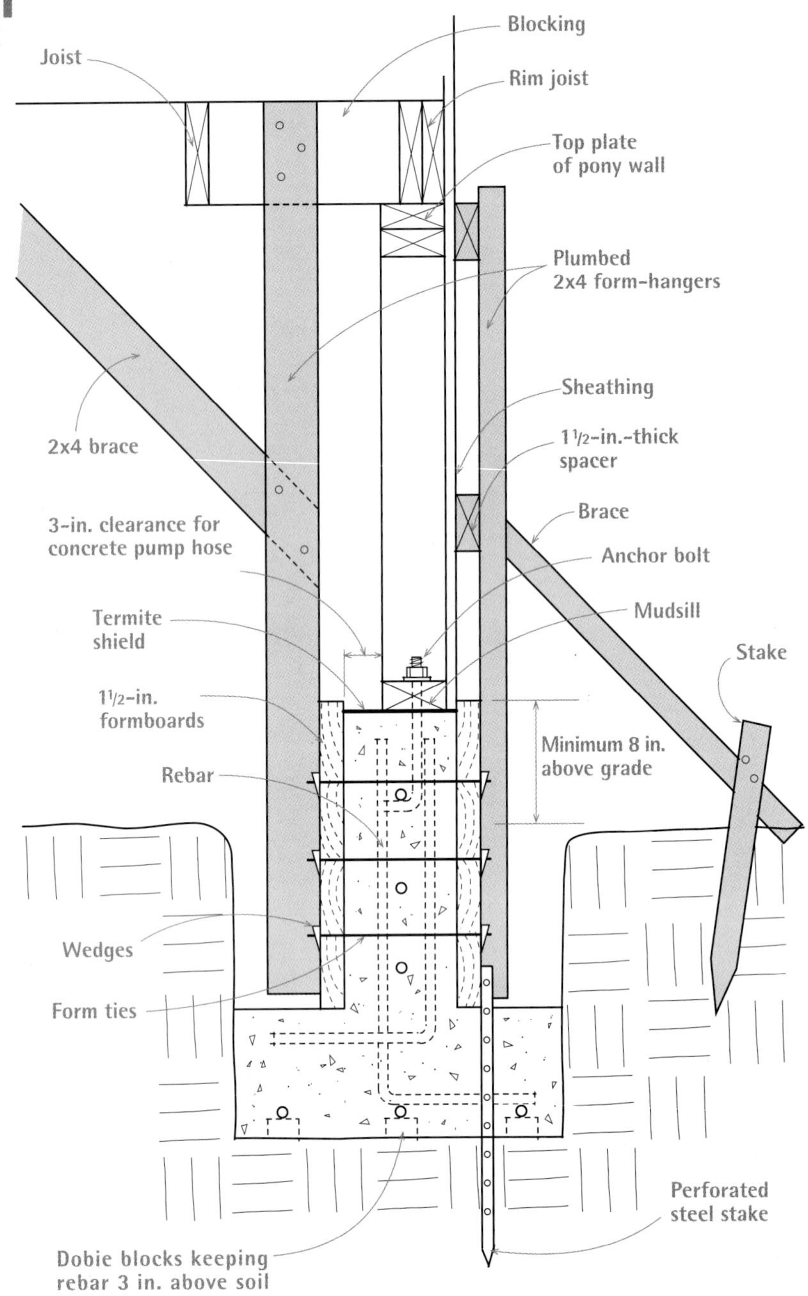

In this example, foundation walls are flush to the sheathing, and the trench walls serve as forms for the footings. Details will vary slightly, depending on the direction of joist (as described in the text) and on other framing particulars.

CONCRETE FORMWORK

Correctly positioning 1½-in.-thick formboards can be tricky, and there are myriad ways to do so. Here's a relatively foolproof method in which you erect the inner (house side) form walls first, by nailing them to 2x4 *form-hangers* nailed to joists. This method also enables easy access for tying rebar, reattaching sills, and the like.

Inner form walls. If floor joists run perpendicular to the foundation wall, start by nailing 2x4 form-hangers into the joists at both ends of the foundation wall section being replaced. The 2x4s should extend down into the foundation trench, stopping 1 in. to 2 in. above the tops of footing forms, if any. Position each 2x4 so its edge is exactly 9½ in. from the outside face of the foundation (8-in.-thick concrete plus 1½-in.-thick formboard). Nail the bottom formboard to the 2x4s, then add 2x4 form-hangers between the first two. Spacing 2x4 form-hangers every 32 in., use two 16d nails to nail them to each joist. Then stack additional formboards atop the first until the top board is slightly above the bottom of the mudsill. As shown in "Concrete Forms for a Shallow Foundation" at left, run diagonal 2x4 braces from joists to the 2x4 form-hangers to stiffen the inner form wall, thereby keeping it plumb and in place (see also the photos on the facing page).

If the joists run parallel to the foundation, first add blocking between the rim joist and the first joist back, across the top of the pony wall. Nail the 2x4 form supports to the blocking, much as just described for perpendicular joists. Once the inside forms are complete, you can cut, bend, and assemble the rebar; attach the mudsill to the pony-wall studs; and insert anchor bolts before building the outside form walls. Even if local building codes don't require steel-reinforced

Positioning
FOUNDATION FACES

Traditionally, the outside face of a foundation wall is flush to the edge of the house framing, allowing sheathing to overhang the foundation 1 in. or so, covering the joint between foundation and framing. However, contractors who install a lot of stucco argue that a foundation face flush to the outer face of the sheathing better protects the sheathing edge and creates a stucco edge that's less bulky—that is, one that sticks out less beyond the foundation wall.

foundations, adding steel is money well spent (see "Adding Steel" on pp. 276–277).

Outer form walls. If your foundation is shallow and the sides of the trenches are cleanly cut, you may not need formboards for footings. But if your footings will have formboards, install them before building the foundation's outer form walls.

If there are no footing formboards, drive 4-ft.-long perforated steel stakes into the footing area to secure the bottom formboards for the outer form walls. Plumb and space these stakes out 1½ in. from the outside face of the foundation to allow for the thickness of the formboards. Use two stakes per formboard to get started. Use 8d duplex nails to attach formboards to the steel stakes. Install this first outer formboard a little higher than the inner formboard initially, then hammer the stakes down to achieve level. You may need several tries to drive stakes that are plumb and accurately positioned because the points of the stakes often are deflected by rocks. Use a magnetic level to plumb the stakes.

Once the steel stakes are correctly positioned and the bottom formboards are nailed to them, add 2x4 form-hangers so you can hang additional formboards above. But first, nail spacers to the pony-wall studs to compensate for the thickness of the 1½-in.-thick formboards. If the pony-wall studs are sheathed, nail 1½-in.-thick spacer boards to the studs so the back face of the form-board lines up with the exterior sheathing. If the studs aren't presently sheathed, nail up 2-in.-thick spacers to accommodate the thickness of the formboards and the sheathing to come. If the outer face of the foundation wall aligns to the face of the sheathing, you can easily cover that often-troublesome joint with siding.

As you install each formboard atop the preceding one, set the form ties that tie together inner and outer formboards. Form ties are designed to space the formboards exactly the right distance apart; they are available in 6-in., 8-in., 10-in., and 12-in. lengths. At the ends of each form tie, insert metal wedges into the slots to keep forms from spreading when filled with concrete. The top formboard should overlap the mudsill slightly.

The outer formboards are braced by the plumbed 2x4 form-hangers, which are in turn supported by diagonal braces running down to perforated steel stakes or to 2x4 stakes driven into the ground. Under the house, diagonal braces run from the inner form-hangers to the joists.

Note: Spray formboards with form-release oil to facilitate their removal after the concrete has been poured. But be careful not to spill the oil onto the rebar, anchor bolts, or old foundation

Use prelooped wire ties to splice lengths of rebar, overlapping rebar sections at least 12 in. Note the cleanly cut sides of this trench, which will serve as forms for the poured foundation footings. An electric demolition hammer with a shovel bit was used to cut this dense soil.

Ready for concrete—the form walls are up and braced. Metal wedges inserted into the ends of the form ties will keep the formboards from spreading. The next step is to pour the footings before pouring the foundation walls.

To use this rebar cutter–bender, you feed rebar parallel with the base arm for bending, as shown, and perpendicular to the base arm when cutting.

because the oil will weaken the bond with new concrete.

ADDING STEEL

Structural steel used in renovated foundations includes *rebar*; *anchor bolts* to attach framing to concrete; *pins* (or dowels), which tie old foundations to new ones; and a plethora of *metal connectors*, including the popular Simpson Strong-Ties, which strengthen joints against earthquakes, high winds, and other racking forces.

Rebar. Rebar in foundations is not specified by all building codes, but it's cost-effective insurance against cracking caused by lateral pressures of soil and water against foundations. Rebar also can eliminate concrete shrinkage cracks. Common sizes in residential construction are No. 3 (³⁄₈ in. in diameter), No. 4 (¹⁄₂ in.), and No. 5 (⁵⁄₈ in.). One common configuration is No. 4 rebar spaced every 16 in. or 32 in. on center.

In footings and foundation walls below grade, place rebar back 3 in. from forms and at least 3 in. above the soil. On the inner side of the foundation walls, rebar can be within 1¹⁄₂ in. of the forms. You should run rebar the length of a foundation, tying the lengths together after overlapping them at least 24 in. Use *prelooped wire ties* to join them. (Wire ties don't lend strength; they simply hold the bars in place before and during the pour.) Use wire ties to attach rebar to the anchor bolts, pins, form ties, and the like. Use a *cutter-bender* to cut and bend bars on small

jobs. When rebar is delivered, store it above the ground—dirty rebar doesn't bond as well.

Anchor bolts. Place ¹⁄₂-in. or ⁵⁄₈-in. anchor bolts no more than 6 ft. apart in one-story house foundations and no more than 4 ft. apart in two-story foundations. In earthquake zones, 4-ft. spacing is acceptable, but conscientious contractors space the bolts every 32 in. There also should be an anchor bolt no farther than 1 ft. from each end of the sills. For maximum grip, use square washers. When pouring a new foundation, use J-type anchor bolts; the plastic bolt holders shown in the bottom photo on p. 260 will position the anchor bolts in the middle of the foundation wall.

When retrofitting bolts to existing foundations, use ⁵⁄₈-in. *all-thread rod* cut to length. Rod lengths will vary according to code specs and sill thickness. For example, a 10-in. rod will accommodate a washer, nut, and 1-in.-thick mudsill and will embed 7 in. in the concrete. You can also buy precut lengths of threaded rod, called *retrofit bolts*, which come with washers and nuts. Drill through the mudsill into the concrete, clean out the holes well, inject epoxy, and then insert the rods and bolts. The procedure is essentially the same for epoxying rebar pins to tie new concrete to old.

Because bolts, all-thread rods, and other tie-ins are only as strong as the material around them, you should center bolt holes in the top of the old foundation and drill them 6 in. to 8 in. deep, or whatever depth local codes require. Use

Masonry anchors:
1. Anchor bolt holder (monkey paw)
2. Anchor bolt holder
3. Simpson SSTB anchor bolt, used with seismic hold-downs
4. J-bolt anchor
5. Square plate washers
6. Concrete screws (high-strength threaded anchors)
7. Lag screw within expansion shield
8. Pin-drive expansion anchor
9. Wedge expansion anchor
10. Expansion shields for machine screws

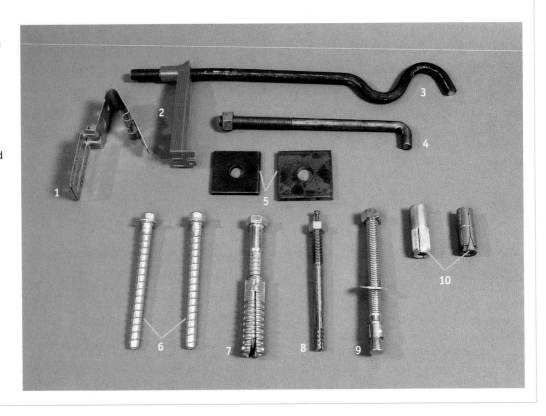

Retrofitting anchor bolts requires drilling in tight spaces. Bolts should be centered in the sill and embedded at least 7 in. into the concrete.

an impact drill if you're drilling concrete. Drill holes ⅛ in. larger than the diameter of the bolt so there's room for epoxy. For example, for ⅝-in. all-thread rod, drill ¾-in. holes; for ½-in. rod, drill ⅝-in. holes. If you make the holes larger than that, the bond probably won't be weaker, but you may waste a lot of expensive epoxy.

Note: To anchor mudsills in retrofits, threaded rod and epoxy have largely replaced expansion bolts. These chemical bonds are almost always stronger than mechanical ones, and epoxy's compressive strength is roughly four times greater than that of concrete.

Pins. Concrete *cold joints* are inherently weak. Cold joints occur when new concrete is butted against old or when separate pours create seams. To keep cold joints from separating, you need to join them with rebar pins. When drilling lateral holes to receive rebar pins that tie old walls to new (or secure a foundation cap), angle the drill bit slightly downward, so the adhesive you'll inject into the hole won't run out and so pins will be less likely to pull out.

Local codes and structural engineers will have the final say on sizing and spacing rebar pins. But, in general, drill ⅝-in. holes for ½-in. rebar to be epoxied; drill holes at least every 18 in. and embed rebar at least 7 in. into the top of foundations, and at least 4 in. into the side of 8-in.-thick

walls. Extend rebar epoxied into the old foundation at least 18 in. into new formwork, and overlap rebar splices at least 12 in.

ATTACHING THE MUDSILL

Builders may attach new mudsills at different stages of form assembly. But here are the essentials. Before nailing up the mudsill (to the bottom of the pony-wall studs), predrill for anchor bolts, as described earlier, making sure that no bolt occurs under a stud. If local codes require metal termite shields, tack them to the underside of the mudsill after drilling it for anchor bolts. Then, using a pneumatic nailer, end-nail the sill to the studs, using two 16d nails per stud. If there isn't enough room to end-nail upward into the sill, jack the mudsill tight to the studs and toenail down from the studs into the mudsill.

Once the mudsill is nailed to the studs, insert anchor bolts into the predrilled holes, screw on washers and nuts, and tie the free ends of the bolts to the rebar. At this point, the nuts should be just snug; you can tighten them down after the concrete has cured. If you end-nailed the mudsill, make sure the studs are tight to the top plates above. Jack up any studs that have separated. You may also need to brace the pony wall if it's

GETTING A GOOD Epoxy BOND

When you're done drilling into the foundation, you need to clean each hole thoroughly because dust clinging to the sides of the hole greatly reduces the epoxy's bond strength. For this, use compressed air to blow dust from each hole and then a hole-cleaning brush in a repeating blow-brush-blow-brush-blow cycle. Before blowing, put on a respirator mask and eye protection. Then fit a tube into the hole and attach a blow nozzle. Continue the cycle until the hole is clean.

To avoid air pockets in the epoxy, fill each hole from the bottom with a long-nozzle injector. The epoxy should flow in easily. As you insert the threaded rod or piece of rebar, twist it one full turn clockwise to distribute the adhesive evenly. If no epoxy

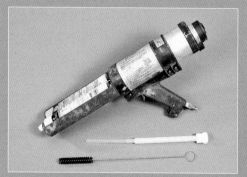

oozes out of the top of the hole when you've fully inserted the threaded rod, you didn't use enough epoxy or the epoxy leaked out. Leave the failed bolt in place, and drill and insert a new one next to it. Or remove the bolt and refill the hole completely with new epoxy.

Rebar epoxied into the old foundation is tied to steel in a perimeter wall of the new foundation. Part of this new wall's footing underpins (flows under) the old concrete. The chalkline immediately above the rebar marks the top of the foundation (see arrow).

This Simpson HD8A hold-down ties framing to the foundation and transfers tension loads between floors.

loose, which is often the case if you have to demolish siding to remove it. Pony walls should be plumb and aligned with the forms.

POURING CONCRETE

After installing the outside formboards, you are almost ready to pour. If you are using a 2-in. hose (interior diameter) to pump the concrete to the site, make sure there is at least a 3-in. clearance between the edge of the formboard and the outside edge of the new mudsill to accommodate the width of the pump nozzle. If necessary, notch the forms so the nozzle can fit. The top of the form should be slightly higher than the bottom of the mudsill.

Fill the footing to the bottom of the wall forms before filling the forms. Some concrete may slop over from the walls onto the footing and bottom formboards, but slopover isn't a problem if you remove the concrete from the form bottoms before it sets up. That will allow easy removal of the forms once the concrete has cured.

When the concrete begins to set but is not completely hard, pull out the perforated steel stakes holding the bottom formboards in place. To do this, remove the duplex nails and then use either a pipe wrench to grip the stakes or a commercial *stake puller*. If you leave the stakes in until the next day, you'll likely be able to pull

them only if you remembered to oil them first. Otherwise, cut the tops of the stakes off and leave the rest embedded in the new concrete.

Hammer the outside of the formboards, then use a *concrete vibrator* to drive out the air pockets. For this, insert the hoselike vibrator into the forms. As the concrete approaches the tops of forms, signal the pump to shut off so that the concrete doesn't spill over the sides. When the forms are full and vibrated, use a trowel to flatten the top of the wall and sponge off any globs on the stakes and forms. Allow the concrete to cure three days at a minimum and seven days for the optimum before removing the forms and shoring, replacing the siding, and tightening down the washered anchor bolts. For further protection against moisture, apply below-grade waterproofing to the outside wall and footing before backfilling.

Capping a Foundation

Capping an old foundation with new concrete is relatively rare but is done when the existing foundation is in good condition and needs to be raised because the house's framing is too close to the ground, allowing surface water to rot sills and siding.

The new cap must be 8 in. above grade. At the very least, that means shoring up the structure, removing the existing mudsill, shortening the pony-wall studs, drilling the old foundation, epoxying in rebar pins to tie the new concrete to the old, and pouring new concrete atop or around some part of the existing foundation. That's a lot of work. So if the existing foundation is crumbling or lacks steel reinforcement, you should replace it altogether.

On the other hand, if the house lacks pony walls and the joists rest directly on the foundation, you have basically two options: (1) grade the soil away from the house to gain the necessary height, which may not be possible if the foundation is shallow, or (2) jack up the house at least 8 in., which means hiring a house mover. Here again, replacing the foundation is usually more cost-effective.

Cubic Yards of Concrete in Slabs of Various Thicknesses*†

AREA (sq. ft.)	1.0 in.	1.5 in.	2.0 in.	2.5 in.	3.0 in.	3.5 in.	4.0 in.	4.5 in.	5.0 in.	5.5 in.	6.0 in.
10	0.03	0.05	0.06	0.08	0.09	0.11	0.13	0.14	0.15	0.17	0.19
20	0.06	0.09	0.12	0.16	0.19	0.22	0.25	0.28	0.31	0.34	0.37
30	0.09	0.14	0.19	0.23	0.28	0.33	0.37	0.42	0.46	0.51	0.56
40	0.12	0.19	0.25	0.31	0.37	0.43	0.50	0.56	0.62	0.68	0.74
50	0.15	0.23	0.31	0.39	0.46	0.54	0.62	0.70	0.77	0.85	0.93
60	0.19	0.28	0.37	0.46	0.56	0.65	0.74	0.83	0.93	1.02	1.11
70	0.22	0.32	0.43	0.54	0.65	0.76	0.87	0.97	1.08	1.19	1.30
80	0.25	0.37	0.49	0.62	0.74	0.87	1.00	1.11	1.24	1.36	1.48
90	0.28	0.42	0.56	0.70	0.84	0.97	1.11	1.25	1.39	1.53	1.67
100	0.31	0.46	0.62	0.78	0.93	1.08	1.24	1.39	1.55	1.70	1.85
200	0.62	0.93	1.23	1.54	1.85	2.16	2.47	2.78	3.09	3.40	3.70
300	0.93	1.39	1.85	2.32	2.78	3.24	3.70	4.17	4.63	5.10	5.56
400	1.23	1.83	2.47	3.10	3.70	4.32	4.94	5.56	6.17	6.79	7.41
500	1.54	2.32	3.09	3.86	4.63	5.40	6.17	7.00	7.72	8.49	9.26
600	1.85	2.78	3.70	4.63	5.56	6.48	7.41	8.33	9.26	10.19	11.11
700	2.16	3.24	4.32	5.40	6.48	7.56	8.64	9.72	10.80	11.88	12.96
800	2.47	3.70	4.94	6.20	7.41	8.64	9.88	11.11	12.35	13.58	14.82
900	2.78	4.17	5.56	6.95	8.33	9.72	11.11	12.50	13.89	15.28	16.67
1,000	3.09	4.63	6.17	7.72	9.26	10.80	12.35	13.89	15.43	16.98	18.52

* This table can be used to estimate the cubic content of slabs larger than those shown. To find the cubic content of a slab measuring 1,000 sq. ft. and 8 in. thick, add the figures given for thicknesses of 6 in. and 2 in. for 1,000 sq. ft.

† Courtesy of Bon Tool Company, © 2003, from Statistical Booklet: Contractors, Tradesmen, Apprentices (see also www.bontool.com).

Concrete Work

Concrete is a mixture of portland cement, water, and aggregate (sand and gravel). When water is added to cement, a chemical reaction, called *hydration*, takes place, and the mixture hardens around the aggregate, binding it fast. Water makes concrete workable, and cement makes it strong. The lower the water to cement ratio (w/c), the stronger the concrete.

POURING A CONCRETE SLAB

Pouring (or placing) a concrete slab is pretty much the same procedure, whether for patios, driveways, basements, or garage floors. Most slabs consist of 4 in. of concrete poured over 4 in. of crushed rock, with a plastic moisture barrier between. In addition, garage floors are often reinforced with steel mesh or rebar to support greater loads and forestall cracking.

To improve the building's energy profile, www.greenbuildingadvisor.com advises installing rigid insulation under slabs. Details will vary regionally, but extruded polystyrene (XPS) panels 2 in. thick are typically recommended. XPS panels are durable, waterproof, and rated R-5 per inch. But using more than 2 in. isn't recommended because too much insulation under a slab will isolate it from the earth's cooling influence during summer months.

This 2-in. (interior diameter) concrete-pump hose is easier to handle than a 3-in. hose. But its smaller diameter requires smaller ³/₈-in. aggregate in the mix. Although a 2-in. hose is much lighter than a 3 in., tons of concrete pass through it—so you'll need helpers to support the hose and move it to the pouring points.

PREP STEPS

As with any concrete work, get plenty of help. Concrete weighs about 2 tons per cubic yard, so if your slab requires 10 cu. yd., you'll need to move and smooth 40,000 lb. of concrete before it sets into a monolithic mass. Time is of the essence, so make sure all the prep work is done before the truck arrives: Tamp the crushed stone, spread the plastic barrier (minimum of 6 mil), and elevate the steel reinforcement (if any) on *dobie blocks* or wire *high chairs* so it will ride in the middle of the poured slab. Finally, snap level chalklines on the basement walls or concrete forms to indicate the final height of the slab—you'll *screed* to that level.

To pour concrete with a minimum of wasted energy, use a 2-in. (interior diameter) concrete-pump hose. A hose of that diameter is much lighter to move around than a 3-in. hose. Another advantage: It disgorges less concrete at a time, allowing you to control the thickness of the pour better. And a 2-in. hose gives easier access to distant or confined locations. *Important:* As you place concrete around the perimeter of the slab, be careful not to cover up the chalklines you snapped to mark the slab height. And as you place concrete in the slab footings, drive out the air pockets by using a concrete vibrator.

Insulating a Slab-on-Grade Foundation

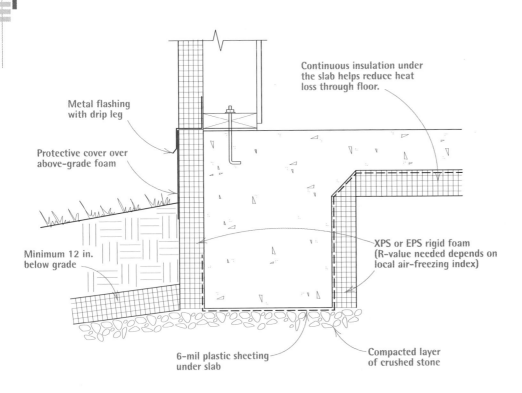

Metal flashing with drip leg

Protective cover over above-grade foam

Minimum 12 in. below grade

Continuous insulation under the slab helps reduce heat loss through floor.

XPS or EPS rigid foam (R-value needed depends on local air-freezing index)

6-mil plastic sheeting under slab

Compacted layer of crushed stone

Establishing screed levels. If the slab is only 10 ft. or 12 ft. wide, you can level the concrete by pulling a *screed rail* across the top of formboards. Otherwise, create *wet screeds* (leveled columns of wet concrete) around the perimeter of the slab and one in the middle of the slab to guide the screed rails. The wet screeds around the perimeter are the same height as the chalklines; pump concrete near those lines and level it with a trowel. This technique is very much like that used to level tile mortar beds, as explained in chapter 16.

The wet screed(s) in the middle should be more or less parallel to the long dimension of the slab. There are several ways to establish its height, but the quickest way is to drive 18-in. lengths of rebar into the ground every 6 ft. or so, and then use a laser level or taut strings out from perimeter chalklines to establish the height of the rebar. In other words, the top of the rebar becomes the top of the middle wet screed. When you've troweled that wet screed level, hammer the rebar below the surface, and fill the holes later.

SCREEDING AND FLOATING

Screeding is usually a three-person operation: two to move the screed rail back and forth, striking off the excess concrete, and a third person behind them, constantly in motion, using a stiff rake or a square-nose shovel to scrape down high spots or to add concrete to low ones. You can use a magnesium screed rail or a straight 2x4 to strike off, but the key to success is the raker's maintaining a good level of concrete behind the screeders, so the screed rail can just skim the crest of the concrete without getting hung up or bowed by trying to move too much material.

Screeding levels the concrete but leaves a fairly rough surface, which is then smoothed out with a magnesium *bull float*, a long-handled float that also brings up the concrete's *cream* (a watery cement paste) and pushes down any gravel that's near the surface. This creates a smooth, stone-free surface that can be troweled and compacted later.

A bull float should float lightly on the surface. As you push it across the concrete, lower the handle, thereby raising the far edge of the float. Then, as you pull the float back toward you, raise the handle, raising the near edge. In this manner, the leading edge of the bull float will glide and not dig into the wet concrete.

FINISHING THE SLAB

After the bull float raises water to the surface, you must wait for the water to evaporate before finishing the concrete. The wait depends on the weather. On a hot, sunny day, you may need to

Ordering Concrete: Be Specific

Concrete has so many different uses (such as floors, foundations, and countertops) and so many admixtures (water reducers, retardants, accelerants, air entrainers, and so on) that the best way to get the mix you need is to specify its use and desired characteristics. That is, when ordering the mix, tell the supplier the quantity you need (in cubic yards), how the concrete will be used (driveway, foundation, patio slab), the loads it will bear, how far it must be pumped, how it will be finished, and other such details.

If you're pouring a slab that will have a smooth finish, you might specify "a 2,500-psi mix but a true five-sack mix," which will be "creamy" enough to finish with a steel trowel. If you specify a 2,500-psi mix but don't describe the finish, the supplier might use four sacks of cement and a water reducer to attain that strength. However, with less cement in it, the mix would be sandier and more difficult to finish.

Where the concrete will be placed can also affect the mix. For example, concrete for a second-story patio far from the street may require a smaller, 2-in. (inner diameter) hose to pump it, so the supplier may specify smaller aggregate (3/8 in. versus 3/4 in.) to facilitate flow. Aggregate size, in turn, affects load-bearing capacity, so a mix with 3/8-in. gravel is often bumped up, say, to 3,000 psi. If the patio slab will also be steel troweled, the mix thus becomes "a 3,000-psi, 3/8-in. aggregate, true six-sack mix." Well, you get the point: Be specific.

Estimating the amount of concrete in cubic yards is straightforward: The calculation is width × length × depth (in feet) of the area you want to cover. You then divide that result by 27 (because there are 27 cu. ft. in 1 cu. yd.). If you're pouring a slab, see "Cubic Yards of Concrete in Slabs of Various Thicknesses" on p. 279 for calculations of cubic yards based on the slab thickness (in inches).

Most concrete-mixer trucks can hold 9 cu. yd. to 11 cu. yd. So if your pour requires more than one truck, ask the supplier to time deliveries 90 minutes apart so you have enough time to deal with each delivery. Finally, don't shave the estimate too close; far better to have too much concrete than too little.

The garage floor is pretty typical of slab-on-grade foundations. Because it has only a single layer of rebar, it can bear the weight of walls only along its perimeter walls.

Fill slab footings first, then vibrate them to drive out any air pockets. Because placing a concrete slab usually entails standing in wet concrete, wear rubber boots.

Screeding is a job for three people: two to level the concrete by moving the screed rail from side to side and a third with a shovel or rake to fill low spots and pull away excess.

Bull-floating the concrete pushes down gravel near the surface and brings the concrete's cream to the surface.

Finishing the slab. Start with magnesium or wood floats, and finish with a steel trowel for a smooth, hard finish. Knee boards distribute your weight as you work.

wait less than an hour. On a cool and overcast day, you might need to wait for hours. Once the water has evaporated, you have roughly one hour to trowel and compact the surface. When you think the surface is firm enough, put a test knee board atop the concrete and stand on it. If the board sinks ¾ in. or more, wait a bit. If the board leaves only a slight indent that you can easily hand float further and then trowel smooth, get to work.

As the bottom right photo shows, knee boards distribute your weight and provide a mobile station from which to work. You'll need two knee boards to move across the surface, moving one board at a time. Then, kneeling on both boards, begin sweeping with a magnesium hand float or with a wood float, if you prefer a rougher finish. Sweep back and forth in 3-ft. arcs, raising the far edge of the float slightly as you sweep away and raising the near edge on the return sweep. The "mag" float levels the concrete. After you've worked the whole slab, it's time for the steel trowel, which smooths and compacts the concrete, creating a hard, durable finish.

As the concrete dries, it becomes harder to work, so it's acceptable to sprinkle *very small amounts* of water on the surface to keep it workable. Troweling is hard work, especially on the back. When the concrete's no longer responding to the steel trowel, edge the corners and then cover the concrete with damp burlap before calling it a day. If the weather's hot and dry, hose down the burlap periodically—every hour, at least—and keep the slab under cover for four or five days. At the end of that time, you can remove the forms. Concrete takes a month to cure fully.

A Mat Slab Foundation

When it comes to cutting-edge technology, a mat slab fits the bill in several respects. Though little known 10 years ago, today mat slabs are a standard solution for building sites with difficult soil conditions. Typically reinforced with a double layer of rebar throughout its pad, a mat slab is capable of bearing loads at virtually any point and distributing heavy loads across an expanse of weak subsoil.

Mat slabs are particularly well suited to level sites, moderate slopes, or sites where you will be cutting the building pad into a hillside. In the project shown on the facing page, mat slabs poured on several levels were a much quicker and more affordable alternative to drilled pier and grade beams because the slab did not require footing excavation and expensive pier drilling. The mat slab only required forms on the outside perimeter, cutting forming cost in half. The engineer determined that a tee footing was not a via-

CREATING A MAT SLAB

2. To create a clean step (a retaining wall) in the foundation, the crew drove in steel stakes, ran horizontal pieces of rebar behind the stakes, put in expanded metal lath behind the rebar, and then backfilled with stone. The yellow 16-mil plastic sheeting is a vapor barrier to prevent moisture from migrating up from the soil to the slab. Dobie blocks atop the sheeting will support the lower level of rebar. To facilitate installing rebar in the retaining wall, formboards are not yet in place.

1. After the building pad was excavated, leveled, and compacted, perimeter forms were set and crushed drain rock was spread. The worker is using a tamper to compact 4 in. of crushed drain rock on undisturbed earth. The stone was held back along the perimeter so that the concrete at the edge of the slab would contain the gravel. Above the compacted stone, there would be 12 in. of concrete; around the perimeter, the concrete will be 16 in. thick.

3. After rebar was installed in the retaining wall, formboards were set and a double layer of rebar was placed in the slab in the foreground. C-shaped rebar spacers connect upper and lower rebar grids, keep the upper layer from sagging, and stabilize the grid during the pour.

5. First pass with the bull float.

4. Because the concrete contains ¾-in. gravel, the mix must be vibrated to be sure there are no air pockets. A boom truck was able to deliver concrete to distant parts of the forms on this 56-yd. monolithic pour.

6. Framing commences on the finished mat slab. The concrete wall along the back edge is a step-down retaining wall between foundation levels.

ble option given the soil conditions and hillside location.

Damp Basement Solutions

To find the best cure for a damp basement, first determine whether the problem is caused by water outside migrating through foundation walls or by interior water vapor inside condensing on the walls. To determine which problem you have, duct tape a 2-ft.-long piece of aluminum foil to the foundation, sealing the foil on all four sides. Remove the tape after two days. The wet side of the foil will provide your answer. Chapter 14 has more about mitigating moisture and mold.

CORRECTING CONDENSATION

If the problem is condensation, insulating basement floors and walls with rigid foam panels can be a complete and cost-effective solution to the problem. See "Curing Basement Condensation" below for more.

Installing foam panels to isolate always-cool concrete surfaces should solve the problem. Piecemeal solutions include insulating cold-water pipes, air-conditioning ducts, and other cool surfaces on which water vapor might condense. Wrap pipes with preformed foam pipe insulation. Wrap ducts and larger objects with sheets of vinyl-faced fiberglass insulation, which is well suited to the task because vinyl is a vapor barrier. Use duct tape or insulation tape to seal seams.

A dehumidifier also can remove excess humidity. For best results, install a model that can run continuously during periods of peak humidity; place it in the dampest part of the basement at least 12 in. away from walls or obstructions. To prevent mold from growing in the unit's collection reservoir, drain it daily and scrub it periodically.

Finally, look for other sources of moisture. An unvented clothes dryer pumps gallons of water into living spaces; vent it outdoors. Excessive moisture from undervented kitchens and bathrooms on other floors can migrate to the basement; add exhaust fans to vent them properly.

DAMPNESS DUE TO EXTERIOR WATER

Position and maintain gutters and downspouts so they direct water away from the house. And, if possible near affected walls, slope soil away from the house.

Besides those two factors, water that migrates through foundation walls or floors is more elusive and expensive to correct. Basically, you have three remedial options: (1) remove water once it gets in; (2) fill interior cracks, seal interior surfaces, and install a vapor barrier; and (3) excavate foundation walls, apply waterproofing, and improve drainage.

Option one: Remove water. *Sump pumps* are the best means of removing water once it gets into a basement. If you don't have a sump pump, you'll probably need to break through the basement floor at a low point where water collects, and dig a *sump pit* 18 in. to 24 in. across. Line the pit with a permeable liner that allows water to seep in while keeping soil out, and put 4 in. of gravel in the bottom.

There are two types of sump pumps. *Pedestal* sump pumps stand upright in the pit. They are water-cooled and have ball floats that turn the pump on and off. *Submersible* sump pumps, on the other hand, have sealed, oil-cooled motors, so they tend to be quieter, more durable, and more expensive. And because they are submerged, they allow you to cover the pit so nothing falls in. A ⅓-hp pump of either type should suffice.

The type of *discharge pipe* depends on whether the pump is a permanent fixture or a sometime thing. Permanent pumps should have 1½-in. rigid PVC discharge pipes with a check valve near the bottom to prevent expelled water from siphoning back down into the pit.

Curing Basement Condensation

Water vapor is everywhere, and the warmer the air, the more moisture it holds. Consequently, when warm summer air comes in contact with the cool concrete of basement walls and floors, it condenses. To "cure" dampness resulting from condensation, you must keep the warm air in living spaces from coming in direct contact with masonry masses by insulating basement floors and walls. Three details will help you do that successfully:

▶ Choose materials that can tolerate moisture and won't be degraded by it, such as *expanded polystyrene (EPS)* panels. Moisture migrates through concrete year-round, so moisture-sensitive materials such as wood, paper-faced drywall, and fiberglass insulation should not be in direct contact with it.

▶ Accept that a small amount of moisture will migrate, and choose materials that allow that. Again, EPS panels are a good choice because they are semipermeable, whereas polyethylene sheet plastic is not. Unable to migrate through plastic, moisture will condense and collect, which could lead to rot and mold.

▶ Make your insulation layer as airtight as possible. Tightly sealed EPS panels will contain conditioned air yet allow minuscule amounts of water vapor to move back and forth.

To see how one builder put these principles to the test in his own basement, see "Installing EPS Panels to Create a Dry Basement" on pp. 450–452.

Option two: Interior solutions. If basement walls are damp, try filling cracks as suggested on p. 263 and applying damp-proofing coatings. (This approach won't work if the walls are periodically wet.) Apply a penetrating sealer such as Drylok Masonry Waterproofer to stop moisture from migrating through. Unlike paint, which adheres to the surface, Drylok penetrates and bonds to the masonry. These coatings can withstand higher hydrostatic pressures than elastomeric paints or gels. Epoxy-based coatings also adhere well but are so expensive that they're usually reserved for problem areas such as wall-to-floor joints.

Impervious plastic vapor barriers were long advocated to block the flow of moisture through foundation walls. But, as explained in "Curing Basement Condensation," a better solution may be to install semipermeable, moisture-tolerant insulation panels against condensing surfaces and thus allow a slight amount of moisture to pass back and forth.

Option three: Exterior solutions. To waterproof exterior foundation walls, first excavate them. At that time, you should also upgrade the perimeter drains, as shown in "Foundation Drainage" on p. 261. Then, after backfilling the excavation, slope the soil away from the house. That is, no waterproofing material will succeed if water stands against the foundation. Before applying waterproofing membranes, scrub the foundation walls clean and rinse them well.

▶ **Liquid membranes** are usually sprayed on to a uniform thickness specified by the manufacturer, typically 40 mil. That takes training, so hire a manufacturer-certified installer. Liquid membranes are either solvent based or water based. Modified asphalt is one popular solvent-based membrane that contains rubberlike additives to make it more flexible and durable. Asphalt emulsions are water based and widely used because, unlike modified asphalt, they don't have a strong odor, aren't flammable, and won't degrade rigid foam insulation panels placed along foundation walls. Synthetic rubber and polymer-based membranes also are water based; they're popular because their inherent elasticity allows them to stay flexible and span small cracks. *Note:* Water-based membranes dry more slowly than solvent-based ones and can wash off if rained on before they are cured and backfilled.

▶ **Peel-and-stick membranes** are typically sheet or roll materials of rubberized asphalt fused to polyethylene. They adhere best on

preprimed walls. To install these membranes, peel off the release sheet and press the sticky side of the material to foundations. Roll the seams to make them adhere better. Peel-and-stick costs more and takes longer to install than sprayed-on membranes, but they're thicker (60 mil, on average) and more durable. Peel-and-stick membranes are often called Bituthene, after a popular W.R. Grace product.

▶ **Air-gap membranes** aren't true membranes because they don't conform to the surface of the foundation. Rather, they are rigid plastic (polyethylene) sheets held out from the foundation by an array of tiny dimples, which creates an air–drainage gap. Water that gets behind the sheets condenses on the dimples and drips free, down to foundation drains. (For this system to work, you must coat the foundation walls first.) Air-gap sheets are attached with molding strips, clips, and nails; caulk the sheet seams.

Pedestal Sump Pump

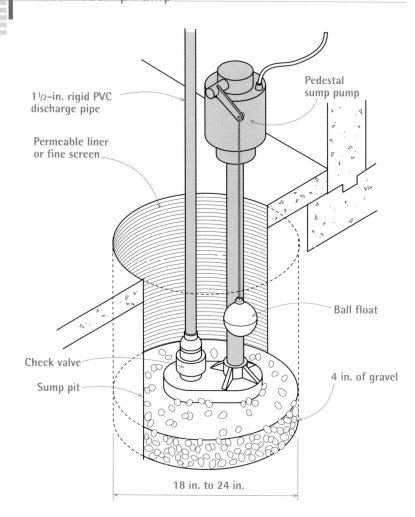

1½-in. rigid PVC discharge pipe

Permeable liner or fine screen

Pedestal sump pump

Ball float

Check valve

Sump pit

4 in. of gravel

18 in. to 24 in.

11

Electrical Wiring

Solid connections are the key to reliable wiring—especially connecting wires to screw terminals on receptacles, switches, and light fixtures. So here's a great tip. Some wire strippers have a small hole near the handle. Insert a stripped wire end, flip your wrist 180°, and—voilà!— a perfect loop.

This chapter is designed to give you a quick overview of the electrical system in your home. Respecting the power of electricity is essential to working safely. Always follow instructions carefully, use appropriate safety equipment, and when in doubt consult a licensed electrician. Before beginning work, check with local building authorities to make sure regulations allow you to do your own work and that you are conforming to local code requirements.

Understanding Electricity

Electricity (flowing electrons called *current)* moves through a wire like water in a pipe. The flow of water is measured in gallons per minute; the electrical flow of electrons is measured in *amperes* or *amps*. Water pressure is measured in pounds per square inch, and the force behind the electrons in a wire is measured in *volts*. The larger the pipe, the more water that can flow through it; likewise, larger wires allow a greater flow of electricity. A small-diameter pipe will limit the flow of water, compared to a larger pipe; similarly, wiring that is too small will resist the flow of current. If that resistance (measured in *ohms*) is too great, the wires will overheat and may cause a fire.

Think of electrical systems as a loop that runs from the generation point (or power source) through a load (something that uses electrical power, a lightbulb, for instance) and back to the generation point. In your home, the main loop, which is the service to your home, is split into smaller loops called *circuits*. Typically, a *hot wire* (usually black or red) carries current from the service panel to one of the various loads, and a *neutral wire* (typically white or light gray) carries current back to the service panel.

WORKING SAFELY

To work safely with electricity, you must respect its power. If you understand its nature and heed the safety warnings in this book—especially shutting off the power and testing with a voltage tester to make sure power is off—you can work with it safely.

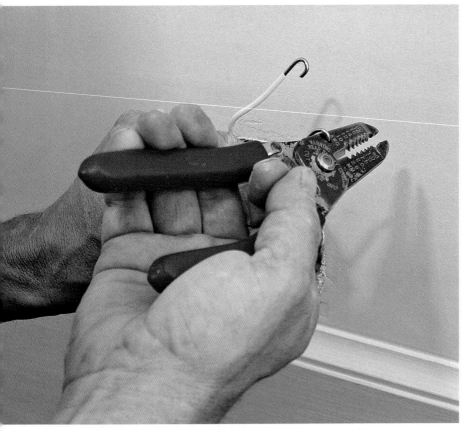

Use a voltage tester to be sure power is off before working on an outlet. If the tester light glows, there is still voltage present at the outlet.

Turn off the power to the circuit at the main service panel before removing receptacle, switch, or fixture covers.

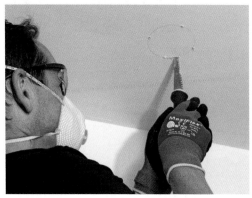

When doing electrical work, wear proper safety equipment, such as gloves to protect hands from the sharp edges of wires, cables, and boxes.

SAFETY ALERT

To work safely on existing circuits, always turn off electrical power at the service panel, and use a voltage tester (see the photos on p. 294) at the outlet to verify the power is off.

▌▌▌▌▌

The cardinal rule of home-improvement projects, which goes double for electrical work, is this: Know your limitations.

Unless you have previous experience doing electrical work and feel confident about your skills, you should leave certain projects to a pro. Working inside a service panel or even removing its cover can be especially dangerous. In most panels, there is an area around the main breaker that remains hot even after the breaker is set to the off position. Also, some older panels don't have a main breaker, and it takes experience to understand the layout of a panel.

Never attempt to remove the cover of, or work in, the main service panel or a subpanel. Call a licensed electrician rather than risk harm.

The safety alerts and safe working practices explained in this chapter will go far to protect you, but the best protection is knowledge. I strongly recommend that readers visit the OSHA/ NIOSH website on electrical safety: www.cdc. gov/niosh/topics/electrical/default.html.

Always wear appropriate safety gear, including rubber-soled shoes, sturdy gloves, safety glasses, and a respirator or dust mask when sawing or drilling overhead. Remember that 120 volts can kill or cause serious shock, so learn and use safe work practices—your life depends on it.

KEY Terms

WATTS. A measure of power consumed. In residential systems, watts are virtually the same as volt-amps.

VOLTAGE. The pressure of the electrons in a system. Voltage is measured in volts.

AMPERES (AMPS). The measure of the volume of electrons flowing through a system (current).

CURRENT. The flow of electrons in a system. Current is measured in amperes (amps). There are two types of current: DC (direct current) and AC (alternating current). AC power is supplied by utility companies to homes.

POWER (VOLT-AMPS OR VA). The potential in the system to create motion (motors), heat (heaters), light (fixtures or lamps), etc. Volt-amps = available volts x available amps (VA).

OHMS. The measure of resistance to the flow of electrons (current) in a wire. The higher the resistance, the lower the flow of electrons.

NEC. National Electrical Code.

SAFETY ALERT

Test the tester first and last. No matter what kind of tester you're using, test it first on a circuit that you know is hot to make sure the tester is working properly, and do the same after you've done the testing. Some testers run on batteries, and those batteries could die after your pretest check and before the voltage test. The post-test check will catch that. And don't think it can't happen—it does.

▌▌▌▌▌

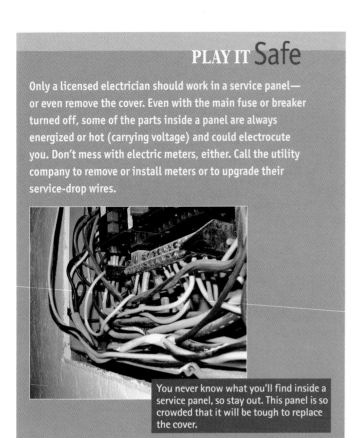
Electricity in Your Home

Power from the utility service is commonly delivered through three large wires, or *conductors*, which may enter the house overhead or underground. Overhead service wires are called a *service drop*. The drop runs to a *weatherhead* atop a length of rigid conduit. When fed underground, service conductors are installed in buried conduit or run as underground service-entrance (USE) cable. Whether it arrives overhead or underground, three-wire service delivers 120 volts (v) to ground and 240 volts between the energized conductors.

Service conductors are attached to a meter base and then to the *service panel*. Straddling the two sets of terminals on its base, the meter measures the wattage of electricity as it is consumed. The service panel also routes power to various *circuits* throughout the house. Increasingly, wireless devices are an important part of a home's electrical capabilities as battery-powered smoke alarms interlock automatically, doorbells with integral cameras show who is at the door, teachable thermostats save energy, and smartphone apps allow you to manage it all even if you're a thousand miles from home.

The utility company will install the drop wires to the building and will install the meter. The homeowner is responsible for everything beyond that, including the meter base and breaker panel, which a licensed electrician should install.

A typical three-wire service assembly consists of two insulated hot conductors wrapped around a stranded bare aluminum wire with an internal steel messenger cable—the bare aluminum wire also serves as the neutral.

The service panel distributes power to circuits throughout the house. Breakers interrupt power if the circuits become overloaded.

A Service Entrance and Panel

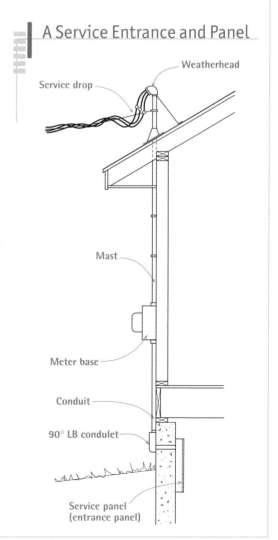

Weatherhead

Service drop

Mast

Meter base

Conduit

90° LB condulet

Service panel (entrance panel)

Wires, CABLES, AND CONDUCTORS

The terms *wires*, *cables*, and *conductors* are often used interchangeably, but they are not exactly the same. A conductor is anything that carries or conducts electricity. A wire is an individual conductor, typically covered in an insulating material, and cable is usually an assembly of two or more wires, protected by an overall plastic or metal *sheathing* (also called a *jacket*). A cable derives its name from the size of the wires within and the type of sheathing it is assembled with. For example, *12-gauge cable* contains 12AWG (American Wire Gauge) wire. More specifically, *12/2 with ground* denotes cable with two 12AWG wires plus a ground wire. NM designates nonmetallic sheathing, MC for metal-clad sheathing, and so on.

Inside the Service Panel

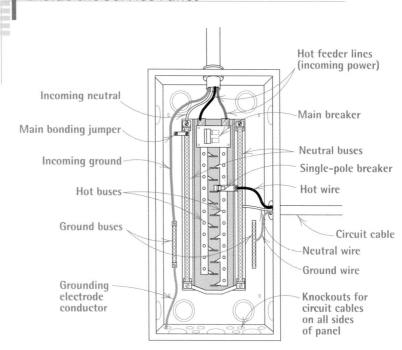

- Hot feeder lines (incoming power)
- Incoming neutral
- Main bonding jumper
- Incoming ground
- Hot buses
- Ground buses
- Grounding electrode conductor
- Main breaker
- Neutral buses
- Single-pole breaker
- Hot wire
- Circuit cable
- Neutral wire
- Ground wire
- Knockouts for circuit cables on all sides of panel

SERVICE PANELS

At the service panel, the two hot cables from the meter base attach to lugs or terminals on the main breaker. The incoming neutral cable attaches to the main lug of the neutral/ground bus. In the main panel, neutral/ground buses must be connected together, usually by a wire or metal bar called the *main bonding jumper*. In subpanels and all other locations downstream from the main service panel, ground and neutral components must be electrically isolated from each other.

In a main fuse box, the hot conductors from the meter attach to the main power lugs, and the neutral cable to the main neutral lug. Whether the panel has breakers or fuses, metal buses run from the main breaker/main fuse. Running down the middle of the panel, hot buses distribute power to the various branch circuits either through fuses or through breakers. The neutral/ground buses are long aluminum bars containing many terminal screws, to which ground and neutral wires are attached.

Each fuse or breaker is rated at a specific number of amps (15- and 20-amp breakers or fuses take care of most household circuits). When a circuit becomes overloaded or a short circuit occurs, the breaker trips or a fuse strip melts, thereby cutting voltage to the hot wire. All current produces some heat, and as current increases, the heat generated increases. If there were no breakers or fuses, and if too much current continued to flow, the wires would overheat and could start a fire. Amperage ratings of breakers and fuses are matched to the size (cross-

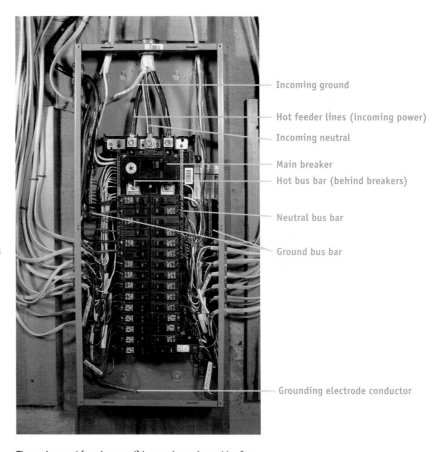

- Incoming ground
- Hot feeder lines (incoming power)
- Incoming neutral
- Main breaker
- Hot bus bar (behind breakers)
- Neutral bus bar
- Ground bus bar
- Grounding electrode conductor

The main panel (service panel) houses incoming cables from the meter as well as the breakers and wires that distribute electricity to individual circuits. At the service panel, neutral conductors (white wires), equipment-grounding conductors (bare copper or green insulated wires), the metal service panel, and the grounding electrode system (grounding rods) must be bonded together.

Stay away from the area around the main breaker. The terminals above the breaker remain hot even when the main is off, presenting a serious hazard if the cover is off.

A meter–main combo, placed outside the house, provides easy access for service or emergencies.

An older fuse box.

sectional area, measured as "gauge") of the circuit wires.

The main breaker. All electricity entering a house goes through the main breaker, which is usually located at the top of a main panel. In an emergency, throw the main breaker to "off" to turn off all power to the house. The main breaker is also the primary overcurrent protection for the electrical system and is rated accordingly. (The rating is stamped on the breaker handle.) If the main breaker for a 200-amp panel senses current that exceeds its load rating, the breaker will automatically trip and shut off all power.

Meter–mains. Increasingly common are meter–mains, which house a meter base and a main breaker service in a single box. Meter–mains allow a homeowner to put the main breaker outside the house, where it can be accessed in an emergency—if firefighters need to cut the power to the house before they go inside, for example. When meter–mains are used, electricians often locate a panel with the branch circuit breakers (called a subpanel) in the garage or another centralized location inside that is easy to access, such as a laundry room.

Fuse boxes. Many older homes still have fuse boxes. Fuses are the earliest overcurrent protection devices, and they come as either Edison-type (screw-in) fuses or cartridge (slide-in) fuses. Edison-style fuses are more common. They have little windows that let you see a filament. When the circuit has been overloaded and the fuse is

blown, the filament will be separated. A blackened (from heat) interior could mean a short circuit—a potentially dangerous situation calling for the intervention of a licensed electrician. The less common cartridge fuses are used to control 240v circuits and are usually part of the main disconnect switch, or serve heavy-duty circuits for an electric range or a clothes dryer.

GROUNDING BASICS

Because electricity moves in a circuit, it will return to its source unless the path is interrupted. The return path is through the white neutral wires that bring current back to the main panel. Ground wires provide an alternative low-resistance path should any of the metal equipment or enclosures become inadvertently energized.

Why is having a grounding path important? Before equipment-grounding conductors (popularly called *ground wires*) were widespread, people could be electrocuted when they came in contact with voltage that, due to a fault like a loose wire, unintentionally energized the metal casing of a tool or an electrical appliance. Ground wires bond all electrical devices and potentially current-carrying metal surfaces. This bonding creates a path with such low impedance (resistance) that *fault currents* flow along it, quickly tripping breakers or fuses and interrupting power. Contrary to popular misconceptions, the human body usually has a relatively high impedance (compared with copper wire); if electricity is offered a path with very low resistance—the equipment grounding conductor—it will take the low resistance path back to the panel, trip the breaker, and cut off the power.

Ground wires (equipment grounding conductors) connect to every part of the electrical system that could possibly become energized—metal boxes, receptacles, switches, fixtures—and, through three-pronged plugs, the metallic covers and frames of tools and appliances. The conductors, usually bare copper or green insulated wire, create an effective path back to the main service panel in case the equipment becomes energized. That allows fault current to flow, tripping the breaker.

The neutral/ground bus. In the service panel, the ground wires attach to a neutral/ground bus bar, which is bonded to the metal panel housing via a *main bonding jumper*. If there's a ground fault in the house, the main bonding jumper ensures the current can be safely directed to the ground—away from the house and the people inside. It is probably the single most important connection in the entire electrical system.

Making Sense of Grounding

Grounding confuses a lot of people, including some electricians. Part of the problem is that the word *ground* has been used imprecisely for more than a century to describe electrical activity or components. The term *ground wire*, for example, may refer to one of three different things:

▶ The large, usually bare-copper wire clamped to a ground rod driven into the earth, or to rebar in a concrete footing. Because the rod or the rebar is technically a grounding electrode, this "ground wire" is correctly called a *grounding electrode conductor*.

▶ Short conductors that connect one piece of electrical equipment to another, to eliminate the possibility of a difference in "potential" between the two, should be called *bonding conductors*.

▶ The bare copper or green insulated wires in all modern circuits, which ultimately connect electrical equipment, such as receptacles and fixtures, to the service panel neutral/ground bar. These connections create a low-impedance fault path back to the service panel. Why? In case the equipment becomes energized by a hot wire touching a metal cover or other part, this conductor allows high current to flow safely and trip the breaker. In this book, the term *ground wire* or *grounding conductor* refers to this conductor. These wires are properly called *equipment-grounding conductors*.

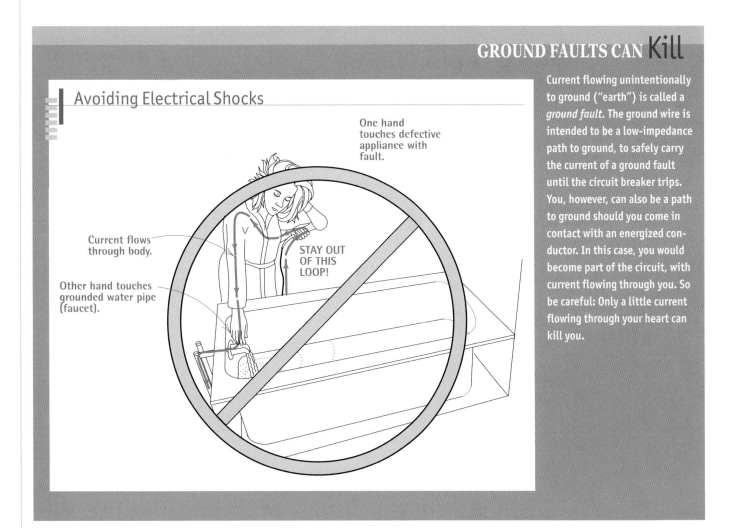

GROUND FAULTS CAN Kill

Avoiding Electrical Shocks

One hand touches defective appliance with fault.

Current flows through body.

STAY OUT OF THIS LOOP!

Other hand touches grounded water pipe (faucet).

Current flowing unintentionally to ground ("earth") is called a *ground fault*. The ground wire is intended to be a low-impedance path to ground, to safely carry the current of a ground fault until the circuit breaker trips. You, however, can also be a path to ground should you come in contact with an energized conductor. In this case, you would become part of the circuit, with current flowing through you. So be careful: Only a little current flowing through your heart can kill you.

Major Grounding Elements

The equipment-grounding system acts as an expressway for stray current. By bonding conductors or potential conductors, the system provides a low-impedance path for fault currents. In a ground fault, the abnormally high amperage (current flow) that results trips a breaker or blows a fuse, disconnecting power to the circuit.

It's required by the NEC to ground metal water piping and metal gas piping in case it becomes energized. An underground metal water pipe can't serve as the only grounding electrode. Otherwise, someone could disconnect the pipe or install a section of nonconductive pipe such as PVC, thus interrupting the grounding continuity and jeopardizing your safety. In new installations, code requires that underground metal water piping be connected to the electrode system and supplemented with another electrode.

Inside this service panel, the smaller copper wire (top) and the insulated white wire feed a subpanel in the house; the thicker copper wire runs to a ground rod.

The main ground wire from the service panel clamps to an 8-ft. grounding electrode (or ground rod) driven into the earth. It diverts outside voltage, such as lightning strikes.

Connections to cold-water pipes and gas pipes prevent shocks should the pipes become inadvertently energized.

Major Grounding Elements

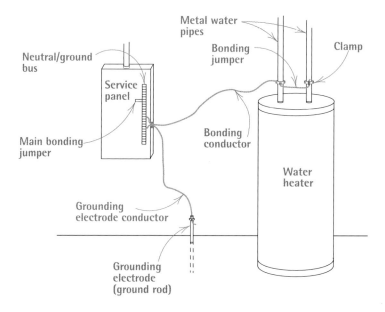

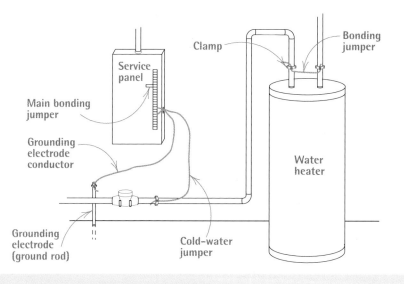

Also attached to the neutral/ground bus in the service panel is a large, usually bare, copper ground wire—the grounding electrode conductor (GEC)—that clamps to a grounding electrode usually either a *ground rod* driven into the earth, or a "Ufer grounding electrode," a 20-ft. length of steel rebar or heavy copper wire poured into the building foundation. The grounding electrode's primary function is to divert lightning and other outside high voltages to the earth before they can damage the building's electrical system. Although the grounding electrode system (GES) is connected to the equipment-grounding system at the service panel, the GES has virtually nothing to do with reducing hazards from the problems in the wiring inside the house. That's the role of the equipment-grounding conductors.

The National Electrical Code (NEC) requires grounding electrode conductor size to be based on the sizes and types of conductors in the service. Typically, residential GECs are size 6 American wire gauge (6AWG) copper. Ground rods are typically ½-in. or ⅝-in. copper-clad steel rods at least 8 ft. long. The Ufer or concrete-encased electrode is the preferred grounding electrode and must be used if new concrete footings are poured.

GFCIS

Ground-fault circuit interrupters (GFCIs) are sensitive devices that can detect very small current leaks and shut off power almost instantaneously. The NEC requires GFCI protection on all bathroom receptacles; all receptacles serving kitchen counters; dishwasher outlets; receptacles within 6 ft. of any sink, tub, or shower stall; laundry-area receptacles; all outdoor receptacles; all unfinished basement receptacles; receptacles in garages and accessory buildings; receptacles in crawlspaces or below grade level; and all receptacles near pools, hot tubs, whirlpools, and the like.

Further, recent code updates require that all new GFCIs have three important features. First, they must be self-diagnostic, with an LED indicator light that will flash or glow red to indicate device failure. Second, should the GFCI fail, it must be self-locking: Current will cease to flow, so that the device must be replaced. Third, GFCIs installed less than 5½ ft. above floor level must be tamper-proof, with plastic shutters that slide across terminal slots to prevent, say, a child from inserting something into a slot.

Wiring a GFCI receptacle is covered on pp. 328–330.

CUTTING POWER AT THE PANEL

Always shut off the power to an outlet before working on it—and then test with an electrical tester to be sure there's no voltage present. In rare instances, a circuit may be mistakenly fed by more than one breaker! Because individual devices such as receptacles, switches, and fixtures can give false readings if they are defective or incorrectly wired, the only safe way to shut off the electricity is by flipping a breaker in the service panel or subpanel.

Turning off the power at a breaker panel is usually straightforward. After identifying the breaker controlling the circuit, push the breaker's handle to the off position. The breaker handle should click into position; if it doesn't, flip it again until you hear a click. (A breaker that won't snap into position may be worn out or defective and should be replaced by an electrician.)

If your home has a fuse panel instead, remove the fuse that controls the circuit, and tape a sign to the panel cover warning that the fuse should not be reinstalled until work has been completed.

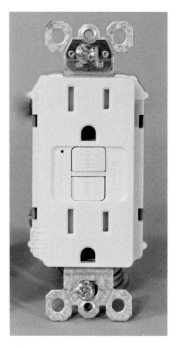

Ground–fault circuit interrupter (GFCI) receptacles detect minuscule current leaks and shut off power almost instantaneously. All bathrooms and kitchens must be GFCI protected, as well as outdoor outlets and garage outlets. Local codes may require additional GFCI protection.

PANEL SAFETY

In a breaker box, simply flip the relevant breaker to the off position.

If you're working on a fuse panel, unscrew and remove the fuse that protects the circuit you're working on.

Once the power is off, lock the panel cover to prevent others from turning the power back on while you're still working.

Always test the tester first on an outlet that you know to be hot, to be sure it is operating properly. Here, the glowing point of a non-contact tester indicates voltage is present.

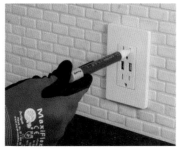

Turn the power off to the circuit. To make sure the power is off, test the narrow (hot) slot first.

Follow up by testing with a probe tester. Insert the tester tips into the receptacle's neutral and hot slots, then into a grounding slot and a hot slot.

Remove the cover and mounting screws. Without touching the sides of the receptacle, test both hot and neutral screw terminals.

(You can also lock the panel cover.) Partially unscrewing a fuse is not a solution because the fuse body is still in contact with the socket and possibly could be jiggled or screwed in enough to reenergize the circuit. Remove the fuse. Likewise, if circuits are controlled by cartridge fuses, pull the cartridge or cartridge block out of the panel.

In any case, once you've cut off the power, shut the panel cover and tape a sign to it, telling others to stay out. Better yet, lock them out. Otherwise, someone not aware of the situation could flip the switch on, energizing the circuit you're working on.

TESTING FOR VOLTAGE

Before doing electrical work, always turn the power off to the circuit. There are several voltage testers you can use to verify that the power is off, but the most common are noncontact testers and probe testers. We recommend using both types.

A noncontact tester is an inexpensive battery-operated voltage tester that is generally reliable and small enough to fit in a shirt pocket. Its plastic tip glows (and it may beep, depending on the model) when the tip is brought close to or touched to a hot (energized) conductor. That is, it can "read" current through a wire's insulation or through a cover plate. Thus you can often detect electrical current at a switch, receptacle, or fixture without removing the outlet cover.

Each time you use a noncontact tester—or any voltage tester—check that it's functioning properly first on a receptacle that you know is hot. After shutting off power at the panel, insert the tester tip into the narrow (hot) slot of the receptacle you intend to work on. If the tester tip does not glow, there is probably no voltage present.

Follow up with a probe tester. Although a noncontact tester is good for a quick initial reading, you should follow up with a probe tester, whose probes make direct contact with conduc-tors. To test for voltage, touch probes to receptacle slots, screw terminals, or wire ends. Because probe testers do not rely on batteries, they are more accurate than noncontact testers.

The probe tester shown above, like many multimeters (p. 304), allows you to choose which electrical function to test. To see if power is present, turn the dial to V (voltage).

Insert the probe tester tips into the receptacle's neutral and hot slots. Next insert the tips into a grounding slot and a hot slot. This test should protect you in case the receptacle was incorrectly wired. The tester screen should indicate no voltage present.

If you need to remove the receptacle—say, to replace it—remove the cover plate *and test one more time.* Being careful not to touch the sides of the receptacle, unscrew the two mounting screws holding the receptacle to the outlet box. (If the box is metal, avoid touching it, too.) Grasp the mounting straps and gently pull the receptacle out of the box. Touch one tip of the tester to the brass screw terminal (hot), while touching the other tip to the silver screw terminal (neutral). If the tester screen does not indicate voltage, it's safe to handle the receptacle and the wires feeding it.

Planning an Electrical Remodel

Being detail oriented is as important to planning as it is to installation. When you plan a wiring project, be methodical: Assess the existing system, calculate electrical loads, check local codes, and draw a wiring floor plan.

If you are only replacing existing devices—changing a light fixture, replacing a faulty switch, or upgrading a receptacle, for example—you seldom need a permit from the local building department. However, if you extend or add any circuit, you must pull (or file) a permit.

Most local electrical codes are based on the NEC. When it's necessary to pull a permit, local code authorities will want to approve your plans and later inspect the wiring to be sure it's correct. Don't short-cut this process: Codes and inspections protect you and your home.

Whatever the scope of your project, if you work on existing circuits, first turn off the power, test to be sure it's off, and tag or lock the panel.

INSPECTING THE FUSE BOX OR BREAKER PANEL

By looking at the outside of the service panel and wiring that's exposed in the basement and attic, you can get a basic overview of the system's condition. If the panel has unused breaker spaces and the wiring insulation is in decent shape, you can probably continue using it and safely add an outlet or two if there is sufficient capacity. However, if the system seems unsafe or inadequate, hire a licensed electrician to open the panel and do a more thorough examination.

Here's what to look for:

Start your investigation at the fuse box or breaker panel. You can learn a lot about the condition of the system by examining the outside of the service box. Examining the inside of a panel or fuse box is best left to a licensed electrician, however.

Rust and corrosion on the outside of a service box or on the armored cable or conduit feeding it can indicate corroded connections inside. Faulty connections can lead to *arcing* (sparks leaping gaps between wires) and house fires, so have a

licensed electrician replace the fuse box or panel. Likewise, if you see scorch marks on breakers or a panel, have a pro examine it.

Melted wire insulation is a sign either of an overheated circuit—usually caused by too many appliances in use at the same time—or of a poor wire connection in which arcing has occurred. In the first case, a homeowner typically installs an oversize fuse or breaker to keep an overloaded circuit from blowing so often, but this "remedy" exceeds the current-carrying capacity of the wire. The wire overheats and melts the insulation, which can lead to arcing, house fires, or—if someone touches that bare copper wire—electrocution.

An oversize fuse may not melt wires where you can see them, but it may have damaged wire insulation in a place you can't see. Have an electrician inspect the electrical system. Installing type-S fuse socket inserts can prevent overfusing.

"Pennying" a fuse is another unsafe way to deal with an overloaded circuit that keeps blowing fuses. In this case, someone unscrews a fuse, inserts a penny or a blank metal slug into the bottom of the socket—a dangerous act in itself—and then reinstalls the fuse. The penny allows current to bypass the fuse and the protection it offers. Here, again, have an electrician examine the circuits for damage to the wire insulation.

Panel covers that don't fit, have gaps, or are missing are unsafe. So, if you see covers that have been cut to fit a breaker, housing knockouts that are missing, bus bars that are visible when the panel cover is on, or mismatched components, hire a licensed electrician to assess and correct those problems. Some older brands of

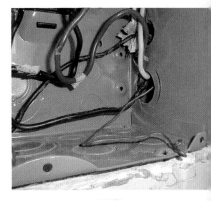

Unsafe! All cables entering a panel must be protected by a cable clamp inserted into a panel knockout. Sharp panel edges could slice the insulation of this unprotected cable. Also, for the cables, the sheathing must extend at least ¼ in. into the box.

This 30-amp main switch and fuses made up the panel for the whole house. It has seen better days. Even if it were safe, it would be dramatically undersized for today's electrical needs.

A 30-Year Pro's Take on Rewiring

If the wiring in an older home appears to be sound and in good repair (see the warning signs below) it's probably OK to continue using it, even though it may not meet code requirements for a new installation. If you are planning to gut the house completely, it makes sense to rip out all the old wiring and completely rewire the house. If you're remodeling only part of the house, leave most of the old wiring in place if it is sound and spend your money rewiring the kitchen, baths, and laundry circuits. That will give you more bang for your buck.

However, you should replace old wiring that's unsafe. If you observe any of these conditions, the wiring should be replaced:

► Circuits that have been extended improperly, as evidenced by loose connections, unprotected splices, or arcing.

► Knob-and-tube wiring whose insulation has deteriorated or that has been damaged. Also, if knob-and-tube wiring in the attic has been covered with loose-fill insulation or insulation batts, that is a serious code violation that could lead to overheating and fire danger.

► Circuits wired with unsheathed wires (other than properly done knob-and-tube) rather than with sheathed cable or conduit.

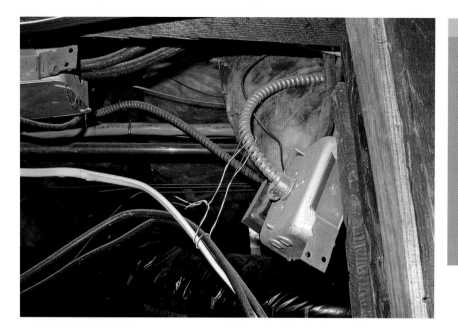

Unstapled cable and unsecured boxes can be inadvertently yanked and stress electrical connections. If you see work this sloppy, suspect substandard wiring throughout the house.

panels and breakers, such as Federal Pacific and Zinsco, have a host of well-documented failures and so should be carefully checked and, where necessary, replaced.

Missing cable connectors or unfilled knockouts enable mice and vermin to enter the panel and nest in it, which can be a fire hazard. Missing connectors also allow cables to be yanked, stressing electrical connections inside the panel. A missing cable clamp may also allow the sharp edge of the panel to slice through thermoplastic cable sheathing, which could energize the panel (if the grounding of the system is not correct) and electrocute anyone who touches it.

A properly grounded panel will have a large grounding wire running from the panel to a grounding electrode (which could be a metal underground water pipe, a ground rod driven into the earth, or a Ufer electrode). For the entire electrical system to be grounded, there must be a continuous ground wire or other effective grounding path running from each device or fixture to the service panel and, by extension, to the grounding electrode. Cold-water and gas pipes must also be connected (bonded) to the grounding bus in the panel.

ASSESSING THE CONDITION OF WIRING

Cables may be visible as they near the service panel and as they run through attics and basements. If there are covered junction boxes, carefully remove the covers and examine the wire splices inside—without touching them. You can also turn off power and pull a few receptacles out to better examine the wires.

Some old houses still have knob-and-tube wiring, which doesn't necessarily need to be replaced. Have it tested to make sure it's still in good shape.

Deteriorated sheathing is a potential shock hazard, so note brittle fiber insulation and bare wire, but avoid touching it. If cable sheathing has been chewed on by mice, rats, or squirrels the cable should be replaced.

NM cable (cable protected by nonmetallic sheathing) must be stapled within 8 in. of single-gang boxes that don't have cable clamps, and within 12 in. of other boxes, and supported by stapling or another method at least every 4½ ft. (54 in.). A cable running through a drilled hole is considered supported. Sagging wire is hazardous because it can get inadvertently strained, jeopardizing electrical connections. Likewise, all boxes must be securely mounted. All NM cable entering metal boxes must be gripped by cable clamps. Single-gang plastic boxes do not require the strain relief of cable clamps, but double-gang (and larger) plastic boxes have integral plastic tension clips that afford some strain relief on cable.

Small-gauge aluminum wiring (10 or 12 gauge) is a fire hazard unless it is correctly spliced to a copper wire with a COPALUM connector or terminated to CO/ALR-rated outlets and switches. If it is incorrectly terminated in a copper-rated only device, the two metals will expand and contract at different rates each time the circuit is under load. This can lead to loose connections, arcing, overheating, and house fires. Aluminum-to-aluminum splices require special splicing techniques, either COPALUM connectors or another listed and approved method.

Wire splices (whether copper to copper, copper to aluminum, or aluminum to aluminum) must be housed within a covered junction box or outlet box. Wires that are spliced outside a box or inside an uncovered box can be a fire hazard because of the dangers of arcing. Loose connections not contained in a cover box can easily ignite combustibles nearby because arcs can reach several thousand degrees centigrade.

Knob-and-tube wiring, although outdated, is usually safe unless individual wire insulation is deteriorated, splices are incorrectly made, or the wiring is overloaded or buried in thermal insulation. Typically, splices that were part of the original installation will not be in a junction box but must be wrapped with electrical (friction) tape and supported by porcelain knobs on both sides of each splice. Nonoriginal splices must be housed in covered boxes. Have knob-and-tube wiring assessed or modified by an electrician familiar with it; it's quirky stuff. The NEC does not allow knob-and-tube wiring to be buried in insulation.

Knob-and-tube wiring lacks an equipment ground (a separate grounding wire), so it offers

Assessing a Circuit's Capacity

To recap briefly, electricity, impelled by *voltage*, flows from the power source. (*Amperes* are the rate of electron flow.) Along the way (at outlets), it encounters resistance and does work. (*Watts* are a measure of power consumed.) It then returns to the power source, its voltage reduced or spent.

Or, expressed as mathematical formulas:

watts = voltage x amperes

amperes = watts ÷ voltage

To determine the capacity of a circuit you want to extend, identify the circuit breaker controlling the circuit and note the rating of the breaker. If it's a general-purpose circuit, the breaker will probably be 15 amp or 20 amp. A circuit controlled by a 15-amp breaker has a capacity of 1,800w (15 amp x 120v); one controlled by a 20-amp breaker has 2,400w.

The total wattage of all loads on the circuit (including the extension) must not exceed these capacities; otherwise, you risk overheating wires. To avoid overloading, it's a good idea to reduce the capacity by 20%. For example, 80% of 1,800w is 1,440w for a 15-amp circuit; 80% of 2,400w equals 1,920w for a 20-amp circuit.

Circuit Capacities*

AMPERES × VOLTS†	TOTAL CAPACITY (watts)	SAFE CAPACITY† (watts)
15 × 120	1,800	1,440
20 × 120	2,400	1,920
25 × 120	3,000	2,400
30 × 120	3,600	2,880

* Safe capacity = 80% of total capacity.
† Amperes multiplied by volts equals watts.

no protection should a faulty appliance get plugged into a receptacle. On the other hand, a knob-and-tube system is run completely on insulators, a plus. The conductors of knob-and-tube wiring were copper, coated with a thin layer of tin (to protect the copper from sulfur in the rubber wire insulation). Uninformed inspectors often mistake the tinned copper wire for aluminum wire.

IS THE SYSTEM ADEQUATELY SIZED?

If receptacles in your house teem with multiplugs and extension cords, you probably need to add outlets. But there are more subtle clues: If you blow fuses or trip breakers regularly, or if the lights brown out when you plug in a toaster or an electric hair dryer, you may need to add new circuits to relieve the overload on existing circuits. This section will help you figure out whether your system has the capacity to do so.

Example of Load Calculation for Single Family Dwelling

CALCULATING GENERAL LIGHTING LOAD			
Type of Load	**NEC Reference**	**Calculation**	**Total VA**
Lighting Load	Table 220.12	2,000 sq. ft. × 3 VA	6,000 VA
Small Appliance Load	Section 220.52	2 circuits × 1500 VA	3,000 VA
Laundry Load	Section 220.52	1 circuit × 1500 VA	1,500 VA
Total General Lighting ▷			**10,500 VA***

Ⓐ CALCULATING DEMAND FOR GENERAL LIGHTING LOAD			
Type of Load	**Calculation**	**Demand Factor (DF)**	**Total VA**
General Lighting	First 3,000 VA × DF	100%	3,000 VA
General Lighting	7,500 ‡ × DF	35%	2,625 VA
Total Lighting, Small Appliances & Laundry ▷			**Ⓐ 5,625 VA**

Ⓑ CALCULATING DEMAND FOR LARGE LOAD APPLIANCES			
Type of Load	**Nameplate Rating**	**Demand Factor (DF)**	**Total VA**
Electric Range	Not Over 12KVA	Use 8KVA	8,000 VA
Clothes Dryer	6,600 VA × DF	100%	6,600 VA
Water Heater	6,600 VA × DF	100%	6,600 VA
Other Fixed Appliances	0 VA × DF	100%	0 VA
Total Load for Large Appliances ▷			**Ⓑ 21,200 VA**

	Total VA (Ⓐ + Ⓑ)		**26,825 VA**
Minimum Service Size	**Total VA / 240V**		**111.77 VA**

The minimum service size is the next standard size above the total VA calculated. Based upon these calculations, the minimum service size is 125 amps.

*Use this to calculate Ⓐ
‡ Total General Lighting Load 10,500 VA – First 3,000 VA = 7,500 VA

USING THIS TABLE

1. Square ft. for general lighting load is for the entire dwelling including habitable basements and attics.
2. NEC requires a minimum of 2 small appliance loads, but it is important to add small fixed kitchen appliances such as refrigerator, microwaves, disposals, dishwashers, large range hood, computers, etc., when calculating this category
3. Minimum of 1 laundry load is required for a single family dwelling.
4. The demand factor calculation is designed to take actual use into account (e.g., it is unlikely that all lights and small appliances will be running at one time).
5. All large load appliances (high wattage) are added at 100%.
6. The final load calculation is the minimum. Often, increasing capacity has little cost impact and is a good practice.

Some appliances require dedicated service because they are heavy energy users. This is a 30–amp, 125/250v dryer receptacle. The breaker for this circuit must also be rated for 30 amps.

The grounding conductor of a cable is connected to a steel box with a special grounding screw, which must thread into a tapped hole.

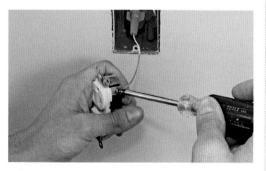

All devices, including receptacles, switches, and lighting fixtures, also must be bonded. Here, a bare ground lead from spliced ground wires in the box is attached to the green grounding screw of a duplex receptacle.

These days, three-wire service feeding a 100-amp service panel is the minimum required by the NEC, and many electricians install 150-amp or 200-amp panels if the homeowners plan to enlarge the house at some point or acquire a lot of heavy energy-using appliances, such as electric ranges and hot tubs. If there are only two large wires running from the utility pole to the house, they deliver only 120v service. A house with two-wire service probably has a 30-amp or 60-amp main fuse or breaker, which is inadequate for modern usage.

Rules of thumb. The only sure way to know if you've got enough capacity to add an outlet or a circuit is to calculate electrical loads, as shown on the facing page. But for the benefit of those who wish that an electrician would just offer an offhand opinion of what works most of the time, here are a few rules of thumb.

▶ **Fuse box service.** If you've got a fuse box with a 30-amp or 60-amp main fuse, the best advice we can give is to upgrade your service. Don't add outlets or circuits until you replace the fuse box with a breaker panel. Fused mains are often abused by people trying to bypass its protection, so insurance companies often charge higher premiums on houses with fuse boxes.

▶ **Adding outlets.** If you have a breaker panel with space to add an additional breaker, you can most likely add a circuit to feed a new outlet or two or more lights. If, for example, you have three-wire service and a 100-amp main, you may have excess capacity.

▶ **Adding a circuit for general use.** If there is an unused space in the panel, have an electrician determine whether the panel can handle another circuit.

▶ **Adding a kitchen or bath circuit.** If you want to add a bath fan or some new light fixtures, and there's space in the panel, have an electrician see if you can add a circuit. Adding a 20-amp, small-appliance circuit to reduce the load on an existing circuit is smart.

▶ **Remodeling a kitchen.** Kitchens are complicated and often full of big energy users. Use the chart on the facing page to help you add up the loads. If there aren't many open spaces for breakers, you may need to upgrade to a larger panel.

▶ **Adding dedicated circuits.** If you need to add dedicated circuits for heavy-use items such as an electric range (50 amps) or a hot tub (60 amps), get out the calculator and do the math.

GENERAL WIRING GUIDELINES

Electricians follow the NEC, as adopted (or amended) by local jurisdictions, which was com-

piled to promote safe practices and prevent house fires. Consider these requirements before you start drawing plans, but be sure to consult local electrical code—it is the final authority in your area.

The guidelines given here apply to all circuits in the house, whether general lighting or heavy-use appliance circuits. Local codes rarely require you to change existing circuits—as long as they are safe—but new electrical work should reflect current electrical code.

Circuit wiring. Wire gauge must be large enough to carry the circuit load and be protected by a comparably sized breaker or fuse at the panel. General-use and lighting circuits are typically 14AWG wire, protected by 15-amp breakers; kitchen, bath, and workshop circuits usually have 12AWG wire, protected by 20-amp breakers.

Acceptable cable. Most circuits are wired with NM cable because the cable is protected behind a finished surface such as drywall or plaster. When circuit wiring is to be left unprotected and exposed, it must be armored cable or in conduit.

Grounding. All receptacles, appliances, and electrical equipment must be connected (bonded) to the service panel via an equipment grounding conductor (ground wire). NM cable contains a separate ground wire, whereas armored cable sheathing and metal conduit provide the path to ground if properly installed.

Boxes. All electrical connections must take place in covered boxes. Boxes where connections are made must be accessible, that is, not buried in a wall or ceiling. Based on local code requirements, boxes may be plastic or metal. If metal, the box must also be connected to the equipment grounding conductor (bonded). If NM cable is used, the ground wire must be connected to a metal box with either a ground screw or a ground clip. If AC cable or metal conduit is used, it must be properly attached to the box to ensure effective bonding. If the box is plastic, it does not need to be (and cannot be) grounded; run a ground wire to the device or fixture only.

GENERAL-USE CIRCUIT REQUIREMENTS

General-use circuits are intended primarily for lighting, but small loads that are connected via a cord and plug, such as televisions, fans, and vacuums, are allowed—as long as the power they draw doesn't exceed the capacity of the circuit.

Lighting and small loads. Although 14AWG wire is sufficient for lighting, electricians often run 12AWG wire on general-use circuits to accommodate additional capacity. Calculate lighting loads at 3w per square foot, or roughly

Arc-fault breakers are designed to detect arcing patterns of current for very short time intervals.

Kitchen LIGHTING BASICS

Kitchen lighting should be designed to utilize natural light during the day and achieve a balance of general light and task lighting at night. Do not be afraid of energy-efficient lighting such as fluorescent or LED. Today's energy-efficient lighting is instant, dimmable, and available in colors that match incandescent light. Kitchen lighting is often highly regulated for energy efficiency. Check with local building officials before you begin your design.

General Lighting

General lighting is meant to illuminate the space generally and can come from recessed cans, surface-mounted fixtures, track lighting, or cove *uplighting*. Consider cabinetry and appliances when laying out new light fixtures. A general rule is 2w incandescent or 1w fluorescent per sq. ft. of kitchen area, but even illumination is the goal.

Task Lighting

Task lighting is meant to provide a higher level of illumination at work areas (sinks, countertops, and islands) and can be achieved with recessed cans, pendants, or *undercabinet* fixtures. If cabinets are over countertops, undercabinet fixtures (T5 fluorescent or LED strips) are by far the best choice and should be spaced for even illumination of the counter surface. For islands and sinks, choose a recessed can with a slightly higher wattage and narrower lamp beam spread, or install pendants with similar attributes.

"lighting") in habitable rooms except in kitchens and bathrooms.

GFCI protection. The NEC requires GFCI protection on all bathroom receptacles; all receptacles serving kitchen counters; dishwasher outlets; receptacles within 6 ft. of any sink, tub, or shower stall; laundry-area receptacles; all outdoor receptacles; all unfinished basement receptacles; receptacles in garages and accessory buildings; receptacles in crawlspaces or below grade level; and all receptacles near pools, hot tubs, whirlpools, and the like.

AFCI protection. Since 2014, the NEC requires arc-fault circuit interrupter (AFCI) protection on all 15-amp and 20-amp receptacles in kitchens and laundry rooms, bedrooms, living rooms, rec rooms, parlors, libraries, dens, sunrooms, and hallways, and on switches serving any of those areas. For more information on AFCI protection, see p. 311.

REQUIREMENTS ROOM BY ROOM

Kitchen and bath appliances are heavy power users, so their circuits must be sized accordingly.

Bathroom circuits. Bathroom receptacles must be supplied by a 20-amp circuit. The NEC allows the 20-amp circuit to supply the receptacles of more than one bathroom or to supply the receptacles, lights, and fans (excluding heating fans) in one bathroom. Receptacles in bathrooms must be GFCI-protected, either by a GFCI receptacle or a GFCI breaker. New or remodeled bathrooms must have a vent fan.

Small-appliance circuits. There must be at least two 20-amp small-appliance circuits in the kitchen serving the kitchen countertops. No point along a kitchen countertop should be more than 2 ft. from an outlet—in other words, space countertop receptacles at least every 4 ft. Every counter at least 12 in. wide must have a receptacle.

Kitchen lighting. Adequate lighting is particularly important in kitchens so people can work safely and efficiently. Lay out a good balance of general and task lighting. Be aware that many jurisdictions have energy-efficiency requirements for lighting in kitchens, so check with your local building authority first.

Bathroom lighting. It is important to illuminate the face evenly in mirrors. Common practice is to place good-quality light sources either above the vanity mirror or on either side of it. Be careful when using recessed cans over the vanity because they can leave shadows across the face. Many jurisdictions also have energy-efficiency requirements for lighting in bathrooms, including the use of high-efficacy lighting and vacancy sensors.

SAFETY ALERT

All bathrooms and kitchens must be GFCI protected. GFCI protection is required for other receptacles, too. Your local building code may have additional requirements for GFCIs.

one 15-amp circuit for every 500 sq. ft. of floor space. When laying out the lighting circuits, do not put all the lights on a floor on one circuit. Otherwise, should a breaker trip, the entire floor would be without lights.

Receptacles. There must be a receptacle within 6 ft. of each doorway, and receptacles should be spaced at least every 12 ft. along a wall. (This is also stated as, "No space on a wall should be more than 6 ft. from a receptacle.") Any wall at least 2 ft. wide must have a receptacle, and a receptacle is required in hallway walls 10 ft. or longer. Finally, any foyer of more than 60 sq. ft. must have a receptacle on any wall 3 ft. or longer.

Outlets. The NEC does not specify a maximum number of lighting and receptacle outlets on a residential lighting or appliance circuit, although local jurisdictions may. Figure roughly nine outlets per 15-amp circuit and 10 outlets per 20-amp circuit.

Light switches. There must be at least one wall switch that controls lighting in each habitable room, in the garage (if wired for electricity), and in storage areas (including attics and basements). There must be a switch controlling an outside light near each outdoor entrance. Three-way switches are required at each end of corridors and at the top and bottom of stairs with six steps or more. When possible, put switches near the lights they control. It should be noted that the light switch may control a receptacle (considered

Dedicated circuits. All critical-use and fixed appliances should, and in most cases must, have their own dedicated (separate) circuits. These fixed appliances include the water pump, freezer, refrigerator, oven, cooktop, microwave, furnace and/or whole-house air-conditioning unit, window air conditioners, and electric water heater. A bathroom heater requires a dedicated circuit, whether it is a separate unit or part of a light/fan. Laundry room receptacles must be on a dedicated circuit; so must an electric clothes dryer.

DEVELOPING A FLOOR PLAN

Drawing a set of project plans can help you anticipate problems, find optimal routes for running cable, minimize mess and disruption, and make the most of your time and money. A carefully drawn set of plans is also an important part of the code compliance and inspection process.

If you're replacing only a receptacle, switch, or light fixture, you usually don't need to involve the local building department. But if you run cable to extend a circuit, add a new circuit, or plan extensive upgrades, visit the building department to learn local code requirements and take out a permit. Your wiring plans should be approved by a local building inspector before you start the project.

Phone first. Call the building department and ask if local codes allow homeowners to do electrical work or if it must be done by a licensed electrician. You may be required to take a test to prove basic competency. This also is a good time to ask whether the municipality has pamphlets giving an overview of local electrical code requirements.

Read up. Make a rough sketch of the work you propose, develop a rudimentary materials list, and then apply for a permit. At the time you apply, the building department clerk may be able to answer questions generated by the legwork you've done thus far. This feedback often proves invaluable.

Inspectors inspect. Inspectors are not on staff to tell you how to plan or execute a job, so make your questions as specific as possible. Present your rough sketch, discuss the materials you intend to use, and ask if there are specific requirements for the room(s) you'll be rewiring. For example, must bedroom receptacles have AFCI protection? Must kitchen wall receptacles be GFCIs if they are not over a counter? Be specific.

Draw up plans. Based on the feedback you've gotten, draw detailed plans. (Pages 302–303 show a set of detailed electrical plans.) They should include each switch, receptacle, and fixture as well as the paths between switches and the device(s) they control. From this drawing, you can develop your materials list. Number each circuit or, better yet, assign a different color to each circuit. When you feel the plans are complete, schedule an appointment with an inspector to review them.

Listen well and take notes. Be low-key and respectful when you meet with the inspector to review your plans. First, you're more likely to get your questions answered. Second, you'll begin to develop a personal rapport. Because one inspector will often track a project from start to finish, this is a person who can ease your way or make it much more difficult. So play it straight, ask questions, listen well, take notes, and—above all—don't argue or come in with an attitude.

On-site inspections. Once the building department approves your plans, you can start working. In most cases, the inspector will visit your site when the wiring is roughed in and again when the wiring is finished. Don't call for an inspection until each stage is complete.

ELECTRICAL NOTATION

Start by making an accurate floor plan of the room or rooms to be rewired on graph paper using a scale of ¼ in. = 1 ft. Indicate walls and permanent fixtures such as countertops, kitchen islands, cabinets, and any large appliances. By photocopying this floor plan, you can quickly generate to-scale sketches of various wiring schemes.

Use the appropriate electrical symbols (p. 302) to indicate the locations of receptacles, switches, light fixtures, and appliances.

Especially when drawing kitchens, which can be incredibly complex, use colored pencils to indicate different circuits. You can also number circuits, but colored circuits are distinguishable at a glance. Use solid lines to indicate cable runs between receptacles and switches and dotted lines to indicate the cables that run between switches and the light fixtures or receptacles they control.

The beauty of photocopies is that you can experiment with different options quickly. As you refine the drawings, refer back to the list of requirements given earlier to be sure that you have an adequate number of receptacles, that you have GFCI receptacles over kitchen counters, that there are switches near doorways, and so on. Ultimately, you'll need a final master drawing with everything on it. But you may also find it helpful to make individual drawings—say, one for lighting and one for receptacles—if the master drawing gets too busy to read.

(continued on p. 304)

Electrical Wiring **301**

Common Electrical Symbols

Symbol	Name
Duplex receptacle	
GFCI duplex receptacle	GFCI
Fourplex receptacle	
240v receptacle	
Weatherproof duplex receptacle	WP
Duplex receptacle, split-wired	
Single-pole switch	S
Three-way switch	S$_3$
Switch leg	
Home run (to service panel)	
Recessed light fixture	ⓡ
Wall-mounted fixture	⊶
Ceiling outlet	○
Ceiling pull switch	Ⓢ
Junction box	Ⓙ
Vent fan	VF
Ceiling fan	CF
Telephone outlet	▶
Two-wire cable	
Three-wire cable	

Lighting and switches layer

🅐 Running 12/2 cable will accommodate the dishwasher circuit.

🅑 Use 14/2 cable for all general lighting home runs (cable runs back to the service panel).

🅒 Verify the dimmer load; dimmers must be de-rated when ganged together.

Power layer

🅐 Wire the GFCI receptacle at the beginning of the run so it affords protection to receptacles downstream.

🅑 Dishwasher/disposal circuit. Install under the sink in the cabinet. Cut the hot (brass) tab on the receptacle to split the receptacle for two circuits. Leave the neutral (silver) tab intact. Be sure to install on a two-pole breaker with a handle tie.

🅒 Single-location GFCI protection.

🅓 Stove is gas, so the receptacle is only for the igniter and clock and is OK to run with the hood. Leave NM cable stubbed at the ceiling, and leave 3 ft. to 4 ft. of slack for termination in the hood/trim. (Note: Never run a stove igniter off a GFCI-protected circuit because it will trip the GFCI every time the stove is turned on.)

🅔 Use 12/3 cable for the home run, so a single cable takes care of the dedicated refrigerator circuit and the countertop receptacle (small-appliance) circuit.

🅕 Home run for counter (small-appliance) circuit 2.

🅖 Oven outlet. Refer to unit specifications to verify receptacle or hard-wired connection.

Electrical Plan: Lighting and Switches Layer

A professional's electrical floor plan may be daunting at first, but it will start to make sense as you become familiar with the symbols used. To make the plans easier to read, they have been divided into two layers: (1) lighting and switches and (2) power, which consists of receptacles and dedicated circuits. (There's some overlap.) The circled letters are callouts that indicate areas warranting special attention. The circled numbers correspond to a master list of lighting fixtures that the electrician developed with the architect. Drawing switch legs and circuits in different colors makes a plan much easier to read.

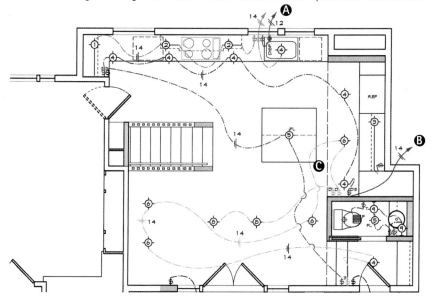

Pendant lights in the dining area are noted by a dotted green switch leg running from the two ceiling boxes to the single-pole switch on the wall.

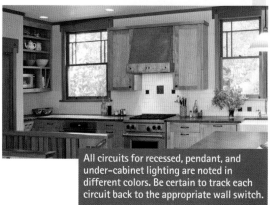

All circuits for recessed, pendant, and under-cabinet lighting are noted in different colors. Be certain to track each circuit back to the appropriate wall switch.

Electrical Plan: The Power Layer

This kitchen remodel is typical in that it has many dedicated circuits (also called designated circuits) and, per code, at least two 20-amp small-appliance circuits wired with #12 cable. Circled letters are callouts that correspond to the lettered notes at left.

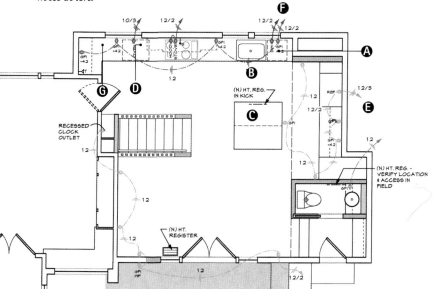

Whether for 120v wall outlets or 240v appliance outlets, each circuit should end with a home run back to the service panel.

Any receptacle that serves the countertop must be GFCI protected. Refrigerators, however, should be run on a non-GFCI receptacle.

Electrical Testers

Testing to see if a circuit or device is energized is crucial to safety and correct wiring. Remember: Always test a tester first to be sure it's working correctly. The first three items discussed below are voltage testers and some perform multiple functions.

Noncontact testers provide a reading without directly touching a conductor. They often allow you to detect voltage without having to remove cover plates and expose receptacles, switches, or wires. Touch the tool's tip to an outlet, a fixture screw, or an electrical cord. If the tip glows red, it means there's voltage present. Noncontact testers rely on battery power and are not fail-safe. They should *not* be used as the final test to determine whether or not a circuit is off and safe to work on. Instead, use a probe tester, as shown on p. 294.

Plug-in circuit analyzers or polarity testers can be used only with three-hole receptacles, but they quickly tell you if a circuit is correctly grounded and, if not, what the problem is. Different light combinations on the tester indicate various wiring problems, such as no ground and hot and neutral reversed. They're quite handy for quick home inspections.

Solenoidal voltage testers (often called *wiggies*) test polarity as well as AC voltage and DC voltage from 100v to 600v. Most models vibrate and light a bulb when current is present. Solenoid testers don't use batteries, so readings can't be compromised by low battery power. However, because of their low impedance, solenoid testers will trip GFCIs. In addition to voltage testers, get a *continuity tester* to test wire runs and connectors for short circuits or other wiring flaws prior to energizing the circuit.

Multimeters Digital multimeters (DMMs) allow you to troubleshoot electrical problems by measuring several aspects of electrical energy, including volts, amps, and ohms. Most DMMs come with probe tips you touch to exposed conductors, though some also have forks or hooks that can measure current running through a cable without exposing the conductors within.

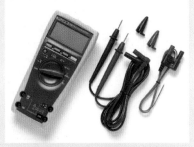

Unusually high voltage readings are a concern because surges can destroy computers, electronic devices, or appliances designed to operate at specific voltages. Conversely, low voltage can also cause damage and may indicate corrosion, loose connections, wiring too small for circuit loads, and so on.

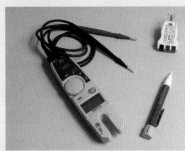

High ohm readings may indicate excessive resistance, which can lead to overheated wires. The cause may be corrosion, faulty connections, cables bundled too closely, or a staple driven so deeply that it has damaged cable sheathing or the wires inside. DMM amp settings measure current flow through wires; amp readings that are too high may indicate a high resistance ground fault, an overloaded circuit, and the potential for overheating and insulation damage.

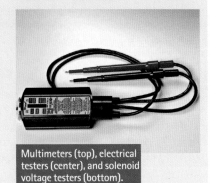

Multimeters (top), electrical testers (center), and solenoid voltage testers (bottom).

If you have questions or want to highlight a fixture type, use callouts on the floor plan. As you decide which fixtures and devices you want to install, develop a separate materials list and use numbered keys to indicate where each piece goes on the master drawing. Finally, develop a list of all materials, so you'll also have enough boxes, cable connectors, wire connectors, staples, and so on. In short, list all you need to do the job.

Tools and Materials

You don't need a lot of expensive tools to wire a house. This section introduces basic tools you'll use most often. Scan chapter 3 for essential safety tools such as safety glasses, work gloves, and respirators, as well as layout and cleanup tools. All materials must be UL- or NRTL-listed, which indicates they meet the safety standards of the electrical industry.

The first test of any tool is that it fit your hand comfortably; the second, that it feels solid and well made. Better tools tend to be a bit heftier and cost more.

HAND TOOLS

All hand tools should have cushioned handles and fit your hand comfortably. Manufacturers now make tools in various sizes, so choose the ones right for you. Don't scrimp on quality.

Pliers and strippers. *Lineman's pliers* are the workhorse of an electrician's toolbox. They can cut wire, hold wires fast as you splice them, and twist out box knockouts. *Needle-nose* (long-nose) pliers can grasp and pull wire in tight spaces. These pliers can loop wire to fit around receptacle and switch screws. A large pair can also loosen and remove knockouts in metal outlet boxes. *Diagonal-cutting* and *end-cutting* pliers can cut wires close in tight spaces; end cutters (sometimes called nippers) also pull out staples easily.

An 18-in. auger bit—also called a ship's auger—bores easily through several studs or wall plates.

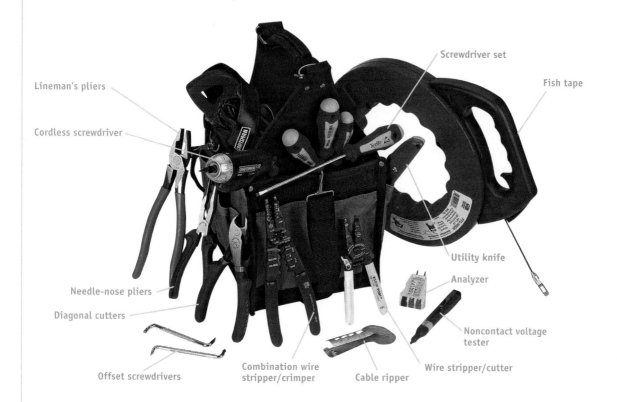

Lineman's pliers

Cordless screwdriver

Needle-nose pliers

Diagonal cutters

Offset screwdrivers

Combination wire stripper/crimper

Cable ripper

Wire stripper/cutter

Noncontact voltage tester

Analyzer

Utility knife

Fish tape

Screwdriver set

A multipurpose or *combination stripping tool* is used to strip individual wires of insulation, cut wire, crimp connections, and quickly loop wire around screw terminals.

A *cable ripper* (see p. 325) strips the plastic sheathing from Romex cable without harming the insulation on the individual wires inside. Many pros use a utility knife to strip sheathing, but that takes practice and a light touch to avoid nicking the insulation of individual wires.

For remodel work, you may need a plaster chisel, flat bar, and a drywall saw.

POWER TOOLS

Buy power tools that are appropriate to your strength and to the task at hand. More powerful tools tend to be heavy and hard to manage, and for wiring, they're often overkill. If possible,

test-drive a friend's power tool before buying your own.

A right-angle drill allows you to fit the drill head between studs or joists and drill perpendicular to the face of the lumber. The pros use drills with ½-in. chucks such as the Milwaukee Hole-Hawg, the DeWalt stud and joist drill, and so on. They're very versatile tools. It's not necessary to get a drill with a clutch; such tools tend to be very heavy and expensive—overkill even for wiring a whole house. A right-angle, D-handle drill and a sharp bit are more than adequate.

Spade bits cut quickly but tend to snap in harder wood. For this reason, most pros prefer auger bits. Self-feeding chipper bits drill doggedly through hard, old wood but won't last long if they hit nails. A ⅞-in. Greenlee Nail Eater bit is a wise buy if your old lumber is nail infested; many companies offer similar nail-eater bits.

Reciprocating saw. A reciprocating saw with a demolition sawblade (see p. 56) is indispensable for most remodeling jobs because it can handle the occasional nail without destroying the blade. You can use a recip saw to cut openings in plaster, but an oscillating tool with a Universal E-cut blade will cut plaster in a more controlled manner.

Cordless power tools. Cordless drills and saws enable you to keep working when the power is off or not yet connected. They don't need an extension cord and so are much more convenient in places with limited access. See p. 60 for more information.

Remodel wiring tools. From right: drywall saw (sometimes called a stab saw), flex bit, bit extension, flex bit steering guide, and reel of fishing tape.

PRO TIP

Cordless drills and screwdrivers reduce the tedium of screwing wires to terminals, attaching devices to boxes, putting on cover plates, and connecting myriad other items. If your screwdriver has a torque clutch, use the lower settings (or a light touch on the trigger) to keep from overtightening or stripping screws. Always tighten cable clamps by hand to avoid over-tightening them and damaging the incoming cable(s).

A wire reel prevents kinks in cable as you pull it.

Box Fill Worksheet*

ITEM	SIZE (cu. in.)	NUMBER	TOTAL
#14 conductors exiting box	2.00		
#12 conductors exiting box	2.25		
#10 conductors exiting box	2.50		
#8 conductors exiting box	3.0		
#6 conductors exiting box	5.0		
Largest grounding conductor; count only one		1	
Devices; for each device, two times the largest connected conductor size			
Internal clamps; one for all clamps based on largest wire present		1	
Fixture fittings; one of each type based on largest wire			

*Table based on NEC 370-16(b) and adapted with permission from Redwood Kardon, Douglas Hansen, and Mike Casey, Code Check Electrical (The Taunton Press).

Gone fishin'. Spring-steel fish tapes or fiberglass fish rods are used to run cable behind finish surfaces. A fish tape is invoked in almost every old wiring how-to book on the market. Today's pros, however, swear by a *pulling grip*, also called a *swivel kellum* (see p. 319).

In most cases, however, it's simplest to use a *flex bit* (flexible bit) to drill through the framing. When the bit emerges, a helper can attach the new cable to a small "fish hole" near the bit's point. Then, using a swivel kellum to keep the new cable from getting twisted, put the drill in reverse and pull the bit (and cable) back through the holes it drilled. No fish tape required.

A 48-in. drill extension will increase the effective drilling length of a flex bit. Use an insulated steering guide to keep the flex bit from bowing excessively.

OTHER USEFUL TOOLS

No two electrician's tool belts look the same, but most contain a tape measure, flashlight, small level, hammer, Speed Square, and a large felt-tipped marker. In the course of a wiring job, you may need several sizes of slot-head and Phillips-head screwdrivers, plus an offset screwdriver and a nut driver.

If you're wiring a whole house, rent a *wire reel*, a rotating dispenser (photo at left) that enables you to pull cable easily to distant points. Reels hold 250 ft. of cable.

Adequate lighting is essential to both job safety and accuracy. If a site is too dark to see what color wires you're working with, your chances of making a wrong connection increase. LED headlamps are fantastic tools.

Sturdy stepladders are a must. In the electrical industry, only fiberglass stepladders are Occupational Safety and Health Administration (OSHA)-compliant because they're nonconductive. Wood ladders are usually nonconductive when dry, but if they get rained on, wood ladders can conduct electricity.

CHOOSING ELECTRICAL BOXES

There is a huge selection of electrical boxes, varying by size, shape, mounting device, and composition. One of the first distinctions to note is that of *new work boxes* and *remodel* or *cut-in boxes*. New work boxes are designed to be attached to exposed framing, as is often the case in new construction and sometimes in renovations where walls and ceilings are gutted. Cut-in boxes are designed for attachment to existing finish surfaces—which frequently involves cutting into plaster or drywall.

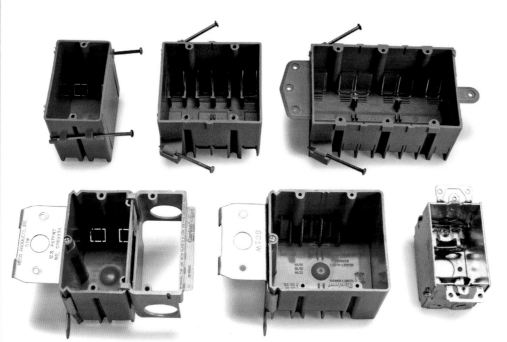

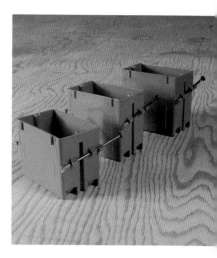

Single-gang boxes come in three sizes: 18 cu. in., 20.4 cu. in., and 22.5 cu. in. In general, bigger is better.

Electrical boxes. Top row, from left: single-gang plastic, double-gang plastic, and triple-gang plastic. Bottom row, from left: single-gang face-nailing adjustable with (orange) snap-on data ring, double-gang face-nailing adjustable, and single-gang metal.

But of all the variables to consider when choosing boxes, size (capacity) usually trumps the others. Correctly sized boxes are required by code and make your job easier because you don't have to struggle to fit wires and devices into a space that's too small.

Box capacity. The most common box type is a *single-gang box*. A single-gang box 3½ in. deep has a capacity of roughly 22½ cu. in., enough space for a single *device* (receptacle or switch), three 12–2 w/ground cables, and two wire connectors. *Double-gang boxes* hold two devices; *triple-gang boxes* hold three devices. Remember: Everything that takes up space in a box must fit without cramping—devices, cable wires, wire connectors, and cable clamps—so follow NEC recommendations for the maximum number of conductors per box.

You can get the capacity you need in a number of ways. Some pros install shallow *4-squares* (4 in. by 4 in. by 1½ in. deep) because they're versatile and roomy. Because of their shallow depth, these boxes can be installed back to back in a standard 2x4 wall. This allows you to keep even back-to-back switch boxes at the same height from one room to the next. Shallow *pancake boxes* (4 in. dia. by ½ in. deep) are commonly used to flush-mount light fixtures.

When you're installing GFCI receptacles or need more room for connectors and devices, use a 4S deep box. Finally, cover 4-square boxes with a *mud-ring* cover.

Metal versus plastic boxes. Metal boxes are sturdy and are available in more sizes than are

Getting BOX EDGES FLUSH

Use an add-a-depth ring ("goof ring") to make box edges flush when an outlet box is more than ¼ in. below the surface—a common situation when remodelers drywall over an existing wall that's in bad shape. Plastic goof rings, being nonconductive, are best. Do not use a steel ring with a plastic box.

Throw a single- or double-gang mud (plaster) ring on a 4-in. box and it's hard to overfill.

Cut-in (remodel) boxes and accessories: 1. Single-gang box with Grip-Lok tab (AKA Madison strap or cut-in strap); 2. Single-gang box with swivel ears; 3. Goof ring (used when a box is set too deep); 4. Round ceiling box with metal spring ears; 5. Double-gang box with swivel ears; 6. Single-gang box with adjustable ears. Box 1 should only be used for switches.

The screw on the side of an adjustable box enables you to raise or lower the face of the box so it's flush with the finish wall.

Adjustable mounting bars (AKA bar hangers). From top: heavyweight bar for new work, where framing is accessible; heavyweight fan–rated remodel bar, which can fit through an opening and be expanded in place; and light new work bar for a light fixture.

plastic boxes. Some metal boxes can be interlocked for larger capacity. Also, metal boxes are usually favored for mounting ceiling fixtures because steel is stronger than plastic. If code requires steel conduit, armored cable (BX), or MC cable, you *must* use steel boxes. All metal boxes must be grounded.

For most other installations, plastic is king. (Plastic boxes may be polyvinyl chloride [PVC], fiberglass, or thermoset.) Electricians use far more plastic boxes because they are less expensive. Also, because they are nonconductive, they're quicker to install because they don't need to be grounded. However, even if a box doesn't need to be grounded, all electrical devices inside must be grounded by a ground wire that doesn't depend on a device for continuity. Box volumes are stamped on the outside of plastic boxes.

Cut-in boxes. The renovator's mainstay is the *cut-in box (remodel box)* because it mounts directly to finish surfaces. These boxes are indispensable when you want to add a device but don't want to destroy a large section of a ceiling or wall to attach the box to the framing. Most cut-in boxes have metal or plastic flanges that keep them from falling into the wall cavity. Where they vary is with the tabs or mechanisms that hold them snugly to the back side of the wall: They could be screw-adjustable ears, metal-spring ears, swivel ears, or bendable metal tabs (Grip-Lok is one brand).

Mounting devices. The type of mounting bracket, bar, or tab you use depends on whether you're mounting a box to finish surfaces or structural members. When you're attaching a box to an exposed stud or joist, you're engaged in new work, even if the house is old. New-work boxes are usually side-nailed or face-nailed through a bracket; nail-on boxes have integral nail holders.

The mounting bracket for *adjustable boxes* is particularly ingenious. Once attached to framing, the box depth can be screw-adjusted until it's flush to the finish surface.

Adjustable bar hangers enable you to mount boxes between joists and studs; typically, hangers adjust from 14 in. to 22 in. Boxes mount to hangers via threaded posts or, more simply, by being screwed to the hangers. Bar hangers vary in

Reading A CABLE

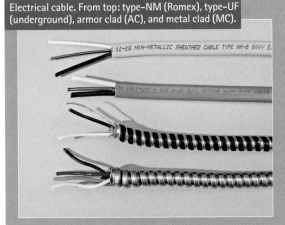

Electrical cable. From top: type-NM (Romex), type-UF (underground), armor clad (AC), and metal clad (MC).

Cables provide a lot of information in the abbreviations stamped into the sheathing. For example, *NM* indicates nonmetallic sheathing, and *UF* indicates underground feeder, which can be buried. The size and number of individual conductors inside a cable are also noted: *12/2 w/grd* or *12-2 W/G*, for example, indicates two insulated 12AWG wires plus a ground wire. Cable stamped *14/3 W/G* has three 14AWG wires plus a ground wire. (The higher the number, the smaller the wire diameter.) The maximum voltage, as in *600v*, may also be indicated.

Individual wires within cable have codes, too. *T* (thermoplastic) wire is intended for dry, indoor use, and *W* means wet; thus *TW* wire can be used in dry and wet locations. *H* stands for heat resistant. *N*, for nylon jacketed, indicates a tough wire that can be drawn through conduit without being damaged.

Finally, make sure the cable is marked *NM-B*. Cable without the final "-B" has an old-style insulation that is not as heat resistant as NM-B cable.

thickness and strength, with heavier strap types (rated for ceiling fans) required to support ceiling fans and heavier fixtures.

CABLE AND CONDUIT

Most modern house wiring is plastic-sheathed cable (Romex is one brand), but you may find any—or all—of the wiring types described here in older houses. Inside cables or conduits are individual wires, or conductors, that vary in thickness (gauge) according to the load (amps) they carry. More about that in a bit.

Nonmetallic sheathed cable (NM or Romex) is by far the most common type of cable. Covered with a flexible thermoplastic sheathing, Romex is easy to route, cut, and attach. Cable designations printed on the sheathing and the sheathing color indicate the gauge and the number of the individual wires inside.

Typically, Romex cable has two insulated wires inside and a ground—which may be insulated or, more often, bare wire. Thus, the Romex used for a standard 15-amp lights-and-outlets circuit will be stamped *14/2 w/grd*. For a 20-amp circuit, *12/2 w/grd* is required. Three-way switches are wired with 14/3 or 12/3 cable, which has an additional insulated wire. Again, wire gauge is rated for the load it can carry, so although you can wire 15-amp circuits with 12-gauge wire, you can't use 14-gauge wire anywhere in 20-amp circuits.

Metal-clad (MC) cable or armored cable (AC) is often specified where wiring is exposed and could be damaged. Some codes still allow armored cable, the older of the metallic cables, but that's increasingly rare. In AC cable, the metal covering of the cable acts as the ground; in MC cable, there is a separate insulated green wire that serves as a ground. To strip either type of metal cable, use a Roto-Split cable stripper; it's vastly superior to the old method of using a hacksaw and diagonal cutters.

Conduit may be specified to protect exposed wiring indoors or outdoors. It is commonly thin-wall steel (electrical metallic tubing, or EMT) or PVC plastic. Metal conduit serves as its own ground. Apart from service entrances, conduit is seldom used in home wiring. When connected with weathertight fittings, conduit can be installed outdoors—and PVC conduit, even underground.

Clamps. Every wiring system—whether nonmetallic, MC, AC, or conduit—has clamps (connectors) specific to that system. Clamps solidly secure cable to boxes to protect connections inside the box, so wire splices or connections to devices cannot get yanked apart or compromised.

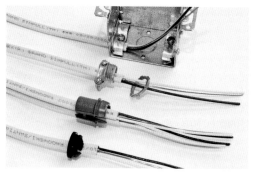

Romex cable connectors. From bottom to top: plastic push-in connector, two-cable hit-lock connector, 3/8-in. NM clamp with locknut, and metal box with internal clamps. Cable connectors are set in box knockouts to prevent wires' insulation from wearing against sharp edges and to protect electrical connections in the box should a cable get yanked.

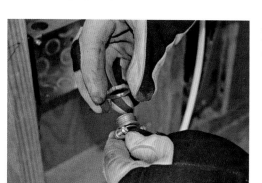

A metal locknut connector consists of two separate pieces. Two screws tighten down to grip the cable.

Plastic snap-in connectors snap easily into place and don't require tightening to secure cable.

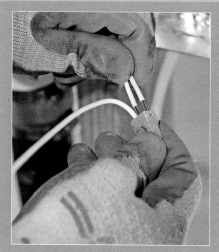

A NEW KIND OF Nut

Splicing with twist-on wire nuts can be problematic because stranded wire tends to slide down solid wire when you join solid wires to twisted-strand fixture leads. Wago Wall-Nuts simplify the task: Strip the wire ends the specified amount, and then push them into nut ports that hold the wires fast. The clear plastic housing allows you to see if the wires are connected, and the ports grasp both stranded and solid wire well. Wagos are also a good solution if the box has very short wires.

Cable clamps in metal boxes also keep wires from being nicked by burrs created when metal box knockouts are removed (see p. 316).

The exception to this rule is single-gang plastic boxes. If framing is exposed and cable can be stapled within 8 in. of the box, code doesn't require cable clamps in a single-gang plastic box. However, two-gang plastic boxes must have cable clamps—typically, a plastic tension clip that keeps cables from being pulled out. And, as noted earlier, all cut-in boxes must contain cable clamps.

Two-piece locknut connectors are still the most common type of clamp, but professional electricians racing the clock swear by *plastic snap-in* cable connectors, which seat instantly and grip NM cable tightly.

Wire connectors. Wire connectors, often called by the brand name Wire-Nut, twist onto a group of wires in a box to splice them together and ensure a solid mechanical connection. The importance of solid connections between spliced wires (or between wires and devices) can't be overstated. If wires work loose, electricity can leap over the gaps (arc) between them and cause a house fire. Wire connectors are sized according to the number of wires and wire gauge they can accommodate; each size is color-coded.

CHOOSING RECEPTACLES AND SWITCHES

Receptacles and switches can differ greatly in quality. Over the life of the device, the difference in price is trivial, but the difference in performance can be substantial. For this reason, buy quality. As shown in the photo below left, cheap receptacles are pretty much all plastic. Their thin metal mounting tabs will distort easily, and they tend to crack if subjected to heavy use.

High-quality receptacles and switches tend to have heavier nylon faces and may be reinforced with metal support yokes that reinforce the back of the devices.

Another indication of quality is how wires connect to a device, whether they are *back-stabbed*—that is, held by a thin metal tension clamp—or solidly secured by screws on the sides of the device or internal mechanical clamps.

If your budget allows, we recommend that you buy specification-grade (spec-grade) or commercial-grade devices from an electrical-supply house. The quality of residential-grade devices sold by home centers varies widely—some of it good, much of it so-so.

Household wiring. Planning your project, ordering electrical supplies, and room-by-room requirements are covered in "Developing a Floor Plan" (p. 301), but here are a few rules of thumb to guide you. General-use and lighting circuits are typically served by 14-gauge (AWG) cable protected by 15-amp breakers, so standard duplex 15-amp receptacles will comprise the bulk of what you buy. However, the NEC requires 20-amp circuits (wired with 12AWG cable) protected by 20-amp breakers for bathroom receptacles; at least two 20-amp circuits for small-appliance receptacles on kitchen countertops; and 20-amp circuits for laundry, garages, workshops, and other accessory spaces.

GFCI protection. To reduce the risk of electrical shock, the NEC requires ground-fault circuit interrupter (GFCI) protection on all 15-amp and 20-amp receptacles located in bathrooms within 6 ft. of sinks, tubs, or shower stalls; in laundry areas; for all kitchen counter receptacles or any other receptacles located within 6 ft. of a sink; for dishwasher receptacles; and for receptacles that are outdoor, in garages, in accessory buildings, or in unfinished basements. GFCI protection may be achieved by installing a GFCI breaker or by

Better-quality receptacles and switches are usually heftier and more reliable than cheaper ones. The quality receptacle on the right has a nylon face, and its back is reinforced with a brass yoke.

Receptacles for different loads. Clockwise from upper left: 30-amp dryer (125/250v), 50-amp range (125/250v), 20-amp duplex, 15-amp duplex, 15-amp tamper-resistant duplex, and 15-amp GFCI.

installing GFCI receptacles. (For more about grounding and ground faults, see pp. 290–293.)

AFCI protection. An arc fault is an explosive discharge of electrical current as it crosses a gap between two conductors, a serious condition because arcs can reach several thousand degrees Fahrenheit. Common causes of arcing in homes include corroded or loose electrical connections or a nail or screw driven through an electrical cable. To prevent house fires caused by arc faults, the NEC requires arc-fault circuit interrupter (AFCI) protection on all 15-amp and 20-amp receptacles in kitchens and laundry rooms, bedrooms, living rooms, rec rooms, parlors, libraries, dens, sunrooms, and hallways, and on switches serving any of those areas. AFCI protection may be achieved by installing an AFCI breaker or by installing an AFCI receptacle at the beginning of a circuit and through-wiring (p. 330) the device to protect receptacles downstream.

Although they protect against different hazards, GFCI and AFCI breakers and GFCI and AFCI receptacles look quite similar. GFCI and AFCI receptacles are also wired similarly (p. 328), so when installing either type of breaker or receptacle, check the label on the device carefully to be sure you are installing the right one. On house circuits that require both AFCI and GFCI protection, you can now buy a dual-function breaker.

Tamper-resistant receptacles. Since 2011, the NEC has required that all receptacles (including GFCI and AFCI receptacles) be *tamper-resistant*, except for those more than 5½ ft. above the floor,

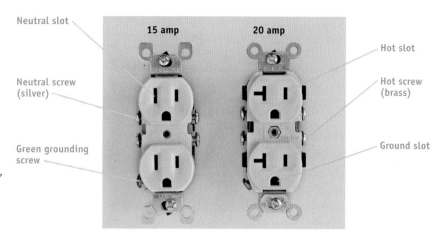

Polarized receptacles. The 20-amp receptacle (at right) has a T-shaped neutral slot so it can receive a special 20-amp plug in addition to standard 15-amp plugs. But 15-amp receptacles cannot receive 20-amp plugs. Both receptacles are also polarized, so that only the large blade of a plug can fit into the large slot of the receptacle.

those behind a not easily moved appliance, those that are part of a light fixture, and nongrounding receptacles used for replacements in nongrounding wiring.

Large appliances, dedicated circuits. Appliances that use a lot of energy, such as electric water heaters, electric ranges and ovens, clothes dryers, central heating and cooling systems, furnaces, whirlpools, and spas, are typically on a dedicated circuit. Cable wire serving such circuits must be matched to the load rating of the appliance, as must receptacles, plugs, and switches (if any).

An exciting new entry to the switch family is the wireless switch, a great boon to renovators because wireless switches can be installed without tearing up walls and ceilings. See "Installing a Wireless Switch" on p. 337.

Matching load ratings. Circuit components must be matched according to their load ratings. That is, a 20-amp receptacle must be fed by 12AWG cable, which is also rated at 20 amps, and protected by a 20-amp breaker or fuse. A 15-amp receptacle or switch should be fed by 14AWG cable, which is rated for 15 amps, and protected by a 15-amp breaker or fuse. (Note that 15-amp receptacles on a multi-receptacle circuit may be installed on a 20-amp circuit as long as the receptacles are properly "pigtailed.") Accordingly, a 20-amp circuit (wired with 12-gauge cable) can be wired with 15-amp receptacles.

NM cable manufacturers color-code the cable sheathing for the commonly used gauges to help correctly match wire size to breakers: White sheathing denotes 14-gauge wire; yellow sheathing, 12 gauge; and orange sheathing, 10 gauge.

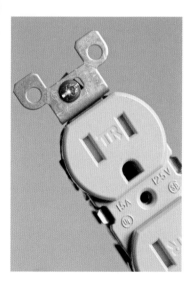

Tamper-resistant (TR) receptacles have internal shutters (the milky plastic visible in the slots) held shut by tension springs. Shutters will not slide open unless both plug prongs are inserted at the same time and with the necessary pressure.

Remodel Wiring Safety Essentials

Before removing box covers or handling wires, turn off the power to the area and use a voltage tester to be sure it is off.

First, remove the fuse or flip off the circuit breaker controlling the circuit and post a sign on the main panel warning people of work in progress. Better yet, if you've got circuit breakers, do as the pros do and install a *breaker lockout* (p. 293) so it will be impossible for anyone to turn it on. Breaker lockouts are available at electrical-supply houses and most home centers.

Testing for power is particularly important in remodel wiring because walls and ceilings often contain old cables that are energized. Here, a noncontact tester is especially useful. Simply touch the tester tip to cable sheathing or wire insulation. You don't have to touch the tester tip to bare wires to get a reading: If a cable, wire, or electrical device is energized, the tip will glow. A noncontact tester can detect voltage through cable sheathing.

Whatever tester you use, test it first on an outlet that you know is live to make sure the tester is working properly.

Always use a voltage tester to test for power before handling cables, devices, or fixtures. Here, a noncontact voltage tester detects voltage through cable insulation.

Rough-in Wiring

Rough-in wiring refers to the first phase of a wiring installation. It is the stage at which you set outlet boxes and run electrical cable to them—as opposed to *finish wiring*, or connecting wires to devices and fixtures.

Rough-in wiring is pretty straightforward when studs and joists are exposed. Whether a house is new or old, running wires through exposed framing is called *new work*, or new construction. If the framing is covered with finish surfaces such as plaster and drywall, however, the job is referred to as *remodel wiring*, or "old work." Remodel wiring is almost always more complicated and costly because first you must drill through or cut into finish surfaces to install boxes and run cable, and later you need to patch the holes you made.

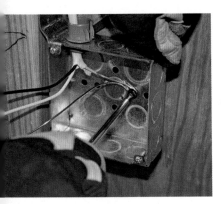

During a rough-in inspection, inspectors demand solid ground-wire splices and, in metal boxes, a ground screw or clip that secures the ground wire.

ROUGH-IN PRELIMINARIES

Wait until rough carpentry is complete before you begin rough-in wiring. Part of an electrician's job is setting boxes so they'll be flush to finished surfaces. Before an electrician starts working, modifications to the framing—such as furring out or planing down irregular studs and ceiling joists—must be complete.

Wait until the plumbers are gone. Waste pipes are large and often difficult to locate, which usually means a lot of drilling and cutting into studs and joists. Once the plumbing pipes are in place, you'll clearly see what obstacles you face and will have more room to move around.

Check your plans often. If there's not a table on site where you can roll out your electrical plans, staple them to a stud—preferably at eye level so you can read them easily. Checking and rechecking the plans is particularly important if you're not a professional electrician.

Be flexible. As you lay out devices, you'll realize that not everything specified on the plans is possible; most plans are developed without knowing exactly what the framing looks like or where obstructions are. Be flexible and choose a solution that makes sense. When running cable around doors and windows, for example, find the easiest path. Sometimes that means running cable in the space above or below. In other words, it may be less work to drill through top or bottom plates.

ORGANIZE YOUR WORK

Perform one task at a time. Each task—such as setting boxes or drilling—requires a different set of tools. So once you have the tools out to do a given task, go around the room and complete all similar tasks. You'll waste less time changing tools and the job goes much faster. In general, the sequence of rough-in tasks looks like this:

1. Walk the room with plans, marking outlet locations on walls and the ceiling.

2. Snap chalklines or shoot laser lines to pinpoint box elevations.

3. Attach boxes to studs and ceiling joists.

4. Drill holes for cable runs.

5. Pull cable through holes and into boxes.

6. Make up boxes—strip wire ends, splice current-carrying conductors, make up (splice) grounds, attach mud rings, and push wires into boxes.

7. Rough-in inspection. (See the box on the facing page and "The Rough-in Inspection" on p. 326.)

After the inspection, finish surfaces are installed. Then, at the *trim-out stage* or finish stage, wires are attached to the devices and fixtures.

ROUGH-IN RECAP: Electrical Code

▶ Circuit breakers, wiring, and devices must be correctly sized for the loads they carry. For example, 20-amp circuits require 12AWG wire. Receptacles can be rated for either 15 or 20 amps. Mismatching circuit elements can lead to house fires.

▶ All wire connections must be good mechanical connections. There must be good pressure between the connectors you are joining, whether wires are spliced together or connected to a device. (The NEC 2017–2020 code cycle requires that installers use a *torque tool*— torque screwdriver or torque wrench—to terminate conductors at electrical equipment.) In general, devices with screw terminals are more reliable than back-wired (stab-in) devices whose internal spring clamps can weaken and allow the wire to become loose.

▶ All wire connections must be housed in a covered box.

▶ Boxes must be securely attached to framing (or the wall, in old work) so that normal use will not loosen them.

▶ Box edges must be flush to finish surfaces. In noncombustible surfaces (drywall, plaster), there may be a ¼-in. gap between the box edge and the surface. But in combustible surfaces, such as wood paneling, there must be no gap.

▶ All newly installed devices must be grounded. Code allows you to replace an existing two-prong receptacle or to replace a nongrounded box that has become damaged. However, if you install a new three-prong receptacle, it must be grounded. The only exception: You can install a three-prong receptacle into an ungrounded box if that new device is a GFCI receptacle. If you extend a circuit, the entire circuit must be upgraded to current code.

▶ In new rough-in work, cable must be supported within 8 in. of a single-gang box without clamps and within 12 in. of any other box and every 4½ ft.

PRO TIP

When marking box locations *on finish surfaces*, use a pencil— never a crayon, grease pencil, or a felt-tipped marker. Pencil marks will not show through new paint. Also, grease pencils and crayons can prevent paint from sticking.

Have materials on hand when it's time to start installing boxes. Electricians often walk from room to room, dropping a box wherever floor plans indicate.

Begin the layout by consulting wiring plans, then walking the room and marking box locations.

A laser level quickly sets box heights around a room.

Here, a laser beam indicates the center of a box. The notations on the edge of the stud indicate that this will be a GFCI receptacle. More commonly, electricians measure box heights from the bottom or top of a box.

Snap a chalkline to indicate box heights if there's no laser on hand.

ORDERING MATERIALS

In general, order 10% extra of all boxes and cover plates (they crack easily) and the exact number of switches, receptacles, light fixtures, and other devices specified on the plans. It's OK to order one or two extra switches and receptacles, but because they're costly, pros try not to order too many extras.

Cable is another matter altogether. Calculating the amount of cable can be tricky because there are many ways to route cable between two points. Electricians typically measure the running distances between several pairs of boxes to come up with an average length. They then use that average to calculate a total for each room. In new work, for example, boxes spaced 12 ft. apart (per code) take 15 ft. to 20 ft. of cable to run about 2 ft. above the boxes and drop it down to each box. After you've calculated cable for the whole job, add 10%.

Cable for remodel jobs is tougher to calculate because it's impossible to know what obstructions hide behind finish surfaces. You may have to fish cable up to the top of wall plates, across an attic, and then down to each box. Do some exploring, measure that imaginary route, and again create an average cable length to multiply. If it takes, say, 25 ft. for each pair of wall boxes and you have eight outlets to wire, then 8 outlets × 25 ft. = 200 ft. Add 10%, and your total is 220 ft. Because the average roll of wire sold at home centers contains 250 ft., one roll should do it.

Many electricians prefer to mark ceiling-fixture centers on the floor . . .

. . . then use a plumb laser to transfer those markings to the ceiling.

Center your hole saw on each ceiling mark.

For convenience and speed, use a hammer length to establish the height of work boxes. It's close to the standard height of 12 in.

ROUGH-IN RECAP: Box Locations

▶ Whatever heights you choose to set outlets and switches, be consistent.

▶ Code requires that no point along a wall may be more than 6 ft. from an outlet. Set the bottom of wall outlets 12 in. to 15 in. above the finished floor surface—or 18 in. above the finished floor surface to satisfy the Americans with Disabilities Act (ADA) requirements.

▶ Place the top of switch boxes at 48 in., and they will line up with drywall seams (if sheets run horizontally), thus reducing the drywall cuts you must make.

▶ In kitchens and bathrooms, place the bottom of countertop receptacles 42 in. above the finished floor surface. This height ensures that each receptacle will clear the combined height of a standard countertop (36 in.) and the height of a backsplash (4 in.), with 2 in. extra to accommodate cover plates.

LAYING OUT THE JOB

With electrical plans in hand, walk each room and mark box locations for receptacles, switches, and light fixtures. Each device must be mounted to a box that houses its wiring connections. The only exceptions are devices that come with an integral box, such as bath fans, recessed light cans, and under-cabinet light fixtures.

Mark receptacle and switch locations on the walls first. Then mark ceiling fixtures. If studs and joists are exposed, use a brightly colored crayon. If there are finish surfaces, use a pencil to mark walls at a height where you can see the notations easily—these marks will be painted over later. Near each switch box, draw a letter or number to indicate which fixture the switch controls.

Once you've roughly located boxes on the walls, use a laser level to set exact box heights for each type of box.

Use the laser to indicate the tops or bottoms of the boxes. Many electricians prefer to determine level with the laser and snap a chalkline at that height so they can move the laser to another room.

To locate ceiling fixtures, mark them on the floor and use a plumb laser to transfer that mark up to the ceiling. This may seem counterintuitive, but it will save you a lot of time. Floors are flat, almost always the same size and shape as the ceilings above, and—perhaps most important—accessible and easy to mark. In complex rooms, such as kitchens, draw cabinet and island outlines onto the floor as well. Those outlines will help you fine-tune ceiling light positions to optimally illuminate work areas.

To install an adjustable box, press the bracket flush against the stud edge and screw it down. The tiny silver screw in the edge of the box is the depth-adjusting screw.

Nail-on boxes cannot be adjusted for depth once installed, so use the depth gauge on the side to ensure that box edges will be flush to finish surfaces.

Boxes with integral brackets have small points at top and bottom that sink into the stud. A mud ring will bring the box flush to the finish wall.

To provide solid support for multigang boxes, first install an adjustable box bracket that spans the distance between studs or use blocking behind the box.

Knockouts on metal boxes require a bit more force, so jab them with a stout pair of needle-nose pliers.

Once the knockout is loose, remove it using pliers.

Use a screwdriver to remove a plastic-box knockout.

INSTALLING NEW WORK BOXES

In this section, we'll assume that framing members are exposed and that boxes will be attached directly to them. Once you've used your plans to locate receptacle, switch, and light fixture boxes on walls and ceilings, installing them is pretty straightforward. Electrical codes dictate box capacity and composition.

Installing wall boxes. In residences, 18-cu.-in. single-gang PVC plastic boxes are by far the most common. This size is large enough for a single outlet or a single switch and two cables. Otherwise, use a 22.5-cu.-in. single-gang box or a four-square box with a plaster ring.

Set each box to the correct height, then set box depth so that its edge will be flush to the finish surface. If you use adjustable boxes, simply screw them to a stud. To adjust the box depth, turn the adjusting screw. Side-nailing boxes typically have scales (graduated depth gauges) on the side. If not, use a scrap of finish material (such as ½-in. drywall) as a depth gauge. Metal boxes frequently have brackets that mount the box flush to a stud edge; after the box is wired, add a mud ring (plaster ring) to bring the box flush to the finish surface.

Multiple-gang boxes mount to studs in the same way. But if plans locate the box away from studs or a multigang box is particularly wide, nail

When the can height is where you want it, fasten its hanger bars to the joists.

Then slide the can to fine-tune its position. Tighten the screws to lock it in place.

Use blocking to get the box in just the right spot.

For box placement midway between joists, use an adjustable bar hanger.

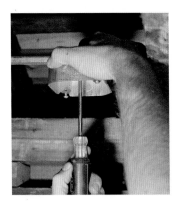

Screw a metal box to the bar hanger. Make sure the box edge will be flush to the finish ceiling.

Alternately, you can nail 2x4 blocking between joists.

blocking between the studs or install an adjustable box bracket and screw the box to it. (The bracket is sometimes called a *screw gun bracket* because a screw gun is typically used to mount it and to attach boxes to it.)

Removing knockouts. Once you've mounted boxes, you'll need to remove the appropriate number of box knockouts and install cable connectors (clamps). Single-gang, new-construction plastic boxes don't need clamps: Simply strike a screwdriver handle with the heel of your hand to drive out the knockout. To remove a metal-box knockout, jab it with the nose of needle-nose pliers to loosen it, then use the pliers' jaws to twist it free.

Installing ceiling boxes. Boxes for ceiling lights are most often 4-in. octagonal or round boxes or recessed light fixtures with integrated junction boxes. Setting ceiling boxes in new work is similar to setting wall boxes, with the added concern that the ceiling box be strong enough to support the fixture. Many electricians prefer to use metal boxes for ceiling fixtures anyway. Ceiling fans require fan-rated boxes.

In many cases, you'll need to reposition the box to avoid obstacles or line it up with other fixtures, but it's quick work if the box has an adjustable bar hanger. To install a recessed can, for example, extend the two bar hangers to adjacent ceiling joists. Then screw or nail the hangers to the joists. Slide the can along the hangers until its opening (the light well) is where you want it, and then tighten the setscrews on the side.

To install a 4-in. box, simply nail or screw it to the side of a joist. If you need to install it slightly away from a joist, first nail 2x blocking to the joist, then attach the box to the blocking. Remember: The box edge must be flush to finish surfaces.

To install a 4-in. box between joists, screw an adjustable hanger bar to the joists, then attach the box to it. Alternatively, you can insert 2x blocking between the joists and screw the box to it.

RUNNING ELECTRICAL CABLE

Once boxes are in place, you're ready to run cable to each of them. It's rather like connecting dots with a pencil line.

Drilling for cable. To prevent screws or nails from puncturing cables, drill in the middle of studs or joists whenever possible. If the edges of any hole you drill are less than 1¼ in. from the edge of framing members, you must install steel nail-protection plates (see p. 318). Always wear eye protection when drilling.

Drill for cables running horizontally (through studs) first. It doesn't matter whether you start drilling at the outlet box closest to the panel or at

PRO TIP

Drilled holes don't need to be perfectly aligned, but the closer they line up, the easier it is to pull cable. Some electricians use a laser to line up drill holes.

When drilling for cable runs, rest the drill on your thigh. This method eases drilling and places holes at a convenient height above box locations. Watch for kickback if the bit binds—it can break a hand or an arm. Wear goggles when drilling.

Identifying Cable Runs

To avoid confusion when it's time to wire devices, identify incoming cables. Use felt-tipped markers to write on the cable sheathing, or write on masking tape wrapped on grouped wires.

Cables running from the panel board to an outlet should be marked "source," "from source," or "upstream." If you're wiring GFCI receptacles, these incoming cable wires attach to GFCI terminals marked "line." Cables running on to the next outlet (away from the power source) are denoted "next outlet" or "downstream"; they attach to GFCI terminals marked "load."

Double- or triple-gang boxes will have a lot of cables entering, so make the cable descriptions specific: "switch leg to ceiling fixture," "3-way switch #2," and so on.

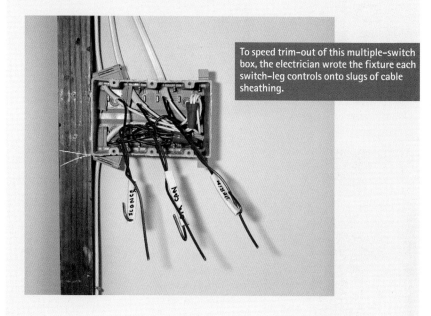

To speed trim-out of this multiple-switch box, the electrician wrote the fixture each switch-leg controls onto slugs of cable sheathing.

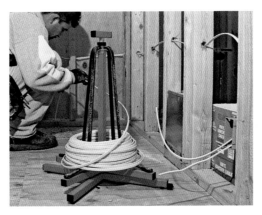

Place a wire wheel near the beginning or end of each circuit.

Staple snugly but not tightly enough to squeeze the cable sheathing. Note the cable spooler on the floor in the background.

For cables within 1¼ in. of a stud edge, install steel plate to protect them from being punctured by a nail or screw.

When bundling two or more cables, use standoffs.

the last outlet on the circuit. Just be methodical: Drill holes in one direction as you go from box to box. However, if you're drilling for an appliance that has a dedicated circuit—and thus only one outlet—it's usually less work to drill a hole through a top or bottom plate and then run cable through the attic or basement instead of drilling through numerous studs to reach the outlet.

If possible, drill holes thigh high. Partially rest the drill on your thigh so your arms won't get as tired. This method also helps you drill holes that are roughly the same height—making cable-pulling much easier. Moreover, when you drill about 1 ft. above a box, you have enough room to bend the cable and staple it near the box without crimping the cable and damaging the insulation.

For most drilling, use a 6-in., ¾-in.-dia. auger bit. Use an 18-in. bit to drill lumber nailed together around windows, doorways, and the like. Using an 18-in. bit is also safer because it enables you to drill through top plates without standing on a ladder. Standing on the floor is a big advantage, as the reaction torque from a heavy-duty drill can throw you off a ladder if a bit binds up.

Pulling cable. For greatest efficiency, install cable in two steps: (1) Pull cable between outlets, leaving roughly 10 in. extra beyond each box for splices, and (2) retrace your steps, stapling cable to framing and installing nail-protection plates if they're required. As with drilling, it doesn't matter whether you start pulling cable from the first box of a circuit or from the last box. If there are several circuits in a room, start at one end and proceed along each circuit, pulling cable until all the boxes are wired. Don't jump around: You may become confused and miss a box.

In new construction, electricians usually place several wire reels by the panel and pull cables from there towards the first box of each circuit.

Running Cable to a New Outlet

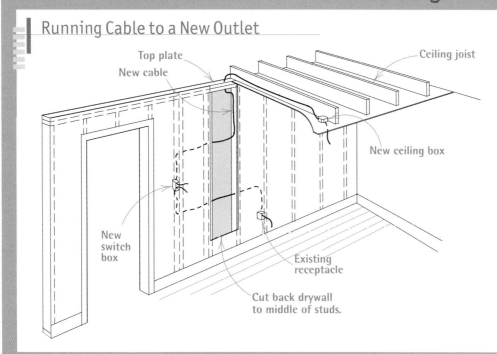

- Ceiling joist
- Top plate
- New cable
- New ceiling box
- New switch box
- Existing receptacle
- Cut back drywall to middle of studs.

Fishing cable to a new outlet can be time-consuming and tedious. Often, it's quicker and easier to cut back sections of drywall to the nearest studs or ceiling joists. With a stud bay exposed, you can pull wire in a hurry and staple cable to studs.

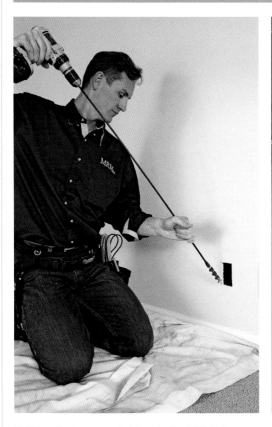

To fish cable to a new outlet, insert a flex bit into the new opening and drill down through the floor or bottom plate.

Pull back after a helper below the floor attaches the cable.

A swivel kellum grip slides over the end of the cable and prevents it from twisting as it's pulled.

PRO TIP

Put a piece of red tape, or even better, tie some yellow "caution" tape around the stud at eye level on the first box in each circuit (the "home-run" box) to ensure that you run cable from it to the panel. This also helps with planning the home-run cable pulls. On a complex job with many circuits, you might run cable between all the outlets in a circuit but forget to install the home-run cable that will energize the circuit. Not something you want to discover after the drywall's up.

REMOVING A WALL BOX

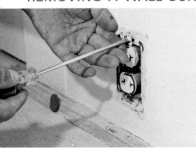

After testing for voltage, remove the cover plate, unscrew the receptacle from the box, and pull it out. Then disconnect the wires.

Remove the old outlet box, drill, fish new cable to the opening, and splice the existing cable to the new. You'll replace the old box with a remodel box.

These circuit segments are called "home runs." Once they've run cable to all the home-run boxes, they move a reel next to each box and continue to pull cable outward until they reach the last box. When doing remodel wiring in a house with a crawlspace, however, electricians often start at the last box and pull cables towards the panel. When they reach the first boxes of several circuits, they will move the wire reels to those locations. From there, they feed, say, three cables down to a helper in the crawlspace. The helper can pull all the cables toward the panel at the same time. This method is much faster than pulling single cables three different times.

Staple cable along stud centers to prevent nail or screw punctures. It's acceptable to stack two cables under one staple, but use *standoffs* to fasten three or more cables traveling along the same path. (Multigang boxes are fed by multiple cables, for example.) Standoffs and ties bundle cables loosely to prevent heat buildup. As you secure cable, install nail plates where needed.

Fishing cable. Most electricians hate fishing wire behind walls or ceilings. It can be tricky to find the cable and time-consuming to patch the holes in plaster or drywall. If you're adding a box or two, try fishing cable behind the wall. But if you're rewiring an entire room, it's probably faster to cut a "wiring trench" in the wall (see p. 323). Before cutting into or drilling through a wall, however, turn off power to the area.

If you'd like to avoid fishing altogether, see "Installing a Wireless Switch" on p. 337.

Fishing cables behind finish walls. If you're adding an outlet over an unfinished basement, fishing cable can be straightforward. Outline and cut out an opening for the new box, insert a flex bit into the opening, and drill down through the bottom plate. (Wear gloves to protect your hands.) When the bit emerges into the basement, a helper can insert one wire of the new cable into the small "fish hole" near the bit's point. As you slowly back the bit out of the box opening, you pull new cable into it. No fish tape required! The only downside is that the reversing drill can twist the cable. This problem is easily avoided by sliding a *swivel kellum* (see p. 319) over the cable end instead of inserting a cable wire into the flex bit hole. Because the kellum turns, the cable doesn't.

Fishing Cable ACROSS CEILING JOISTS

Sometimes there is no open bay to a ceiling fixture location. In that case, you'll have to drill across ceiling joists to run cable to the top plate of the wall. Use a 6-ft.-long flex bit (and a 48-in. extension, if needed) to drill across joists. Flex bits can wander and go off target, so be patient. When the drill bit emerges above the top plate, attach a swivel kellum taped securely to the new cable. Then slowly back the drill bit out, pulling the cable to the opening you cut in the ceiling.

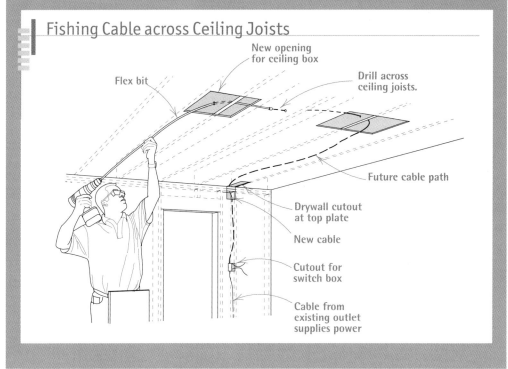

Fishing Cable across Ceiling Joists

New opening for ceiling box

Flex bit

Drill across ceiling joists.

Future cable path

Drywall cutout at top plate

New cable

Cutout for switch box

Cable from existing outlet supplies power

To run cable to a ceiling fixture, start by drilling an exploratory hole with a small-diameter bit.

If you're working alone, jam a long, looped wire into the exploratory hole, then go into the attic and look for it.

Running cable from a ceiling fixture to a wall switch means cutting access holes.

ADDING A REMODEL BOX

1. Position the box and then trace the box outline onto the wall.

2. After you chisel out the plaster within the outline, use a sabersaw or an oscillating multitool to cut lath.

3. After fishing and pulling cable, secure the box to the wall. Here, box ears are screwed to the lath.

Alternatively, you can start by removing a wall box. The closest power source is often an existing outlet. Cut power to that outlet and test to make sure it's off. The easiest way to access the cable is to disconnect the wires to the receptacle and remove it. Then remove the box, which may require using a metal-cutting reciprocating-saw blade to cut through the nails holding the box to the stud. Fish a new cable to the location, and insert the new and old cables into a new cut-in box. Secure the *cut-in box* to the finish surface, splice the cables inside the box, and connect pigtails from the splice to the new receptacle.

Fishing cable to a ceiling fixture. Fishing cable to ceiling fixtures or wall switches is usually a bit complicated. If there is an unfinished attic above or a basement below, run the cable across it, then route the cable through a stud bay to the new box in the ceiling. To run cable to a ceiling light, drill up through the fixture location using a 3/16-in. by 12-in. bit to minimize patching later. Use a bit at least 6 in. long so a helper in the attic can see it—use a longer one if the floor of the attic is covered with insulation. Measure the distance from the

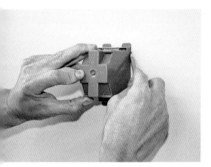

After drilling an exploratory hole and locating the box, trace the box outline onto the drywall.

You'll have the greatest control if you make a series of cuts before using a utility knife or an oscillating multitool to finish the cutout.

To retrofit a ceiling box, mark the box center onto the ceiling, then use a fine-tooth hole saw. Wear a respirator and safety glasses—this is dusty work.

After removing the plaster cutout, drill a hole through the lath so you can pull cable to the box. This worker uses an auger bit, but you could also use a cordless drill with a spade bit.

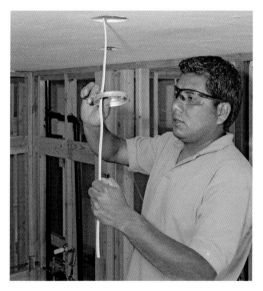

Fish cable through the hole, feed it through a connector in the box, and screw the box to framing or blocking.

bit to the wall; a helper in the attic can use that measurement to locate the nearest stud bay to drill an access hole into. If you're working alone, loop the end of a stiff piece of wire about 1 ft. long and insert it into the drilled hole; friction will keep the wire upright in the hole until you can locate it in the attic.

If there's no access above the ceiling and/or cable must cross several ceiling joists to get from a switch to a light fixture, you'll have to cut into finish surfaces at several points. To access cable in a stud bay, you'll need a cutout to expose the top plate. Using a flex bit may minimize the number of holes you must cut to drill across ceiling joists. But as noted earlier, it may ultimately take less time to cut and repair a single slot running across several joists than to patch a number of isolated holes. Whatever method you choose, make cuts cleanly to facilitate repairs. First, outline all cuts using a utility knife.

INSTALLING REMODEL BOXES

If existing house wiring is in good condition and an existing circuit has the capacity for an additional outlet, turn off the power, cut a hole in the wall, fish cable to the location, and secure a remodel box (cut-in box) to the finish surface (see p. 321). This process is also called *cutting-in* or *retrofitting* a box and, of course, it requires boxes with special mounting mechanisms.

Cutting a wall box in plaster. Hold the new box at the same height as other outlet or switch boxes in the room, and trace its outline onto the wall. Use a stud finder or drill a small exploratory hole to locate studs or wood lath behind. Look for water pipes and other wires. If you hit a stud, move the box. If you hit lath, keep drilling small holes within the opening to find the edges of the lath. If you position the box correctly, you'll need to remove only one or two lath sections.

Use a utility knife to score along the outline to minimize plaster fractures. Remove the plaster within the outline using a chisel. Then cut out the lath, using a cordless jigsaw or, even better, an oscillating tool with a Universal E-cut blade. As you cut through the lath strip, alternate partial cuts from one side to the other to avoid cracking the plaster. Then carefully remove the plaster beneath the box ears so they can rest on lath. Before inserting cut-in boxes, remove box knockouts, insert cable clamps, strip sheathing off the ends of incoming cable, and feed cable into the cable clamps. If more than one cable enters the box, write the destination of each on the sheathing. Secure the box by screwing its ears to the lath.

Cutting a wall box into drywall. Adding a cut-in box to drywall is essentially the same as adding one to plaster. Start by drilling a small exploratory hole near the proposed box location to make sure there's no stud in the way.

There are a number of cut-in boxes to choose from. The most common have side-mounted ears that swing out or expand as you turn their screws.

Hold the box against the drywall, plumb one side, then trace the outline of the box onto the wall. Drywall is much easier to cut than plaster: Simply align the blade of a drywall saw to the line you want to cut and hit the handle of the saw with the heel of your hand.

There is no one right way to cut out the box, but pros tend to cut one of the long vertical sides, then make two or three horizontal cuts across. Then score and snap the last cut. Finish the cutout using the drywall saw and a utility knife.

Retrofitting a ceiling box. As with all retrofits, turn off power to the area and explore first. Follow the mounting recommendations for your fixture. Attach the fixture box to framing.

If there's insulation in an attic above, remove it from the affected area. Be sure to wear eye protection and a dust mask when drilling through any ceiling—it's a dusty job.

Mark the box location, and use a fine-tooth hole saw to cut through plaster or drywall. Place the centering bit of the hole saw on the exact center of the box opening. Drill slowly so you don't damage adjoining surfaces—or fall off the ladder.

If the ceiling is drywall, you're ready to run cable through the hole in the ceiling. If the ceiling is plaster, cut through the lath or leave the lath intact and screw a pancake box through the lath and into the framing. Before attaching a pancake box, remove a knockout, test-fit the box in the hole, and trace the knockout hole onto the lath. Set the box aside and drill through the lath, creating a hole through which you can run cable.

Feed cable to the location, and fit a cable connector into the box. Insert the cable into the connector, slide the box up to the ceiling, and secure it. Strip the cable sheathing and attach the ground wire to a ground screw in the box. Strip insulation from the wire ends, and you're ready to connect the light fixture.

CREATING A WIRING TRENCH

When adding multiple outlets or rewiring an entire room, cutting a wiring trench in finish surfaces instead of fishing cable behind them is much faster. Before cutting or drilling, however, turn off the power to the areas affected. Be sure to wear eye protection and a dust mask.

1. Use a chalkline to mark the top and bottom lines of the wiring trench.

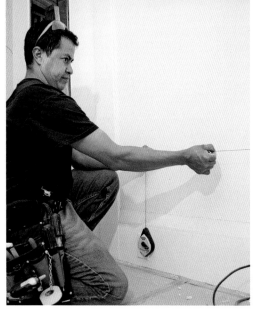

2. To ensure a clean cutline and facilitate patching when you're done, use a utility knife to score along chalklines.

3. Cut along the chalklines, holding the reciprocating saw at a low angle so you don't cut into lath or framing.

4. If walls are drywall, simply pull out the isolated strips. If plaster, use a hammer to break out the plaster and expose the lath.

5. Use a hammer and a flat bar to pry the lath free from the studs. Work slowly to minimize damage.

6. Pull out any nails still stuck in the studs; this will make later repairs easier.

1. For metal boxes, start by removing knockouts and then insert a connector.

2. Screw the box to the side of a stud.

3. For the best appearance, install side-by-side outlets at the same height. The piece being plumbed is a *data ring*, which is essentially a face plate without a box. Data lines have such low voltage that they don't need to be protected in an outlet box.

4. Run cable to each box.

5. Feed cable through cable connectors until you have a generous amount to work with—9 in. or 10 in. is plenty.

6. Staple the cable within 12 in. of each box. Here, the electrician drives staples to the underside of a rough windowsill.

(And before you cut into finish surfaces, see pp. 548–550 for more about lead-paint safety.)

If there are no windows in the walls to be rewired, cut the trench about 3 ft. above the floor so you won't have to kneel while working. If there are windows, cut the trench under the windows, leaving at least 1 in. of wall material under the windowsill to facilitate repairs. If there's plaster, make the trench as wide as two strips of lath. Snap parallel chalklines to indicate the width of the trench. Then use a utility knife to score along each line. Scoring lines first produces a cleaner cut and easier repairs.

Next, use a reciprocating saw with a demolition blade to cut through the plaster or drywall. Hold the saw at a low angle; you'll be less likely to break blades or cut into studs. Using a hammer, gently crush the plaster between the lines. Use a utility bar (flat bar) to pry out the lath strips or drywall section. If you expose any cables in the walls, use an inductance tester to make sure they're not hot. Next, drill through the studs so you can run cable in the trench. Wherever there's an outlet indicated, expand the trench width to accommodate the boxes.

Finally, pull any lath or drywall nails from stud edges. They're easy to overlook because they're small; if you pull them now, patching the trench will go smoothly.

PULLING CABLE TO RETROFIT BOXES

Once you've cut a wiring trench and drilled the holes, installing the boxes and pulling cable is fairly straightforward (photos above). If you're installing metal boxes, remove knockouts and insert cable connectors into their openings. Then screw boxes to studs; screwing is less likely to damage nearby finish surfaces. Be sure that the box will be flush to finish surfaces or, if you'll install plaster rings later, flush to the stud edge. Whenever you install boxes side by side—as with the outlet and low-voltage data ring shown—install them plumb and at the same height.

Installing cable in remodels can be tricky because space is tight and you must avoid bending cable sharply, which can damage wire insulation. Install nail plates wherever the edge of the hole is less than 1¼ in. from the stud edges. Feed cable through the cable connectors into the boxes. Finally, staple the cable to the framing within 8 in. of a single-gang box without clamps, or within 12 in. of other boxes. If there's not room to loop the cable and staple it to a stud, it's acceptable to staple it to other solid framing, such as the underside of a windowsill.

STRIPPING NM CABLE WITH A CABLE RIPPER

Most professional electricians favor utility knives for removing plastic sheathing. But non-pros should use a *cable ripper* to avoid nicking wire insulation. Because the ripper's tooth is intentionally dull (so it won't nick wire insulation), it usually takes several pulls to slit the sheathing completely. Once that's done, pull back the sheathing and the kraft paper and snip off both, using diagonal cutters.

Because cable clamps grip sheathing—not individual wires—there should be at least ¼ in. of sheathing still peeking out from under cable clamps when you're done. If you leave more than ½ in., you make working with the wires more difficult. If you didn't tighten cable clamps earlier, do so now.

If there is only one cable entering a box, just cut individual wires to length—typically, 6 in. to 8 in. If there's more than one cable, you'll add a 6-in. pigtail to each wire group.

SPLICING CABLE

Electricians call the last stage of rough wiring *making up a box*. After removing sheathing from cables, rough-cut individual wires about 8 in. long, group like-wires, and, to save time later, splice all wire groups.

Typically, electricians start by splicing the ground wires, which are usually bare copper. (If they're green insulated wires, first strip approximately ¾ in. of insulation off their ends.) If you use standard wire connectors, trim the ground wires and butt their ends together, along with a 6-in. pigtail, which you'll connect later to the green ground screw of a receptacle.

However, many pros prefer to twist the ground wires together, leave one ground long, and thread it through the hole in the end of a special wire connector (see the center photo on p. 298). If the box is metal, first bond the ground wire to the box, using a grounding clip or a grounding screw.

Splicing hot and neutral wire groups is essentially the same. Trim hot wires to the same length. Strip ¾ in. of insulation off the cable wires and the pigtail, and use lineman's pliers to twist the wires. Screw on a wire connector. If the box will contain two receptacles, create two groups of pigtails.

A cable ripper slits sheathing without damaging individual wire insulation. Use diagonal cutters to cut free sheathing.

see the center photo on p. 298

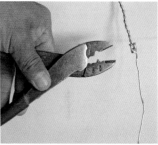

A crimp tool may be used instead of a wire nut to splice ground wires.

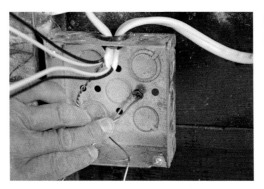

The long wire of the spliced grounds is looped beneath the grounding screw in a metal box.

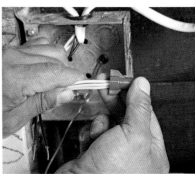

Strip about ½ in. of insulation from wire ends, and splice like-wires together.

For the two receptacles that will be installed in this box, two sets of pigtails are needed.

Fold the wires carefully into the box, and install a mud ring. This box is ready for a rough-in inspection.

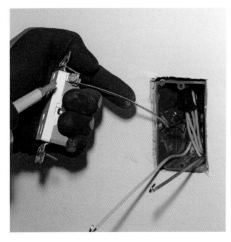

1. After stripping and looping wire ends, attach the ground pigtail to the green ground screw on the receptacle.

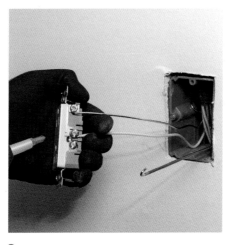

2. Next, connect the neutral-wire pigtail to a silver screw terminal.

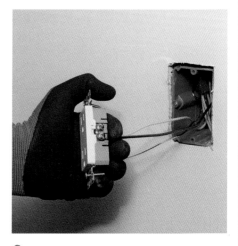

3. Finally, tighten the hot-wire pigtail wire to a brass screw terminal.

Receptacle in Midcircuit

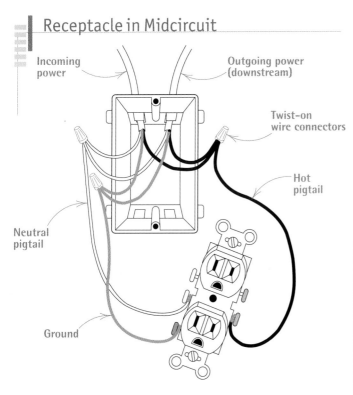

Incoming power

Outgoing power (downstream)

Twist-on wire connectors

Hot pigtail

Neutral pigtail

Ground

By splicing like-wire groups and running pigtails (short wires) to the receptacle in this conventional method, you ensure continuous current downstream.

4. Fold wires into the box so the face of the device will be parallel to the wall, then secure the device to the box.

SAFETY ALERT

The ground screw must compress the ground wire evenly. Never cross the ground wire (or any wire) over itself because the screw would touch only that high spot.

The Rough-in Inspection

At the rough-in stage, inspectors will look for a few key signs of a job well done: cables properly sized for the loads they'll carry; the requisite number and type of outlets specified by code; cables protected by nail plates as needed; neat, consistent work throughout the system; and, above all, ground wires spliced and, in metal boxes, bonded to the box with a ground screw. If grounds aren't complete, you won't pass the rough-in inspection.

At this inspection, only grounds need to be spliced. But since you've got the tools out, it makes sense to splice neutrals in a group, and continuous hot wires (those not attached to switches).

When all splices are complete, carefully fold the wire groups into the box. When you come back to do the trim-out stage, simply pull the wires out of the box, connect wires to devices, and install devices and cover plates.

Connecting circuit wires to a main panel or subpanel is the very last step of an installation. As noted throughout this book, avoid handling energized cables or devices.

Once the wire groups are spliced, gently accordion-fold the wires back into the box until you're ready to wire switches and receptacles.

Wiring Receptacles

Wiring an electrical device is considered part of *finish wiring*—also called the *trim-out stage*—when finish walls are in place and painted. At the trim-out, everything should be ready so that the electrician needs only a pair of strippers and a screwdriver or screw gun.

All work in this section must be done with power off: Use a voltage tester to be sure.

WIRING A DUPLEX RECEPTACLE

The duplex receptacle is the workhorse of house wiring. Receptacles are so indispensable to modern life that code dictates that no space along a wall in a habitable room should be more than 6 ft. from a receptacle and any wall at least 2 ft. wide must have a receptacle.

When a duplex receptacle is in the middle of a circuit, there will be two 12/2 or 14/2 cables entering the box—one from the power source and the other running downstream to the next outlet.

To ensure continuity downstream, all wire groups will have been spliced with wire connectors during the rough-in stage. A pigtail from each splice will need to be connected to a screw terminal on the receptacle. Unless the small tab between screw pairs has been removed (top left photo, p. 331), you need to attach only one conductor to each side of the receptacle.

Loop and install the ground wire to the receptacle's green grounding screw first. Place the loop clockwise on the screw shaft so that when the

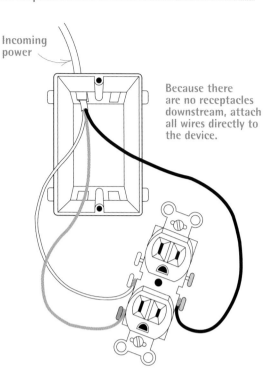

Incoming power

Because there are no receptacles downstream, attach all wires directly to the device.

Two-Slot Receptacles (Nongrounding)

Receptacles with only two slots are ungrounded. Because they are fed by two-wire cable without a ground wire, they are inherently less safe than three-slot receptacles fed with a grounded cable. If existing cables and receptacles are correctly wired and in good condition, most codes allow you to keep using them. Should you add circuits, however, code requires they be upgraded—that is, wired with grounded cable (12/2 w/grd or 14/2 w/grd) and three-slot receptacles.

Replacing a two-slot receptacle with a GFCI receptacle can be a cost-effective way to add protection to that outlet. There will still not be a ground wire on the circuit, but the GFCI will trip and cut the power if it detects a ground fault.

Note: If one slot of a two-slot receptacle is longer, the receptacle will be polarized. That is, a receptacle's brass screw terminal will connect to a hot wire and, internally, to the hot (narrow) prong of a polarized two-prong plug. The receptacle's silver screw terminal connects to neutral wires and, internally, to the neutral (wide) prong of a polarized plug.

Receptacles with two slots (instead of three) are nongrounding types. If the two slots are the same length, the receptacle is also non-polarized and should be replaced with a polarized nongrounding receptacle.

GFCI Receptacle, Single-Location Protection

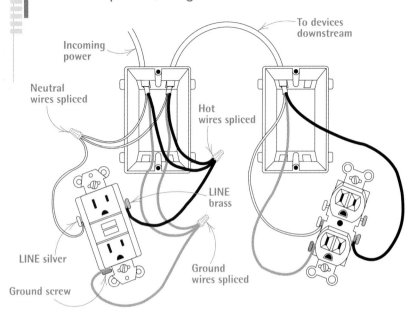

Incoming power

To devices downstream

Neutral wires spliced

Hot wires spliced

LINE brass

LINE silver

Ground screw

Ground wires spliced

This configuration provides GFCI protection at one location —say, near a sink—while leaving devices downstream unprotected. Here, splice hot and neutral wires so the power downstream is continuous and attach pigtails to the GFCI's "LINE" screw terminals. With this setup, receptacle use downstream won't cause nuisance tripping of the GFCI receptacle.

SAFETY ALERT

If you install three-prong receptacles on an ungrounded circuit, the NEC requires that the new receptacles be GFCI-protected and that you label affected receptacles "no equipment ground."

Quality GFCI and AFCI receptacles can be reliably wired by inserting stripped wire ends into terminal holes on the back of the device: Internal clamps grip the wire ends as the screws are tightened down. Or you can loop and attach stripped wire ends directly to screw terminals on each side of the device. Note the "LOAD" and "LINE" descriptors; they are important.

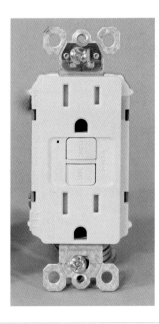

screw is tightened down, the screw head will grip—rather than dislodge—the wire.

Next, loop and attach a neutral conductor to a silver screw terminal. (Some electricians also tighten down the screw that isn't used as a matter of good practice.) Then flip the receptacle over to access the brass screw terminals on the other side. If a looped wire end is too wide, use needle-nose pliers to close it.

Screw down the brass screw so that it grips the hot wire. Pros frequently use screw guns for this operation, but weekend electricians should tighten the screw by hand to ensure a solid connection.

Push the wired receptacle into the box by hand, keeping the receptacle face parallel to the wall. You can hand-screw the device to the box, but if you take it slow and use a fresh bit, a cordless drill/driver is much easier. Finally, install a cover plate to protect the electrical connections in the box and to prevent someone from inadvertently touching a bare wire end or the end of a screw terminal.

When a receptacle is at the end of a circuit—where only one cable feeds an outlet—there's no need for pigtails. Just attach incoming wires directly to the receptacle as shown in the illustration on p. 327. As with pigtail wiring, connect the ground wire first, then the neutral, and then the hot wire.

WIRING A GFCI OR AFCI RECEPTACLE

Ground-fault circuit interrupters (GFCIs) and arc-fault circuit interrupters (AFCIs) offer different but critical code-required protections, so if you are unfamiliar with either term, please review pp. 293 and 310. In brief, GFCIs protect you against electrical shocks, and AFCIs protect against arc faults and reduce house fires. You can achieve GFCI or AFCI protection on a circuit by installing a GFCI or AFCI breaker or, as described below, by installing a GFCI or AFCI receptacle at the beginning of a circuit to protect receptacles downstream.

Note that GFCI and AFCI receptacles look similar, and wiring either type is essentially the same. So before you start your installation, check the label on the device to be sure you are installing the correct one. (For circuits that require both GFCI and AFCI protection, you can buy dual-function circuit breakers.)

When wiring a GFCI or AFCI receptacle, it's important to connect incoming wires (from the power source) to the terminals marked "LINE" on the back of the receptacle. Attach outgoing wires (to outlets downstream) to terminals marked "LOAD." To distinguish line and load

GFCI Receptacle, Multiple-Location Protection

From power source

Spliced ground wires

To devices downstream

Neutral

Hot

*Devices include receptacles, switches, and light fixtures.

A GFCI receptacle can protect devices* downstream if wired as shown. Attach wires from the power source to terminals marked "LINE." Attach wires continuing downstream to terminals marked "LOAD." As with any receptacle, attach hot wires to brass screws, white wires to silver screws, and a grounding pigtail to the ground screw. Note: Here only ground wires are spliced; hot and neutral wires attach directly to screw terminals.

wires during rough-in, write each term on small pieces of the cable sheathing and slip them over the appropriate wires before folding them into the box.

Protecting a single outlet. If the GFCI or AFCI is going to protect users at a single outlet (photos at right), attach wires to only one set of screw terminals. The yellow tape across one set of screws indicates that they are load terminals. If you are hooking up the device to protect only a single point of use, leave the tape in place and connect wires only to the screw terminals marked "LINE." After attaching the ground pigtail, screw down the silver screw to secure the neutral pigtail.

Connect the hot pigtail to the brass screw last, then push the device into the box carefully, hand-screw it to the box, and install a cover plate. *Note:* If you inserted wire ends into holes on the back of the receptacle, you must still tighten down the screw terminals on the sides of the device, which tightens internal clamps that hold them snug.

WIRING A GFCI OR AFCI RECEPTACLE

1. By using pigtails from each wire group, you can wire a GFCI or AFCI receptacle to protect only its outlet and not outlets downstream.

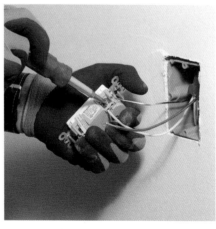

2. Attach the ground pigtail to the green ground screw, then insert the neutral pigtail into a neutral "LINE" terminal hole in the back of the device. Tighten the corresponding screw.

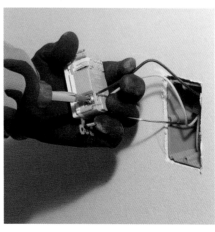

3. Finally, insert a hot pigtail into a hot "LINE" terminal hole on the back of the device and tighten the corresponding screw.

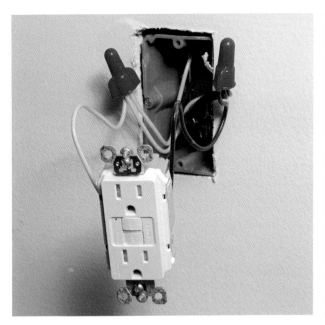 

A GFCI or AFCI receptacle protecting a single outlet has just three pigtails attached. Attach a ground wire to the ground screw, then attach hot and neutral pigtails to "LINE" terminals.

A GFCI OR AFCI receptacle protecting outlets downstream does not use pigtails for hot and neutral wires. Incoming wires from the power source attach to terminals marked "LINE"; outgoing wires attach to terminals marked "LOAD."

Feeding the circuit through a GFCI or AFCI receptacle. If you want to use a single GFCI or AFCI receptacle to protect downstream outlets, feed them from the "LOAD" terminals of the receptacle. That is, connect incoming and outgoing cable wires directly to the device rather than using pigtails. (You should, however, splice ground wires to ensure continuity: Screw the ground pigtail to the receptacle's ground screw.) Again, it's important to connect incoming wires (from the power source) to the terminals marked "LINE" and outgoing wires to terminals marked "LOAD." This is a code-approved way to offer protection downstream without using a breaker.

Attach the ground wire to the ground screw, neutral wires to silver screws, and hot wires to brass screws. If you instead inserted wire ends into holes on the back of the receptacle, screw down all screws on the sides of the device, which tightens internal clamps holding the wires. Be patient as you push the receptacle into the box, folding wires as needed. Install the mounting screws and attach the cover plate.

WIRING A SPLIT-TAB RECEPTACLE

Standard duplex receptacles have a small metal tab between the brass screw terminals. The tab conducts power to both terminals, even if you connect a hot wire to just one terminal. However, if you break off and remove the tab, you isolate the two terminals and create, in effect, two single

receptacles—each of which requires a hot lead wire to supply power.

This technique, known as split-tab wiring, is often used to provide separate circuits from a single outlet, a configuration commonly used when connecting a disposal and a dishwasher. The disposal receptacle is almost always controlled by a switch, which allows you to turn off the disposal at another location. To supply two hot leads to a split-tab receptacle, electricians usually run a 12/3 or 14/3 cable.

To create a split-tab receptacle, use needle-nose pliers to twist off the small metal tab between the brass screws. Next, connect the bare ground wire to the green grounding screw on the device and connect the white neutral wire to a silver screw. If you keep a slight tension on the wires as you tighten each screw, they'll be less likely to slip off.

Flip the receptacle over to expose the brass screws on the other side and connect a hot lead to each brass screw. If you're running 12/3 or 14/3 cable, one hot wire will typically be red and the other black. Finally, push the receptacle into the box, install the mounting screws, and apply the cover plate.

Although this 15-amp split-tab receptacle is fed with 12/3 cable (rated for 20 amps), there's no danger of the load exceeding the rating of the receptacle. Because of the configuration of the slots, the receptacle can receive only a 15-amp plug.

WIRING A SPLIT-TAB RECEPTACLE

1. To convert a standard duplex receptacle into two single receptacles, start by twisting off the small metal tab between the two brass screw terminals.

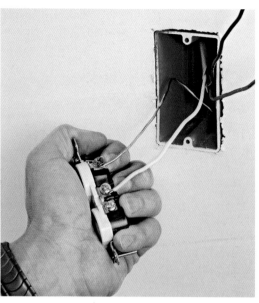

2. Connect the ground wire to the green grounding screw, then the neutral to one of the silver screw terminals.

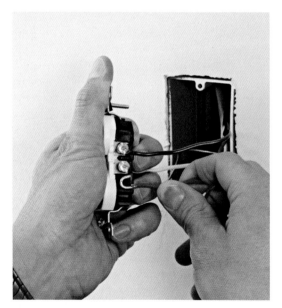

3. Connect the hot wires last. Removing the tab creates, in effect, two receptacles, so each needs a separate hot wire. It's common to use a 12/3 or 14/3 cable for split-tab circuits. Typically, one hot wire of the cable will be black, the other red.

4. A correctly wired split-tab receptacle. Typically, one wire is controlled by a switch.

Wiring Switches

Before connecting or disconnecting wires to a switch, use a voltage tester to make sure that the power to the switch box is off. Test with the switch both on and off to be sure. And never assume an existing switch is correctly wired: After removing the cover plate, test all conductors for power.

Be especially wary of outlets presently controlled by dimmers, as explained on p. 336, "Wiring a Linear Slide Dimmer."

Wiring a single-pole switch. The most commonly installed switch, a single-pole, is straightforward to wire. Spliced together during the rough-in stage, the neutral wires stay tucked in the outlet box. Pull ground and hot-wire groups out of the roughed-in box. Use the hole in the handle of your wire strippers or use needle-nose pliers to loop the conductor ends so they can be wrapped around the screw terminals.

First, attach the ground wire to the green grounding screw on the switch. Orient the wire loop in a clockwise direction—the same direction the screw tightens. When the loop faces the

other way, it can be dislodged as the screw head is tightened.

Next, connect the hot wires to the switch terminals, again orienting wire loops clockwise. One black wire is *hot* (power coming in), and the other is the *switch leg* (power going out to the fixture). With a single-pole switch, however, it doesn't matter which wire you attach to which screw. Generally, pros attach the hot wire last, as they attach the hot wire on a receptacle last.

Once the ground and hot wires are connected, the switch can be tucked into the box. Always push the device into the box by hand until it's flush to the wall. Don't use the screws to draw a device to a box because the device may not lie flat, and it's easy to strip the screw holes in a plastic box. Using a cordless drill/driver is a lot faster than using a screwdriver, but use a light trigger finger on the torque clutch of the driver to avoid stripping the screw head or snapping off the screw.

Wiring a single-pole switch using a "switch loop." This section shows two ways of wiring a single-pole switch when an outlet or fixture box is closer to the power source than to the switch

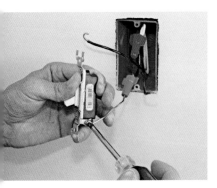

1. When wiring switches, leave neutral wires tucked in the box. Switches interrupt the current flowing through hot wires only. Attach the ground wire first.

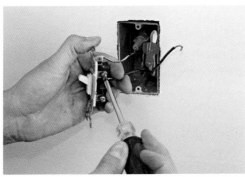

2. Attach hot wires to terminals. One wire is hot (from the power source). The other is the switch leg running to the fixture.

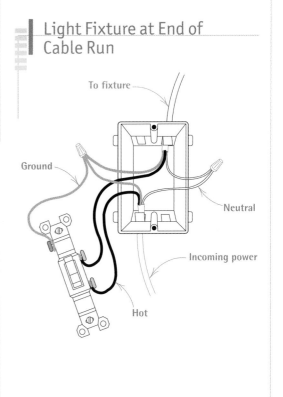

Light Fixture at End of Cable Run

To fixture

Ground

Neutral

Incoming power

Hot

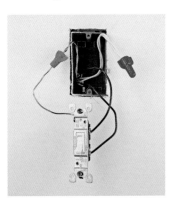

3. Wires are ready to be tucked into the box.

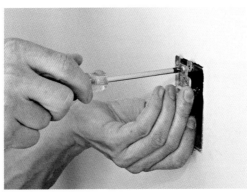

4. Straighten switch(es) in the box, and tighten the screws carefully so you don't strip the box's threads.

Before connecting or disconnecting wires to a switch, use a voltage tester to make sure that the power to the switch outlet is off. Test with the switch flipped both on and off to be sure.

Because switches interrupt only hot wires, you'd think they'd all be easy to wire. As you will see, however, switch wiring can also be quite complex, especially three-way and four-way switches and switches with electronic components. Follow the manufacturer's instructions carefully.

At the switch, start by stripping and looping the wire ends. Next, tape the white wire with black electrician's tape to indicate that it is serving as a hot wire to the back-fed switch. The NEC dictates that the white wire in back-fed wiring is always the hot lead (power coming in). The black wire, on the other hand, is the switch leg that runs back to the fixture.

First, connect the ground wire to the green ground screw on the back-fed switch. Next, connect the switch-leg wire (black), then the hot wire (white taped black) to the switch terminals. To keep looped wire ends snug against the screw shaft as you tighten down the screw, pull gently on wires. Not fumbling with wire ends saves time.

Finally, tuck the wires into the box, screw the switch to the box, and install the cover plate.

box, commonly called a *switch loop*. The first way to wire a switch loop, shown in the drawing below right, can probably be found in 90% of older homes but has been superseded by changes in the electrical code. The second way to wire a switch loop, shown in the drawing on p. 334, conforms to the most recent version of the NEC and should be used for new installations.

Wiring a switch loop: the historical method.
Before the code changed, it was common to run a single length of 12/2 or 14/2 cable as a *switch loop*. This means bringing the power down from the fixture to and through the switch and then back up to the fixture. As such, the white wire taped black in the 12/2 or 14/2 switch loop functions as the incoming hot wire, and the black wire acts as a switch leg to return the power to the fixture. Here, the white wire is actually a hot wire and is marked to identify it as such.

Turn off the power and test to be sure. At the outlet or fixture box, splice all the grounds together. Attach the source neutral wire to the fixture neutral wire. Attach the source hot wire to the white wire (taped black) of the switch loop. Last, connect the switch loop black wire to the black fixture wire.

Note: Here, for convenience, we bend the rule of using a white wire only as a neutral wire and instead wind black tape on each end of the white wire to show that—in this case—the white wire is being used as a hot wire.

Wiring a "Switch Loop": the Historical Method

In the "old school" way of wiring a switch loop, a single length of two-wire cable serves as a switch leg. The white wire in the cable is taped black to show that it is being used as a hot wire. It was a good solution because it conserved copper and was quick to wire, but it has been superseded by the method shown on p. 334.

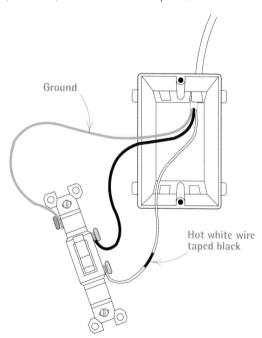

Ground

Hot white wire taped black

A single length of two-wire cable serves as a switch loop

Wiring a "Switch Loop": the Modern Method

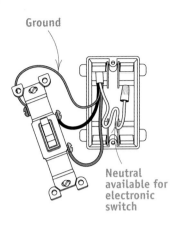

Ground

Neutral available for electronic switch

The NEC requires that there be a neutral in every switch box because some electronic timer switches and other energy-saving controls need a neutral. Thus, if you want to use a switch-loop approach, you must use three-conductor (3-wire) cable, connect the neutral at the power source, and then cap off the neutral in the switch box. Because 3-wire cable has both a black and red conductor, there is no need to re-identify the white conductor as hot, as was done in the old method, because the white wire is a neutral.

1. With the power off, strip sheathing from the 3-wire cable that runs from the fixture box.

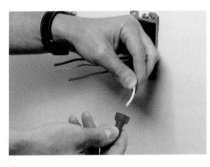

2. If the switch does not need a neutral wire, cap it and fold it into the box.

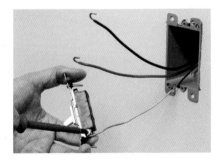

3. First attach the ground wire to the switch ground screw.

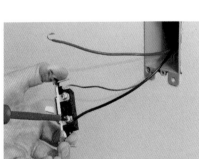

4. Attach the switch leg and the hot lead.

PRO TIP

Unless manufacturer's instructions say otherwise, whenever you splice solid wires with a wire connector, twist the wires together and trim the end before you twist the connector into place. This guarantees a solid connection between the wires should the wire connector come loose. Ideal Industries makes screwdrivers that have a wire nut driver socket in the handle, and this makes it easy to spin the connector on tightly.

▐▐▐▐

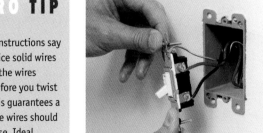

5. Fold the wired device into the box and secure it with mounting screws. Install the cover plate.

Wiring a switch loop: the modern method.
This method of wiring a switch loop reflects recent code changes. Specifically, the NEC requires that there be a neutral in every switch box because some electronic timer switches and other energy-saving controls need a neutral. So if you want to use a switch-loop approach, you must use three-conductor (3-wire) cable. The neutral of the 3-wire cable must be connected to the neutral of the circuit, even if the neutral is not going to be used.

Turn off the power and test to be sure. From the power source at the fixture, run a length of three-wire cable to the switch box. Remove cable sheathing (photo 1 at left) and strip ¾ in. of insulation from the ends of insulated wires. If the switch doesn't need a neutral wire, cap the neutral wire (2) and fold it into the switch box.

If you are attaching switch wires to screw terminals, loop the wire ends. Connect the ground wire to the green ground screw (3) on the switch. Next, connect the switch-leg wire (red), and the hot lead (black), which runs back to the fixture (4). Keep looped wire ends snug against the screws as you tighten them down. When all wires are secured, gently fold them into the box (5), screw the switch to the box, and install the cover plate.

1. Before replacing any device, turn off the power and use a voltage tester to be sure. To test a switch, remove the cover plate and carefully touch the tester to terminals and wires. Here, a voltage tester shows energized wires.

2. When the tester indicates all power is off, unscrew and pull the switch from the box.

Replacing a single-pole switch. You can replace an old single-pole toggle switch with a new one or with a *convertible dimmer* wired as a single-pole dimmer. That is, you can use the existing wires, but first turn off the power to the circuit. Use a voltage tester to see if voltage is present at the switch box (photo 1 at right). If the tester glows, there's power present: Turn it off at the fuse box or breaker panel. Test again. When the power's off, unscrew the old switch and pull it out from the box (2).

Disconnect the switch wires and note their condition. If the cable's fiber sheathing is frayed but individual wire insulation is intact, the wires are probably safe to attach to the replacement switch. If there's debris present in the box, sweep or vacuum it out.

Connect the wires to the new switch (3). There may not be a ground wire to attach to the new switch's ground screw, but code doesn't require grounding a switch if there's no ground wire feeding the box. Once the dimmer's connected, set it flush to the wall, and screw it to the box (4).

Note: Any replacement switch must match the type of fixture it controls, whether line voltage or low voltage. Typically, the dimmer rating is stamped on its face. In this case, the rating specifies, "For permanent incandescent fixtures."

Finally, install the cover plate (5) to protect the connections in the box and to prevent switch users from inadvertently touching the wire ends or dimmer terminals.

3. Check for voltage again, then disconnect the old switch, and examine circuit wires for cracked insulation. If existing wires are in good shape, attach them to the new switch.

4. Screw the replacement switch (here, a dimmer) to the box.

5. Install the cover plate and test the switch.

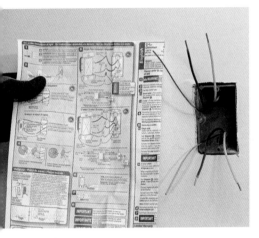

Slide dimmers. From left: Skylark Contour, Luméa, IllumaTech, Diva, Maestro IR (infrared) with remote control. All have a slide bar to preset light levels and a separate on-off switch. The bare-metal flanges around the dimmers are heat sinks.

WIRING A LINEAR SLIDE DIMMER

If you are replacing an existing dimmer, note: Dimmers that do not require neutral wires may be "trickle devices" that allow a minuscule amount of current flow all the time to power the dimmer. Trickle current can be enough to shock you if you work on a fixture without first turning off the breaker (or fuse) that controls the circuit. Always use a voltage tester to test the dimmer and the fixture to make sure no current is present.

All work in this section should be done with the power off.

Slide dimmers have a slide bar that allows you to set the light level and a separate on-off switch so you can turn the light on and off without changing a preset light level. Slide dimmers have largely replaced the old rotary type that combined both functions. Newer slide dimmers offer additional functions, so their wiring has become more complicated, and many sport wire leads rather than screw terminal connections.

For standard single-pole switches, it doesn't matter which screw terminals you connect a switch leg or hot wire to, but it may matter which wire you attach to dimmer leads, so always read the manufacturer's directions. The slide dimmer shown at left is a multiway switch that is convertible: It can be wired as a single-pole or three-way switch, depending on which wires you connect (1). It has a green insulated ground, red and black hot wires, and a red-and-white-striped wire that would be used as a signal wire to a companion dimmer.

Because the convertible device would be used as a single-pole dimmer, we didn't need the red-and-white-striped wire. So we capped it with a wire connector (2). Because this dimmer does not require a neutral, we spliced the neutrals together to feed through to the fixture (without connecting them to the dimmer).

Splice the ground pigtail to the device's ground lead. Then splice the switch leg from the box to the red lead on the device. On devices with wire leads, typically a red lead attaches to the switch leg.

Finally, attach the incoming hot wire to the other hot lead (black) on the device (3). Carefully fold the wires into the box (4) and push the wired dimmer into the box. Screw the device to the box and install the cover plate.

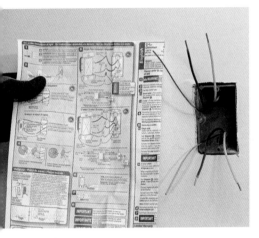

1. Dimmer specifics vary, so read installation instructions before you start: It's critical which circuit wires you attach to dimmer leads.

2. Cap any lead that you won't connect. (Here, a multiway convertible switch was used as a single-pole switch.) Then attach the grounding pigtail to the ground screw.

3. Connect the red lead to the switch leg—the wire that runs to the fixture—then the black lead to the incoming hot wire.

4. Because dimmer bodies tend to be larger than the single-pole switches they may replace, make sure beforehand that the box is large enough.

Installing a wireless switch. Traditionally, adding a light fixture that could be controlled from several locations meant retrofitting three-way or four-way switches—which can turn into a nightmare. Fortunately, today's electricians have another, almost effortless option—installing wireless switches.

Installing three-way wireless capability can be as simple as replacing a mechanical switch (a single-pole toggle, for example) with an electronic *master switch* and locating a *wireless controller* at some distant point. In the photo sequence on this page, we installed a Lutron Maestro Wireless dimmer and a companion Pico Wireless control.

Start by turning off power to the existing (mechanical) switch and use a voltage tester to be sure it's off. Remove the switch cover plate. To be doubly sure the power is off, apply the voltage tester to the switch's terminals and wires. Unscrew the switch from the outlet box, then pull out the switch and disconnect its wires.

Electronic switches are sensitive (and expensive), so follow the manufacturer's installation instructions exactly. As most do, the Lutron electronic dimmer looks like a standard back-wired switch, with a green grounding lead coming off it. Attach the wires per instructions, screw the device to the box, install the cover plate, turn the power back on to the switch, and program the Pico Wireless control via buttons on its face.

Mounting the wireless control—say, at the far end of a hall—is as simple as sticking an adhesive-backed mounting plate to a finish surface. If you want a more permanent mounting, use the screws provided—and expansion anchors if the wall is drywall. The wireless control slips into the mounting plate and is in turn covered by a snap-on plate. The controller needs no wires because it has a tiny battery that's typically good for 10 years. It needs only enough power to "talk" to the master switch.

Even modest wireless devices have a lot of useful functionality. The Pico control also can be clipped to a car visor so that as you approach home, you can turn on the porch light. Inside the house, you can program lights to turn off and on. In a baby's room, for example, you could program a light to dim slowly over a 10-minute period so the baby isn't startled by sudden darkness as he or she drifts off to sleep.

INSTALLING A WIRELESS SWITCH

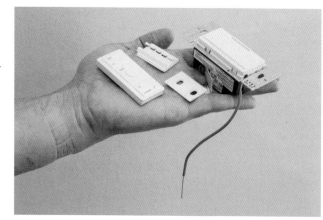

All you need to retrofit wireless switches. From left: a wireless controller, a visor clip, wall-mounting plate for the controller, and an electronic master switch.

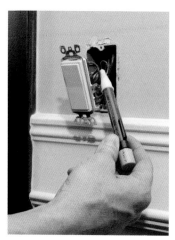

1. Turn the power off, remove the cover plate of an existing switch, and test to be sure the power's off. Disconnect the old switch.

2. Electronic switches are expensive, so read installation directions carefully. Here, existing hot and switch-leg wires are attached to the terminals of an electronic dimmer that serves as a master switch. The green switch lead is connected to a bare copper ground.

3. Tuck in wires, screw the electronic switch to the box, then install the cover plate. After testing the switch, program the wireless controller. The controller will *talk* to the dimmer, telling it to raise or lower the lights—in effect, it is a three-way switch.

4. Down the hall, a small mounting plate is screwed to or stuck on the wall. The controller slides into the plate. No cutting, drilling, or wire-fishing required! Cover plates will make it look like any other switch.

Cover ALL CONNECTIONS

All electrical connections not ending at a switch, fixture, or receptacle must be housed inside a covered junction box so they can't be disturbed. Electricians often use an existing light box as a junction box in which to splice a cable feeding a new fixture. When there's not enough room in an existing box, use a separate junction box to house the splices.

Ceiling FIXTURE ELEMENTS

In this basic setup, the ceiling box mounts to an adjustable bar, which is screwed to ceiling joists. The bracket is attached to the box and the fixture is screwed to the mounting bracket. All metal boxes and brackets must be grounded to be safe. Many electricians use grounding screws in both the box and the bracket, but one ground is sufficient. The metal mounting screws provide grounding continuity to box and bracket. If the fixture has a ground wire, it must be attached to either the ground wire in the box or to the grounding screw on the mounting bracket. The ground wire in the box must be long enough to attach to the ground screw in the box and extend out the box for attaching to the fixture ground wire.

Ceiling Fixture Elements

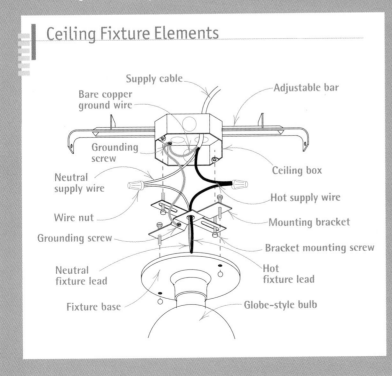

- Supply cable
- Bare copper ground wire
- Adjustable bar
- Grounding screw
- Neutral supply wire
- Ceiling box
- Wire nut
- Hot supply wire
- Grounding screw
- Mounting bracket
- Neutral fixture lead
- Bracket mounting screw
- Fixture base
- Hot fixture lead
- Globe-style bulb

Fixture Wiring

The right lighting fixture can transform a room; it could also cut your energy bills big-time. In this brief section, you'll get an overview of mounting devices, see how to install the most space-conserving fixture—recessed lighting—and learn how to transform an existing incandescent light fixture into one with a more efficient light-emitting diode (LED) light source, thereby reducing energy usage by up to 75%.

MOUNTING FIXTURES TO BOXES

As shown on pp. 308 and 316, there are many mounting options for boxes. The main choice is whether you nail or screw the box directly to a stud or ceiling joist or use an extendable mounting bar to which the box is attached. Either method works fine, but a box that slides along a mounting bar means you can more easily position the light fixture just where you want it.

If mounting screws on all light fixtures were exactly the same diameter and spacing as the screw holes on all boxes, life would be simple and you'd screw the fixture directly to the box. But there are many different box sizes and configurations, and light fixtures vary considerably. Consequently, there are many mounting brackets (shown on the facing page) to reconcile these differences.

Always examine existing outlet boxes before buying new fixtures, and make sure that fixture hardware can be mounted. Otherwise, a routine installation could turn into a long, drawn-out affair with a lot of trips to the hardware store.

Here's an overview of how various fixtures mount to outlet boxes. All metal brackets, boxes, and lamp fixtures must be grounded to be safe. There are special grounding screws (10-32 machine screws, colored green) that ensure a positive connection to metal boxes or plates when installed in a threaded hole. Do not use a wood screw or the like to attach the ground wire to the box; it doesn't provide a good enough connection.

Flat-mounting brackets. Typically, a mounting bracket is screwed to a fixture box, and the fixture is attached to the bracket, either by machine screws or, as is more common for chandeliers, by a threaded post that screws into a threaded hole in the center of the mounting bracket. Brackets can be as simple as a flat bar with screw slots, but some adjust by sliding, whereas others are offset slightly to provide a little more room for electrical connections—and fingers. Ring brackets can be rotated so the slots line up perfectly with outlet-box and fixture screw holes.

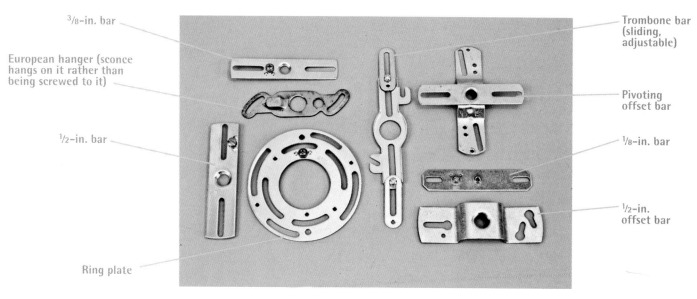

³/₈-in. bar

European hanger (sconce hangs on it rather than being screwed to it)

Trombone bar (sliding, adjustable)

Pivoting offset bar

¹/₂-in. bar

¹/₈-in. bar

¹/₂-in. offset bar

Ring plate

Mounting brackets

No brackets. Some fixtures, such as the recessed lighting fixture shown on pp. 340–341, don't require a mounting bracket. The fixture has its own junction box; once inserted into a hole cut in the ceiling, the fixture is supported by the ceiling. Integral clips and trim pieces pull the fixture tightly to the plaster or drywall ceiling. Recessed cans can be IC-rated (they may be covered with insulation) or may be non-IC-rated (cannot be covered with insulation).

RETROFITTING RECESSED LIGHTING

As the name implies, recessed lighting fixtures fit into the space above the ceiling. Recessed fixtures create a strong cone of light and are frequently used to illuminate work areas or tight spaces. If desired, installing recessed fixtures spaced closely together (half the ceiling height for a 6-in. can with standard baffle trim) will create even lighting rather than "pools" of light.

In retrofit installations, the supply cable to the recessed lighting unit typically comes from an existing ceiling box or nearby switch box. The supply cable feeds to an integral junction box on the fixture. Finding the nearest power source and fishing the wires to the fixture are always an adventure if there's not accessible space above. If the recessed fixture is a low-voltage unit, such as the one shown here, it will come with a transformer, which reduces the 120v current of the supply cable.

Recessed CEILING FIXTURES

Recessed light fixtures vary. The low-voltage model in the photo sequence on pp. 340–341 has a transformer at the end of its assembly to reduce line voltage. This drawing shows a model that runs on line voltage (120v), so it has no transformer. If the unit is watertight, it will have additional trim or lens elements. Closely follow the installation instructions provided with your fixture.

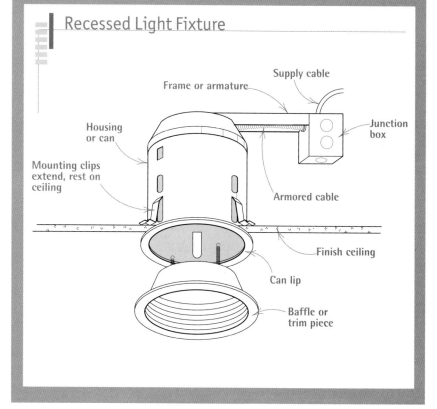

Recessed Light Fixture

Supply cable

Frame or armature

Junction box

Housing or can

Mounting clips extend, rest on ceiling

Armored cable

Finish ceiling

Can lip

Baffle or trim piece

1. There are no absolutes in placing lights, but a single recessed fixture centered in one direction or another usually looks good.

2. Before you commit to a location, drill a small exploratory hole to make sure there's room for the recessed can—i.e., that there's not a ceiling joist, pipe, duct, or wire in the way.

3. As you drill, keep the axis of the bit vertical and the sawblade square to the ceiling. Wear safety glasses.

Cutting a ceiling hole for a recessed fixture.
There's no absolute rule on where to place a recessed light, but in a small space, such as a shower alcove, a fixture centered in one direction or another will look best. In addition, you may want to use a stud finder to avoid hitting ceiling joists above.

Drill a pilot hole to see what's above and to make sure there's room for the can. Make the hole small because if there's an obstruction above it, you'll need to patch it. After drilling the hole, you can insert a 4-in. piece of bent wire and rotate it to see if it hits a ceiling joist.

The small pilot hole will keep the point of a hole-saw blade from drifting. Keep the drill vertical and the circle of the sawblade parallel to the ceiling. There are special carbide hole saws for drilling through plaster. A bimetal hole saw will also cut through drywall, but plaster will very quickly dull the saw in the process. Wear goggles.

If the hole saw is the right size for the can, you won't need to enlarge it. But for the light shown here, the saw was a shade too small, so the installer used a jab saw to enlarge the hole slightly. In a pinch, you can also use just a jab saw.

Test-fit the unit. Although you want the can to fit snugly, the unit's junction box and transformer also need to fit through.

4. If the can fits too snugly in the hole, use a jab saw to enlarge the hole slightly.

5. Test-fit the fixture. The gray box is an integral junction box that will house all wire connections. The black box about to enter the hole is the transformer.

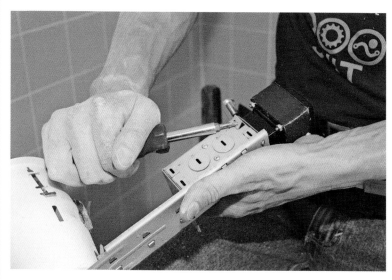

6. After running the cable through the hole you cut in the ceiling, remove a knockout in the fixture's junction box. The cable is not energized at this stage.

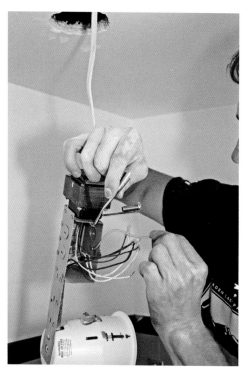

7. Strip sheathing from the cable, feed it into the box knockout, clamp the cable, and splice fixture wires to the supply wires.

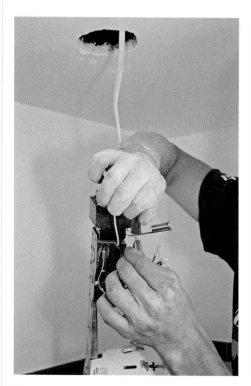

8. Whatever the device, always connect ground wires first.

9. When you have spliced all wires, tuck them carefully into the fixture junction box.

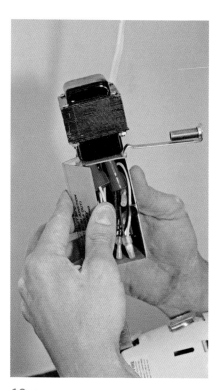

10. Close the junction box cover. Make sure no wires are pinched between the cover and box.

PRO TIP

Recessed ceiling lights that are IC-rated can be covered with insulation. Non-IC-rated cannot. See pp. 425–426 for more information.

PRO TIP

When installing halogen lamps, the pros apply a tiny dab of antioxidant paste to the lamp pins before seating them in a fixture socket to prevent oxidation.

Wiring a recessed fixture. To wire a recessed fixture, remove a Romex knockout from the unit's integral junction box. Inside the knockout, there is a spring-loaded, strain-relief clamp that will grip the incoming cable, so you don't need to insert a Romex connector. Run a length of (unconnected) Romex cable from the nearest power source and feed it into the knockout just removed. Of course, the cable must *not* be energized when you are working on it. (To wire the box with AC or MC cable, remove one of the circular ½-in. knockouts and insert an appropriate connector.)

Inside the fixture's junction box will be two sets of wires that were spliced at the factory. They connect the secondary wires that run from the transformer to the socket. (At the transformer, the current is reduced from 120v to 12v, so polarity is no longer an issue.) There also are three unconnected fixture leads in the box to which you'll splice the supply wires.

Using wire connectors, connect the incoming ground wire to the green fixture lead, the incoming neutral to the white lead, and the hot wire to the black fixture lead.

Tuck the spliced wire groups into the fixture junction box. At the right of photos 8, 9, and 10 on p. 341 is a piece of threaded rod that can be adjusted to support the transformer at the correct height. Snap the junction box cover closed. As with other outlet boxes, code determines the number of wires you can splice in a fixture junction box, based on the cubic inches in the box.

Securing the can. Once the recessed lighting fixture has been wired, push the fixture into the hole, being careful not to bind the Romex cable as you do so. If the fit is snug, use the side of your fist to seat the lip of the fixture flush to the ceiling.

Use a screwdriver to push up the spring-loaded clips that pivot and press against the back of the drywall and hold the fixture snugly in place. To remove the fixture later, pop the clips out.

Insert the bulb (the fixture shown uses an MR16 bi-pin halogen bulb) into the socket. Note that the installer is gripping the lamp's reflector, not the bulb itself. The lamp pins should seat securely. Install the trim piece—this one has a watertight gasket. Snap in the lamp and socket, and push the assembly up into the can. The three arms on the side of the assembly will grip the inside of the can.

11. Feed the cable and the now-wired fixture into the hole.

12. The lip of the can should sit flush to the ceiling; use your fist as needed to seat it.

13. Use a screwdriver to push up the spring-loaded clips that pivot and press against the back face of the drywall to hold the fixture snugly in place.

14. Insert the bulb with a gentle pressure until all pins seat. This is a low-voltage halogen bulb. Wear gloves or hold the bulb by the rim to avoid getting contaminants on the bulb.

15. Install the trim assembly. The black ring in this assembly is a moistureproof gasket.

CONVERTING AN INCANDESCENT FIXTURE TO AN LED

It probably takes more time to write about retrofitting an LED fixture than to do it. Turn off the power to the fixture and use a voltage tester to be sure it's off. Remove the fixture's cover plate and unscrew the incandescent bulb. Screw the threaded adapter into the fixture socket. The other end of the adapter is an orange plastic, quick-disconnect connector that snaps to a matching connector on the LED housing. Snap the two cast-aluminum pieces of the housing and cover together, then screw the cover to the recessed can with three screws.

Similar retrofit kits fit either 3-in. or 4-in. cans. Although all come with some means of dissipating excess heat, the model shown at right, from DMF Lighting, has a heat sink. The 650-lumen lamp is dimmable to 5% and is as bright as a 50-watt incandescent; it consumes less than 12 watts.

Portfolio of Wiring Schematics

The diagrams in this section show a few of the more common circuit wiring variations that you're likely to need when wiring receptacles, fixtures, and switches. (See also the schematics on pp. 326–334.) Unless otherwise noted, assume that incoming cable (from the power source) and all others are two-wire cable with ground, such as 14/2 w/grd or 12/2 w/grd (#14 wire should be protected by no larger than a 15-amp breaker or fuse; #12 wire should be protected by no larger than a 20-amp breaker).

All metal boxes must be grounded. Assume that nongrounded boxes in the wiring diagrams are nonmetallic (plastic) unless otherwise specified. In sheathed cables, ground wires are bare copper. Black and red wires indicate hot conductors. White wires indicate neutral conductors, unless taped black to indicate that the wire is being used as a hot conductor in a switch leg.

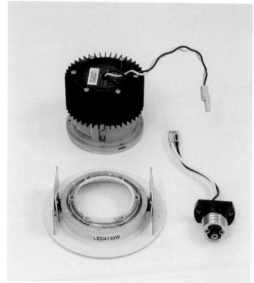

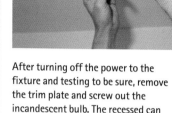

After turning off the power to the fixture and testing to be sure, remove the trim plate and screw out the incandescent bulb. The recessed can stays put.

This simple kit can convert an incandescent ceiling fixture into an energy-saving LED fixture. Clockwise from lower right: a screw-in adaptor, cast-aluminum trim plate and lens, lamp housing, and heat sink.

Snap the housing and trim plate together, screw in the adapter, connect the (orange) quick-disconnect connector, and slide the assembly into the recessed can. Final mounting details will vary, but clips or universal mounting screws will hold the assembly to the can.

Light Fixture at End of Cable Run

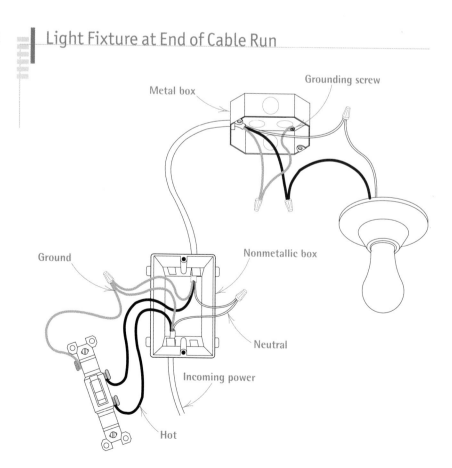

A Simple Switch with a "Switch Loop": the Modern Method

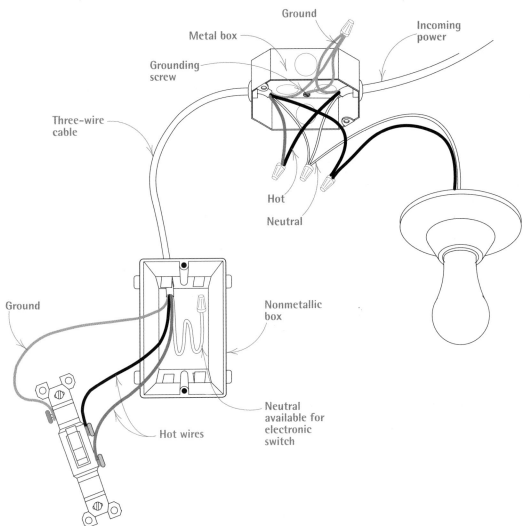

Ground

Metal box

Grounding screw

Incoming power

Three-wire cable

Hot

Neutral

Ground

Nonmetallic box

Hot wires

Neutral available for electronic switch

Close-up: Three-Way Switch

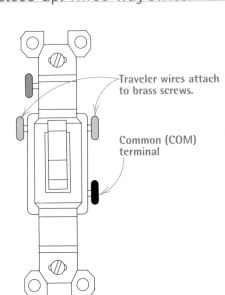

Traveler wires attach to brass screws.

Common (COM) terminal

Three-way switches control a fixture from two locations. Each switch has two brass screws and a black screw (common terminal). The hot wire from the source attaches to the common terminal of the first switch. Traveler wires between the switches attach to the brass screws. Finally, a wire runs from the common terminal of the second switch to the hot lead of the fixture.

Three-Way Switches, Light Fixture Between

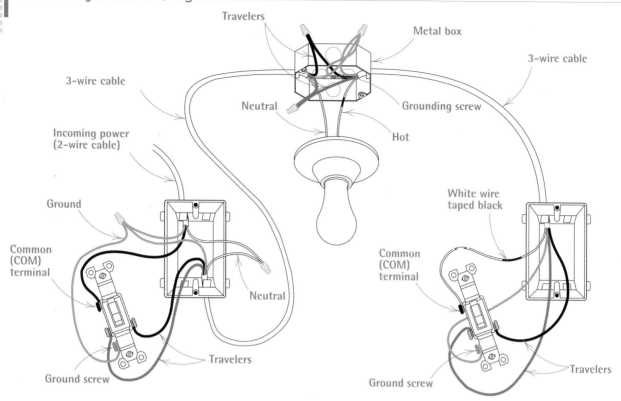

Travelers

Metal box

3-wire cable

3-wire cable

Neutral

Grounding screw

Incoming power
(2-wire cable)

Hot

Ground

White wire
taped black

Common
(COM)
terminal

Common
(COM)
terminal

Neutral

Travelers

Ground screw

Travelers

Ground screw

In this setup, two three-way switches control a light fixture placed between them. Run 3-wire cables between each switch and fixture. Whenever you use a white wire as a switch leg, tape it black to indicate that it's hot. Although this wiring configuration does not meet NEC 2017 requirements for new work, it is very common in older homes and is acceptable as part of existing wiring.

Three-Way Switches: Light Fixture at Start of Cable Run

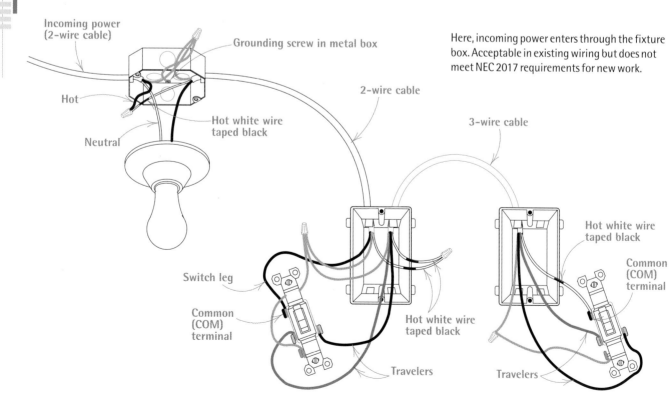

Incoming power
(2-wire cable)

Grounding screw in metal box

Here, incoming power enters through the fixture box. Acceptable in existing wiring but does not meet NEC 2017 requirements for new work.

Hot

2-wire cable

Hot white wire
taped black

3-wire cable

Neutral

Hot white wire
taped black

Switch leg

Common
(COM)
terminal

Common
(COM)
terminal

Hot white wire
taped black

Travelers

Travelers

12 Plumbing

Builders have benefited greatly from the standardization of building materials, and nowhere is this truer than in plumbing. While plumbers once fashioned waste systems from cast iron, oakum, and melted lead, today one needs little more than plastic pipe and solvent-based cement. Better technology enables more people to understand, repair, and install plumbing.

Would-be plumbers should do two things:

▶ **Learn the vocabulary. Some people feel intimidated by the plethora of plumbing terms, especially fitting names. But there's actually a logic to all those names, once you learn what a part does and why it is shaped as it is. Besides, you'll get better service from plumbing-supply clerks if you can speak their language.**

▶ **Consult local plumbing codes before beginning a project. Codes protect your health and that of your neighbors. They spell out when you need permits, what materials you may use, and at what stages the work must be inspected. There is no national code, so most local building departments often follow the Uniform Plumbing Code (UPC) or the International Residential Code (IRC). Get a copy of local plumbing codes from your building department.**

Recommended further reading is Rex Cauldwell's *Plumbing Complete* (The Taunton Press, 2009).

An Overview of Plumbing Systems

A plumbing system is a loop of sorts, created by *supply* (or delivery) pipes that carry potable water to the house and its fixtures and by *drainage, waste, and venting* (DWV) pipes that carry wastewater, effluvia, and sewer gases away from the *fixtures*—sinks, toilets, lavatories, and washing machines.

Flexible PEX tubing is fast overtaking rigid copper and PVC water-supply systems. PEX is faster, easier, and less expensive to install, particularly the soft PEX-A tubing shown, which readily uncoils so it's a snap to feed through holes and snake through walls.

The Water-Supply System

Branch lines

Risers

Cold-water trunk line

Hot-water trunk line

Service pipe

Main supply pipe

Water heater

Originating at a service pipe from the street (or from a well), the main supply pipe splits at a T-fitting, with one leg feeding cold-water trunk lines and the other entering the water heater to emerge as the hot-water trunk line.

These two systems within a system are quite different from each other. DWV pipes are larger and must slope downward so waste can fall freely (by gravity) and sewage gases can rise through vents. Consequently, large DWV pipes can be difficult to route through framing. By contrast, smaller water-supply pipes are easy to run through studs and joists, and they deliver water under pressure, so there's no need to slope them.

THE WATER SUPPLY

The pipe that delivers water to a house (from a city water main or an individual well) is called the *service pipe*. So it won't freeze, a service pipe must run below the frost line and enter a building through the foundation. Typically, a 1-in. service pipe is controlled by a *main shutoff valve* shortly after it enters a building; municipal hookups may enter a water meter first. Plumbing codes also may require a *pressure-reducing valve* if water pressure is more than 80 pounds per square inch (psi).

On the other side of the shutoff valve, the service pipe continues as the *main supply pipe*, commonly ¾ in. in diameter. At some point, the main supply pipe enters a tee fitting where it splits. One leg continues on as a *cold-water trunk line* and the other feeds into the water heater, where it emerges as the *hot-water trunk line*. From the ¾-in. hot and cold trunks run various ½-in. *branch lines* that serve fixture groups. Finally, individual *risers* (supply tubes) run from branch lines to fixtures. Risers are ⅜ in. or ½ in. in diameter and connect to fixtures with threaded fittings. By decreasing in diameter as they get

farther from the trunk lines, supply pipes help maintain constant water pressure.

Before the 1950s, supply pipes were usually galvanized steel, joined by threaded fixtures, but steel pipes corrode and corrosion constricts flow. Consequently, rigid copper piping, which corrodes more slowly, soon replaced galvanized pipe. Joined by *sweat fitting* (soldering), copper was also easier to install and has been the dominant supply piping since the 1950s. Rigid plastic pipe, especially chlorinated polyvinyl chloride (CPVC), has gained market share because it is corrosion resistant, less expensive than copper, and easily assembled with solvent cement. But it may be cross-linked polyethylene (PEX) flexible piping that will finally dethrone King Copper (PEX is discussed later in this chapter).

In contrast to the trunk-and-branch distribution common to rigid piping, flexible PEX tubing systems feature a ¾-in. main water line that feeds into a *manifold*, out of which small supply lines run directly to fixtures. Because it doesn't rely on large pipes to distribute water, this *home-run manifold* system delivers hot water quickly to fixtures, saving water and energy.

DRAINAGE, WASTE, AND VENTING

The DWV system carries wastes and sewage gases away from the house.

Every fixture has a drain trap designed to remain filled with water after the fixture empties. This residual water keeps sewage gases from rising into living spaces. (Toilets have integral traps.) As trap arms leave individual fixtures, they empty into branch drains or directly into a soil stack, which, at its base, turns and becomes the main drain. The main drain then discharges into a city sewer main or a septic tank.

Drainpipes also may be differentiated according to the wastes they carry: Soil pipes and soil stacks carry fecal matter and urine, whereas waste pipes carry wastewater but not soil. Stacks are vertical pipes, although they may jog slightly to avoid obstacles.

Venting is the "V" in DWV. Without venting, wastes would either not fall at all or, in falling, would suck the water out of fixture traps, allowing sewage gases to enter living spaces. Vents admit an amount of air equal to that displaced by the falling water. Thus every fixture must be vented. In most cases, the trap arm exits into a tee fitting whose bottom leg is a branch drain and whose upper leg is a branch vent. Branch vents continue upward, often joining other fixture vents, until they join a vent stack, which exits through the roof.

Because vents must admit enough air to offset that displaced by falling water, vents are approximately the same size as their companion drains. Branch vents and drains are usually 1½-in. or 2-in. pipes, and main stacks and drains are 3 in. Minimums are indicated in "Minimum Drain, Trap, and Vent Sizes" on p. 367. *Important:* Drainpipes must slope downward at least ¼ in. per ft. so that waste will be carried out; vent pipes usually slope *upward* a minimum of ⅛ in. per ft.

DWV pipes may be of any number of materials. An older house may have drain and vent pipes with sections of cast iron, galvanized steel, copper, or plastic. Because some of these materials also are used for supply, let size be your guide:

The DWV System (Drainage, Waste, Venting)

Stack vent

Branch vent

2 in.

Vent stack

1½ in. 1½ in.

2 in.

2 in.

Branch drain

2 in. 2 in.

Soil stack

1½ in.

3 in. minimum

To sewer or septic tank

Main drain or building drain

Drainpipes must slope downward at least ¼ in. per ft. so wastes can fall freely. Vent pipes must slope upward at least ⅛ in. per ft. so sewage gases can rise and exit the building.

A Drain Trap

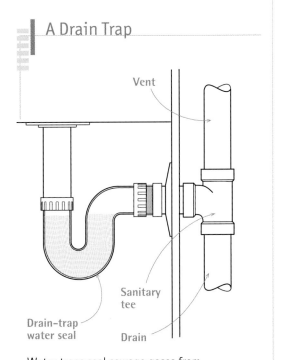

Vent

Sanitary tee

Drain–trap water seal

Drain

Water traps seal sewage gases from living spaces, but they need vents to operate properly. Without incoming air from the vent, falling wastes could suck the water out of traps.

If an existing pipe's diameter is 1¼ in. to 4 in., it's a drain or vent pipe. DWV pipes installed these days are mostly plastic: white polyvinyl chloride (PVC) or black acrylonitrile butadiene styrene (ABS). A host of ingenious fittings enable the various materials to be connected tightly together. *Note:* If sound suppression is an issue, you should insulate plastic pipes or install cast iron.

Planning

If you'll be adding or moving fixtures, you'll need to install pipes, and that will require permits and planning. Start by assessing the condition of existing pipes (see chapter 1), which you can connect to if they are in good shape. Create a scale drawing of proposed changes, assemble a materials list, and then ask a plumbing-supply store clerk or a plumber to review both. If you're well organized, clerks at supply stores will usually be glad to help. However, if you need help understanding your existing system, hire a plumber to assess your system. He or she can also explain how to apply for a permit and which inspections will be required.

IS THERE ENOUGH ROOM?

If you're adding a bathroom, first consider the overall size of the room. If there's not enough space, you may need to move walls. Layouts with pipes located in one wall are usually the least disruptive and most economical because pipes can be lined up in one plane. On the other hand, layouts with pipes in three walls are rarely sensible or feasible unless there's unfinished space above or below in which to run pipes.

FIXTURE ROUGH-IN DIMENSIONS

Once you have a general idea of whether there's enough room, focus on the code requirements for each fixture, which dictate where fixtures and pipes must go. It would be aggravating and expensive if code inspectors insisted that you move fixtures after finish floors and walls had been put in place. So install fixtures and pipes to conform with code minimums. The drawings on pp. 350–351 show typical drainpipe and supply-pipe centers for each fixture and, in most cases, minimum clearances required from walls, cabinets, and the like.

On toilets, the horn—the integral porcelain bell protruding from the bottom—centers in the floor flange. The flange, and the closet bend to which it attaches, should be centered 12 in. from the finish wall for most toilets or 12½ in. from an exposed stud wall. Codes require at least 15 in. clearance from the center of the toilet to walls

or cabinets on both sides: In other words, install toilets in a space at least 30 in. wide. There also must be at least 24 in. of clear space in front of the toilet.

Toilets with 10-in. and 14-in. rough-in dimensions are available to resolve thorny layout issues (such as an immovable beam underneath) or to replace nonstandard toilets. For example, if you replace an old-fashioned toilet with a wall-mounted ceramic tank with a standard (close-coupled) unit, there could be an ugly gap between the toilet and the finish wall. By installing a *14-in. rough toilet*, whose base is longer, you can use the existing floor flange and eliminate the gap behind the toilet.

Install water-supply risers on the wall behind the toilet, 6 in. above the floor and 6 in. to the left of the drainpipe. If there's a functional riser sticking out of the floor, use it. But floor risers are seldom installed today because they make mopping

Minimum Bathroom Dimensions

PIPES IN ONE WALL

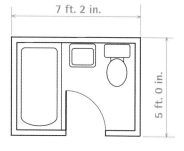

7 ft. 2 in.

5 ft. 0 in.

PIPES IN TWO WALLS

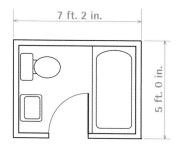

7 ft. 2 in.

5 ft. 0 in.

PIPES IN THREE WALLS

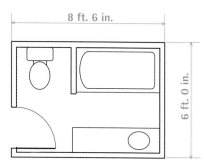

8 ft. 6 in.

6 ft. 0 in.

FIVE FIXTURES

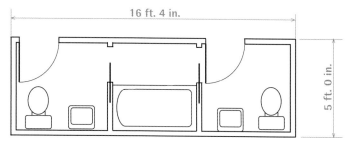

16 ft. 4 in.

5 ft. 0 in.

Toilet Rough-In Dimensions

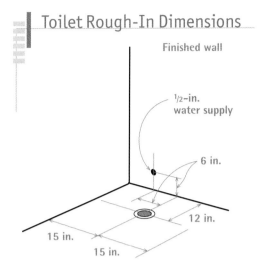

Finished wall

½-in. water supply

6 in.

12 in.

15 in.

15 in.

Center the toilet drain 12 in. from the finished wall behind the unit. Allow at least 15 in. of clearance on both sides of the toilet, measured from the center of the drain.

Lavatory Rough-In Dimensions

½-in. water supply

4 in. to 8 in.

1½-in. drain

18 in.

24 in.

Lavatory centerline

the floor difficult. Clearances around bidets are the same as those for toilets.

Lavatories and pedestal sinks should be a comfortable height for users. Typically, lavatory rims are set 32 in. to 34 in. above the finish floor, but if a family is tall, raise the lav. (If you do, remember to raise drain and supply-pipe holes an equal amount.) Codes require at least 18-in. clearance in front of a sink; 24 in. is better.

Lavatory drains are typically 18 in. above the floor and centered under the lavatory, although adjustable P-traps afford some flexibility in positioning drains. Center supply pipes under the lav, 24 in. above the floor, with holes spaced 4 in. on center. Pedestal sink drains are housed in the pedestal, so tolerances are tight; follow the manufacturer's installation instructions when positioning pipes.

Bathtubs and showers vary greatly, so follow the manufacturer's guides when positioning the pipes. Most standard tubs are 30 in. to 32 in. wide and 5 ft. to 6 ft. long. Codes require a minimum of 18 in. clearance along a tub's open side(s); 24 in. is better.

Freestanding tubs have exposed drain and overflow assemblies, so their 1½-in. drains can be easily positioned to avoid joists and other design constraints. Standard tubs require a hole approximately 12 in. by 12 in. in the subfloor under the tub drain end to accommodate the drain and overflow assembly. If an existing joist is in the path of the tub drain, you may need to

cut through the joist and add doubled headers, as explained further on p. 373.

Positioning supply pipes and valve stems is easier because they're smaller and typically centered on an end wall—although, again, follow the manufacturer's rough-in dimensions for code-required pressure-balancing valves and the like. Place the shower arm 72 in. to 78 in. above the floor so taller users won't need to stoop when taking a shower. Place the tub spout 22 in. high. Tub faucet handles (and mixing valves) are customarily 6 in. above the spout.

Kitchen sinks frequently have double basins, so you can center or offset drainpipes. In a standard 36-in.-wide base cabinet, the drain often is offset so that it is 12 in. from one cabinet sidewall, leaving room to hook up a garbage disposal. To make cabinet installation easier, have the drain exit into the wall rather than the floor. A drain that exits 15 in. above the finish floor will accommodate the height of a garbage disposal (11 in.) and the average depth (9 in.) of a kitchen sink. Sink faucet holes are typically spaced 8 in. on center, so align supply pipes with their centerline, roughly 2 in. above the drain height. Supply-pipe height is not critical because risers easily accommodate varying heights.

SKETCHING LAYOUTS

Make a separate sketch of each floor's plumbing; include the basement and attic, too. Start by creating an accurate outline of the house's footprint, using graph paper and a scale of ¼ in. per 1 ft.

Tub/Shower Rough-In Dimensions

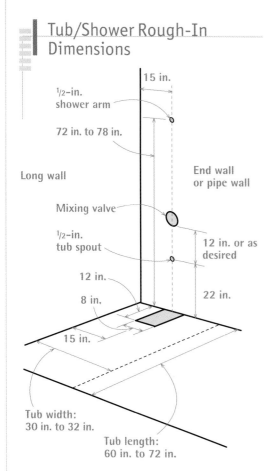

- 15 in.
- ½-in. shower arm
- 72 in. to 78 in.
- Long wall
- End wall or pipe wall
- Mixing valve
- ½-in. tub spout
- 12 in. or as desired
- 12 in.
- 8 in.
- 22 in.
- 15 in.
- Tub width: 30 in. to 32 in.
- Tub length: 60 in. to 72 in.

Kitchen Sink Rough-In Dimensions

- 36-in. cabinet height
- ½-in. water supply
- 12 in.
- 8 in.
- 15 in. or less
- 18 in.
- 1½-in. drain offset

Double sinks are most often installed in kitchens, so the drain is often offset under one sink, as shown in the drawing on p. 380. Braided stainless-steel supply lines are very flexible, so you can rough-in water supply stub-outs at any convenient height; 18 in. is common.

The tub drain and stubs in the end wall should be centered 15 in. from the long wall. Mixing valves are typically set 12 in. above the tub spout, whereas individual valve stems are set 6 in. above the spout, 8 in. o.c., or follow the manufacturer's recommended rough-in specs.

Then use tracing-paper overlays for each floor's plumbing layout. Indicate existing fixtures, drains, supply pipes, water-using appliances, and the water heater. Where pipes are exposed, note the size and dimension of drains and stacks and where the supply pipes exit into the floor above. Especially note the location of 3-in. main drains and vents. If you can cluster fixtures around larger DWV pipes within a room—or from floor to floor—you'll shorten the distance that fixture drains must travel and thus reduce the amount of framing you may have to cut or drill when running the new pipes.

If you're moving or adding fixtures, make separate floor sketches for them, too. By laying tracing-paper sketches of old and new plumbing atop each other, you can quickly see if fixtures cluster and, if you're adding fixtures to an existing system, the closest part of a drain or supply

Framing Considerations

After positioning fixture drains, see if there's a joist in the drain path. If there is, and you can't reposition the fixture, cut through the joist and install doubled headers to redistribute the load. If possible, avoid running larger drainpipes perpendicular to joists and studs because drilling and cutting weaken the framing. But if drill you must, "Drilling and Notching Studs and Joists" on p. 374 shows acceptable hole sizes and locations; "Maximum Sizes for Holes and Notches" on p. 373 also will be helpful.

The trickiest pipe to route is a 3-in. drain, whose outer diameter is 3½ in. If that pipe runs 12 ft. horizontally, sloping ¼ in. per ft., it will drop 3 in. during its run. If it runs between enclosed joists, the pipe will need at least 6½ in. in height—plus the height of any fittings. If floor joists are nominally 2x8s (actually, 1½ in. by 7½ in.), things could get pretty tight. When planning pipe runs, consider pipe dimension, slope, space for fittings, and the actual size of the lumber in your calculations.

Plumber's Isometric Sketch of Three-Fixture Bathroom

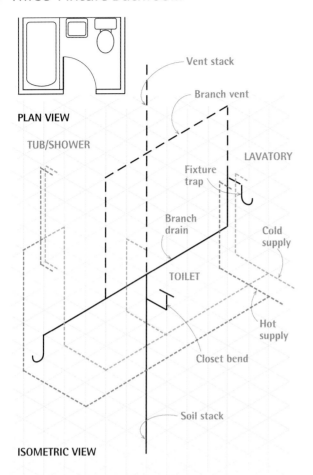

PLAN VIEW

TUB/SHOWER

Vent stack

Branch vent

LAVATORY

Fixture trap

Branch drain

Cold supply

TOILET

Hot supply

Closet bend

Soil stack

ISOMETRIC VIEW

Try to obtain sheets of plumber's isometric paper so you can show bathroom rough-ins in three dimensions. Art- or engineering-supply stores may carry the paper, but the Internet is probably a better bet.

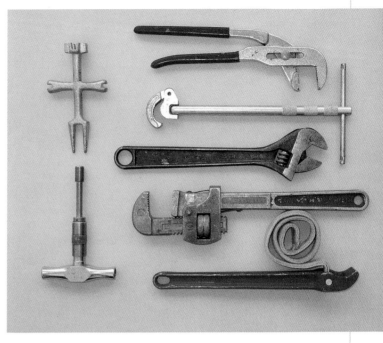

Jaws VII: wrenches and pliers. Clockwise, from upper left: tub-strainer wrench, slide-nut or sliding-jaw pliers, basin wrench, adjustable wrench, pipe wrench, strap wrench (won't mar polished pipe), and no-hub torque wrench.

pipe to connect to and extend from. Plumbers use isometric paper to draw pipe runs, as shown in "Plumber's Isometric Sketch of a Three-Fixture Bathroom" at left, but any to-scale sketch will give you an approximate idea how long pipe runs will be. Sketches also tell you where you'll need fittings because the pipes change direction, connect to branches, or decrease in size.

Tools

With a modest tool collection, you'll be ready for most plumbing tasks.

Pipe wrenches tighten and loosen threaded metal joints, such as ¾-in. nipples (short pipe lengths) screwed into a water heater, galvanized pipe unions, and so on. A pair of 10-in. or 12-in. pipe wrenches should handle most tasks. Get two: Most of the time, you'll need one wrench to hold the pipe and the other to turn the fitting.

Adjustable wrenches (also called Crescent wrenches) have smooth jaws that grip but won't mar chrome nuts and faucet trim. Get several: A 4-in. adjustable wrench is right for the closet bolts that anchor toilet bowls, a 12-in. wrench gives extra leverage for stubborn nuts, and an 8-in. wrench is appropriate for almost everything else.

Strap wrenches aren't a must-have tool but are useful when you need to grip polished pipe without scarring it.

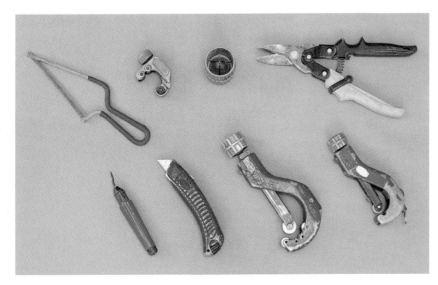

Cutting and reaming tools. Top row, from left: miniature hacksaw, close-quarters cutter, combo chamfer and reamer (cleans burrs from pipe ends after cutting), and aviation snips. Bottom row, from left: reamer, utility knife, large-wheeled tubing cutter (cuts up to 2-in. plastic pipe), and wheeled tubing cutter. The cutting wheels can be changed for different pipe materials.

Slide-nut (sliding-jaw) pliers are good utility tools for holding nuts, loosening pipe stubs, and holding a pipe section while it's being soldered.

Locking pliers (or Vise-Grip pliers) adjust and clamp down on fittings, for example, so you can have both hands free to hold a torch and apply solder.

Basin wrenches are about the only tools that can reach water-supply nuts on the underside of sinks and lavs, where supply pipes attach to threaded faucet stems.

Tub-strainer wrenches tighten tub strainer and tailpiece assemblies.

No-hub torque wrenches tighten stainless-steel band clamps on *no-hub couplings*. Many plumbers use a cordless drill/driver to do most of the tightening, but code requires that final tightening be done by hand.

Pipe cutters (also called wheeled tubing cutters) are the best tools for a clean, square cut on copper pipe. Tighten the cutter so that its cutting wheel barely scores the pipe, then rotate the tool around the pipe, gradually tightening until the cut is complete. Many types have a foldaway deburring tool. Use a *close-quarters cutter* (thumb cutter) where there's no room for a full-size one. If you're installing CPVC plastic supply pipe, use *tubing shears* for clean, quick cuts. A hacksaw works but not as well.

A reaming tool (if your cutter doesn't have one attached) is used to clean metal burrs after cutting copper. Use a *round wire brush* to polish the inside of copper fittings after reaming and *plumber's sand cloth* to polish the pipe ends. If you're cutting plastic pipe, use a *rounded file* to remove burrs—the steel jaws of an adjustable wrench also work well for deburring plastic pipe.

Wide-roll pipe cutters open wide to receive the larger diameters of plastic DWV pipe. *Plastic-pipe saws* have fine teeth that cut ABS and PVC pipe cleanly—and squarely, if used with a miter box. If you need to cut into cast iron, rent a *snap cutter*, also known as a cast-iron cutter. It's the only tool that cuts cast iron easily. Some models have ratchet heads for working in confined places.

A cordless drill and a cordless reciprocating saw are must-haves if you're working around metal pipes that could become energized by electricity and when working in tight, often damp crawlspaces. Old lumber can be hard stuff to drill or cut, so 14.4v cordless tools are minimal. Cordless drills are perfect for attaching plumber's strap, drilling holes in laminate countertops, and so on.

Miscellaneous tools. Clockwise, from left: torpedo level, hammer, 14.4v cordless drill, flint and steel striker (lights torch), gas soldering torch, tape measure, and plumber's sand cloth (used before fluxing pipe).

Water-supply fittings. First column, from top: 3/4-in. 90, 3/4-in. street 90, 3/4-in. street 45, and 3/4-in. 45. Second column, from top: 3/4-in. CxF (copper-by-female) drop-ear 90, 3/4-in. CxF adapter, 3/4-in. CxM (copper-by-male) adapter, and 3/4-in. cap. Third column, from top: 3/4-in. tee, 3/4-in. by 1/2-in. tee, 3/4-in. by 1/2-in. by 3/4-in. tee, and 3/4-in. by 1/2-in. by 1/2-in. tee. Fourth column, from top: 3/4-in. coupling, 3/4-in. by 1/2-in. reducing coupling, 3/4-in. CxF union, and 3/4-in. dielectric union.

If you need to drill 2-in. (or bigger) holes, use a corded drill. Heavy-duty drilling takes sustained power and more torque than most cordless drills have. A 1/2-in. right-angle drill (see p. 61) supplies the muscle you need in close quarters.

Once you have polished metal pipe ends, don't touch them or the insides of fittings with your bare skin; skin has oils that may prevent solder from adhering to the surface. Wear clean, disposable plastic gloves whenever handling, cutting, and soldering pipe.

Propylene gas torches are somewhat more expensive than the propane torches popular with do-it-yourselfers. Either type of torch can solder ½-in. or ¾-in. fittings required for copper pipes, but propylene, being hotter, will do the job faster. That's especially true if there's a small amount of water left in a pipe.

Nonasbestos flame shields protect wood framing when soldering joints. It is also important to have a fire extinguisher nearby.

Your plumbing kit should also include a handful of other tools. *Aviation snips* are used for cutting perforated strap and trimming gaskets, and

a *torpedo level* helps with leveling stub-outs (pipe stubs protruding into a room), sinks, and toilet bowls. You'll also want a hacksaw, screwdriver with interchangeable magnetic bits, utility knife, and a hammer.

Copper Water–Supply Pipe

This section focuses largely on installing rigid copper water pipe. It's strong, easily worked, approved by virtually all codes, and represents more than three-quarters of residential installations. Flexible PEX tubing, which will be discussed later in the chapter, is gaining market share quickly, especially in renovations.

FITTINGS

If you divide fittings into a few categories, their many names start to make sense. Because they do similar things, supply-pipe and DWV fittings often share names.

Fittings join pipes. The simplest fitting is a *coupling*, which joins two straight lengths of pipe. A *reducing coupling* joins different size pipes. A

Plumbing Safety

▶ Get a work permit and a copy of current plumbing codes from your local building department. Follow the codes closely; they're there to protect you.

▶ Get a tetanus shot before you start, and dress for dirty work.

▶ Wear protective eyewear when using power tools, chiseling, soldering, and striking with hammers—in short, for most plumbing tasks. Wear heavy gloves when handling drainpipe and disposable plastic gloves when working with solvent-based cements or soldering. Wear a respirator (not a mere dust mask) when soldering or working around existing soil pipes; P100 filters are the standard protection.

▶ Use only cordless power tools when cutting into supply pipe. If a corded power tool shorts out in that situation, it could be fatal. Before cutting into finish surfaces, shut off the electrical power to nearby outlets, and test with a voltage tester, as shown on p. 294, to be sure power is really off.

▶ Ensure good ventilation when joining pipes because heated solder and solvent-based cements give off noxious fumes. Make sure you have adequate lighting.

▶ When soldering joints in place, place a nonasbestos flame shield behind the fittings to avoid igniting the wood framing. Have a plant spritzer, filled with water, on hand to dampen the wood if you must solder fittings close to framing; make sure there's a fire extinguisher on site. Molten flux or solder can burn you, so be careful.

▶ When connecting to existing DWV pipes, plan the task carefully. Flush pipes with clean water beforehand, and have parts ready so that you can close things up as soon as possible. To avoid weakening nearby joints, be sure to support pipes before cutting them.

▶ Be fastidious about washing thoroughly after handling contaminated waste pipes and chemicals.

▶ If you smell gas, stop working: Running equipment or soldering could spark an explosion. If you can quickly locate the gas shutoff valve outside, shut it off. In any event, clear everyone from the house at once and call the local gas utility.

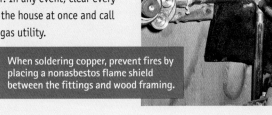

When soldering copper, prevent fires by placing a nonasbestos flame shield between the fittings and wood framing.

Pipe Fitting

Socket depth

Alignment marks

When measuring pipe, allow for socket depths. Also when dry-fitting pipe assemblies, draw alignment marks on pipes and fittings to help you point the fittings in the right direction when assembled. This is particularly helpful when giving plastic fittings one-quarter turn after glue is applied.

Socket Depths of ABS/PVC Fittings*

SOCKET DIAMETER (in.)	SOCKET DEPTH (in.)
1½	¾
2	⅞
3	1½

Fitting sockets vary; always measure depth to be sure.

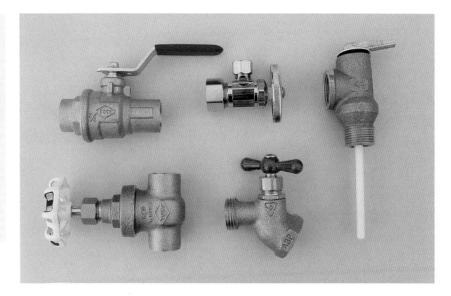

Water-supply valves. Clockwise, from upper left: Lever-handle ball valve, ⅝-in. by ⅜-in. angle stop, TPR valve, female hose bib, and gate valve.

repair coupling has no internal stop midway, so it can slide all the way onto a pipe, then slide back over a new piece of pipe inserted to repair a damaged section. A *union* is a coupling you can disconnect.

Fittings change direction. The most common directional fitting is a *90° elbow*, also known simply as a 90 or an ell. For a more gradual turn, use a *45° elbow*, also called a 45 or a ⅛ bend. A *street ell* is a 90° elbow with one hubless end, which can fit directly into the hub of another fitting. Ditto, a *street 45*.

Tees join three pipes. *Tees* (also spelled T's) allow you to run branch pipes to individual fixtures or fixture groups. *Reducing tees* accept different size pipes. If you want to sound like a pro, "read a tee" by noting its run (length) dimension first (in inches), then its branch leg. If both ends of the run are the same size, mention that number only once, as in ¾ by ½. But if two legs of a tee reduce, cite all three of the tee's dimensions, for example: a ¾ by ½ by ½.

Adapters join different types of pipe. A *sweat/male adapter* has a soldered end and a threaded end. A *sweat/female adapter* has a threaded receiving end. Adapters are also called *transition fittings* because they allow a transition in joining, as just described, or a transition in pipe materials. A *dielectric union* can join galvanized and copper, without the electrolytic corrosion that usually occurs when you join dissimilar metals.

Valves are specialized fittings with moving parts. *Gate valves* are the most common type of *shutoff valve*, although *lever-handled ball valves* are gaining popularity because they are easier to operate. *Hose bibs* have a threaded outlet for attaching a garden hose. *Angle stops* are shutoff valves that control water flow to lavatories, sinks, bidets, and toilets. Temperature- and pressure-relief (TPR) valves are spring-loaded safety valves that keep water heaters from exploding should the water get too hot or the tank pressure too great.

WORKING WITH COPPER SUPPLY PIPE

Type M rigid copper is the most commonly used copper supply pipe in houses, although type L, which is thicker, may also be specified. Type K, the thickest of the three, is usually specified for commercial and industrial jobs.

To cut rigid copper, place a tubing cutter on the pipe so that its cutting wheel is perpendicular to the pipe. Score the pipe lightly at first, until the cutting wheel tracks in a groove. Gradually tighten the cutting jaw as you rotate the tool, until the wheel cuts all the way through. If you tighten the

Measuring and Fitting Pipe

When measuring water-supply or DWV pipe runs, keep in mind that most pipe slides into fitting sockets. The depth of the socket is its *seating distance* (seating depth), which you must add to the face-to-face measurements between pipe fittings. When running pipe between copper fittings with a seating depth of ½ in., for example, add 1 in. to the overall measurement. Rigid ¾-in. copper fittings have a ¾-in. seating depth.

As important, after you dry-fit pipes so fittings point in the correct direction, use a grease pencil or a builder's crayon to create alignment marks on the pipes and fittings. That way you'll be sure the fittings are pointing in the right direction when you make the final connections. Alignment marks are particularly important when cementing plastic pipe because you must turn plastic pipes one-quarter turn after inserting them into fittings; the marks tell you when to stop turning.

CUTTING AND SOLDERING COPPER

1. Hold the tubing cutter square to the pipe and score it lightly at first. Once the cutting wheel tracks in a groove, gradually tighten the cutting jaw as you revolve the tool.

2. Ream the inside of the pipe to remove burrs left by cutting.

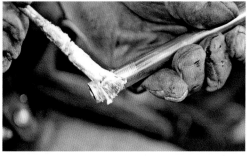

3. Use a strip of plumber's sand cloth to polish the pipe ends slightly beyond the fittings' seating depths. Put on clean disposable gloves after polishing because skin oils can interfere with a solder bond.

4. Use an acid brush to apply flux liberally to the outside of the pipe ends; put a thin, even coat on the inside of pipe fittings, all the way to the bottom of the fitting sockets.

PRO TIP

Quality solder, such as a silver alloy, is easier to work with than standard 95/5 solder: You'll have a wider working temperature range, better void filling, and fewer leaks. Quality solder can cost 50% more than standard types, but it's worth it.

5. Heat the fitting—not the pipe— and apply solder to the lowest fitting hub first. Periodically remove the torch and touch the solder tip to the fitting joint. When the fitting is hot enough, the solder will liquefy and disappear into the joint. (Note: The flame shield behind the fitting was set aside temporarily to get a clearer photo.)

tool too aggressively, you will flatten the pipe or score erratically, thus creating a weak joint. When the cut is complete, clean the end of the pipe with the deburring attachment on the cutter so that you get a good, solid joint. Leftover burrs also increase turbulence and thus decrease flow through the pipe. Use plumber's sand cloth or emery paper to polish both ends of the pipe, and a round wire brush to clean the insides of fittings.

To solder copper pipe, first use a flux brush to apply self-tinning flux (soldering paste) to the outside of the pipe and the inside of the fitting. Then slide the fitting over the pipe. If the fitting is a directional fitting, such as a tee or an elbow, make sure that the fitting points in the correct direction.

Heat the fitting (*not* the pipe), moving the soldering torch so that all sides of the fitting receive heat directly. The flux will bubble. From time to time, remove the torch and touch solder to the fitting seam. When the fitting is hot enough, the solder will liquefy when touched to it. After a few trials, you'll know when a fitting is hot enough. When the fitting is hot, some fluxes change color, from milky brown to dull silver.

Two passes with the solder, completely around the joint, will make a tight seal; more than two passes is a waste. The solder is sucked into the joint, so don't worry if you don't see a thick fillet of solder around the joint. Let solder cool before putting pressure on a joint. After a soldered joint has cooled for a minute, you can immerse it in water to cool it completely, but be careful when handling hot metal.

Soldering in tight spaces can't always be avoided. If a fitting has several incoming pipes—at a tee, for example—try to solder all pipes at the same time. Reheating a fitting to add pipes will weaken earlier soldered joints. Clean and flux the pipes, insert them in the fitting, and keep the torch moving so you heat both ends of the fitting equally. If you must reheat a fitting to add a pipe later, wrap the already soldered joint in wet rags to keep solder from melting. When soldering close to wood, wet the wood first with a plant spritzer filled with water, then use a flame shield to avoid scorching or igniting it.

When space is tight, presolder sections in a vise. Then, when placing the section in its final position, you'll have only a joint or two to solder. If one of the materials being joined might be damaged by heat, solder the copper parts first, allowing them to cool, before making mechanical connections to the heat-sensitive material. For example, if you need to connect copper supply to a pump outtake with plastic or adapters, solder the male (protruding) or female (receiving) adapter to the copper pipe before screwing it in (or on) the pump outtake.

Finally, before soldering pipe to a ball valve or a gate valve, close the valve completely. Otherwise, solder can run inside and keep the valve from closing fully. However, when soldering a shower's pressure-balancing valve or a tempering valve, follow the manufacturer's instructions. Finally, it's helpful to have unions near most valves so that sections can be disconnected without needing to undo soldered joints.

FLEXIBLE COPPER TUBING

Because flexible copper tubing can be bent and run through tight spaces, it is used primarily for short runs to dishwashers (½ in.), ice makers (¼ in. or 5/16 in.), and in similar situations. Chromed copper tubing is commonly used when supply risers will be exposed because it looks good. Flexible tubing is softer than rigid copper pipe, so take pains when you cut it not to collapse the tubing walls by turning a pipe cutter too aggressively. And use a special, sleevelike *tubing bender* to shape it so you don't crimp it, as shown in the top photo on p. 358.

Flexible copper tubing is most often connected either with compression fittings or flared fittings. A *compression fitting* has a ferrule of soft metal that is compressed between a set of matched nuts. A *flared fitting* requires that you flare the tubing ends with a special tool. When using either type of connector, remember to slide nuts onto the tubing before attaching a ferrule or flaring an end. Both types of connection are easy to disconnect, so they are used where repairs may be expected, such as the supply line to a toilet. Don't reuse ferrules, however; replace them if you need to disconnect fittings.

A Compression Fitting

³⁄₈-in. chromed copper tubing

Compression nut

Compression ferrule

Angle stop

Solderless Copper Connections

Traditionally, joining copper pipe and fittings required soldered joints. Solderless copper systems, invented in Europe in the 1950s and introduced to the U.S. in the 1990s, have rapidly gained ground since then and are approved by all major codes and most states.

There are two main types of solderless copper fittings: *push-connect* and *press-connect* (shown below). Both types offer a range of fittings, valves, and specialty items, and both feature rubber O-rings (gaskets) inside to create a positive watertight seal. Most push-connect types can be joined without tools and can be taken apart and reassembled; press-connect fittings, which require a special pressing tool, are permanent.

Prepwork. Prep is the same for both push-connect and press-connect systems. Check pipe ends for damage, because pipes that are dented or out of round may not seat correctly in fittings. Look inside each fitting to make sure its rubber O-rings are intact.

Use a felt-tipped pen to mark the correct seating depth onto the pipe. Use a tubing cutter to cut the pipe square, ream its inside to remove burrs, and chamfer or sand its outside so that a sharp pipe end doesn't nick or slice the O-ring.

Making Connections

For push-connect fittings, push the pipe into the fitting until it seats. If the fitting can be disassembled, its manufacturer will provides a plastic disconnect clip that, when pushed against the fitting, unlocks a gripping ring inside the fitting.

To join press-connect fittings, you'll need a pressing tool, which has different sizes of jaws. Choose a pressing jaw appropriately sized for the fitting (for example, ½ in.). Place the pipe into the fitting, then place the tool's pressing jaws over the bead (hub) of the fitting. Hold the jaws of the tool perpendicular to the tube/fitting and pull the trigger. When the pressing cycle is complete, release the pressing jaws and inspect the joint.

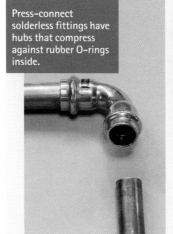

Press-connect solderless fittings have hubs that compress against rubber O-rings inside.

Pressing tools have changeable jaws, which are sized to fittings. Squeeze the tool's trigger and its jaws will close tight, compressing the fitting.

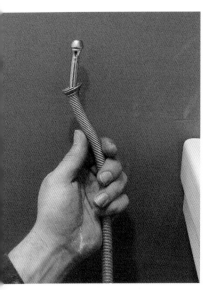

Use a tubing bender to shape chrome supply risers; the wire coils of the bender support the soft tubing and keep it from buckling as you gradually shape it.

In most situations, braided stainless-steel supply lines are a better choice than flexible copper tubing: Braided lines are strong, look good, and can be connected and disconnected as often as needed.

PEX Supply Pipes

PEX is a flexible tubing system that's been used in Europe for radiant heating and household plumbing since the 1960s, but it wasn't widely used in potable-water systems in North America until the 1990s. There are three principal types of PEX, denoted A, B and C, which refer to the manufacturing process for each—not to a grading system.

PEX is short for "cross-linked polyethylene." A cross-link is a bond linking one polymer chain to another. In general, the greater degree of cross-linking, the stronger, more durable, and heat-resistant the tubing will be. PEX-A products typically achieve an 85% cross-linking; PEX-B, 65% to 70% cross-linking; and PEX-C, 70% to 75% cross-linking. *Note*: All PEX systems used in the U.S. must meet stringent ASTM standards for burst pressure, sustained pressure, excessive temperature, bend strength, pipe dimensions, and so on.

PEX is now approved by all major building codes and all states in the U.S., although some local codes still restrict its use—check your local building department to be sure. As PEX tubing, tools, and techniques become better known, its market share is likely to continue growing. There's a lot to like.

PEX ADVANTAGES

Flexibility. Because PEX tubing can bend in short-radius turns, snake easily through walls, and run directly from a distribution point to a fixture, it requires fewer fittings than rigid materials such as copper or CPVC.

Less expensive. PEX's per-length costs are much less than that of copper pipe. (Between 2000 and 2010, the price of copper quadrupled.) Though PEX compression-style fittings may cost more than copper or PVC fittings, far fewer PEX fittings are required. And with fewer fittings to connect, labor costs to install PEX are significantly lower.

Fewer leaks. Leaks most often occur at pipe joints and fittings—fewer connections mean fewer leaks. Corrosive interactions with minerals in the water can cause pinhole leaks in copper. When the water freezes in rigid pipes, PVC and copper piping usually split and leak at the first

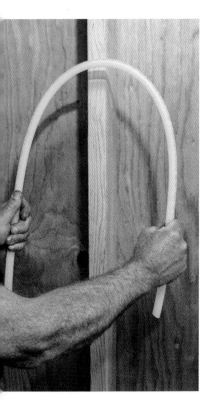

Flexible PEX tubing can bend in a small-radius turn but should never be forced into a sharp 90° turn, which could kink it.

incidence; water-filled PEX can typically withstand a half-dozen freeze-thaw cycles before it ruptures. Drain pipes that will be exposed to freezing temperatures.

Better water pressure at fixtures. With longer runs and fewer sharp turns, there is less in-line turbulence and fewer points at which water flow can be constricted or slowed. Also, hot water will arrive faster at the tap or showerhead because, unlike with metal pipe, there's minimal heat loss through conduction with PEX.

It's quiet. PEX tubing expands slightly, minimizing air hammer—the banging that takes place in rigid piping when taps are turned off suddenly and running water stops abruptly. That ability to expand also means less-pronounced pressure drops, so bathroom users will suffer fewer scalding or freezing showers.

A safer choice. Because most copper systems use a torch to solder joints, fire is a real hazard. With PEX, there's no open flame, no flux, no solder, no pipe cements—in short, nothing toxic to leach into the water supplies. Because it's chemically inert, PEX won't corrode as metal pipe will when installed in aggressive water conditions.

Well suited to renovation. There is a wide range of fittings that connect PEX tubing to existing copper or PVC systems. When PEX lines are connected to manifolds (p. 359), one can simply flip a lever to shut off water and make repairs. If PEX lines kink, they are easy to cut and repair with a coupling; to remove kinks in PEX-A, just heat tubing with a heat gun.

PEX DISADVANTAGES

Proprietary systems. PEX systems tend to be exclusive, with many brands requiring proprietary connectors and crimping tools. If, for example, you use the crimp rings and fittings designed for PEX-B on PEX-A tubing, leaks may develop. In any event, most manufacturers will refuse to honor warranties if you mix and match brands.

Fittings can be costly. PEX fittings, especially compression types, are more expensive than comparable copper or PVC ones. Fortunately, PEX systems require fewer fittings.

Sunlight degrades it. In as little as 30 days, PEX starts to become brittle when exposed to direct sunlight, so store it away from sun, then get it installed and covered quickly.

Flame melts it. Although PEX can withstand high water temperatures, it will melt when exposed to open flame. It must not be directly connected to gas- or oil-fired water heaters and must be kept away from flue pipes, recessed lights, and excessive heat.

Critters can chew through it. PEX is soft, so animals and some insects may penetrate or chew it. Raccoons and rodents may be attracted by the sound of water running through it.

THE ABCs OF PEX

Each type of PEX tubing is slightly different to work with, so research each one online, find out which types are available in your locale, and, of course, talk to plumbers who have worked with the stuff. Our sources tell us:

PEX-A has been used for more than 50 years, the longest of any type. As noted above, it has the highest degree of cross-linking. It is the softest and most flexible material and, given its low to no coil memory, it's easy to pull long, straight runs. PEX-A's bending radius of 6x O.D. (outer diameter) enables short-radius turns, though all types of PEX will kink if forced into a sharp 90° turn. If PEX-A tubing kinks, however, you can use a heat gun to repair it. PEX-A is usually the most expensive of the three. (Major PEX-A makers: Uponor, MrPEX, Rehau.)

PEX-B has the lowest degree of cross-linking of the three types, the highest bursting pressure (the least likely to burst), and, according to one study, the highest resistance to chlorine and oxidation. PEX-B is stiffer than PEX-A and has significant core memory so it can be challenging to work with. If PEX-B tubing kinks, it must be cut and joined with a coupling. PEX-B is the least expensive of the three types. (Major PEX-B makers: Everhot, Viega, Bluefin, Watts.)

PEX-C has less testing data available so it's difficult to compare. In 2013, it was the subject of a class-action settlement because of alleged cracking. It has little coil memory; it is softer than PEX-B but harder than PEX-A. If it kinks, it too must be cut and coupled. (Major PEX-C makers: Roth, Nibco.)

WORKING WITH PEX

Although PEX fittings and connection tools differ, there are basically four types: clamp-ring, press (also called push-fit or slip-in), compression fitting, and cold expansion. It is important to note that fittings and connection tools tend to be proprietary. So although it may be physically possible to attach fittings on a different brand of tubing, doing so may result in leaks and will almost certainly void your warranty.

Submanifold System

PEX tubing may employ manifolds to supply water to individual fixtures, with a shut-off valve for each. There are many ways to lay out PEX plumbing. In a home-run manifold system such as that shown in the photo below, a single 3/4-in. trunk line feeds into a single manifold, from which 1/2-in. tubing feeds all the fixtures. The drawing below shows a submanifold system, which requires less tubing and less drilling to route it.

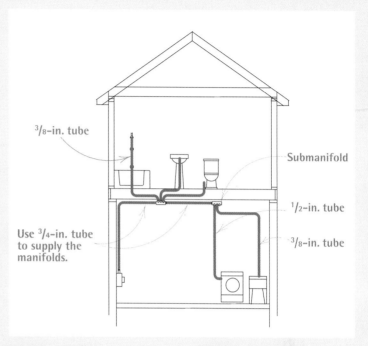

In PEX water-supply installations, central manifolds (home-run manifolds) distribute hot and cold water to individual fixtures or fixture groups. Flexible tubing requires far fewer fittings than do rigid materials.

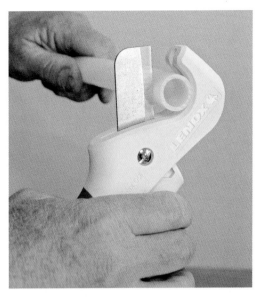

In this photo sequence, a plumber cuts and connects Uponor AquaPEX tubing. Using a simple cutting tool will ensure a square-cut end, which is essential to its seating correctly to a fitting.

Creating a cold–expansion connection in PEX–A requires a special expansion tool that enlarges both the tubing and its ProPEX fitting.

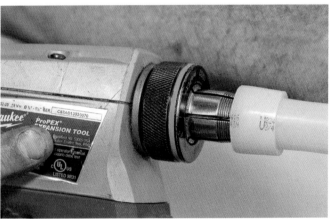

Five or six trigger squeezes will expand the PEX-A tubing and fitting uniformly because tool jaws rotate slightly with each expansion.

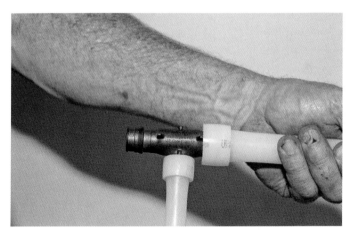

After expanding the PEX components, slide them immediately onto a fitting—here, a tee. The PEX will contract tightly to the fitting, creating a watertight seal.

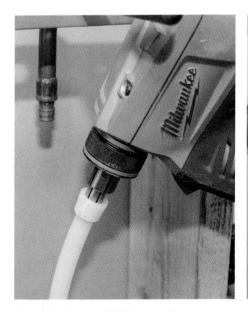

One of the features of PEX tubing is that you can run it directly from a water distribution point (such as a manifold) to a fixture without intervening connections that might leak.

After expanding the PEX-A components, slide them onto the fixture stub-out within 10 seconds. If you wait too long, you may have to re-expand the PEX.

The cold-expansion method of joining PEX-A is generating a lot of buzz in the field; the photo sequence on the facing page shows a plumber connecting Uponor AquaPEX tubing. Following that on this page is a potpourri of clamp-ring and compression fittings connecting PEX-B.

CONNECTING PEX-A

Uponor's system of cold-expansion connection is based upon the cross-linking properties of PEX-A and its tendency to return to its original size and form after being expanded—sometimes referred as its "memory."

Start by using a PEX cutting tool to square-cut the tubing. Remove burrs or debris that might interfere with fittings and slide a ProPEX ring over the end of the tubing.

You can use a hand expander or, as shown, a battery-powered Milwaukee ProPEX expansion tool. Insert the expansion head of the tool into the tubing until it seats, then squeeze the tool's trigger to expand the end of the tubing and the ring fitting. Five or six expansions are enough. With each trigger-squeeze the tool head will rotate about an eighth of a turn to ensure that the PEX is expanded evenly around its circumference.

Remove the tool head and immediately slide the expanded tubing (and ring) over the fitting. You should feel a slight resistance as the tubing slides to the shoulder of the fitting. In 10 or 15 seconds, the PEX tubing and ring will contract and tightly grip the fitting. (When it's cold, the plastic will contract more slowly.) Note: If the PEX stops short of the fitting shoulder, pull the tubing off the fitting and expand it once more.

CONNECTING PEX-B

Clamp ring. After cutting PEX tubing squarely, slide a steel ring (collar) onto the tubing, insert a ribbed brass connector into the flexible tubing, and then use a crimping tool to compress the ring and squeeze the tubing tight to the connector.

Push fittings. Also known as a slip-in fitting, a push fitting employs a mechanism with tiny, stainless-steel teeth inside. Square-cut the end of the tubing, and push it firmly into the fitting until it seats. You then slide a *collet* forward to secure the tubing. This fitting has the benefit of being reusable—just slide the collet and pull out the tubing—though it's advisable to cut a new end if you remove the tubing more than once.

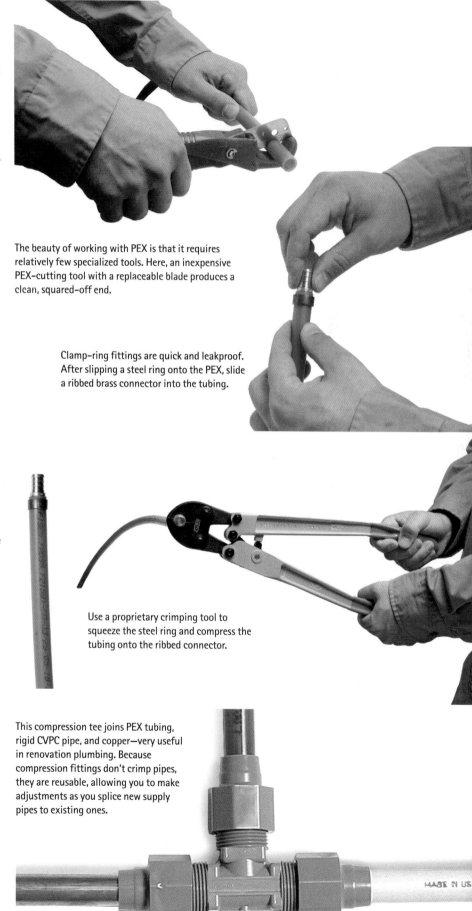

The beauty of working with PEX is that it requires relatively few specialized tools. Here, an inexpensive PEX-cutting tool with a replaceable blade produces a clean, squared-off end.

Clamp-ring fittings are quick and leakproof. After slipping a steel ring onto the PEX, slide a ribbed brass connector into the tubing.

Use a proprietary crimping tool to squeeze the steel ring and compress the tubing onto the ribbed connector.

This compression tee joins PEX tubing, rigid CVPC pipe, and copper—very useful in renovation plumbing. Because compression fittings don't crimp pipes, they are reusable, allowing you to make adjustments as you splice new supply pipes to existing ones.

PEX tubing does require specialized fittings, such as plastic support elbows at tight bends and proprietary clamps where tubing is attached to metal stub-outs.

Compression fittings. Compression fittings are well known to anyone who has ever connected a chrome supply riser to a lavatory or to a toilet tank. Insert the end of the PEX tubing into the fitting sleeve, then tighten the nut of the threaded fitting. Specialized compression fittings can join different pipe materials.

TYING PEX TO EXISTING SYSTEMS

Because PEX is flexible, you need far fewer fittings. But you do need some, especially if you want to tie PEX into existing supply pipes. So, as is the case with copper fittings (see p. 353), there's a plethora of PEX fittings and adapters that enable you to connect PEX to rigid copper or CPVC plastic. Some change pipe materials, size, *and* direction, such as a ¾-in. copper by ½-in. PEX 90 (also called a *reducing ell*). There are also PEX-specific pieces such as the plastic support elbow shown above, which prevents the tubing from collapsing when it must make a sharp turn.

Note: If you do join PEX to a sweated copper fitting, solder the copper parts and allow them to cool before attaching the PEX.

Galvanized Steel Pipe

Galvanized pipe corrodes and constricts, reducing flow and water pressure, so it is no longer installed as water-supply pipe. If your existing system is galvanized and the water flow is weak, replace it as soon as possible. If you're not quite ready to rip out and replace all of your galvanized pipe, you can replace or extend sections with rigid copper. However, you must use a *dielectric union* (see the bottom photo on p. 353) to join copper sections to steel. Otherwise, electrolysis will take place between the two metals, and corrosion will accelerate.

Today, galvanized pipe is largely limited to gas-supply service. Because of safety consider-

PRO TIP

Have a friend help you set plastic DWV pipe. Although plastic is light, it is cumbersome. Once you've applied pipe cement, you have about 30 seconds to position the fittings before it sets. With two people working, one can hold a fitting while the other pushes the pipe and twists it one-quarter turn.

ations and the difficulty of threading pipe without a power threader, *have a licensed plumber install gas-supply service pipes*. A plumber will install gas shutoff valves to *gas-supply stubs*; short lengths of flexible gas-supply pipe run from there to a fixture.

CPVC Supply

Most plumbing codes allow CPVC for hot- and cold-water lines, but check with local authorities to be sure. CPVC is a good choice for hard-water areas because—unlike copper—CPVC won't be corroded by chemicals in the water. *Note:* CPVC is a different material from PVC, which is widely used as drain and waste pipe; PVC may not be used as supply pipe, however, because it releases carcinogens.

Working with CPVC supply pipe is much like cutting and joining plastic DWV pipes, explained at some length later in this chapter. The main difference is that waste pipes are larger. Briefly, here's how to join CPVC: Cut the pipe ends square using a plastic-pipe saw or plastic-pipe cutting shears. Clean the pipe ends as well as the inside of the fitting. Next, apply solvent-based cement to each. Insert the pipe into the fitting, turning either the pipe or the fitting a quarter turn in *one direction only* to spread the cement. Finally, allow the glued joints to set adequately before putting pressure on the line.

Before installing CPVC, make sure that you have the adapters needed to join the new plastic pipes to existing metal pipes and fixtures.

DWV Materials

ABS and PVC plastic pipe are by far the most common materials for drainage, waste, and venting lines, although cast iron is still specified where sound suppression is important. Plastic pipe is strong, the most corrosion resistant of any DWV pipe, and it's easy to cut and assemble using special solvent-based cements. It's light enough for one person to handle, reasonably priced, and extremely slick inside, which ensures a good flow of wastes.

It's by far the favorite DWV material of amateur plumbers—and many pros. But there are a few disadvantages. Many codes prohibit using plastic pipe outside because of durability and UV degradation issues, and if you don't spread the cement evenly or allow it to cure before stressing the joints, plastic can leak.

Cast iron is relatively corrosion resistant, though it will rust in time (decades), and its mass deadens the sound of running water. Although it's heavy to work with, it is still specified by many professionals for high-end jobs, where codes

ELBOWS

Ell or 90

Vent ell

Street ell

Long-sweep ell

22½° ell

45° ell or 45

Closet bend
or 4x3

TEES OR TEE FITTINGS

Sanitary tee

Vent tee

Double tee or cross fitting

Reducing tee

WYE FITTINGS

Wye fitting

Double wye

Combo (or tee/wye)
(drawing exaggerates
sweep)

POTPOURRI OF PARTS

P-trap (required for
all fixtures except a toilet)

Trap adapter has a
slip-nut coupling
with a plastic washer

Closet flange, which
glues to a closet bend

Cleanout adapter

Cleanout plug

Coupling

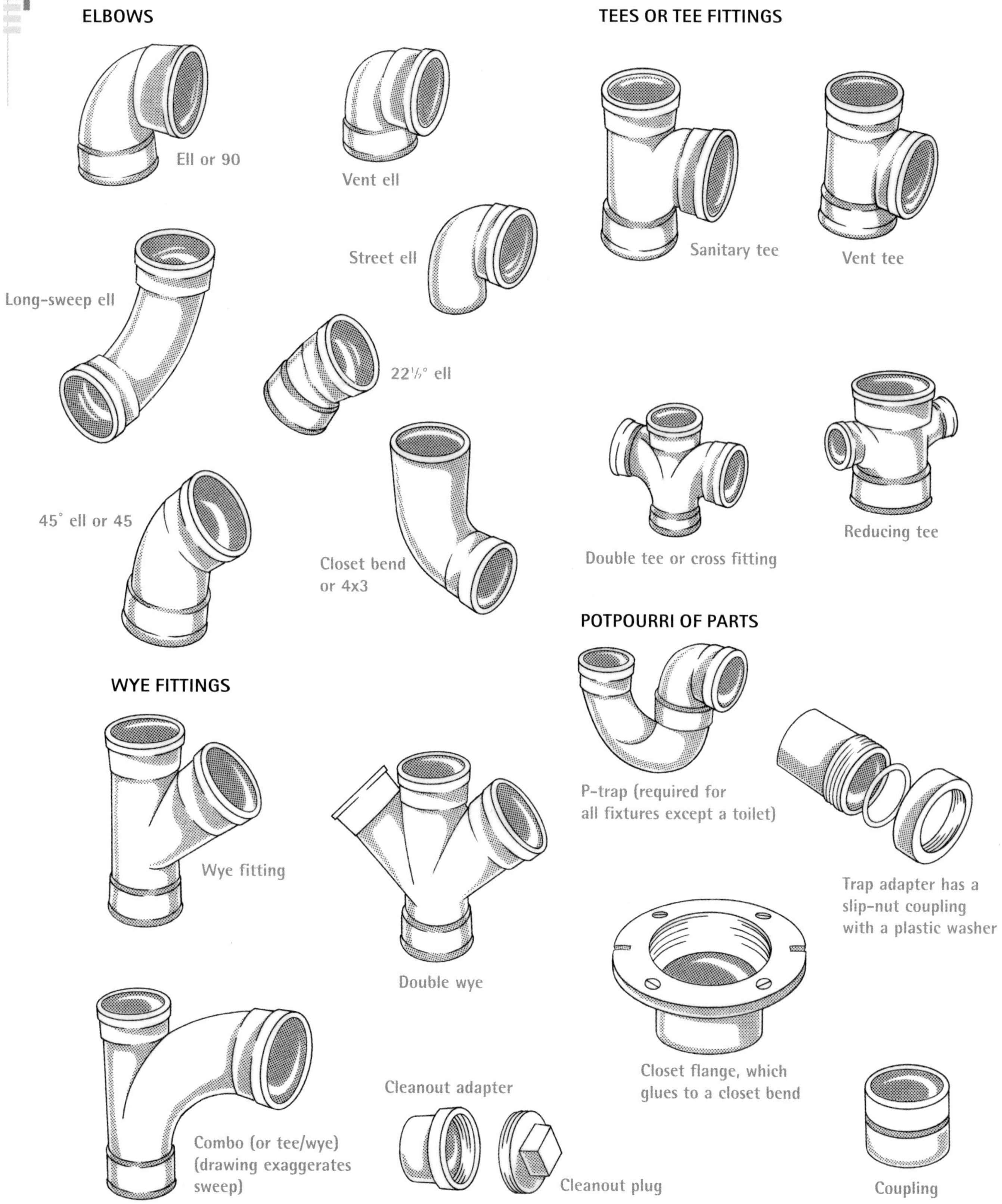

The fittings shown are ABS plastic, but their shapes are essentially the same as those of copper and cast-iron DWV fittings of the same name. Drain fittings—such as the long-sweep ell, the combo, and the sanitary tee—turn gradually so wastes can flow freely, without clogging. Vent fittings have tighter turning radii because they carry only air. Finally, street fittings have one hubless end that fits directly into the hub of another fitting, which is useful when space is tight.

Heat-Recovery DRAINS

As their name suggests, heat-recovery drains (HRDs) capture some of the heat that goes down the drain when you shower, do the dishes, or wash clothes. The heart of an HRD is a special drain wrapped with hollow copper coils that acts as a conductive heat exchanger. As heated water runs down the drain, it preheats clean water running through the coils. That preheated water can then go directly to a fixture, to a traditional water heater, or to a separate storage tank. Already preheated, that water takes less energy and so costs less to bring to the desired temperature. HRD technology is certain to change, but it seems most energy-efficient if its preheated water feeds a tankless, on-demand water heater. Because an HRD creates a near instantaneous heat transfer while someone is, say, showering, there is no need to store the preheated water in a tank, where it would require additional energy to keep its temperature constant.

How an HRD Works

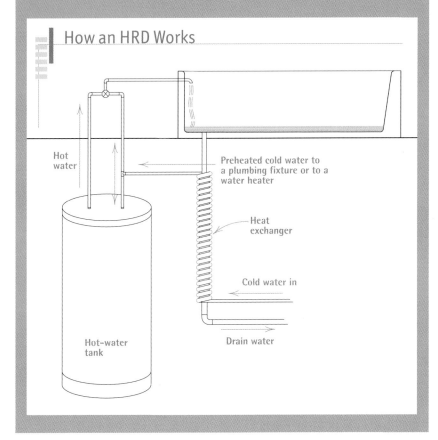

Hot water

Hot-water tank

Preheated cold water to a plumbing fixture or to a water heater

Heat exchanger

Cold water in

Drain water

PRO TIP

Don't use ABS cement on PVC pipe or vice versa. Because solvent-based cements partially dissolve plastic to create a chemical weld, their chemistry is quite specific. Using the wrong cement or joining PVC and ABS pipe is a code violation because it can lead to weak joints and leaks.

allow pipes on building exteriors, and where codes require cast iron in multistory buildings. Ever since *no-hub couplings* replaced lead and oakum, cast iron has been easier to connect, but it still takes skill and strength to cut cleanly and support adequately. Consequently, it's rarely installed by amateur plumbers. Professionally installed cast-iron systems cost, on average, 40% to 50% more than plastic-pipe installations.

Copper DWV pipe is installed mostly on jobs with bottomless budgets. Copper is lightweight, durable, and undeniably handsome. Because its walls are relatively thin, copper DWV pipe is sometimes specified where there are tight turns. However, copper costs two to three times as much as a plastic DWV installation, and it's less corrosion resistant. Compared with cast iron, copper's thin walls don't suppress sound nearly as well.

PLASTIC DWV PIPE

Cut ABS or PVC pipe with a plastic-pipe saw and a miter box, or with a wide-roll pipe cutter. If you use a cutter, gradually tighten its cutting wheel after each revolution. Whatever tool you use to cut the pipe, use a utility knife, a rounded file, or a deburring tool to clean off burrs before sanding the cut lightly with emery paper. Use a clean cloth to wipe off any grit.

Dry-fit the pipes and fittings before cementing them together. Dry-fitting allows you to determine the exact direction you want the fitting to point, as well as the depth of the pipe's seat in the fitting. Pipe cement sets so quickly that there's no time to fine-tune fitting locations. Use a grease pencil or a builder's crayon to draw alignment marks on the pipe and the fitting; a yellow or white grease pencil works well on black ABS pipe, as shown in the bottom left photo on the facing page. Then take apart the dry-fit pieces and apply the cement.

Apply plastic-pipe primer to the outside of the pipe and to the inside of the fitting. Then, using the cement applicator, apply a generous amount of solvent-based cement to the outside of the pipe and the inside of the fitting hub. Immediately insert the pipe into the fitting so that it seats completely. Then turn the fitting (or the pipe) a quarter turn in *one direction only*—stop when the alignment marks meet. When you are finished, the joint should have an even bead of cement all around. Allow the joint to set completely before putting pressure on it.

One-coat, no-primer plastic-pipe cements are new to the market: They seem promising but as yet are unproven for the long haul. Research them carefully before you commit.

After dry-fitting DWV pipes and putting alignment marks on the pipes and fittings, disconnect them and apply solvent-based cement to the outside of pipes and the inside of fittings. Wear plastic gloves to protect your skin.

Insert cemented plastic pipes all the way into the fitting, and give a quarter turn to spread the cement evenly. The yellow crayon lines are the alignment marks.

CAST IRON

To the inexperienced eye, all cast iron looks the same, but it's not. If you lightly rap most cast iron with a rubberized tool handle, you'll hear a muffled thud; old "light iron," however, will reverberate somewhat, with a higher, tinny tone. If you suspect that you have light iron, which was widely installed in the northeastern United States until the 1940s, hire a plumber to have a look and, if necessary, make any cuts you're planning. If you try to cut light iron with a conventional snap cutter, the pipe may crush and collapse.

To cut into cast iron to extend a DWV system or replace a corroded fitting, rent a snap cutter. Many cast-iron joints are hubbed, in which a straight pipe end fits into the flared hub. But increasingly, sections of straight pipe are joined via no-hub couplings with inner neoprene sleeves, as shown in "Splicing a Branch Drain to

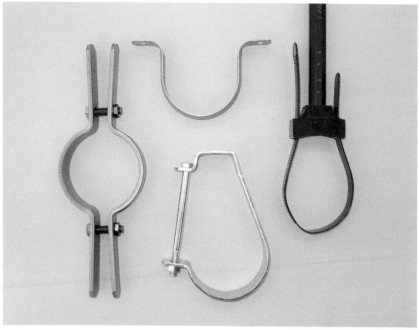

DWV pipe supports. Clockwise from left: riser clamp (stack clamp), steel rigid pipe strap (U-clamp), ABS pipe strap hanger, and J-clamp (often used with all-thread rod).

Here, two horizontal ABS lavatory drains meet at a figure-5 fitting, also called a double combo. An ABS vent rises out of the fitting's top hub, while the drain hub connects to a cast-iron stack via a no-hub coupling. A riser clamp supports the bottom of the cast-iron pipe, and steel nail plates protect the ABS.

A cast-iron snap cutter has beveled cutting wheels along the length of its chain.

No-hub couplings. At the upper left, a transition coupling joins DWV pipes with different exterior diameters, a 2-in. copper vent pipe, and a 2-in. cast-iron takeoff on a 2 by 3 wye fitting. The other couplings join 3-in. no-hub cast-iron fittings.

PRO TIP

If you smear liquid soap on the outside of cast-iron pipes and no-hub fittings, you'll have an easier time pulling the neoprene sleeves onto them.

▌▌▌▌

a Stack" on p. 371. The new pipe–fitting assembly should be 1 in. shorter than the cutout section.

Before cutting into cast iron, however, support the pipe on both sides of the intended cut to prevent movement, which could weaken joints. Use *stack clamps (riser clamps)* if you're cutting into a vertical section of pipe. Use *strap hangers* (see the top right photo on p. 365) if cutting into a horizontal section. Mark cutlines on the cast iron with a grease pencil. Then wrap the snap-cutter's chain about the pipe, gradually tightening until the chain is snug and the tool's cutting wheels align over the cutlines. Crank the cutter's handle to continue tightening the chain until the pipe snaps cleanly. Make the second cut and remove the old pipe section. *Caution:* Wear goggles during this operation.

Once the cuts are complete, slide the neoprene sleeve of a no-hub coupling over each remaining pipe end. You may need to roll each sleeve back on itself, as you would roll up the cuffs of a long-sleeved shirt. Insert the transition fitting or replacement pipe section, unroll the neoprene sleeves onto the fitting or pipe, and tighten the steel band clamps. You can also tie into a cast-iron drain without cutting into it by building out from an existing cleanout, as shown in "Extending a Cast-Iron Main Drain" on p. 372.

Venting Options

Until you expose the framing and actually run the pipe, it's difficult to know exactly how things will fit together—especially vents. Because correct venting is crucial, this section discusses several venting options. But first, here are a few terms to keep straight: A stack is a vertical pipe. If the stack carries wastes, it's a *soil stack*. If the

stack admits air and never carries water, it's a *vent stack*.

Most of the venting options described next are examples of *dry venting*, in which a vent stack never serves as a drain for another fixture. But there are hybrids; for example, if a vent stack occasionally drains fixtures above it, it is a *wet vent*. Wet vents must never carry soil wastes, and many local codes prohibit all wet venting. But when it's legal and the vent is one pipe size larger than normal to ensure a good flow, wet venting can be safe and cost-effective because it requires fewer fittings and less pipe.

BACK VENTING

Back venting (also known as *continuous venting*) is the dry-venting method shown on the facing page, and it's acceptable to even the strictest codes. All the fixtures in the drawing have a branch vent. In a typical installation, the trap arm of, say, a lavatory empties into the middle leg of a sanitary tee. The branch drain descends from the lower leg, the branch vent from the upper. When a branch vent takes off from a relatively horizontal section of drainpipe, the angle at which it departs is crucial. It may go straight up or it may leave at a 45° angle to work around an obstruction. But it must never exit from the

NO-HUB Couplings

No-hub couplings (also known as banded couplings, band-seal couplings, and hubless connectors) consist of an inner neoprene sleeve, which fits over the pipe ends or fittings, and an outer corrugated metal shield, which is drawn tight by a stainless-steel band clamp. No-hub couplings are widely used to join cast-iron pipe and no-hub fittings in new construction, but they are also invaluable to renovators.

For example, if you want to add a plastic shower drain to a cast-iron stack, no-hub couplings can accept either a cast-iron or a plastic no-hub fitting and seal it tightly to the pipe ends once you've cut into the stack. (Support both sides of the section to be cut out so it can't shift during cutting and weaken other joints.)

When joining DWV pipes of different materials, use specialized transition couplings whose neoprene sleeves are sized for incoming pipes with different outside diameters, such as the coupling used to join 2-in. copper and 2-in. cast iron, shown in the bottom photo at left.

Maximum Distance: Trap Arm to Vent*

TRAP ARM DIAMETER (in.)	MAXIMUM DISTANCE TO VENT
1¼	2 ft. 6 in.
1½	3 ft. 6 in.
2	5 ft.
3	6 ft.
4	10 ft.

* Also maximum distance of stack-vented fixture trap arm to stack, based on calculations found in the UPC, T10-1.

Back Venting (Continuous Venting)

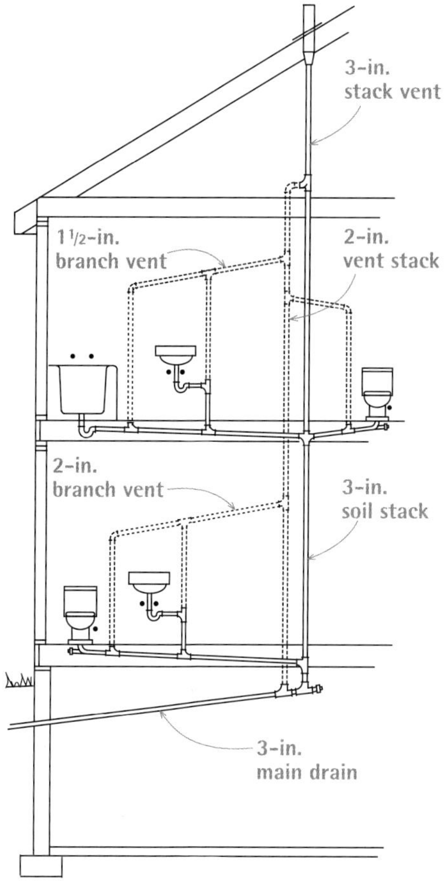

In this illustration of back venting, all fixtures have a dry branch vent—that is, no vent ever carries water. The fixtures on the first floor require a 2-in. branch vent because the toilet's 3-in. drain needs more incoming air to equalize its large waste flow.

side of a drainpipe: If it did, it could become clogged with waste.

Branch vents must rise to a height of at least 42 in. above the floor before beginning their horizontal run to the vent stack. This measurement adds a safety margin of 6 in. above the height of the highest fixture (such as a sink set at 36 in.), so there is no danger of waste flowing into the vent. Since branch vents run to a vent stack, they should maintain an upward pitch of at least ¼ in. per ft., although the UPC allows a vent to be level if it is 6 in. above the flood rim of a fixture.

STACK VENTING

Clustering plumbing fixtures around a central stack is probably the oldest method of venting. In the early days of indoor plumbing, plumbers noticed that fixtures near the stack retained water in their traps while those (unvented) that were at a distance did not. You can vent three bathroom fixtures (lavatory, tub, and toilet) off a 3-in. stack vent, without additional branch vents—if you detail it correctly, as shown in "Stack Venting" on p. 368.

Note: When stack venting, never place a toilet above the other fixtures on the stack: Its greater discharge could break the water seals in the traps of small-dimension pipes. If you must add fixtures below those already stack vented, add (or extend) vent stacks and branch vents. The maximum allowable distance from stack-vented fixtures to the soil stack depends on the size of the

Minimum Drain, Trap, and Vent Sizes

FIXTURE/APPLIANCE	DRAIN/TRAP SIZE	VENT SIZE
Toilet	3 in. or 4 in.	2 in.
Bathtub/shower	2-in. drain; 1½-in. trap	1½ in.
Shower stall	2 in.	1½ in.
Lavatory	1½-in. drain; 1¼-in. trap	1¼ in.
Paired lavatories	2-in. drain; 1½-in. trap	1½ in.
Bidet	1¼ in.	1¼ in.
Kitchen sink (with or without disposal)	1½ in.	1½ in.
Dishwasher	1½ in.	1½ in.
Laundry tub	1½ in. or 2 in.	1½ in.
Clothes washer standpipe	2 in.	1½ in.

Stack Venting

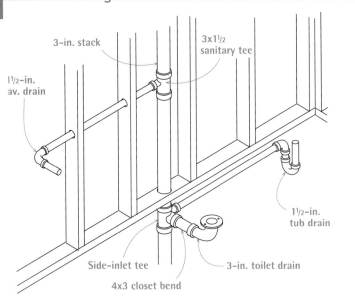

3-in. stack

1½-in. av. drain

3x1½ sanitary tee

1½-in. tub drain

Side-inlet tee

4x3 closet bend

3-in. toilet drain

If close enough to a stack vent of adequate size, fixtures can use it for venting (see "Maximum Distance: Trap Arm to Vent" on p. 367). *Note:* The side inlet serving the tub enters above the toilet inlet; this fixture group must be the highest on the stack.

pipe serving a particular fixture (see "Maximum Distance: Trap Arm to Vent" on p. 367).

VENTING TOILETS

Because they have the biggest drain and vent pipes of any fixture, toilets can be the trickiest to vent. When space beneath a toilet is not a problem, use a setup such as the one shown below in "Venting a Toilet," in which a 2-in. vent pipe rises vertically from a 3 by 2 combo, while the 3-in. drain continues on to the house main. The 3-in.-diameter toilet drain allows the vent to be as far as 6 ft. from the fixture, as indicated in the table on p. 367.

When space is tight, say, on a second-floor bathroom with finished ceilings below, the drain and vent pipes must descend less abruptly (see "Constricted Spaces" below). Here, the critical detail is the angle at which the vent leaves the 3 by 2 combo: That vent takeoff must be 45° above a horizontal cross section of the toilet drain. If it is less than that, the outlet might clog with waste and no longer function as a vent. As important, the "horizontal" section of the vent that runs between the takeoff and the stack must maintain a minimum upward pitch of ¼ in. per ft.

Venting a Toilet

2-in. vent

Closet flange

4x3 closet bend

3x2 low-heel vent or 3-in. tee with 2-in. bushing

3-in. drain

3-in. combo

When there's plenty of space under a toilet—say, in an unfinished basement—the branch drain can descend steeply.

Constricted Spaces

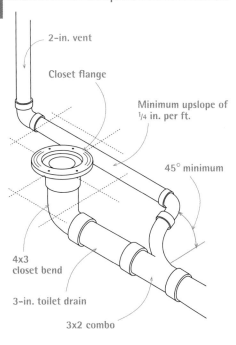

2-in. vent

Closet flange

Minimum upslope of ¼ in. per ft.

45° minimum

4x3 closet bend

3-in. toilet drain

3x2 combo

When a branch drain must travel through a floor platform before reaching a stack, the drain must slope gradually. Here, the angle the vent takes off from the toilet drain is critical—it must not be less than 45°, as depicted in the cross-section drawing on the facing page.

When you've got two toilets back to back, you can save some space by picking up both with a single *figure-5 fitting (double combo)*, like the small one in the bottom right photo on p. 365. From the top of the fitting, send up a 2-in. or 3-in. vent; from the two side sockets use two 3-in. soil pipes serving the toilets; and use a *long-sweep ell* (or a combo) on the bottom to send waste on to the main drain. This fitting is about the only way to situate back-to-back water closets and is quite handy when adding a half bath that shares a wall with an existing bathroom.

OTHER VENTING OPTIONS

Common vents are appropriate where fixtures are side by side or back to back. This type of vent usually requires a figure-5 fitting.

Loop vents are commonly installed beneath an island counter in the middle of a room. The sink drain is concealed easily enough in the floor platform, but the branch vent, lacking a nearby wall through which it can exit, requires some ingenuity. This problem is solved by the loop shown in "Venting an Island Sink" below.

In addition to the fittings shown in the drawing, note these factors as well: The loop must rise as high under the counter as possible and at least 6 in. above the juncture of the trap arm and the sanitary tee to preclude any siphoning of wastewater from the sink. The vent portions may be 1½-in. pipe, but the drain sections must be 2 in. in diameter, and drain sections must slope downward at least ¼ in. per ft.

Air-admittance valves (AAV) are one-way mechanical vents designed to eliminate the need for conventional branch vents for fixtures too far from a wall. As water drains from a sink, it creates a partial vacuum within the pipes, depressing a spring inside the AAV and sucking air in. When the water is almost gone and the vacuum is equalized, the spring extends and pushes its diaphragm up, sealing off outside air once again and preventing the release of sewer gases. Because fixture drains with AAVs don't need lateral vent runs or additional vent-stack penetrations in the roof, they allow greater design flexibility, while saving considerably on labor and materials. Mechanical vents were once intended

Vent-Takeoff Cross Section

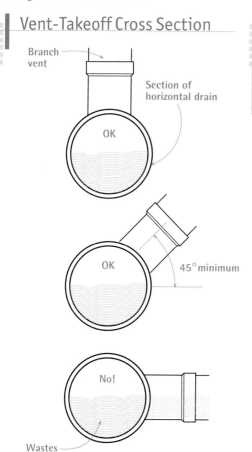

Vents must take off from horizontal drains at a 45° angle minimum. If the takeoff angle is less than 45°, wastes can block the vent.

Venting an Island Sink

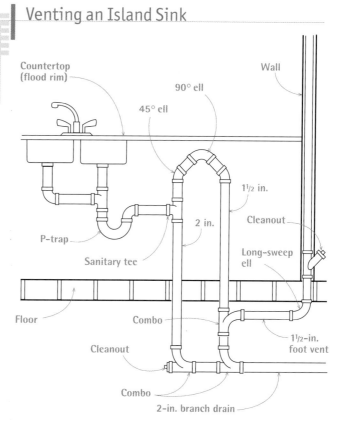

A looped vent is one code-approved way to vent an island sink. The loop should extend as high as possible under the countertop. Loop preassembly makes construction easier. The foot vent must connect to the loop via a combo fitting and slope upward to the stack at a minimum of ⅛ in. per ft.

GOING WITH THE Flow

To optimize flow and minimize clogged pipes, follow these guidelines:

▶ DRAINAGE FITTINGS. Use a long-sweep ell (90° elbow) or a combo when making a 90° bend on horizontal runs of waste and soil pipe and where vertical pipes empty into horizontal ones. Use a standard ell when going from horizontal to vertical. Where trap arms join vent stacks, use sanitary tees. (Long-sweep fittings are not required on turns in vent pipe; regular tees and ells may be used there.)

▶ CLEANOUTS. Cleanouts are required where a building main joins a lead pipe from a city sewer line or septic tank, at the base of soil stacks, and at each horizontal change of direction of 45° or more. Also, install cleanouts whenever heavy flow increases the possibility of clogging, such as in back-to-back toilets. There must be enough room around the cleanout to operate a power auger or similar equipment.

Pipe-Support Spacing

PIPE MATERIAL	HORIZONTAL SUPPORTS	VERTICAL SUPPORTS
Water supply		
Copper	6 ft.	10 ft.
CPVC	3 ft.	10 ft. and midstory guide
PEX	32 in.	Base and midstory guide
DWV		
ABS or PVC	4 ft. and at branch connections	10 ft. and midstory guides if pipe ≤ 2 in.
Cast iron	5 ft.	Base and each story; 15 ft.

When splicing a branch to an existing drain, support both sides of the takeoff fitting. Here, J-clamps are on both sides of a combo fitting. Transition couplings join copper to cast iron; you could use similar couplings to splice ABS or PVC drainpipes to cast iron.

to be only temporary, but their valve mechanisms have been improved so that air-admittance valves are now accepted by major building codes, including the IRC.

VENT TERMINATION

To reduce chances that vent gases will enter the home, stack tops must be at least 6 in. above the upslope side of the roof and at least 3 ft. above any part of a skylight or window that can be opened. A vent stack must be at least 12 in. horizontally from a parapet wall, dormer sidewalls, and the like. Finally, stacks must be correctly flashed to prevent roof leaks.

Roughing-in DWV Pipes

In new construction, pros typically start the DWV system by connecting to the sewer lead pipe, supporting the main drain assembly every 4 ft. and at each point a fitting is added.

Renovation plumbing is a different matter altogether, unless an existing main is so corroded or undersize that you need to tear it out and replace it. Rather, renovation plumbing usually entails tying into an existing stack or drain in the most cost- and time-effective manner. There are three plausible scenarios: (1) cutting into a stack to add a branch drain, (2) building out from the end of the main drain where it meets the base of the soil stack, and (3) cutting into the main drain in midrun and adding fittings for incoming branch drains.

This discussion assumes that the existing pipes are cast iron and that new DWV pipes or fittings are ABS or PVC plastic, unless otherwise noted. If you're adding several fixtures, position the new branch drain so that individual drains can be attached economically—that is, using the least amount of pipe and fewest fittings. Remember, drainpipes must have a minimum downward slope of ¼ in. per ft.

Run clear water through the drains before cutting into them. Flush the toilets several times and run water in the fixtures for several minutes. Then shut off the supply-pipe water and post signs around the house so people don't use the fixtures while work is in progress.

SPLICING A BRANCH DRAIN INTO A STACK

If you're adding a toilet, have a plumber calculate the increased flow, size the pipes, recommend fittings, and—perhaps—do the work. Adding a lav, sink, or tub, on the other hand, is considerably easier and less risky—mostly a matter of splicing a 1½-in. branch drain to a 2-in. or 3-in. stack. The basic steps are clamping the stack before cutting it, inserting a tee fitting into the stack, and joining the branch drain to that fitting.

Let's look at splicing to a cast-iron stack first. Start by holding a no-hub fitting (say, a 2 by 1½ sanitary tee) next to the stack and using a grease pencil to transfer the fitting's length to the stack—plus ½ in. working room on each end. (This gap at each end will be filled by a lip inside the neoprene sleeve.) Install a stack clamp above and below the proposed cuts. Then use a snap

Vent Termination

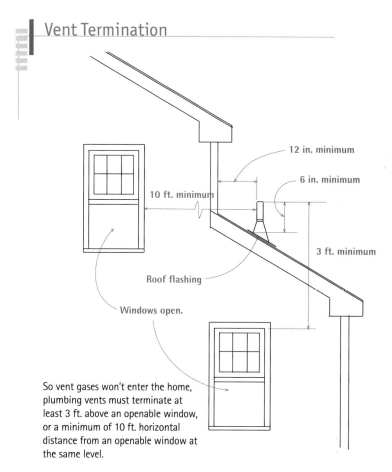

So vent gases won't enter the home, plumbing vents must terminate at least 3 ft. above an openable window, or a minimum of 10 ft. horizontal distance from an openable window at the same level.

Splicing a Branch Drain to a Stack

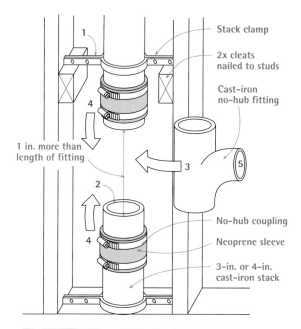

A CAST-IRON STACK

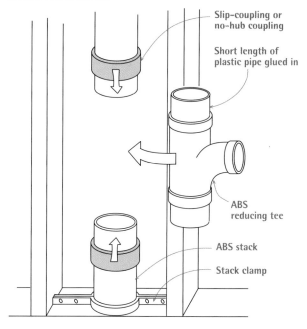

AN ABS-PLASTIC STACK

1. Using stack clamps, support the stack above and below the cuts. Mark and cut the stack. 2. Slide no-hub couplings onto cut stack ends (you may need to roll the neoprene sleeves on first). Insert a no-hub fitting. 3. Slide couplings over fitting ends. 4. Tighten. 5. Connect the branch drain to the no-hub fitting.

Glue two short lengths of ABS pipe to a tee. Mark an equivalent length plus ½ in. on both ends onto the ABS stack to indicate cutlines. (Each ABS slip-coupling has an inner lip that nearly fills the ½-in. space.) Support and cut the stack. Finally, join the pipes by slipping the couplings in place.

cutter to make the two cuts. Drill through studs as needed to run the branch drain. Next, slide no-hub couplings onto both cut pipe ends; in most cases, it's easiest to loosen the couplings, remove the neoprene sleeves, roll a sleeve halfway onto each pipe end, and then replace the couplings.

Insert the no-hub fitting, unroll the sleeves onto fitting ends, slide the banded clamps over the sleeves, orient the fitting takeoff, and tighten the clamps with a no-hub torque wrench. Finally, use a *transition coupling*, which is a special no-hub coupling that accepts pipes of different outer

diameters, to tie the new 1½-in. plastic branch drain to the cast-iron no-hub coupling.

Tying into an ABS or PVC stack is essentially the same, except that you'll use a wheeled cutter to cut the stack. And, instead of using a no-hub coupling, glue short (8-in.) lengths of pipe into the tee fitting, then use plastic slip couplings to join the 8-in. stubs to the old pipe. (The slip couplings also glue on, with an appropriate solvent-based cement.) Use a reducing tee, such as a 2 by 1½. Be sure to support the stack above and below before cutting into it.

Extending a Cast-Iron Main Drain

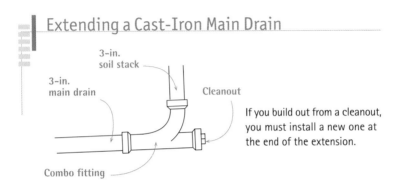

If you build out from a cleanout, you must install a new one at the end of the extension.

3-in. soil stack
3-in. main drain
Cleanout
Combo fitting

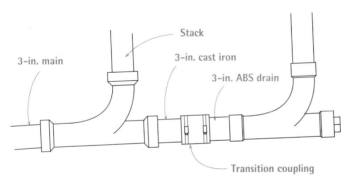

Stack
3-in. main
3-in. cast iron
3-in. ABS drain
Transition coupling

EXTENDING WITH 3-IN. ABS
If the present cleanout is a cast-iron inset caulked with oakum, remove the oakum and the inset and replace it with a short section of 3-in. cast-iron pipe. From there, use a transition (no-hub) coupling to continue with 3-in. ABS plastic.

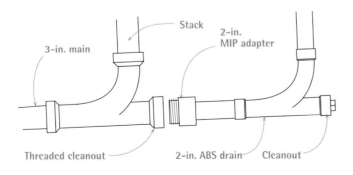

Stack
3-in. main
2-in. MIP adapter
Threaded cleanout
2-in. ABS drain
Cleanout

EXTENDING WITH 2-IN. ABS
If there's presently a threaded cleanout opening and you are adding a tub, lav, or sink—but not a toilet—use a plastic MIP (male × iron pipe) adapter.

BUILDING OUT FROM THE MAIN DRAIN

Extending the DWV system out from the end of a cast-iron main drain—where it joins the soil stack—can be the least disruptive way of tying in a new drain if there's a cleanout at the end of the main drain that you can remove. Before cutting into existing drains, however, support both sides of the section to be cut, using pipe clamps or strap hangers.

The exact configuration of the end run will depend on the size of the main drain, the fitting currently at the base of the stack, the fixtures you're adding, and the size of the drain needed to serve them. If you are not adding a toilet, the drain extension can be 2-in. pipe, which can be attached with a reducing bushing such as the male-threaded adapters shown in "Extending with 2-in. ABS" at left. If you're adding a toilet, however, the extension must be 3-in. pipe, often inserted with a *ribbed bushing* to ensure a tight fit. If it's not possible to insert the 3-in. pipe into an old cleanout leg, you may need to cut out the existing combo and install a no-hub combo to build out from.

Note: If you build out from an existing cleanout at the end of the main drain, you'll need to add a new cleanout at the end of the extension.

TYING INTO THE MAIN DRAIN IN MIDRUN

Before tying into the main drain in midrun, flush the drain and support both sides of the section you'll cut into. Then install strap hangers to support both sides of the 3-in. or 4-in. drain. Tying into a cast-iron or plastic drain is essentially the same procedure as splicing into a stack, but it requires different fittings. With one hand, hold the no-hub combo fitting you'll add next to the drain section, and with the other hand, mark cutlines onto the drain using a grease pencil. The cut marks should be 1 in. longer than the length of the fitting to accommodate the thickness of the *stop lip* inside each no-hub coupling's neoprene

Framing for Toilets and Tubs

You may need to cut through joists to accommodate the standard 4 by 3 closet bend beneath a toilet or the drain assembly under a standard tub. In that event, reinforce both ends of severed joists with doubled headers attached with double-joist hangers. This beefed-up framing provides a solid base for the toilet as well. If joists are exposed, you can also add joists or blocking to optimize support.

Toilets. A minimum 6-in. by 6-in. opening provides enough room to install a no-hub closet bend made of cast iron (4½ in. outer diameter) or plastic (3½ in. outer diameter). The center of the toilet drain should be 12 in. from a finish wall or 12½ in. from rough framing. If joists are exposed, add blocking between the joists to stiffen the floor and better support the toilet bowl, even if you don't need to cut joists to position the bend.

Bathtubs. A 12-in. by 12-in. opening in the subfloor will give you enough room to install the tub's waste and overflow assembly. Ideally, there should be blocking or a header close to the tub's drain that you can pipe-strap it to. To support the fittings that attach to the shower arm and spout stub-outs, add cross braces between the studs in the end wall. To support tub lips on three sides, attach ledgers to the studs, using galvanized screws or nails. Finally, if there's access under the tub, add double joists beneath the tub foot.

Maximum Sizes for Holes and Notches

FRAMING ELEMENT	HOLE DIAMETER (in.)	NOTCH DEPTH (in.)
Bearing studs		
2 × 4	1⅜	⅞
2 × 6	2³⁄₁₆	1⅜
Nonbearing studs		
2 × 4	2	1⅜
2 × 6	3¼	2³⁄₁₆
Solid lumber joists		
2 × 6	1¾	⅞
2 × 8	2½	1¼
2 × 10	3⅛	1⅝
2 × 12	3¾	1⅞

sleeve. (If the main drain is cast iron, use a snap cutter to cut it; if it's plastic, use a wheeled cutter.)

After cutting out the drain section, use no-hub couplings to attach the new no-hub combo fitting. Slide a neoprene sleeve onto each end of the cut drain, insert the no-hub combo, and slide a sleeve onto each end of the combo. Align the combo takeoff so it is the correct angle to receive the fixture drain you're adding. Finally, tighten the stainless-steel clamps onto the couplings.

CONNECTING BRANCH DRAINS AND VENTS

After modifying the framing, assemble branch drains and vents. Here, we'll assume that the new DWV fittings are plastic.

The toilet drain. After framing the toilet drain opening, install the 4 by 3 closet bend, centered 12 in. from the finished wall behind the toilet. Install a piece of 2x4 blocking under the closet bend and end-nail through the joists on both ends. Use plastic plumber's tape to secure the bend to the 2x4. What really anchors the closet bend, however, is the *closet flange*, which is cemented to the closet bend and screwed to the subfloor.

The flange is screwed to the subfloor, yet it will sit atop the finish floor when it's installed. If the finish floor is not in yet, place scrap under the flange so it will be at the correct height. If, on the other hand, the flange is *below* the finish floor, you can build up the flange by stacking plastic *flange extenders* until the assembly is level with the floor. Caulk each extender with silicone as you stack it, and use long closet bolts to resecure the toilet bowl. (Check with local codes first because not all allow extenders.)

Once you've secured the closet bend, add pipe sections to the bottom of the bend, back to the takeoff fitting on the main drain that you installed earlier. Maintain a minimum slope of ¼ in. per ft., and support drains at least every 4 ft. Dry-fit all pieces, and use a grease pencil to make alignment marks on pipes and fittings.

PRO TIP

If the neoprene sleeve inside a no-hub coupling won't slide on easily, it may have a small stop lip inside—sort of a depth gauge to stop the incoming pipe in the middle of the sleeve. Soap the inside of the sleeve to reduce friction. You could use a utility knife to trim off the lip, but that would be more time-consuming and you're likely to puncture the sleeve.

Drilling and Notching Studs and Joists

It's often necessary to notch or drill framing to run supply and waste pipes. If you comply with code guidelines, given in "Maximum Sizes for Holes and Notches" on p. 373, you'll avoid weakening the structure. Although that table is based on the following rules of thumb, remember that local building codes have the final say.

Joists

You may drill holes at any point in the span of a joist, provided the holes are at least 2 in. from the joist's edge and don't exceed one-third of the joist's depth. Notches are not allowed in the middle third of a joist span. Otherwise, notches are allowed if they don't exceed one-sixth of the joist's depth.

Studs

Drilled holes must be at least 5/8 in. from the stud's edge. Ideally, holes should be centered in the stud. If it's necessary to drill two holes in close proximity, align the holes vertically, rather than drilling them side by side. Individual hole diameters must not exceed 40% of the width of a bearing-wall stud, if those studs are doubled and holes don't pass through more than two adjacent doubled studs; hole diameters must not exceed 60% of the width of nonbearing-wall studs. Notch width may not exceed 25% of the width of a bearing-wall stud or 40% of the width of a nonbearing-wall stud.

Edge Protection

Any pipe or electrical cable less than 1¼ in. from a stud edge must be protected by steel nail plates or shoes at least 1/16 in. thick to prevent puncture by drywall nails or screws.

Pipe Slope

DWV pipes slope, so before drilling or notching framing, snap sloping chalklines across the stud edges, then angle your drill bits slightly to match that slope. Drill holes ¼ in. larger than the outside dimension of the pipe so the pipe feeds through easily. Nonetheless, if DWV pipe runs are lengthy, you may need to cut pipe into 30-in. sections (slightly shorter than the distance between two 16-in. on-center studs) and join pipe sections with couplings. That is, it may be impossible to feed a single uncut DWV pipe through holes cut in a stud wall.

Notching and Drilling Limits

JOISTS

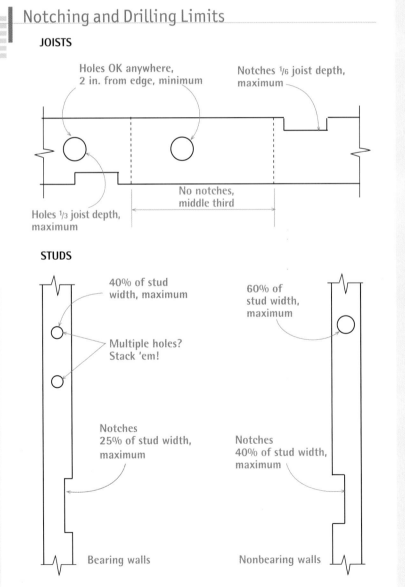

Holes OK anywhere, 2 in. from edge, minimum

Notches 1/6 joist depth, maximum

No notches, middle third

Holes 1/3 joist depth, maximum

STUDS

40% of stud width, maximum

Multiple holes? Stack 'em!

Notches 25% of stud width, maximum

Bearing walls

60% of stud width, maximum

Notches 40% of stud width, maximum

Nonbearing walls

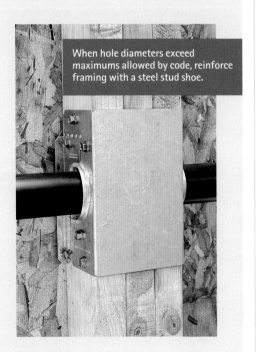

When hole diameters exceed maximums allowed by code, reinforce framing with a steel stud shoe.

The toilet flange (orange ring) will sit atop the finish floor. If the finish floor has not been installed, place scrap under the flange, elevating it to the correct height. This assembly is essentially the same as that shown in "Constricted Spaces" on p. 368.

Support vent stacks in midstory by using plumber's strap to tie stacks to blocking between studs.

Other fixture drains. Run the 1¼-in., 1½-in., and 2-in. fixture drains up from the main drain takeoff. Drains must slope at least ¼ in. per ft., and all pipe must be rigidly supported every 4 ft. and at each horizontal branch connection. Support pipes with rigid plastic pipe hangers (p. 365), or plastic-pipe strap, as shown above. Support stacks at the base and at midstory by strapping or clamping the pipe to a 2x block running between the studs or by using stack clamps.

Run the tub branch drain to the subfloor opening where the tub trap arm will descend. Pipe stub-outs for lavs and sinks should stick out into living spaces 6 in. or so; you can cut them off or attach trap adapters later. All branch drains end in a *sanitary tee*. The horizontal leg of the tee receives the trap arm from the fixture, and the upper leg of the tee is the beginning of the branch vent.

Vent runs. Next, assemble vent runs, starting with the largest vent—often the 2-in. or 3-in. pipe rising from the combo fitting below the closet bend. Individual branch vents then run to that vent stack, usually joining it in an *inverted tee fitting*, typically 4 ft. to 5 ft. above the floor. Support all stacks in midstory with clamps or straps. Horizontal runs of 1½-in. branch vents must be

at least 42 in. above the floor, or 6 in. above the flood rim of the highest fixture, and those runs typically slope upward at least ¼ in. per ft. Continue to build up the vent stack, with as few jogs as possible, until it eventually passes through a flashing unit set in the roof. For code requirements at the roof, see "Vent Termination" on p. 371.

TESTING THE DWV SYSTEM

Once you've assembled all pipes of the DWV system and connected it to the sewer main—but before hooking up fixtures—test for leaks. A common test is to fill DWV pipes with water, after capping the stub-out for each fixture drain and blocking the combo fitting at the foot of the building drain—as described later in this section. Use a garden hose to fill the largest stack: All DWV pipes are interconnected, so you need to fill only one stack to fill all. Should you see leaks, drain the system, fix the leaks, and refill. If you see no leaks, allow the water to stand at least overnight or until the inspector signs off on your system.

There are several types of pipe cap. Reusable rubber caps or plugs eliminate the need for gluing. A *jim cap* fits over the end of a pipe and tightens with a ring clamp. *Test plugs* fit into pipe ends and are expanded by a wing-nut assembly. The most common and least expensive, however, is a *glue-on cap* that fits inside a DWV pipe stub. Allow pipe cement to dry a day before filling pipes with water. When the test is completed, drain the system and cut off the small sections of drainpipe in which caps are glued. Where a stack is several stories high, this is the only type of cap guaranteed not to be dislodged by a weighty column of water.

The linchpin to this pipe-filling test, however, is a *double dynamiter*, a spring-loaded double test plug that fits into the T-Y combo at the foot of

the building drain. As shown in the top photo below, this tool has two rubber balls that can be expanded or contracted by turnscrews on the shaft. Insert the balls so that the forward one lodges in the drainpipe, then expand that ball; the second ball should block the open leg of the combo. To release the water, contract the balls of the double dynamiter in the order in which you expanded them. Loosened, the forward ball will allow the test water to run down the drain; releasing the second ball allows you to remove the tool. Label the respective turnscrews so you don't confuse them: If you release the second ball first, you may get a faceful of water.

If there are finish ceilings in place below new pipes and you don't want to risk wetting them with a failed connection, use an *air-pressure test* in which all openings (including stacks) are sealed. Typically, an inflatable bladder attached to a gauge is inserted into a cleanout at the base of the soil stack, and air is pumped into the DWV system. If the gauge shows no pressure loss over a given period, the inspector signs off.

Roughing-in Supply Pipes

Water-supply pipes are easier to run than DWVs because they're smaller and don't need to slope. Metal supply pipes should be bonded to the house's electrical grounding system (see p. 292).

Run supply pipes to fixtures once hot and cold trunk lines are connected. Run ¾-in. trunk lines, using ½-in. pipe for branch lines serving two fixtures or fewer. Individual supply risers for toilets and lavatories are often ⅜ in. You save some money by using smaller-diameter pipes, but the main reason to reduce pipe diameter is to ensure adequate water pressure when several fixtures are used simultaneously. *Reducing tees*, such as the ¾ by ½ shown in the bottom photo on p. 353, provide a ½-in. branch takeoff from a ¾-in. trunk line.

Support horizontal runs of copper supply pipe at least every 6 ft., but if pipes run perpendicular to joists, plumbers usually secure the pipe every second or third joist. Support vertical runs of copper at every floor or every 10 ft., whichever is less. Support horizontal runs of CPVC supply pipe every 3 ft.; vertical runs should be supported every 10 ft., with clamps or plumber's strap attached to blocking. Support PEX tubing every 32 in. on horizontal runs and every 10 ft. on vertical runs (with midstory guides).

Keep hot and cold pipes apart at least 6 in. They should never touch. To conserve energy, reduce utility bills, and get hot water sooner at fixtures, install closed-cell foam insulation

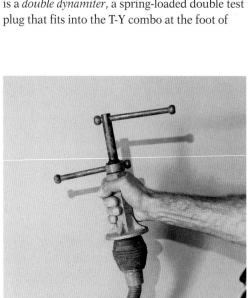

When filling the DWV system for testing, use a double dynamiter to block the combo fitting at the base of the main drain.

Testing plugs. Clockwise, from bottom: 1½-in. test plug, 3-in. test plug, 4-in. test plug, and "jim cap." The first three plugs are inserted into pipes and expand via a wing nut; the jim cap slips over the outside of a pipe and tightens with a band clamp.

sleeves on hot-water pipes. As noted earlier, water-supply stub-outs should protrude at least 6 in. into living space. To hold stub-outs in place, solder them to perforated copper straps nailed or screwed to studs.

Code requires shutoff valves for every fixture riser. Supply pipes to outdoor spigots or unheated rooms should have shutoff valves and unions within the main basement so pipes can be drained. Install *water-hammer arrestors* on branch lines to appliances such as washers or dishwashers, whose solenoid valves stop water flow so abruptly that pipes vibrate and bang against the framing.

To test the supply system before installing drywall, solder caps onto fixture stub-outs and turn on the water. (If you're installing CPVC supply, cement caps onto stub-outs.) If there are no leaks, install steel nail-protection plates over any pipes that lie within 1¼ in. of a stud edge, or use *steel stud shoes* over notched studs. Then install finish surfaces.

Installing Fixtures

Before you can install a new fixture, there's often an old one to remove. If it's necessary to shut off water to several fixtures during installation, capping disconnected pipes will allow you to turn the water back on even if all the new fixtures haven't been installed.

DISCONNECTING FIXTURES

Before disconnecting supply pipes, shut off the valves that control them. As mentioned, code requires a shutoff valve on each fixture riser, but older systems may have only a main valve that shuts off water to the whole house. After shutting off the controlling valve, open the faucets to drain the water.

Lavatory or sink supply pipes may have unions that can be disconnected using two pipe wrenches. Otherwise, *water-supply nuts* (water nuts) will connect the tops of supply risers to threaded faucet stems on the underside of the basin. To loosen water nuts, use a basin wrench, which has a shaft 10 in. to 17 in. long and spring-loaded jaws set at a right angle to the shaft. If the lav is old and you intend to reinstall it, save the water nuts because the threaded faucet stems may be nonstandard.

To disconnect a fixture's drainpipe, use two pipe wrenches to loosen the slip-nut coupling of the P-trap. If older galvanized couplings have seized up, heat them with a propylene gas torch and tap them lightly with a hammer to free the joint. Then try again with wrenches. Be sure to

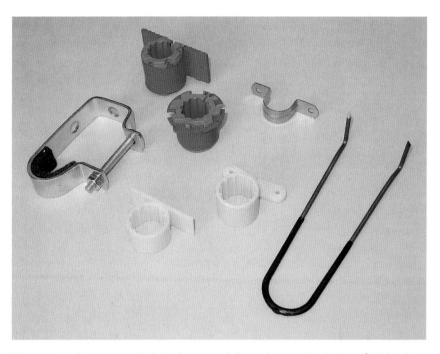

Water-supply pipe supports. Clockwise from upper left: two Acousto-Plumb clamps (which reduce pipe noise by cushioning vibrations), a copper pipe strap, a ³/₄-in. by 6-in. plastic-coated wire hook, two plastic suspension clamps, and a felt-lined J-clamp.

After running DWV branch drains to fixtures, install supply-pipe assemblies and solder stub-outs to perforated strap.

wear a respirator to avoid inhaling smoke from gaskets and such.

Once you've disconnected the drain and supply pipes, lift the lav/sink off its wall hanger, pedestal, or cabinet base and set it aside. An old cast-iron lav can be quite heavy, so lift it with the aid of a helper. Place a plastic bag over the drainpipe stub, and secure it with a rubber band to keep sewer gases at bay. Disconnect fittings carefully if you want to reuse them.

Setting a Pedestal Sink

Installing a pedestal sink takes planning, a lot of adjusting, and two people. For starters, determine well in advance the height of the 2x blocking needed to anchor the sink so you can cut that board into stud walls well before the drywall goes up.

Preattach the sink's hardware before mounting it on the pedestal. Next, level the pedestal base, shimming it as needed. *Ribbed plastic shims* (also called ribbed stability wedges) work well for this task because their ribs keep them from slipping, even if it's necessary to stack wedges on a badly out-of-level floor. Once the base is level, set the sink atop it and check it for level in two directions—front to back and side to side—using two torpedo levels, as shown in the left photo on p. 380. Chances are, you'll need to reset the sink several times to get it level and stable because sinks and pedestals are often not perfectly mated.

Once you're pleased with the sink's placement, use a pencil to mark the locations of the lag-screw holes on the wall. Remove the sink, predrill pilot holes, replace the sink, check for level again, and line up the pilot holes with the mounting holes in the back of the sink. Before inserting the lag screws, however, lift the sink slightly so a helper can slide out the pedestal and attach the drainpipes. That done, slide back the pedestal and reset the sink, then tighten the washered lag screws to secure the sink. But don't overtighten or you'll crack the porcelain. Attach the supply risers top and bottom, test for leaks, and you're done.

It's much easier to attach hardware to the underside of a sink before mounting it. Here, the plumber uses flexible stainless-steel lines to connect hot and cold faucet valves to the spout inlet. The threaded bottom of each faucet tee—one is visible, at right—receives a 3/8-in. water-supply riser and a water nut that holds it tight.

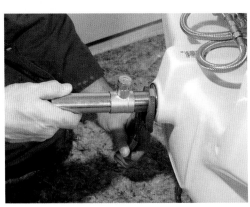

Apply plumber's putty or silicone under the flange of the drain tailpiece, insert it into the drain hole, then use sliding-jaw pliers to tighten the locknut on the underside of the basin.

PRO TIP

When connecting supply lines to fixtures, code requires hot on the left, cold on the right. When your eyes are shut tight against soapsuds and you're fumbling for a faucet, it's reassuring to know which spigot is which.

To remove a toilet, shut off the water by shutting the chrome *fixture stop* near the base of the unit. Flush the toilet and remove the remaining water with a cup or an inexpensive plastic hand pump. Disconnect the tank from the toilet bowl by loosening the bolts that hold the sections together. If the tank is wall hung, use a *wide-jaw spud wrench* to loosen the slip nut between the tank and the bowl. The toilet bowl is fastened to the floor by two bolts that rise from the floor flange; unscrew the nuts capping the bolts on both sides of the bowl. Rock the toilet bowl slightly to break the wax seal on the bottom. Then lift up the bowl and immediately block the drainpipe by stuffing it with a plastic bag containing wadded-up newspapers.

Tub drain assemblies may be hidden in an end wall or they may exit into a hole cut into the subfloor under the drain. The drain and overflow assembly is usually held together with slip couplings, so use a pipe wrench to loosen them. If the drain is a solid piece, cut through it. Supply pipes may be joined with unions or they may be soldered; it's easiest just to cut through supply risers. With those pipes disconnected, you can move the tub.

If it's a standard tub (rather than a freestanding tub), you may need to cut into the finish surfaces at least 1 in. above the tub to expose the tub lip, which is often nailed to studs. If you're discarding the tub and don't care about chipping its enamel, use a cat's paw to pull the nails. If the tub is too heavy or tightly fit to slide out of its alcove, you may need to cut the studs of the end wall so you can slide the tub out.

INSTALLING LAVATORY BASINS AND KITCHEN SINKS

Lavatory basins are supported by pedestals, cabinet counters, legs, wall-mounted brackets, or a combination of these, whereas kitchen sinks almost always attach to base cabinet countertops. Counter-mounted lavs or sinks are particularly popular because of the storage space underneath.

Preassemble the hardware. Before mounting a sink or lav, attach its hardware, including faucets, spout, and the drain *tailpiece*. These connections are easier to make when the fixture is upside down before installation. Insert the threaded faucet stems through predrilled holes in the sink or lav body, and tighten the washered nuts on the underside. Many manufacturers supply a rubber gasket, but when that's lacking, spread a generous layer of plumber's putty between the metal and the porcelain. Don't overtighten. Once the faucets are secure, you can

Lavatory Assembly

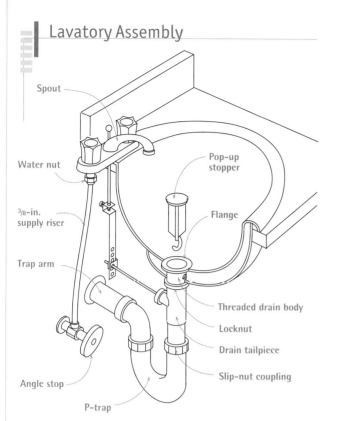

- Spout
- Water nut
- 3/8-in. supply riser
- Trap arm
- Angle stop
- P-trap
- Pop-up stopper
- Flange
- Threaded drain body
- Locknut
- Drain tailpiece
- Slip-nut coupling

connect the risers loosely to the threaded faucet stems, allowing you to reposition them if needed when attaching their lower ends to the angle stops.

Set the unit. Sink (or lav) installations vary, depending on whether the unit is surface mounted, under mounted, flush mounted, wall mounted, or set atop a pedestal. Self-rimming units are among the most common. Once you've attached the hardware, apply a bead of silicone caulk to the sink lip, turn the unit over, and press it flat to the surface (or underside) of the counter. Some sinks need nothing more to secure them, although many have mounting clips similar to those shown at right. Wall-mounted models slip down into a bracket, which must be lag-screwed to blocking attached to studs—preferably *let into* the stud edges. Level the sink front to back and side to side.

Connect the drain. With the sink or lav in place, connect the drainpipe. To the drain stub sticking out of the walls, glue a threaded male *trap adapter*, which will receive a slip coupling. Slide the trap arm into the coupling, but don't tighten it yet. The other end of the trap arm turns down 90° and, being threaded, couples to an adjustable P-trap, which you can swivel so that it aligns to the tailpiece coming down from the lav. The other end of the P-trap has another slip coupling, into which the sink tailpiece fits. When trap pieces are correctly aligned, tighten the slip couplings.

Kitchen sinks are much the same, except that the upper part of a sink tailpiece is threaded to

Sink-Mounting Details

For rimless and self-rimming sinks, first set the mounting device or sink edge in plumber's putty, which will compress.

RIMLESS SINK

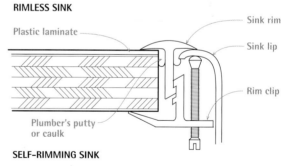

- Plastic laminate
- Plumber's putty or caulk
- Sink rim
- Sink lip
- Rim clip

SELF-RIMMING SINK

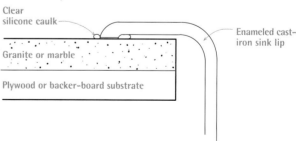

- Clear silicone caulk
- Granite or marble
- Plywood or backer-board substrate
- Enameled cast-iron sink lip

UNDERMOUNT SINK

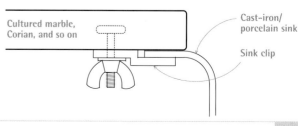

- Cultured marble, Corian, and so on
- Cast-iron/porcelain sink
- Sink clip

It usually takes several tries and some fine-tuning to level the pedestal, level the sink in two directions, and lag-screw the sink to a 2x blocking let into the studs.

Because the slot in the bank of the pedestal is narrow and the wall is close, there won't be enough room to tighten slip-nut couplings on the drain. Instead, after starting the lag screws, lift and support the front of the sink while a helper slides the pedestal forward. After connecting the drain fittings, slide the pedestal back and lower the sink.

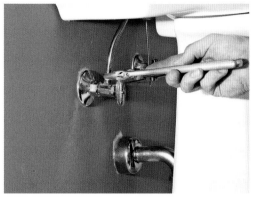

Check the sink for level one last time, tighten the lag screws to secure the sink, and connect the supply risers top and bottom.

tighten to the bottom of a strainer body. To drain double sinks, use the hookups shown in "End-Outlet Continuous Waste," below. Back-to-back lavs or sinks can also share a common drain, by using a figure-5 fitting as shown on p. 365.

Connect supply pipes. To each supply pipe stub-out, attach a shutoff valve, typically an *angle stop* with a compression fitting. Slide the angle stop's ½-in. socket over the stub-out, and tighten the fitting so that the ferrule inside compresses and forms a positive seal. Alternatively, you can sweat ½-in. male threaded adapters onto the stub-outs, wrap Teflon tape on the threads, and screw on a shutoff valve with a ½-in. threaded female opening.

Riser attachments depend on whether you install rigid chromed tubing, which inserts into a compression fitting on the angle stop, or a flexible braided supply line, which has nuts on both ends. Rigid tubing must be shaped with a tubing bender and cut to exact length, whereas braided supply can be easily twisted or looped so it fits.

End-Outlet Continuous Waste

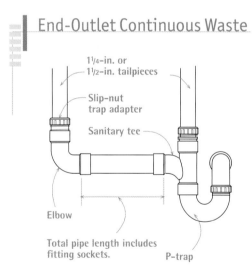

1¼-in. or 1½-in. tailpieces

Slip-nut trap adapter

Sanitary tee

Elbow

Total pipe length includes fitting sockets.

P-trap

A common assembly for double sinks or lavatories.

Factory-Installed Tank-to-Bowl Connection

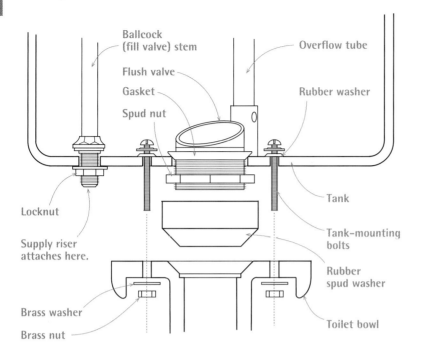

Ballcock (fill valve) stem

Flush valve

Gasket

Spud nut

Overflow tube

Rubber washer

Locknut

Supply riser attaches here.

Brass washer

Brass nut

Tank

Tank-mounting bolts

Rubber spud washer

Toilet bowl

To avoid corrosion, use stainless-steel or brass bolts and nuts.

PRO TIP

Don't overtighten closet bolt nuts or you'll crack the porcelain on the foot. Once the nuts seem snug, gently try to rock the bowl. If it doesn't rock, the nuts are tight enough, though you should return after the toilet's been used for a few weeks and snug the nuts one last time.

INSTALLING A TOILET

When you're ready to install the toilet, remove the plastic bag you inserted earlier in the closet bend to block sewage gases. Place closet bolts in the closet flange if you haven't already done so.

Set the bowl first. Then attach the tank to the bowl. If you're reinstalling an old toilet, as shown in the bottom left photo on p. 382, leave the parts connected, and set the toilet as one piece. But if it's a new toilet, setting the base first is easier on your back. Place the wax ring in the closet flange so that the ring's plastic funnel centers in the flange. Some manufacturers recommend placing the wax ring on the toilet horn and then inverting the toilet bowl, but the wax ring may not adhere and the funnel may not align.

Have help aligning closet bolts to holes as you set the toilet bowl. Don't rock the bowl when setting it, which could excessively compress the wax ring on one side, creating a gap. Instead, press the bowl down evenly, then use a small adjustable wrench to tighten the nuts gradually, alternating sides, until the bowl is secure. Place a torpedo level atop the bowl edge to see if the unit is level side to side and front to back. If the bowl needs shimming, use plastic shims, which can be chiseled or cut flush to the toilet foot so they're

not visible. Don't trim the closet bolts until you've attached the tank and tested the unit for leaks.

Mount the tank. Standard two-piece toilets have tanks that bolt directly to bowls. In addition to bolt holes, tanks have two fittings on the bottom: a *threaded ballcock stem*, which is screwed to the supply riser, and a larger flush valve, which is tightened to a spud nut. Typically, a rubber spud-nut washer covers the spud nut and cushions the tank–bowl juncture to prevent leaks; there may also be a separate, preinstalled sponge-rubber gasket to cushion the tank and bowl. Tighten the spud nut and position the spud-nut washer, insert the washered tank-mounting bolts into the bottom of the tank, and set the tank atop the bowl so that bolts line up with the holes in the bowl. Carefully follow the manufacturer's instructions about caulking mating surfaces because some caulking compounds may deteriorate the gaskets.

To prevent the tank-bolt threads from turning and cutting into the rubber washers or gaskets, hold the bolts steady with a long screwdriver as you tighten the nuts on the underside of the bowl shelf, using an adjustable wrench. Moving from one side to the other, tighten the nuts snugly. Use only brass or stainless-steel bolts and nuts. Connect the water supply, fill the tank, flush the toilet several times, and check for leaks. If there's

PRO TIP

Use only noncorroding (brass or stainless-steel) screws and bolts to secure the closet flange or the toilet bowl; other materials will corrode. To help you align the bolt holes on the bowl with bolts in the closet flange, buy extra-long, 3-in. by $5/16$-in. closet bolts. They'll be long enough to line up easily, even when the wax ring is in the way, and you can trim excess length without difficulty.

INSTALLING A TOILET

Dry-set and shim-level the toilet bowl before centering the wax ring over the closet flange. Once the bowl is placed on the ring, it can't be lifted without replacing the ring.

In most cases, set the toilet base before attaching the tank. Here, an already assembled toilet is being reset onto a new wax ring after a new tile floor was installed. After feeding closet bolts through the bowl, apply even pressure to seat the bowl on the wax ring.

leaking between the tank and the bowl, tighten the nuts. If there's leaking only near the foot of the bowl, the wax ring may have failed: In this case, pull the toilet and replace the ring. If there are no leaks, trim the closet bolts and caulk around the perimeter of the foot.

Toilet-supply connections are essentially the same as sink or lavatory risers. The standard toilet-supply riser is ⅜-in. chrome tubing that attaches (at the top) to a threaded ballcock stem on the underside of the tank and a ⅝-by-⅜ angle stop at the bottom. A better option is a ⅜-in. flexible braided stainless-steel supply line: It won't crimp, attaches to the same fittings, and can be easily disconnected.

Bidets. A bidet is easier to install than a toilet. Although a bidet requires hot- and cold-water connections, only liquid waste (drain water) is produced, so a 1¼-in. drain will suffice. Mount the bidet base securely, but it doesn't need to be seated in a wax ring. In fact, the drain takeoff is similar to that of a tub, which is described next.

INSTALLING A BATHTUB

First and foremost, follow the installation instructions provided with your tub. Failure to do so could void your warranty. For most models, after framing the three-walled alcove around the tub, attach 2x4 ledgers to support the tub lip on three sides. Then cut an opening in the subfloor for the drain assembly, dry-fit the tub, and check for level. For lightweight steel and fiberglass tubs, many installers next remove the tub, apply 30-lb. building paper over the subfloor, place a mortar bed 1 in. to 2 in. thick, and set the tub into it. The mortar bed stabilizes the tub and minimizes flexing, which could lead to leaks around the drain or surface cracking. Of course, place the tub in the mortar before it sets so the mortar will conform to the shape of the tub. If the tub is made of cast iron, on the other hand, don't bother with a mortar bed—cast iron doesn't flex appreciably. But it's famously heavy, so be sure to have at least three workers on hand to move it.

Drain and overflow assembly. Before the final installation of the tub, preassemble the tub's drain and overflow assembly and test-fit it to the tub openings. Slip-nut couplings make adjusting pipe lengths easy. Once the tub back is in the alcove, install the assembly in the tub: Put a layer of plumber's putty between any metal-to-enamel joint that isn't gasketed. While a helper holds the assembly to the end and underside of the tub, hand screw the threaded strainer into the *tub shoe (waste ell)*. Then use a *strainer wrench* to tighten it

Once water-supply and DWV pipes are roughed in and 2x4 ledgers are nailed to the studs to support the tub lip, slide the tub into its alcove.

solidly. The overflow plate inside the tub screws to a mounting flange in the overflow ell. There are many types of drain cap (stopper) mechanisms; follow the manufacturer's instructions.

The last drain connection to be made is a P-trap, which slides onto the tub tailpiece descending from the tee. Adjust the trap so that it aligns with the branch drain roughed in earlier. *Note:* If the drain assembly will be inaccessible, code requires a glued-together drain joint. The only exception allowed is the slip-nut coupling that joins the tub tailpiece to the P-trap.

Water supply. Next, attach the tub's supply pipes. Level and mount a *pressure-balancing valve* to the cross brace let into the end-wall studs. (The valve, also called an antiscald valve, is typically set at a maximum of 120°F to prevent scalding.) Follow the manufacturer's instructions for attaching the pipe to the valve. Although gate and ball valves are usually closed when sweating copper pipe to them, balancing valves may need to be open or disassembled before sweating pipe to the valve bodies.

Like most shower/tub valves, a balancing valve has four pipe connections: one each for hot- and cold-supply pipes, one for the pipe that runs to the shower arm, and one that services the spout. To mount shower arms and spouts, screw brass, threaded female drop-eared ells to the cross braces. Because chrome shower arms and spouts can get marred while finish surfaces are being installed, screw in 6-in. capped galvanized nipples. The faucet stem(s) and the balancing valve are protected with a plastic cover until the finish work is done. Turn the water on and test for leaks. Once you're sure there are none, you are ready to close in the walls around the tub.

Replacing a Water Heater

Most municipalities require a permit to replace a water heater, primarily because they want to ensure that the heater's TPR valve is correctly installed. Even though most local codes allow homeowners to replace water heaters, hire a licensed plumber for this job unless you have a lot of plumbing experience. Plumbers know which brands and hookups require the fewest service calls and can assess the condition of vent pipes and replace them if needed. Besides, thanks to wholesale discounts, plumbers can probably install a new unit for only slightly more than it would cost you if you bought the heater and fittings at retail prices.

A plumber should also be willing to peer through the vent thimble with a flashlight and mirror to check the chimney's interior. All manner of debris can accumulate in the bottom of a chimney—from soot to nests—and that debris can block a chimney, hamper flue draft, and possibly force carbon monoxide into living areas. The National Fire Protection Association suggests annual chimney and flue inspections and whenever a new type of burning appliance is vented into the flue, but inspection is really not a plumber's job. The Chimney Safety Institute of America offers a state-by-state listing of chimney

Tub Drain-and-Overflow Assembly

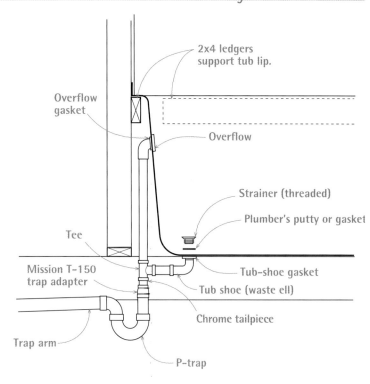

2x4 ledgers support tub lip.

Overflow gasket

Overflow

Strainer (threaded)

Plumber's putty or gasket

Tee

Mission T-150 trap adapter

Tub-shoe gasket

Tub shoe (waste ell)

Chrome tailpiece

Trap arm

P-trap

If the tub drain is not accessible, plumbing codes require that joints be glued together to prevent leaks. To join the tub tailpiece to the trap assembly, use a Mission T-150 trap (1½-in. tubular to 1½-in. pipe).

A shower wall at rough-in. A pressure-balancing antiscald valve (which mixes hot and cold water) is at the bottom of the assembly. The smaller valve above it controls water volume. Note the nail plates to protect the pipes.

Finishing touches. To avoid marring the surface of a chrome shower arm, insert a sliding-jaw pliers handle into the pipe and turn it into final position. When the bathroom is painted and all brightware is installed, remove the protective plastic from the shower walls.

PRO TIP

Don't accept water heaters whose boxes are bashed or torn. Water-heater elements are sensitive. If the box has been handled roughly or dropped, anode rods, liners, or valves may have been damaged.

Turn off the gas first—the stopcock will be perpendicular to the gas line—before removing an old gas-fired water heater. Use an adjustable wrench to disconnect the gas coupling.

Once the gas line is disconnected, attach a hose to the drain at the bottom of the water heater and drain the tank.

A new gas-fired water heater in mid-installation. The red-lever-handled shutoff valve on the cold-water pipe is required by code. After installing the TPR safety valve, plumbers will reattach the vent pipe.

Gas-Fired Water Heater

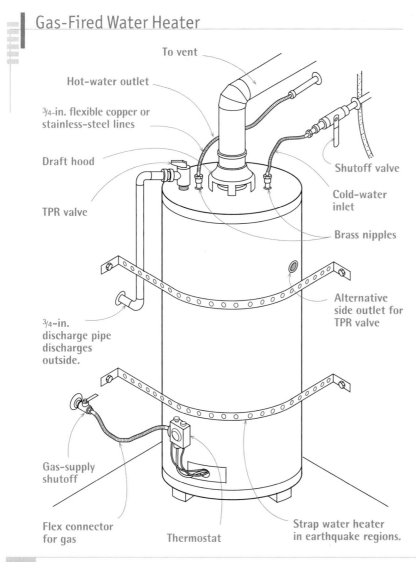

To vent

Hot-water outlet

¾-in. flexible copper or stainless-steel lines

Draft hood

TPR valve

¾-in. discharge pipe discharges outside.

Gas-supply shutoff

Flex connector for gas

Thermostat

Shutoff valve

Cold-water inlet

Brass nipples

Alternative side outlet for TPR valve

Strap water heater in earthquake regions.

services with certified staff (check out its website at www.csia.org).

Above all, installers should follow the water-heater manufacturer's installation instructions closely to ensure a safe installation and to safeguard the unit's warranty should the water heater fail to function properly.

Draining the old water heater. The specifics of disconnecting power or fuel to the old unit will vary, depending on whether the water heater is gas fired, fuel-oil fired, or electric. Once the installer has disconnected the fuel or power source, the water should be shut off and the tank drained. Typically, a hose will be attached to the drain valve at the bottom of the tank. Using a pair of pipe wrenches, unions (if any) on the hot- and cold-water pipes will be taken apart; if there are no unions, pipes will be cut 6 in. to 12 in. above the top of the heater—or a couple inches above the cold-water shutoff valve—by means of hacksaw or a wheeled cutter. *Caution:* Reciprocating saws aren't used because the vibration may weaken nearby pipe joints and cause leaks.

Installing the new heater. If the unit is gas or oil fired, the installer will disconnect the draft hood and vent pipe and either wire them up out of the way or set them to one side. As noted earlier, the plumber should inspect the vent pipes. If the hood or pipes are rusty or corroded, they should be replaced. When the old tank is empty, it can be walked out of the way. Be mindful of sharp edges on the newly cut pipes and the area around the old tank. *Note:* If codes require strapping the unit, steel straps should be bolted to the wall behind before putting the new unit in place. Finally, if the pad beneath the old tank is in poor condition or badly tilted, consider installing a new prefab concrete pad. Then the new water heater can be walked into position.

Making connections. What the installer does next depends on the size and condition of the pipes, what fittings are present, and, of course, what type of water heater it is. There's no single right way to assemble pipes, but the photos on the facing page show a typical installation in progress for a gas-fired unit. There are ¾-in. brass nipples screwed in the tank inlet holes, flexible stainless-steel lines, sweat-to-threaded male adapters, valves, and (at the top) ¾-in. rigid-copper trunk lines exiting to the upper floors. Flexible stainless-steel or flexible copper supply lines are highly recommended for top-of-tank connections: Female nuts on both ends make them easy to disconnect for future repairs.

Installing a TPR valve. Many new water heaters have preinstalled TPR valves. If there is none, the plumber will install a TPR valve into the threaded outlet atop the unit or in a side outlet a few inches down from the top of the tank, lightly coat the TPR valve's threads with pipe compound, and then use a pipe wrench to install the valve. Next, the plumber will install a discharge pipe into the TPR valve's threaded outlet. The pipe may be galvanized or rigid copper—but not plastic—and must slope downward. The discharge pipe should be terminated about 6 in. above the floor, at a safe location where hot water won't scald anyone if it discharges.

Note: Threaded pipe fittings should be coated with pipe compound or wrapped with Teflon tape to ensure a positive seal.

Final steps. The plumber will check the water heater for level, shim the base as needed, and tighten the earthquake straps, if any. When all fittings are connected, turn on the cold water to fill the tank. Open the hot-water faucets to expel air. When the tank is full, water will gush from the faucets. At that point, shut the faucets, and reconnect the fuel or power source as specified in the manufacturer's instructions. *Note:* If the installer disconnected the bonding jumper wires from the hot- and cold-supply pipes, those wires should be reclamped now to ensure proper grounding for the house's electrical system.

PRO TIP

When replacing a water heater, put unions and lever-handle ball valves on both the hot- and the cold-water pipes. Code requires a shutoff valve only on the cold-water pipe, but having them on both can make periodic drainage and repairs easier.

Tankless water heaters are more expensive and more complicated to install than tank–style heaters because tankless models require expertise in plumbing, electrical wiring, and gas fitting. The yellow tubing entering at right is a CSST gas line.

Installing a Tankless Water Heater

Tankless water heaters, also known as *on-demand, instant,* or *flash* water heaters, don't have storage tanks for heated water. When someone opens a hot-water faucet and water begins to flow through a tankless heater, a flow sensor ignites the unit's burner, and a heat exchanger honeycombed with coiled water pipes heats up almost instantaneously. When the faucet or shower valve shuts off and the water stops flowing, the sensor shuts off the burner.

Tankless water heaters are generally more complicated and more expensive to install than a traditional tank heater but, on the whole, are less expensive to operate because there's no energy wasted keeping a tankful of water hot around the clock. In other words, tankless water heaters consume energy only when hot water is needed. About the size of a suitcase, tankless units are a great favorite when space is tight, too.

Their other great draw is that properly sized tankless units never run out of hot water. (Endless hot water is not free, of course: Linger too long in the shower and you'll squander those operating efficiencies.) As a rule of thumb, a tankless unit with 140,000-Btu input will suffice for a one-bath home with one or two users; a 190,000-Btu input should do for a two-bath home with two to four people. (These figures assume a unit with an energy factor of 0.82 or greater.) If

PRO TIP

In areas that lose power periodically, a tank of hot water is a nice thing to have on hand. If you live in an area where the power goes out regularly, think twice about going tankless.

your home has three baths or more, a pair of tankless water heaters located close to points of use may be a better call. Qualified installers can help calculate an optimal size for your home.

LOCATING THE UNIT

If you locate a tankless water heater as close as possible to the fixtures that use the most hot water, you will lose less heat in transmission. In new construction, that's often easy to do. In renovation, however, new tankless units are frequently installed near the site of the old tank-style heater that's being replaced—primarily because pipe hookups are already there.

Before deciding, however, consider what connections your new water heater needs. Most tankless units require ¾-in. water pipes, so if your old heater has only ½-in. water lines, that's one less reason to use that location. If your tankless heater is gas fired (as opposed to electric), it will require a ¾-in. gas line. Tankless units have greater Btu input requirements than tank-style water heaters. Electricity is rarely a problem because cables are easily routed and gas-fired units need only 110v to run fans, sensors, and the like.

Venting can be problematic, though, if you hope to use an existing exhaust vent. Whereas some tankless heaters have separate intake and exhaust pipes, many others have a single concentric *direct-vent pipe*—a pipe within a pipe—to supply fresh air and exhaust combustion gas. That pipe will have an optimal pitch to carry off combustion condensation; see installation instructions. Lastly, tankless models vary according to whether they can be installed inside or out and how close vents may be installed to operable windows, eave vents, and so on.

OUT WITH THE OLD, IN WITH THE NEW

If there's an old tank-style water heater, turn off electrical power and shut off the water supply to the tank. Use a voltage tester to be sure the power is off. Drain the tank, disconnect electrical connections, and cut the hot- and cold-water pipes to remove the old tank. If the old heater had a draft hood and exhaust vent, remove them, too: Most tankless models use concentric vent pipes.

Whether you are replacing an old water heater or installing a new one from scratch, you need five things: (1) a route for a vent pipe and an opening to the exterior, if you are installing the tankless unit indoors, (2) tie-ins to ¾-in. water pipes, (3) an electrical outlet for the unit's power cord, (4) a gas line if the unit is gas fired, and

Plumb and mount the water heater. Just visible to the right of the unit are existing copper water pipes—a convenient place to tie the tankless unit into house lines.

Connecting to the underside of a tankless water heater is easiest before it is mounted. Here, threaded stubs get an application of pipe dope and Teflon tape.

Sweating copper *leads* to valve assemblies is also easier beforehand. In the foreground, the (red) hot-water assembly is being soldered. When soldering, heat the fitting, not the pipe. Valve assemblies (valve kits) save time by combining fittings, including unions, shutoff valves, and drains.

This takeoff atop a tankless unit gives a good cross section of a concentric direct-vent pipe. Air for combustion is pulled in through the outer pipe; exhaust gases are expelled through the inner one.

(5) a sturdy place to wall-mount the unit—whether 2x lumber or, say, a ⅝-in. plywood panel.

Review the installation instructions. If space is tight, do as much work as you can before mounting the tankless unit. For example, install tees on existing supply pipes—so you can tie in the tankless unit's cold-water intake and hot-water outtake pipes—and make a cutout for the unit's vent pipe, if needed. Using flexible corrugated stainless-steel tubing (CSST) for gas pipe and flexible PEX for water-supply pipe will be easier to route in a renovation—if they are allowed by your local building codes.

STURDY MOUNTING, SOLID CONNECTIONS

Test-hang the tankless heater to make sure there are adequate clearance distances and room to make connections. Before mounting the unit, pros often preassemble hot- and cold-water shutoff valve assemblies. Apply pipe dope and Teflon tape to the threaded stubs on the underside of the unit's housing, and screw on a pair of union couplings. If you are working with rigid copper, this is also a good time to sweat (solder) pipe leads to valve assemblies, as shown in the photo at top right. Then attach a TPR valve to the outtake (hot-water) valve assembly.

Plumb the tankless unit and screw it to the wall, being sure to insert a vibration-suppression gasket between the unit's metal housing and the wall. To connect water pipes, start by screwing water valve assemblies to the union couplings on the bottom of the unit, then connect the sweated leads to house supply pipes.

On the other hand, if your water-supply system is flexible PEX tubing, it will employ compression fittings rather than soldered ones. *Note:* PEX may need a transitional fitting to attach it to the water heater. Because PEX will melt when exposed to open flame, it must not be directly connected to gas- or oil-fired water heaters and must be kept away from some types of flue pipe.

After connecting water pipes, flush the lines to dislodge any debris that may have collected during the installation.

Gas connections performed by a licensed gas fitter/installer are similar to water-pipe connections. A ¾-in. union fitting joins the gas line to a threaded stub on the underside of the heater. Galvanized and black steel gas piping need a *dirt leg* (sediment trap), a short leg of vertical pipe just upstream from unit connections, to act as a cleanout. The gas line should also have a shutoff valve. Check all gas-line connections for leaks by

At right center, the gas installer has added a new leg to feed the tankless water heater under the house. To the coupling below the blue shutoff lever, he will attach CSST—a flexible, code-approved material that is more easily routed than galvanized or black steel piping.

On the outside, the end of the vent pipe is covered by a rubber flange.

turning on the gas, spraying soapy water on joints, and looking for active bubbling.

VENT PIPES

Vent-pipe diameter, maximum vent length, pitch, clearances, adapters, and the like, will be specified in the installation instructions, so read them (and local code requirements) closely. If the heater is installed indoors, you may vent it up through a roof or out through an exterior wall. As a rule of thumb, keep vent runs as straight as possible—the fewer bends the better.

In the photo sequence, the tankless water heater was installed in a well-ventilated crawl-space. The location was optimal because a "wet wall" above serviced both a kitchen and a bathroom, but venting a concentric pipe with a 5-in. outer diameter (O.D.) up through the roof would have been impractical. So the installer placed an elbow atop the unit and from it ran vent pipe to the nearest exterior wall, maintaining a ¼-in.-per-ft. pitch *downward* toward the exterior. That slight downward pitch is necessary to allow condensation in combustion gases to run away from the unit and drip outside. (A tankless water heater has a combustion fan that pulls air into the burner; that same fan gently expels exhaust gases, even though the vent pipe has a slight downward pitch.)

At the exterior wall, trace the O.D. of the pipe onto the sheathing, and drill a hole in the center—through the sheathing and siding—to locate the opening outside. Vent kits typically come

The 5-in. concentric pipe shown here pitches slightly downward toward an exterior wall so combustion condensation will drain away from the heater.

with a simple rubber flange that slides over the end of the pipe, which can be caulked to the siding. You can also dress up the termination with a mounting plate. To keep rain from damming up behind a mounting plate, however, honor the First Rule of Flashing: Fit the top of the plate *under* the siding course *above* and *over* siding courses *below*.

Control panel

Hot-water outlet

Union

Temperature- and pressure-relief (TPR) valve

Drain

Water shutoff

CSST gas-supply line

Cold-water inlet

³/₄-in. union

Gas shutoff

Gas valve

A close-up of connections.

Kitchens and Baths

No other rooms are renovated as often as kitchens and bathrooms, in part because we've changed the way we live. In the old days, homeowners regarded kitchens and bathrooms as drab utility rooms, best situated at the back of the house, away from guests. Times change. These days, if you throw a party, everybody hangs out in the kitchen. Bathrooms aren't exactly spartan anymore, either. Today's kitchens and baths contain so many cabinets, counters, fixtures, and appliances that it takes careful planning to make them all fit—and still have room for people to move around. This chapter will help with that.

Kitchen Planning

The best kitchens can accommodate your personal tastes and lifestyles as well as your physical characteristics, such as your height.

WHAT GOES ON IN YOUR KITCHEN?

Start planning by imagining a day in the life of your kitchen, being as specific as possible about the activities—and the actors. Do you want a sunny spot to have coffee, read the paper, and wake up? Will the kitchen table double as a desk for homework? Or will the kitchen be a command center in which you field calls and arrange after-school carpools while tossing the salad?

Keep a notebook in your present kitchen and jot down observations about what goes on—as well as a wish list for what you'd like changed. Many entries will be cooking specific: Is there enough storage and counter space to prep several dishes? Does the cook like company? When you entertain, how large is the crowd? Is there a convenient place for cookbooks? Such consider-

An open floor plan and a granite-topped island allow the cook to visit with guests in the dining room while prepping food and mixing ingredients.

Standard Cabinet Dimensions

REFERENCE*	SPACE	DIMENSION (in.)
A	Height above the finish floor Kitchen countertops Bath vanity countertops	 36 32–34
B	Base cabinet depth	24
C	Height and depth of kickspace	4
D	Wall cabinet distance above: Standard countertop Sink and cooktop	 18 30
E	Depth of wall cabinet	12–15
F	Typical wall cabinet height (8-ft. ceilings)	30
G	Highest usable shelf	80

* Letter refers to "Figuring Dimensions" at right.

Figuring Dimensions

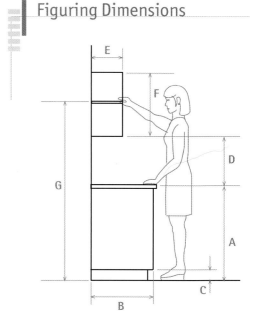

ations will be useful in evaluating your kitchen and establishing priorities for the new one.

CABINET HEIGHTS AND CLEARANCES

Over the years, architects, appliance designers, and builders have adopted a set of physical dimensions that, in theory, make kitchens safer and easier to use. As shown in "Figuring Dimensions" above, these dimensions work for most people but in the end may not suit everybody. As a rule of thumb, a counter is the right height if you can place your palms flat on it, with a slight bend to your elbow. If the standard counter height of 36 in. above the finish floor isn't right for you, lowering or raising it an inch or two may do the trick. However, if you're thinking of selling the house fairly soon, your ideal counter or shelf height may not appeal to the average buyer.

Equally important are the clearances needed to move easily in a kitchen—clearances that homeowners frequently overlook when laying out new kitchens. Be sure to allow enough room to open cabinet doors fully and still walk around them. Traffic lanes through work areas are vital because cooks frequently handle hot, sharp, or heavy objects. Keep a 60-in. minimal clearance if the work area doubles as a corridor. Ideally, though, family traffic should bypass the cooking space. If a kitchen has two or more doors, you may be able to reroute that unwanted traffic by eliminating one of them while gaining counter and cabinet space in the process.

COUNTER AND CABINET SPACE

Meal preparation consists of food prep, cooking, and cleanup, ideally with counter space for each job. Prepping the food—washing, cutting, and mixing—takes the most time, so give as much space as possible to counters near the sink and the cooktop.

Sink counters should be 24 in. wide on each side of the sink, though 36 in. is better, allowing plenty of room for food prep and the air-drying of pots and pans. Because dishwashers are 24 in. wide, they fit neatly under a 24-in. counter. If the kitchen is tiny and there's no undercounter dish-washer, 18-in.-wide sink counters are minimal. Have a backsplash behind the sink.

Cooktop counters should be at least 18 in. to 24 in. wide on both sides of the unit, and at least one side should be made of a heat-resistant material. Placing a stove on an exterior wall keeps exhaust-fan ducts short, but never place a gas stove in front of a window because a draft could blow out burners. The wall behind a stove should be washable.

By the refrigerator, next to the latch side, have a counter at least 15 in. wide so you can place things there as they go in and out of the fridge. If the refrigerator and the sink share a counter, the space should be 36 in. to 42 in. long. Because this counter is typically a food-prep area, you'll need a large surface to store countertop appliances.

Minimum Kitchen Work-Space Clearances

SPACE	DIMENSION (in.)
In front of base cabinet	36
Between base cabinet and facing wall	40
Between facing appliances	48
Work space plus foot traffic	60

COUNTER Outlets

Kitchen counters 12 in. wide or wider must have at least one electrical receptacle to serve them. All points on a counter must be within 2 ft. of a receptacle, and all counter receptacles must have GFCI protection. Chapter 11 addresses this.

Cabinet space has few rules. The best indicator of how much cabinet space you need is the number of appliances, bowls, and paraphernalia you own. Or use this rule of thumb: Figure 18 sq. ft. of basic storage plus 6 sq. ft. for each person in the household.

LAYOUT CHOICES: WORK AREAS

The person preparing and cooking the meal moves primarily in a space bounded by the refrigerator, the stove, and the sink—the so-called work triangle. When laying out such work areas, designers try to keep the distance traveled between the three points within 12 ft. to 22 ft. Three of the layouts shown at right feature a work triangle. The fourth is a single-line kitchen, but the distance traveled should be roughly the same.

U-shaped kitchens are the most practical because they isolate the work area from family traffic. Because cooks spend a lot of time rinsing veggies before dinner and washing pots after, put the sink at the base of the U, with the refrigerator on one side and the stove on the other. If one person preps food or washes while the other cooks, their paths won't cross too often. If possible, place the sink beneath a window so the eye and the mind can roam.

L-shaped kitchens are popular because they allow various arrangements. That is, you can put a dining table or a kitchen island in the imagi-

Recommended Counter Space and Clearances

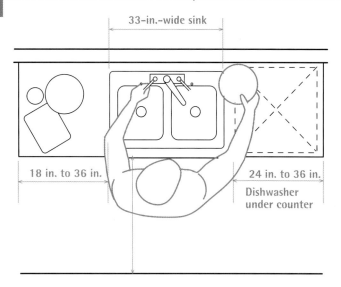

Each work area—food prep, cooking, and cleanup—should have adequate counter space so a cook can work efficiently, with enough clearance to move safely. Counters with dishwashers underneath must be at least 24 in. wide; otherwise, 18 in. is the minimum. See "Minimum Kitchen Work-Space Clearances" at left.

Common Kitchen Layouts

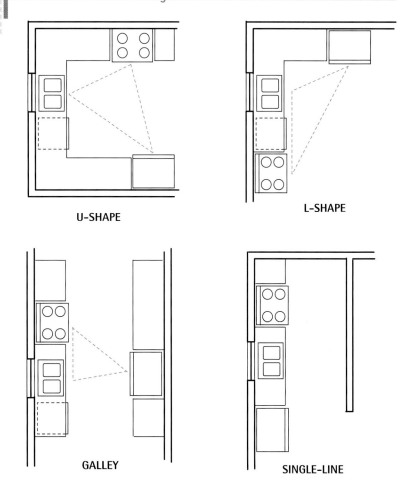

U-SHAPE

L-SHAPE

GALLEY

SINGLE-LINE

Kitchen Lighting Basics

Kitchen lighting should be a combination of natural light (windows), general lighting, and task lighting to illuminate specific work areas. For light that is both warm and efficient, combine incandescent and fluorescent bulbs. Warm fluorescent lights are another option.

General lighting can come from overhead fixtures, recessed ceiling lights, track lighting, or perimeter lighting. Mount ceiling lights 10 in. to 12 in. out from cabinet faces to illuminate kitchen surfaces evenly, while minimizing shadows cast by wall cabinets or by people using the counter. For an average-size kitchen (75 sq. ft. to 100 sq. ft.), ceiling-mounted general lighting should total about 30 watts LED or 200 watts incandescent; if there's recessed ceiling lighting, four 14-LED bulbs should be enough. For larger kitchens, figure 2.8 watts of LED or 20 watts of incandescent light per 10 sq. ft. of kitchen area.

Task lighting over sinks and cooktops should be at least two 10-watt LED or 75-watt incandescent bulbs. Ideally, task lighting should be placed behind a face trim board of some kind so that the bulbs shine more on the work surface than in one's eyes. To illuminate countertops, task lighting is often installed under wall cabinets, hidden by a face board or a cabinet rail. Low-voltage LED or slimline fluorescent bulbs can be shielded by a face board that's only 1¼ in. high. In general, under-cabinet lights should be two-thirds as long as the counter they illuminate. (See also "Kitchen Lighting" on p. 410.)

Adding Cabinets, Refining the Layout

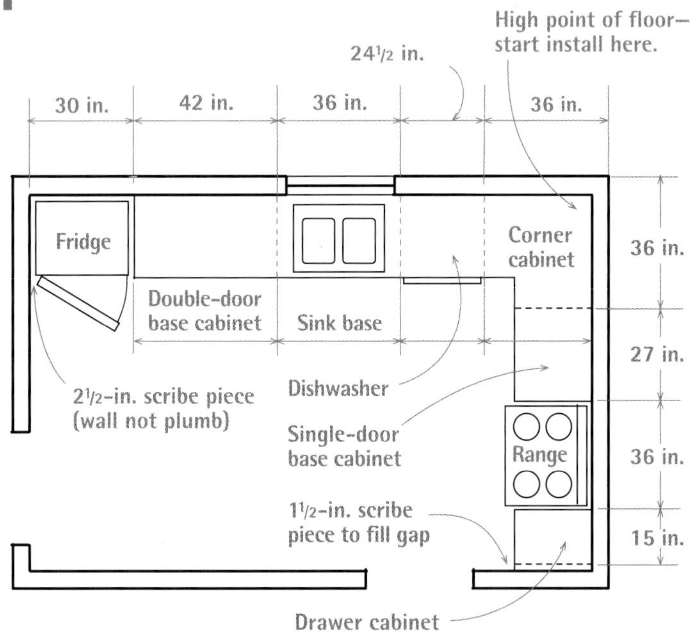

30 in. | 42 in. | 24½ in. | 36 in. | 36 in.

High point of floor— start install here.

Fridge

Double-door base cabinet

Sink base

Corner cabinet — 36 in.

2½-in. scribe piece (wall not plumb)

Dishwasher

Single-door base cabinet

27 in.

Range — 36 in.

1½-in. scribe piece to fill gap

15 in.

Drawer cabinet

Once you choose a layout that works well, use base cabinets to tie appliances and work areas together. On your floor plan, note room irregularities that could affect layout and installation. Using light pencil lines, mark cabinet and appliance locations onto the walls.

nary fourth corner. However, this becomes a somewhat less efficient setup if one leg of the L is too long. Again, position the sink in the middle.

Galley kitchens create efficient work triangles, but they can become hectic if there's through traffic. If you close one end of the galley to stop traffic, the galley should be at least 4 ft. wide to accommodate two cooks. To avoid colliding doors, never place a refrigerator directly across from an oven in a tight galley kitchen.

Single-line kitchens, common to small apartments, are workable if they're not longer than 12 ft. and there's a minimum of 4 ft. to the opposite wall. Compact, space-saver appliances can maximize both floor and counter space.

Islands are great in multiple-use kitchens, for they can provide a buffer between cooking tasks, a place to eat breakfast or read, or a place to sit and talk with the cook. Make the island roomy, with at least 10 sq. ft. to 12 sq. ft. of open counter space, so someone hanging out won't be in the flight path of hot dishes coming and going.

KITCHEN CABINET LAYOUTS

Once you've chosen a work layout that you like, make to-scale floor plans: A ¼-in. to 1-ft. scale provides a good amount of detail for a single room yet still fits on an 8½-in. by 11-in. sheet of graph paper. Include windows, doors, appliances, and cabinets. You may find it helpful to cut to-scale rectangles to represent the refrigerator, sink, and cooktop. If you cut them from different-colored paper or label each piece, you'll have an easier time trying out your layouts.

Basic layout. Refining the layout is a fluid process, but a few spatial arrangements are so common they're almost givens. Place the sink under a window. Don't put a refrigerator and a stove side

Elevations PLEASE!

After you have sketched a floor plan, draw elevations: They will show you how well kitchen components will work when used. When a dishwasher in the corner is being unloaded, for example, will its lowered door make it impossible to open nearby cabinet drawers? When lowered, will the door clear pulls on adjacent drawers and doors? Elevations also show how cabinets will relate to windows and lower cabinets.

by side because one likes it hot, the other cold. In general, place the refrigerator toward the end of a cabinet run, so its big doors can swing free. When the appliances are comfortably situated, fill in the spaces between with cabinets.

Try not to fit cabinets too tightly to room dimensions. If you're fitting cabinets into an older house, it's safer to undersize cabinet runs slightly—allow 1½ in. of free space at the end of each bank of cabinets—so you have room to fine-tune the installation. You can cover gaps at walls or inside corners with scribed trim pieces.

Speaking of inside corners, allow enough room for cabinet doors to open freely.

Cabinet dimensions. Basically, there are three types of stock cabinets: base cabinets, wall cabinets, and specialty cabinets.

▶ **Base cabinets** are typically 24 in. deep and 34½ in. tall so that when a countertop is added, the total height will be 36 in. Base cabinet widths increase in 3-in. increments, as do wall cabinet widths. Single-door base cabinets range from 12 in. to 24 in. wide; double-door base cabinets run 27 in. to 48 in. wide. Drawer

PRO TIP

When cabinets arrive, inspect the packaging for signs of abuse or breakage—crushed corners or torn cardboard—before unwrapping them. Make sure the cabinets and hardware are the styles you ordered, and cross-check your order against the shipping invoice to be sure all parts are there. Report damaged or missing parts immediately.

Cabinet Basics

Cabinets today are basically boxes of plywood, particleboard, or medium-density fiberboard (MDF) panels that are glued and screwed together. Side panels, bottoms, and partitions are typically ¾ in. thick; back panels are usually ¼ in. thick. On custom and semicustom cabinets, you can request thicker stock, but it will cost more.

Cabinet faces are either *frameless* (the edges of the panels are the frame, although they may be veneered or edgebanded) or *face frame* (a four-sided wood frame covers the edges of each box).

Frameless cabinets (also called European style) have fewer elements and a simpler design, so they are easier to manufacture. (The cabinets shown in the installation photos are frameless.) Doors and drawers typically lie flush on the case and overlay the panel edges. Usually, there's ⅛ in. between the door and the drawer edges.

Face-frame cabinets offer more visual variety. You can expose more or less of the frame, vary the gaps between drawers and doors, use different hinge types, and so on. In general, designers who want a more ornamental, less severe, more traditional look often specify face-frame cabinets.

In addition to the elements above, cabinet cases have *mounting rails* (also suspension rails) that you screw through to secure the cabinets to studs. Base cabinets also have *stretchers*—plywood webs across the top—to make boxes more rigid, keep partitions and sides in place, and provide something solid to screw the countertop to.

This frameless base cabinet will have a countertop or its substrate attached to the two plywood webs (stretchers) running across its top. The two mounting rails on the back of the cabinet will be screwed to studs, securing the unit.

Kickspaces

The indentation at the bottom of a base cabinet that provides room for your toes, so you can belly up to the cabinet while prepping food or doing dishes, is called the *kickspace*. Without a kickspace, you'd need to lean forward to work at the counter—a sure recipe for backaches. Custom-made cabinets sit on a separate *rough toekick* (also called a plinth or subbase), which is often assembled on site, whereas most (but not all) factory-made base cabinets arrive with toekicks built in. Toekicks are covered by a *kickface*, or finish toekick, a ⅛-in. plywood strip with the same finish as the cabinets or a vinyl strip; the kickface is better installed after the finish floor.

Scribe Pieces

Cabinet assemblies also include small but important filler strips called *scribe pieces* or *scribe panels*. These typically have a rabbeted back so they can easily be ripped down to fill gaps between cabinets or between a cabinet and a wall. On a frameless cabinet, a separate scribe piece may be attached to a side panel, near its face, whereas on face-frame cabinets, the frame stile (vertical piece) has a rabbeted back edge (for scribing). In addition, many cabinet side panels extend slightly beyond the back panels so those side panels can be scribed to fit snugly to the wall, as shown in the left photo on p. 399. Custom cabinetmakers often create a separate scribe panel to dress up the end cabinet in a run and cover any gaps along the wall.

When installing cabinets, start from a high point in the floor—in this case, at the upper right corner of the photo. As you work out from the high point, add shims as necessary to level the toekick (or the base of the cabinet) in two directions.

Cabinet-Mounting and Edge Details

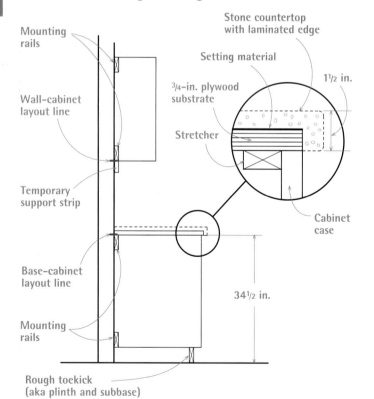

Mounting rails

Wall-cabinet layout line

Temporary support strip

Base-cabinet layout line

Mounting rails

Rough toekick (aka plinth and subbase)

Stone countertop with laminated edge

Setting material

¾-in. plywood substrate

Stretcher

1½ in.

Cabinet case

34½ in.

Better-grade cabinets have mounting rails on the outside of back panels so the rails are not visible inside the cabinet case. The edge detail shown in the enlargment is typical. The countertop substrate—here, ¾-in. plywood—screws to stretchers at the top of the cabinet case. Screw the cabinet bottom to the rough toekick.

cabinets vary from 15 in. to 24 in. wide. Tray units are generally 9 in. to 12 in. wide.

▶ **Wall cabinets** are 12 in. to 15 in. deep, with 12 in. being the most common depth. They vary from 12 in. to 33 in. high. Wall cabinet widths generally correspond to base cabinet widths so cabinet joints line up.

▶ **Specialty cabinets** include tray cabinets, base corner units, corner units with rotating shelves, tall refrigerator or utility cabinets, and wall-oven cabinets. Specialty accessories include spice racks, sliding cutting boards, and tilt-out bins. Specialty cabinet dimensions vary. Base sink cabinets range from 36 in. wide (no drawers on either side) to 84 in., typically in 6-in. increments.

▶ **Other optional features** include plywood vs. particleboard construction and dovetailed drawer boxes vs. boxes simply screwed together. You can also vary the thickness of drawer bottoms, depending upon the size and use of the drawer. A 36-in.-wide drawer full of pots and pans, for example, should have a ½-in.-thick bottom, rather than the ¼ in. that would suffice for a small silverware drawer. Hardware options include full-extension and/or soft-close drawer slides and drawer pulls that are aesthetically compatible with the kitchen's style.

ORDERING CABINETS

When you're satisfied with your kitchen layout, take it to a home center and get an estimate on the cabinets. Or you can go online, where numerous websites will walk you through measuring and ordering. These days, you can order fine cabinets by mail and receive pieces with tight joinery, matched wood grain, and excellent finishes. You can choose from thousands of cabinet cases, doors, drawers, and hardware types. If you've never ordered cabinets before, it's smart to hire a finish carpenter to review your layout before ordering to help figure out exactly what you need.

Installing Cabinets

The key to a successful cabinet installation is leveling the base and wall cabinets and solidly securing them to wall studs and to the floor. As noted earlier, carefully measure and assess the kitchen walls, floor, and corners before you order the cabinets—and review those measurements and conditions again after the cabinets arrive. The photo sequence that follows shows a custom cabinet installation, but most of its advice is relevant to the IKEA cabinet installation on pp. 400–403.

LAYING OUT CABINETS

Using a long level atop a straightedge or a line laser, locate the high point of the floor. It's easier to set a base cabinet (or rough toekick) at the floor's high point and shim up the other cabinets to that level than it is to cut down cabinet bases and toekicks. From the floor's high point, measure up the height of a base cabinet (usually 34½ in. high) and mark the wall. Use a laser level to transfer the base cabinet mark to other walls, creating a level line around the room, which we'll call the *base cabinet layout line*.

Marking off elements. Along the base cabinet layout line, mark off fixed elements, such as the stove, range hood, and refrigerator. Often, a sink cabinet will center under a window. If upper cabinets are to frame a window evenly on both sides, mark the edges of those cabinets. Once the large elements are marked onto the walls, mark off the widths of the individual cabinets. For frameless cabinets, measure from the outside of the side panels. The frames of face-frame cabinets extend slightly beyond the side panels, creating slight gaps between the boxes. Much of the time, the sides of wall and base cabinet units will line up vertically because they are the same width.

Marking wall cabinets. Use the base cabinet layout line to establish the bottoms of wall cabinets, too. Because wall cabinets are normally placed 18 in. above the finish countertop, measure 19½ in. up from the base cabinet layout line to position the bottoms of the wall cabinets; shoot a laser level through that mark and lightly pencil a second level line around the room, which is the *wall cabinet layout line*. Over refrigerators and stoves, the bottoms of the wall cabinets will be higher. If you also are installing full-height pantry or broom cabinets, make sure their tops align with the tops of the wall cabinets; if they don't, raise or lower the wall cabinets until the tops line up. Next, mark off the width of the wall cabinets along the wall cabinet layout line.

Indicating scribe locations. Layout marks should also include *scribe locations*, where you must install a narrow scribe piece (filler strip) to cover a gap between cabinets and an appliance or a space between an end cabinet and an irregular wall. Where cabinets meet at inside corners, 1½-in.- or 2-in.-wide scribes are often needed to offset drawers or doors slightly, so they have room to pull past the cabinet knobs or appliance handles sticking out from the adjacent bank of cabinets.

Marking studs. Finally, find and mark stud centers, to which you'll screw the cabinets. To find studs, use either an electronic stud finder or a really strong magnet, or drive small finish nails

into wall areas that will be covered by cabinets. Whatever works! Use a spirit level to plumb light pencil lines that indicate the stud centers. It's desirable to screw into as many studs as you can to secure wall cabinets, but screwing into only one stud is acceptable for base cabinets and for narrow wall cabinets that don't reach two studs.

INSTALLING BASE CABINETS

Cabinet installers disagree about whether it's easier to install base or wall cabinets first. If you hang the wall units first, you won't need to lean over the base cabinets as you work. If you install the bases first, you can brace the bottom of the wall cabinets off the bases and thereby install the uppers single-handedly. There isn't one right answer, but the photos on p. 407 make a case for hanging the wall cabinets after installing stone countertops. Above all, be patient. Setting cabinets means endlessly checking and rechecking for level, fussing with shims, and so on. So don't begrudge the time it takes. You can't hurry love or cabinets.

Setting rough toekicks. If your cabinets have separate toekicks, install them first, starting at the highest point on the floor—as you did during layout. Make the toekicks as long as possible to minimize joints because joints tend to sag and separate under load. Level the toekicks side to side, front to back, and from section to section. Shimming is an inexact science: As a rule of thumb, shim under the corners and in the middle of a span—roughly every 18 in. to 24 in. A 24-in.-deep base cabinet is typically supported by a 20-in.-deep toekick.

If floors are seriously out of level—say, 1 in. in 8 ft.—construct several *ell supports* such as the one shown in the right photo on p. 396. Screw one leg of each ell to the subfloor, level the top of the toekick, and then screw the side of the toekick to the upright leg of each ell. Ells aren't hard

Leveling cabinet bases and toekicks takes shims and several spirit levels, as well as patience. After leveling each unit in two directions, run a third level diagonally to the adjacent toekick to make sure all are at the same height.

PRO TIP

To make rough toekicks, rip down ¾-in. plywood, which is more durable and water-resistant than particleboard should there be a leak. Don't use 2x4s because they are rarely straight enough to use as a subbase for cabinets. Besides, rough toekicks must be 4 in. high, and a modern 2x4 placed on edge would be just 3½ in. high.

INSTALL THE Toekick FIRST

If finish floors aren't yet installed and you don't want the cabinets dinged up by the flooring installers, then install only the toekick initially, shimming it level and screwing it to the subfloor. This is especially recommended if you'll be laying tile floors because mortar and grout are messy. Then flooring installers can run the flooring snug to the toekick, covering the shims. When the flooring is complete, simply place the base cabinets atop the level toekick and screw them down. When constructing the toekick, increase its height by the thickness of the finish floor so the top of the toekick will be 4 in. above the finish floor. If you're installing tile over a mortar bed (1 in. to 1½ in. thick), make the rough toekick 5 in. to 5½ in. high.

Because you'll be securing both base and wall cabinets to stud centers, use an electronic stud finder to help locate them.

to construct, and they're far more stable than a 1-in.-high stack of shims. Once you've leveled the toekicks, place the base cabinets atop them and see how everything fits together. If this dry run looks good, set aside the cabinets and screw the toekicks to the subfloor.

Setting cabinets with integral toekicks. If your cabinets have integral (built-in) toekicks, be sure to review the preceding section on rough toekicks. Shimming units with integral kicks is similar but more difficult. Basically, you'll shim each cabinet under its sidewalls, front, and back. The difficulty arises because you can't go back and adjust rear shims once you've installed the next cabinet. So take the time to level the top of each base cabinet perfectly. Otherwise, the order in which you install cabinets is the same for either type.

Setting base cabinets. If you're installing a single run of cabinets along one wall, it really doesn't matter where you start. However, if there's a sink cabinet centered under a window, start there. If your cabinet layout is L- or U-shaped, start in a corner because there, where cabinet runs converge in a corner, their tops will need to line up perfectly if the countertop is to be

level in all directions. So take pains to be sure that first corner top is at the right height—in relation to the base cabinet layout line—and level in all directions. Once that corner cabinet is perfectly level, you have a good shot at extending that level outward as you add cabinets.

When you've leveled the corner cabinet in all directions, you can screw it to the toekick and, through its mounting rails, to the studs behind it. But more often, carpenters prefer to "gang" cabinets together, lining up their tops so they're level and, using quick-release clamps with padded jaws, aligning and drawing the cabinet edges or face frames together. Once you've lined up the cabinet edges and frames, use wood screws to join them. Drill pilot holes first with a countersink bit so the screw heads will be flush. If cabinet panels are ¾ in. thick, use 1¼-in. screws to join them so the screw points don't pop through.

After securing the cabinet edges and frames, check the cabinet tops for level and height one last time. Then, depending on the type of cabinet, screw the cabinet bottoms to the toekicks, or screw integral toekicks to the subfloor. Finally, screw the cabinet backs to the studs, through the pilot holes you predrilled. If a wall is wavy, shim

TWO WAYS TO SECURE A TOEKICK

PRO TIP

As you install each cabinet, first transfer the stud center marks to the mounting rails on the back of the cabinet. Then drill through the marks, using a bit that's smaller than the shanks of the mounting screws—or a countersink bit. Drill slowly to avoid splintering the plywood on the inside of the cabinet, or stop the countersink bit just as its point emerges. Finish drilling from the other side.

Once you've leveled all the toekicks in a cabinet run, screw them to the subfloor. If you use square-drive screws, the driver bit is less likely to slip out when the screw meets resistance.

If a floor is badly out of level, avoid using a stack of shims to level a unit because they wouldn't be stable. Instead, use plywood ell supports: Screw one leg of the ell to the subfloor, then screw the leveled toekick to the other leg.

low spots behind the mounting rails; otherwise, screws could distort the mounting rails and possibly misalign the cabinet boxes. Screws should sink at least 1 in. into the studs, so use #8 screws that are 2½ in. or 3 in. long. If your base cabinets have top and bottom mounting rails, drive two screws per stud to anchor the cabinets—in other words, sink a screw each time a mounting rail crosses a stud. To hang upper cabinets, use washer-head cabinet screws (AKA button-head screws).

Setting sink bases. Sink bases with back panels take a bit more work because you must bore or cut through the back panel for pipe stub-outs and electrical outlets, if any. Perhaps the easiest way to transfer the locations of those utilities to the back of the cabinet is to position the cabinet as close as you can to layout marks on the wall, then place a spirit level behind it. Holding the level vertically, place it next to each stub-out, plumb the level, and mark that pipe's position on the wall and on the cabinet's back stringer. Pull the sink base away from the wall, measure how far each stub is below the layout line, and measure down an equal amount on the back of the cabinet. Use a slightly oversize hole saw to bore holes, stopping when the saw's center bit comes

Once you've leveled the toekicks along a wall, start setting the base cabinets on top, and check them for level as well. If the cabinets are in an L- or U-shaped layout, work outward from a corner.

When you're sure that base cabinets are at the correct height and leveled, align their front edges or face frames and use padded clamps to draw adjacent cabinets together. Then sink two wood screws through side panels to secure them.

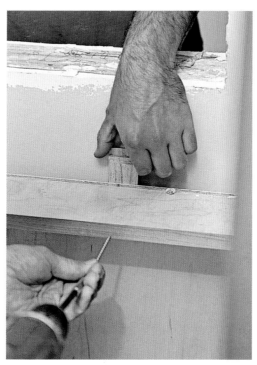

If the walls are irregular—and most are—shim behind the cabinet-mounting rails before screwing them to the studs. Otherwise, back panels and rails could distort.

Screw through cabinet bottoms into toekicks. Predrill screw holes with a countersink bit so screw heads will be flush.

Scribing A BASE CABINET

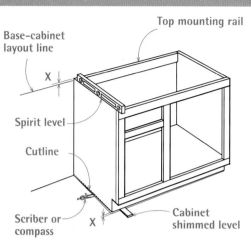

If a base cabinet with an integral toekick sits above the base-cabinet layout line once the unit has been leveled, scribe and trim its bottom to reduce its height. Set the scriber to the amount (X) that the cabinet sits above the layout line.

To cut down a base cabinet (with an integral toekick) whose top is too high, place the cabinet as close as possible to its final position and shim it level. Set a pencil compass to the distance the cabinet top projects above the base cabinet layout line, and scribe the base of the cabinet along the floor. By cutting along those scribed lines with a jigsaw, you'll reduce the height of the cabinet the correct amount. If either side panel is finished, place masking tape along the base of the panel before scribing and cutting. A pencil line drawn on the tape may be more visible, and the tape will keep the metal shoe of the jigsaw from scratching the panel.

through the inside of the cabinet. Finish drilling from the inside of the back panel to avoid splintering it.

Setting islands. Kitchen islands are installed much the same as other base cabinets, except that they can't be screwed to studs. Therefore, the rough toekick must be sturdy and well attached to the subfloor. For that reason, use ell supports

to level rough toekicks or integral toekicks and anchor them to the subfloor. Here, glue and screw the ells to the subfloor after snapping chalklines to show you exactly where the island will sit. Place an ell at least every 18 in. to 24 in., and to further bolster rough toekicks, add crosspieces at the same interval. You can't overbuild a kitchen island, especially if you've got kids who think cabinets are jungle gyms.

HANGING WALL CABINETS

Wall cabinets must be leveled, plumbed, and solidly anchored, so transfer the locations of stud centers to the back of each wall cabinet, and predrill screw holes in the mounting rails, as you did for base cabinets. Remove the doors and shelves so the cabinets will be easier to lift and position next to the alignment marks you drew earlier along the wall cabinet layout line. Before lifting anything, however, use a 6-ft. spirit level to refresh your memory as to where the wall's surface is out of plumb and where it bulges or recedes. Make light pencil notations on the wall.

Supporting cabinets. It's better to have a helper hold wall cabinets in place as you mount them. But if you're working solo, the simplest support is a *temporary support strip*, a straight, predrilled ¾-in. by 1½-in. plywood strip placed immediately below the wall cabinet layout line and screwed to

Bore slightly oversize holes in the sink cabinet so you'll have an easier time lining up pipe stubouts. When the installation is complete, spray expanding foam to fill the gaps.

SCRIBING A PANEL TO AN IRREGULAR WALL

To scribe an end panel (here, a refrigerator panel), first level and plumb its edges. Put painter's masking tape on the panel, and scribe directly onto the tape to make the line more visible. If the wall irregularity is slight, hold a pencil flat to the wall and slide it up and down.

A belt sander held 90° to a scribed line enables you to see how much wood you're removing. Use a 120-grit belt, back-bevel the edge slightly, keep the sander moving, and stop just shy of the line. Finish off with a sanding block or a handplane.

Hanging wall cabinets is easiest with four hands: Two hold and adjust, while two check level and drive screws.

each stud with a #8 wood screw (see "Cabinet-Mounting and Edge Details" on p. 394). With the cabinet bottom sitting atop it, the strip will support the box's weight, freeing one of your hands to screw the top mounting rail to a stud. Use a washer-head cabinet screw (button-head screw) to screw the mounting rail to studs.

Plumbing and shimming. Once that first screw is in, the cabinet should stay put, so you'll have both hands free to shim the cabinet and check for plumb. A cabinet sitting on a leveled strip should have a level bottom and plumb sides—but check to be sure. The front of the cabinet also must be plumb. If it's not, insert shims between the wall and the cabinet. Although you can easily shim behind the top mounting rail, the support strip will prevent shimming from underneath; instead, shim the bottom corners from the side. If the top of the cabinet needs to come forward, slightly back out the screw in the top mounting rail. Once all the cabinet faces are plumb, drive a second screw through the top mounting rail and a third screw through the bottom rail, near the shim point. You will add another screw to the bottom rail later, after you remove the support strip and shim behind the fourth corner of the cabinet box.

Ideally, each wall cabinet should be secured to at least two studs with two #8 wood screws through the top mounting rail and two screws

through the bottom rail. However, many wall cabinets are too narrow to reach two studs. Screwing cabinet boxes to each other lends additional support and spreads the load. But if a cabinet will be heavily loaded or if you're uneasy hanging it on only one stud, cut open the wall and *let in* (mortise) a piece of 1x blocking into at least two studs. You'll need to repair the wall—a rough patch is fine if it's hidden by cabinets—but you'll have plenty to screw to.

Tying cabinets together. Install the wall cabinets in roughly the same order you did base cabinets. If the cabinet layout is L- or U-shaped, start with a corner cabinet and work outward. As you set successive cabinets, place a straightedge or a 6-ft. level held on edge across several cabinet faces to make sure they're flush. You may need to back out screws or drive them deeper to make the cabinets flush. Once they're flush, clamp and screw them together as you did for the base cabinets. At that point, you can remove the support strip. With the strip gone, the space behind the bottom mounting rail will be accessible, so add shims and screws as needed.

Finishing touches. Use trim or a tile backsplash to cover screw holes. Custom cabinetmakers often use a piece of trim with the same finish as the cabinets as a support strip and just leave it in place. Thus if the front of the cabinet has a

If you plumb and level the corner units, it will be easier to level cabinet runs on both walls. The corner strip, held in place with a clamp, covers the gap between the two cabinets. Note: European-style hinges—seen here on the cabinet cases—allow you to remove the doors easily, so there's less weight to lift.

Kitchen Cabinet Space Stretchers

Getting more storage space sometimes means making better use of the space you already have. One great way to do this is by converting existing shelves in base cabinets into *sliding shelves*, which are mounted on drawer glides so they can be pulled out. With this modification, the whole shelf is accessible, not just the front of it. Look for side-mounted, ball-bearing drawer slides rated for 75-lb. or 100-lb. loads because 24-in.-wide shelves (the standard base cabinet depth) can hold a lot.

A number of manufacturers make shelf mechanisms that slide and pivot, thereby improving access into some of the most hard-to-reach places, such as corner cabinets. Cabinet organizers, whether for spice jars, silverware, or pots and pans, enable kitchen users to lay hands on the item they want with less searching.

As the front baskets of these Hafele units pull out and swivel 90°, the back baskets move into the cabinet opening so they can be accessed.

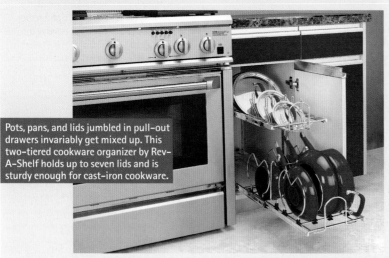

Pots, pans, and lids jumbled in pull-out drawers invariably get mixed up. This two-tiered cookware organizer by Rev-A-Shelf holds up to seven lids and is sturdy enough for cast-iron cookware.

1¼-in.-wide trim piece running along the bottom to hide a lighting strip, use a piece of 1¼-in.-wide trim as a support strip. Cover the cabinet-mounting screws with wood plugs or stick-on screw covers. Finally, install the shelves, drawers, doors, and hardware such as pulls or knobs.

These days, most cabinets use European-style door hinges (shown on the bottom of the facing page), which are easy to remove, reattach, and adjust. Hinges are set into 35mm-diameter holes bored precisely into the door frames. The doors are attached to the cabinet box via baseplates. These hinges easily clip on and off the baseplates without the need to remove any screws. Once the cabinets are installed, clip on the doors and use the adjustment screws to raise the hinges up or down, in or out, until the *reveals* (gaps) between the doors are equal and the doors lie flat.

A Personal Take on IKEA Cabinets

The Swedish retailer IKEA is one of the largest sellers of assemble-it-yourself cabinets in the world. Its cabinets are modestly priced, smartly designed, reasonably durable, and machined so exactly that the average Joe or Jane can put them together. In general, the customer service is good. Finally, there's a wide range of support materials to help customers assemble and install cabinets, including videos, step-by-step photo broadsheets, online help, phone support, and the infamously minimal pictographs shipped with every IKEA order.

Because IKEA has such thorough installation instructions, it makes little sense to duplicate them here. So the comments and the photos below are one person's experience with the IKEA process—or, rather, four people's experiences because My Beloved and I corralled a pair of friends to help us assemble the cabinets one weekend. So here are 12 things four people learned that I hope will be useful if you are thinking of doing likewise.

PRELIMS

1. IKEA's kitchen design service is a great deal. For roughly $150, a representative will come to your house, measure the kitchen, recommend different cabinet configurations, and work up a materials list and estimate. Some of that $150 fee will be applied toward the purchase, if you decide to buy. The service also generates a set of detailed floor plans and elevations of the kitchen, which are invaluable if, say, you buy the countertop separately or hire a contractor to install the cabinets.

INSTALLING IKEA CABINETS

1. Before installing cabinets, survey walls and floors. Note out-of-level floor sections, out-of-plumb wall areas, and low spots that may need to be shimmed. Mark stud locations. Working from your kitchen layout, mark a level line indicating the top of wall cabinets and then a lower line indicating the height of the suspension rail.

2. IKEA wall cabinets hang from a leveled metal suspension rail. Cut it to length and screw it to every stud it crosses. Where the rail crosses low spots of more than 1/4 in., insert shims between the wall and the rail so the rail stays straight. When the rail is secure, slide mounting-bolt heads into the suspension rail. The bolts slide freely so you can align them to predrilled holes in the backs of the wall cabinets.

3. The job will go more smoothly if you get help hanging wall cabinets. If you have a corner cabinet, hang that one first. As one person supports each cabinet, the other person slides the mounting bolts until they stick through holes in the cabinet back. Inside the cabinet, fit a small metal plate and nut over each bolt. Tighten nuts loosely so you can level and adjust each cabinet. Shim cabinet bottoms so they line up to adjacent cabinets.

4. Once cabinet faces are aligned, use quick-release clamps to draw adjacent cabinets together. Clamp jaws should be cushioned or shimmed to avoid marring the cabinets. (Note the mounting hardware to the right of the clamp.)

5. The insides of IKEA cabinets have predrilled holes. Depending upon the cabinet door you buy, drill through the second or fourth holes from the top *and* bottom, using a 3/16-in. bit. Drill through adjacent cabinet sides as shown, insert the connection screws into the holes, tighten them to join the cabinets, and remove the clamps. When you've joined all the cabinets, tighten all the mounting nuts that secure cabinets to the suspension rail.

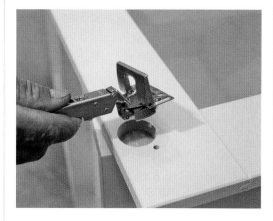

6. To attach doors, insert cup hinges (aka European hinges or German hinges) into the predrilled holes in the door frames. These hinges attach to . . .

. . . receiving hinge parts screwed into predrilled holes along the insides of the cabinets. Hinges can be adjusted in three directions so that doors sit flat to cabinets and align to adjacent doors.

2. Speaking of which, IKEA will, for a fee, install the cabinets—but only if IKEA assembles them, too. In other words, you can't save money by building the cabinets and having IKEA install them.

3. If there is an IKEA outlet reasonably close, pick up the cabinets yourself. Shipping is expensive: A friend in Maine paid $127 for an Akurum cabinet . . . and an additional $124 to have it shipped. If you do pick up the cabinets, get a friend or two to help: Unassembled cabinets are heavy and their boxes are unwieldy.

4. Don't order cabinets until you're almost ready to install them. Even unassembled, cabinets take up a lot of room. Thoroughly vacuum the site before the shipment arrives, especially if you've been renovating. IKEA cabinets' melamine surface can be scratched or chipped by grit, stray nails, and the like.

AT HOME WITH IKEA

5. Once the IKEA shipment arrives, scrutinize the master shipping list to be sure everything's there. Because boxes are large and heavy, this is another two-person job. Contact IKEA immediately if anything's missing.

6. Open each box and check its contents against the parts list enclosed—usually part of an unstapled set of instruction sheets with pictographs. For us, everything was there. By the way, the customer service areas of IKEA stores typically have bins of miscellaneous assembly hardware, free for the taking, if you're short a part.

7. Keeping track of a kitchen's worth of parts takes meticulous organization. Because IKEA assembly hardware is uniform, we assembled all like parts in plastic zip-top freezer bags, labeled them, and put them someplace safe. The large hardware elements, such as drawer slides, we grouped in smaller cardboard boxes. Leave the melamine panels for each cabinet in their shipping box until you're ready to assemble the cabinet. Tempting though it may be, do *not* throw out the large boxes.

D-DAY

8. Invite friends and make a weekend of assembling the cabinets. It's possible for a handy couple to assemble a kitchen's worth of cabinets

Dividing the chores is the best way to install cabinets in a cost-effective and timely manner. Most of us are capable of assembling cabinets and installing interior hardware such as drawer glides. Installing cabinets is a somewhat more difficult task, depending upon how irregular walls and floors are.

Installing base cabinets starts with a review of wall and floor conditions—including the high point of the floor. Measure up 4⁵/₁₆ in. from that high point, and run a level line around the room. This line represents the top edge of a wall strip screwed to each stud it crosses; the wall strip supports the backs of base cabinets.

Lining up and leveling base cabinets is easier than hanging wall cabinets: The wall strips already align cabinet backs, adjustable plastic feet can be screwed up or down to adjust cabinet fronts, and you're not fighting gravity.

Here, the carpenter notes the location of stub-outs for the kitchen sink drain and supply pipes. He will transfer those locations to the back of the sink cabinet and cut holes before installing the sink cabinet.

in two days, but it's easier to maintain momentum with more bodies. If one member of the party has a good collection of tools or has assembled IKEA cabinets before, all the better.

9. Allow plenty of room for assembly and storage once each cabinet is completed. Flattened shipping boxes make an excellent assembly area, protecting both your floors and the melamine surfaces.

10. Keep cool. The pictographs showing installation are minimal, but if you study them carefully you'll find the details are there. More than once I was sure IKEA had screwed up when panel holes didn't line up. But after revisiting the instructions and conferring with my friends—until one of us had an aha moment—the assembly was always correctly machined. Typically, a panel had to be flipped 180°.

11. Never force things. Every bolt or screw has its place, so if it isn't fitting, revisit the instructions. Likewise, resist the temptation to redrill a hole that displeases you. And when tightening screws, don't overdo it: Panel cores are particleboard, which can strip if you apply too much muscle to a screw. If you use a screw gun, make sure it's a variable-speed model, and take it slow.

12. Hiring pros to install the cabinets is not a bad idea if you're pressed for time or not all that comfortable doing it. Aligning cabinets takes skill and patience. You can also leave some of the more complicated IKEA hardware for them to

assemble, such as the sliding shelf mechanism that pulls out of a corner cabinet.

P.S.: A year later the cabinets are operating smoothly and look new. IKEA cabinet hardware has some great options, including drawer dividers that keep even wide drawers well organized and *self-closing mechanisms* that prevent drawers and doors from ever slamming shut.

Countertops for Kitchens and Baths

Most materials in this section require special training and equipment for installation. In fact, solid-surface and stone-polymer countertop makers will sell counter stock only to certified fabricators. Some materials, such as plastic laminates, can be successfully installed if you're handy. But, as you'll see, desirable details, such as wrapped front edges and integral backsplashes (no seams to leak!), are best fabricated in a pro shop. All in all, it's smarter to choose a reputable installer than to try installing a countertop yourself. Tile countertops are discussed in chapter 16.

CHOOSING COUNTERTOPS WISELY

The wide array of countertop colors, materials, and details (such as edges and backsplashes), give you tens of thousands of combinations to choose from. Here's help narrowing your choices.

Plastic laminate, often called Formica after a popular brand, was widely installed in mid-century homes in the U.S. but is now mostly used in utility rooms and work areas. Standard laminate is only $\frac{1}{16}$ in. thick, so it is usually glued to a $\frac{3}{4}$-in. particleboard substrate to lend rigidity and ensure adequate support. Plastic laminate is tough, stain-resistant, economical, and available in extra thick and fire-resistant options. One disadvantage is that the seams can degrade and admit water. And once the substrate is water damaged and the laminate is lifting, the countertop is beyond repair.

Postformed plastic laminate tops. Today, almost all residential laminate countertops are *postformed*—meaning a shop or factory adheres and wraps a single, continuous sheet of laminate to a particleboard substrate, creating a seamless joint between the counter and backsplash. This leak-free joint is the principal reason why contractors order postformed laminate tops. Adhering laminate to a substrate is emphatically *not* a DIY task.

If you opt for a postformed top, the installation becomes a good deal simpler. The counter installer's principal tasks will then be scribing the top of the backsplash so it fits flush to the wall,

Countertop Choices

CHARACTERISTIC	PLASTIC LAMINATE	SOLID SURFACE	QUARTZ COMPOSITE	CERAMIC TILE	STONE	CONCRETE	WOOD
Durability, scratch resistance	Fair	Excellent*	Excellent	Good	Good[†]	Good	Fair[‡]
Ease of cleaning	Good	Excellent	Excellent	Fair[§]	Excellent	Good	Fair
Stain resistance	Good	Good	Excellent	Fair	Fair	Poor	Poor
Water resistance	Good	Excellent	Excellent	Fair	Good	Good	Poor
Heat resistance	Poor	Fair	Good	Excellent	Excellent	Excellent	Poor
Expense	$	$$$	$$ to $$$	$$	$$$	$ to $$	$$

* Material scratches, but scratches are easily sanded out.
[†] Harder stones (granite) wear well; softer stones (soapstone) scratch and stain easily.
[‡] Durability depends on type of wood and finish; wood is generally a bad choice near sinks and water.
[§] Glazed tiles resist stains, water, and heat, but grout joints deteriorate if not sealed and maintained.

THE BEAUTY OF Templates

Templates are useful for transferring accurate measurements to any sheet material or flat surface, whether it's a piece of drywall that needs to be notched beneath an exposed stair or a new door that must be fit to an old, out-of-square door frame. When the going gets rough, the pros make templates. Nowhere is this truer than with granite slabs. Often, countertop suppliers create templates during the "measure date" described in "Ordering Countertops" on p. 406.

Templates are most often made by hot-gluing strips of 1/8-in.-thick plywood (also called doorskin), which is rigid enough to keep its shape yet light enough to transport and position on counter stock.

The more complex the site conditions or cutouts, the more information the template contains. This small section notes the type of granite, edge bevels, thickness, and location of the sink cutout. The template's back edge registers to the window width.

drawing mitered sections together with *draw-bolts*, and securing the counter substrate to the top of the base cabinet. However, as discussed in the next section, the first step in fitting countertops is taking careful measurements.

Solid-surface countertops began with DuPont's Corian, which is still the best-seller in this category. Most brands are polyester or acrylic resins with a mineral filler. Solid-surface countertops have a lot going for them: They're water- and stain-resistant; nonporous and easy to clean and thus great for food prep; and leakproof because they have seamless, chemically bonded backsplashes and integral sinks. Unlike laminates, solid-surface sheets are the same material top to bottom, so if you scratch them or put too hot a pot on them, you can sand out the blemishes. Most brands come with a 10-year warranty. Note that this material is generally expensive.

Quartz composites are a well-established group that includes brands such as Caesarstone, Zodiaq, and Silestone; the blend is roughly 93% quartz and 7% polymers and pigments. Quartz composites have many of the virtues of stone countertops and few of the failings. They are scratch-, heat-, and stain-resistant and are completely nonporous so they're easy to clean. Backsplashes can be bonded so tightly to countertops that there's no seam to collect crud or allow water to penetrate. Composites are unlikely to fracture during transit because mineral particles are uniformly dispersed and free from the imperfections of natural stone. Quartz composites come in a range of colors and, unlike natural stone, their colors are consistent. So the color you choose from a box of samples will be an almost perfect match to the material you get. Because its stone components are, essentially, debris, quartz composite has a good "green" profile.

Ceramic tile is beautiful and durable. Its great variety of colors, shapes, and sizes allows almost unlimited freedom to create your own patterns. Glazed tiles themselves are largely resistant to heat, water, stains, and scratches. Tile can be applied successfully over a plywood substrate or a mortar bed, and a diligent novice can install it successfully. Tile prices vary widely and thus can fit almost any budget. But its grout joints are relatively fragile, easily stained if they're not sealed and maintained, and tend to collect crud. Especially around sinks, tile counters just don't hold up. Standing water will, in time, seep through grout joints between tiles and along backsplashes—especially as grout ages and cracks—and in time degrade plywood substrates.

Natural stone slabs include granite, marble, soapstone, limestone, and slate. Stone is naturally beautiful, but it's also heavy, hard to work, and very expensive. Yet for all its heft, stone is relatively fragile and must be supported by a substrate—usually ¾-in. plywood. Granite, the most popular stone, is available in slabs ¾ in. and 1¼ in. thick, up to 5 ft. wide and 10 ft. long. Section seams are filled with a caulk that's color-matched to the granite, so seams are virtually invisible. To create a thicker edge, fabricators epoxy two layers of stone. When pattern-matched and polished correctly, the seams are almost invisible.

Harder stones such as granite are highly scratch-resistant. Softer stones, such as soapstone, can be scratched, but their softness allows you to easily sand out scratches or buff them out with steel wool. Stone is generally water-resistant, although most stones are somewhat porous and will stain unless protected. Penetrating sealants such as Aqua Mix Sealer's Choice and Miracle Sealants 511 Porous Plus are well regarded in the stone trade. For more recommendations, visit www.stoneworld.com.

Because stone is a natural material with inherent flaws and irregularities, few installers offer a warranty once an installation is complete. In the trade, it's called "a taillight warranty"; in other words, once the installer's truck leaves your driveway, the warranty expires.

Concrete has become très chic as a countertop material because it's tough, can look uptown or rugged, and can be cast into custom shapes. It can be stained and polished to look like exotic stone, factory-cast with colorful glass shards in the bottom of its mold. Or it can be cast directly atop a base cabinet. Concrete weighs roughly the same as stone. However, because a concrete slab needs to be fully 1½ in. thick, you need to overbuild base cabinets to support the additional weight. (Stone tops need to be 1½ in. thick only along their front edge.) Because the bottom of in-place form molds can sag between supports until the concrete sets, it's smart to add cross-counter webs to support the forms. Concrete is notorious for cracking and staining, but you can minimize that by reinforcing the slab adequately with steel or poly fibers and finishing the surface meticulously.

Buddy Rhodes (www.buddyrhodes.com) has a full line of materials specially formulated for concrete countertops, includng concrete mixes, fortifiers, colors, admixes, sealers, tools, and polishing compounds. One of the best books on casting your own countertop is Fu-Tung Cheng's *Concrete Countertops* (The Taunton Press, 2001).

PRO TIP

If you're cutting a finish surface such as plastic laminate, cover the metal shoe (base) of your jigsaw with masking tape to keep the shoe from scratching the countertop.

PRO TIP

Once the laminated countertop is scribed and secured to the base cabinet(s), do the cutout for the kitchen sink. Kitchen-sink cutouts are best done on site because the remaining substrate front and back would likely break in transit if you did the cutout at the shop.

Wood is a sentimental favorite because it's warm and beautiful, but it must be correctly finished and carefully sealed to prevent damage around sinks or along seams where water could collect. There are so many types of wood that it's tough to generalize about traits, but most types scorch, scratch, and stain easily, and water will swell and rot wood unless you keep it sealed. A growing number of fabricators such as Spekva of Denmark are producing oil-finished butcher-block and laminated-wood countertops that will remain durable if you periodically apply a food-safe oil recommended by the manufacturer. These countertops are pricey but, in Spekva's case, the wood is harvested from sustainable, managed forests.

ORDERING COUNTERTOPS

Once you've identified several reputable countertop shops in your area, get bids. For your protection, develop detailed floor plans and, as you proceed, put everything in writing. That way, you'll be sure that quotes from different shops reflect similar details, deadlines, and so on. The following guidelines are adapted, with permission, from Sullivan Counter Tops, Inc. (visit www.sullivancountertops.com).

Visit suppliers in your area. Look at countertops in several different showrooms, and discuss your options with the salespeople. Share any information—samples of wall paint or cabinet finishes, magazine ads showing counter surfaces, photos of sinks or sink fixtures—that will clarify the style and look you're after. A salesperson's willingness to spend time and answer questions says a lot about a supplier.

Get bids. Provide a floor plan, drawing, or sketch to the countertop suppliers, and ask each to generate a written bid. The bid should also specify a completion date, terms of payment, and the scope of the installation. In most cases, quoted prices will not include plumbing, electrical work, or adjustments to the cabinets such as sink cutouts and leveling plywood substrates. Tearing out old countertops is usually extra, too. Based on bids and supplier reputations, choose a supplier.

Schedule the job. Most installations require two visits from the supplier: the "measure date" and the "install date." It's difficult to pinpoint a measure date until the cabinets are installed, but in general, two weeks' to four weeks' notice should be enough. Typically, solid-surface (such as Corian) and plastic-laminate countertops require about one week between the measure date and the install date. Quartz-composite countertops (such as Corian Quartz [formerly known as Zodiaq]) require about two weeks. Most countertop suppliers will allow you to change the installation time without penalty, provided you give them enough notice.

Be prepared for meetings. The measure date is your last chance to give input on details such as underlayment issues, color, edge treatment, and backsplash detail. Whatever you finally decide on, get it in writing. During this meeting, the supplier will review job-site conditions, so the general contractor should be there, too.

Important prerequisites

▶ Cabinets must be set before the job can be measured. In other words, the cabinets must be screwed together and screwed to the walls, not just pushed into place. The cabinets cannot be moved even $\frac{1}{8}$ in. after the countertop supplier has measured because countertops are fabricated to close tolerances.

▶ Cabinets must be set level. As a general rule, cabinets must be set on a level plane within $\frac{1}{8}$ in. over a 10-ft. length. Such stringent requirements are a concern not only to installers but also, in some cases, to manufacturers as a condition of warranty.

▶ All appliances and sinks should be on the job site at the time of the measure. The fitting of sinks and appliances is often critically close. Design or construction issues that could cause problems or delay the installation should be resolved on the measure date. If the sinks and appliances are on site, the supplier can inspect them. If there's a defect, damage, or, say, a sink rim that won't fit the countertop, you'll need time to replace the item before installation.

Installing a Granite Countertop

Once you've looked at granite samples and narrowed your choices to a few varieties, consult the fabricator—the company that will be cutting and installing the stone. For example, if you choose a granite with large crystals, joints between sections may be more obvious than those between finely grained stone. Besides, some richly figured stones are more likely to crack or spall when subjected to everyday use. A fabricator's practical concerns can be a good counterpoint to a kitchen designer's artier point of view. In any event, cabinets and the plywood substrate must be installed before measurements for a stone countertop can be made.

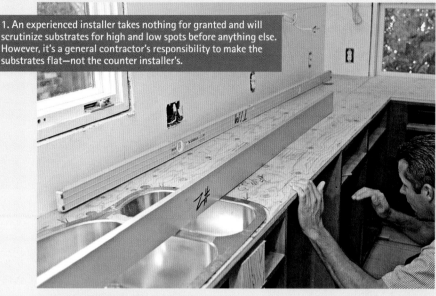

1. An experienced installer takes nothing for granted and will scrutinize substrates for high and low spots before anything else. However, it's a general contractor's responsibility to make the substrates flat—not the counter installer's.

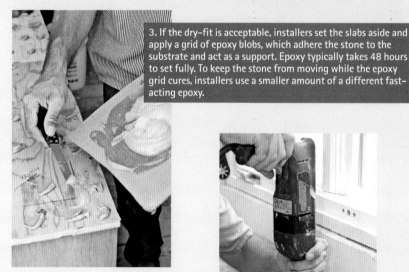

3. If the dry-fit is acceptable, installers set the slabs aside and apply a grid of epoxy blobs, which adhere the stone to the substrate and act as a support. Epoxy typically takes 48 hours to set fully. To keep the stone from moving while the epoxy grid cures, installers use a smaller amount of a different fast-acting epoxy.

2. After ensuring that the substrate is flat, installers typically lift the slabs, stand them on edge, and carefully lower them into position to test clearances, cutouts, overhangs, and so on.

6. To avoid transporting a slab weakened by cutouts or holes, installers usually bore faucet holes on site, using a diamond-tipped hole saw.

4. After caulking the undermount sink with silicone, installers set the slab on the epoxy grid. To avoid spreading the epoxy too thin, installers don't slide the stone in place—they lift and lower it as close to the final position as possible. Suction-cup handles help.

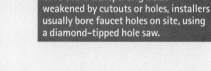

5. After leveling the slabs, pros ensure that the joints are flush by sliding a razor blade back and forth across them. If one side is higher, the razor blade clicks as it hits a high point or falls from a high point to a low one. More sensitive than a fingertip!

Choosing a Range Hood

Cooktops and stoves should be vented by a range hood. In addition to sucking up the smoke of a charred steak, range hoods exhaust airborne grease that might otherwise migrate to a cool corner and feed mold or adhere to woodwork and degrade its finish, as shown in the photos on p. 561.

Range hoods are most often *wall-mounted* directly over a range. Alternatively, there are *downdraft* and *side-draft* vents that pop up from a counter area to suck away fumes. Over island and peninsula ranges, you can install *chimney-type* vents. In general, install the type of vent that will carry exhaust gases outdoors with the shortest and straightest duct run possible. Because heated air rises, wall-mounted and chimney types are inherently efficient, whereas downdraft and side-draft vents pull heated gases in directions they wouldn't go naturally and can even pull burner flames sideways.

Range hoods vary from low-powered and inexpensive to custom-designed units that cost thousands of dollars. When sizing range hoods, the rule of thumb is 100 cfm (cubic feet per minute) per 12 in. of width. So a 300-cfm wall-mounted hood should be adequate to vent the average four-burner, 36-in.-wide range. If that same range is located on a kitchen island, its range vent should draw approximately 400 cfm. In general, bigger is not better when sizing range hoods. For one thing, larger hoods are noisier. Midsize hoods average 3 sones (a measure of noise) to 3.5 sones, which is too noisy to have a conversation near; monster hoods can reach 8 sones. (As a comparison, refrigerators register 1 sone.) Oversize hoods can also expel so much air that they create *backdrafting,* in which negative in-house air pressure draws furnace or fireplace exhaust gases back down the chimney (see chapter 14 for more on backdrafting).

Ideally, a range hood should be slightly wider than the range, say, 3 in. wider on each end, and mounted 30 in. above the range. But follow the hood maker's suggested mounting height; more powerful hoods can be installed higher. Finally, buy a unit with a good-quality filter that can withstand regular washing with soap and water. Most filters are aluminum mesh, but better ones are stainless steel; many can be popped into the dishwasher, which spares homeowners a very greasy and unpleasant task. In general, be skeptical of range hoods that recirculate air through a series of filters rather than venting it outside.

> Range hoods should complement architectural details found elsewhere in the kitchen—in this case, the hood's ogee molding echos shaped shelf edges and cabinet panels.

This enameled cast-iron farmhouse sink is big enough to rinse as large a turkey as you can lift. The faucet's spout swings out of the way for complete access to the sink.

Kitchen Sinks

Because a kitchen is one of the most-used rooms in the house, do a little research before choosing its most-used fixture.

CHOOSING A KITCHEN SINK

Keep four criteria in mind when considering sinks: size and number of bowls; sink-hole compatibility with faucets and accessories; materials; and mounting styles.

Size and number of bowls. Sinks are available in dozens of sizes and shapes. For the record, a 22-in. by 33-in. two-bowl unit is the most popular, perhaps because it fits neatly into "standard" sink cabinets 24 in. deep by 36 in. long. Two bowls allow you to prep food in one and put used bowls and pots in the other during the mad dash to dinner. Typically, there's a garbage disposal on one side. If you've got a dishwasher and don't need to wash dishes in the sink, a large single-bowl sink is best suited to washing pots and pans; 10 in. deep is optimal. If you've got plenty of counter space and two family members who like to cook, install two separate sinks. By the way, three-bowl sinks (45 in. wide) are overkill for most home kitchens.

Most sinks have one to four holes. Typically the spout and faucet handles take up three of those holes, but some single-lever faucets require only one hole. Filling an extra hole is generally not a problem, what with soap dispensers, hot-water dispensers, spray units, filtered-water dispensers, and so on to choose from. Incompatibility more often occurs when a faucet assembly's valve stems have a different spacing than the holes in the sink. Always measure the sink and the faucet assembly to be sure. Likewise, if the sink is undermounted, an installer must drill holes into a countertop, making sure they have the same spacing as the faucet to be installed.

Materials. Sink materials include many of those used for countertops. You can use almost anything that will hold water, but ideally it should be light enough to install without breaking your back, easy to clean, durable, moderately priced, heat-resistant, and stain-resistant. Few materials fill the bill as well as stainless steel.

▶ **Stainless-steel sinks** represent roughly three-quarters of all kitchen installations. Typically, they have a brushed or polished finish; brushed finishes are easier to maintain because water spots don't show as conspicuously. A sink's gauge (thickness) is the real differentiator. Thicker gauges (16 to 18 gauge) are harder to flex and dent and are quieter to use, whereas thinner gauges (20 to 22 gauge) are less expensive, less durable, and more inclined to stain.

▶ **Enameled cast-iron or enameled-steel sinks** are available in numerous colors and provide a classic look that works in modern and traditional kitchens. Enameled sinks have a hard finish, but the enamel can chip, making the metal substrate likely to rust. Abrasive cleaners quickly dull enamel finishes. Cast iron is so heavy that it takes two people to install it and so hard that it's monstrously difficult to drill if you need an additional hole for a water filter or some other accessory. Delicate dishes or glasses dropped on it are doomed.

▶ **Solid-surface sinks** are usually manufactured from the same material as the counter and glued (chemically bonded) to the underside of the counter for a seamless, leakproof joint that won't catch food scraps. As with counters, solid-surface sinks are stain-resistant, nonporous, and easy to clean. And you can sand them smooth if they get gouged or scorched.

▶ **Acrylic sinks** have a lot going for them. They're lightweight, nonporous, and easy to keep clean. They're also stain- and crack-resistant, available in many colors, and reasonably priced. However, compared with solid-surface materials, acrylics are relatively soft, so they should be cleaned with non-abrasive cleaners. They can be damaged by extreme heat and may be incompatible with petroleum-based cleaners and caulks.

Mounting style. There are a number of mounting styles, although almost all require a bead of sealant along their perimeter to keep water from getting under the sink rim. Mounting styles

include self-rimming, undermount, integral, flush mount, and separate rim.

▶ **Self-rimming sinks** are popular and easy to install because the sink rim sits on the countertop—after you've applied a bead of silicone sealant around the perimeter of the sink cutout. Heavier sinks, such as cast iron, are held in place by the adhesion of the sealant and the weight of the sink, whereas lighter sinks, such as stainless steel, require clips on the underside of the counter. When a self-rimming sink is set under a countertop, as shown on p. 410, it may be called a counter-under sink.

▶ **Undermount sinks** are placed under a counter whose sink opening must be finished because it isn't covered by the sink rim. Counters with undermount sinks are easy to keep clean because there's no rim to block food scraps; just sweep them into the sink. Typically, clips attach the rim to the underside of the counter; many contractors also add framing inside the sink cabinet to support the sink when it's filled with water.

▶ **Integral sinks** are bonded to a counter of the same material, creating a seamless, leak-free joint. Integral sinks are common to solid-surface and quartz-composite counters.

▶ **Flush-mount (tile-edge) sinks** have a rim the same thickness as the tile layer when both rest atop a thin-set mortar bed. Such sinks are typically enameled steel or cast iron. You can fill the tile-sink rim joint with grout or silicone sealant, but acrylic latex sanded caulk has the best qualities of both and comes in several colors.

▶ **Separate-rim sinks** are usually stainless steel and employ a separate stainless-steel rim to cover the joint between the small sink rim and the edge of the counter. To prevent leaks, you must seal both the sink side and the counter side of the rim.

INSTALLING A KITCHEN SINK

Because self-rimming sinks are the most common, this section focuses on their installation. If your sink is of a different type, follow the manufacturer's installation instructions.

Lay out the opening. Mark the sink cutout (opening) on the counter or the plywood substrate. Most sinks come with a paper template of the cutout; if yours doesn't, make one of cardboard. To do that, turn the sink upside down onto the cardboard and, with a felt marker, trace its outline. Next, use a yardstick to draw a second outline ¾ in. inside the first on the cardboard. Sink rims are typically ¾ in. wide, so the inner

WILL Sink Hardware FIT?

Before installing countertops, make sure there's enough room for the faucet, spout, soap dispenser, dishwasher air gap, and any other hardware installed near the sink. Move the sink forward if needed to increase space between the sink and the backsplash. Is there room for faucet handles to rotate? Will a tall faucet spout clear a windowsill above the sink?

When you're sure everything fits, cut out the 3/4-in.-thick cabinet stretcher (p. 393) in the area behind the sink where the hardware will be installed. Otherwise, the threaded throats of most deck-mounted faucets will not be long enough to penetrate the countertop, subtop plywood, and the stretcher (usually 2¼-in. total thickness). Make sure cutouts leave enough room for the nuts and washers that hold the hardware in place, as well as room for a wrench. Cutting the stretcher after the sink and countertop are installed is a difficult task that is best avoided.

Cutting sink openings is best left to a pro if your countertop is stone, quartz composite, or any other hard, expensive material. The plywood substrate will also need cutouts, obviously.

Kitchen Lighting

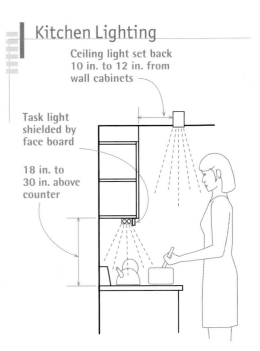

To illuminate work areas without strong shadows, use a combination of task lighting and general lighting.

Sink-Mounting Details

SELF-RIMMING SINK

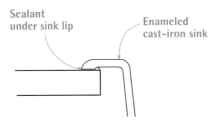

Weight of sink holds it in place.

UNDERMOUNT SINK

INTEGRAL SINK

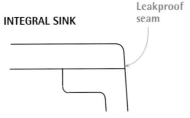

Counter and sink are same material, bonded chemically.

FLUSH-MOUNT OR TILE-EDGE SINK

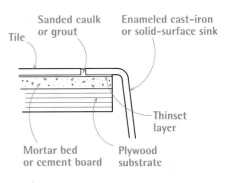

SEPARATE-RIM SINK

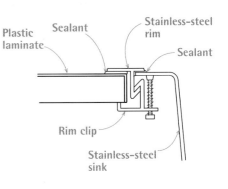

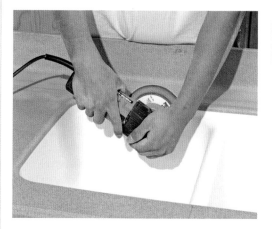

Solid-surface sinks are chemically bonded to a counter of the same material, creating a leak-free seam.

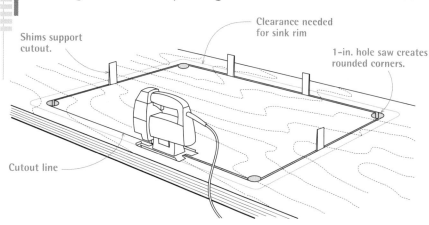

Shims support cutout.

Clearance needed for sink rim

1-in. hole saw creates rounded corners.

Cutout line

Because sink rims are typically 3/4 in. wide, create cardboard templates to show both the clearance needed for the rim and the cutout line needed for the sink body.

line represents the size of the cutout, and the outer line shows how much clearance the sink rim needs. Position the cutout template on the counter so there's clearance on all sides, then use a felt marker to trace around the template.

Cut the opening. Drill a hole at each corner of the cutout for your jigsaw blade; if you use a hole saw to drill the corners, its arc should match the rounded corners of the sink body. As you cut each side of the sink opening, stop just short of the corner hole. Then drive a shim into the sawkerf—from the underside of the counter—to keep the cutout section from falling. (A wood shingle is a perfect shim.) With a few shims in place, finish cutting to the corner holes and lift the cutout section, using the corner holes as finger holds. Use a wood rasp to smooth rough cutlines or splinters.

Install the sink. Wearing heavy work gloves, put your fingers in the sink drain and faucet holes, lift the sink, and lower it into the cutout. Two people should lift and set the sink if it's cast iron; put wood scraps around the cutout to set the sink on so it doesn't crush your fingertips. Check the sink's fit in the cutout—look under the counter as well—before lifting the sink out. Trim the cutout as needed. Then mount the drain basket, faucet assembly, and accessories to the sink; they're harder to attach once the sink is in place. To cushion the sink hardware and create a watertight seal, use the flexible seals or plastic plates provided by the manufacturer. If the unit has a hollow body, put a bead of plumber's putty beneath its lip. *Don't do this if the countertop is stone*; the oil in the putty may stain the stone. Silicone will work, too, but it can make the faucet difficult to remove if you decide to replace it.

Just before installing the sink in its opening, apply a cushion of sealant for the sink rim to rest on. The sealant will be a layer of plumber's putty

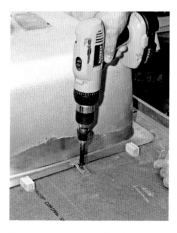

Here, an undermount sink is being clip-mounted to the underside of a solid-surface counter . . .

. . . and then lifted as a unit and placed atop a base cabinet. Supplemental plywood frames inside the cabinet will also support the sink rim.

This self-rimming stainless-steel sink was installed as a counter-under sink. After routing the perimeter of the cutout so the sink rim would be flush to the plywood, the installer set the rim in epoxy. Just before installing the stone, he caulked the top of the rim with silicone.

or, more likely, silicone caulk. Follow the sink manufacturer's recommendations. Make any final adjustments to the sink's position before applying a bead of silicone caulk along the edge of the sink rim, where it rests on the counter. With a moistened finger, compress this caulk line and remove the excess silicone. (If you will install a counter over the sink rim, apply the silicone on top of the sink rim just before the countertop is installed, as shown in the bottom photo on p. 411.)

Fasten mounting clips, if any, and attach the supply risers and the sink drain.

TIP

If there's a window in a shower, it should be fiberglass or metal, not wood. Sustained moisture will swell wood and make the window inoperable.

Bathroom Planning

If a bathroom is comfortable and its fixtures are in decent shape, you might want to add only light fixtures or a vanity. If the bathroom is musty, add a bath fan (see chapter 14). But if a bathroom is drafty, cold, or uncomfortable, you may be wise to tear out finish surfaces, insulate it well, and position the fixtures more efficiently. Here are a few thoughts on what makes bathrooms pleasant.

CREATING BATHROOMS THAT WORK

Here are 12 factors that help make a bathroom comfortable, functional, and easier to clean:

▶ **Adequate room.** Bathroom space isn't efficient if there's not enough room to use the fixtures easily and safely. "Bath-Fixture Clearances" on p. 414 shows minimums, which may be superseded by local building codes.

▶ **Keep it secure and intimate.** Although shared bathrooms should be accessible to the rooms they serve, bathroom users should feel secure once inside. Avoid multiple-door accesses. Keep the room's scale intimate as well: Warm, cozy spaces are best.

▶ **Put private fixtures far from the door.** The most-used and least-private fixture, the lavatory, should be nearest the door so people can pop in and wash their hands quickly. But toilets and tubs should be farther away. Insulate walls and install a tight-fitting door to suppress sound. Cast-iron waste pipes are quieter than plastic ones.

▶ **Use alcoves and half-walls.** Placing fixtures in alcoves and odd spaces around the perimeter of a room maximizes the floor space in the middle. Isolating toilets or tubs with their own doors also makes it possible to share a bathroom during morning rush hours while preserving privacy.

▶ **Take advantage of natural light.** Windows and skylights allow rooms to be small without causing claustrophobia. To block the view of neighbors, install translucent or textured glass, or place windows high on the wall. (In the window trade, glass that allows privacy is called "obscure glass.") Wood windows in showers are generally not a good idea because water sits on windowsills and rots them. Ideally, skylights should open.

▶ **Make sure there's enough artificial light.** For general lighting, plan on 7 watts to 10 watts of LED lighting or 30 watts to 40 watts of incandescent lighting per 10 sq. ft. For fixtures in alcoves, use a 10-watt to 14-watt LED recessed, vaporproof ceiling fixture. To illuminate bathroom mirrors, however, install light fixtures on the walls: one over the mirror and one on each side. Ceiling fixtures alone will make that face in the mirror look ghoulish.

▶ **Choose comfortable fixtures.** "Standard Cabinet Dimensions" on p. 390 suggests counter heights. If they are too high or low for your family, change them so you don't need to stoop or stand on tiptoes. As for tubs, if you can't stretch out in a standard tub or squared tub as you'd like, look into oversize models or slope-back, cast-iron clawfoot tubs in a salvage yard. Shower stalls should be big enough to towel off in—36 in. by 36 in.

▶ **Invest in well-made shower valves,** showerheads, and lavatory faucets with nickel- or chrome-plated finishes. They cost two to three times what bargain home-center accessories do, but they'll last longer. The same goes for towel bars, switch plates, mirrors, and other accessories—buy quality.

▶ **Use appropriate materials.** Water reigns in a bathroom, so use materials that can withstand it. Resilient flooring and tile are great on bathroom floors; wood isn't. Even when wood is face-sealed with a tough finish, its end grain can absorb water. And in time, standing water will cloud most finishes. As explained in "Setting Beds" on p. 490, don't use drywall as a substrate for tile around tubs and showers.

▶ **Ventilate.** Even if there's a window in the bathroom, be sure there's an exhaust fan in the ceiling near the shower. If it's a light/fan combo, the fan switch should have an integral timer so the fan can keep running after the light is turned off. See "Controlling Moisture and Mold" on p. 435 for the whole story.

▶ **Add GFCI protection.** All electrical outlets, including fans, must be protected by GFCIs. Shocks could be fatal in such a moist environment, so the National Electrical Code requires GFCI protection on all bathroom and many kitchen outlets. See chapter 11 for more information.

▶ Choose easy-to-clean details. Countertops with integral bowls are much easier to keep clean because there's no seam for crud to collect in. For the same reason, undermount sinks are preferable to sink rims or mounting clips that sit atop the counter. Nonporous baseboards or backsplashes allow you to swab corners with a mop or sponge without worrying about dousing walls or wood trim. Finally, you can mop bathroom floors in a flash if you have wall-hung toilets.

AN OVERVIEW OF BATHROOM FIXTURES

This short section provides an overview of what's available and a few buzzwords to use when you visit a fixture showroom. Start your search on the Internet, where most major fixture and faucet makers show their wares and offer design and installation downloads. One of the most elegant and informative sites is www.kohler.com.

Lavatories (bathroom sinks) are available in a blizzard of colors, materials, and styles. Styles include *pedestal* sinks, *wall-hung* units (including corner sinks), and *cabinet-mounted* lavatories. Wall-hung lavs use space and budgets economically, but their pipes are exposed, and there's no place to store supplies underneath. Pedestal sinks are typically screwed to wall framing and supported by a pedestal that hides the drainpipes. Counter-mounted lavatories are the most diverse, and they use many of the mounting devices discussed earlier in this chapter for kitchen sinks. Less common are *vanity top* (vessel) basins, which sit wholly on top of the counter.

The tile on the tub surround and floors is Solnhofen limestone, which was formed 150 million years ago in the Mesozoic Era, when warm seas covered present-day Germany. Close up, you can see fossilized sea snails in the stone.

The understated beauty of nature continues in this cabinet's soft, beveled edges and muted finishes. Slip-matched "ropey" cherry doors and drawers are edged with solid wood. Note the fine-grained "absolute black" granite countertop and matte black metal pulls.

Lavatory materials should be easy to clean, stain-resistant, and tough enough to withstand daily use and the occasional dropped brush or blow dryer. Although stainless-steel lavs are tough and stain-resistant, their sleek, polished look is distinctively modern, so they may not look good when matched with traditional porcelain fixtures. *Vitreous china* (porcelain) lavs have a hard, glossy finish that's easy to clean and durable, and is generally as durable as *enameled cast iron*. Hard use can chip and crack vitreous or enameled fixtures. *Spun-glass* lavatories are made from soda lime glass, often vividly colored and irregularly textured. Finally, *solid-surface* lavato-ries are typically bonded to the underside of countertops of a similar material (see "Counter-tops for Kitchens and Baths" on pp. 403–407).

Important: Make sure lavs and lav faucets are compatible. Most lavatories are predrilled, with faucet holes spaced 4 in. (centerset) apart or 8 in. to 12 in. (widespread) apart. There are single lavs for single-hole faucets. And undermounted lavs may have no faucet holes at all; you need to drill the holes in the counter itself.

Toilets and bidets are almost always vitreous china and are distinguished primarily by their types and their flushing mechanisms. Close to 99% of toilets sold are of these three types: tradi-tional *two-piece* units, with a separate tank and bowl; *one-piece* toilets; and *wall-hung* toilets. The other 1% includes composting toilets, reproduc-tions of Victorian-era toilets with pull-chain flushing mechanisms, and so on. As noted in chapter 12, toilet-base lengths vary from 10 in. to 14 in. (12 in. is standard), which can come in handy in a renovation when the wall behind the toilet is too close or too far.

Choosing a toilet is a tradeoff among factors such as water consumption, loudness, resistance to clogging, ease of cleaning, and cost. *Wash-down* toilets are cheap, inclined to clog, and banned by some codes. *Reverse-trap* toilets are quieter and less likely to clog than wash-downs. The quietest and most expensive models are typi-

Bath-Fixture Clearances

Shower stall — 24 in. to 32 in.

Lavatory — 6 in.

15 in. minimum

30 in. to wall

6 in.

Toilet — 36 in. to wall

16 in. minimum

Tub — 30 in. to wall

6 in.

Fixture

Bathrooms with minimal clearances are a tight fit. If you've got room to spare, by all means space fixtures farther than the prescribed minimums.

This palette of colorful, durable materials is rustic yet sophisticated.

cally *siphon-jet* or *siphon-vortex* (rim jet) toilets. Siphon toilets shoot jets of water from beneath the rim to accelerate water flow. Kohler also offers a Power Lite model with an integral pump that accelerates wastes to save water and a Peacekeeper toilet that "solves the age-old dispute over leaving the seat up or down . . . . To flush the toilet, the user simply closes the lid." A Nobel Prize for that one!

Bathtubs and shower units are manufactured from a number of durable materials. And site-built tub/shower walls can be assembled from just about any water-resistant material. Bathtubs are typically enameled steel, enameled cast iron, acrylic, or fiberglass; preformed shower units are most often acrylic or fiberglass. *Steel tubs* are economical and fairly lightweight; set in a mortar bed, they retain heat better and aren't as noisy if you knock against them. *Cast iron* has a satisfying heft, retains heat, and is intermediate in price, but it's brutally heavy to move. For that reason, many remodelers choose enameled steel, acrylic, or fiberglass to replace an old cast-iron tub. *Acrylic* and *fiberglass* are relatively light-weight and are available in the greatest range of shapes and colors; their prices range widely, from moderate to expensive.

Standard tub sizes are 2 ft. 6 in. to 3 ft. wide; lengths are 4 ft. 6 in. to 5 ft. However, if space is tight, you can opt for a compact tub or replace a tub with a shower stall. Shower stalls come as compact as 32 in. sq., but that's a real elbow knocker. Stalls 36 in. sq. or 36 in. by 42 in. are more realistic. One-piece molded tub/shower units don't win beauty prizes, but if properly detailed and supported, they are virtually leakproof.

CHOOSING A LAVATORY FAUCET

If you're buying a new lav and new faucets, make sure they're compatible. As noted earlier, lavatories often have predrilled faucet holes spaced 4 in., 8 in., or 12 in. apart. Most inexpensive to moderately priced faucets with two handles have valve stems 4 in. on center. Beyond that, the biggest considerations are faucet bodies, finishes, handle configurations, and spout lengths.

Faucet bodies will last longest if they're brass, rather than zinc, steel, or plastic. Brass is less likely to leak because it resists corrosion and can be machined to close tolerances. Forged brass parts are smoother and less likely to leak than cast brass, which is more porous. If you spend five minutes operating a nonbrass faucet, it will likely feel looser than it did when you started.

Faucet finishes are applied over faucet bodies to make them harder, more attractive, and easier to clean. The most popular finish by far is polished chrome, which is electrochemically bonded to a nickel substrate; it doesn't corrode and won't scratch when you scrub it with cleanser. Manufacturers can apply chrome plating to brass, zinc, steel, and even plastic. But although chrome plating protects faucet surfaces, their inner workings will still corrode and leak, if they're inclined to.

By the way, brass-finish faucets consist of brass plating over chrome (over a solid-brass base). Brass finishes oxidize, so they should be protected with a clear epoxy coating.

Faucet handles are an easy choice: one handle or two. Hot on the left, cold on the right—who'd have thought you could improve on that? But there's no denying that single-lever faucets are much easier to use. Also consider the valves inside the faucets. Ceramic-disk and brass ball valves will outlast plastic and steel.

Spouts should have a little flare and be long and high enough to get your hands under them to lather up properly. If you're a hand scrubber, look for a spout at least 6 in. to 8 in. long that rises a similar amount above the sink.

Here, *moderne* sink fixtures coexist nicely with the bright colors and strong geometric shapes of the tilework.

Energy Conservation and Air Quality

In the good old days, houses were often drafty and cold, but because energy was cheap, homeowners could make do by cranking up the thermostat. All that changed in the 1970s, when energy costs went through the roof—literally, in houses with uninsulated attics. In response, builders yanked fuel-guzzling furnaces and replaced tired windows and doors. They caulked air leaks and installed weatherstripping to block air infiltration and slow the escape of conditioned air, and fervently insulated walls, floors, and ceilings. The insulated layer between inside and outside was reborn as the *thermal envelope*.

Although tightening the thermal envelope saved energy, it created a whole new set of problems, including excessive interior moisture, peeling paint, moldy walls, punky studs, and a buildup of pollutants that were never a problem when windows rattled and the wind blew free. Today, some houses are so tight that furnaces lack enough incoming air to burn fuel or vent exhausts efficiently. And in others, turning on a bathroom fan or a range hood can create enough negative pressure to pull exhaust gases back down the chimney (*backdrafting*) or suck mold spores up from crawlspaces.

Fortunately, there's help. The rise of resource-efficient and environmentally responsible building—*green building*—has spawned a number of excellent websites that address energy conservation, air quality and related issues. Among my favorites are www.GreenBuildingAdvisor.com (GBA), www.BuildingGreen.com, www.USGBC.org (U.S. Green Building Council), www.BuildItGreen.org, and www.BuildingScience.com. The last site is the home of Joe Lstiburek, whose *Builder's Guide* series on climate-specific building is unrivalled. The *JLC Guide to Energy*

Air-sealing is a crucial first step before insulating. Here, polyurethane foam is used in the attic floor to seal gaps along the top plate of a partition. Penetrations for plumbing pipes and electrical cables may be small but they can add up to big heat loss.

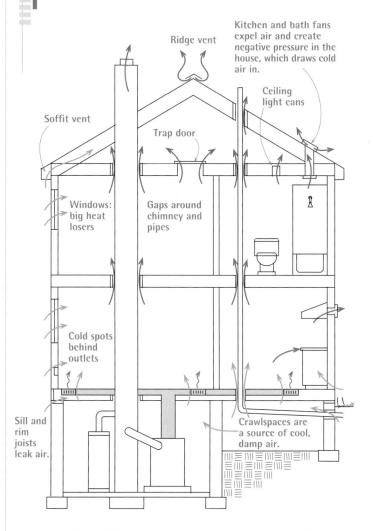

Ridge vent

Kitchen and bath fans expel air and create negative pressure in the house, which draws cold air in.

Ceiling light cans

Soffit vent

Trap door

Windows: big heat losers

Gaps around chimney and pipes

Cold spots behind outlets

Sill and rim joists leak air.

Crawlspaces are a source of cool, damp air.

By insulating living spaces, you create a boundary—a thermal envelope—between inside and outside air. Leaky walls and windows are obvious energy losers, but so are holes and gaps inside the envelope, which allow warm air to flow into unheated attics, creating negative pressure in living areas that draws in cold replacement air.

Efficiency (Hanley Wood, 2011) is very helpful. Lastly, Martin Holladay's GBA blogs are often witty and always well informed (and they have been compiled into the book *Musings of an Energy Nerd,* The Taunton Press, 2017).

A Nine-Step Energy Retrofit

In the last decade or two, the importance of saving energy has hit home. For many homeowners, however, this awareness has come with a price: The nagging feeling that if you don't whip your house into airtight, energy-sipping, good-as-green shape *right now*, you're not doing your part. Fact is, an all-at-once deep energy retrofit that cuts energy use 50% to 90% can cost $100,000 and up.

How Heat Moves

Heat is transferred in three ways: radiation, conduction, or convection. Radiation is the movement of heat through space, in the form of infrared waves; heat and light waves radiate in all directions from a central source, such as the sun or a hot stove. Conduction is the passing of energy from particle to particle, usually between objects touching each other, as when a stovetop burner touches a tea kettle. Convection is the transfer of heat or cold by the movement of air, with warmer air rising and cooler air descending.

Heat transfer in houses is usually a combination of all three mechanisms. For example, sunlight radiates into a room and heats a dark tile floor. A cat lying on the floor receives radiant heat from the sun and conductive heat from the floor. And even in a room where the air seems still, there will be convective loops as warmed air rises from warmed floors and heat vents, and falls near cold windows. As likely, there will also be convective heat loss—that is, air moving through walls can account for 20% to 30% of the total heat loss of an insulated house.

Building materials conduct heat at different rates. The more resistant a material is to heat transference by conduction, the higher its *resistivity value*, or *R-value*. In general, the less dense the material, the better insulator it is and the higher its R-value. And the denser the material, the better it conducts heat or cold and the poorer it is as an insulator. Thus, masonry materials tend to be poor insulators, while fluffy cellulose fibers are excellent. Glass, being very dense, is an excellent conductor but a poor insulator, so enormous sums have been spent to create triple-pane windows with respectable R-values. Recently, heat loss through wood framing—*thermal bridging*—has received a lot of attention and is generally remedied by covering framing with rigid insulation panels.

It's important to note that insulation is a *thermal barrier*—it resists the transfer of heat—because of trapped air pockets within it. Most insulation* does not stop airflow, however, and if air flows through insulation, its R-values plummet. (Air leaks can also carry a tremendous amount of moisture into wall cavities.) Thus, to be an effective thermal barrier, most insulation must be used with an *air barrier* that is continuous and contiguous. Exterior air barriers are usually assemblies of sheathing, housewrap, or building paper and, increasingly, rigid insulation; interior air barriers are usually drywall. In either case, sealing the seams and gaps of materials in the building envelope is the key to making air barriers continuous and airtight.

Spray foam and rigid-foam panels effectively stop airflow and may not require separate air barriers.

GETTING GREEN WITHOUT GOING BROKE

PRO TIP

If the attic is insulated, you'll need to put on gloves and a face mask and move that insulation before you can seal openings in the attic floor. But don't merely cuss those batts; examine them. Fiberglass batting actually filters dirty air, so look for blackened areas on the undersides of batts, where heated air has blown through ceiling cracks into the attic.

The good news is that you don't have to get green all at once. With planning, patience, and paying as you go—rather than going into hock—you can lower utility bills incrementally and feel good about the progress you're making. It's still a green path; it just meanders a bit more.

The nine steps below are a logical progression—which renovations rarely are. So if your home has a pressing need, tend to that first, and as you do so, try to find solutions that will improve the energy profile as well. In the remodel described on p. 45, for example, the homeowners had a drafty house with asbestos siding. After stripping the siding, they built up 2x4 studs into 2x6s, sealed air leaks, and insulated—all from the exterior. This, in turn, meant that interior finish surfaces were undisturbed.

The rest of this chapter more or less follows this sequence—with the exception of upgrading windows, which is covered at length in chapter 6. So search below for a fuller explanation of most steps.

Step 1: Get an energy audit. A thorough energy assessment is essential because it's a specific analysis of your house. Using diagnostic equipment such as blower doors, duct blaster fans, and infrared cameras, audits can pinpoint where heated or cool air is escaping. If mechanical ventilation is needed, an audit will note that, too. After the inspection, you will get a report that includes the home's energy rating, along with an estimation of annual energy use and costs. The report will usually include recommended energy retrofits and their costs, as well as the potential annual savings and probable payback times for each improvement.

Step 2: Seal air leaks. It has been estimated that up to 30% of the heat loss in some houses is due to leaks in the thermal envelope, so air-sealing should always precede insulation. In a "gut" renovation when the sheathing is exposed, it's a straightforward task to seal building seams and openings. In a selective retrofit, an audit is invaluable in tracking down elusive leaks, such as gaps where drywall panels don't quite meet, under or over wall plates, around the perimeter of the attic, and around openings for electrical outlets or plumbing penetrations.

Step 3: Add or upgrade mechanical ventilation. Once you've cut infiltration by sealing air leaks, stale air may build up inside the house. This problem can be remedied inexpensively, however, by adding a *whole-house exhaust fan* not much bigger than a standard bathroom fan

and a few *passive intake vents* (p. 432). The only downside of this solution is that it draws unconditioned air into the house, which will need to be heated (or cooled). The solution to that waste of energy, a somewhat more costly and complicated fix, is incorporating a *heat-recovery ventilator* (HRV) into the house's HVAC ductwork. There's an illustration of an HRV system on p. 435 but, in brief, an HRV uses the heat of stale exhaust air to temper fresh incoming air.

Step 4: Insulate, starting at the top. Because hot air rises, few energy retrofits are more effective than insulating an attic or roof. If you need a new roof, stripping it and installing rigid-foam panels before reroofing is a pricey but premium route to take. If you want to convert the attic to living space, spraying foam to the underside of the sheathing is a viable and less expensive way to insulate from the inside. But if your attic is unfinished and accessible, blowing in cellulose is hands-down the most cost-effective way to go—after you've sealed drywall gaps, pipe chases, electrical penetrations, the framing around the chimney, and so on. Especially in cold climates, more is better: Roughly 17 in. of blown-in cellulose will get you an R-60 attic floor.

Step 5: Insulate inside the basement. Basement walls are an easy energy retrofit because they're accessible. Insulating rim joists and basement walls is worth the money in the Cold Belt; less so in sunny climes unless you want to convert the basement to conditioned living space. In either case, correct moisture problems first. If you can't afford to insulate the walls, do insulate the rim joists. To insulate basement walls, use rigid-foam panels, as described on pp. 450–451. Heat loss through floors is negligible, so insulating them makes little financial sense unless you intend to install a finish floor over the concrete.

Step 6: Upgrade wall insulation. If you're gutting finish surfaces and exposing studs, insulating walls is a romp. Dense-pack cellulose and fiberglass batts are both effective. More often, older homes have some wall insulation, but it has settled or was never installed uniformly. Blower-door and smoke-stick testing and thermal imaging can show cold spots that need air-sealing and supplemental insulation. Typically, insulation is added by prying up siding, drilling holes in sheathing, blowing in dense-pack cellulose or injecting foam, plugging holes, and restoring siding. It can be cost-effective, but takes great skill to do a thorough job that doesn't look piecemeal.

Step 7: Replace or upgrade windows.

Replacing all single- or double-pane windows with high-efficiency, full-thickness (1⅜-in.) triple-pane windows is a huge expense: Installed, new units cost $900 to $1,200 each. Those without tens of thousands of dollars on hand might choose instead to replace one wall of windows at a time or, yet more frugally, to weatherstrip existing windows and then install low-e storm windows. Good-quality storm windows can achieve about half the U-factor of new windows at roughly one-sixth the cost.

Step 8: Upgrade your heating system.

Compared with the cost of replacing windows, installing a 92% AFUE (annual fuel-utilization efficiency) furnace for $4,000 to $6,000 seems like a bargain. It's not higher on the list because without improvements to the thermal envelope you're not making the most of the investment. Further, by tightening the house you can reduce the size of a new furnace or boiler. If you want to save money beforehand, seal leaky ductwork, guided by the Duct Blaster testing of your energy audit. Sealing and repairing ducts is relatively cheap and in winter months can save up to 20% on utility bills.

Step 9: Install renewable energy devices.

A solar panel or some other renewable-energy device may be part of your new HVAC system. More commonly in retrofits, solar hot water or photovoltaic panels that generate electricity can supplement an existing system. Incorporating renewable components into existing HVAC systems can be a costly and complex undertaking, but, presently, there are a number of federal and state incentives for renewables, including energy-efficient mortgages and financing, as explained at right. Passive solar designs reduce energy costs by taking advantage of solar energy. Although the basic ideas are fairly simple, putting them into practice can be complex, requiring the help of a design professional.

Getting an Energy Audit

Improving a home's energy profile begins with a careful diagnosis of where it's losing heat. This diagnosis can be intuitive, based on years of remodeling houses. Most energy loss is due to air infiltration, after all, and old houses leak air in predictable places—attic floors, for example. Moreover, you can feel cold air entering a leaky window on the back of your hand. Yet as building science has become more pervasive, the pressure for rigorous scientific audits has become irresistible. That trend will continue as energy prices continue to rise.

ENERGY Rebates AND INCENTIVES

Although energy-conservation programs change as political winds shift, renovators looking for ways to finance energy retrofits should start at the DOE's website, www.energy.gov.

From there you'll find additional links and descriptions of federal, state, and private programs:

▶ *Approved Energy-Efficient Appliance Rebate Programs:* www.energy.gov/energysaver/energy-saver

▶ *Database of State Incentives for Renewables and Efficiency:* www.dsireusa.org

▶ *Energy-Efficient Mortgages and Financing:* www.dsireusa.org

This last site has additional links to:

▶ *Financing an Energy-Efficient Home:* https://www.energy.gov/energysaver/design/energy-efficient-home-design

This fact sheet from the Department of Energy features an overview of energy-efficient financing programs from mortgages to home improvement loans.

▶ *U.S. Department of Housing and Urban Development: Energy-Efficient Mortgage Program:* https://www.hud.gov/program_offices/housing/sfh/eem/energy-r

The Energy-Efficient Mortgage Program is one of many Federal Housing Authority programs that insure mortgage loans to encourage lenders to make mortgage credit available to borrowers, such as first-time homebuyers, who would not otherwise qualify for conventional loans on affordable terms.

▶ *Energy Ratings and Mortgages:* www.resnet.us/energy-rating

Energy-efficient homes may qualify for mortgages that take into account a home's efficiency. Residential Energy Services Network (RESNET) provides information on home energy rating systems, energy-efficient mortgages, and finding certified energy raters and lenders who know how to process energy-efficient mortgages.

▶ *Refinancing for Energy-Efficiency Improvements:* http://ase.org/home

An overview of refinancing to make energy-efficiency improvements from the Alliance to Save Energy.

Regulations and incentives have also increased the need to verify causes—and remedies—in a scientific manner. From federal tax credits and programs to state or utility incentives, many homeowners have access to incremental rebates, based on the number of energy-saving measures they implement. As of 2010, 25 states had adopted climate action plans to reduce greenhouse gases, so many cities now require energy audits as part of the permit approval process. Typically, energy audits are triggered by some combination of a project's size and valuation—say, a project size of 500 sq. ft. or more and a $50,000 budget.

Lastly, there are emerging state and federal incentives for energy-efficient mortgages and financing, all of which require energy ratings as a condition for lending (see above). It's worth noting that bank loan officers who keep abreast of such programs also smile on energy-efficient mortgages because borrowers with lower monthly

Blower–door testing helps quantify the amount of conditioned air (and hence energy) being lost through leaks. As the blower depressurizes the house, outside air is pulled in through holes in the thermal envelope.

utility bills are better loan prospects and their homes are more highly valued than comparables that waste energy.

Now let's look at diagnostic tests that may be part of a scientific home energy audit, which takes four to six hours to complete.

BLOWER-DOOR TESTING

Because air infiltration is the major cause of energy leaks, auditors first try to quantify the volume of air being lost. After closing exterior doors and windows, turning off combustion appliances such as furnaces and gas water heaters, closing flues and shutting fireplace dampers, the auditor uses a blower door to depressurize the house. This pulls outside air into the house, in effect exaggerating the holes in the thermal envelope.

A blower-door assembly includes an airtight, expandable frame, a calibrated fan, and a manometer—a pressure gauge with two channels. One of the manometer channels measures the difference between inside and outside pressure, while the other measures the difference between the calibrated fan and the inside. The door fan blows air out of the house until the pressure difference reaches 50 pascals (Pa), a standard reference basis. Once pressure readings are steady, the manometer calculates the total flow of all leaks in the house. That airflow can be measured as air changes per hour (ACH) or in cubic feet per minute (cfm).

Getting an aggregate figure for house air leakage will suffice for some audits, but a blower-door fan can also be used in tandem with other devices to help pinpoint leaks, as described below. It's important to note that once the house

envelope has been tightened, blower-door testing is also an essential tool to determine if a house needs mechanical ventilation and to make sure that combustion appliances have adequate air supply and that running bath fans, range hoods, and other fans won't cause *backdrafting*.

SMOKE PENCILS AND THEATRICAL SMOKE

When the house is depressurized, an auditor may use a *smoke pencil* (also called smoke toys or smoke sticks) to help make individual air leaks visible. Smoke pencils emit a chemical smoke, so you can see the air being pulled in through leaks. *Theatrical smoke*, as the name implies, is a dramatic volume of "smoke" generated by a theatrical fog generator, primarily used to indicate duct leaks.

Who requests the audit usually determines how the information is used. When a homeowner gets an audit to satisfy a city planner or a mortgage lender, the smoke demonstration is usually educational. Seeing air leaks in unexpected places often convinces homeowners that they need to spend a little more to get a tighter house. If, on the other hand, an auditor is working with a *home performance contractor*, the blower door, smoke pencil, and thermal images shown here are used to direct the retrofit work. Typically, the blower keeps running as the crew seals air leaks—until a targeted (lower) leakage rate is attained.

THERMAL IMAGING

An infrared camera's ability to see invisible sources of heat loss can inform even a pro, whether he's trying to track down elusive air leaks or develop a retrofit strategy. Thermal images

A smoke pencil helps make air leaks visible.

show not only air leaks but also settled or absent insulation and *thermal bridging* (heat loss through relatively poor insulators, such as framing). It can also help locate and identify moisture issues. The images indicate temperature differences between surfaces, so the camera is most effectively used when there is a temperature difference of at least 10°F from the inside to the outside. Using a blower door in conjunction with the infrared camera can help accentuate and locate problem areas.

The price of thermal imaging cameras has declined dramatically—from thousands of dollars to hundreds—but interpreting their images can be subtle. Generally, darker colors indicate cold air, whereas lighter hues indicate warm air. With a typical color palette, cold air leaks tend to appear as blue fingers. As with any tool, using an infrared camera takes training and practice, without which results can be difficult to interpret.

DUCT TESTING

Duct testing is important because leaky ducts can be responsible for as much as 30% of lost energy, to say nothing of heat-starved rooms or, in some cases, a buildup of carbon monoxide in living spaces.

The auditor starts by temporarily sealing heat registers and supply grilles. He connects a variable-speed duct blower to a central air return, a supply plenum, or the HVAC system's fan cabinet. You can test ducts without pressurizing the inside of the house. However, as energy consultant Donn Davy points out, "If one pressurizes the interior of the house to 25 pascals and then adjusts the duct to equalize that pressure in the duct system, the reading will reflect 'leakage to the outside,' which is more useful for calculating real energy losses, since leakage within the envelope still heats (or cools) the interior, just not as effectively."

To find and fix individual leaks, a theatrical fog generator is often placed next to the duct blower. Ducts fill with "smoke," which then escapes out of duct connections, missing end caps, faulty register-to-boot seals, or leaky connections at the plenum. Occasionally, flexible ducts in crawlspaces get crushed or gashed by clumsy "repairmen," and some types of older flexible ducts degrade over time and develop holes and tears. Predictably, the largest leaks are sealed first, with special attention given to leaks in return-ducts near combustion appliances such as gas water heaters. There, leaking ducts could create negative pressure, pulling dangerous flue gases back into living spaces. Sealing ductwork is covered on p. 428.

Take thermal images in early morning, when the temperature differences (delta T) between inside and outside air are the greatest. The greater the delta T, the greater the color contrasts in the thermal image.

Leaky ducts in a forced hot-air system can squander energy. After sealing heat registers and air intakes, an energy auditor uses a fan to pressurize the ductwork, sometimes blowing theatrical smoke as well, so technicians can see just where ducts are leaking.

TESTING COMBUSTION AND HOUSEHOLD APPLIANCES

Because heating and cooling appliances can affect energy costs, health, and safety, many home audits include those devices, as well as gas-fired water heaters. According to a GreenHomes America spokesman, "We see potentially serious combustion safety and gas leak issues in as many as 25% of the homes we visit before we touch anything. And equipment that has poor or marginal draft before an air-sealing—or any retrofit project—can be made even worse. Carbon monoxide is deadly, and even chronic low levels can make people sick."

Auditors use digital combustion analyzers to see how efficiently your furnace, boiler, or gas water heater is consuming fuel. Such tests can also detect problems within a furnace, such as a cracked heat exchanger, which could allow exhaust gases to enter ducting—and living spaces. To test the draft in flue pipes, the auditor will shut all doors and windows, turn on all exhaust fans in the house in an attempt to create negative pressure, and then insert a digital probe into the flue pipe. A good auditor will also look for telltale signs of earlier backdrafting, such as rust or scorch marks near the draft hood of a furnace or water heater. Audits with health and safety components routinely check gas pipe and meter connections for leaks as well.

A thorough audit will also employ electricity-usage monitors to determine how efficient household appliances are. Kitchens are fertile testing grounds: They account for nearly 30% of the utility bill, and refrigerators and freezers together gobble more than two-thirds of the kitchen total. Ovens, microwaves, cooktops, and dishwashers account for the remaining third.

THE AUDIT REPORT

Reports vary, but those based on a whole-house energy audit such as the widely used HERS-II* (home energy rating system) will typically include:

▶ **An energy-efficiency rating** that places the house along a continuum, with a 0 rating (net zero) meaning that the house's net purchased energy is zero, and a 100 rating indicating that it meets 2008 IRC energy specs. Ratings for most existing homes are considerably above 100, showing a need for improvements. HERS ratings are helpful to homeowners because they can see, at a glance, their home's relative efficiency.

▶ **Scoring individual components.** The aggregate HERS score is broken down to show the efficiency of each house system. Based on the auditor's diagnostic tests, the report includes duct leakage, ACH or cfm numbers, insulation R-values, windows' solar heat gain coefficient (SHGC) ratings and U-values, furnace AFUE scores, and so on. A companion document, usually software generated, explains scores and industry acronyms at greater length and suggests probable savings based on improvements such as, say, increasing roof insulation.

▶ **Auditor's recommendations.** Here, skilled and personally committed auditors can really shine. Having gotten to know you and your house, they can make specific recommendations on where you should spend to improve efficiency, safety, health, and comfort. For example, "If you want to get serious about air quality—which makes sense with a 2-year-old in the house—I would seal the crawlspace with plastic sheeting, 9-mil or better." Above all, an auditor's recommendations should be objective, putting homeowners' needs above anything else.

In California, there are two types of HERS reports: One has to do with Title 24 verification, relating to 14 measures in the energy code. HERS-II is a whole-house verification, projecting how a house will use energy over the years.

HOW TO FIND AN ENERGY AUDITOR

The best way to find a reputable energy consultant, architect, or contractor is to ask a building professional whose opinion you trust. Two companies that have helped develop energy-auditing standards now certify independent auditors and

Digital combustion analyzers see how efficiently a furnace, boiler, or gas water heater is consuming fuel. An audit may also test the draft in flue pipes.

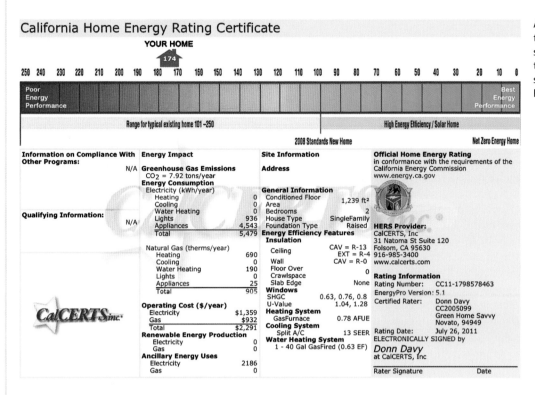

California Home Energy Rating Certificate

YOUR HOME
174

250 240 230 220 210 200 190 180 170 160 150 140 130 120 110 100 90 80 70 60 50 40 30 20 10 0

Poor Energy Performance | Best Energy Performance

Range for typical existing home 101–250 | High Energy Efficiency / Solar Home

2008 Standards New Home | Net Zero Energy Home

Information on Compliance With Other Programs: N/A

Qualifying Information: N/A

Energy Impact
Greenhouse Gas Emissions
CO_2 = 7.92 tons/year
Energy Consumption
Electricity (kWh/year)
Heating	0
Cooling	0
Water Heating	0
Lights	936
Appliances	4,543
Total	5,479

Natural Gas (therms/year)
Heating	690
Cooling	0
Water Heating	190
Lights	0
Appliances	25
Total	905

Operating Cost ($/year)
Electricity	$1,359
Gas	$932
Total	$2,291

Renewable Energy Production
Electricity	0
Gas	0

Ancillary Energy Uses
Electricity	2186
Gas	0

Site Information
Address

General Information
Conditioned Floor Area	1,239 ft²
Bedrooms	2
House Type	SingleFamily
Foundation Type	Raised

Energy Efficiency Features
Insulation
Ceiling	CAV = R-13
	EXT = R-4
Wall	CAV = R-0
Floor Over Crawlspace	0
Slab Edge	None

Windows
SHGC	0.63, 0.76, 0.8
U-Value	1.04, 1.28

Heating System
GasFurnace 0.78 AFUE
Cooling System
Split A/C 13 SEER
Water Heating System
1 - 40 Gal GasFired (0.63 EF)

Official Home Energy Rating
in conformance with the requirements of the California Energy Commission
www.energy.ca.gov

HERS Provider:
CalCERTS, Inc
31 Natoma St Suite 120
Folsom, CA 95630
916-985-3400
www.calcerts.com

Rating Information
Rating Number: CC11-1798578463
EnergyPro Version: 5.1
Certified Rater: Donn Davy
CC2005099
Green Home Savvy
Novato, 94949
Rating Date: July 26, 2011
ELECTRONICALLY SIGNED by
Donn Davy
at CalCERTS, Inc

Rater Signature | Date

A whole-house energy audit, such as the HERS-II (home energy rating system) report shown here, will typically rate a house's energy profile so homeowners can see at a glance how energy-efficient their home is.

---

maintain databases of consultants. They are the Residential Energy Services Network (www.resnet.org) and Building Performance Institute (www.bpi.org). The Department of Energy's Energy Star website, www.energystar.gov, also has a list of certified auditors and contractors. Given the growth in energy audits, a number of contractors have become *home performance contractors* who do both the audits and the remedial work. This niche is so promising that GreenHomes America, which offers energy audits, does retrofit work, and helps homeowners "secure low-interest financing or government… incentives to offset the cost."

The question invariably arises: Can a company doing the work also be an unbiased auditor? Clearly, an independent energy auditor has fewer conflicts of interest. But it has become clear that more energy retrofits are actually completed when homeowners hire a combined auditing-contracting company. Having one entity oversee the retrofit and sift through a slew of government incentives simplifies things for homeowners. So does knowing exactly whom to call if you're not happy how the work was done. Whatever auditing-contracting model you choose, interview carefully, call references, check the company's record with the state licensing board, and get a second opinion if ever you're feeling pressured or uncertain.

Air-Sealing

As noted early in the chapter ("How Heat Moves" on p. 417), most insulation achieves its full R-value only when used with an air barrier. Air leaks allow conditioned air to escape, but they also prevent insulation from doing its job. In new construction—or when gutting finish surfaces in a renovation—installing a continuous air barrier is a straightforward task because access is easy. But in most energy retrofits, upgrading the air barrier means finding air leaks in the building envelope and sealing them one at a time. As tedious as this task can be, it's time and money well spent because, dollar for dollar, sealing air leaks is the most cost-effective energy retrofit you'll undertake—especially if the attic is unconditioned (unfinished and unheated). Air infiltration can account for 20% to 30% of energy losses, especially when a home has fiberglass batts or cellulose insulation.

CAULKS, SEALANTS, AND OTHER AIDS

Because this section is about sealing air leaks, *sealants* is probably the best overall term for different materials that stop air infiltration. Most of us use *caulk* or *caulking* when filling building gaps and cracks, but here let's use "caulk" to denote a flexible sealant that is squeezed out of a tube and doesn't expand to fill the gap or crack.

By the way, sprayed-on foam insulation—sometimes called *flash* foam—is often used to

Careful air-sealing takes a lot of time—and many cans of expanding polyurethane spray foam. Local codes may require the use of fire-rated foams to seal penetrations and gaps that could become fire updrafts.

insulate *and* air-seal building seams but I'll discuss it in the following section on insulating.

Caulk choices. Caulks are covered in chapter 4, so let's keep it short and specific here. Caulks are best used for gaps less than ½ in. wide; for wider gaps, start with a foam *backer rod*, then apply caulk. Surfaces being caulked need to be clean, stable, and dry or the caulk may not stick. Scraping away flaky substrates, blowing or wiping away dust, and drying the surface are important prep steps. *Important:* Caulks have become extremely specific, so choose a type well suited to your task. Here's an overview of general categories:

▶ *Water-based caulks*, which include latex, vinyls, and acrylics, are easy to work with and probably the best choice if you're sealing around interior door and window casings and other moldings. Because they are so easy to clean up, you can use a wet finger to *tool* (shape and compress) caulk joints. They can be painted.

▶ *Silcone-based caulks* are famously tenacious, weather well and stay flexible, adhere to many different substrates, and aren't affected by UV rays. Solvent-based, they can be difficult to work with. Pure silicone can't be painted, but *siliconized acrylics* can be.

▶ *Synthetic-rubber caulks* adhere well and stay stretchy, so they are a good choice for sealing gaps in exterior trim and wood siding. Once cured, they can be painted with latex. *Note:* They are volatile until they cure. Wear a respirator when applying, don't use them indoors, and check local VOC regulations to see if they're allowed.

▶ *Modified silicone polymer caulks* are as easy to work with as latex and as tenacious as silicone. They stay flexible in cold weather, can be painted with water-based paints, don't shrink, and aren't as volatile as synthetic rubbers. They are the most expensive caulks, however, and don't have as long a track record as the others.

Polyurethane foam sealants expand to fill gaps. They are the workhorses of air-sealing, typically applied using 12-oz. to 33-oz. aerosol cans with straw-type applicators. Though their formulas vary slightly, for the most part they are rigid once cured, so they should not be used to fill gaps that will be subject to movement. They should not be exposed to weather or used outside. *Important:* All foam sealants expand, and some so aggressively they can bow window frames. When choosing a foam, first decide whether the task requires low-, moderate-, or high-expanding foam. By naming foam products for their intended uses—"Windows and Doors," "Gaps and Cracks," "Maxfill," "Fireblock"—manufacturers have made choosing a little easier.

Because foam sealants make contact when they're wet and then expand to fill the space, surface prep is not as critical as it is with caulks to ensure adhesion. Ideally, though, the surface should be stable (not loose) and reasonably free of debris. And there's a limit to the width of the gaps they can fill, so if a gap is wider than 1 in. you may want to first pack it with loose fiberglass and then seal it with foam. For the most part, foam sealants are benign, but read the labels and take appropriate safety precautions, including adequate ventilation, turning off pilot lights and not using around open flame, and wearing a respirator and safety glasses.

Weatherstripping, discussed at length in chapter 6, is the sealant of choice when a gap or opening will be subject to movement, such as the perimeter of an attic hatch, a basement's bulkhead door, or a leaky basement window. Tubular stripping made of neoprene or sponge foam can be compressed to seal out drafts.

Foam outlet and switch plate gaskets can be quickly installed beneath the cover plates of electrical receptacles and switches. Typically the gaskets are sponge foam or, in tech-speak, "UL-listed, cross-linked close cell polyethylene insulating foam." Bought in bulk, they're cheap enough to use throughout the house.

Sprayable caulk is intended to seal new-building seams before installing fiberglass batts, but it might be a fast way to seal gutted walls or an addition. There are presently two brands of sprayable caulks—EnergyComplete and EcoSeal—both developed by fiberglass insulation makers. Both caulks are quite thick and must be applied by high-pressure sprayers to the inside of stud bays, as shown in the top photo on the facing page.

Neither product touts any appreciable R-value: They're air sealants, not insulation. Both are low-VOC, so a simple respirator is sufficient

Fire-Blocking CAULKS AND FOAMS

Holes in an attic floor allow heated air to escape from conditioned spaces and so waste energy. In the event of a fire, penetrations in top plates especially can hasten the spread of fire from one floor to the next. To fill such penetrations, local codes may specify fire-barrier (fire-blocking) foams or caulks. Being *intumescent*, these caulks and foams expand when exposed to superheated air and can withstand direct flame.

when applying the stuff. Both solutions are aimed at the professional market—and priced accordingly.

AIR-SEALING THE ATTIC

The "stack effect" of warm air rising makes an unconditioned attic *the* place to start sealing. Informed by your energy audit, suit up (gloves, safety glasses, respirator, and long sleeves), then go up. Wear a headlamp so your hands will be free to work. If attic floor joists are exposed, take along a short plank or two to sit on. Step only on the plank, never on the ceiling between joists. If the floor is covered with loose-fill insulation, take a dustpan to move it aside temporarily as you seal penetrations. Attics heat up quickly as the sun warms the roof, so start early in the morning. Start with the largest openings in the attic floor.

Is the attic hatch insulated? You can buy an insulated cover or build one by gluing two layers of 1-in.-thick rigid-foam panels to ½-in. plywood. Around the edges, use sponge-foam or neoprene weatherstripping as a seal.

A 2-in. gap between the chimney and framing is required by building codes. But you can pack the gap with a noncombustible material such as rock wool, and cover it with sheet metal—metal drip-edge works especially well. Apply a bead of adhesive caulk to the framing around the chimney; place the drip-edge in the caulk, with its other edge snug to the chimney; screw the drip-edge to the framing; then apply a bead of fire-rated caulk to seal the metal to the chimney.

Open framing above soffits and chases can be a freeway for drafts, so cover large openings with rigid-foam panels or drywall panels, apply caulk or expandable foam to seal panel edges, then screw them to framing.

Old bath fan boxes can leak moisture-laden air into the attic. Apply caulk or foam to seal fan-box flanges to the ceiling; use mastic or metal duct tape (not cloth) to seal holes and cracks in the fan box. If any fan exhausts into the attic—rather than to the outside—add a vent pipe that exits through the roof or an exterior wall. Fans that terminate in the attic can soak insulation, encourage mold, and even rot framing.

Recessed ceiling lights (ceiling cans) are big leakers, especially old, non-IC-rated cans that can't be covered with insulation (IC means insulation contact). Replacing non-IC-rated cans is the safest and, ultimately, most cost-effective way to deal with their energy loss. Enclosing non-IC-rated fixtures in a sealed, insulated box is sometimes suggested as a solution. But older incandescent fixtures can get quite hot, and the

When wall surfaces are exposed, it's possible to spray an air-sealing caulk before insulating. Sprayable caulks were developed to improve the efficiency of fiberglass batts by stopping air movement within walls.

The code-required gap between chimneys and combustible materials (framing) can be a freeway for air leaks. Here, a prescored coil of metal stock is used to create dams around a masonry flue. It's nailed to the framing and then caulked with fire-rated caulk to seal gaps. Some builders use leftover metal drip-edge as an air dam.

PRO TIP

If you have knob-and-tube wiring that is still active (energized), do not spray it with foam sealant or cover it with insulation. Instead, call a licensed electrician to replace it.

build-up of heat inside a box could be a fire hazard. Replacing old ceiling cans with IC-rated ones is a safer way to go.

IC-rated cans are either airtight or not airtight. Airtight cans don't need sealing, although some installers caulk can rims to drywall. Non-airtight IC cans are rated for direct contact with insulation. Expanding foam is a type of insulation, so apply spray foam to holes in can bodies to seal them.

If using foam to air-seal IC-rated cans makes you uneasy—or codes don't allow it—there's another option: Cover cans with *insulation protection covers* such as those manufactured by Tenmat. Made of mineral wool, the covers physically isolate the light cans. Tenmat's fire-rated light covers are UL-certified.

Plumbing pipes, electrical boxes, and top plates are simple to seal—if you can find them. Vent stacks are easy because they stick up, but water supply pipes, electrical cables and junction boxes, and the top plates of partitions are usually buried under insulation. There's no substitute for removing insulation and surveying every square

foot when looking for air leaks. Frequently, the underside of batts or loose-fill areas will be discolored by dusty air. Remove insulation, apply spray foam to fill penetrations, replace insulation, and keep looking. Drywall seams and the top plates of partitions are especially easy to overlook; a schematic of the floor below, showing partition locations, may help you find them in the attic.

Sealing the perimeter of an attic with sloping roofs is tricky because the triangular space where roof meets wall is so tight; definitely wear a hard hat. With long arms and the straw applicator on the spray foam can, you should be able to spray where ceiling drywall abuts exterior walls. The top plates of exterior walls are especially big energy losers because wood is a poor insulator.

If the eaves are unvented, butt fiberglass batts snug against the exterior wall, ideally covering top plates in the process. If you'll be insulating the attic floor with loose-fill insulation, place 2-in.-thick foam panels along the attic perimeter so that the rigid foam covers the top plates as shown at left. Caulk panel edges with caulk or expanding foam for a tight seal.

If the eaves are vented, pushing insulation tight to exterior walls may block air coming up from the eave vents. You'll need to install some kind of blocking to contain the insulation, but that blocking may, in turn, keep you from covering the top plates with insulation. Perhaps the best solution in that case is installing a baffle/air chute such as the AccuVent product shown on p. 432.

Lastly, if your finished attic has kneewalls, try to air-seal and insulate the back side of the walls—the side facing the exterior. Because rigid

Placing Rigid Foam along the Attic Perimeter

The roof-sidewall juncture loses a lot of heat. Top plates have low R-values (thermal bridging), and dropped soffits often interrupt the drywall so there's little to stop infiltration. And vented eaves allow yet more air to blow through. Placing 2-in.-thick rigid-foam panels between ceiling joists—and over end plates—partially solves the air leaks, while holding back loose fill from drifting down into vented eaves.

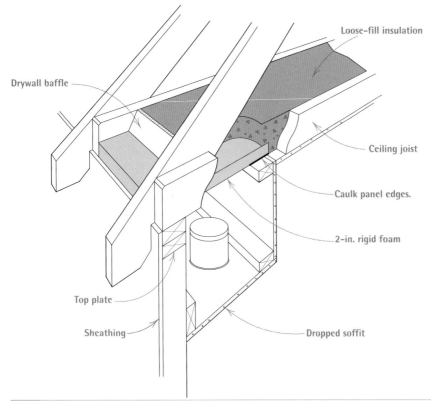

Drywall baffle

Loose-fill insulation

Ceiling joist

Caulk panel edges.

2-in. rigid foam

Top plate

Sheathing

Dropped soffit

foam has higher R-values than fiberglass batts, fitting foam panels into stud bays and sealing their edges with expanding foam will probably reduce energy loss most effectively.

AIR-SEALING WALLS, WINDOWS, AND DOORS

Correct *flashing* (chapters 6 and 7) is an essential first step to airtight windows and doors; weather-stripping (p. 136) is a good second step. Installing storm windows is a cost-effective alternative to replacing windows; it typically halves the energy escaping through old windows. Then, start sealing air leaks you can easily access.

Receptacles and switches on exterior walls can be quickly sealed by unscrewing cover plates and installing a compressible foam gasket. You can also stop a major source of air leaks and save energy by replacing incandescent recessed lights with LED retrofit kits (p. 343), which typically come with an air-sealing gasket.

If you're planning to paint, that's a good opportunity to seal room junctures that may be leaking, notably wall–ceiling joints, along baseboard trim, and around the perimeters of window and door casing on exterior walls. A water-based latex or acrylic caulk is the best material for the job: After applying a thin bead of caulk, you can use a fingertip to shape and compress it into the joint. Once it is painted, you'll never notice it.

Caulking a building's exterior seams, especially where siding meets door and window casing, will also cut air infiltration. Use synthetic rubber or modified silicone caulking. Exterior lights and outdoor receptacles usually have foam gaskets (check to be sure they're still flexible); caulk around outlet boxes to seal gaps in the siding.

Air-sealing window and door jambs is not usually part of an energy retrofit because the gaps between jambs and rough openings are covered by casing. But if you'll be removing casing to fine-tune old windows (p. 156), apply a bead of minimally expanding foam into the cavity around the jambs.

Under sinks and inside cabinets, DWV and water supply pipes often exit through rough cut-outs in the drywall. Look in upper cabinets, too, where exhaust fans are frequently housed. Expanding foam can work wonders around such pipes and ducts. Investigate, too, shared walls with an attached garage. Code requires that shared walls be firewalls constructed of ⅝-in. drywall on both sides—with no through-holes, such as back-to-back receptacles—but builders don't always get the details right. There, use fire-rated caulks or fire-rated foams.

If blower-door or duct testing indicates a leaky fireplace damper or leaky register boots, attend to them, too. More on sealing ducts to follow.

AIR-SEALING BASEMENTS

If your basement is an unconditioned space, or if you have a crawlspace, insulate the ceiling to retain heat in the floors above. Installing open-cell spray foam insulation will air-seal and insulate joist bays simultaneously, while also creating a semipermeable membrane that will allow moisture to migrate. Alternatively, if you don't have the money to spray-foam the basement ceiling, you can insulate it with rigid-foam panels, using cans of spray foam to air-seal panel edges.

Next, air-seal rim joists, which, along with the joint between the mudsill and the top of the foundation, are the biggest energy leakers in the basement. The most cost-effective way to seal rim joists is to cut and fit pieces of rigid-foam insula-

To fill gaps around window and door frames, use a mild–expanding polyurethane foam; more expansive foams can bow frames and cause a window or door to bind in the frame.

Rim joists lose a lot of heat in cold climates. Here, a Connecticut homeowner cut and fitted XPS foam panels into joist bays and sealed edges with expanding foam. Seal any penetrations in basement walls.

tion into each joist bay, sealing the perimeter of each foam panel with expanding spray foam. Spray-foam any foundation holes that go all the way through, too.

To air-seal concrete or concrete basement walls (assuming there are no moisture problems), caulk cracks and gaps, using masonry caulk or a compatible expanding foam. If you'd like to turn the basement into conditioned living space, you can air-seal and insulate basement walls quickly by installing EPS foam panels, as described on p. 450. *Note:* Do *not*, however, install plastic (polyethylene) sheeting as an air or vapor barrier on basement walls. It will trap moisture.

Inoperable basement windows should be sealed shut with expanding foam. Windows and doors that must remain operable for light, ventilation, or egress, in case of emergency, should be weatherstripped with tubular neoprene compression gaskets.

While you've got the spray foam can out, fill holes around pipes and wiring rising out of the basement to forestall air leaks to the floors above. Lastly, if your HVAC system is in the basement, now's the time to seal ductwork, as described below.

AIR-SEALING DUCTWORK

If it's not possible to seal duct leaks while a smoke test is in progress, use a grease pencil or a felt marker to mark leaks so you can go back and seal them later.

Ductwork is joined by various mechanical means. *Rectangular ducts* are fabricated to snap together, so if a rectangular joint is leaking, section ends may have simply separated. With the aid of a helper, push sections together. *Round ducts* with crimped ends are friction-fit, with three self-tapping, sheet-metal screws spaced evenly to hold them together. *Insulated flex ducts* are joined to rigid ducts, floor boots, tees, and so on, by using two *duct straps* (zip ties), which are drawn tight with special *tensioning pliers*. Typically, two ties are needed because insulated flex ducts have an inner liner and an outer jacket to attach.

The size of gaps will determine which materials you use to seal them. For best results, start air-sealing near the main plenum or the cabinet

Rigid-metal duct (left) offers the least resistance to airflow; seal its joints and insulate the runs through unheated areas. Flexible insulated ducts, such as Wire Flex (center), and Alumaflex (right) don't need fittings to make turns, but Wire Flex can be punctured. Alumaflex offers a good balance of strength and flexibility.

Sealing Ducts

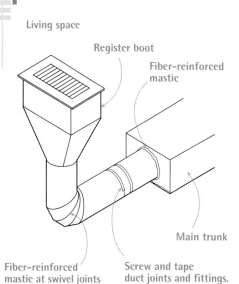

Living space

Register boot

Fiber-reinforced mastic

Main trunk

Fiber-reinforced mastic at swivel joints

Screw and tape duct joints and fittings.

Secure joints between duct sections and fittings with at least three sheet-metal screws. Then wrap the joints with aluminum duct tape—not fabric duct tape. Apply fiber-reinforced mastic to hand-snipped and swivel joints.

To seal duct joints, apply a generous amount of water-based mastic sealant. Correctly applied, the sealant should ooze through gaps; when it hardens, those tiny mastic fingers will hold the seal in place.

housing the system blower, where air pressure will be greatest. Holes less than ⅛ in. wide can be sealed by brushing on mastic. If the hole is inaccessible—say, atop a duct that's close to a subfloor—you can smear on mastic with your hands if you're wearing disposal latex or nitrile gloves. If the gap is wider than ⅛ in., first cover it with self-adhering fiberglass tape, then coat the tape with mastic.

Fiberglass tape and mastic can also be used to seal gashes in insulated flex ducts, but because such ducts "give" somewhat as pressure is applied to them, *oriented polypropylene (OPP) tape* and *foil-backed butyl tape* are often used to patch flex ducts because both tapes stick aggressively. Ironically, you should *never* use cloth-backed duct tape to seal ducts because heat causes its adhesive to dry out and come undone.

VAPOR RETARDERS

Any discussion of air-sealing would be incomplete without mentioning *vapor retarders*, commonly called *vapor barriers*. As essential as water is to life, excess moisture inside a house can lead to big problems—especially as houses become more airtight. Thus, materials that are part of a building's thermal envelope are chosen, in part, on how *permeable* or *impermeable* they are to water vapor.

Permeability is measured in *perms*: The lower the number, the more impermeable the material is and the more it will impede moisture's movement. The higher the perm number, the more easily moisture can pass through it. Accordingly, vapor retarders are classified:

▶ **Class I (impermeable)** vapor retarders have permeance levels of 0.1 perm or less and are rightly called vapor *barriers*. The group includes polyethylene sheeting, rubber membranes, sheet metal, aluminum foil, foil-faced sheathing (insulated and not), and glass.

▶ **Class II (semi-impermeable)** vapor retarders have permeance levels between 0.1 perm and 1 perm. Class II materials include oil-based paint, kraft-paper facing on fiberglass batts, unfaced extruded polystyrene (XPS) panels thicker than 1 in., and traditional stucco applied over #30 building paper.

▶ **Class III (semipermeable)** vapor retarders have permeance levels between 1 perm and 10 perms. The group includes plywood, OSB, #30 building paper, most latex paints, unfaced expanded polystyrene (EPS) panels, and unfaced polyiso (polyisocyanurate) foam panels.

▶ **Class IV (permeable)** vapor retarders have permeance levels greater than 10 perms. Class IV materials include housewraps, unfaced fiberglass insulation, cellulose insulation, and unpainted drywall and plaster.

Vapor barriers are one of the most contentious topics in building science. Air moves back and forth through the thermal envelope as the seasons change. During summer months, when the air outside is warmer, it tends to move inward. In colder months, heated indoor air migrates outward. As water vapor migrates outward along with the heated air, the moisture condenses as it hits cold exterior surfaces. If that condensation is sustained and excessive, it can soak and, in time, rot the framing. (Warm, moist air escaping through ceilings causes the picturesque but damaging ice dams of New England roofs.)

In response to winter condensation problems, builders began installing a polyethylene vapor barrier on the living-space side of insulation to prevent moist air from migrating into wall cavities and condensing there. In winter, vapor barriers work well enough. But in warm months, when moist outside air migrates into wall cavities, it is blocked by the vapor barrier and stays in the walls. Heating and cooling climates have opposite problems when it comes to poly vapor barriers. In hot, humid climates, where air conditioning runs constantly in the summer, a polyethylene vapor barrier installed behind the drywall becomes a condensing surface for moisture.

These days, there's an uneasy consensus that vapor barriers should be used sparingly in regions of extreme cold, such as climate Zones 6 and higher (see the map on p. 439). That is, vapor barriers should *not* be used in moderate or hot climates. Take note renovators: If a home has polyethylene vapor barrier behind the drywall, do not compound the problem by retrofitting foil-faced insulation on the exterior of the building. Moisture in the walls would have nowhere to go, and that's what builders are increasingly concerned about. There must be some drying potential for exterior walls, either to the outside or the inside.

Better understanding of the movement of air and moisture has led to a wider use of class II and class III vapor retarders. Some builders use what's called the Airtight Drywall Approach to stop air leaks and the movement of moisture into walls. And a new generation of exterior latex paints are proving more durable than oil-based paints because latex, being semipermeable, is less likely to be lifted off of a substrate by trapped moisture.

Vented and Unvented Roofs

Vented attics with insulation on the floor are marvels of nature and man. After you air-seal openings in the attic floor and cover it with an appropriate amount of insulation, nature does the rest. Rising as it warms, air flows up from the eaves and out of ridge or gable vents, carrying off moisture that may have migrated from living spaces. In cold climates, insulation and ventilation combine to keep the roof cold, thus minimizing melting snow and ice dams. In hot climes, the same combination rids attics of sultry air and moderates temperatures in the floors below.

Codes require a net-free ventilation area (NFVA) of 1 sq. ft. of vents for each 300 sq. ft. of attic space, equally divided between eave and ridge vents. But building scientist Joe Lstiburek takes issue with that, opting instead for more ventilation at the eaves—say, a 60/40 split, with the eaves getting the greater proportion. Lstiburek reasons that giving eaves more ventilation "will slightly pressurize the attic. A depressurized attic can suck conditioned air out of the living space, and losing that conditioned air wastes money."

As near perfect as vented attics are, though, they have weak points energy-wise. You can pile insulation as high as you like in the middle of the floor to attain required R-values, but the tight spaces where roof slopes meet sidewalls are a problem. Do your best to air-seal and pile insulation over top plates—without blocking soffit vents and the like. Installing rigid insulation over wall sheathing can reduce energy loss in this vulnerable juncture, but it's a prohibitively expensive fix unless you already need to strip siding for some other reason.

A Vented Soffit

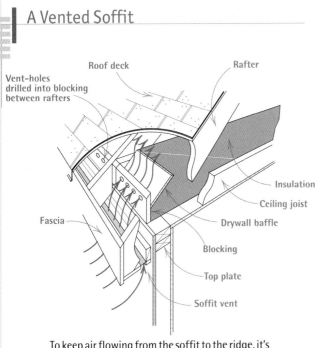

To keep air flowing from the soffit to the ridge, it's important to keep attic insulation from covering the vent to the exterior. Typically, builders use rigid foam to weigh down loose insulation along the sidewalls or install combination baffles/air chutes to ensure airflow.

Insulating under the Roof

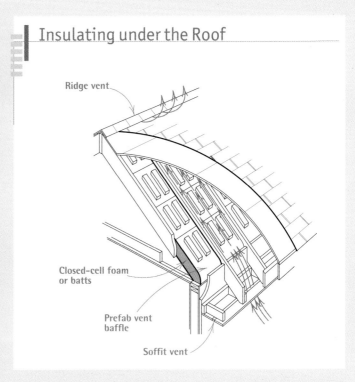

Venting the roof deck. If you want to turn the attic into conditioned space or you've got a room with a cathedral (vaulted) ceiling, insulating and ventilating become more complex. The venting path is essentially the same—air flows up from the eaves and out at the ridge—but creating a vent channel under the roof deck can be challenging. (The IRC requires 1 in. of airspace under roof sheathing; Joe Lstiburek calls for 2 in. of airspace, minimum.) Installing a prefabricated vent baffle (above) is certainly the fastest and perhaps the most cost-effective way to create a vent channel: Staple prefab baffles between rafters and use canned spray foam to air-seal baffle edges.

Insulating under the roof is the conventional way to go and probably the most affordable if the roof is in good shape. Your heating zone, your budget, and the depth of your rafters will decide what type of insulation to use. Filling joist bays with closed-cell spray polyurethane foam will provide the greatest R-value per inch, but it's the most expensive option. If you have 2x10 or 2x12 rafters, you may be able to reach requisite R-values, with some combination of batts between the rafters and XPS rigid foam under them. Furring out rafters to gain additional depth is another option, albeit a labor-intensive one.

Insulating over the roof may be a more attractive option if your rafters aren't deep enough or your roof is worn out and you need to strip it anyway before reroofing. The built-up roof shown in the drawing at right relies on an array of 2x4 *purlins* (horizontal pieces) that capture the foam panels installed over the existing roof deck and provide nailing surfaces for materials installed over the foam. An array of vertical 2x4 *spacers* creates vent channels over the foam and under the new plywood sheathing (a second roof deck) to which the new roofing will be nailed. This is a very complicated assembly, which should be attempted only by seasoned builders. Perhaps the most expensive solution is shown here.

An unvented roof is often the only viable option when roof framing is complicated, such as when there are hips, valleys, dormers, or skylights that would prevent eave to ridge ventilation; when the house has no soffits and hence no soffit vents; or when adding eave vents would clash with the house's architectural style. Unvented roofs also make a lot of sense in high-wind or high-fire areas, where vents might admit drenching rains or embers.

Here again, the choice of what insulation to use and whether to install it over or under the roof deck depends on the condition of the present roof, the depth of rafters, the R-value you must attain, and cost. Cost aside, the most straightforward route is spraying closed-cell polyurethane foam on to the underside of the roof deck. If you need additional R-value, install XPS foam panels to the underside of rafters, and then cover the foam with drywall. Correctly done, this assembly creates a premium air-and-thermal barrier. Again, it's a complex roof that needs to be impeccably detailed by a pro.

Of course, there are caveats. (There are always caveats.) Unvented roofs in the snow belt may still need some type of venting to keep the roof cold and prevent ice dams. And some roofing manufacturers won't honor warranties if their shingles are installed over unvented roofs. For more on this complex and constantly evolving topic, visit Joe Lstiburek's website: www.buildingscience.com.

Insulating over the Roof

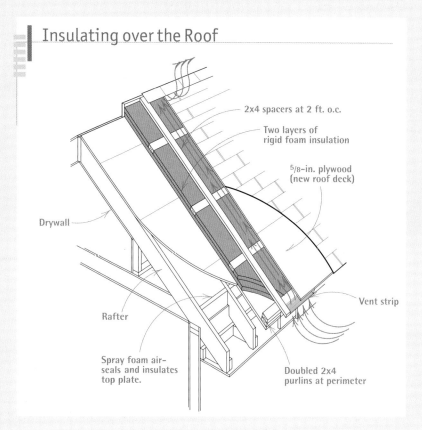

2x4 spacers at 2 ft. o.c.

Two layers of rigid foam insulation

⅝-in. plywood (new roof deck)

Vent strip

Doubled 2x4 purlins at perimeter

Drywall

Rafter

Spray foam air-seals and insulates top plate.

Unvented Roof

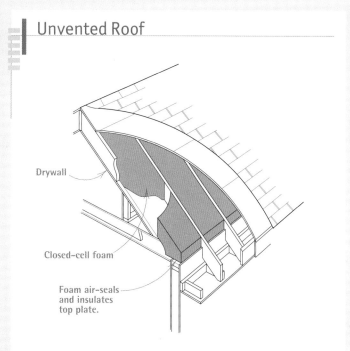

Drywall

Closed-cell foam

Foam air-seals and insulates top plate.

Increasing Controlled Ventilation

After air-sealing the house, you may need to add controlled ventilation—some mixture of mechanical ventilation (fans) and natural convection—to exhaust stale air, remove excess moisture, and avoid furnace backdrafting. To moderate temperatures under the roof and get rid of moisture that has escaped from living spaces, attic and roof vents work like a charm as long as uprising air is allowed to flow freely. Better yet, this venting is free. Looking at mechanical solutions, whole-house exhaust fans and a few passive intake vents don't cost much.

SOFFIT-TO-RIDGE VENTILATION

If air-sealing and insulating attic floors are the first steps to reducing excessive moisture and heat in an attic, increasing ventilation is the second. Nothing exhausts moisture or cools the area under a roof as effectively as soffit-to-ridge ventilation, as discussed here and in chapters 5 and 7. (Gable-end vents help but are usually 1 ft. to 2 ft. below the highest and hottest air.) As a bonus, in winter, cool incoming air can prevent snowmelt and ice dams along eaves. In summer, when unvented roofs can reach 150°F to 160°F,

soffit-to-ridge ventilation can prolong shingle life and make upper-floor rooms appreciably cooler.

Keeping vent channels open from soffit to ridge is essential to keeping air flowing. Continuous soffit vents are typically screened to keep animals and insects out of the house. Inside the attic, it's equally important that as you air-seal the perimeter and insulate the attic floor, you prevent insulation from clogging vent holes (intakes) drilled in the blocking (see "A Vented Soffit" on p. 430) or drifting down into soffit vents. The simplest way to keep insulation where it belongs is to install baffles where attic floor joists meet the rafters.

Standard how-to advice is to construct baffles on site by cutting up rigid-foam panels—without saying exactly where or how you attach the baffles in the tiny triangular spaces under the roof. Moreover, the spaces between rafters almost always vary, so precutting baffles outside the attic won't work. In a 30-ft. by 40-ft. attic, you'll need roughly 80 lin. ft. of baffles, which could take days to fabricate, fit, and seal.

AccuVent, a baffle fabricated out of recycled plastic, seems a good solution to some of the problems just mentioned. Unlike rigid-foam baffles available at many home centers, AccuVents are flexible and less likely to fracture when fit into the cramped triangular spaces where sloping roofs meet walls. Staple the flexible vent's lower end to the inside of a top plate, roll its upper end snug to the roof deck and staple it, and then seal its edges with spray foam. Ridges in the back side of the vent keep it spaced about 1 in. from the sheathing, so air can flow over it. It's both a baffle and an air chute, in other words.

INSTALLING AN EXHAUST FAN

The least expensive mechanical ventilation is a whole-house exhaust fan—similar to a bath fan but rated to run continuously. (This fan is in addition to a bath fan and a kitchen range hood.)

It's hard to imagine an exhaust fan as small as 50 cfm ventilating a whole house, but that size can vent up to 1,500 sq. ft. Exhaust fans made by Delta, Panasonic, Broan, and others offer integral humidity sensors that turn fans on or off, based on the humidity level. Some models also vary fan speed as humidity levels rise or fall.

To avoid depressurizing the house and causing *backdrafting*, it's wise to install two or three *passive intake vents* as well. The cost of fan and intakes is modest—only a few hundred dollars—so spend a little more and have an HVAC specialist size and locate components for you and help you choose a 24-hour automatic timer to control the fan.

A Prefab Baffle/Air Chute

Both air chute and baffle, the AccuVent keeps insulation from getting into soffits and blocking airflow. It may be tricky to install if roofing nails stick out from the underside of the sheathing.

Baffle

Seal baffle edges with expanding foam.

Batt or loose-fill insulation

Bath-Fan BELLS AND WHISTLES

Removing moisture is the primary function of a bathroom fan—good to remember when considering all the extra features you could buy. First, get a quiet fan: 2 sones to 3 sones is tolerable, 1 sone is quiet, and there are even 0.3-sone fans. Next, consider switches. Because fans usually need to continue venting after you leave the room, get an electronic switch with an integral timer so the fan can keep running after the light has been turned off. Or put the light and the fan on separate switches. You can also connect the fan to a humidistat, which is a moisture sensor that will turn off the fan when a preset moisture level is reached.

Noise won't be much of an issue. Whereas cheap bath fans are rated at 3 or 4 sones, some energy-efficient models run at less than 0.3 sone. Remote inline fans are another option: They're larger and noisier, but because they are typically located in attics or crawlspaces, they're out of earshot.

To see how simple whole-house ventilation can be, let's look at installing a bath fan.

Installing a bath fan. Place the fan near the shower—over it if possible—and run a duct from the fan out the roof or through a gable-end wall. Soffit and sidewall vents aren't as desirable because expelled moisture could get drawn up into the attic by a soffit-to-ridge updraft. Keeping moisture out of attics and wall cavities is crucial, and you can help achieve it by caulking the fan housing to the ceiling and sealing each duct joint with foil duct tape, *not* cloth-backed duct tape.

Position the fan on the ceiling. If there is an unfinished attic above, drive a screw up through the ceiling to approximate where you want to put the fan. Make a cardboard template of the fan box (housing) and then go up into the attic. Once you've located the screw (you may have to move insulation to find it), place the fan template next to the nearest joist—most fan boxes mount to ceiling joists—and trace around it. (If the fan box has an adjustable mounting bar, you have more latitude in placing the fan.) Use a jigsaw or reciprocating saw to cut out the opening in the ceiling. To keep the drywall cutout from falling to the floor below, screw to the drywall a piece of scrap wood slightly longer than the cutout.

Mounting the fan. To mount the fan box, you may need to remove the fan-and-motor assembly first. If the housing flange mounts flush to the underside of the ceiling, as shown in the photos at right, use a piece of drywall scrap to gauge the depth of the unit relative to the finish ceiling. But whether the housing flange sits above or below the ceiling drywall, caulk the flange well with

Before screwing a bathroom fan box to a ceiling joist, apply several parallel beads of silicone caulk between the box and the joist to minimize vibration. The scrap of drywall to the right of the box ensures that fan-box flanges will be flush to the finish drywall ceiling.

After mounting the fan box, attach ducting to the fan box's exhaust port. Keep duct runs short, and seal metal duct-fitting joints by wrapping them with self-adhering foil tape.

polyurethane sealant to create an airtight seal between the two materials. To further secure the fan and anchor the edges of the drywall opening, run blocking between the joists—along two sides of the opening—and screw the drywall to the blocking. In some cases, you'll be screwing through the fan's housing flanges as well.

Follow the wiring diagrams provided by the manufacturer. In general, it's easier to run electrical cable through a switch box first because junction boxes inside fan housings tend to be cramped. Bathroom fans should be protected by a ground-fault circuit interrupter (GFCI); see chapter 11 for more information.

Installing the roof vent. Keep duct runs as short as possible to reduce air resistance. After attaching the lower end of the flexible duct to the fan's exhaust port and sealing the joint with metal duct tape, hold the free end of the duct to the underside of the roof sheathing (or gable-end wall) and trace its outline onto the sheathing. Drive a screw through the middle of the circle. Then go outside and locate the screw, which represents the middle of the vent hole you need to cut. Sketch that

circle onto the roof. If the circle would cut into the tabs of any shingle—roughly the bottom half of a shingle strip—use a *shingle ripper* to remove those shingles before cutting the vent hole in the sheathing. Be gentle when removing asphalt shingles and you can probably reuse them. Use a hooked *shingle knife* to cut the circle into any remaining shingles and the roofing paper.

Flash the fan's roof vent as you would any other roof vent: Feed its upper flange under the shingle courses above and over the courses below. Caulk or nail the flange edges per the installation instructions and renail the surrounding shingles. Once the roof vent is flashed, go back under the roof and attach the free end of the duct, also sealing that joint with metal duct tape.

A Bathroom Fan

Roof vent

Flexible metal duct

Metal foil tape

Silicone caulk between flange and drywall

Fan box

Ceiling drywall

To keep moisture from leaking into the attic, apply silicone caulk between fan-box flanges and mating surfaces, such as drywall. To ensure airtight joints, use metal foil tape to seal ducting to the fan's exhaust port and to the roof vent or sidewall vent.

REMOTE Inline FANS

Even a well-made bathroom fan will be relatively noisy if its motor is in the bathroom ceiling 2 ft. from your head. But if you install the fan some distance from the bathroom, you'll reduce the noise considerably. That remote location may mean longer duct and wiring runs, but routing them is rarely a problem. In fact, with a large-enough fan motor and multiple ports, you can vent two bathrooms with one fan. Because longer duct runs can mean greater air resistance, consider installing rigid-metal or PVC ducts, whose smooth surfaces offer less resistance, rather than flexible metal ducts. Alternatively, you could oversize the fan slightly: Remote fans are noisier than standard bathroom fans anyway. Tucked away in an attic or crawlspace, who'll hear them? Better fan makers such as Fantech and American Aldes offer acoustically insulated cases to deaden sound further.

A BALANCED VENTILATION SYSTEM

Adding a whole-house exhaust fan and a handful of passive intake vents will get rid of stale air and admit fresh air for a modest investment. But the cycle still wastes money. Stale though that exhausted air is, it has been heated or cooled at great expense. And fresh intake air still needs to be conditioned. So mechanical engineers next devised a heat-exchange system that would use exhaust air to temper incoming fresh air. At the heart of this balanced system is a heat-recovery ventilator (HRV) in cold climates or an energy-recovery ventilator (ERV) in warmer climates.

Typically, an HRV/ERV system has two fans, one to bring in fresh air and one to expel stale air. Its heat exchanger recovers 75% to 80% of the heat in the outgoing air and uses it to temper incoming air. As both air streams move through the exchanger, it can filter pollen and dust from incoming air and remove excess moisture. And, of course, by equalizing air pressure in tight houses, the system prevents negative house pressure and backdrafting. And the system can be installed cost-effectively because it can be tied into existing duct work. On average, HRV/ERV elements can be retrofitted for $1,500 to $3,000. The principal expense is not the equipment, but the labor required to cut into (and repair) finish surfaces to add ducts.

Controlling Moisture and Mold

Mold needs three things to grow: water, a temperature range between 40°F and 100°F, and organic matter, such as lumber or paper. If your renovation includes gutting finish surfaces, replacing standard drywall with paperless drywall (p. 456) is a smart move in high-humidity areas. Otherwise, the best way to thwart the growth of mold is to exhaust excess moisture—send it outdoors, that is.

Interior moisture generally isn't a problem unless it's excessive and sustained. (Relative humidity indoors should be 35% to 40% during the heating season.) Signs of excessive moisture include condensation on windows, moldy bathrooms or closets, soggy attic insulation, and exterior paint peeling off in large patches. Figure out where the water's coming from, and you're halfway to solving the problem.

SOURCES OF EXCESSIVE MOISTURE

Common sources of excessive interior moisture are (1) air infiltration; (2) poorly installed roofing or incorrectly flashed windows and doors; (3) improper surface drainage; (4) unsealed

Balanced HRV System

Heat-recovery ventilators (HRVs) can be incorporated into existing forced hot air ductwork cost-effectively. HRV systems extract heat from outgoing exhaust air and use that heat to warm incoming fresh air—thus saving energy. In hot climates, an energy-recovery ventilator (ERV) uses outgoing stale air to cool incoming fresh air; ERVs can also temper the humidity of warm, incoming air. Because either system has both intake and exhaust vents, there is no danger of depressurizing the house.

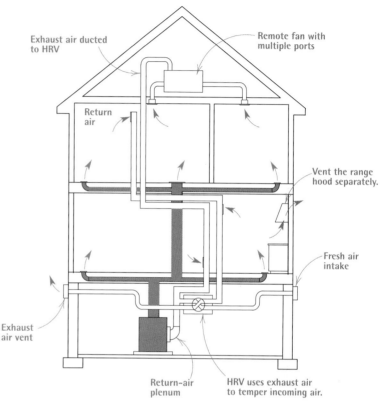

Exhaust air ducted to HRV

Remote fan with multiple ports

Return air

Vent the range hood separately.

Fresh air intake

Exhaust air vent

Return-air plenum

HRV uses exhaust air to temper incoming air.

crawlspaces; (5) inadequately vented bathrooms, kitchens, laundry appliances, water heaters, and furnaces; and (6) leaking HVAC ducts.

By far the largest source of interior moisture comes from kitchens and bathrooms. But as we just looked at installing an exhaust fan and treated the other sources elsewhere, let's next consider sealing crawlspaces.

SEALING CRAWLSPACES

Most crawlspace or basement wetness is caused by improper surface drainage—especially clogged gutters and downspouts. For more on reducing dampness caused by exterior water, see chapter 10.

Crawlspaces are well named: They tend to be dark, dank, dirt-floored areas only a few feet high. To disperse moisture, building codes typically prescribe 1 sq. ft. of screened vents for each 150 sq. ft. of dirt floor, or 1 sq. ft. of vents for every 1,500 sq. ft. of floors covered with a moisture barrier. Problem is, open crawlspaces mean cold floors and heat loss in winter, and in summer, warm, moist air entering through vents invariably condenses on the cooler surfaces of the crawlspace—leading to mold and worse. So it makes more sense to seal and condition crawlspaces. Otherwise, mold spores growing in the crawlspace will be sucked into living spaces by bath and kitchen exhaust fans and carried all the way up to the attic by the *stack effect* of rising heated air.

To seal crawlspaces, first rake the floor to remove debris and sharp rocks, which could puncture a plastic moisture barrier. Heavy sheeting will last longer: 6-mil polyethylene is minimal, but commercial waterproofing firms, such as Basement Systems, use 20-mil polyester cord–reinforced sheeting, which can withstand workers crawling and objects stored on it.

In a 1,000-sq.-ft. rectangular crawlspace without jogs, it typically takes five large sheets of polyethylene to isolate the space: a single floor sheet that runs about 1 ft. up onto walls and four wall pieces that overlap at the corners and the floor by 1 ft. and run up the walls to a height 2 in. to 3 in. below the mudsills. (Leave mudsills exposed so they can be inspected periodically.) Because the sheets are heavy, cut them outside on a well-swept driveway, roll them up, and then unroll them in the crawlspace. Overlap seams roughly 1 ft., caulking each overlap with polyurethane sealant, and then taping the seams with a compatible peel-and-stick tape such as Tyvek tape. Use polyurethane caulk to attach the tops of sheets to the crawlspace walls; if the walls are dirty, wire-brush them first to ensure a good seal.

Because the moisture barrier must be continuous to be effective, sealing floor sections becomes more difficult if there are masonry piers or wood posts present. In that case, use two pieces of polyethylene to cover the floor, with each piece running roughly from the base of a

post to the crawlspace perimeter. Slit the plastic and run it up 6 in. to 8 in. onto each pier; caulk and tape the plastic to the pier. If wooden posts rest directly on masonry pads, jack each post enough to slide a piece of metal flashing or heavy plastic underneath; otherwise, moisture will wick up through the post and eventually rot it.

The wall portions of the sheeting will be less likely to pull loose if you mechanically attach them. If you use sheeting as heavy as a pool liner (20-mil), you can drill holes through it into the concrete and drive in nylon expansion fasteners. Lighter grades of polyethylene can be wrapped several times around furring strips and then attached to walls with powder-actuated nails. (For this operation, eye and hearing protection are essential.) If condensation persists in the sealed crawlspace, insulate the walls with EPS foam panels. Alternately, if basement walls are dry, you can spray foam insulation onto walls in lieu of foam panels. Ultimately, it may be necessary to add a dehumidifier to condition humid crawlspace air.

CLEANING UP MOLD

Mold can't grow without moisture, so first identify and correct the source(s) of the excess moisture before you start cleaning up. Otherwise, the mold can return. If mold is extensive, hire a professional remediation company.

Necessary precautions. Limit your exposure to mold spores by wearing a respirator mask with N95 filters, rubber gloves, eye protection, and disposable coveralls, which you should discard at the end of each day. After assessing the mold's extent, determine the shortest way out of the house for contaminated materials—maybe out a window—to minimize spreading mold spores to

PRO **TIP**

If you're unsure whether basement-wall wetness is caused by moist interior air condensing or ground water seeping through, try this: Wipe dry a section of wall, then duct-tape a 1-ft. by 1-ft. piece of aluminum foil to the dried area. In a day or two, remove the tape and note which side of the foil is wet.

Poor ventilation and a damp dirt floor helped this gaudy fungus blossom on a crawlspace joist. The same conditions encouraged mold to flourish behind the baseboards on the floor above.

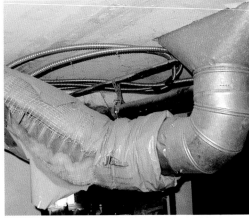

Poorly supported, shedding insulation, and leaking air, this duct should be replaced. Low spots in sagging ducts can become reservoirs for condensation and mold.

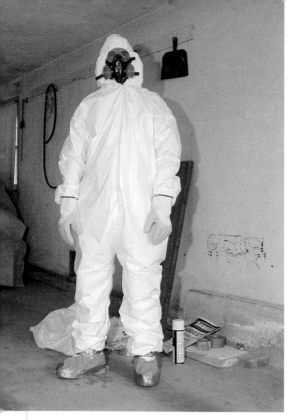

Whether they're removing asbestos or mold-contaminated drywall and framing, pros dress for the job. Here, protective clothing includes a respirator mask with N95 filters, gloves, boot covers, and disposable Tyvek coveralls and hood. Duct tape seals the coveralls at the wrists and ankles.

clean areas. Use sheet plastic to seal doorways and heating registers in affected areas, and turn off HVAC systems until the remedial work is complete. Seal damaged materials in plastic before transporting them from the site. Never sand moldy materials because that will spread spores. Finally, rent a commercial-grade vacuum with HEPA filters; if possible, vent it to the outside.

Assessing the extent. If mold is limited to small areas at the top of a bathroom or exterior wall, it may be surface mold caused by condensation or inadequate ventilation. However, if mold is widespread around windows or doors, bathroom drywall is crumbling, or tiles mounted on drywall are loose, there's probably mold growing in the walls. Start looking at the base of the walls. Turn off the electrical power to the area, remove the baseboard, and use a utility knife or a hole saw to cut small holes in the drywall. If there's no mold, you can easily patch the holes and cover them with the baseboard. More likely, you'll find stained or rotted wall plates and extensive mold colonies.

Throw out moldy drywall. On the other hand, moldy lumber and engineered wood products such as plywood, particleboard, and OSB may

just have surface mold, so probe them with an ice pick or pocketknife to see how sound they are. If they are spongy, replace them. Engineered wood products are particularly susceptible to rot because they contain adhesive binders that fungi feed on.

Remediation. Wash surface mold with soap and water and let it dry well. There's no need for caustic bleaches to kill mold spores (and irritate your lungs) because washing should remove mold. After the surface has dried, paint it with a stain killer such as B-I-N. If mold has caused the drywall's paper facing to roughen or delaminate, cut back the drywall at least 1 ft. beyond the damaged area and replace it.

If your inspection revealed mold growing inside wall cavities, use sheet plastic to seal off the affected area, including the heating registers, then cut back damaged drywall to the nearest stud center on both sides, and cut out damaged framing, if any. If you must replace more than one stud, erect temporary shoring to support the loads above (see chapter 10). To contain spore-laden dust, have a helper hold the hose of the commercial-grade HEPA vacuum near the materials being cut. Using soapy water, scrub the surface mold from the framing, and allow all materials to dry well before installing new drywall—framing moisture content should be 15% to 20% or less. (Borrow or buy a moisture meter to check.) Wrap moldy debris in 6-mil plastic and have it carted away.

Choosing Insulation Wisely

There are dozens of insulating products, which I'll classify into five groups: (1) batts (precut lengths or continuous rolls), (2) loose fill (blown into open spaces) (3) dense pack (blown into confined spaces), (4) rigid-foam panels, and (5) spray foam.

BATT INSULATION

Batts may be made of fiberglass, mineral wool, or recycled cotton, but fiberglass batts are by far the industry leader, accounting for three-quarters of residential insulation sales. Inexpensive and easy to install, batts are favored by DIYers. Faced with kraft paper or foil, batts may be stapled to the face or edges of studs, joists, or rafters; unfaced batts are friction-fitted between framing. In unfinished attics, unfaced batts are placed between floor joists or atop existing batts to improve heat retention.

However, for batt insulation to meet the R-values claimed by manufacturers, batts must be installed precisely so they fill cavities between framing members without gaps or thin spots. One problem is that framing in old houses is frequently irregular, so batts with precut widths often don't fit well. Consequently, if installers don't fill gaps, cut batts a bit short, allow facing flanges to pucker, or don't take time to fit insulation behind pipes and electrical cables, air movement can dramatically reduce their R-value. This is especially true for batts made of fiberglass because fiberglass is less dense than most other insulating materials.

Fiberglass batts. Because it is such a big seller, fiberglass makers are constantly reworking the stuff. Fiberglass is famously itchy, of course. So in response to eye, skin, and lung irritation caused by loose glass fibers, makers now offer batts encapsulated in perforated or woven plastic wrapping—which you can carry bare-handed and install overhead without fibers raining down on you. In some products, the fiberglass has been reformulated so that it's soft, itchless, and formaldehyde free. There are also fiberglass batts of varying density: 3½-in.-thick batts that are rated at R-11, R-13, and R-15; and 5½-in.-thick batts rated at R-21. *Note:* If you compress batts into the cavities, you decrease the insulation's loft and thus reduce its R-value slightly.

Cotton batts (often called *blue jean* insulation) are formaldehyde-free and itchless. Because of their density, they slow air movement, soundproof to a degree, and insulate well. Made from recycled cotton, batts are treated with a natural biostat (borate) to inhibit mold and make them fire-resistant. Wear a disposable paper mask when installing cotton: Although its lint is more benign than airborne glass fibers, avoid breathing it anyway.

Mineral wool (also called *rock wool*) is spun from natural stone such as basalt or made from metal oxides and is the most fire-resistant of any insulation. Rock wool, used to insulate wall cavities early on, is now enjoying a revival among green builders. Relatively dense, it insulates and soundproofs pretty well. Wear a respirator and gloves when working with it, for it has health-related concerns similar to those of fiberglass.

LOOSE-FILL INSULATION

Usually cellulose or fiberglass, loose-fill insulation is most often used to retrofit unfinished attics. Loose fill can be blown in quickly, does a great job of filling irregular and hard-to-reach spaces, and outperforms batt insulation. It's also something a homeowner can install. Smaller blowers can be rented at most home centers, along with a 2-in.- or 3-in.-diameter hose and the bales of insulation that one worker feeds into the blower hopper, while the other worker sprays the loose-fill insulation. The more inches of loose fill you install, the higher the R-value you'll achieve. Depths of 12 in. to 18 in. are fairly common.

Note: Prepping the space is important, so see the earlier section on air-sealing an attic. You'll want to cover any large hole such as chases, keep loose-fill insulation 2 in. away from the chimney,

PRO **TIP**

Standard fiberglass can itch like crazy. So even if you wear a long-sleeved shirt and a pair of gloves, wash well at the end of each day with cold water. Exactly why cold water reduces itching is unclear (shrinks pores perhaps), but it works.

||||

Precut fiberglass batts are by far the most commonly installed insulation. Always read labeling before opening to be sure the batts inside are of the correct length, width, thickness, and R-value for the task at hand.

The explosion of new insulating materials includes these itch-free, environmentally friendly cotton batts, also called blue-jean insulation, which are created from mill wastes.

Recommended Levels of Insulation

Insulation levels are specified by R-value. R-value is a measure of insulation's ability to resist heat traveling through it. The higher the R-value, the better the thermal performance of the insulation. The table below shows what levels of insulation are cost-effective for different climates and locations in the home.

Uninsulated wood-frame wall:
▶ Drill holes in the sheathing and blow insulation into the empty wall cavity before installing the new siding, and:
▶ Zones 3 to 4: Add R-5 insulative wall sheathing beneath the new siding.
▶ Zones 5 to 8: Add R-5 to R-6 insulative wall sheathing beneath the new siding.

Insulated wood-frame wall:
▶ Zones 4 to 8: Add R-5 insulative sheathing before installing the new siding.

Map courtesy of the Department of Energy, www.energystar.gov

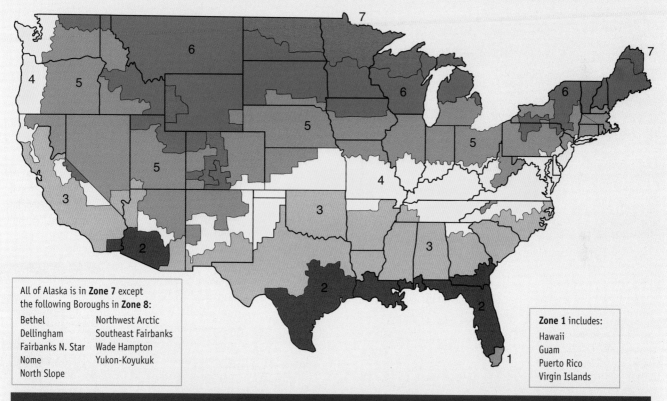

All of Alaska is in **Zone 7** except the following Boroughs in **Zone 8**:

Bethel	Northwest Arctic
Dellingham	Southeast Fairbanks
Fairbanks N. Star	Wade Hampton
Nome	Yukon-Koyukuk
North Slope	

Zone 1 includes:
Hawaii
Guam
Puerto Rico
Virgin Islands

ADD INSULATION TO ATTIC

Zone	Uninsulated Attic	Existing 3–4 in. of Insulation	Floor
1	R30 to R49	R25 to R30	R13
2	R30 to R60	R25 to R38	R13 to R19
3	R30 to R60	R25 to R38	R19 to R25
4	R38 to R60	R38	R25 to R30
5 to 8	R49 to R60	R38 to R49	R25 to R30

and install some kind of baffle to prevent loose fill from blocking soffit vents.

Loose-fill fiberglass has fluffy, spun-glass fibers and an R-value of 2 to 2.7 per in.; to attain an R-49 rating on an attic floor, you'd need to blow in a thickness of approximately 19 in. Not as dense as cellulose, it is more air-permeable. Researchers report that in temperatures below 0°F, convection currents passing through loose-fill fiberglass can reduce its effectiveness by 30%. In the event of a minor roof leak, fiberglass will probably dry quickly; being inorganic, it is also mold-resistant. Loose-fill fiberglass is slightly more expensive than cellulose.

Loose-fill cellulose is mostly recycled newspapers, ground up and treated with borate to make it fire- and mold-resistant. Denser than fiberglass, it is less air-permeable and averages R-values of 3.2 per in.; to attain an R-49 rating, you'd need to blow in about 15 in. of loose-fill cellulose. Its lower cost, higher R-value, and lower tendency to irritate skin give cellulose a slight edge over fiberglass. Cellulose's major downside is that it absorbs water and dries slowly. Were there a major leak, that retained water could soak the drywall ceiling below.

Spray polyurethane foam is rarely used to insulate attic floors because it's prohibitively expensive. But it is occasionally used as a thin *flash coat* to air-seal an attic after it's been prepped. The foam is then covered with loose-fill fiberglass or cellulose until the desired R-value is achieved.

DENSE-PACK INSULATION

Dense-pack insulation uses the same bales of cellulose or fiberglass as loose-fill installations, except that the materials are blown into confined spaces at higher pressures—say, 3 lb./cu. ft.—by more powerful blowers. Dense-pack insulation should be installed only by trained specialists. It is not a job for DIYers.

In renovations where interior surfaces have been gutted (or in new construction), the insulation is confined by a polymer mesh netting stretched taut across studs or rafters, then glued or stapled to framing edges so it stays put. The installer slits small holes in the netting, inserts the nozzle of the blower hose, and fills each bay. It takes skill. If too much insulation is blown in, the netting may bulge so much that drywall can't be installed over it.

More often, renovators use dense-pack insulation when they want to increase the R-value of walls while leaving interior finish surfaces intact. They gain access to wall cavities by removing small sections of siding, drilling holes through exterior sheathing, and blowing in insulation (usually cellulose) until each cavity is filled. Here again, it takes skill. A seasoned installer can tell by the whine of the blower when the correct density is achieved. How long it takes to fill a stud bay is also telling. If a bay fills too quickly, firestops may be blocking part of a bay; if a bay takes forever to fill, it may be open at the bottom and insulation is billowing into the crawlspace. As with any installation, it's important to survey and prep the structure to avoid such problems. An energy auditor's thermal image of cold spots can also help an installer insulate around obstructions in a wall.

At low densities (1.5 lb./cu. ft.), cellulose traps air and is an effective insulator (R-3.5/in.). When it's dense-packed (3 lb. to 4 lb./cu. ft.), that increased density stops most air infiltration, although dense-pack is not an air barrier per se. In fact, some New England contractors report that old houses retrofitted with dense-packed

Insulation Values

TYPE OF INSULATION	R-VALUE PER INCH
Batts and blankets	
Fiberglass	3.2
Cotton	3.5
Mineral wool	3.2
Loose fill	
Fiberglass	2.2–2.7
Cellulose	3.2
Dense pack (3 lb./cu. ft.)	
Fiberglass	2.5–3.5
Cellulose	3.8
Rigid-foam panels	
Expanded polystyrene (EPS)	4.0
Extruded polystyrene (XPS)	5.0
Polyisocyanurate	6.5
Spray-on (contractor applied)	
Closed-cell polyurethane	6.5
Open-cell polyurethane	3.6
Magnesium oxide	3.8

cellulose are as airtight as new houses with polyethylene vapor barriers. For the merits of cellulose versus fiberglass, read the previous section, "Loose-Fill Insulation."

RIGID-FOAM INSULATION PANELS

Rigid-foam panels are less permeable (p. 429) than any insulation material except polyurethane spray foam. Thus, when correctly installed, panels can reduce air and moisture infiltration, eliminate heat loss due to thermal bridging (p. 204), and increase R-value. Retrofitting panels to a building's exterior—over wall or roof sheathing—is invariably a major undertaking, however, so try to coordinate it with other compelling renovation tasks such as stripping a worn-out roof or, say, replacing asbestos siding. Incorporating the additional thickness of rigid-foam panels into exterior assemblies also takes great skill, so make sure a seasoned crew is assigned to it. On the other end of the skill spectrum, panels can be a quick and cost-effective way to create a dry, finished basement.

There are principally three types of rigid-foam panels used in residences, and each has characteristics that suggest where it can be best used. All three types are available in 2-ft. by 8-ft. or 4-ft. by 8-ft. panels with thicknesses from ½ in. to 2 in., and all come unfaced or foil-faced. Facings reduce breakage, make panels less permeable, protect foam cores from degradation, and give panels greater cohesion. Rigid-foam panels have relatively poor UV and fire-resistance, so check with local code authorities before installation.

Expanded polystyrene (EPS), the classic white styrene board, is a closed-cell foam. EPS is the least expensive of the three types, with an R-value of R-4 per in. Unfaced, EPS is semipermeable (5 perms per in.), allowing water vapor to migrate through it, without degrading it. Of late, building scientists favor allowing some moisture migration, so EPS is a good choice for finishing basement walls (see p. 450). Termites and beetles sometimes nest in rigid foam, so if you are thinking of installing it over exterior sheathing, consider a borate-treated variety of EPS, such as Perform Guard. EPS is also the most environmentally benign type.

Note: Installing rigid-foam insulation over sheathing as part of an energy retrofit can be effective, but it's also complicated, not only in detailing around doors and windows but also in making sure you get the right amount of foam for your climate zone.

Extruded polystyrene (XPS) is intermediate in price and R-value (R-5 per in.). Panels thicker than 1 in. have a perm rating of less than 1, so XPS is the most water-resistant of the three foams and thus best suited for insulating foundations. This closed-cell foam also has the highest compressive strength of the three, so it can withstand the roughest usage; it is widely used as exterior sheathing. XPS can be readily identified by its pastel hues: pink, green, blue, and yellow.

Polyisocyanurate (polyiso or PIR), the most costly and most effective insulation, has an initial R-value of 7.4 per in. and a residual R-value of

R-Values of Common Building Materials

MATERIAL	R-VALUE
8-in. concrete (solid)	0.90
4-in. common brick	0.80
3½-in. wood stud (on edge)	4.5
½-in. plywood sheathing	0.63
¾-in. plaster or ½-in. drywall	0.40
Glass	
Single pane	0.89
Double pane	1.91
Triple pane	2.80

Rigid-foam insulation panels such as polyiso cut easily with a crosscut saw. To reduce the movement of air around panel edges, seal them with a compatible canned spray foam or caulk.

How much insulation you need depends on climate, the house's heating system, and which part of the house you're insulating, as shown in "Recommended Levels of Insulation" on p. 439. Or you can use the U.S. Department of Energy's interactive Zip Code Insulation Program, which is free online at: www.ornl.gov/~roofs/Zip/ZipHome.html

To get a detailed insulation plan, type in your zip code, then check the boxes that best describe the house. Major insulation manufacturers offer similar calculators. You should also consult your local building authority, which has the final say and, in many cases, can tell you about tax incentives or rebates that encourage homeowners to insulate and save energy.

If stud walls are exposed, always seal air leaks before you start insulating.

R-6.5 per in. A closed-cell foam, it has the lowest compressive strength of the three—though its foil facing improves its durability. Foil-faced polyiso is commonly used on roofs because of its high-temperature stability.

SPRAY POLYURETHANE FOAM INSULATION

Spray polyurethane foam (SPF) is usually a two-part foam whose components travel through separate, heated hoses to mix at the spray gun nozzle. Sprayed wet onto a surface, the chemicals undergo an exothermic reaction (give off heat), foam, and expand dramatically, then gradually harden as they cure. Because polyurethane foams expand 25 to 100 times the volume of the wet mixture, SPF air-seals and insulates superbly. It adheres well to a variety of surfaces and fills better than any other sealant in hard-to-reach spaces such as the roof-sidewall juncture. *When correctly mixed, applied, and cured*, SPF is inert, non-off-gassing and odorless and thus is safe in both interior and exterior applications.

How much the foam expands determines its density, its R-value, its permeability to water vapor, how much material must be used and thus its cost, how the foam must be applied, and where it may be used. Foam that is more expansive and less dense is called *open-cell* or *"half-pound" foam*; foam that expands less and is denser is *closed-cell* or *"two-pound foam."*

Open-cell foam weighs about 0.5 lb./cu. ft and has an R-value of 3.6 in., roughly the same as fiberglass batts or loose-fill cellulose. Although open-cell foam is an effective air sealant, it is semipermeable to water vapor; thus it must not be used on exterior applications. In interior uses, local building codes may require that it be used with a class II vapor retarder (p. 429). Because of its lower R-value, open-cell foam must be applied thicker, so framing members must be deeper to accommodate it.

Closed-cell foam weighs about 2 lb./cu. ft. and has an R-value of 6.5 per in. With a perm rating of less than 0.8, 2 in. of closed-cell foam is a class II vapor retarder, and thus well suited to exterior applications. Given closed-cell foam's greater density, it may be the better choice on houses framed with smaller lumber. For example, 2x4 walls sprayed with 2-lb. foam can attain an R-value of 26 or higher. Less expansive than open-cell, closed-cell foam is typically installed in several passes, which often results in greater control and less trimming.

A skilled installer is essential. Sprayed polyurethane foam is a remarkable performer if applied precisely. But if its chemical components are not heated to the correct temperature, mixed in the right ratio, and applied in the prescribed thickness, the foam may not cure correctly and could emit noxious fumes for months. If the substrate being sprayed is too moist or too cold, the installation can fail. Research online to find a reputable installer; also consult your state's Contractors License Board, which keeps a file of complaints. A seasoned installer will also mask work areas carefully to avoid spray sticking to floors, windows and doors, and finish surfaces; supply workers with air-fed respirator systems; clean the job site well and advise clients how best to avoid exposure.

Installing Insulation

As part of an overall renovation plan, try to time installing insulation so that it coincides with other major tasks. For starters, it will save money. When the roof is worn out and needs stripping, for example, it is relatively easy to cover roof decking with rigid foam before reroofing. If siding with asbestos must be removed, exposing the sheathing will facilitate retrofitting exterior foam panels or blowing dense-pack cellulose into wall cavities.

Installing insulation could fill many books, so below you'll find representative installations for each of the six major types of insulation.

SPRAYING FOAM ONTO CEILINGS

Good-quality spray foam installations should be preceded by fastidious prep work. Before the insulation installers arrive, remove all tools and materials from the rooms to be foamed and thoroughly sweep all floors. In areas with significant solar gain, radiant barriers should be installed between rafter bays to reflect radiant heat absorbed by the roof and reduce cooling costs. To forestall possible interactions between foam chemicals and plastic pipe, some contractors also cover sprinkler pipes in the ceiling with pipe insulation.

Foam is a messy business, so prep is crucial. Before starting, the installation crew should mask everything they don't want foam to stick to. Typically, this means stapling sheet plastic over windows, doors, subfloors, walls and finish surfaces. Noted one installer, "Closed-cell foam is a monster to remove once it hardens. It is almost as dense as wood, so scraping if off is a lot like cutting lumber. Whatever time it takes to mask off, it's worth it."

To ensure that R-values are achieved, workers should periodically probe newly installed polyurethane foam to determine its depth.

The other crucial prep detail protects installers. Sprayers must wear a hooded Tyvek suit and gloves, with seams taped to keep foam particles out and, most important, a respirator mask connected to an air-fed system. In such systems, fresh outside air is pumped into the mask, creating positive pressure in the mask so that no chemical fumes can enter. Wearing the suit, gloves, and respirator is the safest way to spray foam, period. Masks must fit tight and be well maintained to avoid seal leaks, avoid clogs, and work optimally.

Polyurethane foams vary, so a reputable supplier will recommend a formulation based upon house-specific situations. For example, if rafters are not deep enough to attain the desired

PRO TIP

Because of potential heat build-up, most building codes forbid burying bundles of seven or more electrical wires in foam.

To prevent foam from sticking where it wasn't wanted, the crew stapled sheet plastic over windows, doors, plywood subfloors, finish surfaces, and stud walls to a height of 8 ft.

R-values, a high-density (2-lb.) foam may be required. In the job shown on these pages, 2x10 rafters gave installers a deep cavity to fill with open-cell foam. Building codes may also require fire-retardant additives in the formulation, to delay both the start of a fire and its flame spread. (Speaking of foam characteristics, the sound suppression qualities of open-cell foam are superior to those of closed-cell foam.)

Correctly installed, both closed-cell and open-cell polyurethane foam insulations are chemically inert. Both types are also superb air-sealants, perhaps the single most important reason to use foam: Air leaks are the single biggest waste of heating and cooling dollars. In cold weather, blocking air leaks also prevents the condensation of moisture on the underside of roof sheathing. Once the foam has adhered to a wood surface, there's no room for moist air to condense. (Air that can move through or around fiberglass batts and condense on sheathing or framing leads to lost energy and, in many cases, mold on wood surfaces.) Typically, $2\frac{1}{2}$ in. to 3 in. of open-cell foam is enough to stop condensation, although that may vary depending upon the climate zone, ambient moisture, and so on.

INSTALLING FIBERGLASS BATTS

All batts are installed in basically the same way, so these tips for installing fiberglass batts also hold true for cotton and mineral-wool batts, unless otherwise noted.

Getting started. Carefully seal air leaks before insulating; air currents can dramatically reduce an installation's R-value. Then suit up with the appropriate safety gear to keep fiberglass off your skin and out of your lungs. Wear a respirator, eye protection, long-sleeved shirt, long pants, and work gloves.

To determine how much insulation to buy, measure the square footage of walls, ceilings, and floors and then divide by the number of square feet in an insulation package. Also printed on the packaging is the insulation's R-value and the width of the batts. Because most joists, studs, and rafters are spaced 16 in. on center, 15-in.-wide batts are the most common size. The fewer cuts you make, the faster the job will go; thus many contractors buy precut 93-in. batts to insulate standard 8-ft. walls. (Although 8 ft. equals 96 in., the 3-in. shortfall in batt length anticipates the space occupied by the top and bottom plates.) Batts that long can be a bit unwieldy, though, so you might want to use precut 4-ft. batts in those stud bays where you must fit insulation around pipes and wiring.

When no vapor retarder (p. 429) is required or one will be installed later, many contractors prefer to install unfaced batts. They're quicker to install because there's no facing to cut through, and you can friction-fit the batts. By contrast, kraft paper–faced or foil-faced batts must be cut carefully to avoid tears and stapled to framing. In general, place insulation facing

Pro gear: a Tyvek suit, gloves, and an air-fed respirator mask. Fresh air is pumped into an air-fed system, creating positive pressure in the mask so that chemical fumes cannot enter it.

Here, rafter bays are being filled with BaySeal OCX—OC stands for "open cell" and the X denotes the fire retardant it contains. (Fire-retardant foam typically gives residents about 10 minutes to escape a burning house after the foam first outgasses a distinctive smell.)

toward the side of the building that's usually warmer—in cold climates, place the facing toward the inside of the building.

Cutting and placing insulation. Fiberglass insulation cuts easily with a utility knife, although you'll have to make several passes. Change the blade as soon as it gets dull. Professional insulators use a long-bladed *insulation knife* or hone one edge of a putty knife until it's razor sharp. A common way to cut an insulation batt is to place it on the subfloor, measure off a cutline, and press a 2x4 into the batt to compress the insulation and guide the knife. If you're cutting several pieces to the same length, you can save measuring time by marking the batt length

on the subfloor with masking tape. To cut faced batts cleanly, place the facing side down on the subfloor.

Many insulation contractors prefer to insert batts into stud bays and trim them in place. To trim a batt for a narrow bay, place one side of the batt into the bay and use the stud on the other side as a cutting guide. To get a tight fit, cut the batt about ¼ in. wider than the stud bay. As you place batts, make sure they fill the bays fully. Fiberglass insulates best at "full loft," so before placing a batt between studs or joists, shake it gently to plump it up to its full thickness. If you need several batts to fill a bay, butt their ends together rather than overlapping and compressing them. To avoid compressing the edges of faced batts, staple facing flanges to the edges—not the sides—of studs or rafters. Use a hammer tacker with ⅜-in. staples.

Insulating the attic. First, seal air leaks and add baffles (pp. 430 and 432) to keep batts from blocking soffit vents. If attic floor joists are exposed, place planks across them so you can move safely. (Stepping from joist to joist is a good way to step through the ceiling.) If there is a rough floor and you're not going to finish the attic, you might consider removing floorboards every 6 ft. to 8 ft. and blowing in loose-fill insulation. But batts are easy enough to install if you can pry up floorboards and insulate the floor in sections. Starting at one side of the attic, pry up and stack the floorboards, place batts, and renail floorboards before moving to the next section.

Fitting batts is pretty straightforward. Cut batts to length and place them so they fill the joist bays completely. If there's diagonal bridging, slit batts down the middle, 4 in. to 6 in. from the end, and fit the slit ends around the bridging. As in walls, split batts (see the photos on the facing page) to feed them over and under wires and pipes. *Note:* Keep insulation and other combustible materials back at least 3 in. from masonry or metal chimneys and non-IC-rated recessed light fixtures, as described on pp. 424–425. When adding insulation, place new batts perpendicular atop old ones.

If you intend to make the attic a conditioned space, you can forego insulating the floor and instead insulate the kneewalls and rafters. But because insulated floors soundproof somewhat, you may want to insulate them anyhow. Before inserting batts into the rafter bays, staple *air chutes* to the underside of the roof sheathing. The chutes create a 1-in. space between the sheathing and the insulation so that air can keep flowing under the roof. If the rafters are spaced regularly on 16-in. centers, unfaced friction-fit batts will stay in place until the drywall goes up. But if the rafter spacing is wide (24 in. on center or greater)

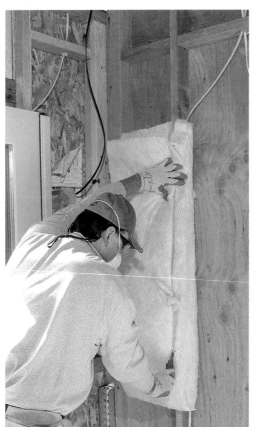

The pros trim insulation in place: It saves time and ensures a tight fit without measuring. Here, an installer leaves the batt folded because his 3–in. knife blade is long enough to cut through a double thickness in one pass.

As you place fiberglass batts in each stud bay, plump them up to full loft, and adjust them so they fill the space completely.

or irregular, friction-fit batts may sag or fall out. Instead, you might (1) staple paper-faced batts to the rafter edges or (2) have foam insulation sprayed to the underside of the roof sheathing after installing air chutes.

Another option: An unvented roof. The options above assume that the roof is vented. Thanks to the air-sealing and insulating properties of spray polyurethane foams, unvented roofs (also called *hot roofs*) have become increasingly popular. Unvented roofs are an especially attractive option when:

▶ *Complicated roof framing* makes eave-to-ridge ventilation all but impossible. Hip roofs, valleys, dormers, and skylights, for example, may preclude continuous ventilation flows to a ridge.

▶ *A house lacks soffit vents*—or lacks soffits altogether. It may be possible to add eave or *drip-edge vents*, but retrofitting ventilation to an old house is rarely easy. At the very least, it will require scaffolding so one can safely and accurately cut vent slots along the eaves or drill through blocking to add vent holes. Moreover, such vents may clash with the house's architectural style.

▶ *Strong winds drive rain into vents* or, in hurricane-prone regions, so pressurize a house that its windows blow out or its roof blows off.

▶ *Rafters aren't deep enough* to accommodate the code-required amount of insulation and an air channel flowing over that insulation.

In any region, eave ventilation can be a two-edged sword, admitting air and moisture into a vulnerable area and one that's difficult to insulate well. A wall's top plates, for example, lose heat through conduction (thermal bridging), a problem made worse by cold air washing over them. For more about vented and unvented roofs and illustrations of each, see pp. 430–431.

INSULATING AN ATTIC WITH LOOSE FILL

Before installing loose-fill insulation, read "Air-Sealing the Attic" (p. 425). Loose-fill materials can blow everywhere, so it's important to cover chases and other large openings so they don't fill with insulation. Equally important is keeping the material away from non-IC-rated recessed lighting cans, chimneys, and eave vents. You can blow loose fill over existing batts, but first hand-pack fiberglass into gaps between batts and joist bays that blown-in insulation might not otherwise fill.

Working in the attic has its risks, especially if it's unfinished. If the joists are exposed, run

To prevent cold spots behind pipes and wiring, split batts in two—sort of like pulling apart a sandwich—so that each piece is roughly half the batt thickness. Slide one half of the batt behind the wiring or pipes, and place the other half in front. Where it would be tedious to split and slide batt ends behind the obstruction in the middle of the stud bay, instead slice halfway through the batt as shown; split the fiberglass at the cut-line and fit the insulation behind the obstruction.

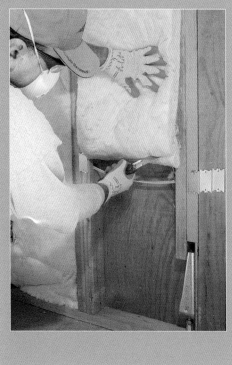

INSULATING AROUND ELECTRICAL Outlets

Electrical outlet boxes in exterior walls can be big energy losers because builders often forget to insulate behind them. To stop air leaks, split insulation batts and slide portions behind the boxes (below left). In very cold climates, you might also want to install airtight outlet boxes. Be careful not to nick cables around boxes.

Before insulating around any outlet (below right), disconnect the electrical power to the area, and use a voltage tester to be sure the power is off.

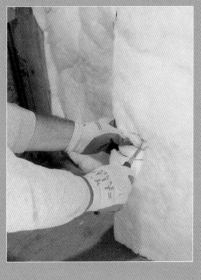

When installing batt insulation, be careful not to block eave-to-ridge ventilation pathways along the underside of the roof sheathing. The holes in the unfilled bays at the right are fascia vents.

planks across them so you can move safely around the attic. Stepping from joist to joist is risky—if you miss, you could step through the ceiling. Having enough light is essential, so set up several work lights and wear a headlamp. Dress the part: respirator mask, safety glasses, work gloves (especially if blowing fiberglass), long sleeves, and long pants. Lastly, determine how deep the insulation should be when you're done, and staple cardboard depth gauges to the side of joists so you'll know when you've reached that goal.

Insulating with loose fill is a two-person job. Down below, spread a plastic tarp and place the blower as close to the attic access door as possible (ideally, outside or in a garage). Loose-fill materials are, well, loose, and they can make a huge mess. Anyhow, the person downstairs (also wearing a mask, etc.) feeds bales of cellulose or fiberglass into the blower hopper. Feeding the right amount of insulation takes practice, but after a while, the machine's whine tells you when you've got a steady flow going.

Up in the attic, the person handling the hose starts by blowing insulation into the farthest reaches and gradually works toward the access door. The trick is applying the loose fill evenly. Angling the hose slightly upward and allowing the insulation to "rain down" seems to disperse the loose fill better than aiming directly at the section you're trying to fill.

An alternative: If the attic has a rough floor of boards, pry them up in the center of the attic to expose the joists. Then blow insulation into the joist bays. It's difficult to blow insulation much farther than 4 ft., so remove boards every 8 ft. or so and feed in about 4 ft. of the blower hose under the floorboards on both sides of the opening, gradually withdrawing the hose as the bays fill.

DENSE-PACKING WALL CAVITIES

There are several ways to dense-pack insulation, as noted on p. 440. In the sequence shown on p. 450, insulation is blown in through holes drilled into the exterior, so interior finish surfaces remain intact. Dense-pack materials are the same cellulose or fiberglass bales used in loose fill, except that here they are blown into confined spaces at higher pressures. Thus dense packing should be done only by trained installers. It's not a job for DIYers.

Before starting, review earlier sections on sealing air leaks, correcting excess moisture, and keeping insulation away from potential ignition sources. Knob-and-tube wiring that's still energized must not be covered with insulation. If such wiring is present, have an electrician

When blowing loose-fill cellulose in an attic, angle the blower hose slightly up so the loose fill falls evenly onto the section you're filling. Aiming the hose directly at the spot you're filling is more likely to create a mound.

INSULATING Floors ABOVE BASEMENTS AND CRAWLSPACES

INSULATING Floors ABOVE BASEMENTS AND CRAWLSPACES

When insulating a floor over an unconditioned crawlspace or basement, you're fighting gravity and moisture.

First, let's deal with gravity. The easiest way to install fiberglass batts without needing three hands is to precut a number of thin wood slats—¼-in. fence lath is light and springy—¼ in. longer than the distance between the joists. As you hold the unfaced batts in the joist bays with one hand, use the other to wedge the slats into place, under the insulation. You can also buy precut wires, "lightning rods," for the same purpose.

If the subfloor area is damp or if there's heavy condensation during warm months, rigid-foam panels or spray polyurethane foam is a better choice than fiberglass batts. Use a compatible construction adhesive or spray foam in a can to glue the foam panels to the underside of the subfloor. If the joists are irregular, trim the foam panels a bit smaller and use expanding foam to fill gaps.

replace it beforehand. Because it's impossible to know what's hidden behind walls, thermal images from an energy audit may be helpful to identify cold spots.

Equipment. Blowing insulation is dusty work, so a tight-fitting respirator mask and safety glasses are essential, as are work gloves when dense-packing fiberglass.

Insulation blowers and hoses can be widely rented, and some suppliers loan the equipment free to any contractor who buys the insulation from them. It takes a powerful machine to dense-pack walls, typically one wired for 240v or 120/240v. Almost all pumping units require two workers, one to feed insulation into a blower hopper and the other to operate the hose. Consequently, most machines have a remote on/off switch so the hose operator can stop the blower as cavities fill. A remote switch also allows one to shut off the blower at the first sign of a clogged hose. Most machines have adjustable gates or air inlets that control the air-insulation mixture. The equipment supplier should explain safe operating procedures to both workers.

Units typically come with 3-in. corrugated plastic hoses whose sections join with steel couplings. When filling wall cavities at low pressure, installers often use a 3-in. by 1-in. reducer to minimize the size of the holes they'll need to patch later. When dense-packing cellulose, some pros prefer to duct-tape a 5-ft. to 6-ft. length of 1-in. clear vinyl tubing to the end of the 3-in. hose so they can see if the insulation is flowing freely. Because narrower hoses are more likely to clog, the pros run an air-rich, low-insulation mix through the vinyl tubing and blow just air intermittently to minimize clogging.

In addition, installers need a drill and drill bit (or hole saw) to drill into exterior sheathing, a flat bar to pry up siding, and a shingle ripper or a reciprocating saw with a metal-cutting blade to cut through nail shanks if there's wood siding.

SAFETY ALERT

Rotating paddles of an insulation blower that are strong enough to break up compressed insulation can also injure hands and arms, so keep them out of the hopper while the paddles are turning. Instead, shut off the machine or use a stick to dislodge balky clumps of insulation.

Vermiculite: A POTENTIAL ASBESTOS RISK

Vermiculite is a small, pebble-like, gold or brown mineral fiber (colored rather like mica) that at one time was widely used for insulation and soundproofing. A loose-fill material, it was often poured in attic floors. But vermiculite often contains asbestos. If you see it in your home, don't disturb it. Only an experienced asbestos-abatement contractor should remove it. Visit the U.S. Environmental Protection Agency's (EPA's) website for more information (www.epa.gov).

If there's vinyl or aluminum siding, one needs a *zip tool* to lift the tops of courses to be removed. If the exterior is stucco, it will take a tungsten carbide hole saw to drill through it.

Important: To be safe, many installers turn off electrical power to the areas to be drilled and filled, and use a voltage tester to make sure the power is off. Or they turn off the service panel's main disconnect and run the blower off a generator.

Insulating wall cavities. The task varies according to the type of siding. It's possible to drill through wood siding and plug it later; but it's preferable to remove the wood siding, drill through the sheathing, and then replace the siding over the plugged sheathing. If there's vinyl or aluminum siding, it must be removed by inserting a zip tool under the top of the course to be removed and then sliding a flat bar under nails holding the siding strip. After the insulation holes are plugged, the siding is replaced. If there's stucco siding, the installer drills directly through

Hoses often clog near the reducer. To clear a clog, remove the reducing nozzle, insert the end of the hose in a trash bag, and turn on the blower. If that doesn't work, rap the sides of the hose and use a stick to dig out whatever you can.

it; plugs will hold well and can be hidden easily by textured patches.

Ideally, holes drilled should be roughly ¼ in. larger than the diameter of the insulation hose. If the holes and the hose are the same size, little air will escape and the wall cavity will become so pressurized that insulation can't be blown in. If holes are too large, insulation will blow all over and patches will be more obvious.

When a wall cavity is almost full, it will become so airtight that it will block the flow of additional insulation, causing the blower to whine. After filling a few cavities, an installer gets a sense of how much insulation is needed to fill each stud bay, and the blower's whine becomes familiar. At that point, the installer can use the remote on/off to shut off the blower and go on to the next cavity. Clear vinyl tubing taped to the end of a hose can be inserted all the way into a stud bay and then gradually pulled back as the cavity fills and the insulation slows.

Plugging holes. Insulation suppliers will have a stock of beveled plugs or corks. After smearing exterior-grade glue around plug edges and driving the plugs flush to the sheathing, an installer can replace the siding, caulk the joints, prime, and paint.

INSTALLING EPS PANELS TO CREATE A DRY BASEMENT

In the last decade or two, building scientists have changed their minds about the migration of moisture through exterior walls. Today's best thinking (www.buildingscience.com) suggests that builders accept that a small amount of moisture will migrate and choose materials that won't

Blown-in installation is typically fed through a 3-in. hose that is reduced into a 1-in. nozzle, which is inserted into holes drilled in the siding. The remote switch draped over the end of the hose enables the operator to shut off the blower quickly should a clog develop.

trap it. Thus, instead of covering basement walls with polyethylene vapor barriers to stop moisture, the wiser course is to choose semipermeable materials such as *expanded polystyrene* (EPS) panels, which allow any moisture that gets into the wall to dry to the inside rather than becoming trapped.

Building on that insight, one "cure" for basement dampness due to condensation (flowing water is another issue altogether) is keeping the warm air in living spaces from coming in direct contact with masonry masses by insulating basement floors and walls. Andy Engel, a former editor of *Fine Homebuilding* magazine, put this theory to the test in his own basement. For the full account, see "The Stay-Dry, No-Mold Finished Basement" (*Fine Homebuilding* issue #169), from which this section has been adapted.

1. Assuming that concrete floors and walls are in good condition and that there is no standing water or running water present, attend to sources of exterior water before remodeling a basement. If the basement floor is not level or in good condition, you can place new concrete over it, after first scarifying it and removing loose material. Do not install a plastic vapor barrier between the new and old concrete.

2. To install a floor, first lay down 1-in.-thick EPS panels directly to the concrete floor, fitting them as tightly as possible and then taping them to keep panels from riding up over each other. EPS's compressive strength is sufficient that you can just cover it with a ½-in. plywood floor—there's no need for sleepers. Apply a low-expanding foam around the perimeter of the EPS, and carefully tape every panel seam using Tyvek tape or a similar product. Use 2½-in. concrete screws to screw the plywood through the EPS to the concrete; you'll need to predrill screw holes first and countersink the screw heads.

3. On walls, use 2-in. expanded polystyrene panels. First, apply an expanding-foam sealant to the walls, then friction-fit the panels in place. As with floor panels, tape joints carefully with Tyvek tape and go around the perimeter of each wall with low-expanding foam to fill any gaps. With concrete floors and walls thus sealed, you can construct 2x4 wood walls with no concerns about wood rot or mold because the concrete has been completely isolated. Placed tight to the EPS, the stud wall will keep the panels from migrating; local codes may also require that you mechanically attach the EPS to foundation walls as well.

4. Finally, cover the stud walls with ½-in. paperless drywall. Because EPS is flammable,

If dense-pack insulation is packed too densely, interior drywall can bulge. So as an installer fills the first few stud bays, a second worker should monitor drywall or plaster surfaces for nail pops, cracks, and bulges. Such excessive pressure isn't common, but it can happen. Second, if a cavity takes forever to fill, either the hose is clogged or the cavity has an opening, and cellulose may be pumping into a nearby cabinet or crawlspace. If insulation is flowing and the cavity's still not filling, an installer should shut off the blower, find the escape hole, and plug it.

Because EPS allows a slight amount of moisture to migrate yet is impervious to moisture damage, EPS panels are an excellent choice for insulating basement walls. But panel seams must be carefully taped to isolate basement walls from conditioned living spaces.

the drywall is a fire-retardant code requirement as well as an aesthetic consideration. By the way, if space is tight in the basement, furring strips or 1⅝-in.-wide metal studs are another option, though furring strips are likely to telegraph irregularities in foundation walls. The greater thickness of 2x4 framing also allows you to run pipes and electrical wires in it.

SPRAYING POLYURETHANE FOAM OVER AN EXISTING ROOF

This last example is more a case history of *why* two building professionals chose a spray-foam roof than a *how-to* of its installation. As suggested on p. 443, the chemistry of polyurethane spray foam is so complex that the best technical advice

I can give is to choose your insulation contractor carefully.

Green reasons for choosing a spray-foam roof. Like many mid-century homes, the 1957 Eichler-designed house had clean lines, lots of glass, roof sheathing that doubled as a finished ceiling, and a flat roof with so little insulation that its inhabitants broiled in the summer and burned through money during the heating season.

When the old tar-and-gravel roof began leaking, the homeowners—a green builder and a landscape architect—started researching replacement roofs that would be energy-conserving, leak proof, cost-competitive, and reasonably green. Ultimately, they chose a closed-cell, sprayed polyurethane foam (SPF) retrofit because:

1. Insulating inside was out. The exposed underside of the 2x6 tongue-and-groove roof deck was classic. Besides, the ceilings were just a shade above 8 ft. and the couple didn't want to lose 4 in. to 6 in. of headroom to insulation.

2. Closed-cell SPF, with an R-value of 6.5 per in., was the perfect antidote to the old roof's dismal energy profile. The old roof assembly, with two layers of ½-in. fiberboard, achieved (maybe) R-3 or R-4. Four inches of SPF would bring it up to at least R-26; 6 in. would bring it to R-39. Spray foam was also particularly well suited to this roof because it had parapets, which defined the edges of foam. (Exterior spray foam only works on flat or low-pitch roofs; for pitches greater than 4-in-12, rigid-foam panels are the over-roof-deck choice.)

3. A spray-foam roof could be applied directly to the old roofing once it was prepped and cleaned up a bit. (See the photos on the facing page.) Not stripping the old roof and sending the debris to the landfill was also a big green plus. Besides, spraying-over would save the labor costs of stripping the old roof.

4. Four inches of closed-cell polyurethane is a class II water retarder, a semi-impermeable membrane that makes leaks all but impossible. As the foam expands and adheres aggressively, it would seal the many pipes and conduits running across the roof of a house with no cavity under the sheathing in which to hide pipes.

5. Compared with the obvious alternatives, an SPF roof was actually less expensive than installing and then insulating either a new torch-down roof or a PVC membrane. And that didn't even include the cost of stripping the old roof. There

Cold-Climate Basement Insulation

Use spray foam to seal rim joist, mudsill edges.

Fill joist bays with EPS pieces.

Notch and seal rigid EPS foam panels around joists.

2-in. rigid foam

Paperless drywall

Construction adhesive

Wood furring strips or flat-framed 2x4 wall

Pressure-treated sole plate

Cementitious board 12 in.

The flat roof of this 1957 Eichler-designed house was a good candidate for a polyurethane spray-foam roof.

were energy-efficiency rebates available for installing an SPF roof, but the couple was so busy that, as the gent put it, "We couldn't afford the time to save money."

6. So was this spray polyurethane foam a green choice? Yes and no, but mostly yes. Since the early days of sprayed foam insulation (the earliest foams were urea-formaldehyde based), the insulation industry has cleaned up its act a lot. Although the two chemical components of SPF are largely petroleum based, once the polyurethane mixture foams and cures, it is chemically inert and doesn't offgas. (Manufacturing PVC, in comparison, involves some particularly toxic chemicals.) Not sending old roofing to the dump is a plus, as is the excellent R-value of the material. (Sometimes it takes oil to save oil.) An SPF roof can also be quite durable when the polyurethane is sprayed with a special acrylic top coating that protects the foam against UV degradation. Lastly, the foam's light color will reflect sunlight so, all in all, it should be a relatively cool roof.

A good deal of prep work must precede the foam spraying. Here, electrical conduits running above the roof had to be raised so they would be accessible after 4 in. of foam was added.

Polyurethane foam can be sprayed right over old roofing—a big selling point—but the old roof must be clean and well attached. Here, a worker installs tabbed roofing nails every 2 ft. to 3 ft. to secure the old roofing.

Spray foam adheres aggressively, so it's important to mask everything you don't want foamed. Here, polyurethane foam is being applied to the chimney. In the background is the masked-off parapet.

Polyurethane foam expands to seal everything sticking out of the roof. Once sealed with a special acrylic topcoat, the polyurethane foam will be protected from UV degradation. Once cured, the closed-cell foam is also strong enough to walk on.

15 Finish Surfaces

People have used plaster in buildings since prehistoric times. Archaeologists have unearthed plaster walls and floors dating to 6000 B.C. in Mesopotamia. And the hieroglyphics of early Egypt were painted on plaster walls. In North America, plaster had been the preferred wall and ceiling surface until after World War II, when gypsum drywall entered the building boom. Although drywall represents a historic shift in building technology, the shift was more one of evolution than revolution because drywall's core material is gypsum rock—the same material used since ancient times to make plaster.

Readers interested in installing drywall should get a copy of Myron Ferguson's *Drywall* (The Taunton Press, 2012), far and away the best book (and video) on the subject.

Drywall

Sometimes called Sheetrock after a popular brand, drywall consists of 4-ft.-wide panels that are screwed or nailed to ceiling joists and wall studs. Sandwiched between layers of paper, drywall's gypsum core is almost as hard and durable as plaster, though it requires much less skill to install. Appropriately, the term "drywall" distinguishes these panels from plaster, which is applied wet and may take weeks to dry thoroughly.

Panel joints are concealed with tape and usually three coats of joint compound that render room surfaces smooth. Each panel's two long front edges are slightly beveled, providing a depression to be filled by joint tape and compound. Each layer of joint compound should be allowed to dry thoroughly before sanding smooth and applying the next coat.

Sturdy scaffolding keeps you safe and allows you to focus on the task at hand. This lightweight setup can be rolled easily from room to room.

Drywall Types, Uses, and Specifications

DRYWALL TYPE AND THICKNESS (in.)	COMMON USES	MAXIMUM FRAMING SPACING	AVAILABLE LENGTHS (ft.)
Regular			
¼	Renovation, covering damaged surfaces; curved surfaces	16 in. o.c., as double layer over framing, single layer over existing plaster, etc.	8, 10
⅜	Renovation, mostly on walls	16 in. o.c.	8, 10, 12
½	Most popular thickness, walls and ceilings	16 in. o.c.	8, 9, 10, 12, 14, 16
⅝	Ceilings, walls needing rigidity	24 in. o.c. on walls, 16 in. on ceiling	8, 9, 10, 12, 14
Moisture-resistant (MR)			
½	High-humidity areas such as bathrooms, kitchens	16 in. o.c.	8, 10, 12 (other sizes special order)
⅝	Bathroom ceilings	24 in. o.c. on walls, 16 in. on ceiling	8, 10, 12 (other sizes special order)
Paperless/Fiberglass Mat			
½	Walls in mold-prone areas such as basements, bathrooms, laundries	16 in. o.c.	8, 9, 10, 12, 16
⅝	Ceilings in mold-prone rooms	24 in. o.c. on walls, 16 in. on ceiling	8, 10, 12, 14, 16
USG Sheetrock UltraLight			
½ non-fire rated	Hang on walls and ceilings: reduced-weight panels won't sag	24 in. o.c.	8, 10, 12, 14, 16
⅝ non-fire rated	Ceiling applications with wet ceiling textures, heavy insulation, high humidity	24 in. o.c.	8, 10, 12, 14, 16
⅝ fire rated	Same uses as type X; check local codes before ordering	24 in. o.c.	8, 10, 12, 14, 16
Fire-resistant (type X)			
⅝	Satisfies 1-hour fire rating	24 in. o.c.	8, 10, 12
Fire-resistant (type C)			
½	Satisfies 1-hour fire rating	24 in. o.c.	8, 10, 12
⅝	Exceeds ASTM standards, OK in multifamily housing units	24 in. o.c.	8, 10, 12

Always check local codes for acceptable type, thicknesses, and framing spacing.

DRYWALL TYPES

Drywall's paper facing and core material can be manufactured for special purposes to make it more flexible, moisture-resistant, fire-resistant, sound isolating, scuff-resistant, and so on. Paperless drywall is also available for use in interior rooms where mold could be a problem because of high humidity.

Several factors will determine the size and type of drywall you choose:

▶ **Where it will be used.** Local building codes may require moisture-resistant (MR) drywall in high-humidity rooms or fire-resistant (type X or type C) panels elsewhere to retard fires.

▶ **Distances it must span.** Because gypsum is relatively brittle, the drywall must be thick enough to span the distance between ceiling joists without sagging and between wall studs without bowing (see "Drywall Types, Uses, and Specifications" on p. 455).

▶ **Skill and strength of installers.** The longer the sheets, the heavier and more unwieldy they are to lug and lift, especially a concern if you're working alone or if ceilings are high. Fortunately, USG Sheetrock UltraLight Panels will lighten the load somewhat, as they weigh one-third less than regular drywall.

▶ **Access to work areas.** Using sheets longer than 8 ft. reduces the number of end joints that need taping. Few installers use anything larger than 12-ft.-long sheets of drywall: 14-ft.- and 16-ft.-long sheets are incredibly unwieldy and difficult to maneuver into most rooms.

Regular drywall comes in four thicknesses (¼ in., ⅜ in., ½ in., and ⅝ in.) and in sheets 8 ft. to 16 ft. long in 2-ft. increments. There also are 4-ft. by 9-ft. sheets. To minimize wall joints when installing drywall horizontally, regular drywall comes in 48-in. and 54-in. widths.

The most commonly used thickness is ½ in., typically installed over wood or metal framing. A sheet that size weighs about 60 lb., still manageable for a strong installer working solo.

To increase fire resistance and deaden sound, you can double up ½-in. panels, but that may be overkill. More often, a single layer of ⅝-in. drywall is used for that purpose. Being stiffer, ⅝-in. panels are harder to damage, so they're also a smart idea in hallways if you've got kids. And they're less likely to sag between ceiling joists or bow between studs.

Renovators commonly use ¼-in. and ⅜-in. sheets to cover damaged surfaces and thereby avoid the huge mess of demolishing and removing old plaster. For best results with this thin dry-wall, use both construction adhesive and screws to attach it. However, neither thickness is sturdy enough to attach directly to studs in a single layer.

Two layers of ¼-in. drywall are routinely bent to cover curving walls, arches, and the like. Attach the second layer with construction adhesive and screws. If the curved area has a short radius (3 ft. to 5 ft.), wet the drywall first (discussed in detail later in this chapter). There's also a ¼-in. *flexible* drywall with heavier paper facings designed for curved surfaces, but this usually needs to be special-ordered.

Moisture-resistant drywall (MR board) is also called *greenboard*, after the color of its facing. Its moisture-resistant core and waxed, water-repelling face are designed to resist moisture in bathrooms, behind kitchen sinks, and in laundry rooms. In general, it is a good base for paint, plastic, or ceramic tiles affixed with adhesives and for installation behind fiberglass tub surrounds.

Although MR board can cover most bathroom walls, it should not be used above tubs or in shower stalls. In particular, it's not recommended as a substrate for tile in those areas because sustained wetting and occasional bumps will cause the drywall to deteriorate, resulting in loose tiles, mold, and water migration to the framing behind. As a substrate for tile around tubs and showers, cementitious backer board is far more durable and cost-effective. Mortar is the premier substrate for tiling, but it takes a skilled hand to do it right.

Paperless/fiberglass drywall was developed in response to concerns about mold. Organic matter such as paper or lumber is a food source for mold, so by replacing paper facing with fiberglass mats, paperless drywall resists the growth of mold, even in basements, kitchens, bathrooms, and other areas of high humidity. Although paperless drywall can be installed with basically the same tools and techniques used to install regular drywall, paperless panels are somewhat more fragile and the dust more irritating. Taping, cutting, and sanding these panels is slightly different, as described on p. 473.

That noted, fiberglass-mat drywall such as DensShield is a much easier tile substrate to work with than a cementitious board, which requires a power saw or a grinder to cut and generates an incredible amount of dust. DensShield has a treated, water-resistant core and fiberglass mats on front and back for strength and moisture- and mold-resistance. DensGlass, another fiberglass-mat gypsum panel, can be used as exterior sheathing and is an acceptable substrate beneath brick, stone, and stucco.

Fire-resistant drywall (type X or type C) is specified for furnace rooms, garages, common walls between garages and living spaces, underneath stairs, and shared walls in multifamily buildings. To make type X drywall panels stronger under fire conditions, manufacturers add glass fibers to the gypsum core. Type X drywall is available in a thickness of ⅝ in., widths of 48 in. and 54 in., and lengths up to 16 ft.

Type C has greater heat-resistive qualities because, in addition to glass fibers in its core, it also contains vermiculite, a noncombustible material. Type C is available in ½-in. and ⅝-in. thicknesses and in the same widths and lengths as type X drywall. Both types of fire-resistant drywall are installed like regular drywall. To achieve a one-hour fire rating for single-family residences, most codes specify ⅝-in. fire-resistant drywall.

USG Sheetrock UltraLight Panels weigh roughly 40 lb. for a ½-in.-thick 4x8 panel, as opposed to the 60 lb. for regular drywall panels of the same size. Less weight also means less sag, so you can confidently hang a ½-in. UltraLight panel on ceiling joists or rafters spaced 24 in. on center—instead of the ⅝-in.-thick drywall usually specified for ceilings. Scoring, cutting, attaching, and finishing these panels is the same as installing regular drywall except, of course, they're easier to lift. For non-fire-rated uses, ½-in. panels are available in standard widths and lengths to 16 ft. Fire-rated ⅝-in. panels also are available.

Other specialty panels are available:

Foil-backed drywall is sometimes specified in the Frost Belt to radiate heat back into living spaces and prevent moisture from migrating to unheated areas.

Abuse-resistant drywall has a high-density Type X core with reinforced-paper or fiberglass-mat faces to prevent surface scuffs, abrasions, and indentations. *Impact-resistant* panels such as DensArmor Plus were designed for commercial-building interiors, but they might be a good choice in, say, a home shop or a utility room that gets heavy use.

Sound-mitigating drywall is typically installed in multiple layers with staggered joints, with special sound-attenuating adhesive between layers.

Blueboard is a base for single- or two-coat veneer plastering and is now widely used instead of metal, wood, or gypsum lath. It is available in standard 4-ft.-wide panels.

Gypsum lath is specified as a substrate for traditional full-thickness, three-coat plastering. Its panels are typically 16 in. by 48 in.

Basic drywall tools. Top: drywall hammer, utility saw, and rasp. Bottom, from left: 6-in. taping knife, spackling knife, utility knife, multibit screwdriver, chalk, and chalkline box.

Cementitious backer board has a core of cement rather than gypsum. (HardieBacker is one popular brand.) Used as a tile substrate, it is installed much like drywall but must be cut with a diamond blade in a power saw or grinder (see chapter 16 for details).

TOOLS

You can install drywall with common carpentry tools—framing square, hammer, tape measure, and utility knife. Still, a few specialized, moderately priced tools will make the job go faster and look better. If you've got high ceilings, rent scaffolding.

Layout tools include a 25-ft. tape measure, which will extend 8 ft. to 10 ft. without buckling; a 4-ft. *aluminum T-square* for marking and cutting panels; a *chalkline* for marking cutlines longer than 4 ft.; a *compass* or a *scriber* to transfer out-of-plumb wall readings to intersecting panels; and a 2-ft. framing square to transfer the locations of outlet boxes, ducts, and such onto the panels.

Cutting and shaping tools may be simple, but they must be sharp. Drywall can be cut with one pass of a sharp *utility knife*, a quick snap of the panel, and a second cut to sever the paper backing, as shown in the top center photo on p. 468. Buy a lot of utility-knife blades and change them often; dull blades create ragged edges. Use a *Surform rasp* to clean up cut drywall edges. The sharp point of a *drywall saw* enables you to plunge-cut in the middle of a panel without drilling, although the edges of the cut will be rough.

A *drywall router* or a *laminate router with a drywall bit* is the pro's tool of choice for quick, clean cuts around electrical outlet boxes, ducts,

For 8-ft. to 9-ft. ceilings, benches can support platform planks and enable workers to raise and attach panels easily.

and the like. With a light touch and a little practice, you can use this tool to cut out boxes already covered by drywall panels, as shown in the bottom photo on p. 468. Typically, a standard ⅛-in.-dia. drywall bit is used on boxes and ducts, whereas a ¼-in. drywall bit is used to trim excess drywall around windows and doors. *Note:* The larger bit creates an exponentially greater amount of dust.

Alternatively, you can use a *utility saw* to cut out the waste portion of a drywall panel that you've run long into a doorway or window opening.

Lifting tools will help get drywall panels up into place so they can be attached. Two lifting tools can be handmade: A *panel lifter* inserted under the bottom of a panel will raise it an inch or so, leaving your hands free to attach the drywall.

Metal panel lifters are not expensive, but 1x2 scraps work almost as well.

The second homemade tool, a *T-support*, temporarily holds a panel against ceiling joists while you attach it. Cut a 2x4 T-support about ½ in. longer than the finished ceiling height so you can wedge it firmly against the panel. Or you can rent an adjustable *stiff arm*, a metal version of a 2x4 T-support.

Adjustable *drywall benches* should enable you to reach 8-ft. or 9-ft. ceilings easily. Alternately, you can lay planks across sturdy wooden sawhorses.

Ultimately, renting a drywall lift and/or scaffolding is the safest way to go, especially if ceilings are higher than 10 ft. If there's no danger of falling off your work platform, you can focus on attaching drywall. Scaffolding is also indispensable during the taping and sanding stages.

Attachment tools are typically a corded screw gun or a drill with a Phillips screw bit to drive drywall screws. A cordless drill is fine for dry-

A panel lifter leaves your hands free to attach the drywall panel.

Drywall router (bottom) and screw gun.

This Senco DuraSpin 14.4v driver employs a clutch to prevent overdriving screws, as well as a collated strip of fasteners that frees up one hand to hold a drywall panel in place.

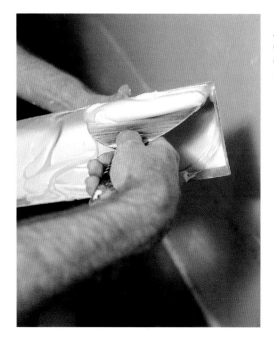

A mud pan holds enough joint compound to cover hundreds of screw holes or several long panel joints.

walling a single room, but pros who have thousands of screws to drive prefer corded screw guns, which have clutches and depth settings that set the screw heads perfectly—just below the surface. That may change, however, as cordless drivers continue to improve battery life and add useful features. The Senco DuraSpin shown above uses collated fastener strips that contain enough screws to install a panel.

Pros also use a drywall hammer for incidental nailing. A standard carpenter's hammer will do almost as well, but the convex head of a drywall hammer is less likely to damage the paper facing of a panel.

A caulking gun is also an important tool. Typically, a caulking gun will be used to apply a sound-attenuating adhesive, such as Green Glue Noiseproofing Compound, between drywall panels to dissipate vibrations. Adhesive is also used when adhering panels to a surface that can't easily be screwed into.

Taping and finishing tools are used to apply joint compound to the seams between panels. The workhorse of taping is the *6-in. taping knife*, perfect for filling screw holes, spreading a first layer of joint compound, and bedding tape.

To apply the successively wider and thinner second and third coats of joint compound, you'll need wider taping knives or *curved trowels*. Taping knives typically have straight handles and blades 10 in. to 24 in. wide; a 12-in.-wide knife will suffice for most jobs. Trowels have a handle roughly parallel to the blade and a slightly curved blade that "crowns" the compound slightly. Trowel blades run 8 in. to 14 in. long.

Applying "mud" (joint compound) takes finesse, so most pros use a *mud pan* or a *hawk* to hold enough mud to tape several joints. As they work, drywallers are constantly in motion: scooping mud, centering it on the knife blade, scraping off the excess, and returning it to the pan or onto the hawk.

Corner knives enable you to apply mud to both sides of an inside corner simultaneously. To finish outside corners (those that project into a room) consider making your own tool. Boil a plastic flat knife, and once it's soft, bend it into the shape you need.

For high-volume jobs, you can rent taping tools that dispense tape and compound simultaneously. See "Mechanical Taping Tools" on p. 473 for more information.

Sponges and a pail of water will keep tools clean as you go. Even tiny chunks of dried compound will drag and ruin freshly applied layers,

Wet-SANDING

Using a large sponge to wet-sand drywall joints will definitely reduce dust, but wet-sanding isn't feasible for a project of any size because you must rinse the sponge and change the water continually. Also wet-sanding soaks the paper facing, sometimes dislodges the tape, and tends to round joint compound edges rather than taper them. That said, if you're drywalling a small room and don't like moving the furniture out of the room, wet-sanding is a cleaner way to go.

PRO TIP

New taping knives and trowels may have burrs or sharp corners that can tear the drywall's paper facing. Before you use a new tool, sand the edges and corners lightly with very fine emery paper. Rinse these tools repeatedly with warm water as you work, dry them when you're done, and store them apart from heavy tools so their blades won't get nicked.

A dust-free sander attached to a shop vacuum cuts dust dramatically, but in unskilled hands it will oversand soft topping coats.

PRO **TIP**

Tape dispensers hold paper or mesh tape and clip to your belt, so your tape is always ready to roll. Bigger dispensers hold a 500-ft. roll.

▮▮▮▮

so rinse tools and change water often. A perfectly clean 5-gal. joint compound container is a great rinse bucket. Use a second one to store and transport delicate trowels and knives.

Sanding equipment starts with special *black carbide-grit sandpaper*, which resists clogging; it comes precut to fit the rubber-faced pads attached to poles and hand sanders. Sandpaper grit ranges from 80 (coarse) to 220 (fine); 120-grit paper is good to start sanding with. Finish-sand with 220 grit or a rigid *dry-sanding sponge* (a special sponge that is never wetted).

A sanding pole with a pivoting head enables you to sand higher—8-ft. ceilings are a snap— and with less fatigue because you use your whole upper body.

Sanding joint compound generates a prodigious amount of dust, so buy a package of good-quality paper *dust masks* that fit your face tightly; they're inexpensive enough to throw out after each sanding session. If you're sanding over your head, lightweight goggles and a cheap painter's hat will minimize dust in your eyes and hair. If you're working with paperless drywall, which has fiberglass mats instead of paper facing, a dust mask is essential, as are a long-sleeved shirt, goggles, and work gloves—fiberglass dust itches.

A shop vacuum with a fine dust filter is a must; vacuum at each break so you don't track dust all over the house. Dust-free sanding attachments are available for shop vacuums (as shown in the photo above). Although they virtually eliminate dust, they'll sand through soft topping coats and expose the joint tape in a flash if you're not careful. For best results, run them at low speed settings and use 220-grit sandpaper.

Finally, tape up sheet plastic in doorways to isolate the rooms you're sanding, especially if people are living in the house. Painter's tape will do the least damage to trim finishes and paint.

MATERIALS

This chapter began with sizes and types of drywall. Now let's look at screws, joint tape, corner

Adjust your screw gun so that it dimples the surface of the drywall but does not tear the paper facing.

Drywall Fasteners

ATTACHING TO...	FASTENER USED	DRYWALL THICKNESS (in.)	MINIMUM FASTENER LENGTH (in.)
Wood studs, ceiling joists, rafters	Type W drywall screws (coarse thread)	$3/8$, $1/2$, $5/8$	1, $1\,1/8$, $1\,3/8$ (penetrate framing $5/8$ in.)
Wood studs only	Drywall nails	$3/8$, $1/2$, $5/8$	1, $1\,1/8$, $1\,1/4$
Light-gauge metal framing	Type S drywall screws (fine thread)	$3/8$, $1/2$, $5/8$	$3/4$, $7/8$, 1

beads, and joint compound before planning and estimating supplies.

Drywall screws have all but replaced nails. Here are the three principal types:

▶ *Type W screws* have a coarse thread that grips wood; screws should penetrate framing at least $5/8$ in. In double-layer installations, attach the first layer with $1\,1/4$-in. screws and the second layer with $1\,7/8$-in. screws.

▶ *Type S screws* have fine threads and are designed to attach drywall panels to light-steel framing and steel-resilient channels. At least $3/8$ in. of the screw should pass through metal studs, so 1-in. type S screws are commonly used for single-ply $1/2$-in. or $5/8$-in. drywall installations. If you're attaching drywall to heavy-gauge (structural) steel, use self-tapping screws.

▶ *Type G screws* are sometimes specified to attach the second panel of a fire-rated, double-layer installation. That is, the first panel is the substrate to which the second panel is screwed and glued with construction adhesive. Ideally, screws also should penetrate framing, so ask building inspectors about installation requirements if your local code specifies type G screws.

Nails are still used to attach corner bead and to tack panels in place. Ring-shank drywall nails hold the best in wood; don't bother with other nail types. Nails should sink $3/4$ in. into the wood.

Construction adhesive enables you to use fewer screws and provides peace of mind for those screws that miss the mark—ones that catch only part of a stud or that are overdriven. You can buy drywall adhesive but, in truth, any adhesive that meets ASTM C557 standards will do just fine.

Joint tape is used to reinforce drywall seams and is available as 2-in.-wide paper tape and $1\,1/2$-in.- or 2-in.-wide *fiberglass mesh tape*. Self-adhesive mesh tape is popular because you can apply it directly to drywall seams and then cover it with joint compound in one pass. With paper tape, you must first apply a layer of joint compound, press the paper tape into it, and apply a coat of compound over that. If there's not enough compound under paper tape, it may bubble or pull loose. That said, some professionals still swear by paper tape because it's less expensive, less likely to be sliced by a taping knife, and won't stretch.

Here's Chip Harley's rule of thumb: "If you're using setting-type compounds (i.e., hot mud), use fiberglass tape. If you're using drying-type compounds (i.e., all-purpose compound), use paper tape. The pros generally use hot mud for taping, but they always finish with topping mud because it's so much easier to sand and finish.

By the way, self-adhesive mesh is great for drywall repair. If you press the mesh over a crack or small hole, you may be able to hide the problem with a single layer of joint compound.

Corner beads and trim beads finish off and protect drywall edges. They're available in metal, PVC, and paper-covered variations. Most attach with nails or screws.

Corner beads are used on all outside corners to provide a clean finish and protect otherwise vulnerable drywall corners from knocks and bumps. (As noted earlier, inside corners are formed with just tape and compound.) Corner beads come in a number of different radii; larger bullnose varieties give you a dramatic curve. Whatever type you choose, though, install it in one piece—that is, as a continuous strip from top to bottom.

J-beads keep exposed ends of drywall from abrading. These beads are typically used where panels abut tile or brick walls, shower stalls, or

Fiberglass-mesh joint tape isn't as strong as paper tape, but it sticks directly to the drywall—without a bed of joint compound—so it's faster to apply. Use mesh tape only with setting-type compound.

PRO TIP

Would-be drywallers unsure about size or spacing of screws should consult the CertainTeed *Gypsum Board Systems Manual,* or US Gypsum's *Construction Handbook.* Both are readily available online, with page after page of rated assemblies: STC and fire rating, thickness of materials, insulation types, fasteners required and spacing intervals.

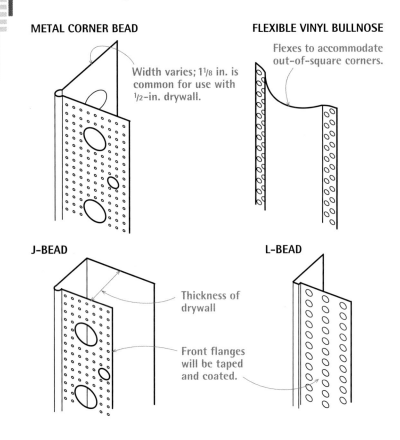

METAL CORNER BEAD

Width varies; 1⅛ in. is common for use with ½-in. drywall.

FLEXIBLE VINYL BULLNOSE

Flexes to accommodate out-of-square corners.

J-BEAD

L-BEAD

Thickness of drywall

Front flanges will be taped and coated.

openings that won't be finished off with trim—in other words, where the edge of the drywall is the finished edge. *L-beads* are similar; they're used where panels abut windows, suspended ceilings, paneling, and so on. In general, L-beads are easier to install on panels already in place. J- and L-beads are sized to drywall panel thickness; some types require joint compound but some don't.

Flexible arch beads often come in rolls that are presnipped, so as you unroll them, they assume the shape of the arch you're nailing or stapling them to. Apply joint compound and finish them as you would any drywall seam.

Joint compounds can be broadly divided into drying types and setting types. They differ in ease of use, setting (hardening) time, and strength.

Drying-type joint compounds are vinyl based and dry as water evaporates from them. They usually come premixed and are easy to apply and sand. Typically, you can apply a second coat 24 hours after the first, if you maintain a room temperature of 65°F. There's little waste with drying-type compound. And, once opened, it will keep for a month if you seal the bucket tightly.

Setting-type joint compounds, which contain plaster of paris, are mixed from powders. They set quickly, allowing you to apply subsequent coats before the compound is completely dry. In general, they bond better, shrink less, and dry harder than drying types. They harden via a chemical reaction, hence their nickname, "hot mud." Depending on additives, they'll set in 30 minutes to 6 hours. Once setting-type compound is mixed, you've got to use it up. It won't store.

Unless you're a drywalling whiz, use a premixed, all-purpose, drying-type joint compound. A 5-gal. bucket will cover 400 sq. ft., roughly, a 12-ft. by 12-ft. room with 8-ft. ceilings. Sometimes you need to thin the compound slightly (in your pan) to get the right consistency. Many pros squeeze a little water from a wet sponge into the compound if it's too thick, then whip the compound into a nice, smooth consistency.

One further distinction: Drying-type and setting-type joint compounds are further formulated as either *taping compounds*—used for the first and second coats—or *topping compounds*, used for the third coat because it feathers out (thins) better and dries faster. Again, all-purpose compound can be used for all three coats, but you might want to experiment with the two types once you've had some practice. Some pros use setting-type compound for the first and second coats and topping type for the third coat.

PLANNING THE JOB

Before estimating materials, walk each room and imagine how best to orient and install panels. These five rules, known to drywall pros, will save you a lot of pain.

Rule 1: Use longer panels to reduce the number (and length) of joints. A 4-ft. by 12-ft. panel is heavier and a bit more unwieldy than a 4-ft. by 8-ft. panel. But hanging larger panels is relatively fast compared with the time it takes to tape, coat, and sand smaller-panel joints.

Rule 2: Think spatially. Running panels horizontally—perpendicular to studs and ceiling joists—can reduce the number of joints and promote stronger attachments. For example, two 4-ft. by 12-ft. wall panels run horizontally will reach an 8-ft. ceiling and create only one horizontal seam to be filled. Two 54-in.-wide (4½-ft.) panels run horizontally will reach a 9-ft. ceiling. Note, however, that 54-in.-wide panels are not stock items, so check on their availability.

Rule 3: Minimize butt joints. Long edges of panels are beveled to receive tape joints, but the short edges (butt edges) are not. Consequently, butt joints are difficult to feather out, and they are likely to crack. So try to minimize the number of butt edges. Where you can't avoid them, position them away from the center of a wall or ceiling. Last, always stagger (offset) butt joints in adjacent panels; never align them. Otherwise, you may need to feather joint compound out 3 ft. to get a barely acceptable joint.

Rule 4: Install drywall that's thick enough. Panels that are too thin may sag between ceiling joists and bow between studs. For example, if you're running panels parallel to ceiling joists spaced 24 in. on center, ⅝-in. drywall is much stronger and less likely to sag than ½-in. panels. Alternatively, you can use lightweight drywall panels (see p. 457), which sag less because they weigh less: a ½-in. UltraLight Panel, for example, can span ceiling joists spaced 24 in. on center without sagging.

Rule 5: Don't scrimp when ordering. Expect a certain amount of waste, especially if you're installing around stairs or sloping ceilings. It's a mistake to try to piece together remnants because that creates a lot of butt joints and looks awful. Likewise, scrimping on screws or joint compound results in weak joints and screw pops.

ESTIMATING MATERIALS

Start by consulting local building codes. They'll specify the type, thickness, and number of drywall layers you must install throughout the house.

It's possible to estimate drywall materials from a set of blueprints, but even the pros prefer to walk the job, measuring walls and ceilings and noting where using longer panels will minimize joints. By using this technique, you get exactly the panel lengths you need.

As you walk through the rooms, record your findings on graph paper with ¼-in. squares. Use one sheet of paper per room, letting each square equal 1 ft. Then, when your materials arrive, you'll know which room gets what.

1. Measure the width and length of each room. In general, think in 2-ft. increments. That is, if a wall is 10 ft. 6 in. long, plan to buy 12-ft. panels and run them horizontally to get the fewest joints. If studs aren't spaced a standard 16 in. on center, note that on your graph paper. (If you're not sure of stud and joist spacing, use a stud finder.) Calculate the number of panels of each dimension you'll need for each room.

2. Note door and window locations and dimensions, but don't deduct their square footage from the room total. Portions of panels cut out for windows and doors will yield a high percentage of nonfactory edges, so it's easiest to discard this scrap or piece it together in out-of-way places, such as closets.

Minimizing Drywall Joints

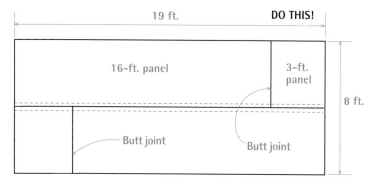

Joints = 27 lin. ft.

Joints = 32 lin. ft.

Panel ripped to 3 ft. wide

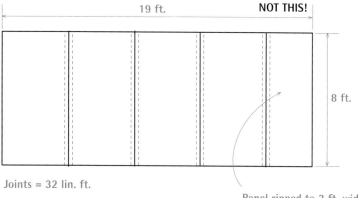

Hanging wall panels horizontally often reduces the number and length of the joints. With studs a standard 16 in. o.c., you could reduce the total lineal feet of taping by running panels horizontally instead of vertically. For the room dimensions shown, however, the upper layout also creates two butt joints, which can be tedious to feather out.

Calculating Drywall Square Footage

PANEL LENGTH (ft.)	SQUARE FEET PER PANEL*		NUMBER OF PANELS		TOTAL SQUARE FEET
8	32	×	40	=	1,280
9	36	×	35	=	1,260
10	40	×	30	=	1,200
12	48	×	25	=	1,200
14	56	×	20	=	1,120
16	64	×	20	=	1,280
Total					7,340

* Calculations assume that all panels are 4 ft. wide.

3. Note the ceiling height. If the ceiling is 8 ft., two standard 4-ft. by 8-ft. wall panels run horizontally will reach it; if they're 9 ft. high, use two horizontal 54-in.-wide panels. If ceilings are higher than 9 ft., you may want to rent scaffolding and a drywall lift.

4. Note the direction and spacing of ceiling joists. Panels run perpendicular to ceiling joists (or rafters, if a cathedral ceiling) are less likely to sag.

5. Consider special rooms. Install moisture-resistant drywall in laundry rooms, behind kitchen sinks, and in bathrooms other than in tub/shower areas. Cover tub/shower areas with cementitious board if they'll be tiled. Install $5/8$-in. type X or type C drywall on shared walls between the garage and living space, and under stairs. If existing plaster is in bad shape and you don't want to tear it out, cover it with $1/4$-in. or $3/8$-in. drywall.

6. Take into account special features, which include arches, curved stairwells, barrel ceilings, odd nooks, built-in bookcases or cabinets, and the like. For curves, you need flexible $1/4$-in. drywall. Expect a lot of waste around complex areas, such as stairways.

7. Think about maneuverability. In older homes, especially those with narrow stairs, there may not be room to maneuver extralong panels. If you're replacing windows, you may be able to have a boom truck deliver panels through window openings, but only if the truck can get close to the house. In short, anticipate how you'll bring drywall into the house. In extreme cases, you may need to use shorter panels for some rooms. Inquire about options with your drywall supplier.

8. Once you've figured out how much drywall you need for each room, figure out where you'll store it. If possible, distribute drywall throughout the house, stacking panels on the floor rather than leaning them against walls, which can damage their edges. In each room, anticipate the sequence of installation—ceilings first!—so the panels you need first will be accessible.

9. Make a master list of all the drywall you need for the complete renovation.

GETTING READY TO HANG DRYWALL

Here's a final checklist before installing drywall:

▶ **The building should be dry and relatively warm (between 60°F and 70°F). Keep temperatures constant. If a room becomes too hot (80°F or higher), joint compound may dry too quickly and crack. If the heating system is inoperable, rent a portable heater. Ventilation also is important for drying: Drying-type joint compound contains a lot of water.**

▶ **Framing lumber must be dry: 15% to 19% moisture content is optimal (see p. 69). Green or wet lumber will shrink as it dries, causing cracks and nail pops in a new drywall job; don't use it.**

▶ **Sight along studs and ceiling joists to see if they are aligned in a plane. To be more**

Estimating Drywall Needs

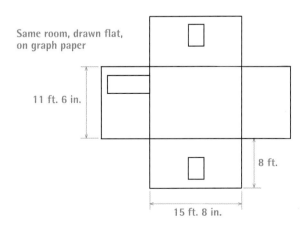

CEILING
Four 12-ft. panels

Joist direction

WALLS
Four 16-ft. panels
Three 12-ft. panels
One 10-ft. panel

11 ft. 6 in.

Window
Two 16-ft. panels

Door

12-ft. panel

10-ft. panel

8 ft.

15 ft. 8 in.

Two 12-ft. panels

Two 16-ft. panels

Same room, drawn flat, on graph paper

11 ft. 6 in.

8 ft.

15 ft. 8 in.

Metal drywall clips can help secure drywall edges in corners where there's insufficient blocking.

Guesstimating Materials

Measuring each room is the only accurate way to develop an estimate of how much drywall you'll need. But a rough estimate of square footage can help you figure about how much joint tape, joint compound, and fasteners you'll need.

Add up the square footage of all rooms to be drywalled. For rooms with 8-ft. ceilings, use a factor of 3.5 multiplied by the room length and room width. For example: 3.5 × 20 ft. × 30 ft. = 2,100 sq. ft. of drywall. For rooms with 9-ft. ceilings, use a factor of 3.85.

For each 1,000 sq. ft. of drywall, you'll need approximately:

▶ 1,000 screws or nails (using construction adhesive reduces this total)
▶ 375 ft. of joint tape
▶ 11 gal. of premixed, all-purpose, drying-type joint compound

Note: If you use construction adhesive to reduce the number of fasteners, a standard 10.5-oz. tube of adhesive will yield 15 lin. ft. of 3/8-in.-wide bead.

---

precise, stretch a taut string or hold a straightedge across the framing members. Running drywall panels perpendicular to framing will help conceal minor misalignments, as will textured finish surfaces. But it's better to adjust misaligned framing, especially if the drywall surface will be brightly lit or otherwise prominent. If a stud is misaligned more than ¼ in., your options include hammering it into line, power-planing it down, and shimming up low spots, as described in chapter 8.

▶ Use a framing square to determine if corners are square, and make sure there's blocking in the corners so you'll have something to attach the panel edges to. Or you can use drywall clips to "float" the corners—and save lumber in the process.

▶ Install steel nail guards to protect plumbing pipes and electrical cables within 1¼ in. of joist or stud edges. Few things are as frustrating as discovering a leak or an electrical short after the drywall is up. Then make a final check of the electrical outlet boxes. They should be securely attached to the framing with their edges flush with the drywall face (see chapter 11).

HANDLING, MEASURING, AND CUTTING DRYWALL

Handling drywall is a bit like waltzing: You can do it by yourself, but it's not all that much fun. For starters, carrying a cumbersome 60-lb. panel around a work site is a two-person job. Both of

you should be on the same side of the panel, same hand supporting the bottom edge, same hand balancing the top. (Imagine ballet dancers in a line.) As you walk, lean the upper part of the panel against your shoulder.

Hanging ceilings is definitely a two-person job, especially if you're hanging long panels to minimize joints. Once the two (or three) of you tack up a ceiling panel, one of you can finish attaching it, while another measures or cuts the next piece.

Measuring isn't difficult, but you need to be aware of framing quirks. For example, if the walls aren't plumb where they meet in the corners, take at least two measurements so you'll know what angle to cut in the end of the panel you're installing. If the two readings vary only slightly—say, ¼ in.—there's no need to trim the end of the panel because taping and filling the joint will take care of the gap. But if readings vary more than ¾ in., trim the panel end at an angle so there's no gap where it abuts the out-of-plumb wall.

Second, if you're running panels perpendicular to the framing, butt ends must meet over joist or stud centers. If framing members aren't evenly spaced, you may occasionally need to trim a butt end to make it coincide with a joist center.

Last, and most important, cut panels about ¼ in. short so you never need to force a panel into place. Forcing will crush an end that you'll need to repair later. If it's a ceiling panel, the gap will eventually be covered by wall panels that fit against it.

PRO TIP

The square factory edge of a drywall panel will not fit correctly when it abuts an out-of-plumb corner. Rough-cut the panel about 1 in. long, position it against the out-of-plumb wall, and use a scriber (a student's compass is fine) to transfer that angle to the face of the panel. Cut along the scribed line, and the panel should fit correctly.

Drywall over Plaster

Drywalling over plaster is a cost-effective way to deal with plaster that's too dingy and deteriorated to patch or too much trouble to tear out. But this requires some important prep work.

▶ If you see discoloration or water damage, repair the leak or find a way to control excessive indoor moisture before attaching drywall.

▶ Locate ceiling joists or studs behind the plaster. Typically, framing is spaced 16 in. or 24 in. on center, but you never know with older houses. If ceiling joists are exposed in the attic above, your task is simple. Otherwise, use a stud finder or drill exploratory holes. Once you've located the joists or studs, snap chalklines on the panels to indicate the centerlines you'll use when driving screws.

▶ Use screws and plaster washers to reattach loose or sagging plaster sections before you install drywall. To minimize the number of fasteners, apply adhesive to the back face of the drywall, and be sure the screws grab framing—not just lath. Plaster washers are shown on p. 477.

▶ For ceilings, use 2-in. type W drywall screws, which should be long enough to penetrate 3/8-in. drywall, 1 in. of plaster and lath, and 5/8 in. into joists. On walls, 1/4-in. drywall is a better choice because drywall shouldn't sag, and thin drywall doesn't reduce the visible profile of existing trim as much. Otherwise, you may either need to build up existing trim or remove the trim and reinstall it over the drywall.

▶ If there's living space above the plaster ceiling, attaching resilient channel may be a good move. These channels bridge surface irregularities and deaden sound. Screw the channels perpendicular to the joists (see the photo on p. 482). Then screw drywall panels perpendicular to the channels.

Trim Considerations

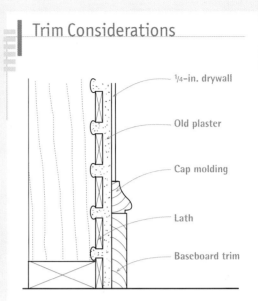

New drywall over old plaster can reduce the visible profile of existing molding it abuts so much that the molding looks undersize. In this case, you have two alternatives: Use molding to build up (increase the thickness of) trim where drywall abuts it. Or remove the trim and reinstall it over the drywall.

Before hanging drywall, check that studs are plumb, corners are square, and (as shown) that studs align in a plane. If not, they must be built up with shims or planed down.

Once you've cut a panel, snap parallel chalklines across the face, indicating stud or joist centers behind the drywall once it's in place. Guided by these lines, your screws or nails will always bed into framing.

Cutting drywall is literally a snap if the blades in your utility knife are sharp. Run the blade along a drywall T-square or a framing square, if that's all you have. In one pass, score the paper covering. Grasp the smaller section and snap it sharply away from the cut, breaking the gypsum core along the scored line. Then cut through the paper on the back side (see photos p. 468).

Cutting is usually easiest if you lean the panel against a wall so you don't need to bend over. But if you've already stacked panels on the floor, you can use them like a workbench, cutting them in place.

Start by tearing off the end papers that join pairs of panels face to face, allowing you to move panels individually. In this manner, you'll cut every other panel from the back.

Most professionals would rather score the front face first, but it doesn't truly matter which side you cut first, as long as your blade is sharp, your snap is clean, and you don't rip or snag the paper on the front face. If the gypsum edge is a bit rough or the panel is a little long, clean up the edge with a drywall rasp. But be careful not to fray the face paper.

Making outlet box, switch, and duct cutouts can be done before or after you hang the drywall. To make cuts before, measure from a fixed point nearby—from the floor or a stud, for example—and transfer those height and width measurements to the panel. A framing square resting on the floor is perfect for marking electrical receptacles. That done, use a drywall saw to punch through the face of the panel and cut out the opening, being careful not to rip the paper facing as you near the end of the cut.

That's one way to do it. Problem is, the cutout rarely lines up exactly to the box.

Using a drywall router is quicker and more accurate. First, make sure the power is off, and push any electrical wires well down into the box so the router bit can't nick them. The ⅛-in. drywall bit should extend only ¼ in. beyond the back of the drywall.

Next, measure from a nearby stud or the floor to the (approximate) center of the box, and transfer that mark to the drywall. Then tack up the panel with just a few nails or screws—well away from the stud or joist the box is attached to. Gently push the spinning bit through the drywall, and move it slowly to one side until you hit an edge of the box. Pull out the bit, lift it over the edge of the box, and guide the bit around the outside of the box in a counterclockwise direction. This method takes a light touch—plastic boxes gouge easily—but it's fast and the opening will fit the box like a glove.

Production drywallers often run panels over door and window rough openings, then use a drywall saw or a drywall router with a ¼-in.-dia. bit to cut the panels flush to the edges of the opening (a ¼-in. bit is less likely to break than a standard ⅛-in. bit). Wear safety glasses and a dust mask whenever you use a router.

ATTACHING DRYWALL

Most professionals use drywall screws exclusively, although some use a few nails along the edges to tack up a panel temporarily. Corner bead is often nailed up, too.

When attaching drywall, push the panel firmly against the framing before driving in the screw. Fasteners must securely lodge into a framing member. If a screw misses the joist or stud, remove the screw, *dimple* (indent) the surface around the hole, and fill it later.

The screw (or nail) head should sink just below the surface of the panel, without crushing the gypsum core or breaking face paper. You will

A Ripping GOOD TIME

Cutting along the length of a drywall panel—ripping a panel—is fast and easy if you know how. Extend a tape measure the amount you want to cut from the panel. If you're right-handed, lightly pinch the tape between the index finger and thumb of your left hand to keep the tape from retracting. Your right hand holds the utility-knife blade against the tape's hook. Using your left index finger as a guide along the edge of the panel, pull both hands toward you evenly as you walk backward along the panel. Remember, the blade needs only to score the paper, not penetrate the core, so relax and keep moving.

later fill the dimple around the head of the fastener with joint compound.

Screw spacing will depend on the thickness of drywall panels, the number of layers, their location on walls or ceilings, the framing interval, and so on, as described in the CertainTeed *Gypsum Board Systems Manual* or US Gypsum's *Construction Handbook*. As noted earlier, both publications are free online. As a general rule of thumb, space screws every 12 in. on-center in the field and 8 in. on-center along edges. If your town requires drywall screwing inspections, ask the building department for code requirements for your area.

Drywall edges are a bit fragile, so place screws back at least ⅜ in. from panel edges. For best gripping, screw heads should be sunk just below the surface of the panel, creating a dimple. Screws driven deeper will tear the facing paper and not hold as well. You can adjust a screw gun's clutch to sink screws at a dimpling depth, but variations in the density of wood will prevent you from doing so uniformly. Consequently, some pros adjust guns to leave screw heads a bit proud, then drive each to a perfect dimple depth using the cordless driver at a slower speed.

PRO TIP

When sizing screws or nails, see "Drywall Fasteners" on p. 461. Fasteners that are too short won't support the panel adequately. On the other hand, fasteners that are too long are more likely to drive in cockeyed or pop the drywall if the framing shrinks.

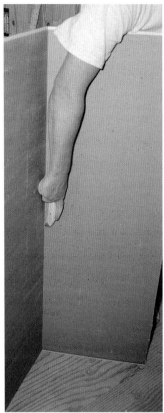

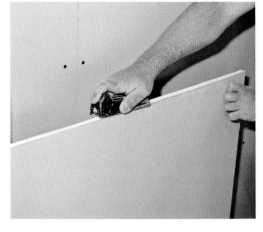

3. If the cut is rough, clean it up with a drywall rasp.

2. Snap the panel sharply away from the face cut (here hidden) to break the gypsum core along the scored line. Then cut through the paper backing along the break.

1. In one pass, score the paper face of the panel using a utility knife guided by a drywall T-square.

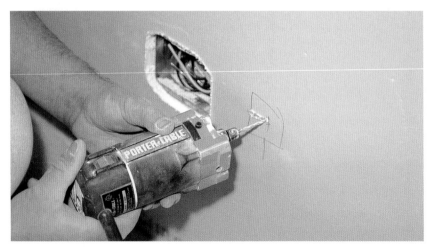

A drywall router quickly cuts out holes around outlet boxes. Beforehand, shut off the electricity to that circuit, and tuck wires well into the box so they can't get nicked.

Nails. Although nails were long used to attach drywall panels, attaching drywall with screws is now the preferred, industry standard method. Because of screws' superior holding strength, panels can be installed with 20% to 30% fewer fasteners (than if you used nails). Screws require less effort to install, hold drywall tighter to framing, and are easier to set properly. They are less likely to break the paper surface, require less effort to fill with joint compound and less likely to visibly "pop" if framing shrinks after finishing.

That noted, the building code requires drywall nails to be spaced every 7 in. for ceiling panels and 8 in. for wall panels along *all* framing members. Double nailing (two nails spaced 1½ in. to 2½ in. apart) is allowed for field nailing. To give a real-world example, it requires 49 nails to attach one 4x8 sheet of drywall to walls framed 16 in. on center and 56 nails for ceilings framed 16 in. on center.

HANGING DRYWALL PANELS

Ceilings. Attach ceiling panels first. It's easier to cut and adjust wall panels than ceiling panels should there be small gaps along the wall–ceiling intersection. Also, wall panels can support the edges of ceiling panels.

First, ensure there's sufficient blocking or metal drywall clips in place to attach panel edges. In most cases, you'll run panels perpendicular to the ceiling joists, thereby maximizing structural strength, minimizing panel sag, and making joist edges easier to see when fastening panels.

The trickiest thing about hanging ceiling panels is raising them. If your ceilings are less than 9 ft. high, drywall benches will elevate workers enough. As you raise each panel end, keep one end low. That is, allow one worker to raise one end and establish footing before the second worker steps up onto the bench. Then, while both workers support the panel with heads and hands, they can tack the panel in place.

If the ceiling is higher than 9 ft., rent a drywall lift. Because the lift holds the panel snugly against ceiling joists, it allows you to have both hands free to drive screws.

Flying solo. If you can't find a drywall lift or helpful friends, you can hang ceiling panels solo by using two tees made from 2x4s. Make the tees ½ in. taller than the height of the finished ceiling. Lean one tee against a wall, with its top about 1 in. below the ceiling joists. Raise one end of the panel up, onto that tee. Being careful not to dislodge the first end, slide the second tee under the other end and raise it until the entire panel is snug against the ceiling joists. (Staple strips of rubberized carpet pad to the tee's top to reduce slippage.) Gradually shift the tees until panel edges line up with joist centers. Be patient and wear a hard hat.

Walls. It's easier to hang drywall on walls than on ceilings. Although one person can usually manage wall panels, the job always is easier and faster with two. Before you begin, be sure there's blocking or drywall clips in the corners to screw panels to. To help you locate studs once they're covered with drywall, mark stud centers on the top plates (or ceiling panels) and sole plates at the bottom.

With a helper and a set of drywall benches, you can safely raise long panels to the ceiling. Lift one end at a time. Whenever possible, run the panels perpendicular to the joists.

If the ceiling is higher than 9 ft., and especially if it's a cathedral ceiling, rent a drywall lift.

Install the top wall panel first, butting it snugly against the ceiling panel. At the same time, level the bottom edge of the wall panel so that subsequent panels butted to it will also be level and correctly aligned to the stud centers.

PRO TIP

Mark joist centers onto the top of the wall plates before you install the first ceiling panel. That will enable you to sink screws into the joist centers when they're covered by drywall. The pencil marks will also help you align screws across the panel, simply by eyeballing from those first screws to the uncovered joists on the other side.

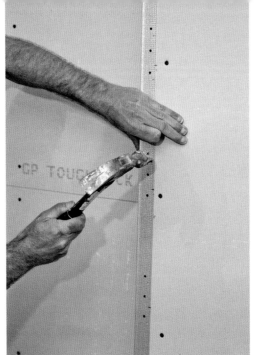

Corner bead protects fragile drywall edges and ensures a clean finished edge.

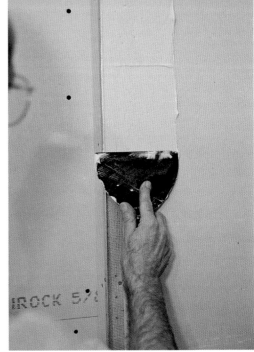

To prevent cracking along an outside corner, this installer covers the edges of the corner bead with mesh tape before applying a setting–type compound.

When hanging wall panels, professionals often start with the top panel, butting it snugly to the drywall on the ceiling. By doing so, they minimize gaps and support ceiling edges better. *Important:* If you're installing wall panels horizontally, the top panel edge must be level and the butt ends plumb. Otherwise, subsequent panels may be cockeyed and butt ends may not be centered over the studs.

Once the upper wall panel is secure, raise the lower panel(s) snug against it. A homemade *panel lifter* is handy because it frees your hands to align the panels and sink the screws. A panel lifter is simply a lever of scrap wood set with a fulcrum in the middle. Pressing down on one end of the lever with your foot raises the other end, which lifts the panel, as shown in the bottom left photo on p. 458.

Doors and windows. Joints around doors and windows will be weak and are likely to crack if panel edges butt against the edges of the opening. That is, run the panel edges at least 8 in. past door or window trimmers, and cut out the part of the panel that overlaps the opening. Pros do this because framing twists and flexes slightly when doors or windows are opened and closed, which will stress drywall joints and cause them to crack.

Expect to waste a lot of drywall when cutting paneling for doors and windows. Old houses are rife with nonstandard dimensions and odd angles, so don't fight it. You can use some of the larger cutoffs in out-of-the-way places such as closets, but remember that the more joints there are, the more taping and sanding. Anyway, drywall panel isn't expensive. So, when in doubt about reusing a piece, throw it out.

Curves. Curved walls are easy to cover with drywall. For the best results, use two layers of ¼-in. drywall, hung horizontally. Apply construction adhesive between the layers. Stagger their butt- and bevel-edged joints. For an 8-ft. panel run horizontally, an arc depth of 2 ft. to 3 ft. should be easy to achieve. Sharper curves may require back-cutting panels (scoring slots into the back so that the panels bend more easily), wetting (wet-sponging the front and back of the sheet to soften the gypsum), or special-ordering flexible drywall panels, which have facings better suited for bending.

Corners. Corner bead reinforces and protects outside corners, uncased openings, and the like. It's available in many materials. For best results, install it in a continuous piece from floor to ceiling. Cut the bead for outside corners about ½ in. short: Push it snugly to the corner and slide it up until it touches the ceiling. The ½-in. gap at the bottom will be hidden by baseboard.

Galvanized metal bead was at one time the only type available, and it's still widely used. To cut it, use aviation snips (also known as tin snips). The metal bead goes on the outside corners before the tape and joint compound are applied. Nail it up, spacing nails 8 in. apart, on both legs of the bead. Then cover it with compound.

Vinyl bead is less rigid than metal and able to accommodate outside corners that aren't exactly 90°. Attach vinyl bead either by stapling it directly to the drywall, spraying the drywall corner with vinyl adhesive before pressing the bead into the adhesive and then stapling, or using a taping knife to press the bead into a bed of joint compound.

Paper-faced beads are embedded in joint compound. One of the best is the Ultraflex structural corner, which comes in varying widths and has a plastic spine that flexes in or out so it can reinforce inside or outside corners. Because they're flexible, such tapes are great for corners of just about any angle.

TAPING AND FINISHING

To finish drywall, seal the panel joints with tape—or cover corners with corner bead—then spread joint compound over them. Typically, three coats of compound are applied in successively wider coats. The first coat, usually a high-strength taping compound, beds the tape. The second coat should be a thin layer of topping compound or all-purpose compound that you feather out to hide the joints. With the third coat, you feather out the compound farther, creating a smooth, finished surface. (See "Joint Compounds" on p. 462 for more about these materials.)

BEDDING THE TAPE

1. Before applying paper tape, cover the seam with a generous bed of joint compound.

2. After using your taping knife to center the tape on the seam, press the tape into the joint compound.

3. After applying a layer of compound over the tape, remove the excess. If the tape moves, you're pressing too hard or your taping knife needs to be cleaned.

4. Use a 10-in. knife to apply the second coat of compound. A quick knife-wipe of each edge will remove the excess from this beautiful "pull."

5. Applied correctly, the third coat should need only hand-sanding. A dry-sanding block is great for light sanding and corners.

First coat. Fill nail holes or screw dimples by applying compound in an X pattern: One diagonal knife stroke applies the compound, and the second diagonal stroke removes the excess.

If you use paper tape on the joints, first apply a swath of taping compound about 4 in. wide down the center of the joint. Press the tape into the center of the joint with a 6-in. taping knife. Then apply compound over the tape, bearing down so you remove the excess.

If you use self-sticking mesh tape, stick it directly over the joint and apply the bedding compound over it. In other words, don't apply a bed of compound first when using mesh tape. Mesh tape must be bedded in setting-type compound, as explained earlier in this chapter.

Allow the first coat to dry thoroughly before sanding it. This will take about a day, if the room temperature is 65°F to 70°F and there's adequate ventilation. Sand lightly with 120-grit to 150-grit sandpaper. Because there are two more coats to come, this taping coat can be left a little rough.

Second coat. The second coat is also called the *filler coat*, and with this one you'll apply the most compound. At this stage, many professionals use a 10-in.-wide taping joint knife and feather out the seams so they are roughly 8 in. to 10 in. wide. After applying the compound, smooth it out with an even wider blade—say, a 14-in. trowel.

When working with joint compound, the lower the angle of the blade and the less pressure, the easier it is to smooth and feather out the mud. The greater the blade angle and pres-

sure, the more compound you'll remove. Because butt-end joints are not beveled, compound spread over the tape would mound slightly at the center. To avoid this, the pros "split the butts." Instead of covering over the tape after it has been bedded, they use a 10-in. knife to apply mud *on each side* of the tape. After that dries, they fill in between those two 10-in.-wide swaths of compound. That way, the joint over the butt ends will be virtually flat.

When this second coat is dry, sand with 150-grit to 220-grit paper. A pole sander will extend your reach and enable you to sand longer without tiring, but don't sand too aggressively or you'll abrade the paper face or expose the tape. Easy does it.

Third coat. The third coat is the last chance to feather out the edges, so use a topping compound, which is easy to thin out and sand because it has a fine consistency and dries quickly. Although premixed compound will be the right consistency, it's OK to add a little water to thin it even more.

Because the third coat is only slightly wider (2 in.) than the second coat, you'll be applying a relatively small amount of compound. Use a 12-in. trowel, with a light touch. Some pros thin this coat enough to apply it with a roller and then smooth it with a trowel. They leave no trowel marks when they're done.

Hand-sand the final coat, using fine, 220-grit sandpaper or a very fine sanding block. Shining a strong light at a low, raking angle on surfaces will highlight the imperfections you need to sand.

The Art of Inside Corners

Use paper tape for inside corners. After applying a bed of compound to both surfaces, crease the tape and place it in the corner. Then use a double-edged corner knife to press the tape into the compound before spreading a layer of compound over the tape.

Some pros snort at corner knives, preferring to use a flat 6-in. taping knife to press tape into compound, one edge at a time. When feathering out joint compound, pros allow the compound to dry on one side of the corner before working on the adjacent surface. In other words, "Never run wet mud into wet mud."

Wrap up. If you intend to texture the surfaces, the third coat doesn't need to be mirror smooth. Even so, don't scrimp on the second coat, or else the joints may be visible through the texture.

To give yourself the greatest number of decorating options in the future, paint the finished drywall surface with a coat of flat, PVA primer. It will seal the paper face of the drywall and provide an excellent base for any kind of paint or wall covering.

DRYWALL REPAIRS

To keep solutions concise, let's divide drywall repairs into four groups: nail pops and surface blemishes; fist-size holes through the drywall; larger holes; and discolored, crumbling, or moldy drywall. Any repair patches should be the same thickness as the damaged drywall. Drywall-repair kits with precut patches are available at most home centers.

Popped nails and screws are generally a quick fix: Drive another fastener 1½ in. away from the popped one to secure the drywall. If it's a popped

Mechanical TAPING TOOLS

Mechanical drywall taping tools are commonly referred to as Bazooka tools, after a popular brand, and they can be rented, usually for two weeks at a stretch. The suite of tools includes a taper that applies tape and compound simultaneously, as well as finishing tools, various head attachments, and flat boxes for taping seams.

These tools are great for working large expanses and creating uniform, flat surfaces that need little sanding. "You can put tape and mud up almost as fast as you can run," notes one pro. Most rental companies supply a video on how to use these tools, but taking a class in addition isn't a bad idea.

Working with Fiberglass-Mat (Paperless) Drywall

Paperless drywall is a great choice for mold-prone areas such as basements and bathrooms because, lacking a paper facing, it denies mold the organic matter it needs to grow. Paperless drywall has a fiberglass-mat surface instead. Consequently, paperless panels must be installed a bit differently.

▶ **Minimize dust.** Airborne fiberglass particles are nasty to breathe or get on your skin, so wear a respirator that fits tight to your face, safety glasses, a long-sleeved shirt, and work gloves. Use hand tools—rather than power tools—to cut panels whenever possible. Use a utility knife to score and snap panels. When making cutouts around door and window openings, use a drywall saw rather than a drywall router.

▶ **Use adhesive *and* screws.** Because there's no heavy paper facing on these panels, it's easy to drive screws too deep—which reduces their holding power. So give the screws a little help by applying a bead of adhesive to stud or ceiling joist edges before attaching panels with screws. By the way, use 1¼-in. drywall screws to attach ½-in. or ⅝-in. panels—the same as you would for regular drywall.

▶ **Use self-adhering fiberglass tape** to reinforce panel joints—rather than paper tape, which mold loves. Use the tape with a setting-type joint compound, and take it easy with the taping knife—it's easy to cut through fiberglass mesh tape.

Paperless drywall has a fiberglass–mat facing that can produce hazardous dust. So when working with paperless drywall, use hand tools to minimize dust, and wear a respirator, safety glasses, long-sleeved apparel, and work gloves.

screw, remove it and fill the hole. Cover the old hole and the new screw with at least two coats of joint compound. If it's a popped nail, don't try to remove it; drive it slightly deeper.

When a piece of drywall tape lifts, pull gently until you reach a section that's still well stuck. Use a utility knife to cut free the loose tape, cover the exposed seams with self-sticking mesh tape, and apply two or three coats of compound—sanding lightly after each. Because the repair area is small, it doesn't matter what type of compound you use, though setting-type compounds are preferred.

To repair drywall cracks, cut back the edges of the crack slightly to remove crumbly gypsum and provide a good depression for joint compound. Any time the paper face of drywall is damaged, cover the damaged area with self-adhering fiberglass mesh tape. Then apply three coats of joint compound.

Small holes in drywall are often caused by doorknobs, furniture, or removed electrical outlet boxes. Clean up the edge of the hole so the surface is flat. Then cover the hole with self-adhering fiberglass mesh tape. (Better drywall-repair kits

have metal-and-fiberglass mesh tape.) Use a light touch when applying the first coat of joint compound, pushing the mud through the tape, but don't press so hard that you dislodge the mesh.

Give the repair three coats of compound, feathering out the edges as you go. Take it easy when sanding. Especially after the first coat, the tape frays easily. By the way, there are precut drywall patches for electrical outlet boxes, which save time.

Repairs in which only tape and joint compound cover the hole are OK just for small holes and gaps.

Large holes should be cut back until you reach solid drywall. Because mesh tape and compound will probably sag if the hole is much wider than 4 in., holes larger than that should be filled with a patch of drywall roughly the size of the damaged area. These patches need to be backed with something solid so they'll stay put.

The easiest backing is a couple of furring strips cut about 8 in. longer than the width of the hole and placed on both sides of the hole. To install each furring strip, slide it into the hole, and while holding it in place with one hand,

Texturing Drywall and Plaster

Joint compound is a marvelous medium for texturing a drywall patch or matching the texture of existing plaster. All that's needed is a little ingenuity.

▶ For a stippled plaster look, place joint compound in a paint tray, thin it with water until it is the consistency of thick whipping cream, and roll it onto the wall or ceiling using a stippled roller. Don't over-roll the compound, or you'll flatten the stipples.

▶ Create a "skip-trowel" texture by adding a tiny amount of sand to thinned mud (in a pan) and then skimming the mixture over the wall. Use a wide knife, which will skip over the slightly raised surface of the sand.

▶ For an open-pore, orange-peel look, use a stiff-bristle brush or whisk broom to jab compound that is just starting to dry. Jab lightly and keep the bristles clean.

▶ To achieve the flat but hand-tooled look of real plaster, apply the compound in short, intersecting arcs. Then knock down the high spots with a rubber-edged Magic Trowel, as shown below.

▶ If you're trying to duplicate a slightly grainy but highly finished plaster surface, trowel on the topping coat as smoothly as possible and allow it to dry. Then mist the surface slightly and rub it gently with a rubber-edged grout float.

You can achieve the irregular, hand-tooled look of plaster by covering drywall with joint compound applied in tight, intersecting arcs. Because of its crack resistance, use 90-minute or 120-minute setting-type compound. It's OK if the drywall isn't completely covered.

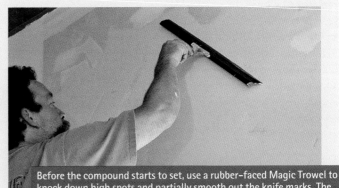

Before the compound starts to set, use a rubber-faced Magic Trowel to knock down high spots and partially smooth out the knife marks. The trowel should glide. Unload excess compound into your mud tray after each pass, and sponge the rubber edge clean every four or five passes.

screw through the drywall into the wood. Screws will pull the furring strips tight to the back of the drywall. Then cut the drywall patch, place it in the cutout area, and screw it to the strapping.

Cover the edges of the drywall patch with self-adhering mesh tape, fill the screw holes, and apply joint compound—three coats in all. Here, a setting-type compound, such as Durabond 90, is a good bet because it dries quickly and is unlikely to sag. The more skillfully you feather the compound, the less visible the patch will be.

For holes larger than 8 in., cut back to the centers of the nearest studs. Although you should have no problem screwing a replacement piece to the studs, be sure to back the top and the bottom of the new piece with lengths of 1x4 furring. The best way to install backing is to screw furring strips to the back of the existing drywall. Then position the replacement piece in the hole and screw it to the gussets, using drywall screws, of course.

Discolored, crumbling, or moldy drywall is caused by exterior leaks or excessive interior moisture. Be sure to attend to those causes before repairing the drywall. Excessive moisture is often the result of inadequate ventilation, which is especially common in kitchens and baths. Leaks around windows and doors are often caused by inadequate flashing over openings.

If the drywall is discolored but solid and if you've remedied the moisture source, wash the area with soap and water, allow it to dry thoroughly, and prime it with white pigmented shellac or some other stain-resistant primer. The same solution works for minor mold on sound drywall.

However, if there's widespread mold and the drywall's crumbling, there's probably extensive mold growing inside the walls. You'll need to rip out the drywall—and possibly some of the framing—and correct the moisture problems before replacing the framing and finish surfaces. In basements where condensation is common in summer months, however, you'll be well advised to install paperless drywall when it comes time to replace finish surfaces.

Chapter 14 covers mold abatement at greater length.

Plastering

This section is limited to plaster repairs because applying plaster takes years to master. The tools needed for plaster repair are much the same as those needed for drywall repair: a screw gun or cordless drill; 6-in. and 12-in. taping knives; a mason's hawk; and a respirator if you'll be removing or cutting into plaster. There's a lot of

PLASTERING

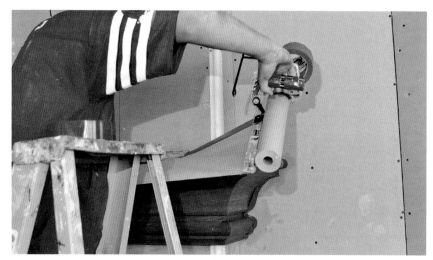

1. If casing, baseboards, or floors are already installed, cover them with paper and tape to protect them from plaster splatters.

2. Mix the plaster to the consistency of soft-serve ice cream before ladling it onto a mason's hawk. For skim-coat plaster, follow the manufacturer's mixing instructions, which typically recommend adding 12 qt. to 15 qt. of water for each 50-lb. bag of plaster.

3. To prevent cracking, cover blueboard seams with self-adhering mesh tape. Load your trowel from the hawk and . . .

4. . . . . sweep the trowel in an upward arc to spread the plaster, then back down to embed it (make it adhere). As you apply plaster, the blade shouldn't lift off the surface. Sweep the blade back and forth, covering bald spots and erasing trowel marks. Use a double-bladed knife on the corners.

grit, so wear goggles, too. The tools and techniques for plaster are similar to those needed for stucco, which is discussed in chapter 7.

ANATOMY OF A PLASTER JOB

Traditional plastering has several steps:

1. Nail the lath to the framing.

2. Trowel a scratch coat of plaster onto the lath. The wet plaster of this coat oozes through the gaps in the lath and becomes a mechanical *key* when it hardens.

3. Trowel on, then roughen the brown coat after it has set slightly.

4. Trowel on a finish, or white coat, which becomes the final, smooth surface.

In the old days, plasterers often mixed animal hair into scratch and brown coats to help them stick. As a result, old plaster that's being torn out is nasty stuff to breathe. The finish coat was usually a mixture of gauging plaster and lime for uniformity. Scratch coats and brown coats were left rough and were often scratched with a plasterer's comb before they set completely so the next coat would have grooves to adhere to. Finish coats were quite thin (1/16 in.) and very hard.

Lath can be a clue to a house's age. The earliest wood lath was split from a single board so that when the board was pulled apart (side to side), it expanded like an accordion. Although metal lath was available by the late 1800s (it was patented in England a century earlier), split-wood lath persisted because it could be fashioned on site with little more than a hatchet. By 1900, however, most plasterers had switched from lime plaster to gypsum plaster, which dried much faster. And

Keying Plaster to Wood

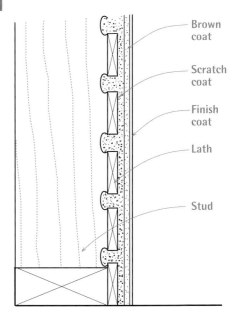

This plaster cross section shows how the scratch coat of plaster oozes through the lath and hardens to form keys, the mechanical connection of plaster to wood.

about the same time, plasterers began using small paper-coated panels of gypsum instead of wood or metal lath. Called *gypsum lath* or *rock lath*, the panels were so easy to install that they dominated the market by the 1930s. But time and techniques march on. After World War II, drywall all but replaced plaster as a residential surface.

SMALL REPAIRS

Small cracks or holes in plaster can be filled with patching plaster or Fix-It-All patching compound, which dries fast and is super strong. Because joint compound is softer, sands easily, and sets slower than patching plaster, it's easier to work with. Setting-type joint compounds are better for patching than drying-type joint compounds, which just aren't as strong. To ensure a good bond, strip paint from the surfaces you're patching and sand the adjacent areas lightly before applying compound.

Small cracks in plaster are repaired by stripping surface paint, cleaning loose plaster, and undercutting the cracks slightly with a knife or a small, sharp-pointed lever-type can opener (also called a "church key"). Undercutting allows the patching material to harden and form a key that won't fall out. Before patching, wet the exposed plaster well and brush on a polyvinyl acetate (PVA)

Repair or Replace Plaster?

To decide whether plaster should be repaired or replaced, first assess how well it is attached to the lath. To do this, near stains, cracks, holes, or sagging sections, press the plaster with your hand. If the plaster is springy, it has probably separated from its lath and must be reattached before you try to repair it.

▶ If the plaster has a few surface cracks and isolated holes but is stable, it can be repaired.

▶ If there's widespread discoloration and cracks wider than 1/4 in. but the plaster's basically stable, cover it with 1/4-in. or 3/8-in. drywall or replace it.

▶ If you see water stains, crumbling plaster, and widespread cracking or sagging surfaces, remove the plaster and replace it with drywall. If there are water stains, of course, find and repair the leak before doing any other work. Widespread sagging suggests that lath has pulled away from framing. Although lath can be reattached, concomitant plaster damage will usually be so extensive that you're better off tearing out the plaster.

bonder such as Plaster-Weld or Elmer's White Glue to bond the patch to the old plaster.

Over the crack, stretch a length of self-adhering fiberglass mesh. Then use a taping knife to spread the joint compound or patching plaster over the tape. Sand the first coat of joint compound lightly, leaving it a little rough so the second coat adheres better. Then apply the second coat, feathering it out to blend in the patch's edges.

When the patch dries, sand it lightly with fine, 220-grit sandpaper, wipe it clean, allow it to dry thoroughly, and prime the patch with PVA primer.

Large cracks often accompany sections of bowed or sagging plaster, which have pulled free from the lath. If the plaster is sound—not crumbling—you can reattach it to the lath using type W drywall screws and plaster washers, which fit under the heads of the screws. A screw gun is good for this operation.

However, before you attach the screws, mark their locations on the plaster, and use a spade bit to countersink a hole ⅛ in. deep for each washer. Sunk below the surface of the plaster, the screw heads and washers will be easy to cover with patching compound.

Place screws and washers every 8 in. to 10 in. on both sides of the crack and anywhere else the plaster seems springy and disconnected from the lath. Once you've stabilized the crack in this manner, scrape, tape, and fill it, as described earlier.

Small holes, the size of an electrical outlet, are easy to fill if the lath is still in place. Remove any

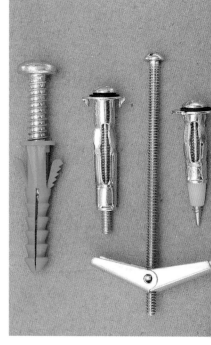

Use these anchors and bolts to attach light loads to drywall and plaster walls. From left: plastic anchor with screw, molly bolt, toggle bolt, and drive anchor.

Patching Cracked Plaster

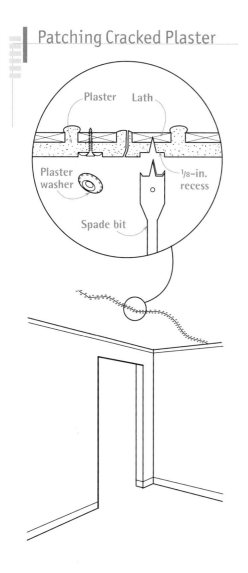

Plaster   Lath

Plaster washer

⅛-in. recess

Spade bit

Cracked plaster often means that it has pulled free from its lath. Use screws and plaster washers to reattach it, countersinking them so they'll be easier to patch.

Patching Holes in Plaster

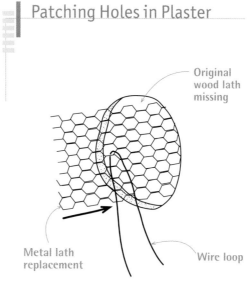

Original wood lath missing

Metal lath replacement

Wire loop

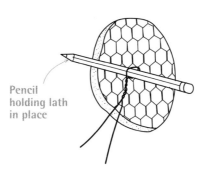

Pencil holding lath in place

If plaster lath has been cut out, replace it before patching the hole. Insert and secure a small section of wire lath with a wire looped around a pencil, as shown. Twirl the pencil to draw the wire lath tight to the back of the plaster, then fill the hole with two coats of patching plaster. Unwind the wire and snip it when the first coat is hard.

loose plaster, brush out the debris, wet the lath and the surrounding plaster well, brush on a PVA bonder, and trowel in patching material. Leave the first coat a little rough, let it dry well, and apply bonder again before troweling in a second, smooth coat.

If the hole has no lath behind it, you'll need to add some. Scrape the loose plaster from the edge of the hole. As shown in "Patching Holes in Plaster" on p. 477, cut a piece of metal lath larger than the hole and loop a short piece of wire through the middle of the lath. Then, holding the ends of the wire, slide the lath into the hole. To pull the lath tight against the back of the hole, insert a pencil into the front of the loop and turn the pencil like an airplane propeller until the wire is taut. The pencil spanning the hole holds the lath in place.

After wetting the lath and surrounding plaster, spread a rough coat of compound into the hole. When the coat has set, unwind the wire, remove the pencil, and push the wire into the wall cavity. The hardened plaster will hold the metal lath in place. Trowel on the finish coat.

Large holes with lath intact should be partially patched with a piece of drywall slightly smaller than the hole. Because a hole with square corners is easier to patch than an irregular one, square up the edges of plaster using an oscillating multitool with an abrasive wheel. (Wear safety glasses.) Be careful not to cut through the lath. Use type W drywall screws to attach the drywall to the lath behind, stretch self-adhering fiberglass mesh tape around the perimeter of the patch, and apply joint compound or patching plaster as described earlier. For best results, the drywall should be slightly thinner than the existing plaster so you have some room to build up and feather out the patch.

As an alternative, pegboard is a dandy substrate for these patches because you can cut it easily with a jigsaw to fit irregular holes. Hold a sheet of ¼-in. pegboard over the plaster, eyeball and trace the shape of the patch through the holes in the pegboard, and screw the pegboard to

Casting Larger Replacement Molding

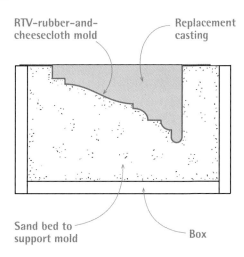

RTV-rubber-and-cheesecloth mold

Replacement casting

Sand bed to support mold

Box

When a repetitive plaster pattern is damaged, replace damaged sections with castings made from sections that are intact.

the lath—textured side out. The patching plaster will ooze through the holes in the pegboard and harden in the same manner that plaster keys into the spaces between lath strips.

RESTORING PLASTERWORK

Restoring damaged crown molding, medallions, or other plaster ceiling ornaments takes patience, hard work, and a lot of skill. You may be better served replacing the originals with plastic or composite reproductions. They are good looking, lightweight, easy to install, and—once painted—indistinguishable from plaster ornaments. Given the hourly rates of a skilled plasterer, this strategy is usually far more cost-effective.

However, if your plaster ornament has a repetitive pattern with only a few damaged sections, you can cast replacement sections by creating a mold from an original, undamaged section.

Removing an ornamental section. If you want to make a casting, you need to remove an undamaged ornamental section. First, support the section that you will remove. To do this, position a 2x4 tee, its head covered with rubberized floor padding, under the ornament to cushion and protect any delicate details. Then use a circular saw with a Carborundum blade or an oscillating multitool with a bimetal blade to cut around the section. The section should include a complete pattern repeat plus 2 in. on each end to allow for some damage when you remove the piece. This is dusty work and you're sure to hit nails, so wear a respirator mask and goggles.

Note: There may well be wires or pipes running through the ceiling, so explore beforehand after turning off the electricity to the area. In many cases, ceiling joists will be exposed in the attic above. Plan cuts so they miss wires and pipes. Replacing the section will be easier if you don't cut through the lath, but that's sometimes impossible to avoid. After cutting around the section, you can often slide a chisel behind it and try to break off the plaster keyed into lath spaces.

Casting replacement sections. To reproduce replacement sections, you must first create a mold. Room temperature vulcanization (RTV) silicon rubber, made by combining two components, is a good choice for molds: It can duplicate fine details and is available from most arts-and-crafts stores and online. After you've removed an original section in good condition, use a toothbrush to remove flaking paint, then repair any small damage with plaster of paris. Shellac the section so its surface will be slick, and let it dry.

To make the mold, paint on a coat of RTV silicone rubber, and allow it to dry. Thereafter, alternate strips of cheesecloth and rubber, allowing each RTV-and-cloth layer to dry before applying the next coat. Three or four layers should give you a mold that's sturdy enough. When the final coat is dry, peel the RTV mold off the original plaster section, and pour a new casting into the mold.

Although plaster is a suitable casting material, it's heavy. If the original object is large—for example, a ceiling medallion—consider casting with a lightweight polymer such as polyurethane or polystyrene, which won't shrink, paints well, and is available in different densities. Or you may need to support the mold in a bed of sand so the new casting material doesn't distort the mold. For larger casts, fill a large enough box with sand, and—before peeling the mold from the original plaster section—press the mold into the sand. Then lift out the mold and peel the rubber carefully from the original plaster. Return the empty mold into the impression it made in the sand. Pour the new plaster (or polymer) and level it off to the top of the mold. When the casting is completely dry, lift it and the mold out of the box and peel off the mold.

Installing new castings. Once you have cast replacement sections, measure both the damaged and replacement sections carefully so the repetitive pattern will match exactly when you install them. Then cut out the damaged ceiling sections, leaving the lath intact and being careful to cut the ends as cleanly and squarely as possible.

Before cutting the new casting to length, minimize fragmentation by first scoring the cutting line with a utility knife and then cutting with a fine-toothed hacksaw.

How you attach the replacement piece depends on its composition. If your casting is plaster, use Durabond 90 quick-setting compound to adhere the plaster ornament to the lath. Dampen the replacement piece so it doesn't leach moisture from the Durabond, and use the 2x4 tee to support it until the compound sets—about 90 minutes. For good measure, predrill holes at a slight angle every 10 in. along the edge of the casting to receive drywall screws with plaster washers. If your replacement casting is a lightweight polymer, you won't need screws; a few beads of construction adhesive or white glue will do the job. Before setting the replacement piece, dry-fit it to make sure it's the same thickness as the old plaster; you may need to build it up slightly. Last, fill the cracks or flaws where the new sections join old before painting the restored ornaments.

Soundproofing

Soundproofing is serious science. Done correctly, it's much more complicated and expensive than just adding fiberglass batts between studs or ceiling joists. Fortunately, there are common-sense solutions that won't cost all that much and will go a long way toward mitigating sound. But first, a little science about how sound travels.

Basically, sound travels through the air or through the structure of a building. Airborne sounds travel in radiating waves through openings of any size. Those sound waves continue until their energy is absorbed—and the more flexible or porous materials they encounter, the sooner that energy is spent. Structure-borne sounds are transmitted through contact between solid materials, and those solids can be as small as a copper supply pipe in contact with a wall plate or a drywall screw sunk in a ceiling joist. To stop structural sounds, you need to isolate them, say, by putting a piece of resilient material between a vibrating appliance and the floor it sits on.

Point being, because *noise*—the sounds we don't want to hear—can travel through tiny openings or be transmitted through structural elements as small as a screw, successful soundproofing is all about paying attention to details. How many details you can attend to will depend on the extent of your renovation. If you're gutting the interior, you can do a lot. But even a room left intact can be improved.

THREE SIMPLE SOLUTIONS

It's finally happened: You're somebody's parent and you're telling your kid to TURN IT DOWN! So next birthday or holiday, get that sonic-busting offspring or your hard-of-hearing mother a pair of really nice headphones, ones so nice and cushy that he or she will want to wear them. You will be spared exploding aliens or the evening news at 60 decibels (db.), and peace will reign. Second, if footfalls in a room overhead have you edgy, a carpet with a substantial pad can be a better sound suppressor than installing insulation between floor joists. Third, heavy curtains, over-stuffed furniture, and rugs will all absorb sound, whereas hard, reflective surfaces such as tile or laminate counters can make living spaces sound as homey as a hospital corridor.

Appliance science. Any appliance with a motor will vibrate and create noise. Washers, dryers, and refrigerators are the biggest offenders because they have the biggest motors. Your appli-ances aren't noisy? Put your ear to the floor when they're running. To mitigate appliance noises in an upstairs in-law unit, one carpenter installed the type of thin foam pad typically used under floating floors, covered that with ¼-in. plywood underlayment, and then installed resilient floor-ing atop that in the apartment's kitchen and laun-dry room. He then mounted the washer and dryer on a plywood platform fastened to the steel-and-neoprene vibration isolators often used to mount commercial cooling units to roofs. Granted, his solution took a few Google searches, but it solved the problem without tearing up the floorboards.

If the floors beneath your appliances aren't adequately supported, they will transmit sound like a drumhead. So consider that when devising solutions.

Ducts, pipes, and wires. You wouldn't want to live without electrical, plumbing, and HVAC sys-tems, but you don't have to live with the racket they make. Where forced-hot-air ducts are acces-sible, wrap them in insulating jackets to conserve heat and suppress sound; where they exit to deliver heat to the house, fill the spaces between ducts and framing with expandable foam. Caulk or foam-fill the holes for plumbing pipes and electrical cables, and you'll simultaneously sup-press sounds and stop drafts.

If your renovation includes gutting walls and exposing these mechanical systems, you can also wrap HVAC ducts and DWV pipes with *dense vinyl jackets*, which incorporate a layer of fiber-glass insulation. ABS or PVC waste pipes are inherently noisy, however, so if you want to silence flushing noises in waste pipes forever, bite the bullet and install cast-iron DWV pipes. Less expensive is making sure that DWV pipes are well secured with pipe straps, hangers, and clamps (see the top right photo on p. 365) and supply pipes are immobilized by Acousto-Plumb clamps (see the top photo on p. 377) and other plastic pipe-support flanges.

Electrical cables and devices are pretty quiet as mechanical systems go, but their routing holes and cutouts for outlet boxes will transmit air-borne sounds—especially when outlet boxes are back to back in a shared wall. Fill holes in fram-ing with expanding foam, caulk around outlet boxes, and, if you want to do an A1 job of stop-ping air leaks and hence sounds, wrap outlet boxes in Moldable Putty Pads MPP+ (facing page). However, do *not* wrap recessed ceiling cans. IC-rated light fixtures (see p. 425) can be covered with loose insulation or insulation batts (but not vinyl jackets). However, non-IC-rated fix-tures *must not be covered with insulation*—or anything else that would cause these fixtures to overheat and cause a house fire.

SOUNDPROOFING BETWEEN FLOORS

Correctly soundproofing between floors may require gutting finished ceilings and exposing joists. Hang two layers of drywall on the ceiling and screw to resilient channels, not to joists, allowing the drywall to float. Cut the ceiling dry-wall short of the walls by ¼ in. and fill the gap with acoustical sealant.

Remember: The fewer screws you use, the bet-ter the soundproofing.

SOUNDPROOFING: A SUCCESSFUL HYBRID APPROACH

This sound-reduction assembly mixes materials from several manufacturers and does an excel-lent job. It features: (1) GenieClip RST sound iso-lation clips, rubber-and-steel mounts that screw to studs or ceiling joists; (2) 25-gauge steel hat channel that snaps into the GenieClips; (3) two layers of ⅝-in. drywall; (4) Green Glue Noiseproofing Compound, applied between lay-ers of drywall; (5) Green Glue Acoustical Sealant, to caulk perimeter gaps around wall and ceiling panels; and (6) Putty Pads, resilient sheets that, when molded to the outside of electrical boxes, stop air and noise leaks. There are many compet-ing products, so choose ones that are effective and easy to install.

Setting up, laying out. Snap horizontal chalklines across stud edges to locate the rows of hat channel. Channels (and thus, the clips that support them) should be spaced 24 in. apart (vertical distance). The bottom row of channels (and clips) should be 3 in. or less from the floor, and the top row of channels (and clips) should be 6 in. or less from the ceiling. Clips should also be placed wherever a section of hat channel ends.

In addition to spacing clips 24 in. vertically, space them 48 in. apart horizontally, as shown in the illustration on p. 482. This is best achieved by installing clips in a zigzag pattern, with clips attached to every third stud for framing spaced 16 in. on center (O.C.) or to every second stud for framing spaced 24 in. on center. Use $1\frac{5}{8}$-in. coarse-thread (Type W) drywall screws to secure sound-isolation clips to studs.

Use a pair of metal-cutting shears to cut hat channel to length; wear sturdy gloves to protect your hands. To attach the hat channel, insert its lower edge into the bottom lip of the GenieClip, then press the top edge of the channel down until it snaps into the top lip of the clip. (This is most easily done with two people.) Where sections of hat channel meet, overlap their ends by 6 in. and join them with two self-tapping sheet-metal screws.

Installing drywall panels. When measuring and cutting wall panels, leave a gap of ¼ in. around the perimeter of each wall section so there is room for acoustical sealant. To elevate the bottom panel(s), place ¼-in. plastic spacers along the foot of the wall and set the panel(s) atop them. Screw the drywall to the hat channel. Typically, one will use $1\frac{1}{4}$-in. fine-thread (Type S) drywall screws to attach ⅝-in. drywall, but size and space screws according to local building codes. When you press drywall that is screwed to hat channels, it will be slightly springy to the touch because it is decoupled (isolated) from the studs. *Important*: Don't use screws that are too long, which could penetrate framing and transmit sound.

Caulk the ¼-in. gap around each wall section with acoustical sealant, and you're ready to install the second layer of drywall. It, too, should have a ¼-in. gap around its perimeter, but stagger panel seams so they do not line up with those of the first layer. This second layer of ⅝-in. drywall will be glued to the first and screwed to the hat channel, using $1\frac{7}{8}$-in. Type S screws spaced according to local building codes.

Green Glue or any other noiseproofing compound should be applied generously to the back

A Putty Pad shaped around an outlet box cuts sound leaks.

Hat channel and Putty Pads are in place as part of soundproofing this ceiling. The GenieClips are also in place, but not visible.

A GenieClip Assembly

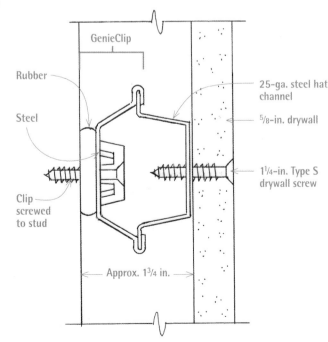

GenieClip

Rubber

Steel

Clip screwed to stud

25-ga. steel hat channel

⅝-in. drywall

$1\frac{1}{4}$-in. Type S drywall screw

Approx. $1\frac{3}{4}$ in.

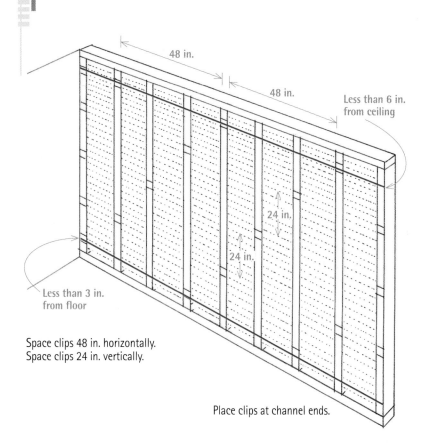

48 in.

48 in.

Less than 6 in. from ceiling

24 in.

24 in.

Less than 3 in. from floor

Space clips 48 in. horizontally.
Space clips 24 in. vertically.

Place clips at channel ends.

This sound studio is insulated on all six sides. Special sound-attenuating drywall will be attached to metal resilient channels, rather than screwed directly to framing. Resilient channels allow the drywall panels to float—further absorbing sound. Window openings were the studio's biggest soundproofing challenge.

of drywall panels; use two 28-oz. tubes of Green Glue for each 4-ft. by 8-ft. panel. Though the adhesive can be applied in a random pattern, it should be distributed evenly. Leave a 2-in. to 3-in. border around panel edges so you can lift it without getting glue on your hands. Noiseproofing compounds, which remain flexible, typically transform the mechanical energy of sound waves into small amounts of heat. Compounds do set, however, so it's important to screw panels to channels within 15 minutes of applying glue.

Caulking and finishing. Caulk with acoustical sealant the ¼-in. gap around the perimeter of the wall section. If there are electrical boxes in walls or ceilings, form Putty Pads to the outside of those boxes to stop air leaks and sound transmission before closing up stud bays with drywall. Spackle, tape, and paint the second layer as you would any drywall.

Soundproofing ceilings is essentially the same—apply Green Glue Noiseproofing Compound between layers of drywall, screw drywall to hat channels secured to ceiling joists, and caulk around the perimeter. And have plenty of help to lift and hold panels in place as you are screwing panels to hat channels.

SOUNDPROOFING: LESSONS FROM A SOUND STUDIO

Stephen Marshall of www.thelittlehouseonthe-trailer.com created a sound studio for a musician/composer client. The studio's design had two functions, which were somewhat at cross-purposes. First, the studio had to be thoroughly soundproofed so instruments could be played without disturbing neighbors. Second, the space had to be pleasant and light-filled during those

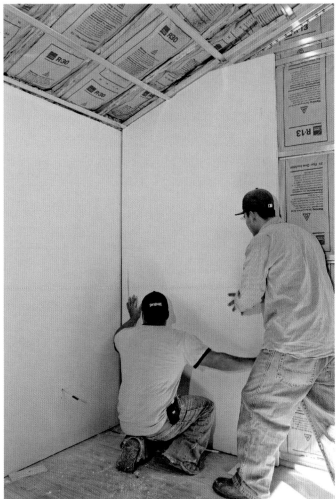

Unlike a typical drywall job, in which ceiling panels are installed first so they can be partially supported by wall panels, this design called for an isolated ceiling. Wall panels were installed first, and ceiling panels were cut ¼ in. *shy* of walls—so the ceiling panels would never touch the walls. The ¼-in. gap was filled with acoustical sealant.

When ceilings are high, use a drywall lift to hold drywall panels snug against resilient channel. Individual screws are less likely to pull through if a panel is uniformly supported.

times when the client was engaged in quieter pursuits.

The studio had fiberglass batt insulation on all six sides and resilient channels attached to walls and the ceiling, to which drywall panels were attached. The drywall was ¾ in. thick, consisting of three pieces of ¼-in. panels laminated with an elastomeric soundproofing compound that never completely sets, so it deadens sound. Panels were staggered so that successive layers covered the seams.

Ceiling and wall panels were screwed to resilient channel, as described in the preceding section. Wall panels were attached first; ceiling panels were then installed so that there was a ¼-in. gap between ceiling and walls. That gap was filled with acoustical sealant—OSI Pro-Series SC175. Additionally, *mass-loaded vinyl* was used

to achieve an STC 60 rating in some of the most problematic areas: electrical boxes and the windows that made the space so pleasant. Electrical boxes were surface-mounted to avoid creating holes in the envelope and wrapped with mass-loaded vinyl jackets. But because the acoustical value of a room is only as strong as its weakest link, the windows were the greatest test. High-performing sound windows would have been prohibitively expensive, so the solution called for sliding window covers, rather like sliding barn doors, which would be constructed from mass-loaded vinyl and rigid insulation. When it was time to make music, the insulated covers would slide over the windows, creating a soundproof cocoon.

16 Tiling

This contemporary tile is based on a Moorish motif 1,000 years old.

Tile surfaces can be beautiful, durable, and—if you're patient—fairly easy to install. Although tile has a hard finish, the ultimate durability of the installation depends on the integrity of what lies beneath.

Choosing Tile

There's a riot of tiles to choose from, including slate, white porcelain hexagonals, ruddy Mexican pavers, tumbled marble, glass mosaic, brick veneer, cast cement, limestone quarry tile embedded with fossils, and so on. You can even paint your own designs on unglazed tiles and then have them kiln-fired. Although some types of tile are better suited to certain uses than others, finding a tile you like is rarely a problem.

SELECTING TILE

Here are seven useful tips for choosing tile appropriate to the job:

▶ Make sure the tile can handle the conditions where it will be installed. Does it need to be waterproof? Will you walk on it? Will you set pots on it?

▶ Sketch the area to be tiled. Include dimensions, fixtures, corners, odd jogs, and adjoining surfaces such as wood flooring or carpet. This sketch is a systematic first step in assembling a materials list.

▶ Choose a tile store with knowledgeable staff that will take the time to answer your questions. Plan to visit the shop on a weekday, which is likely to be less busy and thus a good time to get extra help. High-end tile stores have room mockups and may also have a website showing a wide selection of tiled kitchens and baths. Stores will also display many types of tile in 2-ft.-sq. or 3-ft.-sq. panels. Such visual expanses of tile convey much more than single-tile samples.

▶ If you like a particular tile, have your salesperson determine the manufacturer's specs, which should tell you its suitability for various uses. For example, you wouldn't want to install a wall-rated tile on a floor.

▶ At some point, reconcile the tiles you like with your budget. Some tile is breathtakingly expensive. Also check on availability. Will specially ordered tile arrive in time to meet your renovation schedule?

▶ Determine if trim tile is available for the pattern or type of field pattern you select. Trim tile is used to finish edges and corners and is especially important for counter installations.

▶ Test a tile sample at home. Here, you want to determine its suitability for your intended location by simulating actual use; for example, by scuffing it with shoes, banging it with pots, or dribbling it with water to check for absorption. Does the tile clean easily?

Color and size. In general, smaller tiles are better suited for small areas, such as counters. Larger tiles are more appropriate for larger areas, such as floors. Because light-colored tiles reflect more light, they make a room seem larger. Conversely, dark tiles make a room seem smaller. However, light colors tend to show dirt more readily. Vivid colors or busy designs can provide nice accents, but when used to cover large areas, they may seem overpowering.

Grout is a specialized mortar that seals the joints between tiles. Its color can make a big impact on the overall look. The closer the grout color matches the tile color, the more subdued and formal the surface. The more contrast between grout and tile, the busier, more festive, or more geometric the tile job will appear and the more it will highlight your tiling skill.

Manufactured versus handmade. Mass-produced ceramic tile is popular because it has a clean, classical look, and its uniform size makes an installation more predictable. Most smaller (½-in. to 2-in.) mass-produced tiles come evenly spaced and premounted on paper sheets or fiberglass mesh, simplifying installation.

Handmade tiles offer unique color and a handsome, handcrafted look. But these irregular tiles take greater patience and skill to lay out and install. Handmade tiles often need thinset adhesive applied to each back, as well as a thinset layer troweled onto a setting bed. Because they

These tumbled-marble sheets are attached to a paper backing, which is embedded into the adhesive.

Field and Trim Tile

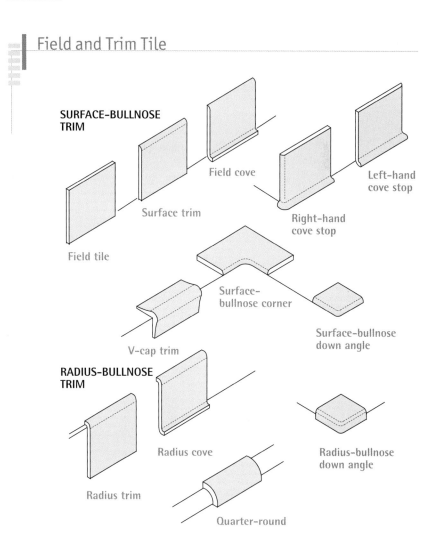

are irregular, handmade tiles may also need to be arranged with plastic spacers to align their edges with the underlying layout lines.

Sheet-mounted tiles. Some tile comes either face mounted or back mounted, typically on 12-in. by 12-in. sheets. To install face-mounted tiles, leave the paper facing on until the adhesive dries, then dampen the paper and remove it. *Disadvantage:* Paper facing obscures tile joints, making them tougher to align. Back-mounted sheets are easier to align, but the bond between the tile and the adhesive may be compromised slightly because the backing remains stuck in the adhesive.

WATER RESISTANCE AND DURABILITY

There are almost as many physical distinctions among tiles as there are tile types, but the most important traits to consider are water resistance and durability.

Water resistance. Here, three of four official categories of tiles include the word *vitreous*, which means glasslike, and suggest how much the tile will resist or absorb water. The categories are *nonvitreous, semivitreous, vitreous,* and *impervious*. Nonvitreous is the most absorptive, and impervious the most water-resistant.

Use nonvitreous tiles on dry areas, such as interior fireplace surrounds and hearths. Use semivitreous or better on shower walls, tub surrounds, backsplashes, and areas that are intermittently wet. Use vitreous and impervious tiles for wet installations such as pools, hot tubs, and outdoor surfaces in rainy climates. In general, the less water a tile absorbs, the less hospitable it will be to bacteria and mold. That's why hospitals and laboratories usually use impervious tiles.

Durability. Tile durability ratings, based on structural strength and surface imperfections, typically assign softer, weaker tile to less demanding areas and harder, impervious tile to heavily trafficked, wet, and outdoor areas. In like manner, tiles are rated for walls, floors, and counters. A reputable tile supplier will give you good advice on appropriate uses and durability and will stand behind the tiles you buy.

Trim TILE

Trim tile is specially shaped to trim or finish off surface edges, corners, and other transitions. Trim tile is further classified as surface (or surface bullnose) and radius (or radius bullnose). Surface trim is essentially flat tile with one rounded edge. Radius trim, also known as quarter-round, curves dramatically to conceal the built-up bed on which it is set. Both types of trim include a range of specialty pieces that finish inside corners, outside corners, and wall joints.

Tools

Many tile suppliers sell or rent tiling tools and offer workshops on techniques and tool use.

Tools and safety. Tiling is deliberate, methodical work and is not as inherently dangerous as some remodeling tasks. Still, it poses hazards, so for starters, note these minimal safety rules:

▶ Use a voltage tester to ensure that power has been shut off to outlets, fixtures, switches, and devices you'll work near. In addition, ensure that bathroom and kitchen receptacles have GFCI protection, as spelled out in chapter 11. Corded power tools should be double insulated and grounded with a three-prong plug. Or use cordless tools instead.

▶ Rubber gloves reduce the risk of electric shock and prevent skin poisoning from prolonged handling of mortar, adhesives, sealers, and the like.

▶ Wear goggles when cutting tiles, whether making full cuts with a wet saw or nibbling bites with a tile nipper. Tile shards can be as sharp as a scalpel.

▶ Wear a respirator when mixing masonry materials, applying adhesive, cutting cementitious backer board, and so on.

▶ Knee pads will spare you a lot of discomfort. Buy a pair that's comfortable and flexible enough to wear all day. Flimsy rubber knee pads won't protect your knees.

Classic vitreous porcelain tile is often used in bathrooms because it resists stains and sheds water.

▶ Open windows and turn off pilot lights on gas appliances when using volatile adhesives or admixtures. Closely follow manufacturer's instructions.

BASIC TOOL KIT FOR TILE

▶ Safety equipment: rubber gloves, goggles, respirator mask, voltage tester, and knee pads.

▶ Measuring and layout: straightedges, framing square, spirit level, pencil or felt-tipped pen, chalkline, tape measure, story pole, and scribe (or an inexpensive student's compass).

▶ Setting and grouting: notched trowel, margin trowel, plastic spacers and wedges, beater board, rubber mallet, grout float, round-cornered sponge, and clean rags.

▶ Cutting: snap cutter, tile nippers, utility knife with extra blades, and wet saw.

▶ Cleanup: sponges, rags, plastic buckets, plastic tarps, and shop vacuum.

▶ Miscellany: hammer and wire cutters.

MEASURING AND LAYOUT

Substrates are never absolutely flat or perfectly plumb, so layout is a series of reasonable approximations. Clean tools give the most accurate readings, so wipe off mortar or stray adhesive before it dries. Here's what you'll need:

▶ A 4-ft. spirit level is long enough to give you an accurate reading. It's indispensable for checking plumb and leveling courses of wall tiles. If a 4-ft. level proves unwieldy on the short end walls of a bathtub, use a 2-ft. level or a torpedo level instead.

▶ A tape measure lets you measure areas to be tiled, triangulate diagonals for square, and perform general layout.

▶ A chalkline allows you to mark tile layout lines before applying adhesive.

▶ Straightedges are useful for aligning tile courses, marking layout lines on substrate, and guiding cuts on backer board and plywood. Professional tilesetters have metal straightedges of different lengths, but wood's OK if it's straight and sealed to resist water.

▶ A framing square is used to establish perpendicular layout lines on floors, walls, and countertops.

▶ A story pole is a long, straight board marked in increments representing the average width of a tile plus one grout joint (see "Storytelling" on p. 500). With it, you can quickly see how many tiles will fit in a given area, as well as where partial tiles will occur.

Straightedges are indispensable for tile layout and installation. They tell you whether surfaces are flat and help you align tile edges.

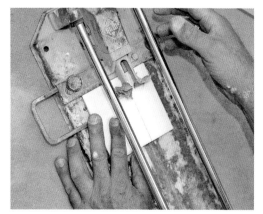

Snap cutters work great for straight cuts on vitreous tile. With this model, you score the tile in one pull and then push down on the tool's wings to snap the tile along the scored line. Here, blue painter's tape keeps the cutter's wings from scratching the tiles.

Wet saws are relatively cheap to rent, and they cut almost any type of tile cleanly. Wear safety glasses and hearing protectors when using one. To extend blade life, change the water often.

Using nippers is more like nibbling an ear of corn than chomping through a hamburger. Twist the nipper handles slightly as you break off little pieces of the tile's edge, and be patient.

▶ A scribe is used to fit sheet materials (cementitious backer board, plywood) to a bowed wall or to transfer the arc of a toilet flange to a tile.

CUTTING

Always wear goggles when cutting or nipping tile, especially when using power tools.

A snap cutter works well on manufactured vitreous and impervious tile. This tool has a little cutting wheel—make sure it's not wobbly or chipped—that should score the tile in one pull. Then reposition the handle so the "wings" of the tool rest on the scored tile, and press sharply to snap it. *Note:* When using a snap cutter, it's tough to get a clean break on nonvitreous tile, tiles with textured surfaces, and floor tiles. For those you'll need to use a wet saw.

A wet saw, which you can rent, is especially useful for cutting nonvitreous or irregular tiles or trimming less than 1 in. off any tile. It cleanly cuts all tile types. To make U-shaped cuts around soap dishes and the like, make a series of parallel cuts with the wet saw before removing the waste with tile nippers.

Tile nippers allow you to cut out sections where tiles encounter faucet stems, toilet flanges, and the like. Nippers require some practice and lots of patience. Take small nibbles—use only part of the jaws—nibbling away from the sides of a cut into the center, gradually refining the cutout. As you approach your final cutlines, go slowly.

A carbide-tipped hole saw is perfect for cutting holes for faucet stems and pipe stubs. To be safe around pipes and water, use a cordless drill with this saw.

A handheld grinder with a diamond blade can cut curved lines in tile (make a series of shallow passes) or plunge cuts for holes in the middle of a tile. Be sure to stop the cuts short of your cutlines and remove the waste with a pair of nippers. Because a grinder is noisy and throws lots of dust, use it only outdoors, and wear a respirator mask and eye protection. An *oscillating multitool* (see p. 57) with a grinder blade is a good alternative.

A utility knife scores cementitious backer board (which you snap like drywall), marks off tile joints in fresh mortar, cleans stray adhesive out of joints, and so on. However, if you've got a lot of backer board to cut, use a handheld grinder with a diamond blade instead.

SETTING AND GROUTING

Setting means positioning and adhering tile to a substrate. *Grouting* means sealing the joints between tiles with a special mortar.

A notched trowel spreads adhesive. Two edges of the tool are flat, designed to spread the adhesive initially. Then, on subsequent passes, use the notched edges to comb a series of parallel ridges, which will spread evenly when the tile is pressed into it. Notch height should be about two-thirds the thickness of the tile.

A margin trowel is a utility tool that's great for mixing small batches of powdered adhesives, cleaning mortar off other trowels, buttering individual tiles with adhesive, and removing excess grout or adhesive that oozes up between tiles.

Plastic spacers and wedges enable you to shim individual tiles so their edges align to your layout lines.

A beater board is just a flat board placed over tile sections and rapped gently with a rubber mallet to seat the tiles in the adhesive. Not all tilesetters use a beater board. Many just press tiles in firmly or use a fist to seat them better.

A grout float (rubber-faced trowel) applies grout in a process that takes at least two passes. Holding the face of the grout float at about 30°, sweep the grout generously over the tile and pack it into the joints. Then, holding the float almost perpendicular to the surface, remove the excess grout, unloading it periodically into a bucket. To avoid pulling grout out of the joints, make your passes diagonally across the tile joints.

Round-cornered, tight-pored sponges are less likely to pull grout out of tile joints. After grout starts to haze over, wipe lightly with a dampened sponge, rinsing the sponge often. Get several types of sponges. Kitchen sponges with scrub

A small group of setting and grouting tools. Clockwise from upper left: grout floats, notched trowels, hand-drill mixing bit, and sponges (of which you'll need a variety).

pads on the back are useful for removing stubborn grout.

A mixing bit in an electric drill can mix large amounts of powdered adhesives or grout. Slow mixing speeds of 300 rpm to 400 rpm work best. Keep the bit immersed to minimize mixing in air, which weakens the batch. Wear a respirator.

Materials

Here's a quick survey of materials you might use to create a durable tiling job.

Think of the job as if it were a layer cake. For example, in floor tiling, the bottom layer (conceptually the table under the cake) would be the floor joists. Nailed to the joists, in most cases, is a plywood *substrate*. For a wet installation such as a shower wall, next comes a *waterproofing membrane*, followed by a setting bed of cementitious backer board or a *mortar bed*. Troweled onto the setting bed is a *setting material*, typically *thinset adhesive* or *organic mastic*. Tiles are applied to the setting bed, and once the bed has hardened, tile joints are grouted. Later, a *sealer* may be applied to make tile and grout more water- or stain-resistant.

Methods and materials are rarely predictable in a renovation, so it's a good idea to survey the back or underside of the surface you're about to tile, both to see how many layers there are and to check if they're in good condition. You can pull out a heating register to see a cross section of the flooring, for example. Or test-drill a small hole in an inconspicuous spot to determine the thickness

Mastering THE MESS

Controlling the mess is a big part of successful tile setting. If you're tiling a tub surround, cover floors with builder's paper or plastic, mask off cabinets, and line the bathtub with a heavy canvas drop cloth before you start. Masonry debris is abrasive, so vacuum as it accumulates. Also, keep a clean 5-gal. plastic bucket (joint compound pails are perfect) full of clean water for sponging mortar, adhesive, or grout off tools. Last, a bundle of clean, dry rags is useful for buffing dried grout haze off tile and soaking up messes.

Because handmade tiles are irregular, they often need to be moved slightly after they've been set in adhesive. Use plastic shims to raise tiles so they're level with others in the course.

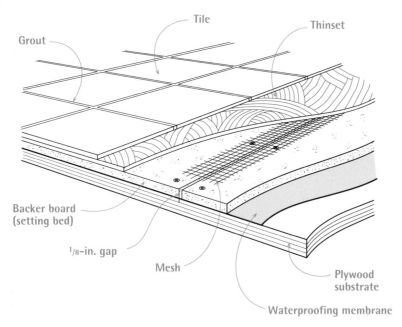

Grout — Tile — Thinset

Backer board
(setting bed)

1/8-in. gap

Mesh

Plywood
substrate

Waterproofing membrane

Applying a mortar bed takes strength and a lot of skill. After installing a curing membrane and attaching wire mesh, you must carefully mix and trowel the mortar on, level or plumb float strips, and screed off excess mortar until the bed is uniformly flat. Then, after it has set a bit, use a wood float to roughen its surface slightly, which improves the adhesion for the thinset to follow.

and composition of an existing wall, floor, or countertop.

SETTING BEDS

Tile can be set on a variety of setting beds. Keep in mind that the substrate below the setting bed must be securely attached to the framing. That assembly must be thick and stable enough to support loads with minimal deflection (1/360 of the span). And when used in damp or wet areas, it needs to remain unaffected by sustained exposure to water.

Mortar beds. Where walls aren't plumb or floors aren't flat, mortar beds are usually the best setting bed. They can easily be screeded level or plumb to create a flat surface. In fact, a mortar bed is the *only* practical choice if you're tiling the sloping floors of a shower stall. But few novices have enough skill to float a mortar bed. If your surfaces are badly out of whack and you're not experienced in floating mud, hire a pro for this job.

Backer board. If walls and floors are reasonably plumb and level, backer board is a durable setting bed for wet and dry installations. It's stable, unaffected by moisture, and easily cut and attached. Backer board is also called cementitious backer units (CBUs), HardieBacker, Durock, and WonderBoard—the last three being popular brands. These products feature a cement-based core, reinforced by fiberglass mesh or integral fibers. Because there's considerable variation among brands, always consult product literature for details on installation.

Glass-mat gypsum board. This board (one brand is DensShield) has a water-resistant core and a heat-cured acrylic coating. It's not as rugged as backer board, but it's an acceptable setting bed in tub surrounds and other light-duty wet areas. Don't confuse this product with drywall. Glass-mat gypsum board is specifically designed as a tile backer.

Unpainted drywall. Drywall is an acceptable setting bed where walls stay dry. *Caution:* Never bond tile directly to drywall in damp or wet installations. Moisture-resistant drywall is an acceptable substrate in damp or wet installations only when it is covered with a waterproofing membrane and then a mortar bed or backer board.

Tile applied directly to drywall in wet installations invariably fails sooner or later, often because people showering bump the walls, compressing the drywall's gypsum core. No longer supported evenly by the core, the grout loosens, water enters and soaks the paper, and—all too often—the framing rots.

Exterior-grade plywood. It's usually a mistake to use plywood, even exterior grade, as a setting bed. Exposed to moisture, plywood tends to swell and delaminate. If you must use it as a setting bed for floor or countertop installations, be sure to cover it with a continuous waterproofing membrane; the base should be at least 1⅛ in. thick (for example, ½-in. plywood underlayment laminated to a ⅝-in. plywood subfloor). On walls, ⅝-in. plywood is the minimum. *Never* use particleboard, OSB, or interior-grade plywood as setting beds.

Self-leveling compounds. Self-leveling compounds (SLCs), such as LevelQuik, have many of the virtues of a mortar bed but require few of the skills needed to float one. Basically, SLCs are fortified mortar powders mixed thin and poured onto out-of-level floors. With a small amount of troweling, they spread across the floor and, within minutes, start to set. Just two hours later, you've got a hard, almost perfectly level mortar-setting bed ready to tile.

That's the short list of common setting beds. For other materials, see "Odd or Problematic Setting Beds" on p. 499.

ADHESIVES

Once you've chosen a suitable setting bed for your wet or dry installation, choose a compatible adhesive. Adhesives vary greatly from brand to brand, so again, always follow the manufacturer's mixing and application instructions exactly. There are three major groups of adhesives: *mastics*, which come ready-mixed; *thinset adhesives*, which are cementitious powders generally mixed on site just before setting the tile; and *epoxy thinsets* that, like most epoxies, require you to mix a hardener and a resin.

Organic mastics are the least expensive of the three adhesive options. Because they come pre-mixed, they're the most convenient option, but they're also the weakest. They are OK for attaching tiles to dry counters or walls—over drywall, for example—but they're inappropriate where there's water, heavy use, or heat. Mastics just don't have the strength of thinsets.

Mastics require a nearly flat setting bed. That is, when they are applied thickly to fill voids, they neither cure completely nor bond thoroughly. Mastic cleans up well with water or solvent if you remove the excess material at once. Opened containers don't keep well, so throw away any left-over mastic after you've set the tiles.

Thinsets have great bonding and compressive strength. Being cement based, they bond best with mortar beds or backer board but are appro-

Here, a tiler uses a margin trowel to apply thinset to a trimmed slate tile. With the correct consistency, thinset spreads easily yet will adhere to a trowel turned on edge. Because polymer–modified thinsets have great bonding strength, you're safer wearing rubber gloves. Also, be sure to clean tools immediately.

priate for virtually all setting-bed materials. Thinsets also are used to laminate rigid setting beds to substrates to create an inflexible substructure for tiling. (Construction adhesive is also used in such laminations, but it is flexible, so it doesn't achieve the rigidity of a thinset lamination.)

Despite the cement ingredients they have in common, thinsets vary widely, depending on their additives. *Water-based* thinsets are the weakest of the group, although they are generally stronger than mastics. *Latex-* and *acrylic-based* thinsets (also known as polymer-modified thinsets) are strong, somewhat water-resistant, and, all in all, the best choice for bonding tile to backer board, mortar beds, SLCs, drywall, and concrete slabs. And they're a close second for bonding almost everything else.

Most thinsets are mixed from powder. After mixing, they have a "bucket life" of about two hours. After being troweled onto a setting bed, they start to set in 15 to 20 minutes.

Epoxy thinsets have excellent compressive and tensile strengths. They bond well and yet retain flexibility when cured. After drying, they are unaffected by moisture, so they're suitable for all situations and substrates. There's a catch, of course: Epoxies are four or five times more expensive than other thinsets and quite temperamental. You must mix the liquid resins and hardeners in exact proportions with the dry ingredients. Setting times are similarly exacting. If directions say 20 minutes, you can set your watch by them. Above all, clean up epoxy before

Waterproofing a Tub Surround

Water-resistant (WR) drywall or greenboard

1/8-in. gap

Waterproof membrane

Fiberglass mesh tape

Thinset

4-in. overlap

Backer board

Tile

This tub surround would be sufficiently rigid without the WR drywall. But installers frequently add a drywall layer to build up the wall thickness when they'll be using bullnose edge trim.

PRO TIP

Manufacturers frequently change their grout colors, so buy 10% more than you need for your current installation. Once the job is done, wrap the extra grout in a plastic bag, label it as to where it was used, and store it in a dry place. While you're at it, buy caulking the same color as your grout; many tile stores carry color-matched caulk, both sandless and sanded.

it sets; some types sponge clean with water, others with solvents.

GROUT

Grout is a specialized mortar that seals the joints between tiles. Most grouts contain sand, cement, and a coloring agent. Grout may also contain additives to stabilize color, increase water and stain resistance, and increase strength and flexibility. Most grouts and premixed additives are sold as a powder, which is subsequently mixed with liquid and allowed to stand (or slake) for 10 minutes before final stirring to the correct consistency.

Use *sandless grout* for joints narrower than 1/8 in. Use *sanded grout* for joints 1/8 in. and wider.

Most tile suppliers carry grout in hundreds of colors. Whatever the color, remember that the greater the contrast between grout and tile, the more obvious the joints and workmanship.

MEMBRANES

Tile, grout, and many setting beds are unaffected by water. But they are porous, so water can migrate through them, potentially damaging plywood substrates or wood framing. To prevent damage in damp or wet areas, install a *waterproofing membrane* first. Even areas that are normally dry, such as entryways, should have a modest building-paper membrane if the floors will be subject to wet mopping and dripping umbrellas.

For walls above the water line in wet areas and for countertops subjected to occasional water, a 15-lb. building-paper membrane is standard, but installing a 60-minute stucco paper (which is impermeable for 60 minutes) makes sense. It's stronger, less likely to tear, and more water-resistant. When installing such membranes, overlap the lower courses and vertical seams of paper by 4 in. Although some builders recommend 4-mil polyethylene as a waterproofing membrane, stucco paper has one big advantage: Unlike plastic, it is semipermeable.

All wet installations need a waterproofing membrane. Here, a tub surround gets two layers of Fortifiber's Super Jumbo Tex 60 Minute stucco paper, which is a fiber-reinforced barrier that's tougher and more water-resistant than regular building paper.

Therefore, it allows water to escape should any get behind the barrier.

Below the water line, such as in a shower pan, you need to protect wood substructure with an impervious membrane. Thus most shower pans are lined with sheet rubber, such as 30-mil, fiber-reinforced chlorinated polyethylene (CPE), whose seams are overlapped and chemically bonded with a solvent. Of course, you don't want to puncture CPE shower-pan membranes with screws or nails. Instead, roll the membrane onto a fresh layer of latex thinset adhesive, and cover it with a mortar bed.

Note: The comments in this section are generalizations. Follow the manufacturer's installation instructions for specific adhesives, membranes, and setting-bed materials.

Getting Ready to Tile

If the substructure beneath the tile isn't sturdy and stable, the job won't last. Likewise, if walls aren't plumb or floors aren't level, tiles may adhere, but they may not look good. Start by assessing the existing surfaces. And that will inform your next steps, which can range from merely sanding finish surfaces to tearing out and reframing with studs and joists. The condition of existing floors, walls, and counters will also determine which setting bed you choose—and whether you should tile at all.

ASSESSING AREAS TO BE TILED

To check whether floors or countertops are level, use a long spirit level or a shorter level atop a perfectly straight board. Take several readings and use a pencil to mark individual high spots and dips. If variations from level exceed ⅛ in. in 10 ft., floating a mortar bed or pouring SLC may be your best bet for establishing a flat setting bed. If the surface irregularities are less than that or the substrate just needs stiffening, adding a single layer of backer board may be all you need.

If room corners aren't square or facing walls aren't parallel, you may need to angle-cut floor tiles around the room's perimeter. This is not ideal, especially in narrow alcoves or hallways, but baseboard trim will partially cover those angled cuts. Similarly, at the back of counters, backsplashes will cover angle-cut tiles.

To check walls for plumb, use a long spirit level or a plumb bob; a taut string is also handy to detect high and low spots. Begin by surveying the entire wall. Unless the tiled corners are plumb, you'll have tapering cuts or mismatched grout lines where the planes converge. To correct out-of-plumb walls, your choices are floating a mortar bed, reframing the walls, or not tiling at

When checking surfaces for plumb and level, take several readings with a spirit level, especially in corners. If they're out of plumb, tile joints won't align, and the mismatch will be glaringly obvious. Correct the condition or don't tile.

Corner out of Plumb

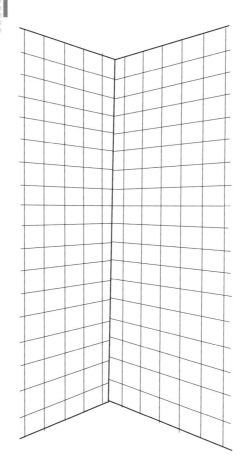

Corner walls too far out of plumb can't be tiled successfully. Their tile joints won't align, and the mismatch will be glaringly obvious.

If you're replacing tub/shower walls or installing backer board, protect chrome gooseneck pipes or threaded spout stubs by replacing them with 6-in. pipe nipples. The nipples are placeholders for the originals, ensuring that pipe stubs line up to holes in the backer board. Here, Foilastic flashing reinforces the waterproofing membrane and later will serve as a dam for silicone caulking.

Removing the toilet lets you reinstall it on top of the new tile. Some people mistakenly leave the toilet in place and as a result are forced to make a lot of unsightly tile cuts around its base, which can also be troublesome to caulk and maintain.

Begin by turning off the shutoff valve to stop incoming water, disconnect the supply line, flush the toilet and remove the remaining water, and disconnect the anchor bolts holding the base to the floor.

Because toilets are heavy, find someone to help you move the toilet out of the way. To block septic gases and keep objects from falling through the closet flange into the closet bend, stuff a plastic bag filled with crumpled newspaper into the pipe (don't forget to remove it before reinstalling the toilet).

With the waste pipe temporarily sealed, take a look at the toilet's *closet flange* atop the closet bend. Ideally, the top of the flange should be the same height as the finish floor. If your tiling increases the height of the floor ½ in. or less, the height of the flange shouldn't be a problem. Just run tiles to within ⅛ in. of the flange. When you apply a new wax ring to the bottom of the toilet horn, the wax will compress and seal the joint

PRO TIP

Weather-resistant barriers and flashing designed for exteriors often double as waterproof membranes indoors, beneath tile. After all, a shower is just indoor rain. Although sold primarily as stucco paper, Fortifiber's Super Jumbo Tex 60 Minute will also keep shower substrates dry even if you take hour-long showers. Polyken Foilastic, a peel-and-stick flashing, does a good job of sealing shampoo niches, pipe cutouts, and troublesome joints around kitchen counters (see the top drawing on p. 507).

all. *Note:* A wall that's plumb in the corner may have a twisted stud elsewhere that throws another section out of plumb.

Finally, survey surfaces for water damage, deflection, and other factors that could affect tile. Examine the bases of bathroom and kitchen fixtures for discoloration, delamination, and springiness, especially under toilets and tubs. Crumbling grout atop a tub often means that water has gotten behind the tile. Open kitchen and bath cabinets and examine the undersides of sinks and countertops. If you see discoloration, probe it with an awl to determine whether materials are solid. Particleboard countertops often deteriorate from sink leaks and dishwasher steam. If there's extensive rot or subfloor delamination, replace failed sections, as described in chapter 8. To test for deflection, thump walls with your fist, or jump on the floors. If you see or feel movement, there may be structural deterioration or, more likely, the substructure may be undersize for the span.

PREPPING THE ROOM

Tiling will go faster and look better if you first remove fixtures and other obstructions so that you can lay a continuous field of tile. This is a good time to relocate or enlarge electrical boxes, install thresholds, or cut a little off the door bottoms so they don't scrape when tiling raises the floor level. For information on disconnecting and installing plumbing fixtures, see chapter 12.

The Thick and Thin of Setting Beds

Thin-bed installations such as latex thin-set adhesive over backer board are, well, thin. Because they offer little depth for adjustment, framing must be exact:

▶ Subfloors and countertops must be level and flat to within ⅛ in. in 10 ft. That is, no high or low points greater than ⅛ in. of level.

▶ Walls must be plumb and flat to within ⅛ in. in 8 ft.

As you might expect, standards are more tolerant for mortar-bed installations, which are thick enough to accommodate less-than-perfect framing. Mortar-bed tolerances are roughly double the thin-bed specs given here. A good place to check on these and other questions is www.tileusa.com, the website of the Tile Council of North America. Each year the site updates its handbook of tile-installation standards, spelling out acceptable materials and structural requirements for each type of setting bed.

adequately. In fact, you can buy extrathick wax rings for such situations.

But if the tiled floor will be more than ½ in. higher than the closet flange (which may result if you install a mud bed or backer-board setting bed), replace the flange and set the new one higher. If waste pipes are plastic, cut off the existing bend-and-flange section and cement on new components to give you the flange height you need. This is easier said than done, however: If there's no room to maneuver new pipes, you may need to cut into flooring or framing. Many plumbers prefer to build up existing flanges by stacking ½-in. plastic *flange extenders* (the same diameter as the flange), caulking each with silicone, and using long closet bolts to resecure the toilet base. But check your local plumbing code to see if this method is allowed.

If drainpipes are cast iron, you may want to hire a plumber to replace flanges that are too low or waste pipes that have deteriorated. There may not be enough room to cut out pipe sections, or, in the case of deteriorated pipes, there may also be rotted framing or subflooring to attend to before replacing pipes and fixtures.

Removing a sink may be advisable before retiling a countertop. The method depends on the sink type: whether countertop, pedestal, or wall mounted. For each, shut off the water, then disconnect supply lines and drainpipes.

Countertop sinks vary in their attachment. If you are retiling a countertop, you must remove a self-rimming sink because its lip rests atop the tile. However, you may want to remove other sink types to better match the color of tiles, upgrade a tired old sink, and so on. Most sinks are held in place with clips on the underside of the counter and sealed with a bed of caulking or plumber's putty between the sink lip and the counter. After disconnecting the pipes, unscrew the clips and, if necessary, break the caulking seal by running a putty knife between the sink lip and the counter. If the new sink is smaller than the old one, you'll need to reframe the opening in the counter.

Remove in-counter faucet assemblies, then tile within ¼ in. of the holes, and caulk the spaces with silicone or plumber's putty. If your installation will involve just thinset and tile, the old valve stems should be long enough to reuse. But if you're building up the setting bed with backer board or mud, buy new faucet assemblies with longer valve stems.

Shower and tub hardware can be masked off with plastic bags if you're not tearing out the shower walls or building up setting beds, but do remove chromework so it doesn't get discolored by mortar or adhesive.

Tile Height at Toilet

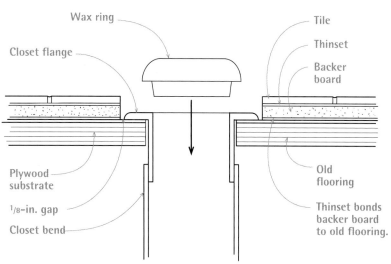

Ideally, the top of the toilet flange should be the same height as the finished floor. If not, consider the options outlined in the text.

To remove a showerhead assembly, gently pry the escutcheon from the wall (it may be seated in plumber's putty). Then wrap a rag around the chrome gooseneck pipe and use a pipe wrench to unscrew it. (The rag prevents the wrench from gouging the chrome finish.) Removing valve handles is slightly more complex because you must first unscrew valve handles from valve stems, and those screws are frequently hidden behind decorative caps. Once you've removed handles and escutcheons, wrap the exposed valve stems with plastic so their threads don't get fouled with mortar.

The last item on the shower wall, the tub spout, can often be unscrewed by hand. If not, you can usually gain some leverage by inserting a rubberized pliers' handle into the spout opening.

Tile to within ¼ in. of the valve stems and pipe stubs, and caulk the gaps with silicone so water can't get behind the wall. Escutcheons will cover the cut tiles.

Build up electrical boxes so they're flush with new tiled surfaces. After turning off electricity to the box—and using a voltage tester to make sure it's off—remove the outlet faceplate, unscrew the device from the box, and screw in a *box extender*. Run tiles to within 1/16 in. of the extender; the faceplate will cover tile cuts. *Note:* All bathroom receptacles and all those within 4 ft. of a kitchen sink must be GFCIs.

Move appliances so the floor they're sitting on can be tiled. Where those appliances are undercounter, anticipate the additional height of the new flooring and raise or alter countertops

PRO TIP

If the thicker new walls reduce the visible profile of existing valve stems, don't panic. Most major plumbing suppliers offer threaded valve extensions to make the stems longer. That's much cheaper than tearing out the walls to replace the valves.

Extending Electrical Boxes

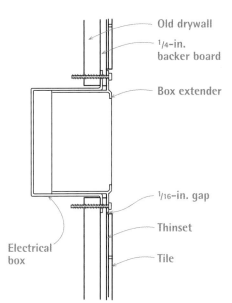

- Old drywall
- 1/4-in. backer board
- Box extender
- 1/16-in. gap
- Thinset
- Tile
- Electrical box

A box extender is usually a plastic sleeve that screws to an existing electrical outlet box so that the box face is flush to a new tiled surface.

PRO TIP

Don't use drywall screws to attach backer board. They're not strong, screw heads can shear off, and drywall screws invariably corrode in damp and wet applications. Use only corrosion-resistant screws or nails. Don't use drywall joint tape either; it's not up to the task.

A handheld grinder with a diamond blade makes a faster, more reliable cut in backer board than a utility knife, but it also produces more dust.

accordingly so appliances can be returned to their nooks.

Cut door bottoms so there's about 1/4-in. clearance between the bottom and the highest point of the tiled floor or the threshold. Do this after the tile and threshold are set because it's difficult to know beforehand exactly how thick the floor will be.

Choose a threshold that reconciles floor heights and materials on either side. For this, you'll need to think through its installation, such as scribing and cutting it to the door jambs and the adhesives or fasteners.

Installing Setting Beds

This section addresses mainly the most common setting beds and mentions only briefly those that are less common or problematic. Backer-board brands vary, so follow manufacturer-specific recommendations about waterproofing, connectors, installation procedures, and so on.

COMMON SETTING BEDS

Here you'll find additional details on backer board, mortar beds, SLCs, drywall, and concrete slabs. Setting tile directly on plywood is not recommended, but it's widely done, so that's addressed, too.

Installing backer board. Backer boards are cementitious backer units. They are strong, durable, and unaffected by moisture—so they are superb setting beds for wet and dry installations. However, because moisture will wick through CBUs, install a waterproofing membrane first in wet applications to protect wood substructures from damage.

Wear a respirator and eye protection when cutting and drilling backer-board panels, which can be scored and snapped much like drywall

To install a backer-board setting bed over an existing substrate, drive 2-in. galvanized roofing nails into the framing. Alternately, use 2-in. corrosion-resistant screws.

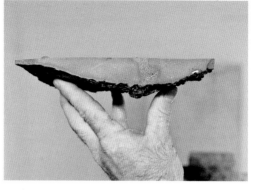

Note how well self-leveling compound levels itself when accidentally dumped onto the ground of a work site. Even its thin, tapered edge is strong. When used to level floors, its optimal thickness is about 1 in.

(many installers score both sides). Although a utility knife can do the job, a dry-cutting diamond blade in a handheld grinder leaves other methods in the dust—literally. Wear a respirator when using this grinder, as well as hearing and eye protection. To drill pipe holes, use a carbide-tipped hole saw.

For most backer-board installations, space galvanized roofing nails or corrosion-resistant screws every 6 in. to 8 in. Screws are more expensive and slower to install, but some tilesetters swear by them; Rock-On cementboard screws cut their own countersink so the heads will be flush. Nail advocates argue that nails are less likely to crush panel edges and are easy to drive flush. To attach 1/2-in. backer-board panels directly to studs, use 11/4-in. screws or nails. If installing panels over drywall or plywood substrates, use 2-in. screws or nails.

Backer-board panels are available in a variety of widths (32 in. to 48 in.), lengths (3 ft. to 10 ft.), and thicknesses (1/4 in., 5/16 in., 7/16 in., 1/2 in., and 5/8 in.). Thinner panels are typically installed over plywood or drywall. Use at least 1/2-in. backer board if you're attaching it to bare studs; otherwise, it will flex too much and crack the tile joints. For a floor rigid enough to tile, install 1/2-in. backer board over 3/4-in. tongue-and-groove plywood, with joists spaced 16 in. on center. For all installations, leave a 1/8-in. gap between the backer-board panels. Cover those joints with 2-in.-wide, self-adhering fiberglass mesh tape before covering the tape with thinset adhesive—the same material used to set the tiles.

Feather out the thinset as flat as possible, but it doesn't have to be perfect because the joints will be covered by adhesive and tile. Finally, leave a 1/4-in. expansion gap where the panels abut the base of walls, tubs, and plumbing fixtures; you'll fill those gaps later with flexible sealant. Keep the bottom edge of backer board 1/4 in. above the tub so water doesn't wick into panels; caulk the gap later with silicone.

Installing the mortar bed. Mortar beds make a superb substrate but are complicated to install. First, attach a *curing membrane* (a waterproofing membrane beneath the mortar) over the framing or drywall, then add *reinforcing wire mesh*. Next, apply two or more parallel mortar columns, and place a wooden *float strip* atop each column. Checking frequently with a spirit level, tap the float strips into the mortar until the floor strips are level or the wall strips are plumb. Then fill between the strips: Dump mortar onto floors between the strips or trowel it onto walls. Flatten the mortar by placing a *screed board* across the float strips and drawing it side to side in a sawing motion. Dump excess mortar into a bucket as the screed board accumulates it.

Once the mortar bed is more or less flat, remove the strips, and fill the float-strip voids with mortar. Then trowel out the irregularities. To help the thinset coat adhere, lightly roughen it by rubbing the surface with a wood float or a sponge float. Allow the mortar to set about an hour before using a margin trowel to clean up the mortar bed's edges. Some veteran tilesetters set tile immediately thereafter, but most mortals should allow the mortar to cure for 24 hours before tiling.

Mixing mortar in correct proportions is an art. Floor mud, or *deck mud*, is dry and rather crumbly: 1 part portland cement, 5 parts sand, and 1 part water. However, once screeded, compacted, and well cured, deck mud can support great loads. Wall mud is wetter and more like plaster because it must be spread onto vertical surfaces; it contains lime to improve its adhesion. Wall mud's proportions are 1 part portland cement, 4 parts sand, 3/4 part lime, and 1+ parts water; use enough water so the mud trowels on easily. Add water slowly because mud won't stick if it's too wet.

Applying leveling compound. SLCs can level isolated low spots or even whole floors. Application requires few skills beyond opening 50-lb. sacks of SLC powder, mixing the powder

Create a level mortar bed by drawing a metal screed board across two float strips pressed into mortar columns. After the mortar has been screeded, the wood float strips are removed and their voids filled with mortar.

with water, and pouring the mix onto a floor. You don't even need to spread it around much. It flows like water, levels itself, and starts to harden in about 15 minutes. Well, that's a bit oversimplified but not by much.

First, be sure the substructure is sturdy enough to bear the weight. Specs for one popular SLC, LevelQuik, recommend at least a ¾-in. exterior-plywood subfloor over joists spaced up to 24 in. on center; use two ½-in.-thick pours to achieve a 1-in. optimal thickness. Wait 24 hours between pours. Whatever the substrate, it should be clean, dry, and free of chemicals—such as

curing compounds in concrete slabs—that might prevent a good bond.

Before pouring, install a waterproof membrane and reinforcing mesh, which is usually wire, although a self-furring plastic lath called Mapelath shows promise. One essential prep detail: *Completely* seal and dam off the section of floor you're leveling, or the free-flowing SLC mix will disappear down the smallest hole and form a heavy mortar pad where you least want one. Pay close attention to board joints, baseboards, and the like; caulk or seal joints with duct tape, pack them with fiberglass insulation—whatever it takes to contain the liquid until it hardens. SLCs are expensive but, in most cases, less expensive than floating a mortar bed. As important, they're great setting beds.

Preparing masonry surfaces. Concrete walls, slabs, and block are good setting beds as long as they've cured for at least a month and as long as they're clean (no chemical residues), dry, free from active cracks, and level or plumb within ⅛ in. in 10 ft. (If it's an out-of-level floor, see "Applying Leveling Compound" on p. 497.)

If you're tiling basement surfaces, the big issue is cracks. If masonry cracks expand and contract seasonally, it's unwise to tile over them because the tiles will crack. Likewise, if one side of a crack is higher than another, it's probably caused by soil movement. (If the crack is inactive and both sides are level, you can vacuum out the crack, dampen it, and fill it with a latex thinset adhesive before applying the thinset setting bed.)

Dry installations. Unpainted drywall is an acceptable setting bed for dry installations. (In damp or wet installations, never adhere tile directly to drywall.)

In dry installations, use at least ⅝-in. drywall if attaching it directly to studs 16 in. on center. Or sandwich two layers of drywall, with a layer of adhesive between, to create a more rigid lamination. But if you install a double layer of drywall, offset the panel edges by at least 16 in.

Stronger INSTALLATION, LESS WASTE

Construction adhesive is often used to join backer board to other substrate layers, but thinset adhesive (applied with a ¼-in. square-notched trowel) creates a far more rigid lamination. When installing backer board, cut pieces lengthwise if possible. For example, if you have 2-ft.-wide countertops, cut a single piece of backer board 2 ft. by 8 ft. rather than into two 2-ft. by 4-ft. pieces. The panel will be slightly stronger, there will be no waste, and you'll have one fewer seam to tape and top.

To avoid cutting handmade tiles on the tub sidewall, the installer laid out tiles on the floor, using plastic spacers to simulate grout joints. She then adjusted the thickness of mortar beds on the tub walls so that the tile assembly fit exactly. (Vacuum floors well before laying tile on them because dust can compromise a setting bond.)

In either case, leave a ⅛-in. gap between panel edges, cover the joints with self-sticking fiberglass mesh tape, and then apply a layer of latex thinset adhesive over the tape. Drywall tools are fine for this application—but *not* drywall joint tape or compound. When installing drywall as a setting bed, you don't need to fill screw holes or feather out joints perfectly smooth because you'll be covering them with thinset and tile.

Plywood beds. Plywood is not recommended as a setting bed, but if you must use it, use only exterior grade and leave ⅛-in. gaps between the panel edges. Plywood substrates for floors and countertops must be at least 1⅛ in. thick—best achieved by laminating a ½-in.-thick plywood underlayment panel to a ⅝-in. plywood subfloor. To prevent squeaking and to stiffen the assembly, apply construction adhesive between the panels. Offset the panels so their edges don't align. In addition to the adhesive, use 1-in. corrosion-resistant nails or screws spaced every 6 in. To secure this laminated plywood to the joists, drive 16d galvanized nails into the joist centers. To avoid high spots that might crack the tiles, sink all screw or nail heads below the surface of the top layer. Sand and vacuum the plywood before notch-troweling on an epoxy thinset adhesive.

ODD OR PROBLEMATIC SETTING BEDS

The beds described next may require special techniques and materials.

Plastic laminate countertops are acceptable setting beds if they're solidly attached. Scuff the surface with 80-grit sandpaper, wipe with a rag dampened with solvent to remove grit and grease, and fill any voids. Then use an epoxy-based thinset to bond the tiles. Alternatively, you can cover the old laminate with ¼-in. backer board, adhering it with an application of epoxy thinset and 1-in. corrosion-resistant screws spaced every 6 in. around the perimeter of the countertop and every 8 in. in the field. Sink screw heads flush, vacuum the backer board, and use latex or epoxy thinset adhesive to bond tiles. *Note:* The recommended 1-in. screw assumes the combined thickness of the countertop materials is at least 1⅛ in.

Tiling over existing tile is a reasonable alternative to ripping it out, as long as the old tile isn't cracked and is well adhered and the substrate is solid. Scuff the tile with a carbide-grit sandpaper. Vacuum the surface well, and wipe with a damp rag. Because tile surfaces are not perfectly regular and grout joints are recessed, first use a flat trowel to spread a layer of epoxy thinset to build up grout joints and level the surface. Wait a day. Then use a notched trowel to apply a setting bed of epoxy thinset.

Two caveats: Because of the risk of leaks, don't tile over tiled shower-stall floors. Rather, tear out the old floor, replace the shower-pan membrane, and tile atop a newly floated mortar bed. Second, don't install tiles 2 in. sq. or smaller over existing tile because they will telegraph the old surface's irregularities. Instead, use large tiles.

Resilient flooring is acceptable if there's a single, uncushioned layer that's well adhered to a stable subfloor. Cushioned or multilayered flooring will flex too much to be a stable base for tile, so, to be sure, use a utility knife with a hooked blade to cut out a cross section of flooring in an inconspicuous spot.

Painted walls are OK as long as the paint is well attached and the wall doesn't flex. Drill a small exploratory hole to determine the composition and thickness of the wall. If it's drywall less than ⅝ in. thick, install a layer of ¼-in. or ⅜-in. drywall over it. If the wall is traditional plaster (hard to drill through), it's probably fine. Prep painted walls by sanding them with 100-grit

sandpaper, and wipe with a damp rag. Use a latex thinset adhesive.

Other situations

▶ Papered walls? Strip 'em! Vinyl wall coverings are supposedly tenacious enough to support tile, but it's risky.

▶ Veneer paneling? Not recommended. Typically ¼ in. to ⅜ in. thick, it will flex, cracking grout joints and eventually dislodging tiles.

▶ Wainscoting or lumber flooring? Not recommended. Solid wood expands and contracts, and all those board seams would be a nightmare to fill.

Tile Estimation and Layout

At this point, we'll assume that the substructure is sturdy and stable and the setting bed is in place. Careful layout is the key to a good-looking job, so don't begrudge the time it takes. The right layout will align tile joints correctly, create a pleasant symmetry, allow you to cut tiles to size beforehand, and—most important—enable you to set tile accurately and quickly while the clock is ticking for that fast-drying adhesive.

ESTIMATING TILE

If you're installing a popular tile that a local supplier has in stock, wait until you've installed the setting bed before estimating tiles. If you order too many, most local suppliers will take back extras, as long as they aren't damaged or returned too long after purchase. Ordering tiles is not complicated unless surfaces to be tiled have a lot of jogs, recesses, odd angles, and obstacles. Using a tape measure and a pad of graph paper, calculate the square footage of the surface to be tiled and add 8% to 10% for waste, damage, and future repairs.

Handmade or exotic tiles are another story. Because they're expensive and must be ordered well in advance, suppliers rarely accept returns. To save money, try to draw layouts that are accurate to an inch so you can count individual tiles. But you should still order extra tiles—say, 5% above your tile count—which is preferable to waiting two months for the next tile shipment from Italy. Besides, tile colors can vary greatly between batches. Your detailed drawings will also help you accurately frame out the area to be tiled.

Finally, when ordering tile, calculate the number of trim pieces separately from your calculations for field tiles. For each distinct piece of tile trim (such as surface cove, V-cap trim), add 15% to the lineal feet of trim indicated by the layout.

PRO TIP

Once you've established a room's two main (perpendicular) control lines, snap as many secondary layout lines as you like. Many tilesetters also snap lines around the perimeter of the room to indicate where cut tiles begin; they set all the full tiles within the lines first, then set cut tiles at the base of walls, cabinets, and fixtures.

▮▮▮▮

Storytelling

A story pole (also called a jury stick) is like an oversize yardstick but is divided into units that represent the average width of one tile plus one grout joint. Story poles give you a quick read on the number of full-size tiles you'll need to get from point A to point B. For these homemade measuring devices, any straight board will do.

Order at least *two* specialty trim pieces for any one-of-a-kind piece (such as *radius-bullnose down angle*). This is also the time to order compatible adhesives and color-matched grout and caulk.

TILE LAYOUT

Most tiles are square or rectangular. So the most common floor-layout dilemma is in imposing a grid that's basically square onto a room that isn't. Laying out walls and countertops is much the same, except that wall layouts are more affected by plumb. Wherever they occur, though, layout lines have the same purpose: They keep tile joints straight. When setting tile, it's easy to obsess about individual tiles and spacers, getting lost in close-up details. Layout lines help you keep the big picture in view and keep tile joints from straying.

FOUR TIME-TESTED TIPS OF TILING

Although the following rules make sense most of the time, bend them when you must.

Use full tiles at focal centers. A focal center is any area that the eye is drawn to: the front edge of a counter; a room's entryway; or a center of activity such as a sink, large window, or hearth. Common sense says full tiles look better than cut tiles, so put full tiles in conspicuous areas. Conversely, put cut tiles where they'll be least noticed.

Cut as few tiles as possible. Cut tiles are extra work, and they don't look as good as whole tiles. To avoid cutting tiles, you might be able to shift the layout a little to the right or left or *slightly* vary the width of tile grout joints.

Make layouts as symmetrical as possible. This rule is both an extension and, occasionally, a contradiction of the two above. Imagine a kitchen-sink counter: The sink is certainly a focal center but may be smack-dab in the middle of a tile field. There's often no way to avoid cutting tiles around the perimeter of a sink. In that case, try to shift the layout (or the sink) so that you can cut tiles an even amount on both sides of the sink (see the bottom drawing on p. 507). The result

Tiling a Floor

Tile layouts impose a grid that's basically square in a room that often isn't. Start by recording the room's dimensions, use a framing square to see which corners are square, and note any obstacles to be tiled around.

FIRST CONTROL LINE

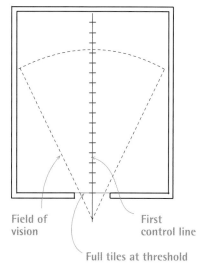

Field of vision

First control line

Full tiles at threshold

SECOND CONTROL LINE

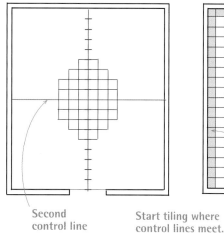

Second control line

Start tiling where control lines meet.

PARTIAL TILES

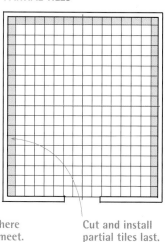

Cut and install partial tiles last.

If the doorway is the focal center of the room, use a story pole to find the tile-joint mark closest to the middle of the doorway. Ideally, the first control line will run through that mark, and there will be full tiles along the threshold.

The second control line is perpendicular to the first, intersecting it roughly midway along its length. Although you can start setting tiles anywhere, it's best to start where control lines meet in the highly visible middle of the room.

Place angle-cut and partial tiles away from the room's focal center. Because cutting tiles takes longer than installing full tiles, most installers cut tiles after the field of full tiles is in place.

will look much better than almost-full tiles on one side and narrow tiles on the other.

This also is a good rule for small counters, which you can see from side to side without turning your head. If you must cut tiles, split the difference at each end.

Don't use tile pieces half size or smaller. They'll look terrible. It's better to adjust the width of tile joints. Or shift the layout so that you have large cut tiles on both ends rather than a row of narrow ones on one end alone.

Tiling a Floor

Using a framing square and a tape measure, check to be sure the room's corners are square and parallel. Make a quick sketch of the room, showing which way corners diverge. Or sketch this directly onto the setting bed.

LINING UP THE TILE GRID

Floor layout begins by identifying the focal point of the room and snapping two chalklines (*control lines*) onto the setting bed perpendicular to each other. In the photos on these two pages, the doorway is the focal point of the room. Thus, the first control line will run through the middle of that doorway and continue across the room to the opposite wall at roughly a right angle.

Full tiles look best in a doorway, so the installers butted uncut tiles against the threshold and continued the tile course until it met an adjacent wall. Then they selected a tile that was roughly in the middle of the doorway, and along one side of the tile, they stretched a taut chalkline that ran across the room to the opposite wall. As two people held the string taut, a third person measured from the string to the adjacent wall to make sure that the string was parallel to the wall.

FLOOR LAYOUT

1. This installer is lightly snapping a chalked control line that will run roughly through the midpoint of the doorway to the room's opposite wall. In the background are dark full tiles he initially positioned outward from the wall to mark the eventual tile joint near the doorway's midpoint.

Applying Thinset Adhesive

Once you've snapped layout lines and vacuumed the setting bed one last time, use the straight edge of a notched trowel to spread thinset adhesive. Then, using the notched edge of the trowel, comb ridges into the adhesive. As you apply and comb adhesive, try to stop just short of the layout lines so you don't obscure them.

In theory, you can start setting tiles anywhere, but it's usually best to start where the control lines meet and work out toward the walls. That way, you know that tile joints in the middle of the room—which are the most visible—will line up. Set several tiles, then pull up one and examine the back, which should be uniformly covered with adhesive. If it isn't, you may have applied too thin a layer of thinset, used a notched trowel with teeth that are too small, or mixed the thinset too thin. If you are using large, handmade tiles such as Mexican pavers, they are often irregular, so you may need to butter additional adhesive onto their back surfaces before setting. For buttering, use the straight edge of a trowel.

Conversely, if adhesive oozes up between tiles, the notched trowel's teeth are probably too large; try one with smaller teeth. Use a margin trowel to remove excess thinset between the tiles before it hardens. Left in place, the thinset between tiles would prevent grout from filling the joints and bonding properly.

2.

Be Consistent when Placing Tiles

When placing tiles in a grid, be consistent. If, for example, you place the *top left* sides of the first tile directly on the intersecting chalklines of your grid, then your unit measurements must include space for the width of grout joints in the *bottom right* sides of that tile. Alternatively, some tilers prefer to set back the top left sides of the first tile one-half the thickness of the grout joint. It doesn't matter which placement you choose, just that you be consistent throughout the grid as you measure out from already set tiles to position new ones.

The First Tile Becomes a Template for the Grid

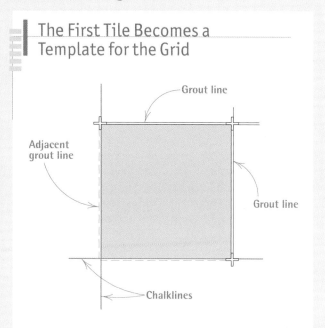

Grout line

Adjacent grout line

Grout line

Chalklines

4. As tile fills each quadrant, have a straightedge nearby to see if tile edges line up, especially if thinset has obscured the chalklines.

3. When corners aren't square, start in the middle of the room—where control lines intersect—and work toward the walls. As you near out-of-square walls, cut tiles to fit. Avoid sliding tiles into place. Instead, align one edge to a control line or grout joint, "hinge" the tile down, and press it into the thinset.

5. After setting the field of full tiles, attend to those that need cutting or special fitting. The installer used a wet saw to cut the two longer lines, then a nipper to finish off the rounded inner corner.

Once it was established that the chalkline was parallel to the adjacent wall, they snapped a chalkline onto the floor. Because it ran through the middle of the doorway, that chalkline became the *primary control line*. Holding a story pole next to the chalkline, the crew then found a tiling-unit point that was roughly midway along the chalkline. Placing a framing square on this point, the crew next snapped a second chalkline perpendicular to the first: It became the *secondary control line*. In this manner, they divided the room into quadrants roughly equal in size. By measuring from these two control lines, using the story pole, and sighting by eye, the crew could make sure that successive courses lined up.

The rest of such a layout is mostly fine-tuning, with an eye to the four time-tested tips of tiling, as outlined earlier: using full tiles at focal points, cutting as few tiles as possible, making layouts as symmetrical as possible, and avoiding tiles less than half size. For this, a story pole is essential.

SETTING THE TILES

Set full tiles before partial tiles. To get a rhythm going, most tilesetters first set all the full tiles, then attend to partial tiles, which take time to cut and set. However, you may want to make complex cuts beforehand, for example, where tiles meet obstacles in the middle of the field, such as a toilet closet flange. By cutting these tiles first, you can set them quickly and install other full tiles around them. Give the adhesive a day to harden, and you're ready to grout the surface.

Note: This description is much condensed. If you're a perfectionist, you'll fuss with the spacing between tiles or between tile sheets and continue making little adjustments until the job is done. Be sure to stand back from time to time for an overview of the layout.

OFFSETTING TILE JOINTS TO ADD VISUAL INTEREST

The layout method just described—using two primary control lines set at right angles—is useful regardless of the tile shape you're working with. When your tiles are rectangular rather than square, however, you may want to create a more interesting pattern by offsetting the end joints, as shown in the photo at left below. (To compare other examples of offset joints, see the brick patterns on p. 239. The *common bond* pattern, with offset end joints, is far more interesting than the *stretcher bond*, in which end joints all line up.)

Offsetting tile end joints requires a slightly greater attention to detail, however, because if you offset every other row of tiles, one of the control lines will be covered half of the time. This is easier to see than to explain. The tiler in the

Offsetting end joints so that they align every other row creates a more interesting pattern—especially when installing rectangular tiles.

Installing an offset tile array is a bit trickier because every other row of tiles will cover one of the control lines. Here, the first tile in the first row sits within the intersecting chalklines, but the first tile in the second row is offset by 8 in., covering the chalkline.

photo sequence is working with 8-in. by 16-in. slate tiles and offsetting each end joint by 8 in. Therefore, end joints line up every other row. As you can see in the bottom right photo on the facing page, the first tile in his first row fits nicely into the intersection of the control lines; the first tile in the *second* row overshoots the control line by 8 in.; the first tile in the third row lines up, and so on.

Not to overthink things, but when one of your control lines is covered half the time, you will probably rely more on the control line that isn't covered up (because it runs parallel to tiles' long sides). Our tiler seems to be doing just that. After setting his first row of tiles all the way to a wall, he uses a straightedge to check tile alignments and then weights down the straightedge so it won't move as he adds and adjusts subsequent rows.

Installing each row is an ongoing process of adjusting and aligning—using a tiler's straightedge, a measuring tape, a framing square—but that's true of any tiling job. And, of course, while you're constantly checking tile positions against the two control lines, you must also keep an eye on the third dimension—how level the tiles are, as illustrated in the photo at right below. This job was particularly exacting because tile thicknesses varied somewhat. So in addition to the layer of thinset he applied to the floor with a notched trowel, he also "buttered" the back of each tile, using the trowel's straight edge.

Countertops

The front edge of a counter is almost always the focal point. The primary control line runs parallel to the front of the counter, and all other layout lines are secondary to it. If the counter will be subject to moisture, install a membrane before installing a setting bed.

STRAIGHT COUNTER, NO SINK

The simplest surface to lay out is a straight counter with no sink because it has only one control line. Begin by using a framing square and a story pole to survey the countertop.

"Counter Layout" on p. 508 assumes the counter edges are finished with V-cap trim, a common choice, and tile joints are 1/8 in. wide. Place several V-caps along the counter edge, then measure back from the edge 1/16 in. from each cap to mark the middle of the first grout joint. Snap a chalkline through these marks to establish a control line. Because the front edge is the counter's focal point, you'll place full field tiles next to the row of V-caps.

Using your story pole, measure the length of the counter to see if you must cut tiles. If one end of the counter abuts a wall and the other is open, plan a row of full tiles along the open end, thus consigning cut tiles to the wall end where they'll

PRO TIP

Thinset adhesive should be moist enough to stick to the tile but not so wet that it slides off a trowel. If thinset skins over while you're setting tile, recomb it with a notched trowel. But if it gets stiff in the pan or it doesn't stick readily to the tiles, discard it and mix a fresh batch. Likewise, if you move a tile after the adhesive has started to set, scrape the thinset off the back of the tile and the setting bed, and apply fresh mortar to both surfaces.

The tiler sets the first row of tiles all the way to the wall, places a straightedge along the edge of the row (at left), and weighs down the straightedge so the first row won't move as he adds successive rows.

Periodically, use a framing square to check to make sure end joints line up. The white spacers between tiles ensure the correct spacing for the grout, which will be applied when the tiling is complete.

When tile thicknesses are slightly irregular, butter the backs of tiles with thinset adhesive in addition to troweling adhesive on the floor. That is, you'll butter the backs of all tiles, but apply varying thickness of thinset.

In addition to checking the alignment of tile joints to the two primary layout chalklines, use a straightedge periodically to make sure tile faces are level.

FROM THE ARCHIVES

Because the front edge of a counter is the most visible, start layout and installation there. Place V-cap trim along the front edge to position successive courses of field tile. As you set each course of tile, use a straightedge to align it.

be less conspicuous. If both ends of the counter are open, and you see that you'll need to cut tiles, move the story pole so that cut tiles will be the same dimension on both ends. That decided, mark positions for the tile units along the control line.

Last, measure to the back of the counter to determine whether the final row of tiles will need cutting. You can precut tiles, but on a counter so simple, you can just measure and cut partial tiles individually after all the full tiles are set, likely giving you more accurate measurements anyway.

STRAIGHT COUNTER, WITH SINK

To tile a straight counter with a sink, the layout is much the same as a counter without a sink, except that here, your main concern is making symmetrical tile cuts (if necessary) on either side of the sink. If you need to cut tiles, move the story pole side to side until the tile joints are equidistant on each side of the sink's rough opening. Then transfer those two marks to the control line. Finally, use a framing square to run lines through those marks, perpendicular to the control line, to the back of the counter.

L-SHAPED COUNTER

On an L-shaped counter, you have in effect two counters at right angles to each other, so you will need two control lines, perpendicular to each other, running along the front edge of each section. Any other layout considerations are subordinate to these two control lines because they determine how the two oncoming tile fields will align.

Use your framing square and a straightedge to establish control lines and to keep the tiles aligned once you've turned the corner. After setting V-cap trim tiles, start tiling where the two control lines intersect. As with straight counters,

MOSAIC TILE

After you've set paper-backed sheets of mosaic tile in adhesive, the paper will start to soften, allowing you to reposition the tiles slightly. To move a row of tiles, place the straight edge of a trowel against them and tap the trowel lightly with a hammer handle.

Use a grout float to seat mosaic tile in the thinset adhesive. Choose knee pads that are comfortable enough to wear for extended periods of time.

Waterproofing Counter Edges

Installers often mistakenly overlook countertop edges when installing the waterproof membrane. A membrane is especially important if you're floating a mortar bed because the moisture from curing mortar is enough to swell unprotected plywood edges and, in time, cause V-cap trim to fall off. At the very least, extend the membrane and wrap it down over the edge of the plywood substrate. However, because building paper folds unevenly and can create a welt that won't lie flat, a better solution is to cover counter edges with a self-sticking flashing such as Polyken Foilastic. It will lie flat, it's impervious to water, and most thinsets will stick to it—but check your thinset's specs to be sure.

The back edge of a counter, where it abuts a wall or backsplash, also is susceptible to water damage if not detailed correctly. Run the waterproof membrane all the way to the wall, then flash the countertop–wall joint using self-sticking flashing. Fold the flashing lengthwise (into an L), adhering one leg of the L to the membrane and running the other leg up the wall at least 1 in. above the finish-tile level.

Countertop Front and Back Edges

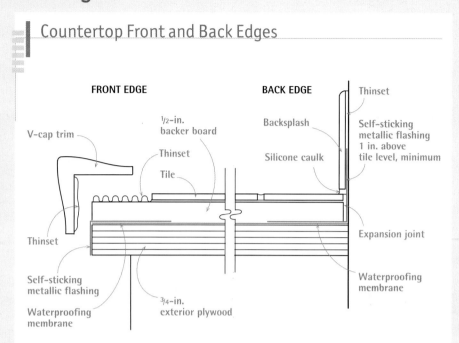

Flash the front edge of a counter, as shown, to prevent the exposed edges of the plywood from wicking moisture from the thinset adhesives and then swelling. Because the back edges of the countertops are also vulnerable to water damage, caulk and flash them, too.

Tiles at the Sink

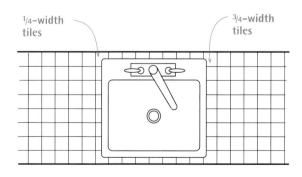

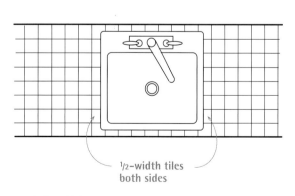

Symmetrical layouts look better. When a layout results in unequal tile widths along the sides of a kitchen sink—a very noticeable spot—either shift the layout or move the sink to create equal tile widths on both sides.

Counter Layout

STRAIGHT COUNTER

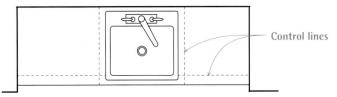

— Control line

A straight counter needs only one layout control line to indicate the first tile joint back from the edge.

COUNTER WITH SINK OR COOKTOP

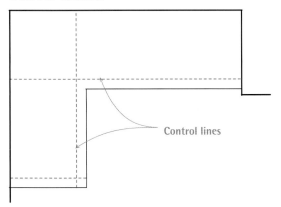

Control lines

If a sink or a cooktop in the counter interrupts the layout and requires tile cutting, mark secondary control lines on either side to indicate where full tiles resume.

L-SHAPED COUNTER

Control lines

An L-shaped counter will have two major control lines, running perpendicular to each other. Add lines as needed to indicate sink placement, open counter edges, and so on.

put full tiles along the front of the counters and work back, relegating cut tiles to the very back to be covered by the backsplash. If you use the same tile for the backsplash, continue the tile joints up the wall so that the backsplash and counter joints line up.

Tub Surround

Never assume tub walls are plumb. Always check them with a 4-ft. level. If walls aren't plumb within ⅛ in. in 8 ft., correct them with a mortar bed or reframe them. Otherwise, tile joints from adjacent walls won't align. Moreover, never

Laying Out a Tub Back Wall

Almost all tubs slope slightly, so use a spirit level to locate the lowest point. From that lowest point, measure up *one tiling unit* (one tile height plus one grout joint) plus ¼ in. and mark the wall. (The ¼ in. allows a gap between the bottom of the tile and the tub; the gap is caulked.) Draw a level control line through that mark, as shown in the illustration below, and extend that level line to all three tub walls. Use a story pole to see if you'll need to cut tiles. If so, lay out tiles so cuts are symmetrical on both ends of the back wall. Draw a plumb line on each end of the wall to indicate where the cut tiles will begin. Finally, through a tile joint along the level control line, draw a plumb control line that roughly bisects the backwall. Start tiling where control lines meet.

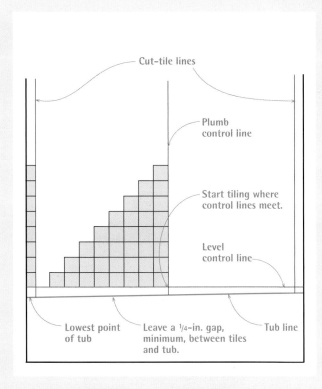

Cut-tile lines

Plumb
control line

Start tiling where
control lines meet.

Level
control line

Lowest point
of tub

Leave a ¼-in. gap,
minimum, between tiles
and tub.

Tub line

TILING A TUB SURROUND

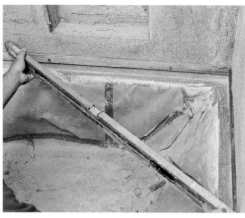

1. After checking tub walls for plumb, use a spirit level to check whether the tub is level along all three sides. If the tub slopes, note the lowest point.

2. At the lowest point of the tub shoulder, measure up one tiling unit (tile width plus one grout-joint width) and mark that onto the wall. Through that mark, draw a level line that extends to all three walls of the surround. Next, nail narrow wooden strips to the underside of that line, as shown.

3. The first course of full, uncut tiles should rest on the wooden strips. After you've installed tiles on all three walls and the setting bed has hardened, remove the wood strips and install cut tiles below.

4. Typically, tilesetters mark a vertical line to bisect the sidewall of a tub. As you trowel on thinset, try not to obscure the line with adhesive. Although this pro is setting a whole wall without interruption, most mortals should set tile a half or quarter wall at a time.

5. Periodically check to see if tile courses are level, inserting plastic shims as needed. Leave them in place until the thinset cures. *Note:* Because walls aren't perfectly regular, you'll often need to use spacers as well as shims. Spacers are uniformly thick; shims are tapered.

6. Use nippers for the curved cuts around pipes and the plastic protective covers over shower valves. Wear goggles when nipping.

7. If it's necessary to cut tiles, place them symmetrically on both ends of the sidewall. Although it's possible to cut all partial tiles at once, measuring each ensures a better fit.

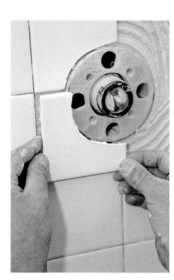

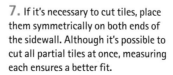

Rejuvenating Grout Joints and Caulking

If your grout is moldy, use a soft-bristle plastic brush to scrub the joints either with household cleaner, a weak bleach solution, or a tile-specific cleaner such as Homax Tile and Grout Cleaner. Wear rubber gloves and goggles, and always brush such solutions away from your face. If the mold returns, try upgrading the vent fan to reduce the moisture in the room. If the grout is intact but dingy, scrub, rinse, and allow it to dry before applying a grout colorant, which will both color and seal the grout. Follow the manufacturer's instructions.

However, if tiles are loose, surfaces flex, or if you see water damage around fixtures at the base of a tub or shower or along the backsplash of a counter, the substrate has probably deteriorated and should be replaced. In other words, you'll need to tear out tiles and substrate.

There's an interim condition, often caused by applying grout that was too thin or by oversponging it, in which tile is intact but grout is worn or crumbling. In that case, use a grout saw or an oscillating multitool (p. 57) to cut out the old grout, taking care not to mar the tile edges. As you'll realize quickly, this job takes patience. Vacuum out the debris, scrub the joints with a cleaning solution, rinse well, and use a grout float to apply polymer-modified grout, which will adhere better. It's possible to regrout only part of a surface, but matching old and new grout color can be difficult, so it's better to regrout the entire surface. Wait 72 hours before sealing the grout joints.

Removing hardened caulk along the tub can be a chore. Chiseling it out is perilous because tub enamel and tile chip easily. Instead, use acetone to dissolve caulk. To use the acetone, cut cotton clothesline to the length of the caulk seam, wet the clothesline with acetone, and place it next to the caulking before covering both with duct tape. Left overnight, the acetone will soften the caulk. *Caution:* Acetone is volatile and thus flammable, and it's nasty to handle and breathe. Leave the bathroom window open while the caulk softens. Don't use acetone around an open flame, such as a pilot light. Wear rubber gloves and a respirator with the appropriate cartridges.

assume that a corner is a good place to start tiling, for it may not be plumb. Instead, establish level and plumb control lines on each wall to guide your layout. Most tilesetters start by laying out the longest wall, which I'll call the back wall. Use your 4-ft. level to determine if the tub is level on all three sides of the surround. If tub shoulders are level, you can start measuring tile courses up from the tub, but in renovation, tub shoulders are rarely level. More likely, the tub will slope. So, from the lowest point of the tub shoulder, measure up one tiling unit and mark it onto a wall. (A tiling unit is a tile width plus one grout joint.) Through that mark, draw a horizontal control line, and extend that line to all three walls of the surround.

Now locate a vertical control line, roughly centered along the back wall. Holding your story pole horizontally, determine whether you need to cut tiles and, if so, where to place them. In most cases, back walls look best if there are symmetrical (equally wide) vertical columns of cut tiles at each end. That decided, choose the joint mark on your story pole closest to the middle of the wall, and run a plumbed line up, bisecting the back wall and the horizontal control line you drew earlier. (This is also a good time to draw plumbed lines at either end of the back wall, indicating where cut tiles begin.)

Next, use your story pole on the sidewalls to see if it's necessary to cut tiles for them and, if so, where to place those tiles. In most layouts, a full column of tiles is placed along the outside edges of sidewalls because they are visually conspicuous; cut tiles are consigned to the corners. Also draw plumb lines to indicate the outside edges of

PRO TIP

About the same time you're sponge-wiping the surfaces of tile, use a margin trowel to remove any grout lodged in the ¼-in. gap where tile meets the tub. Allow a day for the grout to cure, then seal these gaps with an acrylic or silicone caulk. Tile suppliers sell caulk that's either sandless or sanded and color-matched to your grout.

Supporting TILES

It's smart to tape specialty tile pieces in place until thinset has hardened. That's especially true for heavy pieces, such as a soap niche, and for pieces with a relatively small bonding surface, such as bullnose edge trim. *Caveat:* Wait until the field tiles have bonded securely before taping to them.

sidewall tiles. Finally, you may want to draw additional layout lines to subdivide the back wall and anticipate tile cuts around the soap dishes, tub spouts, shower mixing valves, and so on.

As with floor-tile installations, pros often begin setting tub surrounds in the middle of a tile field, where control lines intersect, setting a quadrant of full tiles at a time, then going back later—often, the next day—to cut and set partial tiles and trim pieces. It's also advisable to leave plastic tile spacers in place until the thinset cures. After pulling out the spacers with needle-nose pliers, you're ready to grout.

Getting Grout Right

Most grout is packaged as a powder containing sand, portland cement, colorants, and additives that improve strength and adhesion. Labels on grout bags offer good information, including the correct liquid-to-powder ratio and, on most bags, a chart showing how much grout to buy based on the square footage of the tile area and the width of the grout joint.

Using a margin trowel, mix powdered grout in a clean bucket, starting with three-quarters of the recommended liquid and gradually pouring in the rest. Let the grout stand (slake) for 10 minutes, then stir it again to test its consistency. Grout should be wet enough to stick to the side of a tile but not runny—a consistency rather like hummus or thick toothpaste. Depending on the wetness of the mixture and the temperature of the room, grout starts to set up in about 30 minutes, so mix as much as you can spread in 15 minutes—at which point you should start cleaning up the grout.

If you're grouting a floor or a countertop, just dump the bucket onto the surface. (Grout with the right consistency may need a little help out of the bucket.) Holding the face of the grout float at about 30° to the surface, sweep grout generously over the tile and pack it into joints. After you've packed all joints, remove the excess grout. Holding the float almost perpendicular to the surface, sweep the float diagonally across the tile joints. Periodically unload excess material into a bucket. Diagonal passes are less likely to pull grout out of the joints.

In about 15 minutes, when the grout has begun to set, use a clean, damp (not wet) sponge to clean grout residue off tile faces. Rinse the sponge often. If the sponge pulls grout out of the joints, you can wait a little longer, wring the sponge a little drier, or don't press so hard. Sponge-wiping also smooths out grout joints. So, for best results, use a round-shouldered, tight-cell sponge. Wide-cell sponges will pull grout out of the joints.

In another 15 minutes, use soft, dry rags to rub off any haze that's dried on the tile. Because it's porous, grout can stain. Let it cure for 72 hours before applying a liquid grout sealer or impregnator to the grout alone or to the whole tile surface. Two recommended sealants: TileLab penetrating sealer and TEC sealants.

Holding the grout float about 30° to the floor, pack grout into the tile joints.

In about 15 minutes, when the grout has begun to set, wipe the tile with a clean, damp sponge. Rinse and wring the sponge often. To avoid pulling grout out of joints, sweep the sponge diagonally across tile joints, using a sponge with tight pores.

17 Finish Carpentry

After framing walls, running pipes and wires, insulating, and hanging drywall, it's time to install interior trim. Somewhat like a picture frame, trim is decorative. But it's also functional, concealing gaps and rough edges where walls meet floors, ceilings, doors, and windows. Although finish carpentry is not as fundamental as structural framing or foundation work, it completes the picture and often makes or breaks a renovation project.

Interior trim is often called *casing*, or *molding* if its face is shaped. (In fact, the terms are used more or less interchangeably.) Trim helps establish the character of a room, so it's wise to respect existing trim when replacing or supplementing it. Carefully remove and save existing molding if it's in decent condition. If that type is no longer available, try to locate new molding with a similar feeling. Or you might be able to combine and overlap stock moldings to create a more complex and interesting look. Another choice is using prefab, high-relief synthetics that duplicate large-scale moldings not available in wood today.

Finally, for advice on choosing and installing counters, cabinets, and fixtures appropriate to bathrooms and kitchens, see chapter 13. For sequences of installing doors and windows, read chapter 6.

Tools

Most of the tools for finish carpentry are presented in the basic collection discussed in chapter 3, although upcoming sections address a few specialty tools. Still, by and large, successful trimwork depends more on the hands behind the tools than on the tools themselves. Also, when working with power tools and striking tools, safety glasses are a must, especially when joinery requires close work at eye level.

MEASURING AND LAYOUT

Whenever possible, hold a trim piece in place and use a pencil or a utility knife to mark the cutline. This is usually more accurate than transferring tape-measure readings.

Trim covers gaps between building materials and dresses up a room. Here, baseboards are shimmed 1/2 in. above a concrete basement subfloor so engineered wood flooring or carpet can slide under it. Pneumatic nailers are much faster than hand nailing and far less likely to split or dent trim.

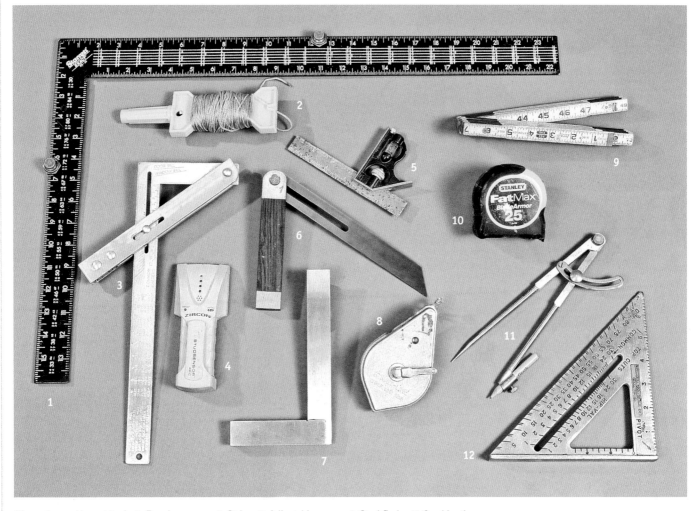

Measuring and layout tools: 1. Framing square; 2. String; 3. Adjustable square; 4. Stud finder; 5. Combination square;
6. Adjustable bevel; 7. Steel try square; 8. Chalkline; 9. Folding rule with sliding extension; 10. Tape measure; 11. Compass;
12. Swanson Speed Square

Tape measures are frequently used to measure trim. Check your tape measure to be sure that the hook at the tape end isn't bent and that the rivet slot hasn't become elongated from the hook's repeated slamming into the case—either of which will give inaccurate readings. For the most accurate readings, start measurements from the 1-in. mark, remembering to deduct 1 in. when taking readings.

A framing square held against a door or window frame quickly tells if it's square or not—a good practice when sizing up a trim job.

A combination square is both a square and a 45° miter gauge, so it can be used to mark square and miter cuts. Because its ruler can be extended from the tool body and fixed with a screw, the square can double as a depth or marking gauge. The combo also has a bubble-level insert for leveling small surfaces such as windowsills.

An adjustable (sliding-T) bevel copies and transfers angles accurately. Because frame corners are rarely 90° exactly, use an adjustable bevel to record the angle needed and then make miter cuts that bisect the actual angle.

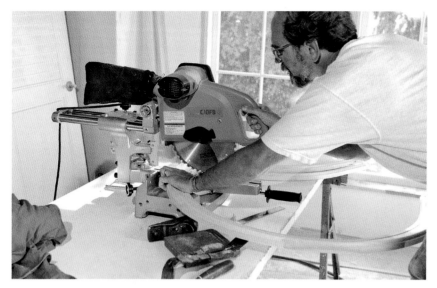

A power miter saw is a must if you're installing a houseful of trim. The 10-in. sliding compound miter saw shown is big enough to make complicated beveled miter cuts in large crown molding.

Domino biscuit joiner and biscuits. The fence on the front of the tool rests on the boards being slotted for glue and biscuits.

versatile. In one stroke, this saw will cut a miter and a *bevel* (an angle cut across a board, in which the sawblade is tilted)—hence the name compound miter. It will also cut through wider stock such as wide baseboards or crown molding.

A tablesaw may be the only table tool you need if you are cutting only miter or butt joints. Tablesaw guides are generally not as accurate or as easy to readjust as the guides on power miter saws, so recutting miter joints will be a bit more work. With a power miter saw, you clamp the stock steady and move the blade, whereas cutting a miter on a tablesaw requires feeding long pieces of trim at an odd angle to the blade.

Still, tablesaws, while costly, can be worth the investment because they can cut stock to length (crosscut) and width (rip cut), prepare edges for joining, and cut dadoes (slots) easily.

A miter trimmer looks like a horizontal guillotine and bolts to a bench. Because its blade is razor sharp, it slices wood rather than sawing it. Although it can shave off paper-thin amounts of wood until joints fit exactly, it's been eclipsed by power miter saws for on-site trim installations.

An oscillating multitool (OMT) (see p. 57) is *the* power tool when you need to cut a piece of trim that's already in place—say, you need to trim the bottoms of door casing. OMTs have a wide range of blades and grinders. A jigsaw also works fine for cutting complex shapes in trim that has not yet been installed.

A Japanese pull saw has an very thin blade. Thus, using a scrap piece of flooring as a height gauge, you can cut the bottom of door casing so exactly that finish flooring will slide snugly under the casing—with no visible gap. For some tasks, a hand tool is just faster.

Use a coping saw to cut along molding profile lines, ensuring a tight fit where molding meets in inside corners. For more, see p. 521.

A plate joiner (biscuit joiner) is a specialized saw with a small, horizontal circular blade that cuts slots into board edges. After slotting boards to be joined, inject glue and insert a football-shaped wooden wafer, called a biscuit, which will swell to create a strong joint with no need for nails or screws.

SHAPING AND SANDING

A block plane and a palm sander are probably all you'll need unless you plan to shape board edges to create complex molding, in which case, get a router.

Block planes are most often used to trim miter joints for a tight fit. If you slightly back-bevel

Four levels—2 ft., 4 ft., torpedo, and a cross-line laser level with plumb—should be part of a finish carpenter's tool chest. Use smaller levels in tight spaces, and use the line laser to run a level line around the room to establish cabinet heights.

CUTTING

Wear safety glasses and hearing protection when operating any of these tools.

A miter box with a backsaw will suffice if you are casing only a doorway or two. A backsaw has a reinforced back so its blade is rigid; it should have 12 teeth per inch (tpi) or 13 tpi, with minimal offset (splay) so it cuts a thin kerf.

Buy a power miter saw if you've got at least a roomful of trim to install. Well worth the cost, a power miter adjusts to any angle for *miters* (angles cut across the face of a board, with the blade perpendicular to the stock). *A sliding compound miter saw*, though more expensive, is more

PRO TIP

For razor-smooth cuts and tight joints, buy an 80-tooth carbide-tipped blade for a 10-in. power miter saw or a 100-tooth blade for a 14-in. saw. If you save such blades for finish work only, they'll last a lifetime.

Handplaning TIPS

When handplaning, clamp the wood securely, and push the tool in the direction of the grain. While holding the shoe of the tool flat against the edge of the wood, angle the tool's body 20° to the line of the board, so that the plane seen from above looks like half of a V. At this angle, the plane blade encounters less resistance and clears shavings better.

center photo on p. 399. *A Festool random-orbital sander hooked to a HEPA vacuum* is a must-have tool if you're rehabbing old trim that may have lead paint. Whatever the sander type, change the paper often; you shouldn't need to lean on a sander to make it cut.

NAILING AND DRILLING

Because most trim is light, it is usually nailed up with finish nails, which have slimmer shanks and smaller heads than other nails. *Trim-head screws* (shown on pp. 81 and 118) are often specified

PRO TIP

Don't bring trim stock to a job site until the drywall joints are dry and the building is heated. Otherwise, trim ends will absorb moisture, swell, and become difficult to install. Never store trim in unheated areas or garages.

miters, the edges of the face will make contact first. Block planes can also shave down a door or window jamb that is too proud (too high above the wall plane), thereby allowing the trim to lie flat. A *power planer* (see the photo on p. 211) can do everything a handplane can but more aggressively, so practice on a piece of scrap and check your progress after each pass. *Caution:* Before planing existing trim, use a magnet to scan the wood for nails or screws, setting them well below the surface before planing.

Rat-tail files and 4-in-1 rasps (see the photo on p. 59) remove small amounts of wood from curved surfaces so that coped joints fit tightly.

Routers are reasonably priced and invaluable for edge-joining, template cutting, mortising, and flush trimming when used with a router table. Router tables vary, but on most you mount the router upside-down, to the table's underside, so the router bit protrudes above the tabletop. A guide fence enables you to feed stock so that the router bit shapes its edges uniformly—much as a large shaper in a lumber mill would.

Before setting up a router table, however, read up. *Fine Woodworking* magazine's website (www.finewoodworking.com) has hundreds of references on routers and router tables. Above all, heed all safety warnings about routers: Their razor-sharp blades spin 10,000 rpm to 30,000 rpm.

Sanders are useful for a variety of jobs. A *palm sander* (or *block sander*) can shape contours and sand in tight places; it is great for light sanding between finish coats. *Orbital sanders* are intermediate in cost, weight, and power. *Random-orbital sanders* sand back and forth and orbitally, so they cut faster and leave fewer sanding marks. If you buy only one sander, this is the one to get. *Belt sanders* have the power to prepare stock and strip old finishes, but use them gingerly. A belt sander can scribe cabinet panels (that is, remove wood from a panel so it fits tightly), as shown in the

Sanders. From left: palm sander, orbital sander, and belt sander.

True Grit: Which Sandpaper for What

COMMON NAME	GRIT NUMBER ("teeth"/sq. in.)	USES
Coarse	40–60	Stripping finishes
Medium	80–120	Sanding down minor bumps
Fine	150–180	Final sanding before finishing
Very fine	220–240	Polish sanding (rarely used)

When it comes to finish nails and nailers, smaller is often better. This pin nailer weighs less than 5 lb. and shoots 1-in. to 1¹/₂-in. nails.

when molding is heavy or complex or when trim pieces will be subject to twisting or flexing, as happens with door frames and stair treads.

A finish hammer has a smaller head than a framing hammer so it's easier to control when trying to avoid denting the trim. Stop when the nail head is almost flush with the wood surface, then use a *nail set* to drive the nail head below the surface. Always set nails before sanding or finishing.

Pneumatic finish nailers have largely replaced hand nailing because pressure settings can be adjusted so the nail goes just below the surface. You don't have to set the nails manually. Finish nailers won't dent trim, and you can nail with one hand while holding joints together with the other. Production carpenters favor pneumatic models with air hoses running to a compressor, but cordless models with spare batteries work well, too, for installing small amounts of trim.

Nailers are designed to shoot specific nail gauges (thicknesses). Standard finish nailers shoot 15-gauge nails, whereas *brad nailers (pin tackers)* shoot 18-gauge to 20-gauge brads. (The higher the gauge number, the thinner the nail shank and the weaker the nail.) Most homeowners should stick with 15-gauge nailers, but brad nailers are great for tacking up trim: Brad holes are tiny, so you can easily pry off and reposition the trim if needed. Brads are also useful for attaching thin cabinet elements such as finish toekicks or cabinet side panels. If you don't have many brads to drive, use a hand brad pusher.

Cordless drill/drivers are the essential tool in most carpenters' belts. As they've grown more powerful, they have become more compact and better balanced, so a ³/₈-in., 18v cordless drill/ driver is now the standard. Typically, they have keyless chucks for quick changing bits and accept either drill bits or screw tips for finish carpentry tasks. From there, the next steps up in power are 20v or 24v *impact drivers*; some models can drive 250 or more 2-in. screws on a single battery charge. (Whatever tool you buy, get a spare battery for recharging while you work with the other

Interior Trim

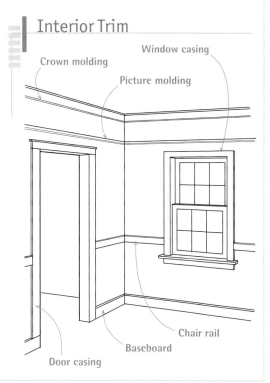

Crown molding

Window casing

Picture molding

Chair rail

Baseboard

Door casing

one.) You'll find more on drills, drivers, and bits in chapter 3.

Materials

Trim materials include custom-milled hardwoods; softwood boards, molding, and stock caps; medium-density fiberboard (MDF); and polymer moldings that replicate detailed historical styles in lightweight, easy-to-install sections.

TRIM OPTIONS

Because trim is costly (especially hardwood trim), buy it from a local shop that mills its own. That way, you'll be more likely to get trim that is straight, knot-free, and stored in humidity- and temperature-controlled warehouses. If you're trying to match existing trim, a local milling shop is also your best bet. You may pay a setup fee but, all in all, the final cost of a room or two of custom trim may be more reasonable than you think.

Talented woodworkers and carpenters sometimes re-create old trim when they need only a few feet of it to complete a renovation. Veteran woodworker Kit Camp uses a tablesaw, a block plane, and sanding blocks, as described in "Site-Made Moldings in a Pinch" (*Fine Homebuilding* issue #206). A homeowner who's good with a router and can find the right bits may be able to do the same. Be advised, though, that the task is very time-intensive: If you need more than 16 lin. ft., it's probably worthwhile to have it milled by a shop.

Stock trim from a lumberyard or home center is often so warped that you must pick through the racks and eyeball each piece to see if it's straight. Discard any pieces that are obviously heavier than the rest—usually a sign of excessive moisture. (Trim is typically dried to 7% to 10% moisture content.) Examine each piece for splits and cupping across the width. Also, sight down the length of each board for excessive twisting. You can force a twisted piece into position by toenailing and clamping it, but the extra stress is likely to open a joint or cause splitting down the road.

If you want the wood grain to show, be even pickier or pay more for a select grade. But if you plan to paint the trim, most surface blemishes can be sanded, filled, or sealed with white pigmented shellac (see chapter 18) or a primer-sealer to suppress knot or tannin bleed-through. *Finger-jointed molding*, which joins short sections of clear softwood, is another option. Although usually painted to conceal its glued finger joints, finger-jointed molding is also available with a wood veneer, which can be stained and clear finished.

Common Molding Profiles

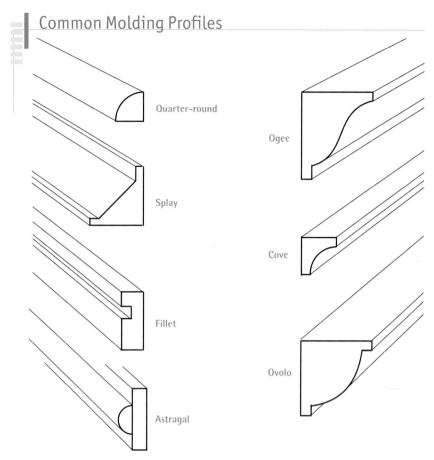

Quarter-round

Splay

Fillet

Astragal

Ogee

Cove

Ovolo

Specialty blocks cover sawcuts, allowing you to join sections of polymer molding without the need for fancy miter cuts.

COMBINING STOCK ELEMENTS

Standard molding is often milled from 1-in. stock (actual size, ¾ in.). You'll find it easy to create more complex trim by combining 1-in. boards with stock molding caps. For example, with baseboards, you might start with a 1x8 and add a quarter-round shoe at the bottom and a cove-molding cap at the top.

However, if you want to dress up a room with complex crown moldings, consider installing polymer millwork instead—whose monolithic casting greatly simplifies installation. Likewise, although you can build them up by hand, mantels and fire-

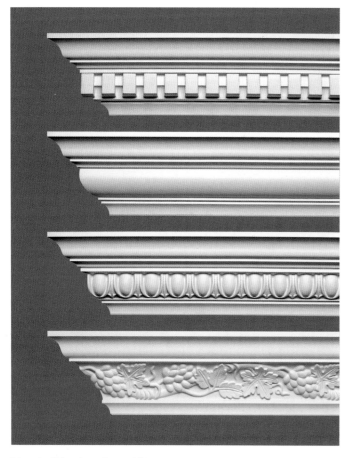

Many traditional cornice-molding types are available in high-density polyurethane. Once filled and painted, they're indistinguishable from wood molding.

closed.) Use a pneumatic nailer to attach it; MDF won't split. Sand it with 150-grit sandpaper, and prime with an oil-based primer (latex roughens the surface). However, MDF does have quirks you need to work around.

Disadvantages. MDF is heavy (a ¾-in. sheet weighs about 100 lb.); lighter versions cost more. It's dusty to cut and shape. Its edges readily suck up moisture. In fact, MDF can swell from ambient moisture, so seal it immediately after cutting or shaping it. Seal cut edges with two coats of shellac-based primer. Then paint all six sides of the panels with an oil-based primer. Perhaps MDF's most annoying quirk is its tendency to mushroom around nail heads. MDF is so dense that it doesn't compress when you nail it; fiber near the nail just bulges up. After setting the nail heads, use a Sandvik carbide scraper to scrape down bumps, then prime it.

Because of MDF's tendency to wick moisture, it's a poor choice for bathroom trim or window installations where condensation is common—no matter how well it's sealed. In those locations, go with wood or PVC trim instead.

POLYMER MOLDINGS

Although many old-house owners prefer wood molding, its supply and quality have been dwindling for decades, leading to a run on third-world forests—now being cut down at an alarming rate. Alternatively, there are polymer moldings (especially polyurethane), which are available in most traditional architectural styles, from simple colonial to elaborate Victorian. Once installed and painted, polymer moldings are virtually indistinguishable from wood trim. The following sections note some of the unique features.

Stability. Unlike wood, polymer molding won't warp, split, rot, or get eaten by termites. Although it does expand slightly (⅜ in. for a 12-ft. piece) in a heated room, special corner pieces "float" over section ends, allowing them to slide freely as they expand. Polymer molding has no grain, so there is no built-in bias to twist one way or the other; there are no splits, cracks, or knots.

Quick installation. Synthetic moldings are less labor intensive. Whereas complex wood moldings are built up piece by piece and their joints painstakingly matched, synthetics come out of the box ready to install. Most polymer moldings are glued up with a compatible adhesive caulk, such as polyurethane or latex acrylic, and tacked up with finish nails or trim-head screws, which are needed for support only until the glue sets. Pieces are so light, in fact, that you can install them single-handedly.

place surrounds also are sold as preassembled units and as kits requiring minimal assembly.

You can combine stock moldings with relatively inexpensive paneling to make wainscoting and frame-over-panel walls. By cutting a piece of paneling in half, you can use two sections, each 4 ft. by 4 ft., topped with a built-up combination of moldings to form a cap. Paneling with vertical, regularly spaced grooves gives the illusion of individual boards. To change the style and create Craftsman-style wainscoting, you could use ¼-in.-thick redwood plywood with the grain running vertically and install redwood strips every foot or two to create detail and cover the seams between sheets. For more, see "Wainscoting" on p. 537.

WORKING WITH MDF

If you want a cost-effective, easily worked material for plain-profile trim, MDF is hard to beat. And you can add visual interest by installing cove, bullnose, quarter-round, or other simple molding along MDF's plain edges.

Advantages. MDF cuts and shapes beautifully. For smooth edge cuts, use a 60-tooth 10-in. blade. Because it has no grain, MDF crosscuts and rips equally well, and its edges can be routed as well, although most MDF trim is simply butt joined. (No need for biscuits to hold the joints

Easy working and finishing. Most polymers can be trimmed like soft pine, using a 12-tpi to 13-tpi saw in a miter box. There's no need for fancy joinery because most systems have corner pieces that cover joints. Touch up holes with plastic wood filler, and caulk field joints on long runs. You also may need a bead of caulk where straight lengths of molding meet existing surfaces that are irregular.

Polymer molding is typically primed white in the factory and could be installed as is, but most homeowners paint it. You paint smooth-surfaced urethanes just like standard wood trim. Some products can be stained, but that gets into the iffy territory of making plastic look like wood.

Basic Skills

Using quality tools and materials matters, but not as much as the skill and judgment of the renovator. This section of tips will help hone your skills in measuring, cutting, and attaching trim.

MEASURING

Accurate measurements are crucial because trim is pricey, and even small discrepancies will stand out. In the following paragraphs, you'll find a few new twists on the old chestnut, "Measure twice and cut once."

Use a sharp point to mark stock. A stubby lumberyard pencil is fine for marking framing lumber. But because the margin of error is small on trim, use a sharp pencil to mark precisely. A utility knife leaves an even thinner line, although it's more difficult to see.

Mark trim in place, if possible. It's almost always more accurate than taking a tape reading and transferring it to stock, especially if your memory's bad.

Change directions. If you normally measure left to right, double-check your figures by changing direction and measuring right to left.

Use templates instead of remeasuring. When you need to cut many pieces the same length, carefully cut one, check it in position to make sure it's accurate, and use that piece to mark the cutline on others. You can also clamp a template to a bench or saw table to act as a stop block. As you cut successive pieces, simply butt a square-cut end against the block, and the blade will cut each in exactly the same place.

When in doubt, go long. If you're not quite sure of the exact measurement and don't want to climb back up the ladder to recheck, cut the piece a little long. You can always make a long board shorter, but reversing the process is quite a trick.

Back-Cutting Trim

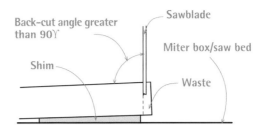

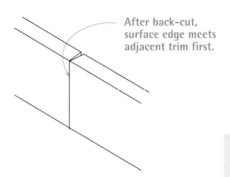

By raising the board's end and keeping the sawblade plumb, you create a back-cut joint whose surface edges can easily be shaved to create tight joints.

PRO TIP

For odd-shaped or complicated pieces, such as winding stair treads, make a template using heavy paper, cardboard, or strips of ⅛-in. plywood hot-glued together. "The Beauty of Templates" on p. 404 has more about templates.

When mitering or coping corner joints, make sure a joint fits well before cutting the other end to final length. Whenever possible, mark the trim in place—that's easier and more accurate than transferring tape-measure readings.

X marks the scrap. As you mark cutlines, pencil a prominent X on the scrap ends of boards. This habit will sooner or later help you avoid wasting trim because you mistook the scrap end for the measured end.

GENERAL CUTTING

Tight trim joints require accurate layouts, sharp saws, and consistent methods.

Recuts are a fact of life. If you're filling and painting trim, slight gaps are acceptable. But if you're using a clear finish, joints must be tight. Before you start cutting trim, always check the accuracy of power-saw miter-stop settings by cutting a few joints from scrap. Then cut stock a hair long so that you can recut joints until they're right.

Cut lines consistently. It doesn't matter whether your sawblade cuts through the middle of a cutline or just past it. What matters is that your method is consistent. For example, moving the width of a sawkerf to one side of the line or the other can make the difference between tight and open joints. Some pros prefer to just "kiss" the inside of the cutline with the sawkerf so that the line stays on the board.

Keep tools sharp. This applies to saws, chisels, planes, and utility knives. Whenever a blade becomes fouled with resin or glue, wipe it clean immediately with solvent. A sharp tool is easier to push and thus less likely to move the stock you're cutting. Likewise, a clean power-saw blade is less likely to bind or scorch wood.

Handsaws usually cut on the push stroke. Start handsaw cuts with gentle pull strokes, but once the kerf is established along the cutline,

emphasize push strokes. (Western-saw teeth are set so that they cut on the push stroke, whereas Japanese saws cut on the pull stroke.) As you continue the cut, keep your elbow behind the saw, which will help you push the saw straight and follow the cutline.

Clamp that stock. If your hands are big enough, it's possible to hold stock against a miter-saw fence with one hand and operate the saw with the other, but it's far easier—and safer—to clamp the stock, using a *spring clamp* or a *quick-release clamp* (see the bottom photo on p. 62). Newer models of compound miter saws have flip-up stops that hold molding against the fence. Finally, use an *outfeed roller* or a sawhorse to support the far end of long pieces so they don't bow or flap as you try to cut them.

CUTTING MITER JOINTS

A miter splits a 90° corner in half, with a 45° cut on each board. With the sawblade set perpendicular to the stock (0° bevel), cut a 45° angle across the face of the trim. When the cut edges are closed together, the boards should form a right angle. Of course, if door or window frames aren't square, corners may be 89° or 91°, requiring that each miter be slightly more or less than 45°, though equal. That is, miter joints should bisect whatever angle is there.

If you'll be painting the joints and the trim stock is relatively narrow (3 in. wide), you can fudge the joints and fill any gaps with spackling. But if you're installing stain-grade molding, especially if it's 5 in. or 6 in. wide, faking a miter joint will look terrible. So if a frame is out of square, take the time to cut and recut joints as necessary so that they bisect the frame's angle.

There are two good reasons to use miters. First, mitering aligns the profiles of moldings so that bead lines and other details join neatly along the joint and sweep uninterrupted around the corner. Second, although flat trim allows you to butt or miter joints at corners, with butt joints you would see the rough end grain of one of the adjoining boards. Even if you sand down the roughness, end grain soaks up extra paint or stain so it often looks noticeably different from adjacent surfaces.

SPLICING TRIM

When a wall is too long for a single piece of trim, you can splice pieces by beveling their ends at a 60° angle and overlapping them (called a *scarf joint*) or by butt joining them and using a biscuit to hold the joint together. If boards shrink, gaps will be less noticeable in a scarf joint because you'll see wood, rather than space, as the overlap

Mitered casing joints look dressy and conceal the end grain of intersecting pieces.

Back-Cutting Miters

Ideally, miter cuts will meet perfectly, creating a tight joint. But back-cutting (also called undercutting) can improve the odds that joints will be tight even if corners aren't perfectly square and frame jambs aren't flush to the surrounding walls. In other words, the front faces of back-cut boards make contact before the backs, so the front edges can be finely shaved to fit. It's far less work to shave the leading edge of a back-cut board with a block plane than it is to recut the joint.

The easiest way to back-cut trim is to shim under it slightly in the miter box or on the saw bed, as shown in the drawing on p. 519. The sawblade is still set at 90° (0° bevel), but the shimmed boards receive a slight bevel because they aren't lying flat. Even a 1/16-in.-thick sliver under the board is enough to give you a decent back cut.

Fussing over a miter joint is probably not worthwhile if you plan to paint the trim because slight gaps can be filled with wood filler or caulk. But open joints are difficult to disguise when wood is to be stained and almost impossible when it is clear-sealed.

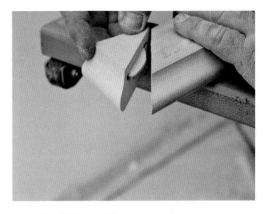

Glued biscuit joints will keep butt joints or miter joints from spreading due to seasonal expansion and contraction. Here, a biscuit joins a mitered window-stool return. Biscuits can also join straight runs of crown molding or baseboard when a wall is too long for a single board.

Two Ways to Splice Trim

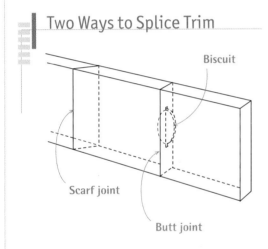

Biscuit

Scarf joint

Butt joint

separates. In general, scarf joints are better suited to flat stock, whereas shaped molding will display a shorter joint line if butted together. (Viewed head on, the joint is a thin, straight line.)

When it's necessary to use several trim boards to span a distance, center end joints over stud centers so you can nail board ends securely to prevent cupping. Where that's not possible, say, where a baseboard butts to door casing, nail the bottom of the baseboard to the wall sole plate, and angle-nail the top of the baseboard to the edge of the casing. Predrill the trim or snip the nail points to minimize splits.

COPING A JOINT

All wood trim shrinks somewhat. Where beveled boards overlap, gaps aren't as noticeable, but shrinkage on some joints—mitered inside corners, in particular—are glaringly obvious because you can see right into the joint. For this reason, car-

Coping a Joint

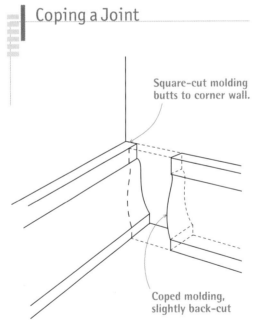

Square-cut molding butts to corner wall.

Coped molding, slightly back-cut

A coped joint is first mitered, then back-cut along the profile left by the miter so that the leading edge of the trim hits the adjacent trim first. That thinner, leading edge can be easily shaved to fit tightly.

To cope a joint, first use a tablesaw or a miter saw to cut the trim at a 45° bevel. Use a jigsaw to cut the straight part of the trim profile, then use a coping saw to back-cut along the shaped part, as shown.

penters often cope inside corners so that their meshing profiles disguise shrinkage. Basically, a coped joint is a butt joint, with the end of one board carefully cut to fit the profile of the molding it butts against.

Cut the first piece of trim square, butt that end right into the adjacent wall, and tack-nail it up. Coping the second piece of trim is a two-step

The Case for Not Leveling Trim

The older a house, the less likely floors and ceilings will be level. Don't make yourself crazy trying to level baseboards and crown molding: You won't succeed, and trim that's level next to a surface that isn't will only emphasize the discrepancy. Interior trim, like politics, is an art of compromise. Trim edges should be roughly parallel to floors and ceilings. As master carpenter Joseph Beals puts it, "Baseboard is effectively floor trim, and the floor plane is the critical reference, level or not."

Mid-wall elements such as chair rails, picture rails, and wainscoting call for yet more fudging. Ideally, chair rails should be level and wainscoting stiles (vertical pieces) should be plumb, but those ideals may clash with existing trim that's neither. In that case, split the difference: Tack up a length of trim that's level. Then raise or lower one end until your eye accepts the compromise.

Trim also can help give the illusion of a level ceiling—helpful, when upper kitchen cabinets must be set level even if the ceiling isn't. So after leveling and securing upper cabinets, install a strip of molding to cover the gap above. (You may need to rip it down at an angle.) If you look for it, you'll see the uneven strip of molding between the cabinets and the ceiling. But if the cover trim matches the cabinet finish, chances are nobody else will notice the difference.

operation. First, cut the end with a 45° miter, as if you were making an inside miter joint. Use a pencil to darken the profile of the edge. Clamp the trim face up. Then, using a coping saw, carefully cut along the profile created by the miter—while slightly back-cutting it. If precisely cut, this coped end will mesh perfectly with the profile of the first piece. If you're not happy with the fit, shave it with a utility knife or recut it from scratch.

Coping requires ingenuity. If the top of the trim curves like quarter-round molding, back-cutting would create a little spur that will probably break off before you can finish the cut, so don't back-cut it. Leave the rounded top as a 45° miter, and chisel a corresponding 45° miter into the top of the piece of molding that you're coping to.

GLUING AND ATTACHING TRIM

To increase holding power and keep joints closed, apply yellow carpenter's glue to mating edges after the casing has been dry-fit and is ready to install. If you allow the glue to tack (set) slightly, the casing ends will slide around less as you nail the trim. But splicing joints with biscuits is by far the better way to keep them from spreading. Use a biscuit joiner to cut slots into mating edges, then inject glue into the slots and spread it evenly on the casing ends. Place biscuits in the slots and reassemble the joints, drawing the joints tight with a single 4d or 6d finish nail angled into butt joints or end-nailed through miter joints. Use a damp cloth to wipe off excess glue.

Finish Carpentry Fasteners

FASTENER	USES	COMMENTS
20-gauge brad	Attach small molding returns.	Glue returns first.
18-gauge brad	Tack-nail trim while adjusting; attach cabinet toekicks and side panels.	Tiny brad holes are easily filled; easy to pry off tacked trim.
4d (1½-in.) finish nail	Attach inside edge of casing to rough jambs (jambs of rough opening).	Snip nail point if worried about splitting casing.
6d (2-in.) finish nail	Attach outside edge of ½-in.-thick casing (through ½-in. drywall) to rough jambs.	Nail should sink at least ½ in. into framing.
8d (2½-in.) finish nail	Attach outside edge of ¾-in.-thick casing (through drywall) to rough jambs; attach baseboard; attach crown molding.	Place nails a minimum of ⅜ in. from edge; snip nail points to minimize splits.
2½-in. to 3-in. finish-head screw	Secure window- or door-frame jambs to rough openings.	Frame jambs twist or flex as doors and windows are operated, so use pairs of screws at each point.

Double–gluing creates strong joints. First, use your finger to rub in a little glue to seal the end grain. When that's tacky, apply a second layer of glue to bond the trim pieces.

Common Interior Glues

COMMON CHEMICAL NAME(S)	PROS	CONS	BRANDS
White carpenter's, polyvinyl acetate	Moderate strength; inexpensive	Runny; poor initial tack; clogs sandpaper	Elmer's Glue-All
Yellow carpenter's, polyvinyl acetate	Strong; good initial tack; sands well		Titebond Original
Polyurethane	Bonds to most materials; sands well; takes stain; fills gaps; water-resistant	Glue expansion can spread joints not tightly clamped; slow tacking time; stains skin	Gorilla Glue; Titebond
Cyanoacrylate	Instant bond; great for nonstructural joints; bonds many materials	Expensive; can't adjust pieces once placed; skin/eye hazards	Super Glue; Krazy Glue
Hot glue	Quick-tack glue to create thin plywood templates	Limited strength, but OK for temporary positioning; low-stress joints	Bostick
Contact cement	Instant bond; resists heat and water; best for attaching plastic laminate and veneers	Can't adjust once sheet and substrate make contact; volatile solvent; needs good ventilation	DAP Weldwood; 3M Fastbond

* Several websites offer interactive product selectors. Specify how and where you'll use the adhesive, and the selector will choose a product (www.titebond.com is particularly good).

Drive nails into framing whenever possible. If framing members are spaced 16 in. on center, nail trim to every stud or ceiling joist it crosses. Where trim runs parallel with the framing, as with side casing, nail the trim at the ends and roughly every 16 in. in the field. Equally important is using the right nail or screw to avoid splitting the wood trim. (The table on the facing page recommends sizes for most trim applications.)

To attach narrow molding such as quarter-round, use a single row of finish nails. On wider molding, use two nails to prevent cupping: Set the nails at least ½ in. from the edge and use a square to line up nail pairs.

It's usually not necessary to predrill softwood trim to prevent splits. If you use a pneumatic nailer, do not nail too close to the edge and don't use too big a nail. However, when nailing hardwood trim or nailing the ends of boards, predrilling is smart. Use a drill bit whose shank is thinner than the nail's. Alternatively, you can minimize splits by using nippers to snip off the nail points, as shown in the top photo on p. 524. It's a bit counterintuitive, but it works.

Before painting, caulk all gaps between the casing and the wall.

Casing a Door

Before casing doors and windows, review "Checking and Prepping the Opening" on p. 120, particularly the remarks on *margining*, centering jambs in relation to a wall's thickness. Then survey the door and window frames to be cased; use a 4-ft. level and a square to see if the frame jambs are plumb, margined, and square.

CASING ELEMENTS

Door casing is trim that covers the gaps around a door frame. It goes on after a door has been hung. Most often casing consists of three pieces: two *side casings* (*leg casings*), which cover frame jambs, and one piece of *head casing*, which goes over the frame head. Six pieces, if you count both sides of a doorway.

There are three common casing joints: *mitered*, preferred for trim that is molded (shaped) because it enables you to match molding profiles as they converge at a corner; *square cut*, made with basic butt joints; and *corner block*, a variation of square cut with discreet blocks at the top corners and sometimes bottom corners as well. (Bottom

Use an end cutter (or nipper) to snip nail points—and wear safety glasses! Nails with blunt points are less likely to split trim because they crush the wood rather than wedge it apart, as triangular points do.

To repair dings from doors or trim, hold a hot, damp cloth over the spot, then apply a steam iron to the cloth until the wood swells slightly. Lightly sand the raised area until it's level. Then use a small artist's brush to apply thinned finish to already finished surfaces.

Door and Window Casings

MITERED

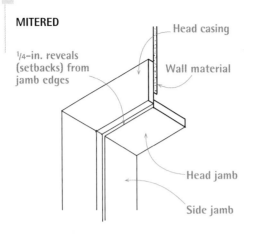

¼-in. reveals (setbacks) from jamb edges

Head casing

Wall material

Head jamb

Side jamb

SQUARE-CUT OR BUTT

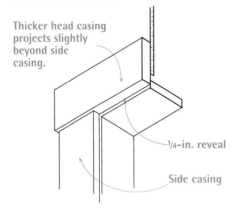

Thicker head casing projects slightly beyond side casing.

¼-in. reveal

Side casing

CORNER-BLOCK

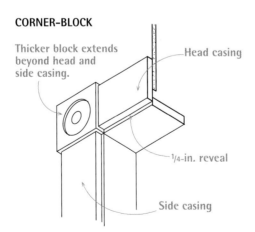

Thicker block extends beyond head and side casing.

Head casing

¼-in. reveal

Side casing

corner blocks are also called *plinths*. Top corner blocks are called *caps*.)

PREPPING THE FRAME

Jamb edges should extend no more than ⅟₁₆ in. beyond finish surfaces. If they protrude more, sink any nails, and then plane down the frame. If frame edges are level or, say, ⅟₁₆ in. below the wall surfaces, leave them alone. If the edges are sunk more than that, build them up with shim strips ripped from stock of the same thickness. Scrape the old frame so that it's flat, glue on the strips, tack with brads if you like, and wipe up the excess glue at once. Also survey the walls around the frame, scraping down globs of joint compound or hammering down (compressing) high drywall spots that would make the trim cockeyed.

MARKING A REVEAL

Boards are rarely perfectly straight, and door and window jambs (and casing stock) are no exception. So instead of trying to nail board edges flush, set the inside edge of casings back ¼ in. from the frame edges. This setback is called a *reveal*: It looks good and will spare you a lot of frustration. Use the rule of your combination square as a depth gauge. Set the rule to ¼ in., and slide along the edge of the frame, making pencil marks as you go. Where head reveals intersect with side reveals, mark the corners carefully.

INSTALLING SQUARE-CUT DOOR CASING

Few frames are perfectly square, so use a framing square to survey the corners. Note whether a corner is greater or less than 90°, and vary your cuts accordingly when you fine-tune the corner joints. Note, too, whether the floor is level because side casing usually rests on the floor.

First, rough-cut the casing. To determine the length(s) of side casings, measure down from the reveal line on the head frame to the finish floor. (If finish floors aren't installed yet, measure down to a scrap of flooring.) Cut the side casings about ½ in. long so you can fine-tune the joints, then tack the casings to the reveal lines on the side frames. (Use 18-gauge or 20-gauge brads.) Next, measure from the outer edges of the side casings to determine the length of the head casing. If the ends of the head casing will be flush to the edges of the side casing, add only ½ in. to the head-casing measurement for adjustments. But if the head casing will overhang the side casing slightly—a ¼-in. overhang is common if head casing is thicker than side casing—add overhangs to your head-casing measurement. Cut the head casing, place it atop the side casings, and tack it up.

Fine-tune casing joints. Because you cut the side casings ½ in. long, the head casing will be that much higher than the head reveal line. But with all casing elements in place, you can see if the head casing butts squarely to the side casings, or if they need to be angle trimmed slightly to make a tight fit. Use a utility knife to mark where the head reveal line hits each side casing, and remove the side casings and recut them through those knife marks. Retack the side casings to the

frame, then reposition the head casing so it sits atop the newly cut side casings. Finally, use a utility knife to indicate final cuts on both ends of the head casing, either flush to the side casing or overhanging slightly, as just discussed.

Glue joints. Butt joints are particularly prone to spreading, so remove the tacked-up casing, and splice the joints with biscuits, as described earlier in this chapter. Spread glue on all joint surfaces, and nail up the side casing. (Insert biscuits, if used.) Draw the head casing tight to the side casing by angle-nailing a single 4d finish nail at each end. To avoid splits, predrill the two nail holes or snip the nails' points. Then remove excess glue with a damp cloth.

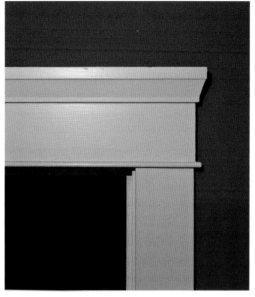

Square-cut casing needn't be plain. Here, square door casing is spiced up with a ¼-in.-thick beaded strip between the head and side casing and a beveled cap molding.

Use a scraper with replaceable carbide blades to shave drywall high spots and hardened joint compound. Use a utility knife to cut back shims still protruding around frames.

Using a combination square as a marking gauge, make light pencil marks on the jamb edges to indicate casing reveal (offset) lines. By offsetting casing and jamb edges, you avoid the frustrating and usually futile task of trying to keep the edges flush.

PRO TIP

Although a reveal is typically ¼ in., sometimes you need to adjust it slightly to give nails a better "bite" going into the edge of a jamb. Jamb stock is only ³/₄ in., so you might, for example, adjust the reveal so it's closer to ⅛ in. Whatever reveal you choose, however, use it consistently throughout the job.

INSTALLING SQUARE-CUT DOOR CASING

Square-cut head casing can easily span ganged windows or, as shown, a double door with windows on both sides. Here, a carpenter tacked up the door side casings before eyeballing the head casing to be sure the joints were flush.

Use a pin-nailer (brad nailer) to tack the head casing until you're sure all joints are tight. When that's done, secure the casing with 6d finish nails.

Slightly back-cut the side casings to ensure a tight fit to the underside of the head casing.

INSTALLING MITERED DOOR CASING

Before installing the casing, first use a framing square to see if the doorway corners are square. If they aren't, use an adjustable bevel to record the angles and a protractor or an angle divider to help bisect them. Then cut the miter joints out of scrap casing until their angles exactly match the frame. You can install mitered casing by first cutting the side casing and then the head casing, as you would with square-cut casing. But some carpenters maintain that the best way to match mitered profiles is to work around the opening. That is, start with one side casing, cut the head casing ends exactly, and finish with the second side casing.

Mark the first piece of the side casing, and then cut it. After marking a ¼-in. reveal around the frame, align a piece of casing stock to a reveal line on a side frame. Where head and jamb reveals intersect in the corner, make a mark on the casing, using a utility knife. (Or make the

utility-knife mark ¼ in. higher if you're not confident you can cut the correct angle on the first try.) Miter-cut the top of the side casing so it matches the bisecting angle you worked out earlier on the scrap. Square-cut the bottom of the casing, and tack it to the frame using 18-gauge or 20-gauge brads.

Cut one end of the head casing in the same bisecting angle, leaving the other end long for the time being. Fit the mitered casing ends together, and align the bottom edge of the head casing to the head reveal line. Then, using a utility knife, mark the head casing where side and head reveals intersect in the second corner. You'll cut through that mark, using the bisecting angle for the second corner (which may be different from the first corners). Again, there's no shame in recutting, so you may want that utility-knife mark to be ¼ in. proud. When its miter is correct, tack up the head casing, then line up the inside edge of the second side casing to the side reveal line. Use a utility knife to locate its cutline. Back-cutting miters slightly can make fitting them easier.

Whether you simply glue mitered joints or biscuit-join them, remove all three pieces of casing before securely nailing them. If biscuit joining, after slotting each piece, reinstall the leg jambs, apply glue and biscuits, and fit the head

Baseboard and Side Casings

SIMPLE BUTT JOINT

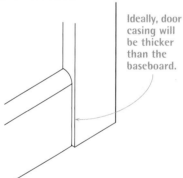

Ideally, door casing will be thicker than the baseboard.

PLINTH-BLOCK JOINT

Plinth block is thicker than door casing or baseboard.

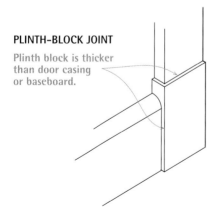

If baseboards are the same thickness as door casings, simply butt them together, or shape the end of the baseboard slightly to reduce its thickness.

Recutting miters is a normal part of installing casing. To adjust for a slight gap at the bottom of the joint, this carpenter is marking the amount he needs to remove from the top of the side casing.

Corner-Block Casing

Casings with corner blocks are a variation of square-cut casing in which you have seven pieces of trim—two plinth (base) blocks, two cap blocks, two pieces of side casing, and one head casing—to measure, cut, and fit.

Start by installing the plinth blocks, which are thicker and wider than the side and head casing. Plinths and cap blocks may line up to reveal lines on door frames, or they may line up with the inside frame edges; be sure to match the detailing of the existing casing.

Tack up plinths, using 18-gauge brads. Then measure from the reveal line on the head frame to the top of the plinth blocks to determine the length of side casings. Tack up the side casings, aligning them to the reveal lines on the side frames. Depending on the detailing of the cap blocks, you may need to recut the tops of side casings. Recut the side casing as needed, then tack up the cap blocks and place a spirit level atop them to see if their top edges align and if they're level.

Finally, measure between the cap blocks to determine the length of the head casing. (Use a rigid folding rule with a slide-out extension for this task.) If the door frame is slightly out of square, cut the head casing 1/16 in. to 1/8 in. long, and back-cut both ends so you can shave them to fit. Once the tack-fit is tight, carefully pry off the tacked-up pieces, cut biscuit slots, glue, insert biscuits, reassemble the pieces, and finish-nail the assembly. Glue all joints even if you don't use biscuits.

casing down onto the biscuits. You can use miter clamps or 18-gauge brads to draw the joint together until the glue dries. Be sure to wipe up the excess glue immediately.

Casing a Window

Casing windows is essentially the same as casing doors, so review earlier sections about prepping frames and installing casing. The main difference is that the side casing of windows stands on a window stool rather than on the floor. Consequently, most of this section describes measuring and cutting the stool, which covers the inside of a windowsill, and the apron beneath the stool. Sills and stools vary, as described in "Windowsills, Stools, and Aprons" on p. 530. The following text focuses on installing replacement stools appropriate to older windows.

Nail window casing to frame jambs and to the rough opening, spacing finish nails every 16 in. Use a combination square to line up each pair of nails.

Window Trim

Rough jamb (trimmer stud)

Frame jamb

Bottom sash rail

Pitched sill

Side casing

Stool horn

Stool

Apron

Rabbeted underside of stool

Interior wall

Exterior sheathing

Use 6d or 8d finish nails to attach casing to rough jambs, 4d finish nails to attach the inside edge of casing to frame jambs, and 4d nails to tie the stool to the apron edge.

MARKING THE WINDOW STOOL

Before you start, decide how far the stool "horns" will extend beyond the side casings and how far the interior edge of the stool will protrude into the room. Typically, horns extend ¾ in. beyond 3½-in.-wide side casings, but use the existing casings as your guide.

To determine the overall length of the stool, mark ¼-in. reveals along both sides of the frame jambs. Then measure out from those reveals the width of side casings plus the amount that the stool horns will extend beyond that casing. Make light pencil marks on the drywall or plaster. Rough-cut a piece of stool stock slightly longer than the distance between the outermost pencil marks.

Next, hold the stool stock against the inside edge of the windowsill, centered left to right in the window opening. Using a combination square, transfer the width of the window frame, from the inside of one jamb to the other, to the stool stock. Use a jigsaw to cut along both squared lines, stopping when the sawblade reaches the square shoulder in the underside of the stool. So you'll know when you've reached that shoulder, lightly pencil the width of the shoulder onto the top of the stool, using a combination square as a marking gauge. Next, cut in from each end of the stock to create the horns. Carefully guide your saw along the shoulder lines, being careful to stay on the waste side of the line. Clean up cutlines with a chisel, if needed.

The cutout section of the stool should now fit tightly between the jambs. Now, rip down the rabbeted edge of the stool so that it will be snug against the window sash, and the stool horns will be flush to the wall (and jamb edges). Push the stool in until it touches the bottom of the sash, then pull the stool back 1⁄16 in. from the sash to allow for the thickness of paint to come. Finally, measure the distance between the stool horns and the wall, which is the amount to reduce the width of the stool, along the rabbeted edge.

Use a tablesaw to rip down the width of the stool, then test-fit it again. If it is parallel to (but 1⁄16 in. back from) the sash rail, you are ready to cut each horn to its final length. If your stool is flat stock, that's the last step. But if the stool is molded (has a shaped profile), miter returns to hide the end grain of the horns, as shown in the top photo on p. 521.

ATTACHING THE STOOL AND APRON

Lightly sand and prime the stool, including its underside and ends, so it won't absorb any moisture from condensation or driving rains. After the paint is thoroughly dry, apply waterproof glue to the underside of the stool, level it, and nail it to

Cutting a Window Stool

TOP VIEW

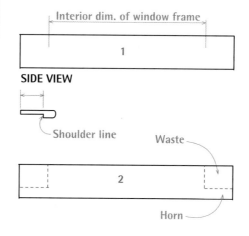

SIDE VIEW

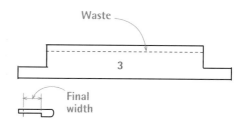

1. After transferring the interior dimensions of the window frame to the stool stock, cut across the stock till the sawblade reaches the stool shoulder.

2. Following the shoulder line, cut in from the ends of the stock to create horns.

3. Rip down the stock so its rabbeted portion butts the inner window sash, less 1⁄16 in.

4. Trim the horns of the stool so they protrude 3⁄4 in. beyond side casings.

Arched Window Casing

There's something inspiring about arched windows. Restoration carpenter Jim Spaulding (shown on pp. 532–533) offers this advice: "Order all the casing from the same shop so that the same knives cut the arches and the legs (side casing). That way, all the profiles will sweep continuously around the frame." Arched casings are different from casings for other windows, so installing them takes some flexibility. After setting the stool and apron, for example, you install its head casing next. Side casings are last.

Although modern window makers offer a limited selection of prefab casing for the arched windows they sell, plan on custom-ordering casing for older arched windows. Correctly determining the radius of the arch is challenging: One method is to tack 1⁄8-in.-thick plywood (also called *doorskin*) to the inside edge of the arched frame head—run it about 1 ft. below the "spring line" of the arch, where the frame becomes straight. Go outside and trace the arch, tracing lightly so you don't bow the plywood. Make templates for each arched window, and take them to a local shop that mills trim.

Note: The inside edge of arched casing must be revealed (set back) from the arch you traced of the frame's inside edge.

INSTALLING A WINDOW STOOL AND APRON

1. Start casing a window by installing its stool. Use a combination-square level or a torpedo level to level it, then 8d finish nails to secure it to the rough sill underneath. However, the stool won't be stable until it is also nailed and glued to an apron under its inside edge.

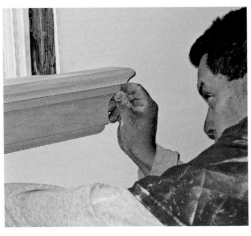

2. Typically, the apron is as wide as the casing above the window stool; the stool horns project ³/₄ in. beyond both.

Windowsills, STOOLS, AND APRONS

Windowsills have both an inner and outer life—one half is interior, and the other is exterior. It makes sense to pitch the outer portion of a sill so it can shed water. But in the old days, windowsills were pitched all the way, front to back, which created a uselessly pitched interior section that had to be covered with a stool piece, as shown in "Window Trim" on p. 528. To fit a pitched sill, usually set at 14° to 20°, the underside of a stool must be partially rabbeted at the same angle so that when the stool is nailed on top of the sill, the top face of the stool will be level. It's an archaic design, but, surprisingly, it survives in some new window designs.

Many modern windows, however, have a sill whose interior portion is flat on top and flush to the insides of the window frame. Consequently, there's no need for a stool or an apron. Such windows may be "picture framed." That is, the casing can be mitered around all four sides of the window frame.

If your windows have traditional stools and aprons, the trickiest part of casing the windows will be fitting the stools. Typically, a stool's outer edge *almost* abuts the inside of the lower window sash (allow a ¹/₁₆-in. space for the thickness of paint), and its interior edge overhangs the apron beneath it. And note that the stool's "horns" extend ³/₄ in. to 1¹/₂ in. beyond the width of the side casings. You can still buy replacement stools for window renovations in older houses.

---

the rough sill using two or three 6d galvanized finish nails. Try not to lean on the only partially supported stool until it's nailed to the top of the apron, which will steady the stool.

Next, cut the apron, which is generally the same casing used for side and head casing. Its thicker edge is butted to the underside of the stool. The apron should be as long as the head casing so that it lines up visually with the outside edges of the side casing. If the apron is molded, cope each end to accentuate its profile or mitercut it and glue on a return. If you'll be painting the casing, caulk along the underside of the stool to prevent drafts. Then butt the apron to the underside of the stool. Nail up the apron, driving 6d finish nails into framing beneath the sill. Finally, nail the stool to the top of the apron, using three or four 6d galvanized finish nails. Set and fill those nails.

Install the side and head casing in the same order that you would case a door: side casing, head casing, and second side casing. The main difference is that window side casing sits on the stool horns. If you're casing side-by-side windows with flat trim, you can run a single piece of head casing over both windows and butt the middle and side casings to the underside of the head casing, as shown in the top photo on p. 526.

Baseboard and Crown Molding

As noted in "The Case for Not Leveling Trim" on p. 522, baseboard and crown molding should follow floors and ceilings rather than level lines projected across the walls. If floors and ceilings are level, fine. Otherwise, leveled trim next to out-of-level surfaces is glaringly obvious.

INSTALLING BASEBOARDS

Install the finish floors first, leaving a slight gap, typically ½ in., between the wood flooring and the walls so that wood strips or planks can expand and contract seasonally. Baseboards cover that gap along the base of the walls. You should also install door casing before baseboards so that baseboards can butt to side casing or plinth blocks. Back-cutting the baseboard slightly yields a tight butt joint against plinth blocks or casing, even if the trim boards are not perfectly square to each other.

INSTALLING MITERED JAMB AND HEAD CASING

1. After installing stop-strips inside the window frame and scribing reveal lines on all three jamb edges, cut and tack up the first piece of casing.

2. Although many carpenters install both jamb casings before measuring the head casing, this carpenter chose to dry-fit the second jamb and head as a pair so he could adjust the miter joint in place. An unglued biscuit held the pieces of casing together as he finessed the joint.

3. Set nails, fill holes, and touch-up sand all surfaces before finishing the wood casing. Generally, there won't be many nails to set if you've used a pneumatic nailer.

Baseboard Strategies

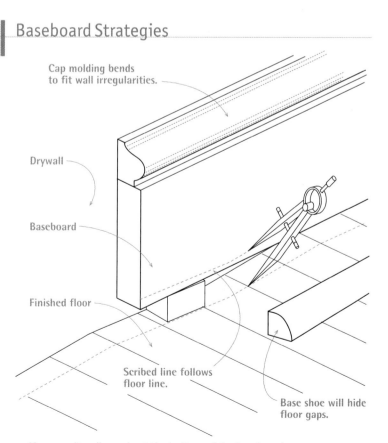

Cap molding bends to fit wall irregularities.

Drywall

Baseboard

Finished floor

Scribed line follows floor line.

Base shoe will hide floor gaps.

If you can't scribe and cut the bottom of the baseboard, use base shoe to cover gaps.

PRO TIP

When ripping down a baseboard, keep the sawblade just off the scribed pencil line. After cutting, clamp the board to a bench and sand exactly to the line, using a belt sander held perpendicular to the board edge. In this case, 80-grit to 120-grit sandpaper is effective because it's not overly aggressive.

Arched windows require complex framing around the arch so you have something solid to nail finish walls and casing to.

1. After scribing a reveal line along the edge of the arched head jamb, tack a finish nail at the apex of the line, and hang (balance) the arched head casing from it. Then level across the "spring lines" of the casing—the points at which the casing springs into its curve.

2. After cutting the arched casing at its spring lines, align the inner edge of the arched casing to the reveal line.

4. Work around the window, nailing the casing every 16 in. The thickness of the casing determines the nail size. In this case, the carpenter used 1-in. brads to nail the inside edge of casing to the frame edges and 6d finish nails along the outside.

3. Install the straight side casing, cutting it a little long on the bottom and then trimming as needed until the casing fits tightly between the arched head casing and the window stool. After dry-fitting the side casing, apply glue and tack it up.

5. If the casing is not wide or thick enough for biscuit joinery, angle 6d finish nails to draw the joint together. To avoid splitting the casing, you first need to snip off the nail points.

Scribe baseboards so they follow the floor. Or, if you're installing the baseboard before the floors are in, shim the trim so it will be a consistent height above the subfloor. Use scrap to cushion the trim from hammer blows, as you tap the trim to align its edges.

PRO TIP

Crown molding must be solidly nailed. So, as you locate studs and ceiling joists, affix a small piece of blue painter's tape to indicate their centers. Place tape sufficiently back from the trim path so you can still see the tape as you position the trim. Using painter's tape is easier than erasing pencil marks, and the tape won't pull off paint.

▐▐▐▐

Locating studs beforehand will make installation easier. If walls have been newly drywalled, look along the base of the walls for screws or nails where drywall is fastened to studs. Otherwise, rap the base of walls with your knuckle until you think you've found a stud. Then drive in a 6d finish nail to locate the stud exactly. Stud finders work, but they are less reliable with plaster walls, whose lath nails meander all over the place.

Scribe the bottom of baseboards to follow the contour of the floor, especially if the floors are irregular. But first, shim the baseboard(s) up about 1 in. above the floor, butt one end of the board to a corner or a door casing, and tack the baseboard to a stud or two to keep it upright. Then run the scribe or compass along the bottom to transfer the floor contour to the baseboard. Cut the scribed line with a fairly rigid sabersaw blade that can cut with the grain yet not wander.

Baseboard joinery employs basic techniques described earlier. Miter outside corners, cope inside ones, and glue all joints before nailing them off. Use two 8d nails (aligned vertically) at every other stud center, and use a single 4d finish nail top and bottom to draw mitered corners tight. Used as baseboard caps, standard moldings, such as quarter-rounds, can hide irregularities between the top of the baseboard and the wall, and they dress up the top of the board. Where baseboards abut door casing and there's no stud directly behind the end of the baseboard, nail the bottom of the board to the wall plate, and angle-nail the top to the side of the casing or plinth block, using an 18-gauge brad to avoid splitting the trim. Finally, set the nails, fill the holes, and caulk all seams before painting.

INSTALLING CROWN MOLDING

Crown molding dresses up the wall–ceiling joint, as do its fancy cousins cornices, which are formed from several boards. Crown molding can be as simple as a single piece of shaped trim angled along the corners of the ceiling, or you can pair it with a *backing trim* to ensure a solid nailing surface, which is not always present in an old houses with irregular framing and springy plaster. Backing trim is, basically, a flat board with a shaped bottom edge.

Start by locating and marking stud and joist centers on the walls and ceilings. Where joists run parallel to a wall or where you can't find framing on a regular nailing interval, install a row of *triangular nailing blocks* along the tops of walls, as shown below. Predrill these blocks to avoid splitting them, and nail them with 8d finish nails to the top plates or studs, spacing the blocks every 24 in. to 32 in. Cut nailing blocks at the same angle as the crown molding, when correctly seated to wall and ceiling.

To determine that angle, cut a short section of crown molding to use as a *seating gauge*. Hold the gauge so that trim edges seat solidly on both the ceiling and the wall. Using this gauge, make a

Crown Molding and Blocking

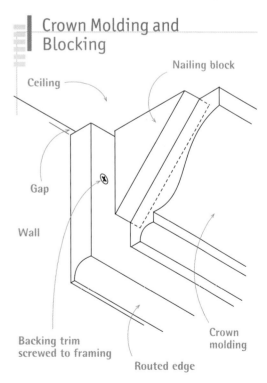

When installing nailing blocks without backing trim, nail the blocks directly to the wall plates. Keep a 1/16-in. space between the back of the molding and the face of the blocks to accommodate wall-ceiling irregularities.

Cut or assemble a small section of crown molding to use as a seating gauge to tell you where the bottom of the molding will meet the wall. A pencil mark every 3 ft. to 4 ft. should do; don't use a chalkline because it could bleed through finish paint.

Outside corners are seldom square, requiring that miters be marked in place. Here, scrap was positioned on both sides of the corner to guide a pencil line along its top edge. At that intersection, the top point of the miter was marked on both scrap pieces. The bottom point of the miter is the corner itself, here being marked.

light pencil mark every 3 ft. to 4 ft. so that when you nail up the molding you'll have reference marks for its bottom edge. (Because crown molding is relatively thin, it easily twists and misaligns.) Nail molding roughly every 3 ft. or to every other 16-in. on-center stud and joist. If walls are too long for a single piece of molding, splice boards over stud centers. Use a nailer to attach crown molding and cornices—hand nailing is too erratic. If you're nailing molding to blocks, use 18-gauge brads to avoid splits; otherwise, use 6d finish nails. Nail the bottom edge first, then the top, keeping the nails back from edges by at least ⅜ in.

To install backing trim, snap a chalkline to line up the bottom edges of the trim. Backing trim is a godsend when you've got level upper cabinets but an unlevel ceiling. Install the backing trim level and the crown molding snug against the ceiling. The amount of backing trim revealed (exposed) will vary, but your eye won't notice it. Use screws to attach the backing trim because they hold better and are less likely to fracture plaster. Before nailing up crown molding, use a seating gauge to mark its position atop the backing trim.

As with baseboard molding, miter outside corners, cope inside ones, and glue all joints before nailing them off. If the first piece of crown molding is long enough to run from inside corner to inside corner, just cut both ends square, pop into place, and cope the ends of adjacent pieces. Miter-cut the crown molding upside down, with its bottom edge up—angled so that the molding's lower edge rests against the back fence of the miter box and the molding's upper edge rests on

the bottom of the miter box. (Inverting the molding in the miter box is the only way to support both of the molding's edges and re-create the same angle the crown molding will have when installed against the wall and ceiling. If you cut the molding right side up in the miter box, the top edge of the molding would be unsupported.) Use the seating gauge you made earlier to establish this angle on your miter saw. Screw a piece of scrap to the saw bed to hold the molding stock in place as you cut it. When in doubt, test the joints by cutting and joining pieces of scrap.

FALSE BEAMS

There are several ways to construct false beams. Two are shown here. The first is to make a ladder frame (imagine a ladder set horizontally) clad with finish boards. This type, shown in the top photo on p. 536, runs perpendicular to ceiling joists so its top board can be screwed to them; end-nail "ladder rungs" to the top board before installing it. Once the top board is secured to joists, attach the bottom board and then the sides. A power nailer is a must because the assembly is shaky until all the boards are on.

The second type, shown in the bottom photos on p. 536, is more correctly called a *box beam* because there's nothing false about the steel I-beam it's disguising. You can order I-beams with bolt holes predrilled, making it easy to bolt plywood nailing blocks to them. The plywood shown was faced with clear fir on three sides and stained to simulate redwood. Because the underside of the box was most visible at eye level, the carpenters took pains to create an even reveal along the bottom of the beam. The gaps along the top of the beam were later covered by the crown molding, shown in the left photo above.

PRO TIP

When installing crown molding, leave the last 2 ft. to 3 ft. unnailed until you've test-fitted all joints. Sometimes the molding needs to come up or down to make a coped joint fit exactly.

Scan left to right on the ceiling, and you'll see the evolution of a false beam. Preassemble the top board (which nails to the ceiling joists) and short nailer blocks. Install the bottom board, then the sides. Use a pneumatic nailer only because hand-nailing will loosen the assembly.

BOXING IN A STEEL I-BEAM

To provide nailing surfaces for the sides of the box beam, first bolt plywood strips to the predrilled I-beam. Attach the bottom panel of the box beam first, then the sides.

Because the underside is the most visible part of the box beam from eye level, measure to be sure the board reveals are consistent. Measuring also tells you exactly where the edge of the bottom board is—so nails don't miss it. Cover gaps along the ceiling with crown molding.

WAINSCOTING

In the old days, when raised-panel wainscoting was constructed from solid wood, fancy joinery was required to accommodate the expansion and contraction of the panels. Today, thanks to the stability of MDF panels and readily available stock molding, you can create good-looking wainscoting with simple joinery. Once painted, this new wainscoting will be almost indistinguishable from that built with traditional materials and methods.

Construct the frame rails (horizontal pieces) and stiles (vertical pieces) from clear, straight 1x4s; if you need more than one board to attain the length you need, use a biscuit joiner (see the bottom photo on p. 514) to splice the board ends. Use this tool to strengthen the butt joints between rails and stiles, too. But first, snap chalklines onto the walls to indicate the position of rails and stiles; if any stiles coincide with electrical outlets, it may be easiest to relocate the outlets so that all the panels along a wall have a consistent width.

Assemble the frame on the floor. After allowing its glued and biscuited joints to cure, tilt the frame upright and fasten it to wall studs with 15-gauge finish nails. To avoid stressing the frame joints, have a helper tilt it up and hold it atop spacer blocks as you nail it. The ¾-in. MDF panels are best routed in several passes to avoid frying the router and scorching the panel edges. Once you're done routing, sand the panel edges lightly and nail them to the wall, leaving an even gap all around, between panel edges and frame elements.

Although you can use any type of stock molding to cover the gaps around the panels, a shaped molding adds visual interest and has a traditional feel—*bolection molding* has a nice profile and a rabbeted back edge that seats neatly against frame edges. Cap the top of the top rail with molding, too, to cover the slight gap between the frame and the wall.

After screwing the preassembled frame to the studs, insert shaped MDF panels, nailing them directly to the wall with 2-in. brads. Leave a 1-in. gap around each panel, which you'll cover with stock molding.

18 Painting

Painting is probably the most popular renovation task because its effects are immediate and
striking. For not much money, you can get a
complete change of scenery and heart. If you
own a few basic tools, your costs will be limited
to the few tools you'll need to rent and the paint
you choose.

This chapter covers both exterior and interior
painting, including trim, doors, windows, and
cabinets. For information on stripping and finishing floors, refer to chapter 20.

Color on the exterior of your home has an impact on its
design. Here the combination provides a cohesive
appearance that enhances the house style.

Essential Prep Work

If you want painted surfaces to look good and
last long, the substrate—such as drywall, plaster,
and wood—must be stable and dry before you
start. Prep work is crucial for a good paint job.
Whether you'll be painting a building's interior or
exterior, follow these guidelines:

▶ **Correct structural or moisture-related
problems.**

▶ **Scrape or sand down paint that's poorly
adhered or applied excessively.**

▶ **Sand surface irregularities.**

▶ **Choose primer that will adhere well and
be compatible with the finish coats.**

▶ **Follow instructions on paint cans,
including optimal temperatures for painting.**

▶ **Sand lightly between coats for better
adhesion.**

▶ **Wear an appropriate mask or respirator.**

▶ **Test for lead paint in houses built
before 1978.**

Choosing Paint

Manufacturers frequently reformulate their
primers, paints, and stains, so look for a reputable supplier who keeps up with changes. Before
buying paint, examine the surfaces to be painted
and think about the conditions it must endure.
Then ask the following questions.

FOUR TELLING QUESTIONS

▶ Interior or exterior paint? Beware of
paint labeled "interior/exterior." It's probably
cheap. Quality exterior paints contain
additives that repel moisture, block UV rays,

and discourage mold. These additives are not substances you'd want to inhale indoors while the paint is curing. In general, don't use exterior paints indoors and vice versa.

▶ Has the surface been painted before? Surfaces should be primed if they (1) have never been painted, (2) have been extensively scraped or sanded, or (3) are "chalky" or poorly prepped. However, if existing paint is well adhered, priming probably isn't necessary; just paint over the old coat.

▶ What type of finish (sheen) do you want? Topcoat finishes range from *flat* (also called dull or matte) to *semigloss* (aka eggshell, velvet, satin) to *gloss*. Glossier finishes tend to be more durable and easier to clean, so they're favored in high-use areas such as bathrooms and kitchens. Enamel, which dries to a hard, durable finish, is best for window sashes, doors, and casings.

▶ Is the surface unusual? Specialty paints are available for masonry; hard, nonporous surfaces; high-humidity areas; and nonslip surfaces. If you have an unusual surface to paint, ask your supplier to recommend a finish.

OIL BASED VERSUS LATEX

Whether you're painting exterior or interior surfaces, latex paints are probably the best bet. The next sections explain why.

Two essential definitions.

▶ Oil-based paints and stains may contain linseed oil, tung oil, or synthetic resins called alkyds. Because alkyds are the most common "oil" in oil-based paints, professionals often use the term *alkyd* instead of *oil based*. However, oil based is the broader, more inclusive term for products that must be thinned and cleaned up with solvents.

▶ Latex paints, also called "acrylics," are water based. They can be thinned with water and cleaned up with warm, soapy water. In recent decades, acrylic latex paints have improved so dramatically that they now meet almost every painting need. Basically, you no longer need to use oil-based paint.

Oil based: advantages and disadvantages. Oil-based paints are durable and tenacious, adhering even to glossy or chalky surfaces. Thus, many pros still insist on an oil-based exterior primer, even if they'll be applying latex topcoats. Many old-school painters also favor oil-based paints for interior trim because they dry slowly and level well, thus minimizing brushmarks.

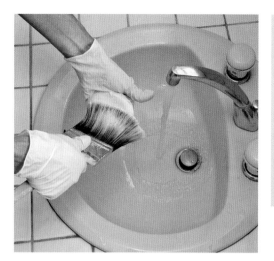

One big advantage of latex paint is that you can easily clean brushes and your hands with soap and water.

Colorful LANGUAGE: PRIMERS, PAINTS, AND STAINS

▶ **PRIMER.** An important base coat, primer is applied to substrates such as raw wood, drywall, plaster, or previously painted surfaces. Above all, primers must stick to the substrate; they may also contain stain blockers, preservatives, pigments, or other additives to hide flaws and ensure more uniform topcoats of paint.

▶ **PAINT.** If primer's job is adhesion, paint's is protection—protecting the primer and substrate from moisture, mild abuse, and (if it's exterior paint) UV rays. Paint must also hold color, dry smoothly, and withstand weather, so its pigments, solvents, and additives must be carefully blended and held together by a binder, or resin.

▶ **BINDER.** Binders determine a paint's penetration, adhesion, drying rate, flexibility, and durability. In relation to pigment, the more binder a paint has, the shinier and more durable its finish will be. Glossy paints tend to have high binder-to-pigment ratios.

▶ **PIGMENT.** Color is determined by pigment. The more pigment a paint has, the more intense its color and the better it hides what's beneath.

▶ **VEHICLE.** A paint's liquid component, the vehicle, is needed to suspend the pigments and binders. Oil-based vehicles (linseed oil, tung oil, or modified oils called alkyds) thin with mineral spirits, also known as paint thinner. Latex paints suspend polymer particles (plastic) in water.

▶ **STAIN.** Penetrating or semitransparent stains are most often pigmented oils that soak into wood and form a thin film on the wood's surface; there are also water-based stains. You can see wood grain through stain. Although stains may contain water repellants, preservatives, and some UV blocking, they don't protect wood as well as paint does and so must be reapplied periodically—say, every two to four years.

▶ **SOLID-COLOR STAIN.** Despite the name, solid-color stain, a fast-growing group of exterior coatings, is more like thinned paint than stain. It's popular because wood texture (but not wood grain) remains visible. However, solid-color stains have only about half the life span of painted surfaces. Acrylic latex solid-color stains are the most durable.

Inside, latex is the only paint to use on drywall because it won't raise the paper facing of panels. Exterior latex is colorfast, durable, and easy to apply. However, its quick-drying characteristic can be a problem if you're painting an exterior in 90° F heat, which causes the paint to dry on the brush. In that case, additives such as Flood Floetrol will slow drying time and extend its brushability.

TIPS FOR THE PAINT STORE

Before heading for the paint store, calculate the square footage of surfaces to be painted. Then compare those figures with the coverage figures listed on the paint containers. Unless a wall is mostly glass, don't bother to subtract the square footage of windows and doors from your total. You'll need the paint someday for touchups.

Predicting the coverage of stains is more difficult, especially if the wood is untreated. Add 15% to 25% if you need to special-order the stain and must wait more than a day or two for delivery. That way, you'll be ensured of having enough stain to finish the job.

If the job is big, you'll save money by buying 5-gal. rather than 1-gal. containers. However, if you need only a few 1-gal. cans, ensure uniform color by mixing their contents in a clean, empty 5-gal. bucket. This way, you'll avoid finishing one can in the middle of a wall and resuming with a noticeably different hue.

Have the paint store mechanically shake the paint for you—unless, of course, the manufacturer's instructions indicate it shouldn't be shaken. For example, polyurethanes and varnishes trap air bubbles when shaken.

The problem is that oil-based paints never completely cure. Rather, they oxidize and, over the years, erode and crack. Any siding, especially wood, expands and contracts as temperatures fluctuate, so the inflexibility of oil-based paints leads to cracking and more commonly to *chalking*, a powdery residue of oxidized oil and pigment. In addition, mold feeds on the organic compounds in oil-based paints. But the biggest problem is their solvents: noxious, volatile, polluting, and tedious to clean off tools and equipment. Availability may also be an issue: Because of more stringent regulations on volatile organic compounds, or VOCs (such as in California), oil-based paints are increasingly harder to get.

Latex: advantages and disadvantages. Acrylic latex has almost everything a painter or a substrate could want: As the paint or stain dries, its water base evaporates with minimal odor, leaving a flexible coat of polymer particles (plastic) that rarely cracks, as oil-based paints often do. Latex also is semipermeable, so moisture generated inside the house can migrate, through the paint, to the outdoors. Because latex is synthetic, it's inhospitable to mold. Finally, latex cleans up easily and dries quickly.

Although some artists can hand-tint 5-gal. quantities to match existing paint, the rest of us should rely on paint-store mixologists. They have color charts, recipes, and accurate measuring tools. The blue object at the lower right is a power-drill paint mixer.

Tools and Equipment

Ladders and scaffolding are essential for many painting jobs. For more on them, see chapter 3.

MASKS AND RESPIRATORS

The level of respiratory protection you need depends on how harmful the particles, vapors, or chemicals you will be exposed to are. So read the fine print on the label, which will list the chemicals present in the paint, or refer to the Material Safety Data Sheet that comes with the finish or is available from the manufacturer. I will use *mask* to denote a lightweight, inexpensive, usually disposable paper device and *respirator* to denote a more serious apparatus that fits snugly, protects against chemicals and vapors, and may have replaceable filter cartridges. Following OSHA guidelines, filters are color-coded according to the substances they protect against.

Dust masks will do for most woodworking, sanding, and general cleanup tasks, but if you are sanding drywall—especially paperless drywall, which has a fiberglass-mat facing—wear a *sanding and fiberglass respirator*. It's advisable to wear a mask or respirator even when the chemicals in a paint—say, latex—are largely benign. Consequently, you can wear a relatively inexpensive *latex paint and odor mask* when using any product with "nuisance-level vapors." You could also choose a lightweight respirator. Point being, concentrates of any vapor can give you headaches or cause more serious health problems in chemically sensitive people, so you need good ventilation and some respiratory protection.

As you'll learn later in the chapter, for any task involving leaded paint, wear a respirator with HEPA cartridges (P100 or N100). For most paint applications, a half-face respirator with replaceable cartridges will be adequate. Masks vary, so find one that fits snugly. To test the fit, cover the cartridge openings and inhale: No air should enter the mask.

Respiratory cartridge service life varies according to the chemical you're using, the type of filter specified for that chemical, ventilation in your workspace, and the care you take to keep cartridges "alive" in storage. Change the cartridges whenever you smell fumes or whenever it becomes difficult to breathe through the respirator. Manufacturers suggest their own change schedules. For example, when filtering epoxy-based paints (very toxic) or urethanes, filters can become loaded in 4 to 8 hours, whereas filters for latex-based paints typically have a longer change schedule. OSHA's website offers three ways to calculate when to change filters.

Protect your lungs whenever you sand, scrape, or strip paint—even if there's plenty of fresh air. While heating old paint for scraping, this worker is wearing a lightweight respirator with P100 particulate filters.

Respirator Filters

Black	Organic vapors
Yellow	Acid gases and organic vapors
Olive	Multigas (several gases and vapors)
Purple	Any particulates: P100
Orange	Any particulates: P95, P99, R99, R100
Teal	Any particulates free of oil: N95, N99, NR100

GLOVES AND GLASSES

Here again, even when working with relatively benign substances, wear gloves to protect your skin and safety glasses to protect your eyes. Safety glasses are especially important when spray painting or rolling paint over your head. Disposable gloves will keep chemicals from irritating or drying out your skin. Latex-free *nitrile plastic gloves* are inexpensive, durable, and so flexible that you can pick a stray brush hair off a painted surface while wearing them. A box of 100 pairs is very inexpensive.

BRUSHES

The bristles of high-quality brushes are *flagged*, meaning the bristle ends are split and of varying lengths, enabling them to hold more paint. As you shop for brushes, pull lightly on bristles to make sure they are well attached to the metal ferrule on the handle. Then, when you gently press the bristles as though painting, they should spread evenly and have a springy, resilient feel. Avoid brushes with stiff bristles.

Bristle types. Bristles are either *natural* (hog bristles, for example) or *synthetic* (usually nylon). Use natural bristles for oil-based paints, varnishes, shellacs, and solvent-thinned polyurethanes. Use synthetic bristles for latexes. Nylon bristles may dissolve in oil-based paints, and natural bristles tend to swell and clog when used with water-

based finishes. Although some synthetic bristles work with either painting medium, once you've used a brush for a particular type of paint, continue using it for that type.

The width of the brush should depend on the area of the surface to be painted and the amount of paint to be applied. Because rollers and spray guns are best for large surfaces, spend your brush money on smaller, better brushes. Many pros praise the 2½-in. *angled sash brush* as the most versatile brush in their arsenal; it's wide enough to smooth out paint on baseboard trim yet slim enough to "cut an edge" at corners and along window casings. If you own only one brush, this should be it. But if you'll be painting many narrow window muntins, also buy a 1½-in. sash brush.

Brush care. Brush care begins with proper use. Don't stab bristles into tight spots. Instead, stroke the paint on. When you take a coffee break, leave a moderate amount of paint on the bristles so they don't dry out.

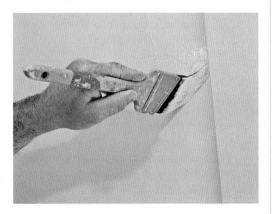

Quality bristles fan out as you apply paint. Here, the tapered bristles of a 2½-in. angled sash brush cut a straight edge where walls meet trim.

PRO TIP

Before storing brushes, wrap them in stiff paper to protect the original shape of the bristles. Never wrap damp brushes in foil or plastic bags: Damp natural bristles can rot, and paint thinner remaining in the bristles can dissolve plastic bags.

Clean brushes immediately after you finish painting for the day. Remove excess paint from the bristles by drawing them over a straightedge, not over the edge of the paint can. Clean the brush in an appropriate solution: paint thinner for oil-based paints, soap and water for latexes.

Wear disposable gloves during brush cleaning, and use your fingers to work the solution into bristles and all the way to the ferrule. After giving each brush a thorough initial wash, rinse it in a fresh batch of solution. When the brushes are clean, rinse with warm water, use a *brush spinner* to remove excess moisture, comb the bristles, and put the brushes into their cardboard sleeves. Don't use hot water when cleaning brushes because it will swell and split the bristles.

ROLLERS, PANS, POLES, AND PADS

Rollers enable you to paint large areas quickly and evenly. In addition to the familiar 9-in. cylinder type, there are also 6-in. "hot-dog" rollers for tight spaces, such as inside cabinets, and beveled corner rollers that resemble a pointed wheel. There are also textured rollers, including stippled, faux finish, and distressed.

Choosing a roller cover. The surface and paint should determine the type of roller cover, also called a sleeve. For example, if you're painting smooth walls, use a short-nap cover (¼ in. to ½ in.), whereas concrete block and stucco need a long nap (1 in. to 1½ in.). Most covers are synthetic and work either with oil-based or latex paints. However, for fine finishes with glossy oil-based paints, use a fine-nap natural-fiber cover. For an ultrasmooth finish when rolling enamels, varnish, or polyurethane, use a fine-nap mohair cover.

Don't buy cheap cardboard-backed roller covers unless you intend to paint a single room with latex and then throw the cover away. (Never use cheap roller covers with oil-based paint. The oil will pluck the fibers from the cylinders and leave them sticking to your wall.) Quality roller covers have plastic sleeves that will survive repeated cleanings. As you do with brushes, use a roller cover for only one type of paint, whether oil based or latex.

Roller pans. Ramped metal or plastic roller pans are routinely sold in packages that include a roller frame and a cover or two, but pros rarely use roller pans. Occasionally, a pro will use a pan to hold a small amount of paint for decorative painting. But when pros roll multiple rooms, they prefer an *expanded metal ramp* inside a 5-gal. paint bucket about half full. This ramp gives you room to load the roller and remove

Heavy paper or cardboard covers help bristles keep their shape.

Expanded metal ramps allow you to load paint quickly and roll excess into the bucket. Note the building paper protecting the floor from paint spatters.

Before washing roller covers, use this painter's 5-in-1 tool to remove the excess paint.

As you load your brush, dip it only ½ in. to 1 in. into the paint before tapping the tip sharply against both sides of the pail to remove the excess.

Smart painters recycle brushes. This one began life as a finish-coat brush. Then, as it got tired and splayed, it was used for primer coats. When its bristles became too crusty and its handle separated from the ferrule, it became a duster.

excess paint quickly—so you can keep painting, rather than repeatedly filling a roller pan.

Paint pads. Pads for paint are about the size of a kitchen sponge and have a short nap. Generally, they are used to paint hard-to-reach spots such as insides of cabinets. They're also used for applying clear finishes such as polyurethanes to flat surfaces.

Extension poles. Whether sectioned or telescoping, extension poles are indispensable for reaching ceilings and upper parts of walls with rollers or pads. Because the poles tax mainly your shoulders and back, rather than your wrist and arm, they enable you to work longer with less fatigue. Another advantage: By painting with an extension pole, you don't need to stand immediately under the drizzle, known as "paint rain."

Roller-cover care. If you buy quality roller covers, clean them as soon as you finish a job. For this, wear disposable gloves. Before washing a cover, use a 5-in-1 painter's tool (see the bottom photo on p. 543) to remove excess paint. Then slide the cover off the metal roller frame and wash the cover in the paint-appropriate cleaner, working out the paint with your gloved fingers. Repeat the procedure with fresh cleaner, then wash with soap and water. Blot the excess moisture with a paper towel or a clean rag. Air-dry the cover by sliding it onto a wire clothes hanger; don't let it lie on its nap while drying. Store the dried cover or pad in a paper bag or foil. If a cover or pad wasn't cleaned properly and has become crusty, throw it away.

Painting Basics

A quality paint job takes preparation, patience, and experience. In addition, professionals also learn how to streamline their moves. As one pro put it, "Any time you eliminate a move in painting, you save time."

PAINTING SAFELY

Almost all paints, including latex, contain VOCs, which are hazardous. The following advice is pertinent for all kinds of paints and stains.

Read the label. There's valuable information on all paint containers: drying time, coverage, thinner data (what to use and how much), and safety instructions. Should an emergency arise—say, a child swallows some paint—the guidance you need may later be concealed under paint drippings. So read up before you open the can or, better, remove and save the label.

Don't breathe paint fumes. Breathing paint or solvent fumes can make you dizzy, impair your judgment, and, over a sustained period, damage your brain, lungs, and kidneys. Set up a fan to blow fumes away from your work area, and always wear a half-face respirator with replaceable cartridges. Rule of thumb: If you can smell fumes while wearing a snug mask, change the cartridge.

Ventilation is a particular problem when chemically stripping paints because the chemicals are strong and because heat guns, sanders, and scrapers increase airborne particles. The now banned lead-based paints are especially dangerous when inhaled or ingested, so if you suspect that you will be stripping lead paint, always test it before disturbing it. See "Lead-Paint Safety" on p. 548 for more.

Protect your eyes and skin. Although most water-based paints are innocuous, oil-based paints can be extremely irritating. In most cases, flush your eyes with water if you get paint in them, and visit a doctor immediately.

There are a few things you can do to protect your skin. Before you start, apply lotion to your

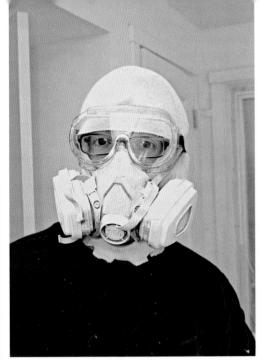

The right protective gear can keep paint mist off your skin and out of your lungs. This includes a respirator with replaceable vapor filters, as well as a "spray sock" over your head, safety goggles, and disposable gloves.

skin to reduce irritation and speed cleanup. Also, wear gloves, even when using latex, because any paint will irritate skin over time. Gloves are a must for oil-based paints.

When it comes to cleanup, painters have traditionally used paint thinner or turpentine to clean their hands and then washed with hot, soapy water. However, hot water opens skin pores, causing them to absorb more solvent. Instead of volatile solvents and hot water, it may be safer to use waterless hand cleaner and wipe it off with paper towels.

Store paint safely. Store paint where children can't reach it. Solvents such as paint thinner, turpentine, and all paints—including latex—should be considered toxic and stored out of reach of children. (Some "green" brands of latex such as Benjamin Moore Natura and AFM Safecoat contain no VOCs, but you still wouldn't want kids to drink them. So store these products safely, too.) Also, store paint where temperatures are moderate because freezing ruins their bonding ability and heat increases their volatility. Close all containers completely so the paint doesn't dry out and contaminants can't get in. Never store rags or steel wool dirty with solvents because of the danger of spontaneous combustion. Dispose of such articles safely: Most paint-can labels carry disposal suggestions, and many municipalities have annual curbside pickups of such materials.

BRUSH BASICS

The pros work steadily and methodically and note what works and what doesn't. The following tips will help you keep the job moving and get great results.

Acclimate a new brush. Stand a new brush in 1 in. of oil-based paint for five minutes. After absorbing a bit of paint, the new bristles will release paint more readily when you start to work, whereas thirsty new brushes may drag at first. It's not necessary to acclimate brushes when using latex, which works into bristles within 20 to 30 seconds.

Avoid overloading your brush. Pros have only ½ in. to 1 in. of paint in the bottom of a paint pail when edging—and the same amount on the tip of the brush. With this small amount, you'll cut a cleaner paint line and keep paint off the brush handle and your hands. Should the bucket tip over, you'll have less mess to clean up.

Retrieve loose bristles. If a bristle comes loose and sticks to the surface, pick it out by dabbing lightly with the tip of the brush. Quality brushes rarely lose bristles.

Paint with gravity. This is close to an absolute rule. Paint, as a liquid or mist, always falls or drips downward, so it's better if it lands on unpainted surfaces—rather than painted ones. Best sequence: ceilings, walls, trim, baseboards.

Paint with the grain. When painting trim, brush paint in the direction of the wood grain. Painting cross-grain doesn't help paint adhere better and will look terrible if brushstrokes dry quickly.

Steady hand, straight paint. Few pros use masking tape to achieve straight lines when brushing paint onto trim, window casing, and the like. Pros feel that tape takes too much time to apply and sometimes lets the paint seep under, leaving a ragged line. Besides, during removal, tape can pull off paint. Patience and a steady hand work better. With a little practice, it's easier than you might think.

ROLLER BASICS

Acclimate roller covers to paint. Work paint into a new roller cover before using it. Load the cover with paint. Then roll it up and down the paint ramp to work the paint down to the base of its nap, and remove the excess.

Roll upward, after loading a roller with paint. If you roll downward instead, you'll be more likely spray excess paint onto walls and floors. Instead, with an initial upstroke, excess falls back onto the roller cover.

Roll paint in a zigzag. Roller covers contain the most paint during the first three to five passes, so first roll a W or an N to distribute "fat paint," which you can then reroll to spread the paint evenly.

Lighten up once the paint is spread on the wall. Especially on outside corners (corners that project into a room), don't bear down on the roller. Too much pressure can make the roller skid or leave roller-edge marks.

Spray Painting

Spray painting is a great way to get a lot of paint on a building fast. Spraying is most appropriate when you've got large expanses to paint, where there's ornate trim, or if the surface is multifaceted (shingles), textured (stucco), or otherwise difficult to cover with a brush or roller. Spraying can also speed a job along when you need to apply numerous thin, even applications, as on cabinet doors.

SPRAYING: NOT A CUT-AND-DRIED DECISION

Although it has some advantages, spray painting is not effortless, nor is it right for all situations:

▶ **Spray painting can be complicated. An even application of paint requires the right tip, the correct paint consistency and gun pressure, good technique, and a feel for**

vagaries of heat, humidity, and wind. Plus, the equipment can be balky if it's not scrupulously cleaned and maintained.

▶ As each section is sprayed, the paint should be immediately worked in with a roller or a brush—at least for the first coat. *Back brushing or back rolling* helps the paint penetrate and adhere better and evens out areas where paint has been applied unevenly. This two-step procedure is best done by two people.

▶ Because paint mist is diffuse, you must mask adjacent areas that you don't want painted. Done well, masking takes a lot of time. Moreover, you shouldn't spray in any wind stronger than a slight breeze.

▶ Consequently, spray painting is less appropriate when you don't have much to paint or if the area to be painted has a lot of doors, windows, or other obstructions that will need to be masked.

Spray-painting equipment. In recent years, spray-painting equipment has become much easier to operate and maintain. The key to successful spray painting, as with any painting, is thorough prep work. That is, begin by correcting moisture problems, removing loose paint and dirt, caulking and filling holes and gaps, and priming unfinished substrates.

Renting a Paint Sprayer

Don't rent spraying equipment that isn't well maintained. A first-rate rental company will size the pump to your job, recommend spray tips, and explain how everything works. For good measure, ask for an operator's manual, too.

Make sure the equipment is clean when you pick it up. Also, ask whether the last paint used in the sprayer was oil based or latex. If you are applying latex and the last paint through was oil, it is not unreasonable to reject the sprayer and ask for another. Or you can ask the renter to spray a small amount of latex to make sure there's no oily residue still in the system.

Finally, make sure you know how to use the equipment, even if you're a bit macho and don't want to admit that you've never operated any sprayer before. One face-saver is, "Say, run a little water and show me how to pressurize this, would you?" Don't leave the rental yard without understanding how the equipment works.

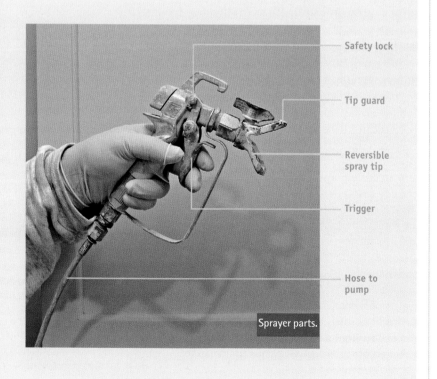

Safety lock

Tip guard

Reversible spray tip

Trigger

Hose to pump

Sprayer parts.

Spray-painting safety begins with a respirator with two replaceable organic-vapor filters. If you'll be spraying exteriors, a half-face respirator should be adequate. For interiors, where paint concentrations build up quickly, wear a full-face respirator, gloves, a spray sock to keep paint mist off your head, and disposable coveralls. The greater the concentration of paint mist, the sooner filters will clog and cease filtering.

Airless sprayers have largely replaced conventional spray equipment with its bulky compressors, pressurized paint pots, and two hoses. Instead, single-hose, airless sprayers force paint through the tip at up to 3,000 lb. of pressure, creating a very fine spray. High-volume airless sprayers can apply coatings of varying viscosity, from extrafine to heavy—and they transfer to the surface a higher percentage of the paint (70% to 75%) than conventional high-pressure paint guns (35%) or high-volume, low-pressure (HVLP) sprayers (60%).

Spray tips control paint volume and pattern, or fan. Fixed-size tips are coded with three-digit numbers: The first digit (2–9) indicates in inches half the width of the paint fan when you hold the tip 1 ft. from the surface being painted. The next two digits (00–99) indicate the size of the tip opening in thousandths of an inch. So a no. 518 tip will spray a 10-in.-wide fan (at 1 ft. away) and has a 0.018-in. opening.

   A few manufacturers make adjustable spray tips, but they're a specialty item. If, for example, you'll be at the top of a tall ladder, need several different spray patterns, and don't feel like climbing down the ladder to change tips, use an adjustable tip. But most of the time, fixed-size tips are the way to go: They cost less and maintain a precise aperture longer.

SPRAY-PAINTING BASICS

Never touch a spray tip while it's spraying: It will inject paint into your skin (and bloodstream), which requires immediate medical attention, including removing the affected skin.

Carefully mask off everything you don't want painted—from ceilings to windows to shrubs. To protect large expanses, use 1½-in. painter's masking tape to attach high-density plastic sheeting. To cover baseboards, windowsills, and the like, apply 12-in.-wide masking paper. Careful masking takes time, but it's crucial to ensure crisp spray-painted edges.

Before turning on the pump, make sure the spray-gun trigger is locked, the pump's pressure control is turned to low, and the priming lever is turned off.

Moving with a Spray Gun

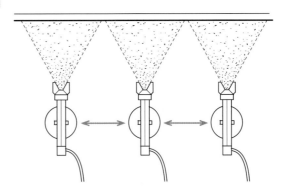

To achieve uniform spray applications, move with the tool, as shown, rather than remaining in one spot and swinging the spray gun in an arc.

Test the sprayer on an inconspicuous area first to make sure it's operating correctly and to familiarize yourself with its fan pattern and volume. If the pressure is correct, paint will stick when the spray tip is 12 in. from the wall. But if paint bounces back at that distance (coating the sprayer and your gloved hand), reduce the pressure.

If the sprayer clogs often, the paint may be contaminated and need straining. Buy strainer bags at a paint store, or, in a pinch, strain paint through an old pair of panty hose or nylons.

Keep the sprayer moving in long, straight strokes. Hold the spray tip 12 in. away from the surface, and overlap passes 30% to 50%. To prevent uneven paint buildup, move the spray tip parallel to the surface. If the surface you're painting has distinct edges, start spraying just before the edge and don't release the trigger until the paint fan is past the far edge. If you must start in the middle of a wall, begin moving your arm before pulling the trigger.

Use a hose that's long enough so you can move freely around the work site. For exterior jobs, you'll need a 100-ft. hose; for most interior jobs, use a 50-ft. one. As you spray, hold a loop of the hose in your free hand to keep it out of your foot path—and away from freshly painted surfaces. Sprayer hoses come in 50-ft. lengths with couplings on both ends.

Start at the top and work down as you spray. Because a fine paint rain falls when you spray, it's better that it falls on unpainted surfaces—which will be painted over for a uniform finish.

Use a cardboard painting shield to keep paint overspray off adjacent surfaces you have already painted or won't paint at all, as shown in the top photo on p. 568.

Keep spray tips clean because they are easily clogged. So when you're done painting for the day, soak them in the appropriate vehicle (solvent for oil-based paint, hot water for latex). If the tip is really gunky, soak it in lacquer thinner.

Replace the spray tip if, after cleaning it, its spray pattern is still blotchy or the spray unit seems to be guzzling paint. Solids in the paint actually abrade the inside of a tip, enlarging it over time.

Keep filters clean and replace them often. Using a paint-appropriate thinner, clean filters every time you change paint colors, at the end of each day, or if you're pumping a lot of paint—say, 40 gal.—by lunchtime. A 100-mesh filter is considered minimum; coarser filters will admit debris that can clog the tip.

Lead-Paint Safety

In 2010, the Environmental Protection Agency (EPA) issued the Renovation, Repair, and Painting (RRP) ruling, which required professional contractors working on houses built before 1978 to take extensive precautions to protect their crews and homeowners from the potential presence of lead dust. Although the EPA exempts homeowners working on their own homes, lead poisoning doesn't exempt anyone. So it's something everyone should take seriously.

UNDERSTANDING LEAD PAINT

Lead-based paint adheres to almost any surface and weathers well, so it's not surprising that it can be found in 90% of houses built before 1940. However, as lead's health hazards became known

Safe lead abatement is two-pronged. First, contain dust by attaching mechanical scrapers or sanders to industrial-strength vacuums with HEPA-rated filters. Second, wear disposable coveralls, booties, and gloves taped shut at all openings, as well as a respirator with replaceable HEPA filter cartridges. A half-face mask and eye protection should be adequate in most cases.

When to Call in a Lead-Abatement Specialist

The EPA's 2010 RRP Rule is complex, and this brief discussion is no substitute for knowing the law. You can read it at www.epa.gov/lead/pubs/renovation.htm.

Broadly stated, the rules are designed to make people aware of the hazards of lead, require renovation firms to be certified in lead-safe work practices, present ways to limit potential contamination, and, in effect, direct those doing the abatement to *capture* lead-contaminated materials, treat them as toxic wastes, and dispose of them safely.

The law is quite specific about how waste is to be captured—requiring, for example, that exterior trim painted with lead paint be individually wrapped in sheet plastic before being carted off site. Containing the limited amount of debris generated by, say, a kitchen remodel is probably within the capabilities of most contractors or

well-prepared homeowners. But if the renovation project involves a sizable modification of an exterior that has tested positive for lead—hire lead-abatement specialists. Complying with particulars of the RRP Rule will be their headache, not yours.

In theory, containing lead-painted debris is straightforward, but in practice, it is a nightmare. Most job sites are already a chaotic stream of delivery trucks, subs, inspectors, and crew coming and going. Add to the mix nervous neighbors—thanks to extensive notifications of lead abatement in progress, posted throughout the neighborhood. If your region is windy, containment may also involve erecting scaffolding and a plastic envelope around the house. It can be a huge undertaking.

in the 1950s, paint manufacturers began to phase it out. It was banned altogether in 1978 by the U.S. government.

Because lead is a neurological toxin, it is particularly damaging to children ages six years and younger, who seem drawn to it because it's slightly sweet. Breathing or eating it can cause mental retardation in children. In people of all ages, lead can also cause headaches, anemia, lethargy, kidney damage, high blood pressure, and other ailments.

Because of its durability, lead paint was commonly used on exteriors, glossy kitchen and bathroom walls, in closets (which rarely get repainted), and as an enamel on interior doors, windows, stair treads, and woodwork. So each time a swollen window or sticking door was forced open, it ground lead paint into flakes and dust. Roof leaks, drainage problems, and inadequate ventilation added to the problem because excessive moisture causes the paint to degrade and detach sooner.

If you have an older home, the presence of lead may not be a dire problem if your home is well maintained and old paint is well adhered. However, lead becomes most dangerous when it becomes airborne, especially during sanding or heat stripping, for then it can be inhaled and easily absorbed into the bloodstream. Therefore, if you are considering renovating your old house, be methodical. Postpone any demolition or paint removal until you test to see if lead paint is present. If it is, develop a plan for dealing with it.

STEP ONE: TEST FOR LEAD

If you suspect the house has lead paint, test it to be sure. There are a number of test kits available at home centers, but at this writing only one satisfies EPA criteria for false positives and false negatives: LeadCheck.

Testing is pretty straightforward. Using a utility knife, cut away a small amount of surface paint to get at earlier layers underneath. Rub the test swab on the exposed area: the brighter the swabbed area, the greater the concentration of lead. Test several areas, such as door and window trim and siding. A LeadCheck kit contains 16 swabs and costs less than $50.

If sample areas test negative, it's probably safe to work on the house, although you should of course still wear appropriate respirators and minimize exposure to paint dust. If renovation is extensive, however, you'd be well served to order a lead-testing kit from an accredited testing lab, take additional samples, and have the lab validate the results you got from the inexpensive kit. *Note:* Some states consider home

Houses built before 1978 should be tested for lead paint. Cut out a small chunk of surface paint to expose older paint layers, and rub the exposed area with a test swab. This bright-red LeadCheck reading indicates a high concentration of lead.

lead tests to be inconclusive. Instead, they require *in situ* lead-testing with an XRF gun (handheld XRF spectrometer).

STEP TWO: COVER UP

Wear a respirator with N100 or P100 HEPA filters and snug-fitting goggles; a full-face respirator is not a bad idea but may be too hot to work in. Disposable coveralls, booties, and gloves are also a must; tape shut neck, wrist, and ankle openings. At the end of day, shed contaminated work clothes at the job site: Wearing them home can endanger your family. *Note:* You can find disposable respirators for about $10, but a reusable respirator with replaceable filter cartridges is a better buy if the job extends beyond a few days.

STEP THREE: CONTAIN THE MESS

Interior containment. Tape shut all HVAC registers and turn off the furnace/AC so it won't start while you're working in the room. To contain lead-paint dust and debris inside, use 6-mil sheet plastic and painter's tape to isolate and contain the mess. Tape plastic across door openings to seal off work areas from living spaces, and cover walls inside the work area with plastic as well so they can be wiped down and vacuumed when you're done. Protect floors with rosin paper or heavy cardboard, and cover that with sheet plastic taped to baseboards to keep it in place. When cleanup is complete, you will roll up and discard the plastic.

Before beginning lead abatement indoors, use painter's tape and sheet plastic to seal off heating registers and doors. Also cover interior walls and floors out from work areas approximately 6 ft. to capture dust and debris. Vacuum and wipe everything when you're finished.

Containing lead-contaminated materials outdoors is challenging. A containment area extending 10 ft. from contaminated surfaces should be covered with 6-mil black plastic. All debris must be bagged or wrapped in plastic before leaving the area. The black plastic also must be thoroughly cleaned with a HEPA-filtered vacuum system before it is folded in on itself and safely discarded.

Exterior containment. It's impossible to predict the movement of air or neighbors, so containing the mess of an exterior renovation is complicated. Start by taping sheet plastic over doors and windows to keep dust out of the house. Next, use 6-mil black plastic to cover the ground to a radius of 10 ft. from any lead-bearing surfaces that will be disturbed. To keep dust, paint chips, and debris inside this 10-ft. containment area, construct a box around it—rather like the sides of a sandbox—using lumber set on edge. Staple the plastic to the box. Lastly, create a cordon of rope or plastic warning tape in a 20-ft. radius from lead-bearing surfaces to keep people out of the work area. If strong winds are an issue, a plastic envelope supported by scaffolding may be required, as noted in the sidebar on p. 548.

STEP FOUR: MINIMIZE DUST

An essential part of confining lead contamination is minimizing dust. Consequently, any sanding, grinding, or scraping tools must be attached to a HEPA-filtered vacuum system. Noxious fumes are equally taboo, so avoid using heat guns, which generate temperatures exceeding 1,100°F. Setting dry, old wood ablaze is another hazard to avoid. So if strip you must, chemical strippers are safer, especially environmentally friendly ones (see p. 560).

Before scraping, stripping, or removing any lead-painted materials, mist the surfaces to suppress dust. During cleanup phases, mist plastic wall and floor coverings too to keep down dust. A gardener's pump sprayer is a good tool for this operation.

STEP FIVE: DECONSTRUCT, DON'T DEMOLISH

Carefully dismantling materials will release less dust than demolishing them. Scoring along the edges of painted trim, siding, and other building materials will reduce debris when removing them. When you pry such materials, do so gradually along their length to minimize breakage. If it's necessary to cut tainted materials to a more manageable size, use a power saw attached to a HEPA-filtered vacuum system. Remember, *every piece* of contaminated building material must be bagged in extrastrength plastic disposal bags—or wrapped in plastic and vacuumed—before removing it from the containment area.

STEP SIX: CLEAN THOROUGHLY

Outside. After bagging or wrapping everything sizable, use a HEPA-filtered vacuum to capture dust and small debris. Mist the plastic to keep dust down, detach the plastic anywhere you stapled or nailed it to the house exterior or the containment frame, then fold the plastic in on itself to capture any remaining residue and bag the plastic. While still in the containment area, vacuum your coveralls and booties, then remove and discard them.

Inside. Once you've bagged, wrapped, or removed any interior debris, wipe down the plastic covering the walls. (Wet- and dry-cleaning cloths can be purchased at most home centers.) Then wipe plastic floor tarps. Roll each plastic sheet in on itself to trap remaining debris, then bag it. Vacuum coveralls and booties and discard them. Finish up by wiping down all surfaces with a wet-cleaning cloth; each cloth is good for cleaning about 40 sq. ft.

Painting the Interior

If you see water stains, widespread peeling, mold, or large cracks that suggest structural movement, attend to the underlying causes first.

GETTING READY TO PAINT

When painting interiors, it's best to move the furniture out. If that's not possible, group it in the center of the room and cover it with a plastic tarp.

Highlight all blemishes for filling later by circling them with a pencil or attaching scraps of painter's tape near them, as shown.

Remove drapes, wall hangings, and mounting hardware, and fill holes. Turn off electric power to the room—use a voltage tester to be sure it's off—and remove the cover plates of electrical outlets and switches. Light fixtures or hardware left in place should be masked off or wrapped in plastic. Finally, cover the floor with canvas drop cloths—plastic is too slippery to work on.

Previously painted surfaces don't need much preparation if the surface is intact: Sand lightly with 150-grit sandpaper or a sanding block. If paint is flaking or loose, remove it with a paint scraper or spackling knife. Then sand rough paint edges with 120-grit to 150-grit sandpaper. Use spackling compound to fill holes, and sand it when dry. Apply a bead of paintable caulk (acrylic latex) to fill gaps where the trim meets walls, smoothing it with a moist finger. Caulking makes the finished paint look much better.

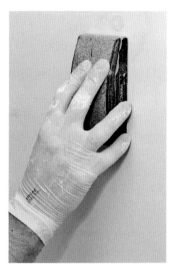

Use a sanding block or fine sandpaper to lightly sand all fills and patches before priming.

After sanding, dry-mop or vacuum surfaces to remove dust. Then sponge-wash them with a mild detergent solution and rinse with clear water. If the walls are especially greasy (kitchen walls, for example), use a more aggressive cleaner like trisodium phosphate (TSP). After rinsing, allow walls to dry thoroughly before painting. A good test of dryness is to check whether a piece of transparent tape will stick or not.

Unpainted drywall and plaster must always be primed. You can prime drywall as soon as the

On Priming Painted Surfaces

Primers bond to substrates and provide a stable base for finish coats. It's wise to prime previously painted surfaces in the following situations:

▶ You're switching paint types—say, applying latex over oil based.

▶ The old paint is flaking, chalking, stained, or otherwise in poor condition.

▶ The old paint is glossy and thus would prevent the new paint from adhering well.

Before applying primer, scrape, fill, sand, wash, and rinse the surface, and allow it to dry thoroughly.

In general, like bonds best to like. That is, latex paint bonds best to latex primer, oil based to oil based. But a quality acrylic latex primer is a good all-purpose choice because it bonds well and suppresses water stains, crayon marks, smoke, rust, and creosote. However, if you get severe bleed-through, switch to an oil-based sealer-primer instead.

Stickability Test

Here's how to test old paint before selecting a new paint to ensure that the new paint will stick:

▶ Bend a paint chip that's coming off. If it cracks, it's oil based; if it flexes, it's latex.

▶ Duct-tape a wet sponge to the wall and wait 15 minutes. If there's paint on the sponge or you can rub any off the wall, it's latex.

▶ Put a few drops of latex solvent such as Goof Off on a painted windowsill; if the paint bubbles, it's latex.

PRO TIP

Amateurs mistakenly paint from a 1-gal. paint can instead of a painter's pail. Consequently, they dip into too much paint, which they then scrape off on the can lip. It's better to dip into less paint and with two flicks of the wrist, dip-tap-tap the brush on both sides of a pail. This removes excess paint that might drip but leaves most of the paint on the brush, letting you paint farther.

final topcoat of joint compound has dried and has been sanded. Some pros still prefer an oil-based primer for drywall, but today's acrylic latexes seal as well and are far easier to clean up.

Plaster surfaces must be cured thoroughly before painting. Although latex primers can be applied as soon as the plaster is dry to the touch, it's better to wait three to four weeks. Latex paint allows some migration of moisture, so plaster can continue to "breathe off" water vapor. Restorers familiar with plaster recommend diluting latex primer 15% with water so the coating is thinner and even more permeable.

Oil-based paints are another story. Because the alkali in plaster can remain "hot" for up to three months, wait that long before using oil-based paints. Otherwise, free alkali in the plaster will attack the paint. Alkaline-resistant primers formulated for new plaster may shorten your wait somewhat, but they must be special-ordered. Before ordering, make sure that the primer will be compatible with the paint.

PAINTING CEILINGS AND WALLS

Before painting, read the earlier sections of this chapter on equipment (especially respirator and masks), safety concerns, and painting basics. All offer tips that can save you hours and keep you safe.

As you paint, be methodical so you won't need to touch up missed areas. Paint top to bottom: Do ceilings, walls, trim, and baseboards before doing doors and windows. Paint back to front. Many painters go to the deepest recess of a room—often, a closet—and work methodically toward a door. Paint inside to out. If you start painting in the backs of built-ins and cabinets, your final brushstrokes on the outermost edges will be clean and crisp.

Once you've prepped the surfaces, masked off baseboards, and spread drop cloths, it's time to paint.

Painting the ceiling begins by using a brush to cut in a 2-in. to 3-in. border where the ceiling meets the walls and near all moldings. This cut-in border reaches where a roller can't and thus allows you to roll out the rest of the ceiling without getting paint on the walls. Later, as you roll within ½ in. to 1 in. of the ceiling–wall intersection, you'll cover the brushmarks, so the paint texture will look uniform. This operation goes much faster if one painter on a step bench cuts in, while the second painter rolls on paint, using an extension pole to reach the ceiling.

To avoid obvious lap marks, paint the ceiling in one session, working across the narrowest dimension of the room. Roll out paint in 3-ft. by 3-ft. sections—about one roller-load of paint.

LINGERING PAINT Smells

As paint dries, it outgases (gives off gases), releasing water vapor or mineral spirits and additives into the air. The warmer the room and the better the ventilation, the sooner the smells will dissipate. Labels on paint cans indicate drying times. Typically, in a room that is 60°F or warmer, acrylic latexes will be dry enough to recoat in two to four hours. Oil-based paints can be recoated in 24 hours. However, odors may linger because the paints need longer to cure: 8 to 10 days for latex, 28 to 30 days for oil-based paints.

First, roll the paint in a zigzag pattern, which distributes most of the paint in three or four strokes, then go back and roll the paint evenly. When the roller is almost unloaded, slightly overlap adjacent areas already painted. Keep roller passes light, and don't overwork an area. Once the paint is spread evenly and starting to dry, leave it alone so its nap marks can level out.

For a smooth finish, use a standard 9-in. roller cover with ⅜-in. to ½-in. nap. Thanks to the extension pole, you can reach the ceiling easily, without needing to stand directly under the roller and its fine paint rain. To minimize mist and

Use a brush to cut in a 2-in. to 3-in. border, then follow up with a roller. You can quickly paint the vast expanse of wall—and cover brushmarks—without getting wall paint on the crown molding.

Primers and Paints*

SURFACE	PRIMER AND PAINT	COMMENTS
Drywall		
Unpainted	Acrylic latex primer and paint	Don't sand between coats.
Painted with oil-based semigloss or gloss	Oil-based (alkyd) or latex	To switch to latex: sand oil-based paint, vacuum, prime with acrylic latex primer.
Painted with latex	Acrylic latex	Sand lightly before first new coat; not needed thereafter.
Plaster		
Unpainted	Acrylic latex primer and paint	Plaster must be cured before painting; dilute primer coat.
Painted with oil-based semigloss or gloss	Oil-based or latex	To switch to latex: sand oil-based paint, vacuum, prime with acrylic latex primer.
Painted with latex	Acrylic latex	Lightly sand before painting.
Interior trim		
Doors, unpainted	Oil-based or latex primer and paint; semigloss finish	Oil-based paint soaks into wood, dries harder, resists abrasion; sand between coats.
Unpainted	Clear finishes, such as polyurethane and varnish	Always seal bare wood or it will become grimy and dull.
Painted with oil-based semigloss or gloss	Oil-based paint	Sand between coats.
Painted with latex	Acrylic latex	Not as durable as oil-based paint.
Exterior		
Siding and trim, unpainted	Acrylic latex primer and paint	Latex stays flexible, allows some moisture migration.
Painted with oil-based semigloss or gloss	Oil-based paint, or acrylic if properly primed	Unless you strip siding, stick with oil-based paint.
Painted with latex	Acrylic latex	

* Oil-based *here is synonymous with alkyd, now mentioned on most containers of paint and stain. Alkyds are synthetic resins that have largely replaced the traditional petroleum-oil base.*

PRO TIP

If there are two painters, divide the job: The first painter leads the way with a brush to "edge" the corners, trim, and other hard-to-roll areas. The second painter follows, rolling over the edging to hide brushmarks, thereby giving the wall a uniform texture. The first painter should edge out 2 in. to 3 in. from the trim and corners, and the second painter rolls to within ½ in.

Use a hot-dog roller to fill those tight spaces over doors, around windows, inside cabinets, and the like.

Rather than masking windowpanes or attempting to cut in clean paint lines, paint slightly (¹/₁₆ in.) onto the glass. After the paint dries, use a razor to cut a clean line.

PRO TIP

Before using a razor to clean paint from window glass, wash the windows well. Otherwise, grit or sanding residue could scratch the glass as it is pulled along the surface by the razor. Those scratches will be all too visible when the sun shines through the glass. For best results, change razors often.

drips, run the roller up and down the bucket ramp several times when loading. But don't fret about small, stray spots on walls because you'll cover them later when you roll the walls.

Painting the walls is nearly the same as painting ceilings—cutting in with a brush and rolling larger areas—except that you can load more paint on the roller. To reduce spatter, roll up on the first stroke; the excess will fall back to the roller. Continue rolling in a zigzag pattern to unload the roller before rolling out the paint. A 6-in. hot-dog roller can paint the areas over doors and windows that are too narrow for a standard 9-in. roller. Rolling is always faster than brushing because you don't need to dip a roller in paint continually. If you're careful around electrical outlets, you can also use a hot-dog roller there.

With more paint on your roller, you can cover slightly larger expanses of wall, say, 3 ft. by 4 ft. If you start at the top of the wall and work down, you'll roll over any drips from above. Cover brushmarks by rolling within ½ in. to 1 in. of the ceiling; this is important when applying darker hues because rolled-on paint reflects light differently than paint that's been brushed on. Slightly overlap adjacent sections. To avoid unloading excess paint along outside corners, lighten up as you roll.

Finally, sand lightly between coats when you apply oil-based paint, especially enamels on cabinets or trim. On walls, use 220-grit sandpaper or a dry sanding block. It's not necessary to sand

After unloading most of the paint on the roller in a zigzag pattern, spread it out evenly, top to bottom.

latex paint, unless you've waited several weeks between coats; until latex is 100% dry, new coats adhere easily.

Painting interior trim begins with preparation tasks. Prepare the trim, window sashes, and doors by filling nail holes with nonshrinking wood filler, priming bare wood, caulking gaps with acrylic latex caulk (letting it dry overnight), lightly sanding all trim with 180-grit sandpaper, and vacuuming dust and debris. Enamel paint—which dries to a hard, glossy finish—is best for trim, window sashes, and doors because it's the most durable. By the way, there are both oil-based and latex enamels.

Painting straight edges requires a quality brush and a steady hand. If you can develop a steady hand, you won't need to use masking tape, which is time-consuming to apply. In most cases, all you need is a 2½-in. or 3-in. angled sash brush, unless your baseboards are exceptionally wide. Start with crown (ceiling) molding, proceed to door and window trim, and finish with the baseboards. Always paint with the grain, cutting trim edges first, then filling in the field with steady back-and-forth strokes. To avoid lap

Faux-Painted Walls

Faux (pronounced *foe*) is French, meaning "false" or "imitation." It's used here to describe various advanced painting techniques that create layered finishes, sometimes to imitate stone or wood or simply to allow underlying layers of paint to show through. Typically, two accent colors are applied to already painted walls and ceilings. If the surfaces are irregular—say, rough plaster—all the better, for colors will look more varied and unpredictable. In general, thin coats of paint allow you to see layers underneath. The photos here show a few faux-painting basics; if this whets your appetite for more, there are lots of good books on the subject.

Faux Texturing

In the sequence shown here, the painter was trying to achieve an old look, as if an imperfectly plastered wall had been painted many times. There's no single right way to apply a faux finish, so you can experiment with methods and materials until you get a look you like. Then just try to re-create that look consistently throughout the room.

1. After pouring a small amount of the first accent color into a shallow container, load your stippling brush lightly. Then quickly jab the bristles at the wall to create a stippled effect. Follow with a dry rag, lightly patting the just-applied paint to flatten the stipple points. Just flatten the points; don't remove the paint itself.

2. Apply the second accent color with a sponge, while the first coat is still wet. Tip: Because dipping a sponge will load too much paint onto it, instead brush the paint onto one side of the sponge, as shown.

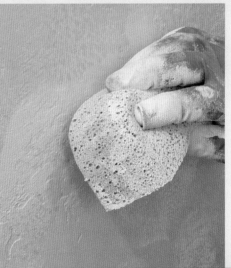

3. Pat the second accent color on. Then flip the sponge to its dry side and gently move the color around, spreading paint, not removing it; the wet side of this sponge would have streaked the paint. Stand back periodically to check whether the faux effect is consistent from wall to wall.

marks, paint about 3 ft. of trim at a time, overlapping adjacent sections while they're still wet. If the paint is drying too fast, add Flood Floetrol to latex paint or its Penetrol to oil-based paint.

If trim edges are thinner than ³⁄₁₆ in., they'll be difficult to cut in without spreading trim paint on the wall. In that case, overlap the wall paint onto the trim edge so that it covers the edge completely, producing a clean, straight line. In other words, the thin edge of the trim will be covered with wall paint, not trim paint, but your eye won't notice.

Windows sashes vary greatly in design. But as a general rule, paint them from the inside of the sash out. That is, if sashes are divided into multiple panes by *muntins* (narrow wood sections between panes of glass), paint the muntins first. Then paint along the insides of sash rails and stiles where they meet glass. Finally, paint the faces of sashes. To develop a rhythm, paint all the vertical muntins—one side at a time—then the horizontal muntins. By painting similar window elements at the same time, rather than jumping around, you'll be less likely to miss elements and the work will go faster.

Don't worry about cutting in clean edges at the glass. Instead, paint slightly onto the glass (¹⁄₁₆ in.), even if unevenly, thus creating a tight seal. After the paint dries, use a razor to cut a clean line on the glazing.

Open windows to paint their edges. When painting double-hung windows, follow the steps at left. If you are repainting the exterior of the house at the same time, go outside and paint the accessible parts of the window. Slide the window sashes back to their original position, and finish painting. To prevent binding, move the sashes as soon as the paint is dry.

Painting a door is easiest if you pull the hinge pins and lay the door across a pair of sawhorses. (If that's not possible, shim beneath the door so it can't move.) For the best-looking results, remove all door hardware except hinge leaves, especially if you're spray painting. Cover the hinges with masking tape. If you prefer not to remove the old latch mechanism and escutcheons, carefully mask them, too.

If you're brush painting a flush door (flat surface), divide it into several imaginary rectangles, each half the width of the door. Apply paint with the grain and overlap the edges of adjacent sections. Work from top to bottom. Painting panel doors is similar, but work from the inside out: Paint the insides of the panels first, next the rails (horizontal pieces) top to bottom, and finally the vertical stiles.

Painting cabinets is faster if you remove and spray paint drawers and doors, and brush paint cabinet frames. You'll need a spray room isolated from the house (a clean garage is ideal); a drying rack for doors; and a sprayer, which you can rent. Be sure to read the earlier sections on painting safety and spray painting, which emphasize ventilation and wearing a respirator.

Start by washing doors and drawer fronts, especially those near the kitchen stove. If your cleaner isn't cutting the grease, try TSP or dena-

Painting a Double-Hung Window (Interior View)

1. Before painting, make sure both sashes are operable. Raise the inner sash, as shown, so you can paint it completely, including the tops and bottoms of its rails (horizontals).

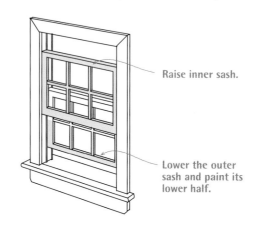

Raise inner sash.

Lower the outer sash and paint its lower half.

2. Reverse the position of the sashes and paint the rest of the upper sash.

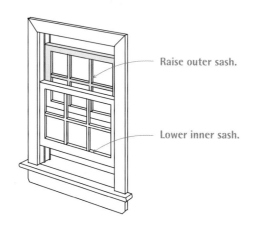

Raise outer sash.

Lower inner sash.

3. Once the paint is dry enough to handle, lower both sashes completely and paint the upper half of the jambs. When that's dry, raise both sashes, and paint the lower half, and then the window trim.

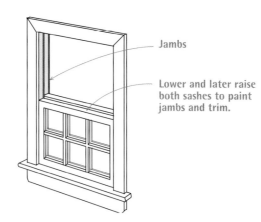

Jambs

Lower and later raise both sashes to paint jambs and trim.

PRO TIP

To paint both sides of a door without waiting for the first painted side to dry, drive a pair of nails into the top and bottom of the door, then rest those nails on a pair of sawhorses. You and a helper can grab the nails and flip the door over. When both sides are dry, pull the nails, then fill and paint the holes.

||||

PRO TIP

Most amateurs fold sheets of sandpaper in half, then in quarters. But sheets will last longer if you fold them in thirds. Folding sandpaper in fourths places abrasive surfaces face to face, causing premature wear.

||||

To spray paint a door, first place it across sawhorses at a comfortable working height. Mask hinges to keep paint off and ensure that the doors will hang correctly when reinstalled. Then move the sprayer smoothly and continuously, maintaining a constant distance from the surface. Overlap preceding passes by about a third. On each pass, begin spraying before the spray tip is over the door, and don't release the trigger until the paint fan is past the far edge. Use your free hand, as shown, to keep the hose out of the way.

tured alcohol; make sure you wear goggles and gloves. That done, examine the cabinet parts and their hardware, and plan to replace the doors or drawer fronts that are warped or not repairable, as well as hardware that's broken or outdated. Before disassembling cabinet parts for spraying, assign each door and drawer a number. Write these numbers just inside the cabinet frame, where they won't be covered by paint. Number lower cabinet doors on bottom edges and upper doors on top edges (the least visible locations), or number them behind the hinges.

Tape over door-hinge mortises if you'll reuse the hinges. Otherwise, paint buildup in the mortise may misalign the hinges and thus the doors. Either cover the mortises with tape or leave the hinges on the doors and mask off the hinges.

Remove hardware before prepping the doors. If existing paint is flaking or the doors are dented, start with 100-grit sandpaper in a random-orbit sander, wipe off dust with a damp rag, and fill cracks and holes with nonshrinking wood filler. Repeat the sequence as needed, ending with a 220-grit sanding by hand. However, if the old paint is in good condition, a single pass with 220-grit paper and a damp rag is all you'll need to prep before painting.

When spray painting only the face of a drawer, mask off the rest.

Homemade Drying Rack

This freestanding drying rack is constructed from 8-ft. lengths of ½-in. galvanized-steel electrical conduit, 2x2 frames lag-screwed together, and two pieces of ¾-in. plywood. The plywood base is roughly 30 in. by 30 in.; the plywood top can be smaller, say, 18 in. by 18 in.

Drilling the holes in the 2x2s slightly larger than the conduit diameter allows quick disassembly. To keep the conduit from getting dinged during transport, store it in 3-in. plastic DWV (drain, waste, and vent) pipe with capped ends; wrapping blue tape around the conduit prevents the metal from marring newly painted cabinet doors. To avoid tipping, load the rack from the bottom, unload it from the top, and balance the weight carefully side to side.

Stains and clear finishes are thinner than paint and more inclined to run, so mask off adjacent areas before starting prep work.

For the most durable surface, apply a coat of primer-sealer to unpainted surfaces. (A painted surface in good condition does not need a primer.) Follow this with three coats of enamel, which will hide well, even if you're applying light paint over dark. Use acrylic-latex paint for the primer and finish coats, even if the cabinets are presently covered with oil-based enamel. Top-quality latex enamel is almost as tough as any oil-based enamel, it dries faster, and it's much easier to clean up. To minimize runs, keep the doors horizontal during spraying and drying. Between coats, sand lightly with 320-grit garnet sandpaper. Painting drawer faces is essentially the same, except that you should mask off the drawer sides. Paint cabinet frames from the inside out, finishing with long, vertical strokes on the frame faces.

Stripping and Refinishing Interior Trim and Wood Paneling

Natural wood can be handsome, but stripping layers of old paint or a dulled finish is an enormously tedious, messy job. The following questions and tests may give you easier options.

SIX QUESTIONS BEFORE STRIPPING

▶ What kind of paint? Trim paint in houses built before 1978 likely contains lead, which becomes hazardous if you sand it or heat-strip it. Yet it may be perfectly safe if it's intact and well maintained. Analyze a paint sample, as explained on p. 549. Also, the more paint layers, the bigger the mess.

▶ What kind of wood? Builders often used plain or inferior-grade softwood for trim they intended to paint. Test-strip a small section to see if the wood is worth stripping. Common pine or spruce and badly gouged wood probably aren't.

▶ How thick is the wood? If wood paneling is a ¹⁄₁₆-in. veneer, it may be too thin to sand, let alone scrape and strip. After turning off the electrical power, move panel battens (vertical pieces) or electrical outlet covers to see the edge of a panel.

▶ Will washing do it? Clear finishes that have become dull and grimy may just need a good washing. Using a damp rag, rub Murphy Oil Soap onto a small section and wipe it dry quickly. If that clouds the surface, stop; but if it brightens the woodwork, keep going.

▶ Scuff-sand and touch up? If a clear finish remains dull after a test-wash or is worn looking, scuff-sanding and a new application of the old finish may do the trick.

▶ Will new coats adhere? If painted or clear finishes are cracking, peeling, or otherwise coming off, new coatings won't stick. To test adherence, use a utility knife to lightly score a 1-in. by 1-in. area into nine smaller squares (like a tic-tac-toe array). Press a piece of duct tape onto the area and pull up sharply: If two or more little squares pull off, you should strip the paint or finish.

What's That Finish?

To identify a finish, rub on a small amount of the test solvents in this list, starting at the top of the list (the most benign) and working down until you've got your answer. When applying solvents, wear rubber gloves, open the windows, and wear a respirator.

▶ **Oil.** If a few drops of boiled linseed oil soak into the woodwork, you have an oil finish: tung oil, linseed, Watco, or the like. If the oil beads up on the surface, the woodwork has a hard finish, such as lacquer, varnish, or shellac. Keep investigating.

▶ **Denatured alcohol.** If the finish quickly gets gummy, congratulations! It's shellac, which will readily accept a new coat of shellac after a modest sanding with an abrasive nylon pad or 220-grit sandpaper. Older woodwork with an orange tinge is often shellac-coated.

▶ **Mineral spirits** (paint thinner). This will dissolve wax immediately. Dampen a rag and wipe once. If there's a yellowish or light brown residue on the rag, it's definitely wax. If your woodwork finish has an unevenly shiny, runny appearance, suspect spray-on wax.

▶ **Lacquer thinner.** This solvent dissolves both varnish and shellac, so try denatured alcohol first. If alcohol doesn't dissolve the finish but lacquer thinner does, it's varnish.

▶ **Acetone.** This one will dissolve varnish, too, in about 30 seconds. But if acetone doesn't affect the finish, it's probably polyurethane.

In this 1920s house, the homeowner wanted an older look for the cabinets. So after spraying three coats of oil-based enamel, the painters rehung the doors and rolled on a final coat . . .

. . . which they then lightly tipped off with a dry brush, intentionally leaving very faint brushmarks.

MOTHER NATURE'S Strippers

Solvent-based paint strippers, which usually contain methylene chloride (dichloro-methane) are effective, exceptionally smelly, and (according to the Centers for Disease Control and Prevention) probably carcinogenic. Not surprisingly, the plethora of green building products now include a range of solvent-free chemical strippers. None of this new batch of strippers removes finish as aggressively as solvent-based strippers, but if you're patient, repeated applications will get the job done. In general, gel strippers work more effectively because they adhere better—especially to vertical surfaces—and don't dry out as quickly. *Note:* Many of these eco-strippers tout their safety for indoor use, but make sure there's adequate ventilation anyhow. Some names to look for: Smart Strip, Ready-Strip, Citristrip, Multi-Strip, Fastrip, and Lift Off.

Solvent-free paint strippers are typically less aggressive than solvent-based varieties—and a lot less smelly. Wear a respirator, safety glasses, gloves, and long sleeves when using any paint stripper.

STRIPPING METHODS

Test-strip small sections of woodwork to see which method—or combination of methods—works best for you.

Metal scrapers with straight edges work well on flat surfaces without too many layers of paint or clear finish. A scraper with changeable heads enables you to scrape varying contours. For best results, hold the scraper head roughly perpendicular to the surface and pull the tool toward you. *Caution:* Sharp scraper heads can easily gouge wood, especially softwoods like fir and pine, whose contours may be obscured by thick paint.

Heat guns soften paint so you can scrape it off. Heat guns can remove many layers of paint, but stay alert when using them. Maintain a constant distance from the surface you're stripping, and keep the gun moving so you don't scorch one spot. Using a heat gun on shellac and varnish gets tricky because they have low kindling temperatures and tend to burn when heated; first, try stripping them with metal scrapers.

Never use a heat gun next to glass—for example, on window muntins—because you could crack the glass. Heat guns can also ignite dry materials within walls, so stop using guns well before the end of the workday so woodwork can cool. Before you leave for the day, sniff around for smoke or anything that smells hot.

For chemical strippers, a rule of thumb is the stronger and smellier the chemical, the faster it will strip paint or finish. Methylene chloride (dichloromethane), for example, will soften multiple layers in 10 to 15 minutes. Fortunately, there is an emerging family of solvent-free strippers that, although not as aggressive, will strip effectively given enough time and reapplications. They are easier on the environment and your lungs, but adequate ventilation is still important. Follow application recommendations on the container label. By the way, gels and semipaste strippers are best for vertical surfaces. Even when brushed on thickly, they won't run.

Chemical strippers require patience and care. Use a rag to cover the cap before opening the container so stripper won't splash on you. Pour stripper slowly into a work pail, and close the container immediately so it won't spill if the container is bumped or knocked over. As you apply stripper, brush away from yourself. To avoid tracking stripper throughout the site, replace plastic tarps as they become fouled with softened paint. Or lay down newspaper, which is cheap and easy to roll up before stuffing it into a garbage bag.

STRIPPING SAFELY

Before stripping woodwork, read "Painting Safely" on p. 544 and "Lead-Paint Safety" on p. 548. Many of the concerns when stripping are the same as those when painting. Most important, wear a respirator with replaceable filters. Also wear rubber gloves, goggles, and a long-sleeved shirt. Lay down plastic tarps (or layers of newspaper) to protect floors and capture paint debris, mask off areas you're not stripping, and make sure you have adequate ventilation. Even if a chemical stripper is relatively odorless and claims to be eco-friendly—keep it off your skin and out of your lungs! Read instructions for all stripping chemicals before using them, and if you're using a heat gun, have a fire extinguisher close by.

PRO TIP

If you're using water-based finishes, use damp rags to wipe dust off surfaces, rather than tack rags. Because tack rags are typically pieces of cheesecloth treated with varnish to make them sticky, they leave a faint, oily film that water-based coatings may have trouble adhering to. Tack rags are fine, however, if followed by oil- or solvent-based finishes.

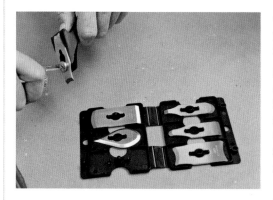

This hand scraper comes with six interchangeable stainless-steel blades, which will fit most contours you're likely to encounter.

Once your tarps or newspaper are in place, brush on stripper liberally. A ⅛-in. to ¼-in. coating of stripper should stay wet long enough to soften all the layers of paint or finish. To make sure that slower strippers stay moist, press a sheet of lightweight plastic (polyethylene) right onto the stripper-coated woodwork; the stripper won't dissolve the plastic. Periodically lift an edge of the plastic and try scraping off the paint. Be patient: Remove the plastic only when the softened paint scoops off easily. Until then, leave the plastic on.

Although renovators typically use a wide spackling knife or a putty knife to scoop off softened paint, a wooden spatula with a beveled edge is a near-perfect tool because it won't gouge the chemically softened wood. Whatever tool you use, unload sludge from your scraper after each pass. Use a toothbrush, a nylon potato brush, or a handful of wood shavings to dislodge softened paint from detailed or hard-to-reach areas. Only occasional spots should need additional stripper.

When the woodwork is bare, scrub off the stripper residue with a solvent recommended by the manufacturer. For solvent-based strippers, typically mineral spirits is applied with a nylon abrasive pad, then blotted dry with paper towels. Follow that with a dilute solution (5% to 10%) of household cleaner in warm water, and wipe that off with paper towels. Allow the wood to dry thoroughly—at least a day or two—before filling holes or sanding. *Note:* Don't use steel wool to scrub stripper or remove paint. Otherwise, steel particles can stick in the wood and then rust, marring the new finish.

REFINISHING WOODWORK

Once your stripped woodwork has dried, patch it with wood putty that dries to the same color as the unfinished wood. (Putty lightens as it dries.) Test a number of putty colors, allowing each to

dry well before test-staining or finishing. When a patch is so hard that your thumbnail can't gouge it, the putty's dry enough to sand.

Sanding. If the woodwork is in good shape and doesn't need filling, just *scuff-sand* it (sand it lightly) with 220- or 320-grit sandpaper before applying a clear finish. More likely, you'll need to use several grades of sandpaper, starting with 80 grit or 100 grit to sand down tool marks or dings, moving on to 150 grit, and ending with 180 grit or 220 grit. Always test-sand an area in an out-of-the-way place, and always use the least abrasive paper that will still be effective.

Sand sections completely with one grit before switching to another, even if you think an area is smooth enough. If you switch from 120 grit to 150 grit while sanding a baseboard, for example, it may have two different shades when you stain or finish it.

If there's a lot of woodwork to sand, use a palm-size power sander (also called a block sander) or a random-orbital sander for the first three sandings, and finish up by hand-sanding with the wood grain, using 180-grit or 220-grit garnet paper. Wrap sandpaper around a standard blackboard eraser or a scrap of 2x4 to hand-sand flat areas; sandpaper wrapped around a dowel works well on concave areas. After sanding, wipe or vacuum the surfaces to remove the dust.

APPLYING A CLEAR FINISH

Clear finishes can be brushed on like paint or wiped on and off with a rag or pad—or some combination thereof. Brushed-on finishes tend to be thick and shiny, whereas wiped-on finishes are thinner and less shiny. As with oil-based paints, use natural-bristle brushes to apply oil-based clear finishes. Use synthetic bristles for water-based finishes.

Don't shake containers of clear finish. If you do, you'll entrap air bubbles. Instead, stir them well by hand until the thick, flattening agent at the bottom of the can is evenly distributed. Clear finishes tend to "skin over," so pour out small amounts into a painter's pail and replace the can lid promptly. If finish builds up around the rim, the lid won't seat well. To prevent buildup, use an old slotted screwdriver to punch slots around the rim's recess. These slots will let excess finish drain into the can, allowing the lid to seat tightly into the recess. If the finish skins over, strain the finish or discard it.

Polyurethanes are favored for wood in kitchens, bathrooms, hallways, and other busy areas. Once cured, they're tough and water-resistant. And they're easy to apply. Oil-based and water-based

COMMON CLEAR-FINISH PROBLEMS

Orange peel, often seen near kitchens, is caused by airborne cooking oils.

Weeping is wood sap excreted over decades.

Wax buildup is characterized by uneven, shiny sections where sprayed-on wax has run.

You can usually apply oil-based penetrating stains over previously varnished surfaces, but they can be tricky to work with. Test-stain an inconspicuous section. If the penetrating stain is compatible with the old finish, it should dry hard overnight.

After brushing stain on, use a clean, dry rag to remove any excess. You may need to apply several coats—over several days—to match existing stains, so be patient.

Staining and Sealing: A Sampler

There should be a Ph.D. in stains. There are all-pigment stains that won't fade (a good choice for window trim), water- and oil-based stains, liquid stains, penetrating stains that both stain and seal, and gel stains that won't run on vertical surfaces. Plus, there are wood conditioners, sanding sealers, presealers, and a plethora of putties and wood fillers. If you want to learn more, excellent resources are Michael Dresdner's *Painting and Finishing* (The Taunton Press, 2002) and *Finishing Wood* (The Taunton Press, 2017).

To achieve an even stain on softwoods such as fir, preseal them with a thinned coat of whatever the clear finish will be, say, 1 part oil-based polyurethane to 4 parts mineral spirits. Presealer soaks into the softer parts of the wood and seals them slightly. But once dry, the surface should still feel like wood.

Oil-based polyurethane over water-based stains is OK, but do not use water-based polyurethane over oil-based stains. The polyurethane won't stick.

polyurethanes are equally durable but require slightly different application methods. For both types, apply at least two coats, preferably three.

▶ Water-based polyurethanes: Because water-based polyurethanes dry clear, use them if you want light-colored wood to stay light. Seal woodwork before application, using a diluted finish. Thereafter, brush on full strength. Follow manufacturer recommendations for drying times. Some water-based polyurethanes dry quickly, allowing multiple coats in a day. There's no need to sand between coats unless a week passes, in which case, use a fine nylon abrasive pad. If you want to add a wax finish, follow recommendations on both the polyurethane and the wax cans—both for compatibility and drying times.

▶ Oil-based polyurethanes: Oil-based polyurethanes impart a rich, amber color to wood. So use them if you favor dark wood or a historic look. They don't need a preliminary sealer coat, but they will flow on better if you thin each coat with 10% mineral spirits. Oil-based polyurethanes dry slowly, so apply only one coat per day, unless the manufacturer recommends otherwise. Here again, sanding between coats is not imperative unless you wait a week between coats—or you need to sand down imperfections. Before waxing an oil-based poly, be sure to wait a week after the last coat dries.

Shellac doesn't have the water resistance of polyurethane, but it dries quickly, has a wonderful old-fashioned sheen, and adheres well to earlier shellac coats, so it can be touched up repeatedly. If wood is new or recently stripped, apply a sealer coat of thinned-down shellac. If you'll be staining the wood, first brush on a sealer of 1 part shellac to 4 parts denatured alcohol, allowing it to dry. Otherwise, brush or wipe on a coat of 1 part shellac to 2 parts denatured alcohol, wiping off the excess and allowing the coat to dry for two hours before sanding lightly with 320-grit sandpaper. Thereafter, apply two or three coats, thinned with 10% denatured alcohol. If the surface is smooth, there's no need to sand between coats. Wait one day between coats and three days before waxing. Because shellac dries so quickly, don't attempt to rebrush it.

Oil finishes include boiled linseed oil, tung oil, and the so-called Danish oils like Watco. Using a nylon pad or a rag, rub a generous amount of oil onto the wood. Let that soak in for 10 to 15 minutes before rubbing off the excess with a clean, dry cloth. With each coat, the wood will darken slightly. Allow each coat to dry for 24 hours, then reapply the oil until you get the look you like; typically, two or three coats do the trick. Oil finishes offer the least protection but are easiest to reapply.

Painting the Exterior

Exterior paint jobs can last 10 years or more if you're fastidious about prep work and attentive to water-related building details. Key factors include proper flashing of windows, doors, and roof junctures; maintaining gutter systems; caulking gaps in exterior siding; and adequately venting excess moisture from interior spaces.

PRO TIP

Shellac mixed from dry flakes is a terrific finish with better water resistance and clarity than premixed shellac. Also, it's a good idea to check the label for an expiration date. Old shellac in the can won't dry properly.

Ladder SAFETY

If you're painting a whole house, the job will go much faster if you rent scaffolding. However, if you decide to use an extension ladder or two, follow these safety rules:

▶ Don't place ladders near incoming electric service lines. When the air is moist, electricity can arc to nearby objects or people, so keep your distance.

▶ Securely position the ladder. Never ascend a ladder that lists to one side. On uneven ground, use a ladder with adjustable leveling feet, as shown on p. 52.

▶ Place the bottom of the ladder out from a building no more than one-quarter the ladder's height.

▶ Wear hard-soled shoes so your feet won't tire quickly on the ladder rungs.

Sheet peeling is caused by excessive moisture migrating through a wall—in this case, an unvented bathroom wall.

WHY PAINT FAILS

Before you sand or scrape anything, figure out why the paint is failing . . . and where.

Blistering is usually caused by painting over damp wood or an earlier coat of paint that isn't dry. Blisters often contain water vapor, although "temperature blisters" are largely hot air, caused by painting a surface that was too hot. Scrape and sand blisters, allow the wood siding to dry thoroughly, and spot prime.

Peeling off in sheets is blistering on a grand scale—sometimes an entire wall. This is most common on older homes lacking vapor or air barriers and occurs especially on siding outside bathrooms or kitchens, when excessive moisture migrates through the wall. If your old house has been retrofitted with insulation and a vapor barrier, peeling may indicate moisture trapped inside the walls and, possibly, rotted framing. At the very least, add ventilator fans to exhaust water vapor. And on outside walls, drive thin plastic wedges behind the lap siding to help moisture escape.

Intercoat peeling, a new coat of paint separating from the old, is a classic case of poor prep work. Typically, chalky old paint was not scrubbed or sanded and thus new paint could not adhere. Or, less often, the painter waited too long between the prime and finish coats. Scrape failed paint, sand, and wash the surface well, letting it dry before repainting.

Wrinkling is caused by applying paint too thickly, painting an exterior that is too cold, failing to thin paint sufficiently, applying paint before earlier coats are dry, or letting the paint get rained on before it cured adequately. Use a power sander to even out the surface before repainting.

Alligatoring, or cross-grain cracking, is caused by too many layers of paint, usually old, oil-based paint. The thicker the paint, the less it can flex as siding expands and contracts. So the paint cracks—sometimes all the way down to bare wood. Alligatoring may also be caused by painting over an undercoat that didn't dry completely. In either case, it's big trouble because you'll need to strip the paint down to bare wood and seal it with a primer-sealer before repainting. It may be easier to replace the siding.

Chalking is a normal occurrence and isn't a problem unless it's excessive, usually the result of cheap paint. Because new paint won't adhere well to a powdery residue, you must scrub and rinse the old surface and allow it to dry before repainting.

Rusty nail stains are common where siding nails are not galvanized. The fastest fix is sanding each stain lightly and priming with a rust-inhibiting primer like Kilz stain blocker. For a longer-lasting repair, sand until you expose each nail head, use a nail set to sink each one 1/8 in. below the surface, prime with stain blocker, and fill with wood filler. Then sand and spot-prime before painting.

PRO **TIP**

Although white pigmented shellacs such as B-I-N are terrific for blocking stains on interior surfaces, they're not advisable for priming exteriors. Hot sun softens and degrades shellac.

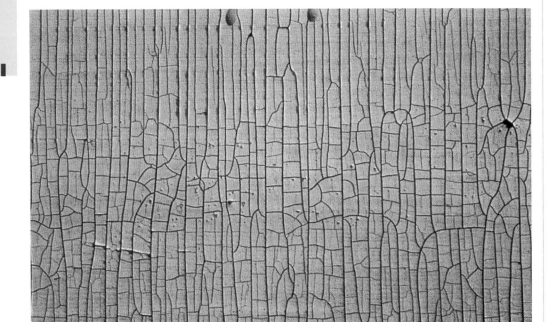

Alligatoring, or cross-grain cracking, is caused by too many layers of old, inflexible paint.

Hand-scrape nooks and crannies that power tools can't reach or could damage. Here, heat guns and chemical scrapers also make sense.

To maximize adhesion, spot-prime seams and gaps before caulking them with an exterior-grade, paintable acrylic or multipolymer caulk. Paint won't stick to pure silicone caulk.

Where a random-orbit sander won't fit, use a palm sander, as shown, with 60-grit or 80-grit sandpaper. A palm sander is also handy for scuff-sanding old paint in good condition so new paint will adhere better.

Tannin bleed-through, a widespread brown staining, occurs when waterborne resins in woods, such as cedar and redwood, bleed through porous latex primers. Scrub the surface well, and prime it with one or two coats of an oil-based (alkyd) primer such as Kilz or Benjamin Moore Fresh Start. Alternatively, you can use an alcohol-based primer, which is also good for covering and sealing fire- and smoke-damaged wood—and its smell. Once you've sealed the bleed-through problem, paint what you like—latex or oil based—over that.

Graying wood is a natural response to sunlight when siding is left unfinished or has been sealed with a clear finish that degrades. Never leave wood siding exposed to the elements—both because bare wood quickly degrades and because paint or clear finishes don't adhere well to degraded wood. Sand and wash the surface, and reapply a clear finish with a UV-blocking agent. Clear finishes need to be reapplied every two to four years. If that sounds like too much maintenance, prime the siding with an oil-based primer, and switch to paint.

Mold and moss are common in humid climates, on north-facing and foliage-shaded walls, and where lawn sprinklers hit the house. Siding that's constantly damp can lead to structural rot. For starters, cut back foliage and adjust the sprinklers. Scrub moss off by hand, using a wire brush. Remove mold by applying a cleaner/mildewcide such as Zinsser Jomax before scrubbing or power washing the surface (but see the cautions later in this section). After the exterior has dried for about a week, prime as needed, and repaint with a paint containing a mildewcide.

PREPARING THE EXTERIOR

Houses built before 1978 may contain lead paint, so before starting prep work, be sure to review

A mechanical scraper with a vacuum attachment is the tool of choice when you've got to strip exterior paint. But set the nail heads first, or you'll chew up expensive scraper blades. Eye protection is a must.

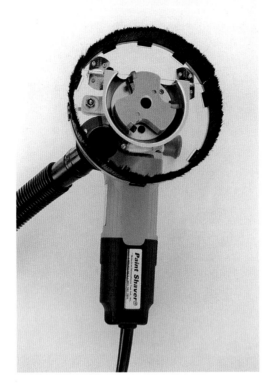

this chapter's earlier sections on equipment and lead-paint safety—and test for lead (see p. 549).

Even though you're working outside, wear a respirator when sanding or scraping paint and when working with solvents or paints of any kind. Likewise, when applying cleaning solutions (which can be caustic), wear rubber gloves and eye protection.

Prep work is prodigiously messy, so spread drop cloths out 8 ft. from your house to protect your lawn and shrubs. Otherwise, you'll be picking paint scraps out of your grass for years or—even worse—exposing kids and pets to lead-based paint. If it's hot and sunny, don't cover your plants with plastic drop cloths; the plants will cook. Instead, use fabric drop cloths, which shade plants, won't tear, and are far less slippery to walk on. But if tests indicate lead paint, capture the debris in heavy 6-mil plastic tarps, which you can roll up and discard at the end of each day.

Washing surfaces. Before installing new wood siding, apply primer-sealer to the front and back faces and to all edges. When that coat is dry, scuff-sand it with 100-grit sandpaper, and dust it off with a whisk broom. Then install the siding before applying the topcoats.

However, if the siding is already painted—even if the paint is in good shape—wash it first. Start by applying a house-cleaning solution, using a garden pump sprayer. A dilute solution of TSP is often recommended, but urban houses may need a cleaning agent with a degreaser that will cut soot, automobile exhaust, and the like. As noted earlier, if there's mold present, use a cleaner with a mildewcide. Once the cleaner has had time to work, rinse it off with a garden hose and allow it to dry thoroughly for a week or so before painting. That's a minimal wash.

To thoroughly wash and rinse an exterior, rent a pressure washer, which has a small boiler and a high-pressure electric pump. The rental company will recommend a detergent suitable for the unit and explain how to use it safely. For most cleaning jobs, 1,800 psi to 2,500 psi is specified—for softwoods such as cedar or redwood, use 1,000 psi to 1,500 psi. Tip sizes range from 0° (concentrated pressure that can easily damage siding) to 40° (a wider fan of water favored for light cleaning). In general, start with a low-pressure setting until you are familiar with the machine, and keep the spray wand moving. *Note:* Always spray downward if you're cleaning lapped siding. Otherwise, you'll force water underneath.

Scraping, sanding, spot-priming, caulking. Once the siding has dried, hand-scrape the loose paint. For this, use a large scraper with a 3-in.-wide blade, preferably one with a forward grip over the blade. Be sure to scrape the lower edges of the clapboards and beneath the windowsills. For hard-to-reach areas when siding abuts trim or where trim is curved or intricate, use a hand scraper with interchangeable blades. If paint doesn't come off easily, that's a good sign—it's well adhered. However, if isolated areas of paint are too thick or obscure ornamental details, use a chemical stripper or a heat gun to remove more paint.

After scraping loose paint, use 80-grit sandpaper to feather out the edges of the remaining paint, smooth uneven surfaces, and scuff up surfaces so new paint will adhere well. For this work, an electric palm sander or random-orbital sander is a good tool, powerful yet light enough to use all day. If you're prepping painted stucco, brick, or concrete, instead use a wire brush. When you're done, brush off the dust with a hand broom.

If the paint is largely intact, you may not need to prime it, but you should spot-prime all areas you've scraped down to bare wood, plus exposed nail heads and cracks, gaps, and holes you intend to fill or caulk. Spot-priming blocks nail stains, seals wood from moisture, and provides a better surface for filler or caulk to adhere to. Use either an exterior-grade polyurethane, a paintable acrylic, or a multipolymer caulk; don't use silicone caulk because paint won't stick to it.

This is also a good time to set and fill nail heads. Because wood filler shrinks as it dries, slightly overfill the holes. When the filler is dry, sand it flush.

Where wood is badly deteriorated, you should replace it. If the trim has only localized rot and would be difficult to replace, scrape the loose matter away and impregnate the remaining area with an epoxy filler, such as the one shown in the photos on p. 175.

Stripping exterior paint. Stripping exterior paint is a nasty job. Fortunately, only a few paint conditions require stripping. One of those conditions is alligatoring, in which many layers of old, cracked, oil-based paint resemble the skin of an alligator (see the photo on p. 564). Unless the siding has deteriorated and requires extensive replacement, however, stripping siding will be much less expensive that replacing it. But if you decide to strip, wear a respirator, eye protection, and other apparel related to lead safety.

Basically, stripping exterior paint becomes a choice between mechanical scrapers and chemicals. Sandblasting is too dirty, damaging, and dangerous to be done by anyone but an expert. Sanding and hand scraping a whole house is impossibly slow. And whereas heat guns are OK for small areas, the snail's pace of stripping a house and the real risk of starting a fire make it a distant third option. So, in the end, some combination of mechanical scraping, chemicals, and limited hand scraping will probably serve best.

Mechanical scrapers such as AIT's Paint Shaver are serious, two-handed tools. Mechanical scrapers look somewhat like angle grinders but have carbide-tipped rotary cutterheads that shave paint from the clapboard faces and edges. The better models have vacuum attachments that collect most of the debris. Nonetheless, have tarps in place before starting, and to prevent cutter damage, set siding nails well below the surface. Thus, as you strip, you'll need a hammer and a nail set to set the nails you missed. Also, be sure to use scaffolding so you can focus on the tool and not your footing.

To minimize damage to the siding, first set the tool's depth adjustment so you need several passes to strip a surface. Finding the right cutting depth is largely a matter of trial and error, so first test the tool on an inconspicuous section. Beyond that, the real trick to mechanical scraping is keeping the steel shoe/guard flat to the surface so the tool strips evenly. Where a Paint Shaver won't reach, use hand scrapers, a chemical stripper, or

small mechanical scrapers like the Metabo Lf724S. When you've finished stripping, use a palm sander or a random-orbital sander with 50-grit to 80-grit sandpaper to smooth out the rough spots before washing, caulking, dusting, and priming.

Chemical strippers are most appropriate where trim is intricate or where you want to remove lead paint without dispersing particles into the air and soil. Strippers vary in strength, environmental impact, working time (4 to 48 hours), and method of application. See the sidebar on p. 560 for a list of solvent-free strippers. Typically, chemicals are brushed thick—roughly 1/8 in.— and allowed to work.

Given enough time, chemical strippers should remove all paint layers in one application, although solvent-free strippers may take repeated applications. Methods of application vary widely, so follow the instructions on the label carefully. One example: To keep its stripper from drying out while working, Dumond Chemicals' Peel Away system comes with plasticized paper that's pressed directly onto coated surfaces. Peel Away's active agent is lye, which is extremely caustic, so after stripping surfaces, you'll need to apply a special neutralizing solution before priming or painting.

APPLYING EXTERIOR PAINT

Prep work done, it's almost time to paint. Even if you've read this entire chapter, you might want to scan it one last time for tips about paint quality, tools, basic techniques, spray painting, and so on.

Here's a quick review of factors mentioned in greater detail earlier in this chapter: Regarding exterior coatings, the more opaque the finish, the better it will protect wood siding. Clear, oil-based sealers help siding shed water without obscuring the wood grain, but they must be reapplied every two to four years because they offer relatively little protection from UV rays. A second option, semitransparent stains resemble thinned-down paints and represent a compromise that adds UV protection but reduces siding visibility. That is, you can see wood texture but not its grain. Third, there's paint that completely hides and thus protects wood the best, if correctly applied. Which brings us to acrylic latex.

Acrylic latex is king. What's not to like? Simply called latex, it's durable, flexible, virtually odorless, and cleans up with soap and water. Use good-quality latex primer-sealer and paint on all exteriors, whether covering existing paint or unfinished siding. Oil-based primers are justified

PRO TIP

Although power washing is widely used to clean stucco, brick, aluminum, and vinyl siding, it's not appropriate for wood siding. It can gouge and even shred wood, force water into gaps around doors and windows, soak insulation inside walls, and inject water into wood that will take weeks to dry. Certainly, never use power washers to strip paint: They'll scar wood and scatter paint flecks to the ends of the earth.

▶ Spray exterior trim first. Mask the siding, spray the trim, and let it dry before removing the first masking. Then mask over (cover) the trim so you can paint the siding.

▶ To mask off windows, apply double-sided masking tape around the perimeter of the window. Press sheet plastic to the tape and trim off the excess plastic.

▶ To keep paint off building elements you don't want painted, use a cardboard shield to block the spray, as shown at left.

▶ Don't spray when it's windy. Even if the air is calm, move cars and lawn furniture away from the house or cover them. If your house is close to a neighbor's, ask the neighbors to do the same or, better, do it for them.

▶ After you spray a section, immediately brush the paint into the surface. Back brushing helps sprayed paint adhere better, look great, and last longer. But because latex dries so quickly, back brush before spraying the next section.

Use a cardboard shield to prevent overspray onto adjacent areas.

After spraying a section of siding, immediately brush the paint into the wood—and into the building seams—using a 4-in. brush.

only if your siding is raw redwood or cedar, or if you've had problems with tannin bleed-through and so need to block stains before repainting. Latex is also best for masonry, stucco, aluminum, and vinyl siding because it's the only coating flexible enough to expand and contract as siding heats and cools.

Optimal conditions. Check the weather before you start painting. Ideally, wait until several consecutive dry days are forecast. If possible, wait a week after a rain. Also allow morning mists to evaporate before painting. Humidity near 90% is risky because it doesn't allow the paint to cure. The best temperatures for curing range from 60°F to 85°F. Don't paint when temperatures are 90°F or above because surfaces that are too hot can cause paint to blister. Finally, stop painting two hours before sundown if nighttime temperatures could drop below 40°F.

If possible, don't paint in strong sun. Paint the west and south faces of the house early in the morning; the north face at noon; and the east face and any part of the south face still remaining in the afternoon.

Latex paint should flow on easily and dry slowly enough that brushmarks level (disappear). To slow the rate of drying, add a dash of Flood Floetrol to latex, as indicated on the container. To slow the rate of drying of oil-based paint, add Flood Penetrol.

A painting sequence. Paint the house from top to bottom. To minimize overlap marks on clapboards, paint horizontal sections all the way across, until they end at window or door trim or at the end of a wall. After painting large sections, go back and paint the trim, windows, and doors, top to bottom. Last, paint gutters, porches, and decks. If you can remove shutters, doors, screens, and the like, do so; they are much easier to paint if placed across sawhorses. Don't bother to mask trim or windows unless you're spray painting. Again, you can scrape stray paint from window glass with a razor tool later.

Using a brush, paint the bottom edges of horizontal siding before applying paint to the flat face of each board. To distribute paint evenly along siding, load the brush with fresh paint, then partially unload it by dabbing every foot or so. Spread out the dabs, brushing the paint in and smoothing it with the grain. For exteriors, a 4-in. brush is the workhorse for the big spaces. A 3-in. angled sash brush is handy for cutting in trim edges and corners. For a sequence of painting double-hung window parts, see p. 556.

If the house has a stucco exterior or some other flat expanse, roll the paint on—after using a brush to cut in the edges—or spray it. Exterior rolling is much the same as interior rolling: Roll on fat paint in a zigzag pattern, before rolling it out evenly. To minimize spatter, roll the first stroke up. In general, the smoother the exterior surface, the shorter the roller cover nap.

Apply primers and topcoats full strength, except when the paint seems to be drying too quickly. Latex dries quickly and adheres well, so you don't need to sand between coats unless you wait several days or more. Consult the label on the paint container for drying times and maximum intervals between coats. For a lasting paint job, apply one coat of primer and at least two topcoats. (It's generally not necessary to prime existing paint that's in good shape and well prepped.) However, prime all siding that has been sanded or begun weathering down to bare wood.

Oil-based stains and semitransparent stains do not need thinning. Apply them full strength to bare wood. If you've stripped the house of its original paint and want to switch to stain, test a small section first. Because a clear stain will probably look uneven, a semitransparent stain is likely the better bet. To apply stain, a paint pad will hold more stain than a brush but requires a little practice to avoid runs.

As with paint, apply stain to the undersides of the shingle or clapboard courses first. To avoid getting stain on your skin, wear a long-sleeved shirt, rubber gloves, a hat, and safety glasses.

PRO TIP

It's impossible to tell from a manufacturer's small sample swatch how a paint color will look on a house. Instead, prime and paint a sheet of plywood—and repaint it until you find a color you love. By the way, paint stores often are willing to tint primer to match your eventual topcoat, which helps topcoats cover better and look truer.

Avoiding Direct Sun

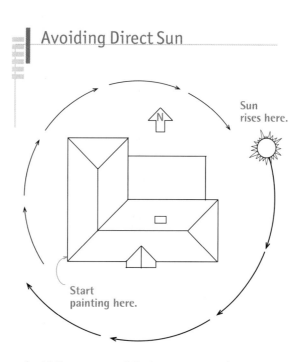

Sun rises here.

Start painting here.

Avoid the sun around the house as you paint so that you apply paint in the shade if possible. Paint applied in full sunlight is more likely to blister later.

19 Wallpapering

Wallpaper has been popular for centuries, particularly in formal rooms, where there's less danger from little ones' grimy fingers and rowdy ways. Paint has long been favored for rooms that get the most use and abuse because it's durable and easy to clean and apply.

Today, wallpaper is available in so many materials—including grasses, bamboo, rice paper, foil, and cork—that they are collectively called wallcoverings, and some of them are stronger, more durable, and easier to maintain than some paints. Many are backed with paper or cloth, with cloth being stronger. But by far the most popular are vinyl-coated wallcoverings, which are washable and grease-resistant.

Most wallcoverings come prepasted. This makes them easier to apply and—years later—easier to strip from the wall. Because the manufacturer has already pasted the covering, all you need to do is unroll it and soak it briefly in a water tray. As you'll see in this chapter, many pros avoid the drips and mess of a water tray by rolling a prepaste activator onto the backing.

Selecting Materials

Choose a wallcovering whose pattern and color are appropriate for the room. There are no hard and fast rules on what works. But in general, lighter colors make rooms look larger, and darker colors make rooms look smaller. Smaller, sub-dued patterns are better for quiet rooms. Splashy floral prints tend to serve better in places such as front halls that bustle with activity. Delicate tex-tiles or grasses are best reserved for rooms with

To hang wallcovering, start by unfolding the top half of a pasted strip and carefully align it to a plumbed line or to the plumbed edge of the preceding strip.

Stringcloth

Cork

English fabric

Grass

Foil

Lincrusta

Vinyl-coated with decorative border

little traffic and little risk of bruising. Also, consider the age and energy level of the people in the room. For example, vinyl-coated murals of rock stars or sports themes will appeal to kids and endure abuse.

Washable wallcoverings can be sponged clean occasionally with a mild soap or cleaning solution. Scrubbable coverings can take a vigorous scouring with a nylon-bristled brush or a pad, as well as stronger cleaning agents. There are also strippable and peelable wallcoverings, discussed at greater length in "Stripping Wallpaper" on p. 577.

For do-it-yourselfers, it's smart to choose a covering that's easy to hang. That is, textured coverings can be fragile and difficult to handle. The condition of existing walls also should affect your decision. For example, heavy coverings can conceal minor wall flaws, whereas lightweight

papers will accentuate flaws and won't conceal underlying bold paints or vivid patterns. If walls or trim are badly out of square, avoid coverings with large, bold patterns because slight mismatches along their edges will be more obvious than if patterns are subdued.

WALLCOVERINGS

Wallpaper, actual paper, is most appropriate for historically accurate restoration and wherever you want fine detail. Although vinyl coverings are increasingly hard to distinguish from paper, vinyls tend to look glossier. Although paper may have an aesthetically pleasing flat finish, it is more vulnerable to grime and abuse.

Vinyl is today's workhorse, available in a dizzying range of patterns and in finishes ranging

Papering OVER DIFFERENCES

Almost everybody still refers to wallcovering as "wallpaper." But in this chapter, wall-covering, the noun, will be the general term for both paper and nonpaper coverings. However, when used as a verb, wallpaper refers to the act of hanging either type of covering. Last, you can assume that advice on papering or prepping walls also holds true for ceilings.

from flat to glossy. Vinyl is especially suitable for areas with traffic and moisture. Most vinyls are washable, and cloth-backed vinyls are usually strippable—that is, they are easily removable when you want to change them. Although no wallcovering is intended to conceal major cracks and irregularities, heavier vinyls can conceal minor flaws.

Fabric coverings include cotton, linen, silk, stringcloth, and wool. They're often chosen to match or coordinate with colors and textures in drapes and fabric-covered furniture. They come paper-backed, acrylic-backed, or unbacked (raw). And the backing largely determines the method of installation. Avoid slopping adhesive or water onto the fabric facing because some fabrics stain easily; delicate fabrics are usually dry hung, in which paste is applied to the wall and the dry wallcovering is smoothed onto it.

Natural textures, such as rice paper, grasses, and bamboo, tend to be expensive, temperamental, and delicate. And because the thinner coverings reveal even minor flaws in wall surfaces, you first need to cover the walls with a lining paper. Still, natural textures are evolving, with vinyl-coated versions that are relatively durable and easy to install. Besides their beauty, most natural textures have no pattern that needs matching.

Foils and Mylar also vary greatly in appearance and ease of handling. For example, heavier, vinyl-laminated foils are durable and easy to install. However, some uncoated metallic coverings retain fingerprints, so you should wear disposable nitrile gloves when hanging them, or perhaps avoid them altogether. That said, foils are well suited to small rooms because they reflect light, thus making the space appear larger.

Lincrusta, an embossed wallpaper similar to a fine cardboard, is making a comeback. The modern version of Victorian lincrusta is called *anaglypta*.

Cork and wood-veneer wallcoverings are finely milled and manufactured to use cork and rare woods efficiently. Typical veneer dimensions are $1/64$ in. thick, 1 ft. to 4 ft. wide, and 12 ft. long. Such specialty coverings may be available through suppliers of professional paperhangers.

Borders are thin strips of wallcovering that run along the edges of walls where they meet ceilings, wainscoting, and trim. They can be installed over wallcovering or directly to drywall or plaster. The surface determines the adhesive.

Choosing the Right Adhesive

Most of the adhesives described here come premixed, unless otherwise noted.

▶ **Clay-based adhesives** dry quickly and grip well. Use them to install heavy vinyls, Mylar, foils, or canvas-backed coverings or to adhere wall liners to irregular surfaces such as concrete block or paneling. *Caution:* Clays stain delicate materials. They may attack paint substrates, and they probably dry too quickly for amateurs to use successfully.

▶ **Clear adhesives** may be the best all-around pastes. They're strippable, grip almost as well as clays, and won't stain. Clear adhesives are frequently classified either as standard mix (good for most lightweight coverings) or as heavy duty (for weightier coverings).

▶ **Cellulose** has the least grip of any adhesive in this listing, but it's strong enough for delicate papers—especially for fine English wallpapers and unbacked murals. It won't stain but is somewhat less convenient because it comes as a powder to be mixed with water.

▶ **Vinyl-to-vinyl adhesive** is recommended for adhering vinyl borders over vinyl wallcoverings or new vinyl wallcovering over old. It's so tenacious that it can't be stripped without destroying the substrate, so wipe up stray adhesive immediately. It's also used to adhere wall liners, Mylar, and foil.

▶ **Prepaste activator** makes prepasted wallcoverings easier to install. It improves adhesion, while letting you avoid the mess of water trays. Prepaste activators are rolled on, which conveniently increases slip time, the time in which you can adjust wallcovering after hanging it.

▶ **Seam adhesive** typically comes in a tube, reattaches lifted seams and tears, and is compatible with all wallcoverings. After applying seam adhesive, roll the seam.

A Wall OF YOUR OWN

Thanks to technology, you can have wall coverings fabricated with virtually any pattern or image you want, including historical documents or wall-size photos of family members. The cost has come down a lot. Make sure these special coverings are treated with a protective coating such as a clear acrylic for easier cleaning.

Ordering Wallcovering

Start by calculating the square footage of your walls and ceilings. Once you've determined the overall square footage, subtract 12 sq. ft. for each average-size door and window. To determine the total number of rolls you'll need, divide the square footage by the square footage listed on the rolls. If you're using American single rolls (see "How Much on a Roll?" below), you could instead divide by 36 (the number of square feet on each roll). But dividing by 30 gives you an allowance for waste.

PRO TIP

When you pick up a shipment of wallcovering, check the code number and run number on the label packed with each roll. Code numbers indicate pattern and color. Run numbers tell what dye lot you're getting. The dyes of different runs can vary considerably and will be especially noticeable side by side. So if you must accept different runs to complete a job, use the smaller quantity in a part of the room that isn't as conspicuous. Patterns tend to be current for at least two years.

The wallcovering type determines which type of paste you need. Pastes come premixed, as shown, or as powders that are mixed with water. Many wallcoverings are prepasted and require that you either roll prepaste activator onto their backing or soak them in water.

LIFE ON THE Edge

The edges of most wallcoverings are pretrimmed at the factory, allowing you to butt them together after matching the patterns. If the edges aren't pretrimmed, do it yourself with a razor knife and long straightedge. Untrimmed edges are called selvage.

If the edges of a pretrimmed roll are frayed, refuse that roll. Similarly, refuse vinyls with edges that have become crimped in shipping or storage because they cannot be rolled flat. To avoid damaging the edges yourself, always store the rolls flat—rather than on end.

PASTES

Like wallcoverings, pastes have evolved. Probably the best advice is to follow the manufacturer's paste specifications, usually printed on the wallcovering label, along with the code and run numbers. If the paste isn't specified, ask your supplier to get that information from the manufacturer.

Wheat pastes were the standard for centuries, but that changed in the 1960s and 1970s, with the introduction of vinyl wallcoverings. Trapped behind an impervious skin of vinyl, wheat paste was an ideal medium for mold. Moreover, wheat paste wasn't strong enough to bond many of the newer, thicker materials. Wheat pastes are occasionally specified for delicate wallpaper, but clay- or starch-based adhesives with additives that increase grip and discourage mold have largely supplanted them.

Pastes come premixed or as powders to be mixed with water. Premixed pastes are generally stronger, more consistent, and more convenient. Once opened, however, they have a relatively short life. In general, the thicker the paste, the quicker it dries and the greater the weight it can support.

How Much on a Roll?

Wallcovering rolls (also called *bolts*) are available in American single rolls, Euro rolls (metric), and commercial widths. At this writing, Euro rolls dominate the market.

▶ American single rolls are 18 in. to 36 in. wide. (A 27-in. width is comfortable for most people.) The wider rolls generate fewer seams but are much more difficult to handle. Whatever the width of an American single roll, it will contain 36 sq. ft. of material.

▶ Euro rolls are 20½ in. to 28 in. wide and are generally sold as double rolls (twice as long). Typically, there are 56 sq. ft. to 60 sq. ft. on a Euro roll.

▶ Commercial coverings are typically 48 in. to 54 in. wide, beyond the skills of most nonprofessionals to hang.

Wall Coverings, Adhesives, and Application

WALL COVERING	ADHESIVES AND APPLICATION	COMMENTS
Prepasted	Paste already on backing; soak in water tray or machine according to manufacturer's recommendations.	To avoid mess of water tray, roll prepaste activator onto backing, which allows longer work time.
Lightweight vinyl	Standard clear adhesive; kitchen and bath adhesive with mildewcide in high-humidity areas	To avoid stretching vinyl when smoothing, use rubber squeegee or plastic wall-covering smoother.
Heavy vinyl	Heavy-duty clear adhesive; clay-based adhesive	If surface to be covered is rough or textured, install liner first.
Vinyl border over vinyl covering; new vinyl covering over old vinyl	Vinyl-over-vinyl adhesive	If adhesive gets on facing, sponge off immediately.
Vinyl border over flat paint	Standard clear adhesive; vinyl-over-vinyl adhesive	Sponge paste off walls quickly.
Wallpaper (including delicate English papers)	Cellulose adhesive; standard clear adhesive	Smooth with soft-bristle brush; if paste gets on facing, blot off with damp sponge.
Natural fibers (grasses, rice paper, bamboo)	Follow manufacturer's recommendation, usually clear adhesive.	Fibers vary greatly; some are vinyl coated and durable; dry-hang delicate types.
Paper-backed fabrics	Cellulose adhesive; standard clear adhesive	Dry-hanging usually best, test-hang small sample to be sure; avoid getting adhesive on fabric face.
Paper-backed upholstery, drapery, or other heavy fabric	Heavy-duty clear adhesive	Requires stronger bond; roll adhesive onto backing, but test-hang small sample to be sure.
Raw (unbacked) fabric	Cellulose adhesive; standard clear adhesive	Dry-hang; avoid getting adhesive on fabric face; test-hang sample to be sure.
Foils; Mylar	Clay-based adhesive; vinyl-over-vinyl adhesive	Not strippable; often dry-hung
Paper-backed murals	Follow manufacturer's recommendation; cellulose adhesive; standard clear adhesive.	Follow manufacturer's recommendations; may require liner and/or dry-hanging.
Canvas backed	Clay-based adhesive	Prime wall first or adhesive and covering may not be strippable.
Lincrusta, anaglypta coverings (embossed, often of heavy paper)	Clay-based adhesive	Roll adhesive onto back of covering; smooth with brush; don't roll seams.
Cork, thin wood veneer	Clay-based adhesive	Run liner perpendicular to finish covering; wipe paste off face immediately.
Wall liner; lining paper	Clay-based adhesive; vinyl-over-vinyl adhesive	Prime walls first; smooth coat or fill textured walls, block, and paneling.

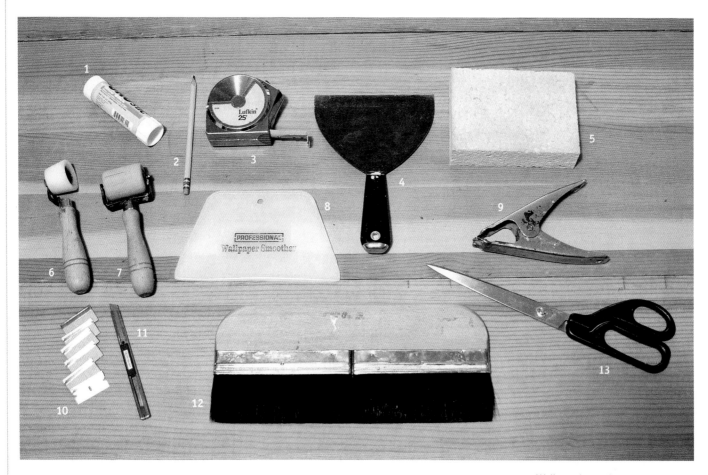

Wallpapering tools:
1. Glue stick for touchups
2. Pencil
3. Tape measure
4. 6-in. taping knife
5. Sponge
6. Beveled seam roller, used close to trim and in interior corners
7. Standard seam roller
8. Smoother-scraper
9. Spring clamp to hold wallpaper while pasting
10. Single-edge razor blades
11. Razor knife with snap-off blades
12. Smoothing brush
13. Shears for rough-cutting strips

If the room has numerous recesses, difficult corners, or a lot of trim to cut around, order an extra roll or two. Also, if the pattern is large, you'll waste more because you'll need to match patterns along seams. On the back of most wallcovering, you'll find the pattern repeat, usually stated in inches: The larger the pattern repeat, the greater the waste. Also, order an extra roll or two for repairs. You never know when a roof will leak or a child will ding a wall.

Equipment

You'll need some special tools and work surfaces to apply your wallcovering.

▶ A spirit level will tell you whether walls and wallcovering edges are plumb. Be sure to plumb the leading edge of the first strip of wallcovering. A 4-ft. level with metal edges can double as a straightedge when trimming selvage (manufactured edges).

▶ Your pasting table should have a washable top about 3 ft. by 6 ft. Avoid covering it with newspaper because newsprint may bleed. To protect the top from scarring during cutting, cover the tabletop with hardboard or use a zinc cutting strip. (If you don't have a

suitable table, lay a sheet of smooth, void-free plywood over sawhorses.)

▶ Have a 16-ft. retractable tape measure for measuring and marking.

▶ A razor knife with replaceable blades gives the cleanest cuts. Don't be stingy about replacing blades during use because dull blades can rip wallpaper. A professional may use 200 to 300 blades on a big job. Some pros prefer single-edge razor blades, though knives with snap-off blades are popular, too.

▶ Shears help you rough-cut from a roll.

▶ Paste brushes spread pastes on backing— or on walls, in some cases.

▶ A roller and pan are needed to spread vinyl paste, which is too heavy to brush on. Ask your supplier how long the nap of the roller cover should be.

▶ A smoothing brush, with soft bristles, will smooth out the wallcovering paste.

▶ A wallpaper smoother smooths vinyls, liners, and other heavy materials. It is also handy for flattening the occasional paste lump.

▶ A seam roller spreads glue along the edges of the strips to ensure that seams will stick well. *Caution:* Seam rollers are not

Primer-Sealers for Wallcoverings

PRIMER-SEALER TYPE	USES	COMMENTS
Pigmented acrylic	Seals all surfaces, including existing wallpaper, vinyl covering, and latex paint; suitable base for all wall coverings	Also known as universal primer-sealer; cleans up with water; protects drywall when coverings are stripped; add pigment to hide existing wallpaper patterns.
Clear acrylic	Same uses as for pigmented acrylic, but can't bond latex paint; suitable base for all wall coverings	Cleans with water; won't protect drywall; can't hide patterns.
Heavy-duty acrylic	Mostly for weighty vinyl coverings used for commercial installations	Soaks into raw drywall, so won't protect it when covering is stripped away.
Alkyd/oil-based	Seals all surfaces except existing wallpaper or vinyl coverings; fast drying (2 to 4 hours); suitable base for all wall coverings	Thin with paint thinner to improve bond with existing paint; protects drywall; can be tinted.
Stain sealer; pigmented shellac	Hides or contains stains from water and smoke, wallpaper inks, grease, crayons, and more	Not a primer-sealer; when dry, apply acrylic primer-sealer topcoat.

LOOKING Sharp

This clever magnetic bracelet keeps a single-edge razor blade handy. A sharp blade is essential, especially if you're working with prepasted wallcoverings soaked in water. Wetted paper will snag a dull blade and rip easily. Blades are far cheaper than wasted wallcovering.

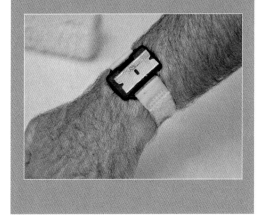

generally recommended for delicate or finely textured papers or grasses.

▶ A 6-in. taping knife, also called a joint knife, is useful for filling low spots and scraping off high spots in a wall. Also use this knife to press the wallcovering snugly against the trim before cutting away excess paper.

▶ A sponge and water pail are handy for wiping excess paste off the pasting table, trim, and most wallcovering surfaces. In general, the sponge should be just damp. Change the water in your pail often. To avoid creating a sheen along the seams, wipe entire strips, not just the edges.

▶ Other useful equipment includes a sturdy fiberglass stepladder; a long, straight board for detecting irregularities in walls and ceilings; and plenty of clean, soft rags. If you use a prepasted wallcovering, you'll also need a water tray in which to soak the strips to activate their adhesives.

Preparing Surfaces

Surface prep determines how well coverings adhere, hence how good the job looks and how long it lasts. Ideally, surfaces should be clean, dry, flat, and stable. Before hanging wallcoverings, assess existing surfaces. Replace or repair them as necessary, then prime and seal them. Sealing surfaces improves adhesion and, just as important, allows you to remove coverings later without destroying the underlying drywall or plaster.

In the old days, a wall was sized, or brushed with a glutinous mixture to improve adhesion. But sizing is rarely done today because it's chemically incompatible with many pastes, causing them to crystallize, lump, and bubble, creating voids where the covering is unattached. Instead, professionals use one of the primer-sealers described in "Primer-Sealers for Wallcoverings" above left.

Before you start prep work, remove furniture from the room or move it to the center of the room and cover it with a tarp. That will allow you to work faster and be safer. And remove cover plates from electrical outlets and switches, which you'll replace after you've papered. Speaking of safety, always use a voltage tester to be sure electrical power is off when you trim wallpaper around an electrical outlet or switch. A single-edge razor blade is recommended here, and you wouldn't want the blade to touch a live conductor. Also, set up ladders and scaffolds so they don't wobble. Wear a respirator when applying chemicals, and clean up waste as you go.

PREPPING PAINTED SURFACES

Before you wallpaper painted surfaces, figure out what kind of paint you've got and what shape it's in. In general, oil-based paints are stable surfaces for wallcovering because they aren't water-soluble. Yet some primer-sealers can stabilize even latex paint. You could scrape off a small patch of paint and have a paint store analyze it, but two simple tests should suffice.

Hot towel test. Soak a hand towel in hot water, wring it well, and rub the paint vigorously for 20 to 30 seconds. If paint comes off on the towel, you've probably got latex. Alternatively, you can use duct tape to hold a moist sponge next to a painted surface for 15 minutes before removing the sponge. If you see paint on the sponge, it's latex.

X-tape test. If paint didn't come off on your towel or sponge, it's probably oil based. To see how well it's bonded, use a razor blade to score a 1-in. X lightly in the surface (don't cut into the drywall or plaster). Press masking tape over the X, and pull it off quickly: If there's no paint on the tape, the paint is well adhered. If paint does come off, scrape and sand it thoroughly before proceeding.

If the existing paint is a well-adhered glossy or semigloss oil-based one, sand it lightly with fine sandpaper (150 grit to 180 grit), using a sanding block or an orbital sander. Then use a sponge mop, dampened with water, to remove the sanding dust. Alternatively, spray or wipe on paint deglosser to dull glossy and semigloss paint surfaces.

If the paint is a well-adhered flat oil-based paint, you can begin hanging wallcovering. Simply rinse the surface with a mild detergent solution to remove grime, rinse with clear water, and allow the surface to dry.

Latex paint should be prepared by scraping lightly and sanding. You needn't remove the entire coat of paint; just sand it enough so the primer-sealer can bond. Avoid gouging or ripping the surface underneath, especially if it's drywall. When you're done sanding, wipe the wall clean and apply a coat of pigmented acrylic primer-sealer.

NEW WALLPAPER OVER OLD

You can wallpaper over an existing covering if:

▶ **It is not highly textured, as lincrusta, stringcloth, and bamboo are.**

▶ **There are no prominent seams.**

▶ **There are no more than two layers already on the wall.**

▶ **The old wallpaper is well bonded.**

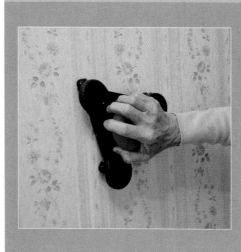

Scarifying TOOL

Some pros recommend using a scarifying tool, such as PaperTiger, to perforate wallpaper so steam can penetrate. But these tools also can perforate drywall and scar the surface. Damage also can occur if you use metal-edged scrapers. So use tools like these only as a last resort when the paste is especially tenacious, and use them with great care. Instead, try steaming the wallpaper before scarifying or scraping it.

Check edges and seams first: If they're peeling or poorly adhered, strip the wallcovering. But if there are only a few isolated loose spots, use seam adhesive to reattach them. Or use a razor blade to cut out the loose seams. Then fill voids with spackling compound, allow it to dry, and sand lightly with 150-grit sandpaper.

Another potential problem is bleed-through from metallic wallpaper inks. To test, dampen a cloth with diluted ammonia (1 part ammonia to 4 parts water) and gently rub the old wallpaper. If inks change color (usually, they turn blue-green), they'll bleed. To prevent this, seal the old wallpaper with pigmented shellac or a similar stain killer. Allow the sealer to dry thoroughly before painting surfaces with a universal primer-sealer.

Otherwise, if existing wallpaper is well adhered, wipe it with a damp sponge, let it dry, and then paint surfaces with pigmented acrylic primer-sealer. Tint the primer-sealer to match the background color of the new wallcovering.

STRIPPING WALLPAPER

Strip existing wallcovering if it is tired or grimy and can't be washed, poorly adhered or damaged, puckered or lumpy because there are too many layers, water stained, moldy, strongly textured, or you want to paint the walls. Before you apply new wallcovering, repair and prime the finish surfaces.

Before stripping, reconnoiter. Peel up a corner of the wallpaper in an out-of-the-way place. Then determine whether the walls are plaster or drywall. Plaster is harder and can survive a lot more steaming and scraping than drywall. Next, test how easily the wallpaper strips. Peelable and strippable types should be relatively easy to remove if the wall was properly sealed before it was papered. But if surfaces were not sized or

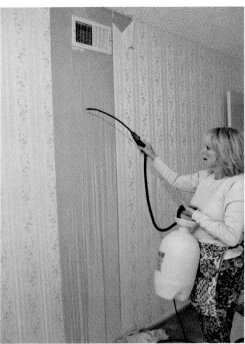

1. Stripping takes patience. Start at one end of a strip, pulling slowly and steadily so that the strip comes off in a single piece, if possible. This covering was peelable, meaning that its facing peeled off, but its paper backing remained stuck to the substrate.

2. To remove the paper backing, spray it with a solution of hot water and a wallpaper stripper such as DIF, which breaks down the paste. Allow the solution to soak in for three to five minutes. Because spraying is messy, place old towels or a tarp at the base of the wall.

3. Applying steam will hasten the penetration of the stripper solution and soften the paste, making removal easier. On this wall, heat from a nearby heat register had baked the paste hard.

Strippable VS. PEELABLE

Strippable wallcovering can be dry stripped (no spraying or steaming needed); both the facing and the backing material come off easily, with only a faint paste residue left on the wall. Peelable coverings usually require spraying with a wallpaper-removing solution or steaming to remove the wallpaper facing. If the paper backing is in good shape, it can stay on the wall as a wall liner for the new wallpaper. Otherwise, remove it, too.

primed first, you're in for some work. Moreover, if the unsealed substrate is drywall, stripping the wallpaper may destroy the paper face of the drywall. In theory, you can patch damaged drywall with a coat of joint compound and then seal it with a PVA primer, but removing the damaged drywall or covering it with ¼-in. drywall will yield far better results.

Stripping wallpaper is messy, no matter what method you use. You'll need painters' tarps or old towels to protect floors from stripping solutions, water that forms as steam condenses, and sticky wallpaper. Canvas tarps, or even old towels, are better than plastic tarps, which tend to be slippery. Have trash bags handy for stripped paper.

As noted earlier, turn off the electricity to areas you're stripping, and use a voltage tester to be sure the power is off.

Use the least disruptive stripping method. Start stripping at the top or bottom of a strip. Use a putty knife or a plastic scraper to lift an edge. Then gently pull off the wallpaper in the largest strips possible. This takes patience.

If you can't pull off the covering or if it begins tearing into small pieces, try spraying a small area with a wallpaper-removing solution such as Zinsser's DIF, which is also available as a gel that you brush on. A time-tested alternative is 1 cup vinegar per gallon of hot water; sponge on or apply using a spray bottle. Allow either solution to soak in for 5 to 10 minutes before trying to pull off the paper. If this method doesn't work, chances are the paper is vinyl coated and the solution is not penetrating. In this case, try a wallpaper steamer. Hold the steamer pan against the wallcovering long enough for the paste to soften—usually a minute or two—then pull or scrape the covering free.

If your wallcovering is peelable, chances are its facing layer will strip off, leaving the paper backing stuck to the wall. To remove it, spray or sponge on wallpaper-removing solution, then apply steam. The backing should release easily;

4. A plastic wallpaper smoother–scraper will scrape off steamed paper backing without damaging drywall beneath. When all paper is off, use a soft–bristle nylon scrub brush to gently remove paste residue.

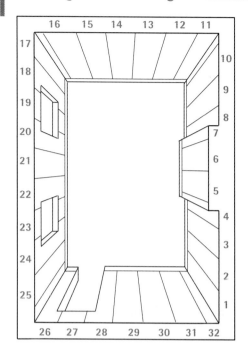

Because a room's final strip of wall-covering usually needs to be trimmed narrower to fit, try to end a job where the final strip is inconspicuous. Here the job begins and ends in a corner.

otherwise, use a plastic scraper or smoother to remove the backing. When the walls are stripped, wash them with a mild cleaning solution. Then rinse and let them dry thoroughly before applying a primer-sealer. If paste lumps remain, remove them with a nylon-bristle scrub brush.

PREPPING RAW DRYWALL OR PLASTER

Newly installed drywall must be sealed before you hang wallcoverings. Otherwise, the drywall's paper face will absorb paste, making it impossible to remove the covering at a future date without damaging the drywall. To seal drywall, apply a coat of universal primer-sealer, which is an excellent base coat for papering and painting.

Allow gypsum plaster to dry two to three weeks before applying either wallcovering or paint. Uncured plaster contains alkali that is still warm, causing the paint and paste to bubble. "Hot plaster" has a dull appearance, whereas cured plaster has a slight sheen. Curing time varies, but keeping the house warm will hasten it. Cured plaster can also be coated with universal primer-sealer. (By the way, old lime plaster, which was used until the early 1900s, took a year to cure.)

Mold and Mildew

Mold is a discoloration caused by fungi growing on organic matter such as wood, paper, or paste. Mildew is essentially the same, but it usually refers to fungi on paper or cloth. Without sufficient moisture and something to eat, mold (and mildew) can't grow. Mold can grow on the paper and adhesives in drywall, but, technically, mold can't grow on plaster because plaster is inorganic. If your plaster walls are moldy, the fungi are growing on grease, soap, dirt, or some other organic film on the plaster.

If walls or ceilings are moldy, first alleviate the moisture. To determine the source, duct-tape a 1-ft.-sq. piece of aluminum foil to the wall. Leave this up for a week. Upon removal, if there's moisture on the back of the foil, the source is behind the walls. If the front of the foil is damp, there's excessive moisture in the living space, which you might reduce by installing ventilator fans or a dehumidifier for starters.

If the drywall or plaster is in good condition, clean off the mold by sponging on a mild detergent solution. (Although widely recommended, diluted bleach isn't any more effective than a household cleaner.) The sponge should be damp, not wet. Rinse with clean water, and allow the surface to dry thoroughly. Then paint on a universal primer-sealer. If you'll be wallpapering the room, use a mildew-resistant, kitchen-and-bath adhesive.

If mold or water stains are widespread and the drywall has deteriorated, there may be mold in the walls. Remove a section of drywall to be sure. If the framing is moldy, it may need to be replaced—a big job and one worth discussing with a mold-abatement specialist, especially if family members have asthma or chronic respiratory problems (for more, see chapter 14).

Before hanging wallcovering, check the trim for plumb and level. For example, if the crown molding isn't level, you may need to raise or lower the wallpaper strips so molding doesn't chop off the top of a prominent pattern.

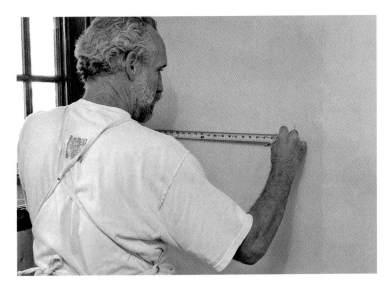

To determine exactly where seams will occur, mark off intervals the same width as your wallpaper.

After determining your starting point, draw a plumb line to indicate the leading edge of the first strip of wallcovering.

Laying Out the Work

Before ordering materials, check that walls are plumb and woodwork is plumb and level. Use a long, straight board or a 4-ft. spirit level to check walls for bumps or dips, and mark them with a pencil so you can spot them easily later. You can hang almost any wallcovering type or pattern, but if door or window casings are markedly out of plumb, small patterns make this less obvious than large, loud patterns.

STARTING AND FINISHING POINTS

To decide where to hang the first strip, figure out where you want to end. The last strip of wall covering almost always needs to be cut to fit, disrupting the strip's full-width pattern. So, avoid hanging the last strip where it will be conspicuous.

A common place to begin is one side of a main doorway. There, you won't notice a narrower strip when you enter the room. And when you leave the room, you'll usually be looking through the doorway. The last piece is often a small one placed over a doorway. Another common starting point is an inconspicuous corner.

To determine exactly where wallcovering seams will occur, mark off intervals the same width as your wallcovering. Go around the room, using a ruler or a wrapped roll of wallcovering as your gauge. Try to avoid trimming and pasting very narrow strips of wallcovering in corners; this usually looks terrible, and the pieces don't adhere well. You may want to move your starting point an inch or two to avoid that inconvenience.

However, if the pattern is conspicuous, you might start layout with a strip centered in a conspicuous part of a wall—over a mantel, over a sofa, or in the middle of a large wall. Choose a visual focal point, and mark off roll-widths from each side of the starting strip until you have determined where the papering will end—preferably in an inconspicuous place.

You may want to center the pattern at a window, if that's the visual center of the room. With a picture window, the middle of the strip should align with the middle of the window. If there are two windows, the edges of two strips should meet along a centered, plumbed line between them unless the distance between the windows is less than the width of a strip. If the distance is less, center a strip between the windows.

Basic Papering Techniques

Before you start hanging wallcovering, turn off the electricity to affected outlets, switches, and fixtures, and check with a voltage tester to be sure the power is off.

Room Focal Points

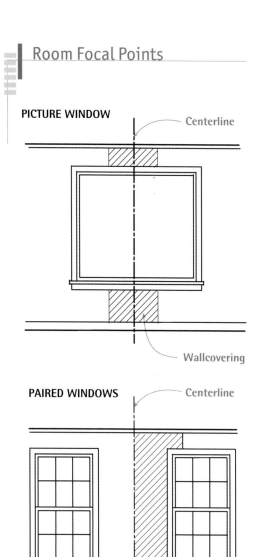

PICTURE WINDOW — Centerline, Wallcovering

PAIRED WINDOWS — Centerline, Wallcovering

Where windows are the focal point of a room, position the wallcovering accordingly. If the window is a large picture window, center the middle of the strip on the middle of the window. If the focal point is two windows, center the edges of the two strips as shown.

Use shears to rough-cut strips, leaving extra at each end for trimming and pattern matching. This pasting and layout table is a professional model, strong yet light, easily transported from job to job.

Three Types of Seams

You can join strip edges in three ways: butt seam, overlap seam, or double-cut seams.

▶ **The butt seam** is the most common. Its edges are simply butted together and rolled with a seam roller.

▶ **An overlap seam** is better where corners are out of square or when a butt seam might occur in a corner and not cover well. Keep the overlap as narrow as possible, thereby avoiding a noticeable welt and patterns that are grossly mismatched.

▶ **Double-cut seams** (also called through-cut seams) are the most complex of the three. They are used primarily where patterns are tough to match or surfaces are irregular; for example, where the walls of an alcove aren't square.

Gently pull cut strips over the edge of the table to counteract the tendency of wallpaper to curl.

How to Cut Seams

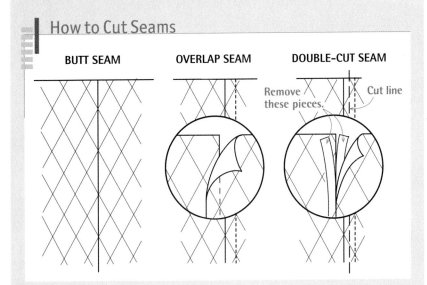

BUTT SEAM	OVERLAP SEAM	DOUBLE-CUT SEAM

Remove these pieces.

Cut line

Measure out from a corner or a jamb casing, if that's where you'll begin, and draw a plumb line that will become the leading edge of the first strip. If the casing is out of plumb, allow the trailing edge of the strip to overlap the casing enough to be trimmed with a razor knife without creating a space along the casing. If the casing is plumb, simply butt the trailing edge to the casing. As you proceed around the room, continually check for plumb.

CUTTING STRIPS TO LENGTH

Measure the height of the wall and cut several strips to length, leaving extra at each end for trimming and vertically matching patterns. Cut the first two strips extra long. Slide the first strip up or down the wall until most (or all) of its pattern shows near the ceiling line. Don't show less than half the pattern. The pattern along the baseboard will be less visible and less important.

On your cutting table, place the second strip next to the first, and align the patterns along their edges with two strips side by side. You'll have a sense of how much waste to allow for pattern matching. (A pattern-repeat interval is often printed on the label packaged with the wallpaper.) Depending on the size of the patterns, each succeeding strip can usually be rough-cut with an inch or two extra at each end and then trimmed after being pasted.

Do the rough-cutting at the table using shears. Do the trimming on the wall using a razor knife. Patterns that run horizontally across the face of a covering are called straight match. Patterns that run diagonally are called drop match and waste somewhat more material during alignment.

Unless you are working with a delicate covering, cut several strips at a time. But be careful not to crease them. Flop the entire pile of strips facedown on the table so the piece cut first will be the first pasted and hung. The table must be perfectly clean; otherwise, the face of the bottom strip could become soiled.

PASTING

Unless you're experienced, buy premixed adhesive. But if mix you must, try to achieve a mixture that's slightly tacky to the touch. Add paste powder or water slowly: Even small increments can change the consistency radically. Finally, mix thoroughly to remove lumps.

As you work, keep the pasting table clean, quickly sponging up stray paste so it won't get on strip faces. Some coverings, such as vinyl, are not marred by stray paste on the face, but many others could be. Although the batch of paste you mix should last a working day, keep an eye on the

Prepasted Papers and Water

Many wallcoverings come prepasted. Typically, manufacturers specify that individual strips be soaked for 30 seconds in a tray filled with lukewarm water. But follow the directions printed on the back or supplied by the retailer. After soaking, pull each strip out of the tray and onto the worktable, book (fold) it, and allow it to expand before hanging it on the wall. Precut the pieces before placing them in the water tray. Otherwise, if you try to trim soaked strips, they'll snag or tear.

Many professional paperhangers will hang prepasted wallcoverings but hate water trays because (1) water and diluted paste drips everywhere; (2) the water in the tray must be changed often; (3) a thin film of paste also ends up on the front of the wallcovering; and (4) if the strips are soaked too long, they may not adhere well. Instead, these pros roll prepaste activator onto the back of strips, just as you'd apply standard paste. Rolling on an activator reduces mess and ensures good adherence to the wall. Last, pros sometimes roll thinned-down paste instead of activator. That may be OK, but first ask the supplier if the two pastes will be compatible.

As you apply paste, roll it out from the center toward the edges. Hold down the wallcovering with one hand to keep it from sliding as you roll.

3. Once the strip is correctly positioned, smooth it onto the wall, smoothing the upper end first and brushing out from the center toward the edges.

1. After pasting the back of the strip, fold both ends in, as shown, so they meet in the middle. Be careful not to crease the folded ends. This folding is called booking the paper.

2. To hang the first strip, unfold the top half and carefully align it to the plumbed line you marked earlier, leaving an inch or so extra at the top. Gently slide the strip into place. Align each subsequent strip to the leading edge of the preceding one.

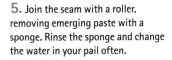

4. With a taping knife pressed against the woodwork, trim the joint with a single-edge razor blade or razor knife. After trimming excess paper and smoothing out the strip one last time, sponge away any paste smeared on the woodwork.

5. Join the seam with a roller, removing emerging paste with a sponge. Rinse the sponge and change the water in your pail often.

6. Use a plastic wallpaper smoother to flatten any lumps of paste. This tool is also useful for smoothing out heavier vinyls and the like.

consistency. Paste should glide on, never drag. Rinse the paste brush or roller when you break for lunch and when you quit for the day.

Until you become familiar with papering, apply paste to only one strip at a time. In other words, hang each strip before pasting another one. Using a roller, apply paste in the middle of the strip, toward the top. Spread the paste to the far edge and then to the near edge. For good measure, run the roller over strip edges twice because it's often hard to see if the paste along the edges is evenly spread.

BOOKING STRIPS

For convenient handling, create loose folds of pasted covering that you can carry to the wall, unfold without mess, hang, and smooth. The most practical folding method is known as booking, perhaps because the folds resemble the folds of a book dust jacket. When folding strips, do so loosely; avoid creasing them.

Booking a strip (typically, for five minutes) also allows it to absorb the moisture in the paste, expand, and contract slightly to its final width. If you do not allow the wallcovering time to expand and contract before hanging it, it will do so on the wall, either buckling or gapping at the seams. Booking times vary: Pros who know their pastes and papers will sometimes cut, paste, and book several strips at once, placing them into a plastic trash bag so the paste doesn't dry out.

However, if you paste several strips at once, keep track of the order in which you pasted them. And hang them in the same order.

HANGING PASTED STRIPS

To hang a strip of wallcovering, unfold the booked upper fold (leaving the lower one folded) and align the edge of the strip to your plumb line. Position the upper end of the strip an inch or so above the ceiling line. Smooth the upper end of the strip first by running a smoothing brush down the middle of the strip and out toward the edges. Working from the center outward, brush air bubbles, wrinkles, and excess paste from the middle to the edges. Align subsequent pieces to the leading edge of each preceding strip, checking periodically to make sure the strips are plumb.

If the upper half of the strip is adhering well, simply unfold the lower fold and smooth the paper down, again brushing down the center and out toward the edges with small strokes.

If a butt seam doesn't meet exactly, you have three choices:

▶ **Move a strip slightly by raising one of its edges and—palm on paper—using your other hand to slide the strip toward or away from the**

seam. Raising one edge of the strip reduces the grip between paper and wall.

▶ **Pull the strip off the wall, realign its patterns along the seam, and brush it down. But you've got to move quickly: Don't wait much more than a minute to pull the strip off.**

▶ **Pull off the strip, quickly sponge-clean the wall, and hang a new strip. Don't try pulling just one edge of the strip, however. At best, it will stretch, draw back when it dries, and open the seam. At worst, you'll pucker or rip the strip.**

SPONGING

It's impossible to overstate the importance of gently wiping paste off wallcovering faces and adjacent surfaces. If paste dries on a painted ceiling, it can pull the paint off. (If you see a brown crust along a ceiling–wall intersection, that's dried paste.) Paste will even pull the finish off wood trim. Vinyl-on-vinyl and clay adhesives are especially tenacious, so sponge off the excess immediately.

Equally important: Change your sponge water often so diluted paste doesn't accumulate. Warm water is best. And wring the sponge almost dry before wiping. When you've wiped the surfaces clean, come back with a soft, dry rag. But apply only light pressure so you don't move the wallcovering, disturbing the seams.

Don't rub delicate wallpapers. Instead, blot them clean with a just-damp sponge. Before you commit to any wallcovering, ask your supplier if it can be wiped (or blotted clean) with a sponge. If not, consider other materials.

PRO TIP

To reach your ceiling, you'll probably need an elevated platform. The safest option is to rent scaffolding. In a pinch, sturdy planks running between two stepladders will do.

Fixing Three Small Flaws

▶ **A paste lump under the covering.** First, try to flatten it with a plastic wall smoother. This may take several gentle passes. (Don't use a metal blade because it would snag on the lump and tear the wallpaper.) If there are many lumps, the paste is unevenly mixed. In this case, pull the strip off the wall and sponge the wall clean. Then adjust or replace your paste mix and start with a new strip.

▶ **Air bubbles that you can't brush out.** This is a common problem with vinyl wallcoverings. Use the point of a razor knife to cut a small slit. As you gently force out the air with a smoothing brush, the slit will flatten out and then become unnoticeable.

▶ **Edge not adhering.** Pull it away from the wall slightly and dab on paste with a small brush. Avoid stretching the covering, especially if it's vinyl.

To avoid a noticeable welt of overlapped paper in an inside corner, pull back one strip and trim off the excess paper. If corners are out of square, plumb the leading edge of each wall's first strip.

TRIMMING AND ROLLING

Where a strip of wallcovering meets a border, such as woodwork, a ceiling line, or a baseboard, use a 6-in. taping knife to press the edges of the covering snug. Cut off the excess by running a razor knife along the blade of the taping knife. To ensure that strips fit tightly against a door or window casing, rough-cut them a little long. Then, using your taping knife, tuck the wallcovering snugly against the casing and trim it more precisely. For clean cuts, razor blades must be sharp.

Conventional wisdom suggests rolling seams 10 to 15 minutes after the strips are in place—that is, after the paste has set somewhat. But the master craftsman shown hanging wallpaper in the photos here prefers to roll the seams before he brushes out the paper. If you position the strips correctly, roll the seams, and then smooth the covering, he asserts, you'll be less likely to stretch the wallpaper. Also, if seams don't align correctly, you want to know that sooner rather than later so you can adjust or remove the strip before the paste sets up.

In any case, rolling may cause paste to ooze from the seams. So be sure to sponge wallcovering clean as you work, unless you're installing delicate or embossed wallcovering, which shouldn't be rolled or wiped at all. Finally, use a moderate pressure when rolling. After all, you're trying to embed the wallcovering in the paste, not crush it.

Complex and Special-Care Areas

Installing wallcovering would be a snap if there were no corners, doors, windows, and electrical outlets, where you need to use extra care.

TILTING TRIM AND COCKEYED CORNERS

In renovation work, door and window casings and corner walls are rarely perfectly plumb, but strips of wallcovering must be. If your first strip begins next to an out-of-plumb casing, for example, overlap it by the amount the casing is off plumb. After brushing out the wallpaper, trim the overlapping edge. Thus the leading edge of that strip will be plumb, as will the leading edge of the next strip. Still, always double-check for plumb before hanging subsequent strips.

Inside corners. If an inside corner is cockeyed, a strip of wallcovering wrapping the corner will be out of plumb when it emerges on the second wall. First, use your spirit level to determine which way the walls are leaning. Then trim down

As you wrap an outside corner with wallpaper, relief-cut the top of the strip at the corner, as shown. Otherwise, the paper won't lie flat on both walls.

the width of the strip so it is just wide enough to reach the second wall—plus a ⅛-in. to ¼-in. overlap. (Save the portion you trim off: If it's wide enough, you may be able to paste it onto the second wall, thus attaining a closer pattern match in the corner.)

Now, hang a strip of wallcovering on the second wall, plumbing its leading edge to a plumbed line you've marked on the wall first. Tuck the trailing edge of the strip into the corner so that it overlaps the first strip. There will be a slight mismatch of patterns, but in the corner, it won't be noticeable. If you don't like the small welt that results from the overlap, use a razor knife to double-cut the seam. However, if your walls are old and undulating, they'll make it tough to cut a straight line. Ignoring a slight welt may spare you a lot of frustration. In any event, don't butt-join strips at corners because such seams almost always separate.

Outside corners. Outside corners project into a room and are very visible. So when laying out the job, never align the edge of a strip to the edge of an outside corner. These seams look terrible initially, and they often fray or separate over time. If

Ceiling Folds

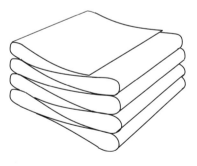

An accordion fold is easiest to unfold as you paper a ceiling and helps keep paste off the face of the wallcovering.

FITTING OVER OUTLETS AND FIXTURES

Before hanging paper over an electrical outlet, switch, or fixture, turn off power to that outlet and check that it's off by using a voltage tester.

Cutting option 1: Loosely hang the paper, locate the outlet, and cut a small X over the center of the outlet, extending the X until the paper lies flat.

Cutting option 2: Loosely hang the paper, and cut around the outside of the outlet box. When the cutout is complete, brush the paper flat.

the edge of a strip would occur precisely at a corner, cut it back ½ in. and wrap the corner with the edge of a full strip from the adjacent wall. Relief-cut the top of the wallcovering where it turns the corner, as shown in the photo on p. 587, so the top of the strip can lie flat. Remember to plumb the leading edge of the new strip.

PAPERING AROUND ELECTRICAL OUTLETS AND FIXTURES

Turn off electricity to the affected outlets and fixtures, and confirm that it's off by using a voltage tester, as shown on p. 294. Remove the cover plates and other hardware from the outlets so the hardware protrudes as little as possible.

For an outlet relatively flush with the surface, simply position the strip over it. Then, over the center of the outlet, cut a small X in the strip. Gradually extend the legs of the X until the strip lies flat. Even though the outlet's cover plate will cover small imperfections in cutting, cut as close as you can to the edges of protruding hardware or the electrical box. Smooth the strip with a smoothing brush, and trim any excess paper. If the edges of the cutout aren't adhering well, roll them with a seam roller.

It's preferable to remove fixtures such as wall sconces, but that's not always possible. For example, sometimes mounting screws will have corroded so badly that you would damage the fixture trying to remove them. In that case, after matching the wallcovering pattern, cut the strip to the approximate length. Then measure on the wall from the center of the fixture in two directions—say, from the baseboard and from the edge of the nearest strip of wallcovering. Transfer those dimensions to the strip you will hang. If you apply paste after cutting a small X, avoid fraying the edges of the cut with your paste brush or roller.

Hang the strip, and gradually enlarge the X until it fits over the base of the fixture. Smooth down the entire strip, trim closely around the fixture, and wipe away any paste that smeared onto the fixture.

PAPERING CEILINGS

People rarely paper ceilings anymore, unless they're trying to replicate a historical look. Even in papered rooms, ceilings are usually painted. Even professionals find papering the ceiling challenging and time-consuming. So get a helper if possible, and paste only one strip at a time until you get the knack of it. Cover ceilings before walls because it's easier to conceal discrepancies with wall strips.

Because shorter strips are easier to handle, always hang across the ceiling's shorter dimension. Snap a line down the middle of the ceiling and work out from it. Cut strips for the ceiling in the same manner described for walls, leaving an inch or two extra at each end for trimming. However, folding the covering is slightly different. It's best to use an accordion fold every 1½ ft. or so, which you unfold as you smooth the strips across the ceiling. (Be careful not to crease the folds.)

With your smoothing brush, sweep from the center of the strip outward. Once you have unfolded the entire strip, make final adjustments to match seams and smooth well. Roll seams after the strips have been in place for about 10 minutes.

ARCHES AND ALCOVES

Consider painting the inside of arches: It will be easier and look better than papering them. However, papering curved sections isn't difficult, provided you allow enough extra wallcovering for overlaps and trimming and for making pattern adjustments.

Before papering an arch, try to position wall strips so their edges don't coincide with the vertical (side) edges of the arch. Just as it's undesirable to have wallpaper seams coincide with an outside corner, seams that line up with an archway corner will wear poorly and look tacky. When hanging strips over an arch, let each strip drape over the opening, then use scissors to rough-cut the paper so it overhangs the opening by about 2 in. Make a series of small wedge-shaped relief cuts in the ends of those strips, and fold the remaining flaps into the arch. Then cover the flaps with two strips of wallpaper as wide as the arch wall is thick. Typically, these two strips meet at the top of the arch in a double-cut seam. If possible, match patterns where they meet.

Double-cutting is also useful around alcoves or window recesses, where it's often necessary to wrap wall strips into the recessed area. Problem is, when you cut and wrap a wall strip into a recess, you interrupt the pattern on the wall. The best solution is to hang a new strip that slightly overlaps the first, match patterns, and double-cut through both strips. Peel away the waste pieces, smooth out the wallpaper, and roll the seams flat.

Papering an Arch

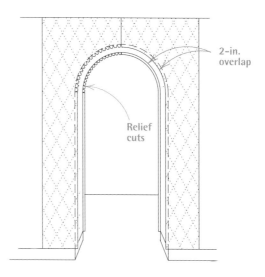

2-in. overlap

Relief cuts

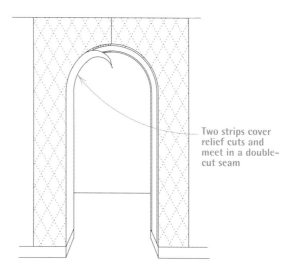

Two strips cover relief cuts and meet in a double-cut seam

Touch-Ups AND REPAIRS

▶ **SMUDGES AND STAINS.** Clean washable wallcoverings by rubbing them gently with soap and water. A commercial cleaning dough removes stains when rubbed lightly over a soiled spot. As the dough gathers grime, fold the dough in onto itself, exposing clean dough surfaces. You can blot (not rub) most nonwashable coverings with commercial, stain-removing solvents. Your wallcovering supplier can suggest one.

▶ **GASHES.** First, try to repaste the torn flap. If that doesn't look good, rip—don't cut—a patch from a spare roll of the same covering. The ragged edge of a ripped piece will be less obvious than straight edges cut with shears or a razor knife.

▶ **DENTS AND CRACKS.** Repair dents and cracks in the wall behind, and gently tear free any unpasted paper around the gash. Paste the back of the ragged-edge replacement, carefully aligning its pattern with that of the existing covering before smoothing down the patch.

20 Flooring

In the old days, flooring was the last building material to be installed and the first to show its age, as it was crushed by footsteps, swollen by moisture, and abraded by dirt. Foot traffic is as heavy and gritty as ever, but today's crop of engineered flooring and floor finishes is far more durable—and varied.

Flooring is only the top layer of a system that usually includes underlayment and subflooring, as well as structural members such as joists and girders. If finish floors are to be solid and long lasting, all parts of the flooring system must be sized and spaced correctly for the loads they will carry. Also, although some flooring materials can withstand moisture better than others, all will degrade in time if installed in chronically damp locations. In other words, correct underlying problems before installing new flooring.

You can rent many of the specialized tools needed to refinish or install flooring. Before you leave the rental company, have a salesperson explain how to operate the equipment safely and, in the case of sanders, how to change the paper.

CHARACTERISTIC	SOLID WOOD	ENGINEERED WOOD	LAMINATE	RESILIENT FLOORING	BAMBOO	PALM	CORK	TILE & STONE	CONCRETE
Durability	Good	Very good	Excellent	Excellent	Very good	Very good	Good*	Excellent	Excellent
Required maintenance	Sweep regularly	Sweep regularly	Damp mop	Damp mop	Sweep regularly	Sweep regularly	Sweep or damp mop	Wet mop	Wet mop
Water resistance†	Poor	Fair	Good	Good to excellent	Poor to fair	Poor to fair	Poor	Excellent	Excellent
Comfort underfoot	Flexes	Flexes	Flexes	Soft	Flexes	Flexes	Soft	Hard	Hard
Green creds	Yes	Mixed	No	Lino, yes; vinyl, no‡	Yes	Yes	Yes	Yes	Mixed
Cost§	$ to $$$	$ to $$	$	$	$ to $$	$$	$$ to $$$	$$ to $$$	$ to $$

* Durable but deforms if furniture sits on same spot for too long.

† Correct persistent moisture problems before converting any area to living space.

‡ Linoleum and recycled rubber are considered green; vinyl is not.

§ Costs do not reflect installation charges.

This chapter begins by introducing some of the more exciting flooring choices. Then it explains how to strip and refinish wood flooring and how to install wood flooring, resilient flooring, and carpeting. Tile floors are covered in chapter 16.

Flooring Choices

These days, choosing flooring is almost as complicated as buying a car. The old standbys such as solid wood, tile, and linoleum have been joined by hundreds of ingenious hybrids, from snap-together laminates that mimic wood or tile to bamboo planks to prefinished maple the color of plums. To make them tougher, floor finishes may include ceramics, aluminum oxides, or titanium.

Basically, you should try to choose flooring that's right for the room. Some factors to consider:

▶ Compatibility with the house's style or historical period

▶ Ease of installation

▶ Ease of cleaning and maintenance

▶ Scratch and water resistance

▶ Durability

▶ Comfort underfoot

▶ Sound absorption

▶ Anti-allergenic qualities

▶ "Green" practices for wood flooring, such as sustainable-forest harvesting

▶ Cost

WOOD FLOORING

The revolution that produced engineered lumber has also transformed wood flooring. In addition to solid-wood strips and planks, there are laminated floorings, some of which can be sanded and refinished several times. There's also a wide range of prefinished flooring.

Solid-wood flooring is solid wood, top to bottom. The most common type is tongue-and-groove (T&G) strip flooring, typically ¾ in. thick by 2¼ in. wide, although it's also available in widths that range from 1½ in. to 3¼ in. Hardwood *plank flooring* is most often installed as boards of varying widths (3 in. to 8 in.), random

These HomerWood hand-scraped planks have a cinnamon cherry finish. Prefinished flooring such as this spares you the effort of sanding, the stink of noxious fumes, and a week of waiting for the floors to dry.

Here's a typical cross section of solid-wood tongue-and-groove strip flooring.

The solid-wood wear layer of this engineered flooring can be sanded and refinished several times. The finishes shown are (left to right) white ash, vintage chestnut, and cherry.

This array of Natural Cork flooring is sealed with a UV-cured acrylic finish that keeps out moisture. The cork wear layer is bonded to a high-density fiberboard.

lengths, and thicknesses of ⅜ in. to ¾ in. *Parquet flooring* comes in standard ⅜-in. by 6-in. by 6-in. squares, though some specialty patterns range up to 36-in. squares.

Because red and white oak are attractive and durable, they account for roughly 90% of hardwood installations. Ash, maple, cherry, and walnut also are handsome and durable, if somewhat more expensive than oak. In older homes, softwood strip flooring is most often fir, and wide-plank floors are usually pine. If you know where to search, you can find virtually any wood—old or new—which is a boon if you're restoring an older home and want to maintain a certain look. On the Internet, you can find specialty mills, such as Carlisle Wide Plank Floors, that carry recycled wood that's often rare or extinct, such as chestnut salvaged from barns or pecky cypress pulled from lakes. There's also new lumber made to look old, such as the hand-scraped cherry shown in the photo on p. 591. It's not surprising that wood flooring is a sentimental favorite. It's beautifully figured, warm hued, easy to work, and durable.

Disadvantages: Wood scratches, dents, stains, and expands and contracts as relative humidity varies. And, when exposed to water for sustained periods, it swells, splits, and eventually rots. Thus, wood flooring needs a fair amount of maintenance, especially in high-use areas. In general, solid wood is a poor choice for rooms that tend to be chronically damp or occasionally wet.

Engineered wood flooring is basically an upscale plywood with a top layer of solid hardwood laminated to a three- to five-layer plywood base. Most types are prefinished, with tongue-and-groove edges and ends. This flooring is typically sold in cartons containing 20 sq. ft. of 2½-in. to 7-in. widths and assorted lengths.

Engineered wood flooring can be stapled to a plywood subfloor, glued to a concrete slab, or *floated* above various substrates. There are also glueless, snap-together systems. (Because floating floor systems can accommodate so many substrates, building environments, and design options, " 'Floating' an Engineered Wood Floor" is discussed on pp. 610–614.) More dimensionally stable than solid wood, engineered wood flooring is especially well suited to areas where humidity fluctuates, such as kitchens or basement rooms.

There are many price points and quality levels of engineered wood flooring, and you get what you pay for. Better-quality flooring has a thicker *wear layer*—the top veneer layer, which can be sanded. Wear layers range from 3/32 in. to ¼ in. thick. In general, a wear layer that was *dry sawn* has grain patterns more like solid-wood flooring, whereas layers that were *rotary-peeled* or sliced from a log tend to look like plywood.

Disadvantages: The wear layer of engineered wood flooring is relatively thin. Although manufacturers contend that a wear layer $\frac{5}{32}$ in. thick can be refinished two or three times, that seems optimistic given the condition of most rental sanding equipment.

Prefinished wood flooring is stained and sealed with at least four coats at the factory, where it's possible to apply finishes so precisely—to all sides of the wood—that manufacturers routinely offer 15-year to 25-year warranties on select finishes. Finishes are typically polyurethane, acrylic, or resin based, with additives that help flooring resist abrasion, moisture, and UV rays. To its prefinished flooring, Lauzon says it applies "a polymerized titanium coating [that is] solvent-free and VOC and formaldehyde-free." Harris Tarkett coats its wood floors with an aluminum oxide–enhanced urethane. Another big selling point: These floors can be used as soon as they're installed. There's no need to sand them or wait days for noxious coatings to dry.

As tough as prefinished floors are, however, manufacturers have very specific requirements for installing and maintaining them, so read their warranties closely. In many cases, you must use proprietary cleaners or "refreshers" to clean the floors and preserve the finish. Also, board ends cut during installation must be sealed with a finish compatible with that applied at the factory.

OTHER NATURAL FLOORINGS

The materials in this group—bamboo, coconut palm, and cork—are engineered to make them easy to install and durable. And their beauty is 100% natural.

Bamboo flooring sounds implausible to people who visualize a floor as bumpy as corduroy. However, bamboo flooring is perfectly smooth. It is first milled into strips and then reassembled as multi-ply, tongue-in-groove boards. Available in the same widths and lengths as conventional hardwood, bamboo boards are commonly $\frac{3}{8}$ in. to $\frac{5}{8}$ in. thick. Bamboo flooring can be nailed or glued. But if you glue it, allow the adhesive to become tacky first so the bamboo doesn't absorb moisture from it.

Bamboo flooring comes prefinished or unfinished and can be sanded and refinished as often as hardwood floors. It's a warm, beautiful surface, with distinctive peppered patterns where shoots were attached. Bamboo is hard and durable, with roughly the same maintenance profile as any natural wood product, so you must vacuum or mop it regularly to reduce abrasion. Avoid installing it in chronically moist areas.

Coconut palm flooring, like bamboo, is plentiful and can be sustainably harvested. Its texture is fine pored, reminiscent of mahogany. Because coconut palm is a dark wood, its color range is limited, from a rich, mahogany red to a deep brown. And it is tough stuff: Smith & Fong offers a $\frac{3}{4}$-in.-thick, three-ply, tongue-and-groove strip flooring called Durapalm, which it claims to be 25% harder than red oak. Durapalm is available unfinished or prefinished. One of the finish options contains space-age ceramic particles for an even tougher surface. So if you're thinking of installing a ballroom floor in your bungalow, this is definitely a material to consider.

Laminated bamboo, such as amber horizontal Plyboo, is about as hard as oak. Thanks to its extensive root system, bamboo can be harvested repeatedly without replanting.

Cork flooring is on the soft end of the hard-soft continuum. Soft underfoot, sound deadening, nonallergenic, and long lasting, cork is the ultimate "green" building material. Cork is the bark of the cork oak, which can be harvested every 10 to 12 years without harming the tree (some cork trees live to be 500 years old).

Traditionally sold as 3/16-in. by 12-in. by 12-in. tiles, which are glued to a substrate, cork flooring now includes colorfully stained and prefinished squares and planks that interlock for less visible seams. Cork flexes, so many manufacturers use a flexible coating such as UV-cured acrylic to protect the surfaces and edges from water. Cork's resilience comes from its 100 million air-filled cells per cu. in., so it's a naturally thirsty material. Wipe up spills immediately and avoid soaking a cork floor when mopping it: Use a damp mop instead and periodically refresh its finish.

Disadvantages: Avoid dragging heavy or sharp-edged objects across it because it will abrade. Chair and table legs can leave permanent depressions.

Typically, engineered cork flooring has a three-ply, tongue-in-groove configuration. The surface layer is high-density cork, the middle layer is high-density fiberboard with precut edges that snap together, and the underlayment layer is low-density cork that cushions footsteps and absorbs sound. First developed in Europe, snap-together panels float above the substrate, so owners can easily replace damaged planks or, when it's time to move, pack up the floor and take it with them. Many snap-together floors are glueless, but floors requiring glue usually need it to bond planks together, not to glue them to a substrate.

PRO TIP

In many of the adobe houses he designed, Albuquerque legend Nat Kaplan continued his tile floors up the wall by using 12-in.-sq. Mexican paver tiles as baseboards. With no fear of drenching wood baseboards or drywall, you can freely swing that mop as you swab the floors.

LAMINATE FLOORING

Here, the term "laminate" refers *not* to fusing thin wood layers but to a group of floorings whose surface layers are usually photographic images covered and protected by a clear *melamine* (plastic) layer. The photographic images often show wood grain, tile, or stone. Although plastic-laminate "wood" flooring may be a hard sell to traditionalists, the stuff wears well and every year captures a larger share of residential flooring. Moreover, as this category increases in popularity, manufacturers offer more and more colors and textures, including many that don't mimic natural materials and are quite handsome on their own.

Developed and first adopted in Europe, laminate flooring most commonly consists of snap-together planks that float above a substrate, speeding installation, repairs, and removal. Of all flooring materials, laminate is probably the most affordable, and as noted, it's almost indestructible. It's also colorfast, dimensionally stable, and easy to clean—though many manufacturers insist that you use proprietary cleaning solutions.

Disadvantages: Laminate flooring dents, exposing a fiberboard core; you can't refinish it, although damaged planks can be replaced; and you shouldn't install it in high-humidity areas or over concrete basement floors because it tends to delaminate.

RESILIENT FLOORING

Vinyl and linoleum are the two principal resilient materials, available in sheets 6 ft. to 12 ft. wide, or as tiles, typically 13 in. or 12 in. square. Linoleum is the older of the two materials, patented in 1863. It may surprise you to learn that linoleum is made from raw, natural materials, including linseed oil (*oleum lino*, in Latin), powdered wood or cork, ground limestone, and resins; it's backed with jute fiber. (Tiles may have polyester backing.) Because linoleum is comfortable underfoot, water-resistant, and durable, it was a favorite in kitchens and baths from the beginning, but it fell into disfavor in the 1960s, when it was supplanted by vinyl flooring, which doesn't need to be waxed.

However, linoleum has proven resilient in more ways than one by bouncing back from near extinction, thanks to new presealed linoleums that don't need waxing. In addition, linoleum (sometimes called Marmoleum, after the longest continuously manufactured brand) has antistatic and antimicrobial qualities. It's also possible to

These colors are a small sample of the hundreds available from Marmoleum, one of the oldest makers of linoleum.

custom-design linoleum borders, which are precisely cut with a water jet. Flooring suppliers can tell you more.

Vinyl has similar attributes to linoleum, though it is a child of chemistry. Its name is short for polyvinyl chloride (PVC). Vinyl flooring also is resilient, tough to damage, stain-resistant, and easy to clean. It comes in many grades, principally differentiated by the thickness of its top layer—also known as its wear layer. The thicker the wear layer, the more durable the product. The more economical grades have designs only in the wear layer, whereas *inlaid* designs are as deep as the vinyl is thick. If you're thinking of installing vinyl yourself, tiles are generally easier, although their many joints can compromise the flooring's water resistance to a degree.

STONE AND TILE

If stone and tile are properly bonded to a durable substrate, nothing outlasts them. However, handmade tiles or stones of irregular thickness should usually be installed in a mortar bed to adequately support them. And leveling a mortar bed is best done by a professional. Tile or stone that's not adequately supported can crack, and its grout joints will break and dislodge. Chapter 16 has the full story.

Tiles are rated by hardness: Group III and higher are suitable for floors. Slip resistance is also important. In general, unglazed tiles are less slippery than glazed ones, but all tile and stone— and their joints—must be sealed to resist staining and absorbing water. (Soapstone is the only exception. Leave it unsealed; most stone sealers won't penetrate soapstone and the few sealers that do penetrate make it look like it's been oiled.) Tile and stone suppliers can recommend sealants; you'll find several on p. 405. If stone and tile floors are correctly sealed, they're relatively easy to clean with hot water and a mild household cleaner.

CARPETING

Carpeting is favored in bedrooms, living rooms, and hallways because it's soft and warm underfoot and because it deadens sound. In general, the denser the pile (yarn), the better the carpet quality. Always install carpeting over padding; the denser or heavier the pad, the loftier the carpet will feel and the longer it will last. Wool tends to be the most luxurious and most expensive carpeting, but it's more likely to stain than synthetics. Good-quality polyester carpeting is plush, stain-resistant, and colorfast. Nylon is not quite as plush or as colorfast, though it wears well. Olefin and acrylic are generally not as soft or

Porcelain 12-in. by 12-in. tile (top); natural cleft slate (bottom).

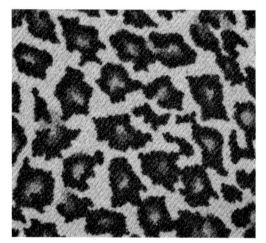

Not grandma's wall-to-wall carpet, unless she lived in Hollywood. This faux-leopard, cut-pile carpet is part of Shaw Carpet's Wildebeest series.

durable as other synthetics, although some acrylics look deceptively like wool.

Disadvantages: Carpeting can be hard to keep clean, and it harbors dust mites and pet dander, which can be a problem for people with allergies. In general, wall-to-wall carpet is a poor choice for below-grade installations that are not completely dry because mold will grow on its underside. Far better to use throw rugs in finished basement rooms.

Refinishing Wood Floors

Wood floor refinishing can be as simple as lightly sanding an existing finish and applying another coat of the same finish or as extensive as stripping the floor finish completely and sanding it several times before applying a new finish. If the floor is just grimy and dull from too many coats of wax, it may just need a thorough washing. If washing doesn't do the trick, try to determine what the existing finish is before you rent a sander.

Refinishing Floors Safely

Refinishing floors is not inherently dangerous. Nonetheless, there are safety issues to consider.

▶ **Electrical.** Before renting sanders, examine their electrical cords and plugs, rejecting any that are frayed or appear to have been sanded over. If you don't have a heavy-duty extension cord, rent or buy one; lightweight household cords could over-heat and start a fire. User's manuals or labels on big sanders indicate minimum cord specs. Household circuits must be adequately sized for the equipment: 220-volt drum sanders often require 30-amp circuits; 110-volt sanders typically require 20-amp circuits. In most cases, a drum sander's 30-amp plug will fit a home's 30-amp dryer receptacle.

▶ **Volatile chemicals.** Finish manufacturers have reduced the volatility and strong odors of their products, but you should always limit your exposure to them by wearing an organic-vapor respirator, long sleeves, and gloves when sanding old fin-ishes or applying new ones. Even fumes from water-based polyurethane are unhealthy to breathe, so as soon as finishes are dry to the touch, open windows to let vapors disperse. And sleep elsewhere until they're completely dry.

▶ **Fire and explosion hazards.** Sparks or open flames can ignite chemical fumes or dust. So before you start sanding or applying finish, turn off pilot lights for water heaters, ranges, and furnace. Also tape light switches down so they can't gen-erate a spark. To dispose safely of rags soaked with finish, put them in a bucket of water for a few days to let volatile elements disperse. Never ball them up and store them in an airtight bag or container while still volatile; they might spontaneously combust.

▶ **Lead paint and asbestos.** Floors painted before 1978 may contain lead-based paints, so don't sand them until you've had the paint tested, as suggested in chapter 18. Lead paint is generally not a problem until the dust becomes airborne or it flakes in an area where small children might eat it. Old linoleum floors may have been adhered with asbestos adhesive, which wasn't banned until 1977. Here again, asbestos is usually harmless if undisturbed, so first consult a local health depart-ment to get the name of a test lab.

PRO TIP

If you're not sure what finish was used on floors, first examine old paint and finish cans in the basement, garage, or workshop. The contents of the cans will almost certainly be useless, but their labels may tell you what's on your floors.

Most rental companies offer drum sanders because their paper clamping slots make changing sandpaper easy. Before accepting a rental drum sander, however, inspect slot lips for nicks or metal spurs, which could damage wood floors. Before leaving the rental company, learn how to insert sandpaper so it's tight to the drum.

THREE TESTS TO DETERMINE A FLOOR FINISH

Wood floors installed in the 1960s or earlier were usually finished with some combination of wax, shellac, and varnish. After that, they were most likely finished with a penetrating oil or oil- or water-based polyurethane. (For more on finishes, see p. 603.) Here are a few tests to help figure out what's there and what to do next.

Test 1: Wax. Place several drops of water on the floor. If the surface turns white in 10 to 15 min-utes, there's wax on the floor. If the water doesn't leave a white spot, try test 2. If the finish is dull, try cleaning it with a wax cleaner. If scratches and scuffs are limited, wax will be reasonably easy to remove by applying wax stripper or min-eral spirits and wiping up the residue. Then apply a new coat of wax. However, if floors are badly abraded and can't be buffed out, sand the floors, refinish them with a penetrating stain, and then wax them. Even if you sand wax-sealed floors down to bare wood, wax clinging to board edges may prevent a nonwax finish from adhering properly. (Get a second opinion from a profes-sional floor refinisher.)

Test 2: Shellac or varnish. Find an area where the finish is in poor shape and scrape it with your thumbnail or a penny. If the finish flakes off, it's shellac or varnish, which were good in their day but should now be sanded off completely and replaced with polyurethane. If the finish doesn't flake, try test 3. If the abraded areas are small, try restoring damaged areas by lightly sanding them, vacuuming and dust-mopping them well, then applying a new coat of finish. If there's not too much sanding to do, you may not need to rent a sander: A random-orbital sander with 100-grit or 120-grit sandpaper should do the job. Use a natural-bristle brush or a lamb's wool pad to apply shellac or varnish.

Test 3: Polyurethane. In an inconspicuous place, brush on a small amount of paint stripper. If the finish bubbles, it's polyurethane. If it doesn't bubble, the floors were probably sealed with a penetrating oil finish. If floor damage is limited, you may be able to touch up the pene-trating oil with a similar substance, testing small areas until you find a good color match. If the finish is polyurethane, which is a surface finish, sand the floor lightly if it is in good shape to help the new coating adhere. Fortunately, polyure-thane will stick to other polyurethane even if one is oil based and the other is water based. As long as the base coat is dry, it doesn't matter whether you apply oil-based urethane over water-based urethane or vice versa. Of course, if the finish is in bad shape, you should sand down the whole floor to bare wood and then refinish it.

RECAP: WHEN TO REMOVE FLOOR FINISHES

Sand floors to bare wood when:

▶ **Floor finishes are gouged, pitted, or showing bare spots.**

▶ **Stains go below the surface, such as mold stains beneath potted plants.**

▶ **Floorboards are irregular or uneven.**

▶ **New finishes won't adhere to the flooring.**

▶ **You need to repair rotted or split boards.**

▶ **The floor is thick enough to withstand a sanding.**

IS IT THICK ENOUGH TO SAND?

To avoid splintering wood floors when sanding them, keep at least ⅛ in. of solid wood above the tongue of T&G flooring. The easiest way to check the floor's thickness is to remove a forced-hot-air floor register and look at the exposed cross section of flooring. If that's not possible, pull up a threshold or a piece of trim and bore a small hole to expose a cross section, or drill in a closet, where no one will see the hole. If you've got engineered flooring, its wear layer (top veneer layer) won't be very thick—5/32 in. is typical—so start sanding with a less aggressive sandpaper, as suggested in "Floor-Sanding Materials" at right. Manufacturers contend that you can sand all

engineered wood floors at least once or twice, and thicker wear layers three to five times.

EQUIPMENT

Most sanding equipment can be rented. Be sure to have a knowledgeable person at the rental company explain how to operate the machines safely, how to change sandpaper and adjust wheels and drive belts, and what size circuit breaker or fuse each tool requires. Finally, inspect each piece of equipment. Sander drums and edger disks should be smooth and free of nicks or metal spurs that could scar floors. Check to see that sander wheels roll freely and that electrical cords aren't frayed or swathed in tape because they have been run over by the sander.

A large floor sander does most of the heavy sanding. Most professional refinishers favor large *belt sanders*, as shown in the photo on p. 590, because their belts are continuous, whereas rental companies usually rent *drum sanders* because the paper is somewhat easier to change. Typically, a special wrench or *key* turns a nut at the end of the drum, which opens a paper-clamping slot on the face of the drum. Drum sandpaper must be tight or it will flap and tear: Use old pieces as templates for new ones.

Caution: A drum sander is a powerful machine that can gouge even the hardest wood, so always keep the machine moving when the sanding drum is down. A lever on the handle lowers or raises the drum. Start the machine only when the drum is up. Then, as you walk, gradually lower the drum.

PRO TIP

Don't use chemical paint stripper to remove a floor finish, even if the floor is painted. Strippers are caustic to wood and hazardous to users, and even the smallest residue—between boards, for example—can create adhesion problems for the new finish.

▐▐▐▐

PRO TIP

Empty sander bags when they become about one-third full. As bags fill up, they become less efficient fillers, and more dust will stay in the air or on the floor.

▐▐▐▐

How Deep Can You Sand?

ENGINEERED FLOORING

Wear layer (top veneer layer)

Flooring nail

SOLID-WOOD FLOORING

Tongue

Flooring nail

Subfloor

You can sand only the top veneer layer of engineered flooring. Solid-wood, tongue-and-groove (T&G) flooring is a lot thicker, but you can sand only to the top of its tongue. If you sand lower, you'll hit flooring nails. T&G nail heads should be just flush, as shown.

Floor-Sanding Materials*

TYPE OF MATERIAL	GRIT SIZE	WHEN TO USE
Sandpaper belts for large floor sander; disks for edger	36	Aggressive; use on first pass if boards cupped, uneven
	36 open coat	Use on first pass if floors coated with wax, paint
	60	Try for first pass; switch to 36 if not enough cut
	100	Second or third pass
Buffer screen (use with backer pad)	100	Final screen before applying finish
	220	Smooth between coats of finish
Sandpaper strips (attach to buffer backer pad)	180	Smooth between coats of oil-based finish
	220	Smooth between coats of water-based finish

* Consult finish manufacturer's specs for sanding requirements.

PRO TIP

When sanding floors, follow the physician's creed, "First, do no harm." It can take hours to repair a trough cut by paper that's too coarse. In fact, you may have to replace the damaged section. So start with the least aggressive sandpaper grit that will do the job, whether it's removing old finish or leveling uneven boards. If that proves too gentle, you can switch to a more aggressive grit.

IIII

Worth a look: The U-Sand, a four-headed random-orbit floor sander, comes highly recommended by Charles Peterson, a hardwood flooring expert and a consultant for the National Wood Flooring Association. Peterson notes that the U-Sand "is aggressive enough to take down floorboards, yet gentle enough for light abrading between finish coats." Its orbiting heads are configured so that you can sand right up to walls, thus eliminating the need to rent an edger. The U-Sand can be rented from Home Depot, True Value, and other chains.

An edger (disk sander) goes where drum or belt sanders can't—along the perimeter of floors and into tight nooks. (Large floor sanders should not be used within 6 in. of walls.) Edgers may be smaller than floor sanders, but they can still gouge flooring quickly. So practice on plywood. The edger's paper is held in place against a rubber disk by a washered nut. To prevent gouging the floor with the edger, many professionals leave three or four used disks beneath the new one, which cushions the cutting edge of the sandpaper somewhat.

A buffer is a versatile tool. With abrasive *buffer screens*, it can lightly sand floor finishes you want to restore or fine-sand a floor that you've stripped down to bare wood. Its slow, oscillating movement is perfect to scuff-sand between finish coats. Or, when the final coat is down, you can put a lamb's wool *buffing pad* on the buffer to bring up the sheen of a finish; thus it's often used to buff out a new wax coat.

Hand scrapers and sanding blocks reach corners, flooring under cabinet toekicks, and other places edgers can't reach. Hand scraping is tedious, but it goes more quickly if you periodically use a fine metal file known as a mill file (bastard file) to sharpen the scraper blade.

Other hand tools you'll need include a nail set to sink nail heads below the surface of the wood before you begin sanding, a hammer, and wide-blade spackling knives or metal squeegees to apply wood filler. If you cut your own edger disks, you'll need a pair of heavy scissors. Brushes and applicators should be matched to

Thanks to orbiting heads in all four corners of the U-Sand sander's business end, you can sand tight against walls, so there's no need for an edger. Nor do you need to "sand with the grain" as you must with a drum sander.

When renting sanders, be sure to get any specialized tools they require, such as the T-wrench needed to change sandpaper on this edger.

specific finish types. You'll find those tools discussed and paired with finishes in "Finishes, Cleaning Solvents, and Applicators" on p. 603.

Personal safety equipment is not optional. Get a close-fitting respirator with organic vapor filters. During the sanding phases, wear eye goggles with side vents; vented goggles admit a bit of sanding dust, but they won't cloud up with water vapor. Drum sanders and edgers are noisy and tiresome; wearing hearing protection will keep you alert longer, so you'll be less likely to gouge the floor because you're punchy with fatigue. Wear disposable plastic gloves when applying finishes or wood filler. If you can, buy latex-free nitrile plastic gloves, which auto mechanics, gardeners, and postal workers swear by. Nitrile gloves are tough enough to withstand automotive solvents and garden grit, yet thin enough to sort mail with. You can find nitrile gloves at auto parts stores, typically sold in boxes of 50 to 100 in sizes ranging from small to extra-large.

Edging and hand scraping are hard on knees, so get knee pads comfortable enough to wear all day.

Rent a heavy-duty vacuum, since there's no point in frying a home vac that's really not up to the task. Ideally, the vacuum should have a HEPA filter to capture dust rather than recirculate it into the room, but not all rental companies carry them. A backpack vac, shown in the bottom photo at right is less likely to bash woodwork and has no wheels to compact sawdust, but most rental companies offer only wheeled canister types.

SANDPAPER AND BUFFER SCREENS

Sandpaper and buffer screens are rated according to the concentration of grit per square inch and the size of the abrasive particle. The lower the grit number, the larger, coarser, and more widely spaced the grit particles. Lower-grit papers cut more aggressively, whereas the higher the grit number, the finer and more closely spaced the grit. Consequently, as you sand floors, each grit should be slightly finer than the preceding one, smoothing out scratches of the previous grits, until you arrive at the grit level specified on the label of your floor finish. *Always read the finish manufacturer's sanding requirements before renting equipment and buying sanding material.*

If you're sanding floors to bare wood, you'll typically need to make two or three passes with a large floor sander and an edger and one pass with a buffer with an abrasive screen before floors are smooth enough to apply finish. (Vacuum after each pass.) Get 36-grit, 60-grit, and 100-grit paper for the floor sander and the

Buffer screens are held on with friction. Use them to fine-sand a floor that's been stripped or to sand between finish coats.

Backpack vacuums are less likely to gouge flooring or bash woodwork, but their capacity is generally less than that of floor models. Empty vacuums when they're one-third full because the fuller they get, the less efficient they become.

Before you start sanding floors, cover cabinets, air registers, and other fixed elements with plastic sheeting. To seal edges, use blue painter's tape to avoid lifting off paint or cabinet finishes, but remove the tape as soon as possible.

credit you for unused paper when you return the equipment.

PREPARING FOR SANDING

After testing floors to determine the finish, empty rooms of all movable items (don't forget window blinds and shades). Then use 3-mil polyethylene sheeting to cover the cabinets, radiators, smoke detectors, doorways, and heating or air-conditioning openings, using painter's tape to avoid pulling paint off the walls and trim. Dust migrates through the smallest openings, so use painter's tape to seal the perimeters of closed doors and keyholes. Because baseboards will get bumped by edgers, remove them—though often that's not possible. Alternatively, you can use a metal shield to protect trim, as shown in the top left photo on p. 602. Vacuum the floor so you can survey it closely for nails sticking up and floor-boards that are split or uneven. Use a nail set to sink nails below the floor surface. If boards are uneven or cupped, you may be able to sand them down evenly if they are solid wood. If any boards are split or splintered, replace them now.

To prevent sanding over the cord, keep the excess looped over your shoulders. Drum sanders sand whether you're going forward or backing up, so as you back up, continually pull the cord away from the path of the sander.

edger; both use the same grit on each pass. To screen the floor before finishing, buff with 100-grit screens backed by a nylon backing pad. To smooth between coats, use a 220-grit screen or hook-and-loop sandpaper strips that attach to the buffer pad.

Note: If floors are coated with paint or wax—which gum up sandpaper quickly—*use open-coat sandpaper* for the first sanding pass. You can use regular closed-coat sandpaper (most sandpaper is closed coat) for subsequent passes.

If you're simply recoating a finished floor, you probably won't need a drum sander and an edger; a buffer with a nylon pad and two grades of screen (100 grit and 220 grit) should do the job. Again, check your floor finish's label to see what grit sandpaper to use between coats. Finally, sandpaper wears out quickly, so get more than you think you'll need. Most rental companies will

Replacing a Floorboard

To remove a damaged board, drill holes across it so you can pry it out in splinters, using a hand chisel. Or you could cut into the damaged board by using a circular saw set to the depth of the flooring and then pry out pieces with a flat bar. To make this pocket cut, rest the heel of the saw on the floor, pull back the saw guard, and slowly lower the front of the saw sole until the turning blade engages the wood. Be careful: Holding a blade guard back is never advisable if you can avoid it, and the saw may jump when it engages the wood. Let the blade stop before you lift the saw.

Find a replacement board that's similar in color and grain: Try to pull a board from a nearby closet or from a floor section that's usually covered by an appliance. Hold the board next to the hole and use a utility knife to mark off the appropriate length. To make the replacement fit more easily, slightly back-bevel its lower edges on a tablesaw. If the stock is tongue-and-groove, use a tablesaw to cut off the lower leg of the groove. Apply construction adhesive to the underside of the new board, then drive it into the opening using a piece of scrap to cushion the hammer blows. It's not possible to nail the board through its tongue, so predrill and face-nail two 6d-finish nails at either end. Use a nail set to drive the nails below the surface. Fill the holes with wood putty.

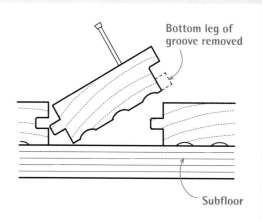

To insert a replacement board into an existing tongue-and-groove floor, use a tablesaw to remove the bottom of the groove. Slightly back-cut the ends of the new board so it will slide in easier.

SANDING FLOORS

A quick review: Shut off all pilot lights, seal off doorways, open windows for ventilation, wear a respirator and ear protection, and start with the least aggressive sandpaper. If that grit isn't cutting it, you can switch to a more aggressive sandpaper. Remember: Lower the sander drum only when the machine is moving.

Start sanding with the drum sander. Sanding with the direction of the wood grain cuts less aggressively and minimizes scratches that must be sanded out later. However, if there are high spots that need to be sanded down or if the floor is painted, sand diagonally to the wood grain on the first pass, then with the grain on all subsequent passes. (The diagonal angle should be 15° to 30° from the direction of the floorboards.) If you must sand the first pass diagonally, use the same grit on the second pass, as you sand with the grain. Because a parquet floor has grain running in various directions, sand it diagonally on the first pass, too.

Using an edger. After completing each sanding pass with the drum sander, use the edger to sand along the perimeter of the room, as close to the base of the walls as you can get. Use the same grit sandpaper that you just used on the drum. When it's upright, the sander disk moves in a clockwise direction, so work from left to right, keeping the edger moving constantly to avoid scour marks. You don't need to press down on

Overlap Sanding Passes

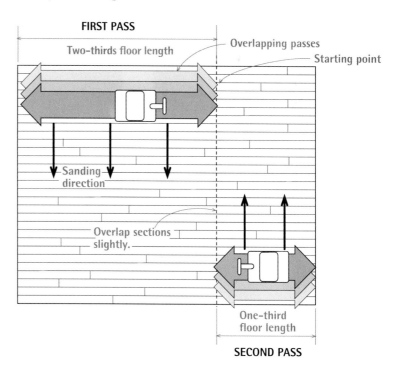

Start along a wall and sand about two-thirds the floor length. Sand up and back. Then raise the drum, and roll the sander over so the next pass overlaps by roughly half a drum-width. Sand until you reach the opposite wall. Then turn the sander 180° and sand the remaining third of the floor.

Edgers sand right up to the base of a wall, but they are aggressive sanders with plenty of torque. To avoid scuffing baseboards and casing, cover the edger bumper with masking tape or, if possible, have a helper shield the woodwork, as shown.

Hand-scrape the areas the edger can't reach. Scrape with the wood grain, before sanding lightly with a sanding block.

the edger to make it work. If the edger is sanding too aggressively, switch to the next finer grit. If it's not sanding aggressively enough, change its sanding disks more often. Again, try to sand with the wood grain as much as possible.

Finish up by hand. Hand scraping and sanding take care of the areas the edger can't reach, such as in corners and under cabinet toekicks. It's hard, tedious work, but fortunately there's not much of it. A sharp scraper will speed the job, scraping with the grain to remove the old finish. Then use a sanding block to smooth out the semicircular edger marks. (*Note:* If there are a lot of edger marks, use a random-orbital sander to feather them out.) Once you've drum-sanded, edged, and hand-scraped the room, vacuum it well before switching to the next-finer grit and sanding the floor a second time. If you need to fill holes or gaps in the floor, do it before the second sanding.

SCREENING FLOORS

After you've drum-sanded and edged the floor with 100-grit sandpaper and vacuumed it well, use a buffer with an abrasive screen to smooth out any remaining marks. Use a 100-grit or 120-grit buffer screen, which is held onto the buffer pad by friction. Because the buffer rotates slowly and the screen is flexible, you can buff right next to the base of the wall. Start along a wall, moving the buffer from side to side (it rotates in a counterclockwise direction). As you did with the drum sander, overlap passes about one-half the width of the buffer pad. Buffer screens wear out quickly, so replace them when you've screened one-third to one-half the floor. Save at least one used screen so you can fold it and use it to hand-screen the corners where the buffer couldn't reach.

Between finish coats, screen the floor to improve adhesion, using a 100-grit or 120-grit screen. Before applying the next coat, vacuum and dry-mop the floor well with a tack rag.

To achieve an even smoother finish, vacuum the floor and wet-sponge it with clear water the night before screening it (the moisture will raise the grain slightly). The next day, when the wood is dry, screen it smooth. Wetting the wood and then screening it is called *popping the grain*. Popping is optional but strongly recommended if you'll be applying a water-based finish. After screening the floor and touching up corners by hand, vacuum the room thoroughly and use clean tack rags to remove dust from any horizontal surface. (A tack rag is a slightly sticky cheesecloth pad that attracts dust.) Finally, dry-mop the floor, wrapping the mop in a clean cloth lightly dampened with the same solvent you used to thin the floor finish.

SELECTING A FLOOR FINISH

Floor finishes are often divided into two categories: *penetrating sealers (penetrants)* and *surface finishes*. Penetrating sealers usually contain plant-based oils, such as tung oil or modified linseed oil, and soak into wood fiber. In time, they harden to seal and protect the wood. Because penetrating sealers form a hard outer shell, they can be easier to touch up by sanding lightly and adding more sealer if wood becomes scuffed or scratched—touched-up areas won't be obvious.

Wood stains penetrate and don't really seal. When they have dried, penetrating sealers are often waxed to make them more durable. But waxed floors are durable only if they're regularly maintained, which takes time. Thus most floors today are sealed with surface finishes, which don't require waxing.

Surface finishes, as their name implies, form a tough exterior shell to resist scuffs, scratches, and moisture. The earliest surface finishes were shellac and varnish, which have been largely replaced by oil-based and water-based polyurethanes. Shellac has poor water resistance and chips easily, and varnishes tend to be strong smelling and slow to dry. Besides, both are extremely flammable. Although surface finishes require less maintenance, their disadvantage is that they can't be touched up when they become worn, so you must refinish the whole floor. Surface-finish sheens range from matte (little shine) to satin to semigloss to gloss. In general, glossier finishes are harder, more durable, and more water-resistant.

As you'll see, polyurethanes vary greatly in ease of installation, drying time, and durability. *Water-based polyurethanes* are the stars of the show these days: tough, nonyellowing, relatively mild smelling, and fast drying (two to six hours).

 TIP

If you're using a water-based floor finish, apply a water-based sealer first. Sealers help topcoats adhere, deter stains, and seal the surface so you can apply thinner, more uniform topcoats. Follow the manufacturer's guidelines to make sure the sealer is compatible and that you use the correct grit between coats.

Finishes, Cleaning Solvents, and Applicators

FINISH	CLEANS WITH	APPLICATOR	COMMENTS
Water-based polyurethane	Soap and water	Synthetic brush; pad; round applicator	Probably best all-around finish for nonpro; tough, water-resistant finish; easy cleanup, low smell; work fast, overlapping edged areas before they dry
Oil-based polyurethane	Mineral spirits	Natural-bristle brush; round solvent-resistant applicator	Tough, durable finish; favored by pros because it dries slower than water-based; slightly stronger smell while drying
Penetrating sealers (tung and modified linseed oils)	Mineral spirits	Lamb's wool applicator, natural-bristle brush	Slow to dry; strong odor; scratches easily but can be touched up with new finish over old; usually waxed
Stain-sealers	Mineral spirits	Varies: natural-bristle brush to clean rags	Same profile as penetrating sealer; finish must be waxed to protect wood
Varnish	Turpentine or paint thinner	Natural-bristle brush; lamb's wool pads	Volatile; strong smelling; slow to dry; hard, amber finish gives historical appearance; often used to match older finish
Shellac	Denatured alcohol	Natural-bristle brush; lamb's wool applicator	Poor water resistance; flammable; chips; rarely used for floors anymore
Various (acrylic-impregnated; acrylic-urethane; UV cured)	Proprietary solvents	Computer-monitored sprayer	Factory applied in highly controlled environment; durable; water-resistant coatings on prefinished wood flooring

Start each coat by cutting an edge around the floor's perimeter and along cabinet bases, using a brush or a paint pad. Then use a T-bar applicator to overlap the edged borders while they're still wet.

For newer, water-based polyurethanes, use a round synthetic T-bar applicator like this one. Pour the finish onto the floor, then spread it in broad sweeps—like window washing with a squeegee.

Use a lamb's wool applicator to apply penetrating-oil finishes.

And brush cleanup is easy with soap and water. Because they contain lower levels of VOCs, water-based polyurethanes also are safer to use. Despite their volatility, *oil-based polyurethanes* are often applied by professional refinishers because they're slower to dry and thus allow more time to even out coats. Although solvent based and stronger smelling, oil-based polys are more durable and water-resistant, and they turn a handsome amber with age. Let them dry for 24 hours before recoating or walking on them. *Moisture-cure polyurethanes* are the most durable of the lot, but because they must absorb moisture from the air, they're temperamental to apply, slow to dry, and best left to professionals.

Finally, there are surface finishes that require highly controlled environments and thus are factory applied to prefinished flooring. This group of finishes may include aluminum-oxide, titanium, or ceramic additives to resist abrasion and may require UV curing, rather than heat curing. Some of the toughest of these finishes infuse acrylic into the wood cells in a modern version of a penetrating sealer.

APPLYING A WATER-BASED FLOOR FINISH

Applying a water-based polyurethane does not differ much from applying an oil-based finish and, as noted in the preceding section, water-based finishes are more benign. Although "Finishes, Cleaning Solvents, and Applicators" on p. 603 offers general guidance, always follow the manufacturer's instructions.

Cutting the edge is the first step when applying any type of floor finish. Use a brush or paint pad to apply a 6-in. swath of finish around the perimeter of the room, along cabinet bases—in short, any place that would be difficult to edge with a large applicator. Pour finish into a sloping paint tray with a replaceable liner so you can easily reload the paint pad, brush, or large applicator.

If you're applying a slow-drying finish, you can edge the whole room before switching to a large applicator. However, because water-based finishes dry quickly, it's best to edge one section of a wall at a time so you can use a T-bar applicator to overlap edged borders while they're still wet. Maintaining a wet edge is the key when applying water-based finishes: Edges that dry before they're overlapped have a distinct lap mark.

Working the floor with a T-bar applicator is like a ballet with chemicals. After cutting the edge, pour a thin puddle of finish along one wall, parallel to the floorboards, but stop the puddle 3 ft. to 4 ft. shy of the far wall. Holding the applicator pad at a slight angle—somewhat like a snowplow blade—pull the applicator through the finish. Angle the applicator so that excess finish

Stirred, NOT SHAKEN

Don't shake clear floor finishes to mix them as you do paint. Shaking will entrap air bubbles and leave blemishes—popped bubbles—when the finish dries. Instead, stir finishes thoroughly from the bottom of the can. Don't thin finishes. If stirring doesn't dissolve the finish "skin" or other solids, strain the finish through a paint strainer.

flows toward the inside of the room so you can spread it out on the return pass. The ballet comes at the end of each turn, as you sweep the applicator pad 180°, spread the finish evenly, and set up for the next pass. It's easier to do than to describe.

Periodically pour more finish in a long puddle to maintain a wet edge. Having a second person to pour the finish and touch up missed spots is helpful but not essential. If you see a missed spot after the finish has started to set, let it dry and be sure to coat that area the following day, when you apply the next coat. Once each coat is dry, screen-sand it lightly, vacuum, and dry-mop it with a tack rag over the mop. Then apply the next coat.

In general, don't walk on the floor until the final coat has cured at least three days, and—because this finish is water based—do not damp-mop it for a month.

Installing Strip Flooring

T&G strip flooring, ¾ in. thick and 2¼ in. to 3¼ in. wide, is by far the most commonly installed wood flooring. Installing it requires few specialized tools and, with a modest amount of prep work, it goes down fast and lasts a long time.

PREP STEPS

Wood absorbs water and swells, so don't bring hardwood flooring on site until the building is closed in and "wet work" (such as plumbing, tiling, drywalling, plastering, and painting) are complete. Allow paint, plaster, or joint compound several days to dry. If necessary, turn on the heating or air-conditioning so indoor conditions will be close to normal (60°F to 80°F) for a week before installing flooring. Open the bundles of wood flooring and allow them to acclimate indoors for 72 hours before installing them.

Filling HOLES AND GAPS

Flooring stores carry color-matched spot fillers and trowel fillers. Spot filler is basically woodworker's putty, applied with a spackling knife to fill nail holes and obvious cracks. Trowel filler, which is thinner, is poured onto the floor in small amounts and worked into the narrow gaps between floorboards, using a large squeegee or a smooth-edge trowel. Done on your knees, applying trowel filler is hard work, requiring pressure to force the filler into gaps and to scrape off excess. Consequently, although spot-filling is common, trowel-filling is not. Note: If you've got wide pine planks, which expand and contract seasonally, don't fill the gaps between them. Brag about their rustic charm instead.

PRO TIP

As you install strip flooring, use wood from several different bundles or cartons to ensure a varied mix of color and grain. If strips are noticeably lighter or darker, distribute them throughout the floor to avoid obvious, odd-colored sections. Stagger board ends in adjacent rows by at least 6 in., because random joint patterns will be visually less intrusive.

Square-edge wood-strip flooring is face-nailed, so use a straightedge to line up the nails for a neat, professional appearance. Because tongue-and-groove flooring is nailed through the tongues, those nails are hidden.

Subflooring AND UNDERLAYMENT

Subflooring is usually CDX plywood or OSB panels whose long edges run perpendicular to floor joists—although in older houses, subflooring may be 1-in. boards run diagonally. To allow for expansion and to minimize squeaks, leave 1/8-in. gaps between square-edge panels, nailing the panels to joists every 6 in.; ring-shank or spiral nails hold best. (T&G panels have integral expansion gaps, so butt their edges tight.) Undersize subflooring often sags between joists, creating high spots over the joists and floors that are springy and squeaky. Adding blocking between the joists may stiffen and quiet floors.

Underlayment is a layer placed over the subfloor and under the finish floor. There are several types of underlayment. Rigid types such as particleboard or hardboard can stiffen the subfloor and level out irregularities—especially important when thin flooring such as linoleum would telegraph gaps or an uneven substrate. Other types of underlayment act as a cushioning layer, such as the padding used beneath carpeting or the foam layer specified beneath engineered wood flooring. Felt paper is sometimes used as underlayment in dry, above-grade applications. Underlayment intended for below-grade use often combines a vapor barrier with some kind of foam cushioning.

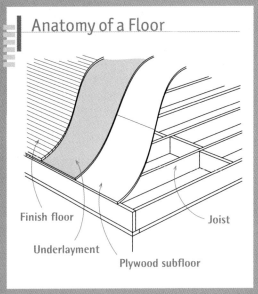

Anatomy of a Floor

Finish floor

Underlayment

Plywood subfloor

Joist

Use a moisture meter to check interiors if your region has high humidity. Home centers and electronics stores carry reliable, inexpensive meters. Ambient humidity indoors should be 35% to 55%; if readings are higher, consider installing a dehumidifier. Also check the moisture content (MC) of wood subfloors and flooring, using a moisture meter with probes. Typically, wood flooring's MC is 6% to 10%. The subfloor's MC should not vary more than 4% from that of the flooring's.

If you're installing floors over a basement or crawlspace, check the humidity of that area, too. If it's too high, correct any contributing factors before installing wood floors; high humidity also encourages mold. Crawlspaces with dirt floors should be covered with plastic and sealed to limit moisture and air infiltration, as described in chapter 14.

Survey subfloors to make sure they're solid, flat, and clean. If floors are excessively springy, stiffen them by adding blocking between the joists, adding plywood or OSB panels over existing subfloors, or sistering new joists to old ones, as described in chapter 8. In older houses, floors are rarely level; so if they're solid, it's more important that they be flat—within 1/2 in. per 10 ft. Use a rental edging sander or a woodworker's belt sander with coarse sandpaper to lower high spots; use strips of building paper (15-lb. felt paper rather than rosin paper) or wood shims to build up low spots. In general, masonry floor-leveling compound is too inflexible to use beneath wood flooring because flooring nails will fragment it and board flexion will fracture it.

If you notice protruding nail heads, not enough nails, or squeaky spots, correct these conditions now. Squeaks can usually be silenced by screwing down subflooring to joists near the squeak or by nailing it down with ring-shank or spiral nails. Vacuum and sweep the floor well. If the floor is over an occasionally damp basement or crawlspace, staple 15-lb. building paper to the subfloor, overlapping roll seams by 6 in. However, don't bother with building paper if the subfloor areas are dry or if the floor is on an upper story.

Finally, remove the baseboard molding if you can do so without damaging it. Baseboards hide the expansion gap between the perimeter of the flooring and the base of the wall. At the very least, install a piece of quarter-round shoe molding to cover the gap if you can't remove the baseboards. If door casings are already installed, undercut (trim the bottoms of) each side jamb by the thickness of the flooring; an *undercut saw* is specially designed for this task. An oscillating

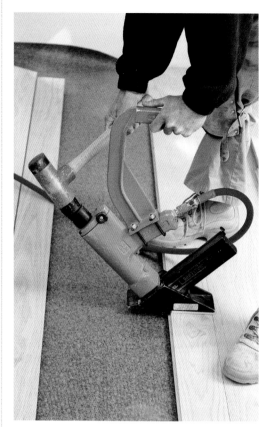

A pneumatic floor nailer will drive nails or staples at the correct depth all day long, once you've calibrated its pressure. You don't need to hit the rubber strike cap hard to make the nailer fire.

Floor Nailing Schedule*

SIZE AND TYPE FLOORING	SIZE NAIL TO USE	SPACING
T&G strips[†] (¾ in. × 1½ in., 2¼ in., 3¼ in.)	2-in. barbed flooring cleat,[‡] 7d or 8d flooring nail, or 2-in. 15-gauge staples with ½-in. crowns[‡]	10 in. to 12 in. apart; 8 in. to 10 in. preferred
T&G[†] strips (½ in. × 1½ in., 2 in.)	1½-in. barbed flooring cleat or 5d cut-steel or wire-casing nail	10 in. apart
T&G strips (⅜ in. × 1½ in., 2 in.)	1¼-in. barbed flooring cleat or 4d bright wire casing nail	8 in. apart
Square-edge strips[§] (⁵⁄₁₆ in. × 1½ in., 2 in.)	1-in. 15-gauge barbed flooring brad	2 nails every 7 in.
Square-edge strips[§] (⁵⁄₁₆ in. × 1⅓ in.)	1-in. 15-gauge barbed flooring brad	1 nail every 5 in. on alternate sides of strip
Planks (4 in. to 8 in.)	2-in. barbed flooring cleat,[‡] 7d or 8d flooring nail, or 2-in. 15-gauge staples with ½-in. crowns[‡]	8 in. apart

*   Adapted from the NWFA: The National Wood Flooring Association, all rights reserved © 2004.
[†] Tongue-and-groove (T&G) flooring is blind-nailed on the tongue edge, with face-nailing required on the starting and finishing runs.
[‡] NOFMA Hardwood Flooring must be installed over a proper subfloor. Use 1½-in. fasteners with a ¾-in. plywood subfloor on a concrete slab. A concrete slab with sleepers every 12 in. on center does not always require a subfloor.
[§] Square-edge flooring is face-nailed.

multitool with a wood blade is even faster. Remove doorway thresholds if they're nailed down. But if they're glued down or set in mortar, simply butt the flooring to them.

EQUIPMENT

Sawdust or debris trapped under a board can mean uneven, loose, or squeaky floors later on, so be obsessive about keeping subfloors clean as you install flooring.

Installation tools include safety glasses, hearing protection, knee pads, chopsaw or small table-saw, hammer, nail set, tape measure, chalkline, flat pry bar to remove trim, large flat-bladed screwdriver to draw board edges tight to each other, flooring mallet, and a manual or pneumatic flooring nailer. For the little bit of face-nailing to be done, use a pneumatic finish nailer; if you haven't got one, use a ¹⁄₁₆-in. bit to predrill holes for the face nails. You'll need white glue to secure floorboards under toekicks and in other odd spaces where it's difficult to reach with any nailer. Finally, rent a shop vacuum if you don't own one. And be sure to have a good-quality broom and a dustpan.

Pneumatic flooring nailers are more expensive than manual nailers, but they don't depend on your strength to drive flooring nails to the correct depth. Nailers aren't foolproof, though. Take a sample of the flooring to the rental company to ensure that the pneumatic nailer will correctly engage the flooring edge profile. That is, the tool may need an adapter to avoid damaging the boards' tongues. On site at the start of the job, calibrate the nailer's pressure by nailing a "practice row" of flooring to the subfloor. Typically, pneumatic nailers are set at 70 psi; adjust the pressure up or down until the tool sets nails correctly, as shown in "How Deep Can You Sand?" on p. 597. Once the setting is correct, pull up the practice row.

LAYOUT, STARTER ROWS, AND BEYOND

There are two places to install a starter row. The first and most obvious place is along a long wall.

Adding a spline, as shown on the left, creates a tongue-and-groove board with two tongues so you can nail outward from that board in two directions. Use a spline when you want to start an installation in the middle of a room.

Framing

Baseboard

Drywall

Flooring

Subfloor

½-in. expansion gap

When walls are out of square, use baseboards or shoe molding to cover the ½-in. gap wood floors require.

Using a piece of scrap to avoid damaging the tongues, drive the boards snugly together before nailing them. The friction between the tongues and the grooves will usually hold them during nailing.

The second place is down the center of the room, which is recommended when rooms are wider than 15 ft., when rooms are complex, when several rooms converge, and when walls are out of parallel by 1 in. or more.

Flooring usually runs parallel to the length of a room, so start by measuring the width of the room at several points to see if the walls are parallel. If the walls aren't parallel, split the difference eventually by ripping down the final row of boards on both sides of the room.

Use baseboards or shoe moldings to conceal the expansion gaps as shown in "Concealing Floorboard Edges" at left.

Installing the starter row along a long wall.
At both ends of the room, measure out from the wall the width of a floorboard plus ¾ in. for expansion. Snap a chalkline through those two points so you'll have a straight line to align the starter row to. Place the groove edges of the first row toward the wall, so the boards' tongues face into the room. If you pick straight boards for the starter row, successive rows will be more likely to stay straight. Face-nail the boards in the starter row, driving pairs of 6d or 8d nails every 10 in. to 12 in. and placing them in 1 in. from the boards' edges.

If you use a pneumatic finish nailer to face-nail the boards, you'll be unlikely to split them. If you hand-drive the face nails, use a ¹⁄₁₆-in. bit to predrill for 6d spiral nails. In either case, sink the nail heads below the surface of the wood, and eventually fill holes with wood putty. Next, use the pneumatic flooring nailer to *blind-nail* (nail through tongues) boards every 10 in. to 12 in. To further avoid splits, don't nail within 2 in. to 3 in. of a board's end. Once the starter row is secured, blind-nail subsequent floorboards until you reach the opposite wall and run out of room to use the pneumatic nailer.

Installing the starter row in the middle of the room. Measure out from both long walls to find the approximate center of the room. If walls aren't parallel, the centerline should split the difference of the measurements between the two walls. Snap a chalkline to indicate the centerline; line up the starter row to it. Because you *don't* face-nail a starter course in the middle of a room, screw temporary blocks—scrap flooring is fine—along the chalkline to keep starter-row boards in place. Otherwise, they could drift as you drive nails through the tongues. Nail down five or six rows before removing the temporary blocks.

Next, add wood *splines* (also called slip tongues) to the grooves of starter-row boards,

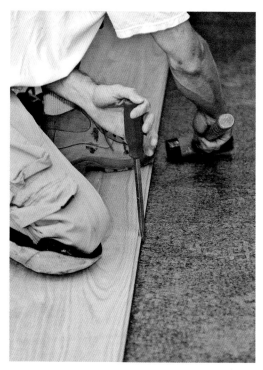

If boards are slightly warped or tongues and grooves are a bit swollen, use a thick screwdriver as a lever to draw them together. Hammer the screwdriver point into the subfloor to get some traction.

Glue the last row of boards, especially those at the base of a cabinet. Only one edge will engage the board next to it, and there's often not enough room to drive nails.

The first and last rows of tongue-and-groove floorboards are usually face-nailed, here with a pneumatic finish nailer. Draw boards tightly together with a flat bar.

which most flooring stores carry. Adding splines allows you to blind-nail toward the opposite wall as well. Glue splines to board's grooves, using scrap flooring to drive the splines snug without damaging them.

Installing the rest is straightforward. To speed the installation and ensure that board ends are staggered at least 6 in. between rows, have a helper rack (spread out) floorboards so you can quickly tap boards into position with the flooring mallet and nail them down. Floorboards come in regular lengths from 12 in. to 36 in. So to create a random joint pattern, use board remnants with irregular lengths to start rows.

Continue blind-nailing boards every 10 in. to 12 in., checking periodically to make sure the rows are straight. If milling irregularities or warping prevents boards from seating correctly, use a large flat screwdriver to lever the boards snug, as shown in the top left photo. Set aside boards that are too irregular to use; professional installers typically order 5% extra to allow for warped or poorly milled boards and waste.

As you approach within a foot or two of the opposite wall or next to a base cabinet, you won't have room to use the pneumatic flooring nailer, so switch to a pneumatic finish nailer. Nor will you have room to swing your flooring mallet, so use a pry bar to draw the boards' edges snug. As

Whenever there's a tricky transition or a complex shape to cut, either scribe the shape or create a template to transfer the shape to the floorboards, as this installer is doing.

Wood Flooring over Concrete

Before purchasing wood flooring for installation over concrete slabs or in below-grade areas such as basements, check the warranty to see if such installations are allowed. When allowed, most require that excess moisture conditions be corrected beforehand. In general, engineered wood flooring is a better choice than solid-wood flooring in such locations because engineered flooring is more dimensionally stable.

When solid-wood flooring is installed over concrete, the slab is typically covered with a 6-mil polyethylene vapor barrier, with 2x4 sleepers (on-face) spaced every 12 in. to 16 in. over the plastic; powder-actuated fasteners then attach the sleepers to the slab. Although it's possible to nail ¾-in. T&G flooring directly to sleepers, it is likely to deflect. Most flooring manufacturers specify a ¾-in. plywood subfloor, with panels run perpendicular to the 2x4s, a ⅛-in. gap between panel edges, and a ½-in. expansion space around the perimeter of the subfloor.

Alternatively, some building scientists argue against using plastic vapor barriers and for covering concrete surfaces with semipermeable, *moisture-tolerant* insulation panels, as discussed on p. 284. They suggest placing 1-in.-thick expanded polystyrene (EPS) panels on the concrete floor. EPS's compressive strength is sufficient that you can then cover it with a ½-in. plywood floor—there's no need for sleepers.

Engineered wood and plastic-laminated flooring can be nailed or stapled to wood subflooring, glued directly to a slab, or "floated" over it. Floating systems typically call for floorboards to be glued or attached to each other, rather than to the subfloor or slab. Many systems feature proprietary underlayments that block moisture and cold and cushion the flooring so you feel less like you're walking on concrete. Cosella-Dörken's DELTA-FL underlayment is a dimpled plastic membrane, and other systems employ foam or felt pads. Floating systems can be used over slabs with radiant heating, as well.

you get within 6 in. to 8 in. of the wall, measure the distance remaining, including ¾ in. for an expansion joint. In most cases, you'll need to rip down the last row of floorboards. If they are less than 1 in. wide, first glue them to the next-to-the-last row and install the two rows as a unit. Or, if you're installing floorboards of varying widths, rip down a wider board. The last row of boards should be face-nailed and glued to the subfloor as well. Finally, install prefinished transition pieces such as thresholds, reducer strips (strips that taper to accommodate differing floor heights), and so on. When you've sanded and finished the flooring, reinstall the baseboards.

Store extra flooring in a dry location. If the flooring has a warranty, file it in a safe place, along with the flooring's code number and floor care information.

"Floating" an Engineered Wood Floor

As explained earlier in the chapter, engineered wood flooring can be thought of as a specialized type of plywood, usually five to seven plies thick, with a hardwood veneer face layer—also called the *wear layer*. Because it is more stable dimensionally than solid wood, engineered flooring is little affected by shifts in temperature or humidity. When used with an appropriate underlayment layer, it has been installed over a wide range of substrates, including plywood, particleboard, concrete, resilient flooring, and even ceramic tile.

This flooring's wide usage has led to diverse methods of installing it, including gluing it to a substrate, nailing or stapling it, and *floating it*—in which individual boards are snapped together or glued together but that, as an assembly, are not attached to the subfloor. This section looks at one way to float an engineered wood floor. Floating a wood floor is an attractive option for many reasons:

▶ It is relatively simple to install and requires few specialized tools.

▶ It is the preferred method for installing a floor below grade—say, in a basement. It is also arguably the simplest way to install flooring over concrete.

▶ It is one of the few acceptable ways to install wood flooring over particleboard—which makes it an ideal choice to upgrade a floor that was previously carpeted.

▶ It is the preferred installation method over subfloors with radiant heat because there are no flooring nails or staples to puncture radiant tubing or electrical cables.

Every box of boards has variations in color and grain, so take boards from several boxes as you install them. Stagger end joints at least 12 in. to 24 in. in adjacent rows. If end joints line up every third or fourth row, it won't be too noticeable.

Heed the maker. The usual warnings about buying quality and reading the manufacturer's installation instructions are doubly true when it comes to engineered wood flooring. The integrity of the product, the thickness of the wear layer, and the durability of the finish all argue for buying premium flooring from a reputable company. Wear layers vary from 1/16 in. to 1/4 in. thick. Supposedly, wear layers 1/8 in. to 1/4 in. thick can be refinished two to five times, though it would take a very light touch to do so successfully.

ORDERING, ACCLIMATING, AND STORING FLOORING

After calculating the square footage of the room(s) to be floored, add 5% to 7% for waste. If you will be installing the flooring in a diagonal pattern, add 10%. It's helpful to have a carton on hand for repairs later, in the event that the product line is discontinued.

Have flooring delivered after drywalling and painting are complete. Ideally, your order should arrive five to seven days before you plan to install it so it has time to acclimate. Flooring makers recommend room temperatures of 60°F to 80°F, with relative humidity of 35% to 65%. If there's new plaster or concrete present, it should have cured for at least 60 to 90 days.

Checking the order. When the shipment arrives, set aside cartons that are damaged, damp, or delaminating. Open them at once and look for crushed edges, cupped boards, water stains, blistered finishes, and so on. If you see any of these conditions, reject the shipment. If cartons look OK, don't bother opening them until you are ready to install the flooring. But if you reject a shipment, you must do so *before* installation. You can't do so afterwards, piecemeal.

Wood is irregular, however, so expect to cut off occasional sections with small knots. One other thing to note when opening cartons: Most boards should be 4 ft. to 8 ft. long. Too many short boards will yield many more end joints, which won't look as good.

Storing flooring. When the flooring arrives, stack cartons flat in the middle of the room. Do *not* stand cartons on ends or sides, which may damage the tongue-and-groove edges. Do not store cartons in an unheated garage, directly on concrete, or in direct sunlight.

TOOLS AND MATERIALS

You need only two specialized tools to install engineered wood flooring: a hard-plastic *tapping block* and a metal *pull bar*. Both are available from online flooring suppliers. Here are the other necessary items:

▶ **Layout:** measuring tape, chalkline (taut string), and straightedge (a flooring board will do).

▶ **Cutting:** chopsaw with a fine crosscut blade for cutting across boards, tablesaw with a fence for ripping boards, sabersaw for complex cuts around pipes and the like, and undercut saw or an oscillating multitool for trimming the bottoms of door casings.

▶ **Miscellaneous:** pry bar, hammer, staple gun or hammer tacker, utility knife, carpenter's square, pencil, safety glasses, work gloves, and broom or vacuum. Optional: finish nailer.

▶ **Materials:** approved adhesive or glue for joining boards, cardboard shims, sponge and rags for wiping up excess glue, construction adhesive for affixing thresholds, and foam or felt paper underlayment. Replacement baseboards or trim if originals are destroyed removing them.

PREPARING THE ROOM

If you *float* engineered wood flooring, you can install it over pretty much any substrate—above or below grade—as long as that substrate is stable, sturdy, dry, and flat. Again, follow the manufacturer's specs that come with your flooring to protect the warranty. This section describes floating engineered wood flooring over particleboard or plywood, typically 5/8 in. or 3/4 in. thick, nailed to 2x10 floor joists spaced 16 in. on center.

Thorough prep work is essential. The substrate must be flat, well attached, and clean. Pry off baseboards, and trim the bottoms of door casings so flooring can slip under them. Here, the installer uses a scrap of flooring as a depth gauge beneath an undercut saw. You could also use an oscillating multitool.

Remove obstructions. If the room is carpeted, remove it and its padding, then pry up nailing strips and any staples stuck in the subfloor. Because the new flooring will run to within ½ in. of finish walls, also remove baseboards, shoe molding (if any), and doorway thresholds, as well as heat registers, baseboard heater end caps, pipe escutcheons, and so on. You will also need to trim the bottoms of door casings so flooring can slide under them. The best tool for this last operation is an undercut saw, as shown on p. 611; an oscillating multitool with a wood-cutting blade (see p. 57) is even faster.

Secure the subfloor. After vacuuming it well, walk the subfloor (or floor) and note squeaks. Using ring-shank nails, nail down squeaky spots, or, better, countersink wood screws into noisy boards. That done, use a straightedge to survey the floor for high and low spots. Shaw Floors' specs call for the substrate to be flat within ⅛ in. of a 6-ft. span. Sand down high spots or fill low spots, using a cement-based leveling compound such as Ardex Feather Finish, which is self-drying and has excellent compressive strength.

Apply the underlayment. Again, follow the manufacturer's specs. Below-grade installations over concrete, for example, can be quite exacting, employing a seam-sealed 6-mil poly film vapor barrier used in tandem with an approved foam pad layer. Above grade, tar paper is often used as underlayment, although quality installations favor a foam layer because it cushions floating floor systems, mitigates substrate irregularities, and soundproofs to a degree. When stapling foam underlayment, use as few staples as possible and drive them flush to the surface.

GETTING THE FIRST ROW RIGHT

Lay the first row of floorboards along the longest, straightest wall, which in most cases will be an exterior wall. Because engineered wood flooring expands and contracts slightly, there must be a ½-in. expansion gap between that first row and the wall. The gap must be maintained all the way around the perimeter of the room.

Layout tips. Use a chalkline to mark the expansion gap—and to note if the longest wall is, in fact, straight. Hold the chalkline ½ in. from each corner of the wall, snap it, and then move along the line with a measuring tape, noting places where the wall is not straight—and by how much. Write measurements on the subfloor so that later you can shim out "low spots" so that the first row of flooring will be straight even if the wall is not.

Next, measure out from the chalkline to the opposite wall to determine if the room is square. If the room is not square, you will need to scribe the floorboards along one wall—preferably along the least visible wall. In the photo sequence, the room was square, so there was no need to scribe floorboards.

Lastly, measure the room's width—from the starting wall to the opposite wall—and divide the room's width by a board's width. Engineered floorboards vary in width, but the most common are 6 in. or 7 in. wide. If yours are that width and the last row would be less than 2 in. wide, split the difference. In other words, trim down the first row so that both the first and last rows will have roughly the same width.

Laying row one. Place cardboard shims every 2 ft. to 3 ft. along the wall to establish a ½-in. expansion gap and, as needed, to create a straight surface for the first row of boards. To determine the length of that first row, measure the length of the wall and subtract 1 in.—to allow ½-in. gaps at each end. Place boards with the grooved edges toward the wall; tongue edges will face out into the room. Apply glue to boards' butt-end grooves, fit boards together by hand, and snug them against the shims along the wall. Tap board ends together, using a pull bar. Then place a straightedge along the tongues to make sure boards are lined up straight.

Note: Place shims behind end joints in the first row, too.

Don't overglue. Board edges are machined to fit tightly and are glued only to add stability. Apply glue to the bottoms of grooves, filling them only halfway. Along long edges, start and stop gluing 2 in. from board ends. When gluing butt ends, stop and start 1 in. from corners. If excess glue oozes out of joints, use a damp rag to wipe it off immediately.

INSTALLING THE REST OF THE FLOOR

As you continue adding rows, take boards from several cartons so there will be a random mix of colors and shades. Many installers spread out enough boards for four or five rows so they can easily choose board lengths that will stagger end joints 12 in. to 24 in. between adjacent rows. Ideally, end joints should never line up exactly across a floor. As you progress, use a tapping block to gently join long edges; periodically use a pull bar to close up end joints. If you install boards in the same direction—say, left to right—you will be less likely to loosen previous joints. Periodically use a straightedge to make sure the tongues of successive rows continue to be straight.

Cover the substrate with an approved foam underlayment. Start installing engineered wood flooring along the longest wall, using shims to establish a 1/2-in. expansion gap. It's important to get the first row of boards straight: Place shims every 2 ft. to 3 ft. and behind end joints.

Floating flooring systems vary. Some are glueless, in which boards snap together, whereas others use glue to join boards to each other—but *not* to the substrate. If your system requires glue, apply it sparingly to board grooves. Half-fill grooves. Along the sides, stop glue 2 in. shy of board ends, and when applying glue to butt ends, stop 1 in. shy of each corner.

After applying glue, use a hammer and a hard-plastic *tapping block* to tap boards together. Be gentle. As the first row is installed with the groove toward the wall, you'll be tapping the tongue edge, which can be crushed if you strike it too hard.

Use a *pull bar* to draw boards together end to end. As you measure and install the end boards in each row, allow room for a 1/2-in. expansion gap along each wall. Without expansion gaps, flooring could buckle because of shifts in temperature and humidity.

If engineered flooring is correctly milled, there should be no gaps between boards. Installers periodically run their fingertips over joints to make sure joints are tight because sometimes your fingers can see what your eyes can't. Sawdust or debris is the most common cause of slight gaps—and another reason to keep the job site clean. (Cut boards in another room or outside.) If a damaged edge causes a gap, remove the board.

The last row of flooring must often be ripped down so it will fit. Hold the board to be cut against the wall (leave a 1/2-in. gap), and mark the amount to be removed from the tongue edge. Because there won't be room for a tapping block, use a pull bar to pull the last boards snug. Whenever a board end meets a threshold or, say, an obstruction such as a vent, be sure to leave room for an expansion gap.

As you approach the opposite wall, there won't be room for the tapping block; use the pull bar instead. Invariably, the last row of flooring will get ripped on a tablesaw. So when measuring the width of that last row, don't forget the 1/2-in. expansion gap along the wall.

Use painter's tape to hold the threshold in place until the construction adhesive dries.

After allowing floorboard glue to dry for 12 hours, pull shims and install baseboards to cover expansion gaps. (Don't toenail through flooring edges when attaching baseboards or shoe molding!) Replace any other trim removed during prep.

In doorways, threshold trim covers gaps in flooring and creates a transition between rooms. To avoid splitting thin threshold stock, use a flexible construction adhesive instead of nails. *Note:* Apply construction adhesive to *one side only* where floors meet in the doorway, as shown in the photo. In other words, one side of a threshold must be able to move freely as flooring expands and contracts.

PRO TIP

The part of walls covered by baseboards often isn't painted. So before replacing baseboard, note where the old trim was nailed to studs. Lightly pencil those stud locations onto the new baseboard and you won't have to search for studs.

Allow at least 12 hours for the glue to dry before walking on the floor. Then replace baseboards and any other trim removed earlier.

Resilient Flooring

Resilient flooring surfaces, such as vinyl and linoleum, bounce back from use and abuse that would gouge or crush harder, less flexible materials. However, vinyl and linoleum are relatively thin, so their durability depends on a subfloor that's thick enough and an underlayment layer that's smooth, stiff, and flat.

Resilient flooring is installed either as tiles or as sheets; both require underlayment. Tiles are generally easier to install—their layout is similar to that for ceramic floor tiles, as described in chapter 16, but are poorly suited to high-moisture areas because of their many seams. Resilient sheets are better for kitchens and bathrooms, as suggested in the kitchen installation shown here.

CHOOSING AN UNDERLAYMENT

Because resilient materials are thin—between 1/16 in. and 3/16 in. thick—they will telegraph subsurface irregularities, such as board joints, holes, and flooring patterns. So underlayment materials must be uniformly flat (no holes or voids), smooth, stiff, and dimensionally stable. Few materials fit the bill. *Note:* It's possible to adhere

Because resilient flooring is flexible and easy to cut, you can fit it after the cabinets are installed, no matter how complex their shapes. Given their wide range of colors and textures, linoleum and vinyl flooring can complement almost any decor.

Creating a Paper Template

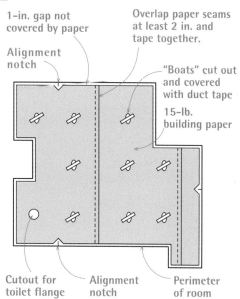

1-in. gap not covered by paper

Alignment notch

Overlap paper seams at least 2 in. and tape together.

"Boats" cut out and covered with duct tape

15-lb. building paper

Cutout for toilet flange

Alignment notch

Perimeter of room

By scribing the perimeter of a room onto felt building paper, you create a full-size template that you can transfer to the resilient flooring. Taping over boat-shaped cutouts keeps the template from moving.

resilient flooring directly to concrete slabs, old resilient flooring, and wood flooring, but that often requires a lot of prep work to make such surfaces perfectly smooth. So in addition to creating a more durable, smoother resilient floor, underlayment speeds up its installation.

Plywood. Plywood is universally acceptable if it's correctly installed and is exterior-grade APA-rated CDX underlayment. It will be stamped "underlayment" or "plugged crossbands." Also, it should be at least ¼ in. thick and have a *fully sanded face* (FSF)—not plugged and touch-sanded (PTS). Type 1 lauan plywood, which has an exterior glue, is also specified by many resilient flooring makers; panels should be at least ¼ in. thick. Three face-grades of lauan are acceptable as underlayment: BB, CC, and OVL. Type 2 lauan is not acceptable. APA Sturd-I-Floor plywood is another option. It's a structural plywood that serves both as subfloor and underlayment. Sturd-I-Floor panels range from ¹⁹⁄₃₂ in. to 1⅛ in. thick; span and loads dictate the thickness.

Hardboard, particleboard, and OSB. Hardboard, a very dense fiberboard, is generally an acceptable underlayment for resilient flooring in dry locations, but it should not be used in kitchens and bathrooms because its joints tend to swell when they get wet. *Particleboard* also swells along its edges when it absorbs moisture. OSB underlayment panels are more stable, but surface roughness can telegraph through resilient flooring. In dry locations, most of these materials are acceptable underlayments, but check your flooring manufacturer's recommendations to be safe. Those specs will also include nail lengths and spacing, as well as acceptable filler materials.

INSTALLING UNDERLAYMENT PANELS

Follow panel and flooring manufacturer specifications for the length and spacing of fasteners and acceptable filler materials. In the installation shown here, the installer attached ⁵⁄₁₆-in. underlayment panels, using 1⅛-in. staples spaced every 4 in. to 6 in. in the field and every 1 in. to 1½ in. along the panels' edges. Stagger underlayment joints so they don't align with subfloor joints. Before filling panel joints and irregularities with a patching compound, use a wide spackling knife or drywall-taping knife to scrape off splinters. If the blade clicks against a nail or staple, use a nail set to sink the fastener below the surface.

Most resilient flooring makers specify a portland cement–based patching compound, which may contain a latex binder. If you use any underlayment other than hardboard, fill and level the panel joints and surface imperfections. But don't fill nail holes because if nails work loose, they'll raise the patching compound as well, creating a bump in the flooring. Apply one or two coats of compound, feathering it out along the edges of the seam. If you're careful, you won't need to sand the compound.

CREATING A TEMPLATE

Bring resilient flooring sheets onto the job site at least 24 hours before working with it so it has time to acclimate to room temperature (at least 68°F). Resilient materials are more pliable when they have warmed and are less likely to crimp or crease. As you roll and unroll resilient sheets, be careful not to crimp the material, which could crease its surface and be visible forever after.

There are several ways to transfer a room's dimensions to resilient flooring sheets but none so accurate as creating a template, especially if there are refrigerator alcoves or base cabinets to work around. Create the template with 15-lb. felt paper, which is inexpensive and, being stiffer than rosin paper, is not likely to tear as you transfer the room's outline to the resilient flooring. Using a utility knife, rough-cut pieces of the paper so they approach within 1 in. of all walls, cabinet bases, and the like. Beyond that, don't agonize about fitting the paper too accurately. That is, the paper doesn't need to butt against walls and cabinets because the scribing tools will span small gaps between the edge of the paper and the perimeter of the room. If the jaws of the scribing tool are 1½ in. wide, they will scribe a guideline onto the paper that is uniformly 1½ in. away from the base of walls, cabinets, etc.

As you roll out individual pieces of paper, overlap their edges about 2 in. and use duct tape to join them. Once you've covered the floor with felt paper, use a utility knife to cut small (2-in. by 5-in.) boat-shaped holes in the paper every 3 ft. to 4 ft., as shown in Step 1 on the facing page. As you cut each boat-shaped hole, cover it with duct tape, which adheres through the holes to the subfloor. This will keep the paper from moving as you scribe the perimeter of the room.

Many installers use a scribing tool or a compass set at about 1½ in. to trace the shape of the room and cabinets onto the paper to create a template. But the installer shown in this photo

MAKING A PAPER TEMPLATE FOR RESILIENT FLOORING

In this installation overview, a felt-paper template, to the left of the island, is about to be scribed to record the room and cabinet outlines. To the right, newly cut linoleum is dry-fit to see what adjustments need to be made before gluing it down.

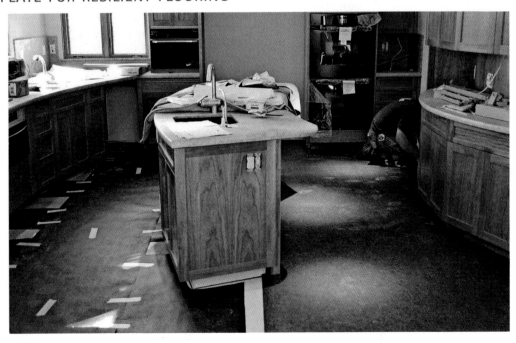

1. Cut boat-shaped openings in the felt-paper template, then cover the holes with tape to anchor the template to the underlayment. To record the perimeter accurately and ensure precise cuts on the flooring material, it's vital that the template stays put.

2. You can use an adjustable compass or, as shown here, a pin scribe and a small framing square to trace the room's perimeter onto the template. By holding one edge of the square against walls and cabinets and scribing along the other edge, you ensure a scribed line that is a uniform distance from the perimeter.

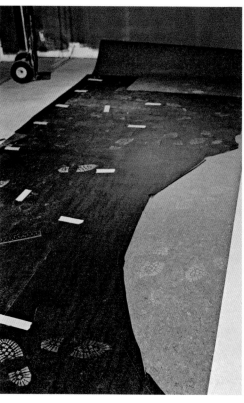

3. After scribing the template, carefully lift it off the underlayment, roll it loosely, and unroll it onto the resilient flooring. Press the taped "boats" onto the flooring so the template won't move.

4. To transfer the outline of the room and the shape of the cabinets to the flooring, align one edge of the framing square to the scribed line. Then use a utility knife to score along the other (outer) edge of the square. But score only one-third the thickness of the flooring.

5. Using the scored line as a guide, use a hook-blade knife to cut all the way through the flooring.

6. After cutting the flooring, carefully roll it with its backing out so it will lie flat when you unroll it onto the underlayment.

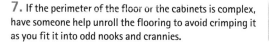

7. If the perimeter of the floor or the cabinets is complex, have someone help unroll the flooring to avoid crimping it as you fit it into odd nooks and crannies.

sequence preferred a *small framing square* and a *pin scribe*. The 1½-in. width of a framing square's blades (legs) ensured a uniform scribing distance, and the square fit easily under the cabinet toekicks. Holding one edge of the square flush to the wall, he ran the point of the pin scribe along the other edge—scribing a light line in the felt paper 1½ in. away from the wall. Because the square's blades are straight, along curved surfaces he moved the square often, making a number of scribe marks to indicate the arc of the curve.

It's OK to use several sections of felt paper if a room is large or unusually complex. In this case, be fastidious about marking section edges so you can reassemble and tape them to the resilient sheeting before transcribing the room outline. On a room of any size, you'll probably need to use several sheets of resilient flooring as well. (Sheet widths vary from 6 ft. to 12 ft.) *Important:* When you're done scribing, gently lift the template—but leave the duct tape stuck to the paper. Loosely roll up the template and carry it to the room where the resilient sheet has been unrolled on the floor, face up, to warm and flatten. Line up template edges to trimmed flooring edges. Then carefully unroll the template so it lies flat atop the resilient material (Step 3). Press down the "boats" so the duct tape sticks to the resilient flooring, anchoring the template.

CUTTING AND FITTING RESILIENT SHEETS

To transfer the outline of the room to the resilient flooring, place a blade of the framing square on the scribed line and run a utility knife along the *outside* edge of the blade, as shown in Step 4 on p. 617. The mark made by the utility knife—1½ in. beyond the scribed line—represents the cutline you'll make in the resilient flooring. But the utility knife should score the flooring only about one-third deep. After you've scored with the utility knife, use a hooked knife to cut all the way through, with the scored line guiding the hooked knife. Hold the hooked knife at a slight angle so it undercuts the edge. At some point, you'll also trim off the flooring's factory edge, which protects the material in transit.

Once you've cut the outline, remove the paper template and loosely roll the flooring with its back facing out. Carry it to the room to test its fit. If you need to retrim the flooring to make it fit exactly, you're in good company. Professional installers always assume they'll trim because no template measurement is ever 100% accurate. When you're satisfied with the flooring's fit and final position, use a pencil to draw *set marks* on the underlayment so you'll know exactly where

the sheet edge should be when you *lap the sheet* (roll it back on itself) to apply adhesive.

ADHERING AND SEAMING THE FLOORING

Some flooring materials are adhered only along the edges (*perimeter bond*), whereas others are completely glued down (*full-spread adhesion*). Flooring secured by full-spread adhesion is less likely to migrate or stretch and hence is more durable. Follow the installation instructions that come with your flooring. Be sure your supplier provides the manufacturer's instructions on adhesion and seaming methods that may be unique to your resilient flooring.

Full-spread adhesion. After the resilient flooring is final-trimmed in place, it is typically lapped back halfway, exposing roughly half the area underneath. Using a square-notched trowel, spread a compatible adhesive on about half the floor. Unroll the lapped portion down into the adhesive, and immediately use a 100-lb. roller on the material to spread the adhesive and drive out bubbles. Roll across the material's width first, then along its length. Next to the walls, use a hand seam roller to seat the material in the adhesive. Repeat the process with the second half of the sheet. If you get adhesive on the face of the flooring, clean it off at once, using a cleaner recommended by the manufacturer. (Most glues clean up with water.)

Seaming edges. If one sheet of flooring doesn't cover the entire floor, you'll have at least one seam edge. Here, be sure to follow the manufacturer's instructions on seam spacing and adhesive applications between individual sections, called *drops*. For example, manufacturers indicate how far back from the edges to apply adhesive and whether the seams should be butted together or overlapped and double-cut through both layers.

Although manufacturers tell you to butt *vinyl* seams tightly, you'll need to leave a hairline gap between the sections of Marmoleum, which contracts along its length and expands across its width. Also, linoleums tend to expand slightly because of the moisture in the adhesive.

Wall-to-Wall Carpeting

Basically, there are two types of wall-to-wall carpeting. Conventional carpeting is laid over a separate rubber or foam padding and must be stretched and attached to *tackless strips* around the perimeter of the room. *Cushion-backed carpeting*, which has foam bonded to its backing, doesn't need to be stretched—it's usually glued down—so it's generally easier to install. However, it must be destroyed to remove it. Consequently, better-quality carpeting is almost always conventional, and that's the focus here.

Carpeting doesn't ask much of subflooring, which can be slightly irregular as long as it is dry, solid, and of adequate thickness (⅝-in. plywood is typical). Carpeting can be installed over existing wood, tile, resilient flooring, or concrete floors, but check the manufacturer's recommendations regarding subgrade installations, acceptable gaps in the substrate, padding thickness and type, and so on. Don't scrimp on padding; buy the densest foam or heaviest rubber padding you can.

CARPETING TOOLS

You can rent most of the specialized tools. To install conventional carpeting, you'll need the following:

▶ A power stretcher stretches carpeting taut across a room, so it can be secured to tackless strips along opposite walls. Cross-room stretching eliminates sags in the middle of a room. You simply add and adjust stretcher sections to extend the tool.

▶ A knee-kicker is used in tandem with the power stretcher to lift the carpet edges onto strips, to stretch the carpet in a closet, or to draw the seams closer together before you hot-glue them.

▶ The stair tool drives carpeting into the spaces between tackless strips and walls and between stair risers and treads.

▶ A seam iron heats the hot-melt carpet seam tape that joins the carpet sections.

▶ A row runner, or row-running knife, cuts between the rows of loop-pile carpeting. Many installers use a large, flat-bladed screwdriver to separate the rows first.

▶ An edge trimmer trims the carpet edges so they can be tucked behind tackless strips; under baseboards; or under transition pieces such as metal carpet doors, which are used in doorways or where dissimilar flooring materials meet.

▶ Seam rollers can be either star wheeled (spiked) or smooth; they press the carpet edges onto the hot-melt seam tape to ensure a strong bond. Use a smooth roller for cut-pile carpets and a star roller for loop-pile carpets.

▶ Miscellaneous tools include a utility knife with extra blades, aviation snips to cut tackless strips, a hammer, a stapler if you're applying padding over plywood, a notched trowel if you're installing padding over concrete, heavy shears, a chalkline, a tape measure, and a metal straightedge to guide utility-knife cuts.

Carpet Layout

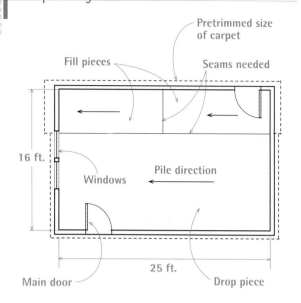

An ideal layout minimizes waste and seams, positions seams away from traffic, and orients carpet pile so that someone entering through the main entrance looks into the pile. Carpet comes in 12-ft.-wide rolls (also called bolts), so 41 RF (running feet) of carpet would allow enough extra for trimming edges in this 16 ft. by 25 ft. room.

PRO TIP

Both knee-kickers and power stretchers have a dial on their head, which adjusts the depth of the tool's teeth. Adjust the teeth so they grab the carpet backing but not the padding.

Carpet Pile and Layout

Most carpeting consists of yarn loops stitched through a backing material. The upper face of the carpet is called the *pile*. When those yarn loops are uncut, the carpeting is called *loop pile*. When the loops are cut, it is called *cut pile*. *Sculpted pile* usually is a mixture of looped and cut, which frequently creates a pattern.

Because carpeting is stored and transported on a roll, its pile gets pressed down in a direction that it retains thereafter. By stroking a carpet's pile, you can determine its pile direction. When you look into the pile, the carpet color looks richer and when the *pile direction* points away from you, the carpet appears lighter. When you install carpet, the main entryway of the room should look into the carpet pile so that it appears as rich and luxuriant as possible.

Also, where it's necessary to join two pieces of carpet, the pile of at least one piece should lean into the seam, thus overlapping and concealing it to some degree. Finally, if you have to use more than one piece of carpet in a room, all pile should point in the same direction. Otherwise, the sections will appear to have different hues.

ESTIMATING CARPET

Carpeting comes on factory rolls whose standard width is 12 ft.; a handful of carpet manufacturers offer widths of 13 ft. 6 in. or 15 ft. Plan on covering most of the room with a large piece of carpet 12 ft. wide, then covering the remaining spaces with smaller pieces joined to the large piece with hot-melt seam tape. Professional installers call the full-width piece of carpet the *drop*; the smaller pieces are called the *fill*. Joining carpet seams is time-consuming, so choose a layout that minimizes seams.

Start by measuring the room's width and length at several points, then make a sketch of the room on a piece of graph paper. A ¼-in. to 1-ft. scale is a good size to work with. On the sketch, include closets, alcoves, base cabinets, floor registers or radiators, stairs, doorways, and other features. Also note the location of doors and windows, particularly the main entrance into the room. Carpet pile should slant toward the main entrance, so that a person entering the room looks into the carpet pile.

Carpet seams and edges must be trimmed, so factor that into your estimate. Add 3 in. for seamed edges, and allow 6 in. extra for each carpet edge that runs along a wall. Stair carpet pile should slant toward a person ascending the stairs. Try to carpet stairs with a single piece 3 in. wider than the stair treads to allow for tucking along both sides of the tread. If the carpet has a repeating pattern, determine how often it repeats and add that amount plus 2 in. to the lengths of smaller pieces that must be seam-matched to the large piece of carpet covering most of the room. Given a detailed layout sketch, a flooring supplier can refine the estimate and order the correct amount of padding, tackless strips, and other supplies.

INSTALLING TACKLESS STRIPS

Wear heavy gloves when handling tackless strips. Nail the strips around the perimeter of the room, leaving ¼-in. gaps between the strips and the base of walls so you can tuck the carpet edges into those gaps. As you nail down each strip, try not to hit the angled tack points sticking out of the strip. (The tacks should always slant toward walls, away from the center of the room.) Because it's difficult to grasp the strips without pricking a finger, use aviation snips to shorten them.

Each strip must be nailed down with at least two nails. Nail the strips *in front of* radiators or built-in cabinets because it would be difficult to nail the strips or to stretch carpeting behind or under such obstacles. To anchor carpet edges in doorways, you can use a *metal carpet bar*, which has angled barbs like a tackless strip, as shown in "Carpet Transitions" on p. 624. Or you can install a *hardwood threshold* to provide a clean edge to butt the carpet to, after first anchoring it to a tackless strip, or folding the carpet under about 1 in. and nailing down that hem.

INSTALLING PADDING

Padding, which is usually 6 ft. wide, should run perpendicular to the carpet to prevent padding and carpet seams from lining up. If the padding has a slippery side, face it up so carpet can slide over it as you position it. Once you've rough-cut the padding, carefully position the pieces so they butt to each other. *Do not overlap padding* sections, which could create a raised welt under the carpet; likewise, padding should not overlap tackless strips. Trim the padding so it butts to the edge of the strips. Then use duct tape, or something as strong, to tape the padding seams together so they can't drift.

If there's a plywood subfloor, staple the padding every 6 in. around the perimeter of the floor and every 12 in. in the field using ⅜-in. staples. If the substrate is concrete, sprinkle a latex-based carpet adhesive such as Parabond M-4259 Solv-Free around the perimeter of the room and spread it out with a notched trowel; it's not necessary to adhere the entire padding. Finally, because concrete is slippery, tape the padding edges so they can't ride up onto tackless strips.

ROUGH-CUTTING CARPET

When rough-cutting carpet, it's helpful if you can unroll it completely. If you don't have enough room to do so indoors, unroll it on a clean, dry sidewalk or driveway. Sweep the area well beforehand, and make sure there are no oil stains on the ground.

Cut the carpet to the overall room dimensions on your sketch—plus the extra you included for seams and trimming along walls. At this stage, don't cut out carpet jogs, such as along cabinets and around doorways, because you'll cut those later, when the carpet is spread out on the floor you're carpeting. *Important: Before making any cuts,* note the direction of the carpet pile, especially if you have more than one piece of carpeting to cut.

Transfer the room dimensions to the carpet. Use a felt-tipped marker on the edge of the carpet to indicate the cutline across its width, or use a utility knife to notch the carpet's edges at each end of the proposed cut.

LAYING WALL-TO-WALL CARPETING

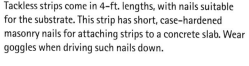

Tackless strips come in 4-ft. lengths, with nails suitable for the substrate. This strip has short, case-hardened masonry nails for attaching strips to a concrete slab. Wear goggles when driving such nails down.

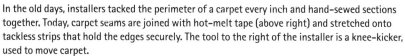

In the old days, installers tacked the perimeter of a carpet every inch and hand-sewed sections together. Today, carpet seams are joined with hot-melt tape (above right) and stretched onto tackless strips that hold the edges securely. The tool to the right of the installer is a knee-kicker, used to move carpet.

In general, cut loop-pile carpet from its pile side (its face), and cut cut-pile carpet from its backing side. To cut carpet with *cut pile*, flop the edge of the carpet over so its backing is up, and snap a chalkline between the two notches you made in the carpet's edges. Cut along the chalkline with a utility knife or a double-edged flooring knife, using a long, metal straightedge to guide the blade. If the carpet is *loop pile*, cut it with the carpet facing up: Locate a pile row near the cutline and run the blade of a large, flat screwdriver between them to separate the rows. Then push a row runner, as shown in the bottom photo at right, steadily down the row to cut through the backing. The carpet pile will guide the cut.

POSITIONING THE CARPET

Carpet is heavy. So get help rolling it up and carrying it. Unroll it in the room where you'll install it. If you measured and cut properly, the edges of the carpet should curl up about 3 in. at the base of the walls. To adjust the carpet slightly once you've unrolled it, lift a corner about waist high. Then, as you stand with one foot on the carpet and one behind it, raise the foot that was on the carpet and, with the side of that foot, kick the carpet sharply. *Note:* By using the side of your foot, rather than your heel or toe, you'll be less likely to stretch or tear the carpet.

Many installers don't cut fill pieces of carpet until they've positioned the drop piece and measured from the drop piece to the wall. This allows

When installing padding over a concrete floor (which is somewhat slippery), tape the padding edges to the edges of tackless strips to keep the padding from riding up onto the strips. Tape just the edges of the strips so the tack points remain exposed.

Adjust the row-runner blade so it cuts just through the carpet backing and not into the padding. Change blades often so they cut cleanly.

This row runner is cutting a looped-pile carpet. Note that the padding seams run perpendicular to the carpet seams to prevent them from lining up.

1. After unrolling the carpet, use the side of your foot, in a kicking motion, to adjust its position, as shown.

2. Many pros prefer to measure from the leading edge of the drop piece before cutting fill pieces. If you find that walls are not parallel, rough-cut the edges of the fill pieces at a slight angle so you'll have enough extra to trim.

3. Before hot-taping carpet seams, help the carpet lie flat by notching carpet edges where they abut doorways, cabinets, and other jogs.

4. Straighten carpet seams before hot-taping them. Using just a knee-kicker, lift the carpet edges onto the tackless strips to create a slight tension. After the hot-melt tape seams have cooled, you can use a power stretcher to reposition the edges more securely to the strips.

5. Once the seam iron is hot enough, place it onto the tape and flop the carpet edges onto the tool so that only its handle is exposed.

them to double-check the size of the fill piece(s) needed. As with finish carpentry, "measure twice, cut once" is good advice, especially if the walls aren't parallel. Once you've cut the fill pieces and positioned them next to the edge of the drop piece, go around the perimeter of the floor and loosely notch the carpet where it butts against door jambs and corners so the carpet will lie flat—but don't trim the carpet edges yet. First, you need to join the carpet sections, using hot-melt seam tape.

HOT-MELT SEAMS

Join carpet seams before stretching and trimming the carpet. Correctly installed, hot-melt seams are strong enough to withstand stretching without separating.

Use the knee-kicker to draw the seam edges together. After lining up the carpet sections, roll back one section slightly and slip a piece of hot-melt seam tape under the carpet edge so that the tape will run exactly down the middle of the seam. The tape's adhesive-coated side should face up. Plug in the seam iron, and let it heat up. Once it's hot, place the iron on top of the seam tape at the start of the seam and let the carpet flop down on both sides, covering all but the handle of the iron. Most irons take about 30 seconds to heat the tape at a given point.

Once the adhesive has melted, move the iron farther along the tape. Then use a seam roller to embed the carpet backing in the melted adhesive. This operation isn't difficult, but you must make sure that carpet edges butt together over the tape, rather than overlap each other. (Back-cutting the carpet's edges slightly with the row runner helps.) If it's a cut-pile carpet, use a smooth roller to press the carpet backing into the seam tape; if it's loop-pile carpet, use a star roller. After you roll a section of seam, weight it down as shown in Step 7. Continue along the seam—in roughly 12-in. increments—until the whole seam is bonded. Allow the adhesive to cool for 20 to 30 minutes before stretching the carpet.

STRETCHING CARPET

Once the carpet seams have cooled, stretch and attach the carpet to the tackless strips around the perimeter of the room, using the knee-kicker, power stretcher, and stair tool. Actually, if you must join carpet seams, you already will have used the knee-kicker to draw the carpet edges taut so the glued seam will be straight.

Typically, stretching begins in a corner, using a knee-kicker. Place the knee-kicker about 1 in. from the wall and rap the cushion of the tool quickly with your knee. When done properly,

6. As you heat sections of hot-melt tape, move the iron forward and roll the seam with a star roller. The spikes of a star roller penetrate the loop-pile carpet and press the backing into the tape adhesive at many points.

7. After seam-rolling each section, weight it down until the tape has cooled.

8. Once the hot-taped seams have cooled, they'll be strong enough to be stretched with a power stretcher and won't pull apart. When extended between opposite walls, the power stretcher's head typically moves the carpet about 2 in. as the lever is depressed.

9. After stretching a carpet, use an edge trimmer to cut off the excess around the perimeter of the room. Then use a stair tool to tuck carpet edges behind tackless strips or under baseboards.

PRO TIP

Before joining carpet seams, apply a bead of seam sealer to the edge of each section of carpet. Seam sealer fills any voids created when you trimmed carpet edges and keeps the pile from unraveling. Seam sealer looks like white glue and is usually latex based.

Carpet Transitions

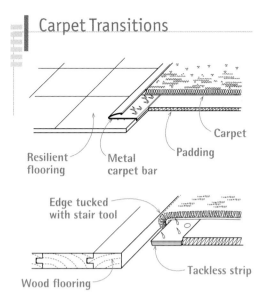

Resilient flooring
Metal carpet bar
Padding
Carpet

Edge tucked with stair tool
Wood flooring
Tackless strip

Tape the edge of the padding to keep it from riding up onto the tackless strips when stretching the carpet.

this maneuver carries the carpet forward and onto the points of the tackless strips. Then use the stair tool to pack the carpet into the small gap between the strips and the wall, also to securely lodge the carpet on the tack point. Knee-kicking takes a little practice. But, as more of the carpet is attached and there is some tension on it, the task becomes easier. As you work along an edge of the carpet, rest your hand on the section just attached. The extra weight will prevent the section from being dislodged by successive knee-kicks.

After securing a corner 2 ft. to 3 ft. along each wall, assemble the power stretcher to stretch the carpet to an adjacent corner across the room. To do this, place the tail end of the power stretcher on a 2x4 block resting along a baseboard of the corner you just attached; the 2x4 prevents the tail end of the stretcher from damaging the finish wall. Unlike the knee-kicker, which rebounds, the stretcher has a lever that extends the tool and holds it there until the lever is released. Before you engage the stretcher's lever, place the stretcher head 5 in. to 6 in. from the wall you're pushing the carpet toward. Once you push down on the lever, the stretcher head should move the carpet about 2 in. forward, leaving you plenty of room to use the stair tool and secure the carpet to the tackless strip.

That's pretty much it. Once the second corner is attached, alternate using the knee-kicker and the power stretcher, somewhat as shown in "A

Carpet-Stretching Sequence" on the facing page. There's no one always-best sequence. Keep an eye on what the carpet is doing, and use the tool that seems right. Seen from above, your carpet-stretching movements would resemble something between a tennis match and icing a cake. Go back and forth, fine-tuning as you go.

TRIMMING CARPET

Once the carpet is secured to the tackless strips and the transition pieces in the doorways, use an edge trimmer to remove the excess; trimmers also tuck the edge of the carpet to a degree. After trimming the edge, go around the perimeter with a hammer and a stair tool, tamping the carpet between the tackless strips and the wall or under the baseboard trim. It's important that the trimmer's blade be sharp, so change razors whenever the tool drags, rather than cuts.

CARPETING STAIRS

If possible, use a single strip of carpet on stairs, eliminating seams. Stair padding can be many pieces because it will be covered by carpet. For the best-looking job, carpet pile on treads should slant toward you as you ascend the stairs. On risers, the pile should, therefore, point down. If you use several pieces of carpet on the stairs, for durability and appearance, carpet seams should always meet in riser–tread joints.

Estimating and ordering. First, determine the width of the runner. On closed stairs (which have walls on both sides), carpeting usually runs from wall to wall. On open stairs (with balusters on one or both sides), carpeting should run to the base of the balusters. In either case, each side of the carpet should be tucked under 1¼ in. to hide the cut edge and prevent its unraveling. Thus measure the width of the stairs and add 2½ in. to that width. So if your stairs are 36 in. wide, you'll be able to cut only three runners from each 12-ft. width of carpeting, allowing for tuck-unders and waste.

To determine the overall length of the stair runner, measure from the edge of the tread nosing to the riser, and from there up the riser to the nosing of the step above. Add 1 in. to this measurement to accommodate the padding over the step and multiply this by the number of steps. Add 6 in. to that total for adjustments at the top and bottom of the stairs.

That's a formula for straight-run stairs. If yours have bends and turns, create a paper template for each step that turns. Each template should cover a tread and the riser below. As noted above, add 1 in. for the distance the padding sticks out around the nosing. However, some old hands at laying carpet feel that stairs should be covered only with continuous pieces, with extra carpet tucked behind riser sections.

Installing carpet on stairs. Nail tackless strips on the stair risers and stair treads so that tack points on risers point down and those on the treads point in, toward the risers. The tackless strip on each tread should be about ⅜ in. away from the riser so there's a gap to tuck the carpet into. The tackless strip on each riser should be about 1 in. above the tread. Tackless strips should be 2½ in. shorter than the width of carpet with tucked-under edges. That is, each edge is 1¼ in. wide and does not attach to the tackless strip. So the strips must stop short by that amount. Instead, tucked-under edges of carpet will be tacked with a 1¼-in. tack on each side, driven into the riser–tread joint.

Padding pieces are as wide as the tackless strips are long. Butt the padding to the edge of the strips. To keep the pad from being seen from the open side of a stairway, cut the riser portion of the padding at an angle, as shown at right. Staple the padding every 3 in. to 4 in. To tuck the carpet edges, snap chalklines on the backing, 1¼ in. from the edges. Use an awl (not a utility knife) to score lightly along the lines, fold the edges under, and then weight them down for a few minutes to establish a slight crease.

Secure the bottom of the stair runner first, overlapping the carpet about ¼ in. onto the floor; push the carpet into the tackless strip at the base of the first riser. (The tackless strip on the first riser should be ¼ in. above the floor.) Press the carpet onto the strip points with a stair tool. Tack the bottom of the rolled edges onto the riser with a 1¼-in. tack at each end. Then pack the extra end of the carpet into the ¼-in. gap between the tackless strip and the floor.

To cover the first step, stretch the carpet with a knee-kicker, starting at the center of the tread. As you push the knee-kicker, use a stair tool to tamp the carpet into the riser–tread joint. Work out from the center of the step, until the carpet is attached to tackless strips along the entire joint. At the end of each side, secure the tucked-under edge with a single 1¼-in. tack. Continue up the stairs, using the knee-kicker and stair tool. If the width of the carpet varies (sometimes a tucked-under hem slips), insert an awl point in the hem and jimmy the tool to move the hem in or out.

A Carpet-Stretching Sequence

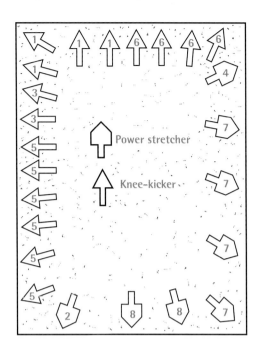

To stretch carpeting, alternate using a knee-kicker and a power stretcher. Typically, use a knee-kicker to secure a short section of carpet to tackless strips nearby, before using a power stretcher to stretch carpet across the room and secure it to strips along an opposite wall.

Stair-Carpeting Details

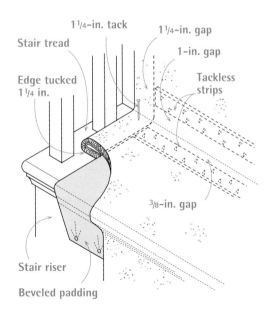

Along open stairs, allow an extra 1¾ in. to tuck under the carpet. Angle-cut padding so it won't be visible from an open side of the stairs.

Glossary of Building Terms

This glossary contains many commonly encountered building terms. You'll find somewhat more specialized terms defined in context in their respective chapters. For a comprehensive resource, get a copy of Francis D. K. Ching's *A Visual Dictionary of Architecture* (John Wiley & Sons). Many of the energy-related definitions were adapted from Energy Star's *Thermal Bypass Checklist* (Department of Energy).

AFCI. Arc-fault circuit interrupter. An electrical device that detects minute current fluctuations associated with arcing and instantaneously cuts power to circuits, thus preventing fires caused by loose or corroded connections, punctured cables, and so on.

Air barrier. Any material that restricts airflow. In wall assemblies, the exterior air barrier is often a combination of sheathing and either building paper, housewrap or board insulation. The interior air barrier is typically drywall.

Air-seal. To fill holes, gaps, or cracks in an air barrier.

Anchor bolts. Bolts used to secure a wood sill or plate to concrete or to masonry flooring or walls.

Apron. The inside trim of a window, installed against the wall immediately beneath the stool.

Arcing. Electrical current jumping a gap between conductors.

ASHRAE. American Society of Heating, Refrigeration, and Air-Conditioning Engineers.

Back-cut. To cut a board or a piece of molding at a slight angle so that more material is removed on the back face of the stock. Allows joints to fit tightly (see "Back-Cutting Trim" on p. 519).

Back-drafting. A state of depressurization (negative pressure) within a house in which combustion gases are pulled back into living spaces instead of exiting through a flue.

Backfill. The replacement of excavated earth in a trench around and against a basement foundation.

Balusters. Usually, small vertical members in a railing between a top rail and bottom rail or stair treads.

Balustrade. An assembly of balusters, top rail, and sometimes bottom rail. Used on the edge of stairs, balconies, decks, and porches.

Base (baseboard). A board that finishes (or trims) the joint between the bottom of the wall and the floor.

Base molding. Molding used to trim the upper edge of a baseboard.

Base shoe. On interior baseboard, molding abutting the floor. Also the bottom plate of a frame wall.

Batter board. At each corner of an excavation, one of a pair of horizontal boards forming an L, fastened to stakes driven into the ground. Stringlines stretched between batter-board assemblies may indicate elevation or the outlines of foundation walls.

Bay. The space between any pair of rafters, studs, or joists.

Bay window. Any window structure projecting outward from the walls of a building, either rectangular or polygonal in plan.

Beam. A structural member, usually horizontal, that supports a load.

Bearing wall. A wall that supports a vertical load in addition to its own weight.

Below grade. Below the surface of the ground, such as a basement floor.

Blind-nail. To fasten a board through its edge so the nail head isn't visible on the face of the work. For example, wood-strip flooring is usually blind-nailed through the tongue of tongue-and-groove boards.

Blocking. Dimension lumber installed to bolster framing and, in the case of floor framing, stiffen joists and reduce flexion. Also see *Bridging*.

Blueboard. Specialized drywall panel used as a base for single- or two-coat veneer plaster.

Brace. An inclined piece of framing lumber that stiffens a wall or floor. Also a temporary wall support removed after framing and sheathing are complete.

Brad nailer. A pneumatic nailer often referred to as a pin nailer.

Brick veneer. A facing of brick laid against and fastened to sheathing of a frame-wall or tile-wall construction.

Bridging (cross-bridging). Small wood or steel members installed diagonally between the top and the bottom edges of the floor joists to brace the joists and spread the action of loads. By contrast, solid bridging (blocking) consists of short lengths of dimension lumber (at least 2x6s) butted perpendicularly between joists for similar effect.

Building paper. A general term for asphalt-impregnated felt paper as well as rosin papers. This term does not include plasticized materials such as Tyvek, usually referred to as housewrap.

Built-up roof (BUR). Low-pitch roofing composed of several layers—today, typically modified bitumen (MB) and fiberglass-reinforced interplies. In the old days, three to five layers of asphalt felt were laminated with coal tar, pitch, or asphalt.

Butt joint. The junction where square-cut ends of two boards meet, or where a square-cut end abuts the face or edge of another board.

Cant strip. A triangular piece of lumber used at the junction of a flat deck and a wall to prevent cracking of the roofing applied over it and to aid water runoff.

Cap. The topmost member of a column, pilaster, door, or window molding, and so on.

Cap flashing. Metal or vinyl flashing over the cap (head casing) of a window or door; crucial to divert water around the unit. Also called head flashing. (shown on pp. 130, 143).

Carriage. See *Stair carriage*.

Casing. Molding of various widths and thicknesses, used to trim door and window installations.

Caulk (caulking). Various flexible sealants intended to stop air or water penetration along building seams. Sealant is the term preferred by manufacturers to distinguish better-quality, more durable materials from caulk, a term for older, cheaper, less durable materials.

CDX. The grade of plywood most often used for roof and wall sheathing and subflooring; the X in CDX indicates that is has exterior-grade glues.

Checking. Cracks that appear with age in many exterior materials and paint coatings.

Chopsaw. Jargon for a power miter saw, most of which can make compound cuts.

Collar tie. Member connecting opposite roof rafters, stiffening the roof structure and keeping rafters from spreading. Also called collar beams (see "Reinforcing a Roof" on p. 15).

Column. See *Post*.

Condensation. In a building, water beads or frost that accumulates on the inside of the exterior covering of a building when warm, moisture-laden air from the interior reaches the point at which it can no longer hold moisture.

Conditioned air. Air that has been heated or cooled by an HVAC system; air within the thermal envelope.

Control lines. Especially useful in tiling, these are the primary layout lines that divide space into quadrants or smaller workable areas. They help in aligning joints and creating visual symmetry. Sometimes called layout axes.

Convective airflow. Airflow that occurs in gaps between insulation and the air barrier due to temperature differences in and across the gap, resulting in a stack effect or driving forces from more to less heat.

Coped joint. In finish carpentry, a joint that allows two pieces of shaped molding to meet in an interior corner (see "Coping a Joint" on p. 521).

Corbel-out. To build out one or more courses of brick or stone from the face of a wall, either for decorative effect or to support elements above.

Corner bead. A strip of formed sheet metal installed on drywall or plaster corners to act as reinforcement.

Corner boards. Vertical trim along the exterior corners of a house. The ends of the siding often abut corner boards.

Corner braces. Diagonal braces at the corners of a frame structure, installed to stiffen and strengthen a wall.

Counterflashing. Two-piece flashing, consisting of base and overlapping cap pieces that keep water from entering joints around chimneys and skylights.

Cove molding. A molding with a concave face, used as trim to finish interior corners.

Crawlspace. A shallow space below the living quarters of a house without a basement, normally enclosed by foundation walls and frequently dirt floored.

Cricket. A small drainage-diverting roof structure of single or double slope, placed at the junction of larger surfaces that meet at an angle, such as above a chimney. Sometimes called a saddle (see "Chimney Flashing" on p. 100).

Crosscut. A sawcut across the grain of a board.

Crown molding. A usually complex molding at the top of an interior or exterior wall.

Cut-in brace. A diagonal brace notched into studs. Also see *Let-in brace*.

d. Indicates nail size. See *Penny*.

Dado. A rectangular groove, usually across the width of a board or plank.

Dense-pack. Insulation typically made from fiberglass or cellulose that is blown under pressure—typically, 3 lb./ cu. ft.— into wall cavities or, on occasion, between rafter bays.

Dew point. The temperature at which a vapor begins to condense as a liquid.

Dimension lumber. Usually 2-in.-thick lumber such as 2x4s, 2x6s, and 2x8s, but not thicker than 5 in. (Wood thicker than 5 in. is called timber.) Most joists, rafters, studs, and planks are dimension lumber.

Door jamb, interior. The surrounding frame of a door, consisting of two vertical pieces called side jambs and a horizontal head (head jamb).

Dormer. A roofed structure that projects from a sloping roof, offering a vertical front wall suitable for windows.

Downspout. Vertical pipe that carries rainwater from roof gutters.

Drip cap. A molding placed on the exterior topside of a door or window frame, causing water to drip beyond the frame.

Drip-edge. A metal edge projecting along the edge of a roof that causes water to drip free of the building, rather than travel, by capillary motion, behind the siding or exterior trim (shown on p. 96).

Drip kerf. A groove under a sill that causes water to drip free from the sill rather than cling to and run down the face of a house.

Drywall. Wall and ceiling interior covering material, usually gypsum panels. Commonly called Sheetrock, after a major brand.

Ducts. Round or rectangular pipes that deliver air to and from a heating plant or air-conditioning device. Also channels that vent exhaust air driven by a kitchen or bathroom fan.

Eaves. The lowest parts of a sloped roof, overhanging a wall.

Edgebanding. A solid-wood band on the edge of a laminated panel, such as plywood.

Elastomeric. A material that stays pliable during the life of its installation and is thus able to expand and contract with temperature changes.

Enamel. Any paint that dries to a hard, usually glossy finish. Because enamels are quite durable, they are often specified for trim, bathrooms, kitchens, and other high-use areas.

End-wall studs. Partition studs that abut the studs in an exterior wall or another partition.

Energy retrofit. Any effort to improve the energy profile of an existing structure, whether by air-sealing, upgrading insulation, installing a more energy-efficient furnace, and the like.

EPS. Expanded polystyrene, the classic white styrene board; a closed-cell rigid foam insulation.

Face-nail. To nail into the face of a board. Also called direct nailing.

Fascia. A flat board fastened along the eaves ends of roof rafters (shown on pp. 170, 173).

Female end. The receiving end of a pipe or socket.

Field. Any relatively flat, unobstructed expanse of building material.

Finish, natural. A transparent finish that largely maintains the original color or grain of wood.

Fire-stop. Located in a frame wall, usually 2x4 or 2x6 horizontal blocking between studs intended to impede the spread of fire and smoke in the event of a house fire.

Flash foam. A thin layer of spray foam whose primary function is to air-seal building seams in advance of insulating them with batt insulation.

Flashing. Sheet metal or other material used in roof and wall construction to shed water and protect a building from water seepage.

Flat paint. An interior paint that contains a high proportion of pigment and dries to a flat, lusterless finish.

Flue. A fire-clay (terra-cotta) or stainless-steel liner in a chimney through which smoke, gas, and fumes ascend. A chimney may have multiple flues, each dedicated to a different combustion source.

Footing. A masonry section, usually concrete, in a rectangular form wider than the bottom of the foundation wall or pier it supports.

Foundation. The supporting portion of a structure below the first-floor construction, or below grade (ground level), including footings.

Frieze. On a building exterior, a traditional trim board between the top of the siding and the soffit.

Frost line. The depth of frost penetration in soil.

Furring. Strips of wood or metal applied to a wall or another surface to even it and, normally, to serve as a fastening base for finish material.

Gable. The peaked portion of the roof above the eaves line of a double-sloped roof.

Gable end. An end wall that has a gable.

GFCI. Ground-fault circuit interrupter. A highly sensitive electrical device that detects minuscule current leaks and shuts off power instantaneously, thus greatly reducing the likelihood of fatal shocks.

Girder. A large, or the principal, beam of wood or steel used to support concentrated loads, such as joist ends, along its length.

Gloss. A paint that contains a relatively low proportion of pigment and dries to a sheen or luster. Most enamels are glossy.

Grade. Ground level.

Grain. The direction, size, arrangement, appearance, and quality of the fibers in wood.

Greenboard. Nickname for water-resistant drywall, named for the green coloring of its paper facing. It has a water-resistant core and water-repellent face. Also called WR board.

Grout. In masonry, a specialized mortar that seals the joints between tiles. The term is also applied to a watery mortar that is thin enough to flow into the joints and cavities of masonry work, filling them.

Gutter (rain gutter). A shallow channel of metal or wood installed below and along the eaves of a house to catch and carry rainwater to a downspout.

Hardiebacker. A cementitious panel used as a substrate for tiling. Hardiebacker and WonderBoard are brand names often used to refer to cement-based backer boards in general.

Header. A beam placed perpendicular to joists, to which joists are nailed, whether in framing for chimneys, stairways, or other openings. Or a beam above a door or window opening.

HEPA filter. High-efficiency particulate air filter. Specified for respirator-mask and vacuum filters when working with hazardous materials such as asbestos, lead paint, extreme cases of mold infestation, and the like. There are also HEPA furnace filters.

HERS. Home energy rating system. An aggregate measure of a home's energy efficiency.

Hinge gain. A shallow slot cut into a door stile or jamb to receive a hinge plate.

Hole saw. A cylindrical saw-blade that mounts to a power drill. Capable of drilling large-diameter holes such as those needed for doorknobs.

HVAC. Heating, ventilation and air-conditioning. Whole-house systems that cool or heat a house.

I-beam. A steel beam with a cross section resembling the letter I.

Infrared imaging. Heat-sensing camera which helps reveal thermal bypass conditions by exposing hot and cold surface temperatures revealing unintended thermal flow, air flow, and moisture flow. Darker colors indicate cool temperatures, while lighter colors indicate warmer temperatures.

Insulation contact (IC). Rating for recessed lights allowing insulation to be placed directly over the top of the fixture. IC is also used to denote "insulation covered."

Insulation contact, air-tight (ICAT) lighting fixture. Rating for recessed lights that can have direct contact with insulation; constructed with airtight assemblies to reduce thermal losses.

Insulation, thermal. Any material high in resistance to heat transmission that reduces the rate of heat loss when placed in the walls, ceilings, or floors.

Interior finish. Material used to cover the framing of walls and ceilings or other interior areas. Drywall and plaster are the most common interior finishes.

IRC. International Residential Code.

Jack rafter. A rafter that runs less than the distance from the ridge to a wall plate. For example, a rafter that spans the distance from the wall plate to a hip, or from a valley to a ridge.

Jack stud. A stud that runs less than the distance from the top plate to the bottom plate. (AKA trimmer stud.)

Jamb. The elements of a door or window frame, excluding the sill; typically two side jambs and a head jamb.

Jigsaw. A portable electric saw with a thin saberlike blade that can make curved or detailed cuts in boards or panels 1 in. thick or less. Also called a sabersaw.

Joint. The juncture of adjacent surfaces, whether sheet materials or framing members.

Joint compound. Generally, a ready-to-apply compound spread onto joints between drywall panels to bed joint tape, fill recessed areas, and—after several applications—hide joints so that the wall appears monolithic.

Joist. One of a series of parallel beams, such as 2x6s or 2x8s, placed on edge to support floor and ceiling loads and supported, in turn, by larger beams, girders, or bearing walls.

Kerf. A slot. A sawkerf is the thickness of its blade. A drip kerf is cut into the underside of a sill to cause water to drip free.

Laminating. Applying a plastic laminate to a core material. In framing, laminating is the nailing or bolting of two or more pieces of lumber together to increase load-carrying ability.

Lath. A thin building material of wood, metal, gypsum, or insulating board fastened to the frame of a building to serve as a plaster base.

Lav. Abbreviation for lavatory, a bathroom sink.

Ledger strip. A strip of lumber nailed along the side of a girder for joist ends to rest on. In deck construction, a critical horizontal frame member bolted to the house. A board that supports a fixture edge, such as a tub lip.

Let-in brace. Metal or wood braces notched into studs. Also called a cut-in brace. To stiffen wall framing, let-in braces run diagonally. By contrast, let-in boards, which support wall-hung sinks, run perpendicular to studs.

Light. The space in a window sash for a single pane of glass. Also a pane of glass.

Lintel. A horizontal structural member that supports the load over an opening such as a door or window. Also called a header.

Loose-fill. Insulation typically made from fiberglass or cellulose that is blown loosely to insulate an area (usually an attic floor) or to augment insulation that is already there.

Louver. An opening with a series of horizontal slats pitched to permit ventilation but exclude rain, sunlight, and unwanted viewing from outside.

Lumber, dressed size. The dimensions of lumber after it has dried and been milled. A 2x4's dressed size is actually 1½ in. by 3½ in. Also see *Nominal dimension*.

Male end. The inserted end of a pipe, duct, or fitting.

Mantel. The shelf above a fireplace.

Margin. To center a window or door frame within the thickness of a wall so that frame edges are flush with finish surfaces on both sides or so that frame edges protrude equally. Centering a door or window frame in an opening is called "margining the frame."

Masonry. Stone, brick, concrete, hollow tile, concrete block, gypsum block, or a combination of these materials bonded together with mortar.

MDF. Medium-density fiberboard.

Mfr. Abbreviation for manufacturer.

Mid-run. Midway through a pipe run, duct run, and so on.

Migration. In engineering, the unwanted movement of a jack or structural member due to the loads on it. Dangerous migration can occur if a jack is not plumb.

Millwork. Generally, any building material made of finished wood and manufactured in millwork plants and planing mills. This term applies to molding, trim, doors, window and door frames, blinds, porchwork, mantels, and so on—but normally not flooring or siding.

Mineral spirits. Solvents used to thin solvent-based paints, clean up solvent-based adhesives, and so on.

Miter joint. The joint of two pieces set at an angle that bisects the joining angle. For example, the 90° miter joint at the side and head casing of a door opening is composed of two 45° angles.

Mold. A fungal growth that feeds on organic materials such as wood and paper—as distinguished from mildew, black stains caused by mold.

Molding. Shaped wood or polymer strips used to finish a room, whether to cover gaps or add visual interest. Also called trim. (See "Common Molding Profiles" on p. 517.)

Mortise. A rectangular cavity or slot cut into the face or edge of a wood member. In wood joinery, a projecting tenon of the same size fits into the mortise. When installing door hardware, a door edge is often mortised to receive a lock case.

Mudsill. See *Sill*.

Mullion. A vertical bar or divider in the frame between windows, doors, or other openings.

Muntin. A small member that divides the glass or openings of a window sash or door.

Nail-protection plate. A metal plate at least 1/16 in. thick, nailed to a stud or joist edge to prevent screws or nails from puncturing cables or pipes running within 1¼ in. of the edge. Also called a nail plate; required by code in many instances.

NEC. The *National Electrical Code*, a book of periodically updated standards adopted by most local code authorities (published by the National Fire Protection Association).

Newel. A post to which the end of a stair railing or balustrade is fastened.

Nominal dimension. The stated size of a piece of lumber before it was dressed (milled) and shipped to the lumberyard. Thus a nominal 2x4 or 2x6 has actual dimensions less than its nominal dimensions: a 2x4 is actually 1½ in. by 3½ in.

Nonbearing wall. A wall supporting no load other than its own weight, also called a curtain wall.

Nosing. The projecting edge of a stair tread.

On center (o.c.). The measurement of spacing for studs, rafters, joists, and so on in a building from the center of one member to the center of the next one.

OSB. Oriented strand board.

Outrigger. The extension of a rafter beyond the wall line. Usually a smaller rafter nailed to a larger one, forming a roof overhang.

Paint. A combination of pigments with suitable thinners or oils. Used as a decorative and protective coating.

Paint roller. A paint-rolling hand tool whose cylindrical rotating arm receives a napped roller cover.

Parting bead. A small wood piece in the middle of both side jambs of double-hung window frames. The beads separate upper and lower sashes and guide the sashes as they are raised and lowered. Also called a parting stop or strip.

Partition. A wall that subdivides spaces within any story of a house.

Penny. As applied to nails, originally indicated the price per hundred. Term is now a measure of nail length and is abbreviated as the letter d.

Pier. A column of masonry used to support other structural members.

Pin nailer. Nickname for a pneumatic brad nailer.

Pitch. The slope of a roof, or the ratio of the total rise to the total width of a house. Roof slope is expressed in inches of rise per 12 in. of run. A roof that rises 8 in. per 12 in. of run has an 8:12 (or 8/12) pitch.

Plaster grounds. Strips of wood used as guides or strike-off edges around window and door openings and around the base of walls.

Plate. Sill plate: a horizontal member anchored to a masonry wall. Sole plate or shoe: the lowest horizontal member of a frame wall. Top plate: the topmost horizontal member of a frame wall, which supports ceiling joists, rafters, or other members. Also see *Nail-protection plate*.

Plow. To cut a lengthwise groove in a board or a plank.

Plumb. Perfectly vertical.

Pointing. In masonry, using a jointer or striking iron to shape and compress fresh mortar. Also called striking a joint.

Polyiso. Polyisocyanurate, or PIR, a rigid foam insulation commonly used on roofs because of its high-temperature stability.

Post. In engineering, a vertical structural compression member that supports loads acting in the direction of its longitudinal axis. Also called a column.

Prehung door. A door that comes prefitted to a frame, with hinges mortised to a jamb.

Preservative. Any substance that prevents the action of wood-destroying fungi or insects.

Primer. The base coat of paint, which seals the surface and minimizes absorption of subsequent coats. Back-priming denotes applying primer to the back sides of siding and exterior trim.

Proud. To cut a building material slightly long (or wide), for the purpose of fitting it tightly or trimming it later. The opposite of cutting a material shy. Also, material that protrudes or bows outward and needs to be trimmed back.

PVA. Polyvinyl acetate (glue).

PVC. Polyvinyl chloride is the basic compound in numerous building materials, ranging from plastic pipes to resilient (vinyl) flooring.

Quarter-round. A small molding that has the cross section of a quarter circle.

Rabbet. A rectangular, longitudinal groove cut in the edge of a board or a plank.

Rafter. One of a series of structural members of a roof designed to support roof loads.

Rafter, valley. A rafter that forms the intersection of an internal roof angle. Valley rafters are usually doubled 2-in.-thick members.

Rail. Horizontal structural members of panel doors, window sashes, cabinet frames, or wainscoting. Also the sloping or horizontal members of a balustrade or staircase extending from one vertical support (such as a post) to another.

Rake. On a sloping roof, trim or framing members that run parallel to the slope and define the outer edges. Rake boards usually project beyond end walls (see "Building Terms" on p. 8).

Repointing. In masonry, repairing deteriorated mortar joints by cutting back old mortar and then packing and pointing (shaping) new mortar into the joint. Also called tuck-pointing.

Respirator mask. A nose-and-mouth mask with replaceable cartridge filters capable of filtering out noxious fumes, harmful chemicals, and so on. As distinguished from a mere paper dust mask.

Reveal. In finish carpentry, the amount (usually, ¼ in.) that a casing edge is set back from a jamb edge. In addition to creating visual interest, reveals spare carpenters the frustrating task of aligning board edges that are rarely perfectly straight.

Ridge. The horizontal line, usually at the top of the roof, where sloping roof planes meet.

Ridgeboard. The board placed on edge at the ridge of the roof, to which the upper ends of rafters are fastened.

Rigid insulation. Insulation typically made from polystyrene, polyurethane, or polyisocyanurate, manufactured into 2-ft. x 8-ft. or 4-ft. x 8-ft. panels of various thicknesses. As an exterior sheathing material, rigid insulation provides a complete thermal break assembly and can effectively shift the dew point outside of the exterior wall construction assembly.

Rim joist. Outermost joists, which create a kind of box around the joist grid. Rim joists sit on a mudsill.

Rip. To cut with the grain of a board, thereby reducing the board's width. Reducing the width of any board is called ripping it down.

Rise. In stairs, the vertical height of a step or flight of stairs.

Riser. Each of several vertical boards used to close spaces between stair treads.

Roll roofing. Roofing material composed of asphalt-saturated fiber, supplied in 36-in.-wide rolls, with 108 sq. ft. of material.

Roofover. Applying a new layer of roofing over an existing one.

Roof sheathing. Fastened atop rafters, boards or panels on which shingles or other roof coverings are laid.

Rough opening (RO). A framed opening in a wall, roof, or floor platform.

Run. In stairs, the front-to-back depth of an individual stair or the horizontal distance covered by a flight of stairs.

R-value (resistivity-value). The ability of a material to resist heat flow through it. The higher the R-value, the greater its resistance to heat flowing through it.

Saddle. See *Cricket*.

Sash. A frame containing one or more lights of glass.

Sash balance. A device, usually operated by a spring or a weight, designed to counterbalance the weight of a window sash so it doesn't come crashing down.

Screed. A straight wood or metal strip drawn across concrete, plaster, or a mortar bed to create a flat surface. Also strips embedded in those materials and acting as thickness gauges.

Scribing. Using a compass or scribing tool to transfer the outline of irregular shapes onto woodwork or onto sheet materials such as drywall or resilient flooring.

Sealer. A finishing material, clear or pigmented, usually applied directly over uncoated wood to seal the surface. Distinct from sealant, which is the term manufacturers use for their better grades of seam and crack sealants, commonly called caulk.

Semigloss paint. A paint that has some luster but that isn't particularly glossy.

Set. To sink a nail below a surface.

Shake. A thick, hand-split shingle.

Shear wall. A wall reinforced to withstand sideways (shear) forces from wind, soil loads, or earthquakes.

Sheathing. Structural panels such as plywood or boards, attached to framing members to strengthen the structure and provide a base for roofing, siding, and flooring.

Shy. To cut something slightly short.

Siding. The exterior cladding of a house, whether made of wood clapboards or shingles, stucco, metal, or vinyl lap siding, and so on.

Sill. In framing (here, a mud-sill), the lowest member of the frame of a structure, which rests on the foundation and supports floor joists and frame walls. Also, the bottom of a door or window opening or unit. The exterior portion of a doorsill or windowsill is often pitched to shed water.

Sleeper. Usually, a wood member embedded in or placed on concrete, to which subflooring or flooring is attached.

Soffit. The underside of an overhang, especially at the eaves of a roof (shown on pp. 170, 173).

Soil stack. In plumbing, the vertical main of a system of soil, waste, or vent piping.

Sole (sole plate). See *Plate*.

Span. The distance between structural supports, such as walls, columns, piers, beams, girders, and trusses.

Speed Square. Another one of those brand names—this one owned by the Swanson Company—that must be driving its trademark lawyers crazy. That is, virtually everybody who uses Swanson's tool or a look-alike calls their tool a speed square. This small framing square with reinforced edges is in every carpenter's tool belt—or should be. A great tool.

SPF. Spray polyurethane foam.

Splash block. A small masonry block placed beneath a down-spout to carry water away from the building.

Spray foam insulation. Insulation available in both open- and closed-cell configurations that is typically made from polyurethane. It is sprayed into construction assemblies as a liquid that expands to fill the surrounding cavity. Once dry, spray foam functions as both an air barrier and thermal barrier. Closed-cell spray foams are more dense and may also function as a vapor retarder.

Square. A standard unit of measure—100 sq. ft.—usually applied to roofing materials. In measurement, two adjacent pieces that join in a 90° angle.

Stair carriage. Supporting member for stair treads. Usually of dimension lumber run diagonally, on edge, and notched to receive the treads.

STC. Sound transmission co-efficient, a measure of a material's ability to reduce noise.

Step-flashing. L-shaped pieces of metal flashing, typically interwoven with shingle courses, to deflect water from roof dormers, chimneys, skylights, and the like (shown on p. 99).

Stile. A vertical structural member in a panel door, sash, cabinet frame, or wainscoting.

Stock. The basic materials from which a building element is fashioned. For example, stair carriages may be cut from 2x12 stock, or window cap flashing may be cut from 26-gauge aluminum stock.

Stool. A flat, horizontal molding at the bottom of a finished window, the part often adorned with small flower pots. Usually rabbeted on the underside and fitted over the inside edge of a windowsill.

Story pole (story board). A straight board marked off in uniform increments, to aid in layout. Story poles are used to mark siding intervals (see the drawing on p. 176) and to help establish tile and joint intervals (see the photos on pp. 177, 500).

Striking a joint. See *Pointing*.

Stringers. In stairs, the supports on which stair treads rest. More or less synonymous with carriage.

Strip flooring. Solid-wood flooring that is typically ¾ in. thick by 2¼ in. wide, with matched tongue-and-groove edges.

Strong. To cut something a bit long. Also see *Proud*.

Strong-ties. Various steel lumber-connectors such as reinforcing plates, straps, clips, hangers, brackets, and so on. Pioneered by the Simpson Company, Strong-Tie Connectors have become the generic name for the category.

Stub-outs. Plumbing pipes that protrude into an unfinished room.

Stucco. An exterior plaster made with portland cement as its base.

Stud. One of a series of wood or metal vertical structural members in walls and partitions. Wood studs are usually 2x4s, though 2x6s are also used.

Subfloor. Plywood panels or 1-in. boards installed over joists to support a finish floor. A sub-floor may also be covered with underlayment.

Suspended ceiling. A ceiling system hung from overhead structural framing.

Tack rag. A slightly sticky cheesecloth pad that adheres dust and is used between sandings.

Taillight warranty. Any warranty that expires when the contractor's truck leaves your driveway.

Tail pieces. Pipe ends.

Termite shield. A shield, usually of corrosion-resistant metal, placed in or on a foundation wall or other mass of masonry or around pipes to prevent termite migration.

Thermal barrier. Term used to describe when flow of heat is restricted or slowed. Accomplished through insulation.

Thermal bridging. Accelerated thermal flow that occurs when materials that are poor insulators displace insulation.

Thermal bypass. The movement of heat around or through insulation. This typically occurs when gaps exist between the air barrier and insulation or where air barriers are missing.

Thermal bypass checklist. Comprehensive list of building details for Energy Star Qualified Homes addressing construction details where air barriers and insulation are commonly missing.

Thermal envelope. The building envelope, as seen through its ability to retain or lose conditioned air.

Thinset mortar. A thin, cement-based setting material (adhesive) troweled onto a mortar setting bed or substrate to adhere tiles. Thinset formulations vary.

Threshold. A strip of wood or metal with beveled edges installed over the sill of an exterior door or over the gap between finish flooring and the doorsill.

Tie-down. A metal connector used to keep lumber joints from separating during hurricanes, high winds, or other conditions of excessive stress. Also called hurricane tie. Also see *Strong-ties*.

Toenail. To drive a nail at an angle other than 90°, thereby increasing its resistance to pull out.

Tread. A horizontal board in a stairway that receives foot traffic.

Trim. Essentially the same as molding. Also, to install door and window casing, baseboards, crown molding, and so on.

Trimmer. In a rough opening, any structural member fastened side by side to a like member, thus doubling it for added strength. Hence trimmer studs, trimmer joists, and trimmer rafters.

Truss. A frame or jointed structure of smaller elements designed to span long distances. Roof trusses, for example, can be engineered to support great loads.

Tuck-pointing. See *Repointing*.

UBC. Uniform Building Code.

U-factor. An aggregate measure of how well non-solar heat flows through a window's glazing and frame. Simply put, U-factor is the inverse of R-value (1 divided by the R-value), so the lower the U-factor, the better.

Underlayment panels. Specified for resilient flooring, a layer—over the subflooring—designed to add rigidity to the subflooring and provide a smooth surface for the thin finish layer that follows.

Unconditioned space. Space outside the thermal envelope whose temperature and moisture level is essentially the same as outdoor air.

UPC. Uniform Plumbing Code.

Utility knife. A knife with sturdy, razor-sharp, retractable, replaceable blades and a hollow handle for storing them.

Valley. The internal angle formed by the junction of two sloping sides of a roof.

Vapor barrier. Impermeable material such as polyethylene sheeting used to retard the movement of water vapor into walls and thus prevent condensation within them.

Vapor retarder. Any material that restricts the flow of moisture. Vapor retarders are commonly classified into four groups, based upon the degree of permeance (permeability) to water vapor. Class I vapor retarders (impermeable) are rated at less than 0.1 perms and are considered vapor barriers. Class IV vapor retarders with a permeance of 10 perms and above are considered permeable.

Vent. A pipe or duct that allows air to flow in or out.

VOC. Volatile organic compound; typically, a noxious solvent in solvent-based finishes or adhesives. You should wear a respirator mask with VOC-rated filters when working with VOCs.

VSR. Variable-speed reversible (drill).

Wallcovering. Formerly wallpaper, the collective term for any decorative sheet material adhered to walls, including paper, vinyl, foil, cloth, cork, and bamboo.

Weatherstripping. Narrow lengths of spring metal, vinyl tubing, or other materials designed to prevent air and moisture infiltration around windows and doors.

Wick. To draw moisture by capillary action.

Wind baffle. An object that serves as an air barrier for the purpose of blocking wind washing at attic eaves.

Wythe. The width of a brick and one mortar joint.

XPS. Extruded polystyrene; a closed-cell, rigid foam insulation that can be readily identified by its pastel hues of pink, green, blue, and yellow.

Index

A

AAV (Air-admittance valves), 369–70

ABS (Acrylonitrile butadiene styrene), 349

ABS pipe, 362, 363, 364, 365, 371–73

Abuse-resistant drywall, 457

AC (Air-conditioning) systems, 20

AccuVent (Baffle/air chute), 432

Acetonc, 510, 559

Acoustic sealants, 482, 483

Acrylic paint. *See* Latex paint

Acrylic sinks, 409

Acrylonitrile butadiene styrene (ABS), 349

Adapters, pipe, 355, 372, 379

Add-a-depth ring (Goof ring), 307

Additions, 39–41, 45–47, 101

Adhesives, 86–88
  cellulose, 572, 574
  clay-based, 572, 573, 574
  clear, 572, 574
  construction, 461, 468, 498
  interior, 522–23
  for resilient flooring, 618
  seam, 572
  selection of, 87
  solvent-free vs. solvent-based, 88
  sound-attenuating, 459
  for tile work, 489, 491–92
  vinyl-to-vinyl, 572
  for wallcoverings, 572, 573, 574, 582, 583, 585
  See also Thinset mortar/adhesive

Adjustable hinges, 127

Adjustable square, 54

Adjustable wrenches, 352

Admixtures, 235

AFCI (Arc-fault circuit interrupter), 18, 300, 311, 328–30, 626

AFUE (Annual fuel-utilization efficiency) furnaces, 419

Aggregate, 235, 281

Air-admittance valves (AAV), 369–70

Air barrier, 417, 626
  See also Vapor retarders

Air-conditioning (AC) systems, 20

Airflow, convective, 417, 432, 627

Air-gap membranes, 285

Air leaks, testing for, 376, 418, 419–23

Airless sprayers, 547

Air-pressure test, 376

Air quality, 416, 417, 422

Air-sealing, 423–29
  attics, 418, 425–27
  basements, 427–28
  definition of, 626
  doors, 427
  ductwork, 419, 428–29
  electrical boxes, 424, 426, 627
  materials for, 423–25, 429, 440
  pipes, 426, 427
  spray foam insulation for, 423–26, 440
  walls, 427
  windows, 427

Alcoves, wallcoverings for, 589

Allen wrenches, 62

Alligatoring, 564, 567

Aluminum gutters, 190

Aluminum wiring, 297

American Society of Heating, Refrigeration, and Air-Conditioning Engineers (ASHRAE), 626

Amperes (Amps), definition of, 287

Anchor bolts, 260, 274, 276–77, 626

Angled-end joints, 213

Angle-dividers, 55

Angled sash brush, 542

Angle stops, 355

Annual fuel-utilization efficiency (AFUE) furnaces, 419

Appliances
  countertops and, 406
  electrical circuits for, 300, 311
  energy-efficient, 422
  load calculations for, 298
  soundproofing, 480
  tile floors under, 495–96

Apron, 528, 529–30, 626

Arc-fault circuit interrupter (AFCI), 18, 300, 311, 328–30, 626

Arched windows, 529, 532–33

Arches, wallcoverings for, 589

Architects, 29, 33, 34–35

Architectural (Laminated) shingles, 103, 104, 108

Arcing, 295, 297, 299, 311, 626

Armored (Metal-clad) cable, 309

Asbestos, 21, 207, 498, 596

ASHRAE (American Society of Heating, Refrigeration, and Air-Conditioning Engineers), 626

Asphalt shingles, 10, 101, 103–10

Astragal molding, 517

Attic hatches, 425

Attics
  air-sealing, 418, 425–27
  inspection of, 14–15, 444
  insulation of, 446–48

Aviation (Tin) snips, 63

Awning windows, 145, 146

B

Back brushing, 546

Back-cut, 519, 521, 626

Back-drafting, 420, 432, 626

Backer board, 490, 494, 496–97, 498

Backer rod, 424

Backfill, 261, 278, 282, 285, 626

Backing trim, 534

Backsplash, 507

Baffle/air chute (AccuVent), 432

Bahco by Snap-On carbide scrapers, 59

Ball valves, 355, 357, 385

Balusters, 217–18, 626

Balustrade, 626

Bamboo flooring, 591, 593

Bar clamps, 62

Base (Baseboard)
  definition of, 626
  installation of, 530, 534
  joints for, 527
  removal of, 606
  scribing, 531

Base cabinets, 393–94, 395–98

Basements
  air-sealing, 427–28
  damp, 284–85, 436
  EPS panels for, 450–52
  inspection of, 16–17, 418
  insulation of, 284, 450–52
  See also Foundations

Basement windows, 428

Base molding, 626

Base shoe, 531, 626

Basin wrenches, 352, 353

Bathrooms, 389, 412–15
  countertops for, 403–407, 413
  electrical circuits for, 299, 300
  exhaust fans for, 425, 433–34
  greenboard for, 456, 464
  inspection of, 16
  isometric sketch of, 352
  layout of, 349, 412
  lighting for, 300, 412
  moisture from, 435
  pedestal sinks for, 350, 378
  planning for, 25, 27, 349, 412–15
  receptacles for, 18, 310
  renovation case histories of, 38
  ventilation for, 16, 412
  See also Plumbing; Showers; Toilets

Bathtubs
  drywall for, 464
  hardware for, 495
  installation of, 382, 383
  planning for, 412
  removal of, 378, 379
  rough-in for, 350, 351, 373
  sizes of, 415
  tub surrounds for, 494, 509–11
  types of, 415

Batter board, 626

Batt insulation, 437–38, 440

Bay, 203, 211, 626

Bay window, 626

Bazooka tools, 473

Beams
  definition of, 626
  engineered, 72, 73–74
  exposed, 222
  false, 535–36
  hidden, 222–25
  for jacking and shoring, 263–64
  needle, 267, 273
  span of, 270
  supports for, 272
  See also Steel I-beams

Bearing walls
  definition of, 195, 626
  demolition of, 206
  determination of, 222
  vs. nonbearing walls, 26, 27–28, 222
  remodeling of, 222–25
  shoring, 206, 212, 221

Beater board, 488

Below grade, 259, 261, 276, 278, 626

Belt sanders, 515, 597

Bevel gauges, 55, 513

Bevel joints, 172

Bidets, 382, 414

Bids, 36

Bifold doors, 118

Binder, 539

Biscuit joiner, 514

Biscuit joints, 521, 522, 525
Blind-nail, 607, 608, 609, 626
Blisters, roofing, 114
Blocking
    in bearing walls, 211
    for crown molding, 534
    definition of, 626
    load path and, 224
    sole plate over, 214
Block planes, 59, 514–15
Blower-door testing, 420, 427
Blueboard, 457, 475, 626
Blue jean (Cotton) insulation, 438
Board feet, calculating, 71
Bolection molding, 537
Bolts, 84–85
    anchor, 260, 274, 276–77, 626
    carriage, 84
    closet, 381
    machine, 84
Bond beam course, 257
Bond patterns, 239, 504–505
Boral TruExterior siding and trim, 171
Border wallcovering, 572
Boxes (Electrical)
    adjustable, 308
    air-sealing, 424, 426, 627
    in ceilings, 317, 323
    cut-in (remodel), 306, 307, 321, 322
    extending, 495, 496
    flush edges for, 307
    grounding, 299, 308, 343
    installation of, 316–17
    junction, 207–208, 296, 338
    for lighting fixtures, 308, 343
    location of, 315
    metal vs. plastic, 299, 307–308, 343
    mounting devices for, 308–309
    plastic, 299, 307–308, 310, 343
    removal of, 320, 321
    requirements for, 299
    screw-on vs. nailed, 306
    selection of, 306–308
    single-gang, 310
    size of, 306, 307
Box nails, 78
Braces
    corner, 627
    cut-in, 627
    definition of, 626
    for jacking and shoring, 264
    let-in, 242, 629
Bracket hangers, 191
Brad nailer, 516–17, 526, 533, 626
Braided stainless-steel supply lines, 358, 380
Breakers, 289–90, 293, 311
    See also Service panel
Brick, 237–44
    adding openings in brick walls, 244
    bond patterns for, 239
    cleaning & sealing, 242–44
    cutting, 238, 239

inspection of, 14
laying, 240–41
painted, 243
refractory, 251
repointing, 241–42
terminology for, 237
types of, 238
    See also Chimneys
Brick cutters, 236, 238
Brick sets, 236
Brick tongs, 236
Brick veneer, 14, 253, 626
Bridging (Cross-bridging), 200, 211, 213, 626
Brushes
    angled sash, 542
    paint, 542–43, 545, 552, 553, 554, 556–57
    paste, 575
Bubble diagrams, 31, 32
Buffers & buffer screens, 598, 599–600, 602–603
Building codes, 30
    for chimneys, 244, 245, 247, 249
    for concrete block work, 254, 256, 257
    for electrical work, 287, 293, 301, 313, 318, 328, 629
    for foundations, 259, 260, 272, 276, 277
    International Residential Code, 628
    for plumbing, 346, 347, 354, 383
    Uniform Building Code, 632
Building departments, 30–31, 35, 195
Building materials, 68–89
    adhesives, caulks and sealants, 86–88
    metal connectors, 78–86
    R-value of, 417, 441
    steel framing, 73, 74–75
    structural and nonstructural panels for, 75–78
    See also Lumber
Building paper (Felt paper)
    description of, 94, 96, 626
    installation of, 169
    under siding, 168
    for stucco, 185–86
    See also Weather-resistant barriers
Built-up roof (BUR), 113, 626
Bullnose molding, 518
Bullnose planes, 59
Butcherblock, 406
Butt hinges, 127
Butt joints (Square cut), 523–24, 525–26, 527, 626
Buying a home, 7–8, 9

C
Cabinets
    air-sealing, 427
    base, 393–94, 395–98
    face-frame, 393
    frameless (European style), 393
    IKEA, 400–403

installation of, 394–400
layouts for, 392–94
ordering, 394
overview of, 393
painting, 556–58, 559
planning for, 391
standard dimensions of, 390, 393
storage space in, 400
wall, 394, 395, 398–400
    See also Countertops
Cable
    bundling, 318
    definition of, 289
    fishing, 319, 320–22
    identification of, 317
    load calculations for, 311
    metal-clad, 309
    NM, 297, 299, 309, 311, 325
    quantity calculations for, 314
    running, 317–22, 324–25
    splicing, 325, 334
    stapling, 318
    stripping, 325
    types of, 308, 309
    See also Wiring
Cable clamps, 309–10
Cable ripper, 305, 325
Cant strip, 627
Cap, definition of, 627
Cap flange, 142
Cap flashing, 142, 143, 149, 167, 169, 627
Cap shingles, 110, 113
Carpenter ants, 215
Carpeting, wall-to-wall, 122, 595, 618–25
Carriage. See Stair carriage
Carriage bolts, 84
Casement windows, 145, 146
Casing
    definition of, 512, 627
    for doors, 122, 130–31, 523–28, 606–607
    removing, 64
    for windows, 142, 150, 524, 528–30
    See also Interior trim
Casing nails, 78
Casting plasterwork replacements, 478, 479
Cast-iron pipe, 362, 364, 365–66, 371–73
Cast-iron sinks, 411
Cat's paw, 65
Caulk and caulking, 86, 87, 88–89
    for air-sealing, 423–25
    application techniques, 88
    for chimney flashing, 99, 100
    definition of, 627
    for doors, 129, 130, 131
    for glazing, 158
    for roof-wall connections, 101
    sprayable, 424–25
    for tile joints, 510
    types of, 88–89, 424
    for windows, 142
Caulking gun, 459

C-clamps, 62
CDX plywood, 76, 78, 627
Cedar Breather (Synthetic mesh), 93, 111, 180
Cedar siding, 182
Ceiling ornaments, 478
Ceilings
    drywall on, 466, 468–69
    electrical boxes in, 317, 323
    fishing cable through, 320, 321–22
    insulation of, 342, 443–44
    leveling, 214, 215
    lighting fixtures in, 315, 316, 338–45, 425–26
    painting, 552–54
    removal of, 210
    skylight openings in, 163
    soundproofing, 480
    suspended, 631
    textured acoustic, 21
    wallcoverings on, 585, 587, 588–89
Cellular PVC plastics, 171
Cellulose adhesives, 572, 574
Cellulose insulation, 437, 440
Cement
    contact, 523
    description of, 235
    fiber-cement siding, 182, 183–85
    fiber-cement trim, 171
    hydraulic, 263
    pipe, 364
    Portland, 235
Cementitious backer board, 457
Chalking paint, 564
Chalklines, 55
    for roofing, 107
    for shingling, 178
    for tile work, 487, 502, 504
Channellock slip-joint pliers, 63
Chases, 214
Checking, 170, 627
Chemical strippers, 544, 560–61, 567, 597
Chimney crown, replacement of, 245–47
Chimney fires, 12, 245
Chimney pad, 262
Chimneys, 244–50
    backdrafting, 416, 420
    building codes for, 244, 245, 247, 249
    cleaning, 245
    flashing for, 10, 11, 99–101
    frame-chimney gap and, 425
    inspection of, 11–12, 14, 244, 245
    liner of, 11–12, 249–50
    rebuilding, 247–49
    relining, 249–50
Chlorinated polyvinyl chloride (CPVC), 347, 362, 376
Chopsaw (Miter saw), 65, 514, 627
Circuit analyzers, plug-in, 304
Circuits (Electrical)
    capacity of, 297, 311

dedicated, 301, 311
general-use, 299–300
load calculations for, 297–300
planning for, 301–304
requirements for, 300–301
Circular saws, 56–57, 183–84
City planners, 30–31, 34
Cladding (Curb caps), 164–65
Clamp-ring fittings, 359, 361
Clamps, 62, 309–10
Clapboards, 179–85
Clay-based adhesives, 572, 573, 574
Cleanouts, 370
Cleanup, 52, 54, 237
Clean zone, 206–207
Clear adhesives, 572, 574
Clear finishes, 540, 558, 561–63
Clips
drywall, 86, 465, 468, 469
H-clips, 86
sound isolation, 480–81, 482
Closed-cell spray foam insulation,
430, 442, 444, 452
Closed-cut valleys, 97, 98, 109, 113
Closet bolts, 381
Closet flange, 373, 381, 494
Coconut palm flooring, 591, 593
Code. See Building codes
Cold expansion fittings, 359, 360,
361
Cold joints, 277
Cold-water trunk line, 347
Collar tie, 15, 627
Collins, Rick, 230–33
Columns. See Posts
Combination square, 54
Combustion analyzers, 422
Comfort, planning for, 24
Common bond pattern, 239, 504
Common nails, 78
Common vents, 369
Compass, 55
Compression fittings, 357, 359,
361, 362
Concrete
air-sealing, 428
brick or stone facade on, 253
cubic yards of, 279, 281
definition of, 235
forms for, 258, 274–78
ordering, 281
pouring, 278
pouring vs. placing, 259
quality of, 261
reinforcement of, 235
screeding and floating, 280
steel reinforced, 260, 274–75,
276
tile work on, 498
See also Concrete slabs;
Foundations
Concrete anchors, 82, 83, 84
Concrete block (Concrete masonry
units, CMUs)
building codes for, 254, 256, 257
cracks in, 262
laying, 254–57

Concrete breakers, 67
Concrete countertops, 405
Concrete-cutting saw, 269, 273
Concrete floors, 269, 591, 610
Concrete forms, 258, 274–78
Concrete pads, replacement of,
269–70
Concrete slabs
finishing, 281–82, 284
mat slab foundation, 259, 282–
84
pouring, 279–82
Concrete vibrator, 278, 280
Condensation, 284, 627
Conditioned air
blower door testing and, 420
definition of, 627
escape of, 416, 420, 423
vented attics and, 430
whole house exhaust fans and,
418
Conduction, 417
Conductors, 289
Conduit, 309
Connectors
metal, 78–86, 225, 260, 264, 309
Simpson Strong-Tie, 83, 85–86
Snap-in, 309
wire, 309–10, 334
wood construction, 84, 85–86
Construction adhesives, 461, 468,
498
Construction set (Working
drawings), 34–35
Contact cement, 523
Continuous (Dry) venting, 366–67
Contractors
foundation, 263, 264, 266, 268,
273, 276
general, 29–30, 34–35, 36
roofing, 91
Control lines
for countertops, 505–506, 508
definition of, 627
first (primary), 501, 502, 504
intersection of, 502, 503
offset joints and, 504–505
room focal point and, 502
secondary, 500, 501, 504
for tub surrounds, 508, 511
Convective airflow, 417, 432, 627
Cooktops, 390
Cooler nails, 78
Coped joints
for crown molding, 535
definition of, 627
filing, 515
techniques for, 521–22
Coping saws, 58, 514
Copper gutters, 190
Copper pipe
cutting, 355–56
for DWV systems, 364
fittings for, 354
inspection of, 19
PEX connections to, 362
soldering, 354, 355, 356–57

solderless fittings for, 357
supporting, 376
for water supply lines, 347, 353,
354–55
Copper tubing, flexible, 357
Corbel-out, 251, 252, 627
Cordless tools, 60–61, 305, 353,
458–59, 516
Corian, 405
Cork flooring, 591, 594
Cork wallcovering, 571, 572, 574
Corner beads, 461, 462, 470, 627
Corner blocks (Plinths), 523–24,
527, 528
Corner boards, 174–75, 627
Corner braces, 627
Corner knives, 459, 472
Corners
drywall for, 471–72
framing of, 200, 204
repair of, 231
shingling, 174, 178–79
wallcovering around, 586–88
Cornice molding, 518
Cotton (Blue jean) insulation, 438
Counterflashing, 99–101, 627
Counter space, 390, 391
Countertops
bathroom, 403–407, 413
concrete, 405
granite, 405, 407, 413
plastic laminate, 403, 405, 499
stone, 405, 411
thinset mortar/adhesive for, 506,
507
tile, 405, 495, 499, 505–508
waterproofing edges of, 506
wood, 406
Cove molding, 517, 518, 627
CPVC (Chlorinated polyvinyl
chloride), 347, 362, 376
Craftsman house, renovation of,
37–38
Crawlspaces, 435–36, 627
Creosote, 245
Cribbing, 264, 265–66
Cricket, 100, 627
Cripple walls, 195, 222
Crosscut, definition of, 627
Crown molding, 478, 530, 534–35,
627
Cup (European-style) hinges, 400,
401
Curb caps (Cladding), 164–65
Current, definition of, 287
Curved trowels, 459
Curves, drywall for, 464, 470
Cushion-backed carpeting, 618–19
Custom trim, 517
Cut-in (Remodel) boxes, 306, 307,
321, 322
Cut-in brace, 627
Cutters, 63
Cyanoacrylate glue (Gorilla Super
Glue), 87, 88, 523
Cylinder locksets, 135–36

D

d. (Penny), 78, 627, 630
Dado, definition of, 627
Danish oils, 563
Dead bolts, 127, 133–35, 136
Dead load, 195
Deck-head screws, 82
Dehumidifiers, 284
Demolition and deconstruction, 71,
206–11
Denatured alcohol, 559
Dense-pack insulation, 418, 440–
41, 448–50, 451, 627
Design development drawings, 34
Designer-builders, 30
Dew point, 627
Diagonal-cutting pliers, 304
Dielectric unions, 377
Dimension lumber, 69, 70, 627
Dimmers, 334, 335, 336–37
Direct-vent pipe, 386, 387, 388
Disclosure statements, 7
Dishwashers, 357, 390, 409
Disk sander (Edger), 598, 601–602
Dobie blocks, 280
Door jambs, 116–18
air-sealing, 427
definition of, 627, 629
head jambs, 116, 123, 128
hinge jamb, 116–18, 121–22,
123, 128, 131–32, 140
latch jamb, 117, 123, 128, 129,
133
oversizing, 119
weatherstripping, 126–27, 136–
39
Doors, 116–42
air-sealing, 427
bifold, 118
casing for, 122, 130–31, 523–28,
606–607
dimensions of, 117
drywall around, 470
energy efficiency of, 126
entry, 126–27
exterior, 117, 118, 124–31, 141
fiberglass, 126
fixing problems with, 141
flashing, 139, 142, 143–44, 167,
168–69
framing of, 222
French, 117, 118, 131–33, 170
hanging, 140
hinges for, 119, 127
inspection of, 13–14
installation of, 119–33
inswing vs. outswing, 138
interior, 117, 119–24
interior trim for, 522–23, 606–
607
locksets for, 118, 119, 127, 133–
36
overview of, 116–19
painting, 129, 556, 557
pocket, 118, 141
precased, 128
prehung, 117, 119–31, 630

removal of, 209
reveal for, 122–23, 124, 125, 128, 130, 525
right- vs. left-handed, 118
rough opening for, 118, 120–21, 124, 131
sealing underneath, 137–38
steel, 126
tile floors under, 496
weatherstripping, 124–25, 126–27, 136–39
wood, 126
Doorsills, 125, 126, 131
Dormer, 90, 96, 100, 101, 107, 627
Double combo (Figure-5) fitting, 365, 369
Double dynamiter, 376
Double-hung windows, 144, 145, 146, 156–59, 556
Dovetail saw, 58
Downspouts
cleaning, 168
damp basements and, 284
definition of, 627
inspection of, 12, 14
installation of, 192, 193
Drainage, waste, and vent (DWV) system, 362–76
air-sealing, 427
branch drain-stack connection for, 373–74
extending the main drain, 372–73
framing considerations and, 351
inspection of, 14, 19
minimum sizes for, 367
overview of, 346–47, 348–49
pipe fittings for, 363, 370
planning for, 28
plastic pipe for, 349, 362, 364–65, 371–73
rough-in for, 370–76
slope for, 374–75
supports for, 365, 370
testing, 375–76
types of pipe for, 362–66
venting options for, 366–70
Drains, heat-recovery (HR), 364
Drain-traps, 348, 367
Drawer faces, 558
Drawer-front screws, 82
Drawings, 23, 26–28, 30, 31–35
of bathrooms, 349, 412
bubble diagrams, 31, 32
of cabinets, 392–94
design development, 34
of electrical systems, 301–304
elevation plans, 392
isometric sketches, 352
of kitchens, 391–92
working set of, 34–35
Dremel variable-speed rotary tools, 58
Dressed size lumber, 629
See also Dimension lumber
Drill bits, 61–62, 305, 489

Drills and drilling, 61–62
cordless, 61
for drywall work, 458–59
for electrical work, 304, 305, 317–18
for finish carpentry, 516
hammer, 67
right-angle, 66, 305
for tile work, 488, 489
variable speed, 632
Drip cap, 138, 627
Drip-edge flashing, 9–10, 94, 96–97, 112, 627
Drip kerf, 142, 627
Driven steel pilings, 259, 260
Drum sanders, 590, 597–98, 601
Dry (Continuous) venting, 366–67
Drying racks, 558
Drying-type joint compounds, 462
Drywall, 454–74
abuse-resistant, 457
anchors & bolts for, 476–77
carrying, 465
on ceilings, 466, 468–69
cutting, 323, 457–58, 466–67, 468
definition of, 627
estimating materials for, 463, 465
fasteners for, 461–62, 467–68
fiberglass-mat (paperless), 456, 473, 541
fire-resistant, 457
foil-backed, 457
hanging, 468–71
length of, 455, 456
measuring, 465–66
moisture-resistant, 165, 456, 464, 628
over plaster, 456, 466
painting, 551–52, 553
planning for, 462–63
for plaster repairs, 478
prep work for, 464–65
removal of, 210
repairs to, 473–75
sanding, 459, 460, 541
sound-mitigating, 457
soundproofing, 480–82
taping & mudding, 459–60, 471–73
texturing, 474
tile work over, 456, 490, 498–99
tools for, 457–60, 473
types & specifications, 455, 456–57
wallcoverings on, 579
water-resistant, 165
Drywall benches, 458
Drywall clips, 86, 465, 468, 469
Drywall routers, 457–58, 468
Drywall screws, 461, 465, 467, 468, 496
Duct Blaster, 419
Ductwork
air-sealing, 419, 428–29
definition of, 627

drywall around, 467
inline fans for, 434
soundproofing, 480
testing for air leaks, 421, 427
Dumpsters, 67
Duplex nails, 78
Dust collectors, 51
Dust masks, 460, 541
DWV. See Drainage, waste, and vent (DWV) system

E
Eaves, 172–73, 426, 447, 627
Edgebanding, 627
Edger (Disk sander), 598, 601–602
Edge trimmer, 619
Eichler-designed houses, 48, 452
Elastomer, 87, 169
Elastomeric coatings, 187–88, 244, 263, 627
Electrical boxes. See Boxes
Electrical circuits. See Circuits
Electrical notation, 301, 302, 304
Electrical systems, 286–345
building codes for, 287, 293, 301, 313, 318, 328, 629
demolition and, 207–208
floor plans for, 301–304
grounding basics for, 290–93, 299, 343
inspection of, 18, 21, 295–97, 326
layout of, 315
load calculations for, 297–300
overview of, 286
planning for, 28, 294–304
rough-in wiring for, 312–14, 315, 326
running cable for, 317–22, 324–25
safety for, 52, 286–87, 288, 291, 312, 327
service entrance and panel, 288
shutting off power to, 293–94
terminology of, 287
tools for, 304–306
voltage testers for, 207–208, 287, 294
See also Cable; Grounding and ground wires; Receptacles; Wiring
Electrical testers, 207–208, 287, 294, 304
Electric demolition hammer, 268
Elevation plans, 392
Enameled sinks, 409
Enamel paint, 539, 558, 628
End-cutting pliers, 304
End-wall studs, 202, 628
Energy audits, 418, 419–23
Energy efficiency and conservation, 416–23
appliances and, 422
doors and, 126
nine-step plan for, 417–19
rebates for, 419
retrofits for, 44, 417–19, 628

skylights and, 159
windows and, 147, 148, 149
Energy-efficient mortgages, 419–20
Energy rebates, 419
Energy recovery ventilator (ERV), 435
Engineered lumber, 68, 71–72, 171
Engineered wood flooring, 591, 592–93, 610–14
Entry doors, 126–27
Epoxies
for anchor bolts, 277
for foundation cracks, 263
mixing, 87
for rotten wood, 175, 216
for screw holes, 139
thinset mortar/adhesive, 491–92, 499, 500
for tile work, 491–92
uses for, 88
EPS (Expanded polystyrene)
for basements, 428, 450–52
for damp conditions, 284
definition of, 628
description of, 441
ERV (Energy recovery ventilator), 435
European-style (Cup) hinges, 400, 401
Exhaust fans
bathroom, 425, 433–34
inline, 434
installation of, 432–35
whole house, 418, 432, 435
Expanded metal ramp, 543–44
Expanded polystyrene. See EPS
Expansion joints, 497
Extension poles, 544
Exterior doors, 117, 118, 124–31, 141
Exteriors, 166–93
painting, 553, 563–69
water-resistant barriers for, 168–70
See also Siding
Exterior trim, 166, 170–75
door casing, 122, 130–31
engineered, 171
installation of, 171–72
painting, 170–71, 172
repair of, 175
siding around, 176, 187
water table, 173–74, 176, 179, 182
window casing, 142, 150
Extruded polystyrene (XPS), 279, 427, 441–42, 632
Eye protection
for blowing insulation, 449
equipment for, 53
for masonry work, 235, 236
for nailing, 77
for plumbing work, 354

F
Fabric wallcovering, 571, 572, 574, 585

Face-frame cabinets, 393
Face-nail, 79, 628
Factory-finished siding, 182
False beams, 535–36
Fascia, 172–73, 628
Faucets
  handles for, 415
  holes for, 407
  for kitchen sinks, 408–409
  for lavatories, 412, 414, 415
  removal of, 495
  space for, 410
  spouts of, 410, 415, 495
Faux-painted walls, 555
Felt paper (Building paper)
  description of, 94, 96, 626
  installation of, 169
  under siding, 168
  for stucco, 185–86
  See also Weather-resistant
    barriers
Female end
  definition of, 628
  fittings with, 352, 355
  for shower connections, 382
  for sink connections, 380
  soldering, 356
  sweat-female adapters, 355
  for water heater connections,
    385
Fiberboard, 171
Fiber-cement (FC) siding, 182, 183–
  85
Fiber-cement trim, 171
Fiberglass doors, 126
Fiberglass insulation
  batts, 437, 438, 444–47, 449
  dense-pack, 440–41
  installation of, 418, 444–47
  loose-fill, 437, 440
Fiberglass-mat (Paperless) drywall,
  456, 473, 541
Fiberglass mesh tape
  for backer board, 497
  for drywall repairs, 474
  for drywall seams, 461, 472, 499
  for ductwork, 429
Fiber-reinforced flashing strips,
  129, 169–70
Field, definition of, 628
Figure-5 (Double combo) fitting,
  365, 369
Figuring dimensions, 390
Files
  flat, 59
  rat-tail, 59, 515
Fillet molding, 517
Finances, 36
Finger-jointed molding, 170, 517
Finish carpentry
  adhesives & glues for, 522–23
  cutting for, 520–22
  fasteners for, 522
  materials for, 517–18
  measuring for, 519–20
  tips for, 524
  tools for, 512–16
  See also Interior trim

Finishes
  clear, 540, 558, 561–63
  for flooring, 596–97, 603–605
  for interior trim, 524
  natural, 628
  oil-based, 559, 563, 567, 569
  polyurethanes, 540, 561–63,
    596–97, 603–604, 605
  stripping, 544, 558–61, 567
  water-based, 562, 603–604, 605
  See also Paint
Finish floors. *See* Flooring
Finish hammers, 63
Finish nails, 78
Firebricks (Refractory bricks), 251
Fireplaces, 20, 247, 250–52
Fire-resistant drywall, 457
Fire-stop, definition of, 628
Fittings. *See* Pipe fittings
Fixed windows, 145, 146
Fixtures. *See* Lighting fixtures;
    Plumbing fixtures
Flange extenders, 373, 495
Flared fittings, 357
Flash foam, 423–24, 628
Flashing, 96–101
  cap, 142, 143, 149, 167, 169, 627
  chimney, 10, 11, 99–101
  definition of, 628
  direct-vent pipes, 388
  doors, 139, 142, 143–44, 167,
    168–69
  drip-edge, 9–10, 94, 96–97, 112,
    627
  for French doors, 170
  galvanic action and, 95
  inspection of, 9–10
  purpose of, 166
  ridge, 101
  roof, 90, 96–101
  for roof-wall connections, 101
  self-sticking, 507
  for skylights, 101, 159, 163–65
  step-flashing, 159, 164, 165, 631
  terminology of, 169
  for tile work, 494
  for valleys, 97–98
  vent-pipes, 98–99
  for windows, 139, 142, 143–44,
    167, 168–69
  wood shingles, 111
Flashing strips, fiber-reinforced,
  129, 169–70
Flashing tape, 168, 169–70
Flat files, 59
Flat paint, 539, 628
Flat roofs, 11, 113–14
Flexible arch beads, 462
Flitch plates, 75, 213
Floating a wood floor, 610–14
Floating concrete, 281
Flooring, 590–625
  bamboo, 591, 593
  coconut palm, 591, 593
  cork, 591, 594
  crowned, 262
  door height and, 122

edges of, 608
engineered wood, 591, 592–93,
  610–14
filling holes in, 605
finish for, 596–97, 603–605
laminate, 591, 594, 610
layout of, 607–10, 612
prefinished, 593
prep work for, 600, 605–6, 611–
  12
protection of, 54
refinishing, 595–605
removal of, 210–11
replacing a board, 601
resilient, 500, 591, 594–95, 614–
  18, 624, 632
sanding, 597–603
screening, 602–603
solid-wood, 591–92, 605–10
stone, 591, 595
strip, 591–92, 605–10, 631
tile, 16, 502–505, 591, 594, 595
tools for, 597–99, 607, 611
types of, 591–95
underlayment for, 606, 612, 614–
  16
vinyl, 21, 594–95, 614–18
wood, 210–11, 591–93, 595–610,
  624
Floor plans, 26–27, 30, 32
  for bathrooms, 349, 412
  for cabinets, 392–94
  for electrical systems, 301–304
  for kitchens, 391–92
Floors
  concrete, 269, 591, 610
  inspection of, 15–16
  insulation of, 449
  springiness in, 262, 270
  subfloor under, 606, 610, 612,
    631
Floor trusses, 72
Flue, 244, 245, 246–47, 249–50, 628
Flush-mount (Tile-edge) sinks, 409,
  410
Foil-backed butyl tape, 429
Foil-backed drywall, 457
Foil wallcovering, 571, 572, 574,
  585
Folding rule, 55
Footings, 254, 259, 628
Forced hot-air (FHA) heat, 20, 28,
  421
Formaldehyde, 76
Form-hangers, 274
Formica (Plastic laminate)
  countertops, 403, 405
Form-release oil, 275
Foundation contractors, 263, 264,
  266, 268, 273, 276
Foundations, 258–85
  building codes for, 259, 260, 272,
    276, 277
  capping, 278
  concrete forms for, 258, 274–78
  cracks in, 262, 263, 285
  definition of, 628

drainage for, 261
fixing damp, 284–85
footings for, 259
inspection of, 14, 16–17
mat slab, 259, 282–84
minor repairs to, 269–72
removal of, 273
replacement of, 272–78
sheathing overhang of, 274
slab-on-grade, 279–82
steel reinforced, 260, 274–75,
  276
symptoms of failure, 261–62
types of, 258–60
Frameless (European style)
  cabinets, 393
Framing, 195–206
  building in place, 203–206
  of corners, 200, 204
  for doors, 222
  layout for, 198–99
  lumber for, 69
  for skylights, 160–62
  smarter way to build, 204–205
  standard stud length for, 199
  steel, 73, 74–75
  terminology of, 199
  for windows, 222
Framing angles, 85
Framing screws, 82, 83
Framing square, 54
Freezers, 422
French doors, 117, 118, 131–33,
  170
French drains, 261
Frieze, 172–73, 628
Frost line, 254, 259, 262, 628
Full-body harnesses (Personal fall-
  arrest systems, PFAS), 92
Fungi, 216
Furnaces, 20, 419
  See also Heating system; HVAC
Furring, 212, 213, 214, 272, 628
Fuse boxes, 290, 293–94, 295, 299
Fuses, 289–90

G

Gable, definition of, 628
Gable end, 100, 101, 102, 628
Galley kitchens, 391, 392
Galvanic action, 95, 193
Galvanized metal bead, 470
Galvanized pipe, 19, 362
Galvanized steel gutters, 190
Gasketed threshold, 138
Gas lines, inspection of, 20
Gate valves, 355, 357
GEC (Grounding electrode
  conductor), 293
General contractors, 29–30, 34–35,
  36
General-use circuits (electrical),
  299–300
GenieClip RST sound isolation
  clips, 480–81, 482
GES (Grounding electrode system),
  293

GFCI (Ground-fault circuit
    interrupter)
  for bathroom fans, 434
  in bathrooms, 412–13
  boxes for, 307
  definition of, 628
  in kitchens, 310
  requirements for, 18, 293, 300,
    303, 310–11, 313
  wiring, 328–30
Girders
  adding, 270
  definition of, 195, 628
  within foundations, 260
  replacement of, 271
  supports for, 272
Glass, cutting, 158, 159
Glazing, 118–19, 158–59
Gloss paint, 539, 628
Gloves, disposable, 542
Glues, 87, 522–23, 613, 618–19
  See also Adhesives
Glulams (Glue-laminated timbers),
    73
Goof ring (Add-a-depth ring), 307
Gorilla Glue (Polyurethane glue),
    87, 523
Gorilla Super Glue (Cyanoacrylate
    glue), 87, 88, 523
Grade, 261, 276, 278, 280, 628
Grade beams with concrete piers,
    259–60
Grain, 603, 628
Granite countertops, 405, 407, 413
Grass wallcovering, 571, 572, 585
Great room, planning for, 43, 44
Greek Revival house, 228–29
Greenboard (WR board), 165, 456,
    464, 628
Green building, 26, 36, 38, 46, 416
Green Glue noiseproofing
    compound, 480, 481–82
Grinders, 242, 488, 496
Grounding and ground wires
  basics of, 290–93, 299, 343
  definition of, 291
  elements of, 292
  for light fixtures, 338, 341, 342
  to metal pipes, 376
  for metal vs. plastic boxes, 299,
    308, 343
  receptacles, 296
  requirements for, 313
  safety and, 291
  splicing, 325
Grounding electrode conductor
    (GEC), 293
Grounding electrode system (GES),
    293
Grout
  application of, 511
  on countertops, 405
  definition of, 235, 628
  rejuvenating, 510
  sanded vs. sandless, 492
  for tile work, 485, 488–89, 492
Grout float, 488, 506, 511

Gutter hangers, 191, 192
Gutters (Rain gutters), 189, 190–93
  damp basements and, 284
  definition of, 628
  inspection of, 12
  installation of, 192–93
  materials for, 190
  repair of, 193
  role of, 166–67, 190
Gypsum board, glass-mat, 490
Gypsum lath, 457, 476

H
Hacksaw, 58
Hammer drills, 67
Hammers, 59, 63, 196, 459, 516
Hammer tackers, 63
Handmade tile, 485–86, 489
Handrails, 16
Handsaws, 514, 520
Hand screws, 62
Hand sledges, 63
Hand tools, 55–56, 58, 59, 62–64
  See also Tools
Hardboard, 14, 75, 171, 615
Hard hats, 53
Hardiebacker, 490, 628
Hazardous materials, 21–22
H-clips, 86
Headers
  definition of, 195, 628
  framing, 199–202
  insulation of, 200–201
Head jambs, 116, 123, 128
Hearing protection, 53
Heat guns, 560, 562
Heating system
  annual fuel-utilization efficiency,
    419
  forced hot-air, 20, 28, 421
  upgrading, 419
  See also HVAC
Heat-recovery (HR) drains, 364
Heat-recovery ventilator (HRV),
    418, 435
Heat transfer, 417
HEPA filters, 541, 542, 628
HEPA vacuums, 54
  for lead paint removal, 156, 515,
    550, 551
  for mold removal, 437
  for refinishing floors, 599
HERS (Home energy rating
    system), 422, 423, 628
Hidden hangers, 191
Hinged single doors, 118
Hinge gain, 140, 628
Hinge jamb, 116–18, 121–22, 123,
    128, 131–32, 140
Hinge knuckle, bending, 133
Hinges
  adjustable, 127
  butt, 127
  for doors, 119, 127
  European-style, 400, 401
  fixing problems with, 141, 142
  loose, 139
Hold-downs, 85

Hole saw, 488, 628
Home energy rating system
    (HERS), 422, 423, 628
Home inspections, 6–7, 8–20, 418,
    444
Hopper windows, 145, 146
Horizontal slider windows, 145,
    146
Hose bibs, 355
Hot-melt carpet seam tape, 619,
    621, 622, 623
Hot-melt glues, 88, 523
Hot-water trunk line, 347
Houses
  buying, 7–8, 9
  Craftsman, 37–38
  Eichler-designed, 48, 452
  Greek Revival, 228–29
  inspection of, 6–7, 8–20, 418, 444
  jacking and shoring, 263–69
  in-law unit in, 39–41, 271
  leveling, 273
  renovation case histories of,
    36–50
  whole house renovation of,
    48–50
Housewrap, 168, 169
HRV (Heat-recovery ventilator),
    418, 435
Humidifiers, 20
Humidistats, 433
Hurricane ties (Twist straps), 85
HVAC (Heating, ventilation and
    air-conditioning)
  asbestos in, 21
  definition of, 628
  energy efficiency and, 419
  inspection of, 20, 21
  planning for, 28
Hydraulic cement, 263
Hydraulic jacks, 264

I
I-beams. See Steel I-beams
IC (Insulation contact), 424, 628
Ice makers, 357
IGUs (Insulated, tempered, double-
    pane glass), 48
I-joists (TrusJoists), 72–74
IKEA Cabinets, 400–403
Impact drivers, 61, 81, 516
Impermeable vs. permeable
    barriers, 429
Infrared imaging, 420–21, 628
In-law units, 39–41, 271
Inline fans, 434
Insect infestations, 214–15
Inspections, home, 6–7, 8–20, 418,
    444
Insulated, tempered, double-pane
    glass (IGUs), 48
Insulated flex ducts, 428, 429
Insulation, 437–53
  of attics, 446–48
  of basements, 284, 450–52
  batt, 437–38, 440
  ceiling, 342, 443–44
  cellulose, 437, 440

  of cold spots, 447
  on concrete floors, 610
  cotton, 438
  for damp basements, 284
  dense-pack, 418, 440–41, 448–50,
    451, 627
  for energy conservation, 418
  extruded polystyrene, 279, 632
  fiberglass, 418, 437, 438, 440–41,
    444–47, 449
  of floors, 449
  of headers, 200–201
  installation of, 443–53
  loose-fill, 438, 440, 447–48, 629
  mineral wool, 438
  recommendations for, 437, 442
  roof, 430–31
  R-values of, 437, 440, 630
  under slabs, 279, 280
  thermal, 417, 628
  types of, 437–43
  of walls, 418, 448–50
  of water supply lines, 376–77
  of wiring, 443, 447
  See also Rigid foam insulation;
    Spray foam insulation
Insulation blowers, 448, 449
Insulation contact (IC), 424, 628
Insulation contact, air-tight (ICAT)
    fixture, 424, 628
Intake vents, 102–103
Integral sinks, 410
Interior doors, 117, 119–24
Interior plywood, 77
Interiors
  painting, 551–58
  stripping and refinishing, 558–63
  See also Drywall; Wallcovering
Interior trim, 512–37
  corner blocks for, 523–24, 527,
    528
  custom, 517
  cutting, 514, 519, 520–22
  door casing, 523–28, 606–607
  fasteners for, 522
  finishing tips for, 524
  gluing & attaching, 522–23
  leveling or not leveling, 522
  materials for, 517–19
  measuring for, 519–20
  miter joints for, 520, 521, 523,
    524, 527–28, 531
  painting, 539, 553, 554, 556
  splicing, 520–21
  square cut, 523–24, 525–26, 527
  stripping, 558–61
  terminology of, 516
  tools for, 512–16
  for windows, 524, 526, 528–30
  See also Molding
Interior windows, 118
Internet access, 54
IRC (International Residential
    Code), 628
Island counters, 369, 392, 398
Isometric sketch, 352

J

Jack flashing (Vent-pipe flashing), 98–99
Jackhammers, 273
Jacking a house, 263–69, 273
Jack rafter, definition of, 628
Jacks
  hydraulic, 264
  post, 264
  roofing, 92, 107
  screw, 264
  types of, 64, 65
Jack studs
  definition of, 629
  elimination of, 205
  lumber for, 195
  marking for, 198
  for window & door openings, 199, 201, 202, 222
Jambs, 117–18
  air-sealing, 427
  definition of, 627, 629
  head jambs, 116, 123, 128
  hinge jamb, 116–18, 121–22, 123, 128, 131–32, 140
  latch jamb, 117, 123, 128, 129, 133
  oversizing, 119
  weatherstripping, 136–39
  See also Door jambs
Japanese pull saw, 58, 514, 520
J-beads, 461–62
Jigsaws, 56, 57, 629
Jim cap, 375–76
Job sites, 4
Joint compound
  application of, 470, 471–73
  definition of, 629
  estimating, 465
  sanding, 460, 470, 472
  tools for, 459–60
  types of, 462
Jointers (Striking irons), 236
Joints
  angled-end, 213
  bevel, 172
  biscuit, 514, 521, 522, 525
  butt, 523–24, 525–26, 527, 626
  coped, 515, 521–22, 535, 627
  definition of, 629
  expansion, 497
  miter, 520, 521, 523, 524, 527–28, 531, 629
  scarf, 520–21
  splicing, 522
  wood construction connectors for, 85–86
Joint tape, 461
Joist hangers, 84, 85, 224
Joists
  air-sealing, 427
  jacking and shoring, 267, 268
  notching & drilling, 374
  reinforcing, 213
  sistered, 213, 216
  span of, 270
  springiness in, 262

supporting, 195, 270
TrusJoists, 72–74
Junction boxes, 207–208, 296, 338

K

KD (Kiln-dried) lumber, 69–70
Kerf, 126, 142, 629
Kerf-in weatherstripping, 136
Keyhole saw, 58
Kickspaces (Toekicks), 393, 395–96, 398
Kiln-dried (KD) lumber, 69–70
Kitchens, 389–412
  cabinet layouts for, 392–94
  counter space in, 390, 391
  electrical circuits for, 299, 300, 303
  energy-efficient appliances in, 422
  floor plans for, 391–92
  galley, 391, 392
  greenboard for, 464
  inspection of, 16
  Islands in, 369, 392, 398
  lighting for, 300, 392
  L-shaped, 391–92, 396, 397, 399, 506, 508
  moisture from, 435
  planning for, 25, 389–94
  range hoods for, 408
  receptacles for, 18, 310, 391
  renovation case histories of, 37, 38
  single-line, 391, 392
  U-shaped, 391, 396, 397, 399
  ventilation of, 16, 369, 408
  work triangle in, 391
  See also Cabinets; Countertops
Kitchen sinks, 389–90, 408–12
  base cabinets for, 397–98
  cutouts for, 406
  faucets for, 408–409
  installation of, 378–80, 409, 411–12
  Island counters for, 369, 392, 398
  mounting style of, 409, 410
  rough-in for, 350, 351
  types of, 408–409
Knee-kicker, 619, 624
Knee pads, 53
Kneewalls, 204, 206, 426–27
Knob-and-tube wiring
  assessment of, 17, 18, 296, 297
  insulation around, 448
  removal of, 295

L

Lacquer, 540
Lacquer thinner, 559
Ladder jacks, 65
Ladders, safety for, 52, 91, 563
Lag screws, 85
Laminated architectural glass, 159
Laminated (Architectural) shingles, 103, 104, 108
Laminated veneer lumber (LVL), 73, 171
Laminate flooring, 591, 594, 610

Laminate trimmers/routers, 58, 457–58
Laminating, 202, 221, 222, 223, 231, 629
Laser levels, 55, 65, 176
Laser plumbs, 65
Laser tape measures, 65
Laser tools, 65
Latch jamb, 117, 123, 128, 129, 133
Latex bonders, 187
Latex caulk, 89
Latex paint
  enamel, 558
  for exterior work, 567, 569
  for interior work, 554
  vs. oil-based, 539–40, 552
  primer, 552
Lath, 186, 457, 476, 629
Laundry rooms, 464
Lavatories (Lav), 413–14
  definition of, 629
  faucets for, 412, 414, 415
  installation of, 378–80
  pedestal sinks in, 350, 378
  planning for, 412
  removing, 377
  rough-in dimensions for, 350
  types of, 414
Layout tools, 54–55, 512
L-bars, 64
L-beads, 462
Lead-abatement specialists, 548
Lead paint
  history of, 22
  inspection of, 15, 207
  removal of, 544, 549–51, 558, 596
  safety for, 548–51, 565–66
  sanding, 515
  testing for, 548
  on windows, 156
Lead pipes, 22
Ledger strip, 629
LED lighting, 343, 412
Let-in brace, 242, 629
Levels, 55–56, 514, 575
Lever-handled ball valves, 355
Lifts & lifting tools, 74, 458
Light (pane of glass), 117, 629
Light framing lumber, 69
Lighting, task, 392
Lighting fixtures
  for bathrooms, 300, 412
  boxes for, 308
  ceiling, 315, 316, 338–45, 425–26
  insulation contact, air-tight, 628
  for kitchens, 300, 392
  LED, 343, 412, 427
  load calculations for, 298, 299–300
  planning for, 302, 303
  recessed, 339–42, 425–26
  wallcoverings around, 588
  wiring & switches for, 332–34, 343–45
Light steel framing, 74–75
Lincrusta wallcovering, 571, 572, 574

Line blocks, 236
Lineman's pliers, 63, 304
Linoleum, 594–95, 596, 614–18
Linseed oil, 563, 594, 603
Lintel, 195, 629
  See also Headers
Liquid membranes, 285
Live load, 195
Load-bearing vs. nonbearing walls, 26, 27–28, 222
Load path, 224
Loads (Electrical), 195, 197, 211, 221
Locking pliers, 353
Locksets, 118, 119, 127, 133–36
Log buildings, 42–44
Loop vents, 369
Loose-fill insulation, 438, 440, 447–48, 629
Louver, 629
Low-e coatings, 160
L-shaped kitchens, 391–92, 396, 397, 399, 506, 508
LSL (Laminated strand lumber, TimberStrand), 74
Lumber, 68–74
  board feet calculations for, 71
  dimension, 69, 70, 627, 629
  dressed size, 629
  engineered, 68, 71–72, 171
  framing, 69
  grades of, 68–69
  handling, 195–96
  kiln-dried (KD), 69–70
  laminated strand, 74
  laminated veneer, 73, 171
  moisture content of, 69–70
  nominal dimension of, 629
  parallel strand lumber, 73–74
  pressure-treated, 70
  salvage, 71
  sizes of, 70
  species of, 69
  standard, 68–71
  terminology of, 69
LVL (Laminated veneer lumber), 73, 171

M

Machine bolts, 84
Main breaker, 290
Main panel. See Service panel
Main shutoff valve, 347
Main supply pipe, 347
Male end
  adapters, 355, 372
  definition of, 629
  for sink connections, 380
  soldering, 356
  trap adapters, 379
  for water heater connections, 385
Mallets, 59
Manifolds, 347–48, 359
Mantel, 629
Margin (Center)
  definition of, 629

for door frames, 121, 122, 123, 125, 128
for French doors, 131
Margin trowels, 488, 491, 511
Masonry, 234–56
anchors for, 276
bond patterns for, 239
cleaning & sealing, 242–44
definition of, 629
preparation for, 236–37
repointing, 630
steel framing with, 74–75
terminology of, 235
tile work over, 498
tools for, 235–36, 242
See also Brick; Chimneys
Masonry paint, 188
Mason's hammers, 236
Mason's levels, 236
Mason's string, 54
Mass-loaded vinyl, 483
Mastics, 87, 429, 489, 491
Mat slab foundation, 259, 282–84
MB (Modified bitumen), 11, 113, 114
MC (Moisture content), 69–70
MDF (Medium-density fiberboard), 75, 518, 537, 629
Measuring tools, 54–55, 512–14
Mechanical scrapers, 567
Mechanical ventilation, 418, 432–35
Medallions, 478
Medium-density fiberboard (MDF), 75, 518, 537, 629
Mesh, reinforcing, 498
Mesh tape, fiberglass, 429, 461, 472, 474, 497, 499
Metal boxes, 299, 307–308, 343
Metal carpet bars, 620
Metal-clad (Armored) cable, 309
Metal connectors, 78–86, 225, 260, 264, 309
Metal roofing, 11
Metals, galvanic action and, 95, 193
Metal studs, 213–14
Metal tension strip (V-bronze), 136
Metal thresholds, 138
Meter–mains, 290
Mfr (Manufacturer), definition of, 629
Mid-run, definition of, 629
Migration, 258, 264, 266, 629
Millwork, 517, 629
Mineral (Rock) wool insulation, 438
Mineral spirits, 559, 629
Mission-barrel tiles, 115
Miter box & backsaw, 514
Miter joints
coped, 521
cutting, 520
definition of, 629
for door casing, 523, 524, 527–28
for window casing, 531
Miter saw (Chopsaw), 65, 514, 627
Miter trimmer, 514

Modified asphalt membranes, 285
Modified bitumin (MB), 11, 113, 114
Modified silicone polymer caulks, 424
Moisture, 432–34, 435–37
See also Water damage
Moisture content (MC), 69–70
Moisture-cure polyurethanes, 604
Moisture meter, 606
Moisture-resistant (MR) drywall, 165, 456, 464, 628
Mold
control of, 435–37
definition of, 629
on drywall, 475
on grout, 510
on paint, 565
signs of, 22
on structural elements, 216
wallcoverings and, 579
Molding
astragal, 517
base, 626
bolection, 537
bullnose, 518
combining elements, 517–18
cornice, 518
cove, 517, 518, 627
crown, 478, 530, 534–35, 627
definition of, 512, 629
fillet, 517
finger-jointed, 170, 517
ogee, 517
ovolo, 517
polymer, 518–19
profiles of, 517, 518
quarter-round, 517, 518, 522, 534, 630
removing, 64
splay, 517
for wainscoting, 537
Mortar
definition of, 235
for firebox rebuilding, 251–52
mixing, 239
pointing, 235
tools for, 242
types of, 237, 240
See also Thinset mortar/adhesive
Mortar bed, 489, 490, 497, 498, 595
Mortar joints
inspection of, 14
repointing, 241–42, 243
striking, 241
Mortgages, energy-efficient, 419–20
Mortise, 629
Mortise locksets, 133–35
Mosaic tile, 506
Moss, 112, 565
MR (Moisture-resistant) drywall, 165, 456, 464, 628
Mud pan, 459
Mudsills
definition of, 631
installation of, 277–78
replacement of, 267, 272, 273

Mullion, 629
Multimeters, 294, 304
Muntin, 117, 556, 629
Muriatic acid, 244
Mylar wallcovering, 572, 574, 585

N

Nailing schedule
for exterior trim, 172
for flooring, 607
for structural carpentry, 79, 197, 198
Nail-protection plate, 317, 318, 465, 629
Nails & nailing, 79–81
clapboards, 181
for drywall, 461, 468
for exterior trim, 172
for finish carpentry, 522
galvanic action and, 95
hammer use for, 196
pneumatic framing, 80–81
removing nails, 198
roofing, 94, 96, 104
sizes of, 78, 80
specialty nails, 80
stains from, 564
for structural panels, 77
types of nails, 78, 80
for wood shingles, 178–79
NEC (National Electrical Code), 287, 293, 313, 318, 328, 629
Needle beams, 267, 273
Needle-nose pliers, 304
Neoprene collars, 98–99
Net-free ventilation area (NFVA), 430
New-construction windows, 149, 154–56
Newel posts, 16, 218–19, 629
NFVA (Net-free ventilation area), 430
Nibblers (Shears), 184
NM (Nonmetal sheathed) cable, 297, 299, 309, 311, 325
No-hub couplings, 364, 366, 372, 373
No-hub torque wrenches, 352, 353
Noiseproofing compound, Green Glue, 480, 481–82
Nominal dimension, 70, 629
Nonasbestos flame shields, 354
Nonbearing walls, 195, 211–16, 222, 629
Noncontact testers, 304
Nonmetal sheathed (NM) cable, 297, 299, 309, 311, 325
Nonpenetrating sealers, 603
Nonvitreous tile, 486
Nosing, 216, 219, 629
Notched trowels, 488

O

Oak flooring, 592
Ogee molding, 517
Ohms, definition of, 287
Oil-based paints
for interior work, 554

vs. latex, 539–40
primer, 552
safety for, 544–45
stripping, 567
thinning, 541
Oil-based polyurethanes, 562–63, 603, 604
Oil finishes, 559, 563, 567, 569
OMT (Oscillating multitools), 57, 514, 606–607
On center (o.c.), 197, 198, 199, 629
Open-cell spray foam insulation, 442, 444, 445
Open-coat sandpaper, 600
Open valleys, 97, 112–13
OPP (Oriented polypropylene), 429
Orbital sanders, 515
Oriented polypropylene (OPP), 429
Oriented strand board (OSB), 72–74, 75, 77, 615, 629
OSB (Oriented strand board), 72–74, 75, 77, 615, 629
Oscillating multitools (OMT), 57, 514, 606–607
Outlets
adding new, 299
air-sealing, 424, 426, 627
drywall around, 467, 468
grounding, 296
insulation around, 448
requirements for, 300
shutting off power to, 293–94
wallcoverings around, 588
See also Receptacles; Switches
Outrigger, definition of, 629
Ovolo molding, 517
Owner–builders, 29, 35

P

Padding, for carpeting, 620
Pads
concrete, 269–70
knee, 53
paint, 544
putty, 480–81
Paint
color of, 569
definition of, 539, 630
enamel, 539, 558, 628
for exterior doors, 129
failure of, 564–65
flat, 539, 628
gloss, 539, 628
latex, 539–40, 552, 554, 558, 567, 569
masonry, 188
oil-based, 539–40, 541, 552, 554–55, 567
oil-based vs. latex, 539–40, 552
peeling, 564
primer, 539
selection of, 538–40, 553
semigloss, 539, 631
storage of, 545
stripping, 544, 558–61, 567
terminology of, 539
thinning, 541
tiling over, 500

types of, 539, 553
wallcoverings over, 577
See also Lead paint
Paint brushes
for cabinets, 556–57
care of, 542–43
for cutting-in, 552, 553, 554
for doors, 556
overview of, 542–43
use of, 545
Painter's tape, 208
Painting
basics of, 544–46
cabinets, 556–58, 559
ceilings, 552–54
doors, 129, 556, 557
drywall & plaster, 551–52, 553
exteriors, 553, 563–69
exterior trim, 170–71, 172
interiors, 551–58
interior trim, 539, 553, 554, 556
prep work for, 538, 565–67
safety for, 544–45, 551
sequence for, 552, 569
spray, 546–48, 556–58, 568
stucco, 567, 569
techniques for, 545–48, 552–54, 556–58, 567–69
tools & equipment for, 541–44
walls, 554
windows, 556
Paint pads, 544
Paint rollers, 543–44, 545–46, 552–53, 630
Palm nailers, 213
Palm sanders, 515
Panel lifters, 458, 470
Panels, structural and nonstructural, 75–78
Paper-faced beads, 471
Paperless (Fiberglass-mat) drywall, 456, 473, 541
Parallel strand lumber (Parallam, PSL), 73–74
Parquet flooring, 592
Particleboard, 75, 615
Parting bead, 147, 630
Partitions, 195, 199, 213, 630
Passive intake vents, 418, 435
Passive solar designs, 419
Passive ventilation, 102
Paste brushes, 575
Pasting tables, 575
Pedestal sinks, 350, 378
Peelable wallcovering, 578–79
Peel-and-stick membranes, 285
Penetrating sealers, 603
Penny (d.), 78, 627, 630
Permeable vs. impermeable barriers, 429
Permits, 30, 31, 33–34, 195
See also Building codes
Personal fall-arrest systems, PFAS (Full-body harnesses), 92
PEX, 28, 358–62
connections for, 359–62
existing supply pipes and, 358, 362

fittings for, 358, 360, 361
manifolds for, 347–48, 359
overview of, 347–48
pros and cons of, 358–59
supporting, 376
types of, 359
for water heaters, 387
PEX-A, 359, 360, 361
PEX-B, 359, 361–62
PEX-C, 359
Piers, 259–60, 282, 630
Pilot holes, 79, 84
Pin nailer, 630
See also Brad nailer
Pins, rebar, 277
Pipe
ABS, 362, 363, 364, 365, 371–73
air-sealing around, 426, 427
cast iron, 362, 364, 365–66, 371–73
cement, 364
galvanized steel, 19, 362
insulation around, 447
lead, 22
measuring and fitting, 355
nail guards over, 465
ordering, 370
plastic, 347, 349, 362, 364–65, 371–73
soundproofing, 480
supports for, 365, 370, 377
See also Copper pipe; PEX
Pipe cutters, 352, 353
Pipe fittings, 354–55
adapters, 355, 372, 379
clamp-ring, 359, 361
cold expansion, 359, 360, 361
compression, 357, 359, 361, 362
for copper pipe, 353, 354–55
for copper tubing, 357
for DWV systems, 363, 370
female end, 352, 355, 356, 380, 382, 385, 628
flared, 357
male end, 355, 356, 372, 379, 380, 385, 629
for PEX, 358, 360, 361, 362
press-connect, 357
push-connect, 357, 361
Pipe-frame scaffolding, 64–65
Pipe nipples, 494
Pipe wrenches, 352
PIR (Polyiso), 630
Pitch, 90, 92, 96–97, 115, 630
Planes (Tool), 59, 65, 514–15
Plank flooring, 591–92
Plans & planning, 23–50
for bathrooms, 25, 27, 412–15
building professionals and, 29–31
case histories of, 36–50
for comfort, 24
for drywall, 462–63
for electrical remodels, 28, 294–304
for finances, 36
for kitchens, 25, 389–94

for lighting fixtures, 302, 303
for the long term, 23, 24
for plumbing, 27, 28, 349–52
process of, 31–35
See also Drawings; Floor plans
Plaster
anchors and bolts for, 476–77
cutting, 322
demolition of, 207
drywall over, 456, 466
history of, 454
mixing, 478
painting, 551–52, 553
removal of, 210
repairs to, 475–79
replacement of, 476
texturing, 474
wallcoverings on, 579
Plaster grounds, definition of, 630
Plasterwork, restoring, 478–79
Plastic boxes, 299, 307–308, 310, 343
Plastic laminate (Formica) countertops, 403, 405, 499
Plastic pipe
for DWV systems, 349, 362, 364–65, 371–73
for water supply lines, 347
See also PEX
Plastic spacers, 488
Plates
definition of, 630
flitch, 75, 213
marking and layout for, 196, 198–99, 200
nailing, 197, 198, 201
rough openings and, 201
sill, 202, 226, 230, 630
sole, 201, 202, 211–12, 214, 630
See also Top plates
Pliers, 63, 304–305
channellock slip-joint, 63
diagonal-cutting, 304
end-cutting, 304
lineman's, 63, 304
locking, 353
needle-nose, 304
slide-nut (sliding-jaw), 352, 353
slip-joint, 63
Vise-Grip, 63
Plinths (Corner blocks), 523–24, 527, 528
Plow, definition of, 630
Plug-in circuit analyzers, 304
Plumb (Vertical), 55, 56, 630
Plumb bobs, 55, 65
Plumbing, 346–88
framing considerations and, 351
inspection of, 19
overview of, 346–49
planning for, 27, 28, 349–52
rough-in for, 349–50, 351
safety for, 354
terminology of, 346
tools for, 352–54, 360, 361
See also Drainage, waste, and vent (DWV) system; Pipe; Pipe fittings

Plumbing codes, 346, 347, 354, 383
Plumbing fixtures
installation of, 377–82
rough-in for, 349–50, 351
See also Bathtubs; Sinks
Plunge routers, 58
Plywood, 76–78
CDX, 76, 78, 627
composition of, 75
hardwood, 77–78
nailing, 197
roof sheathing, 93
structural, 77–78
subfloor, 615
for tile work, 491, 499
Pneumatic air chisels, 242
Pneumatic framing nails, 80–81
Pneumatic nailers, 66
for fiber-cement, 184
for finish carpentry, 66, 516
for flooring, 607, 608
framing, 66, 81
for roofing, 91, 92, 106
for structural panels, 77
Pocket doors, 118, 141
Point failure, 260
Pointing mortar, 235, 241–42, 243, 630
Point loads, 195
Polyiso (Polyisocyanurate, PIR), 630
Polymer molding, 518–19
Polyurethane foam sealants, 424
Polyurethanes
caulks, 88, 89
clear finish, 540, 561–63
for flooring, 596–97, 603–604, 605
for foundation cracks, 263
glue, 87, 523
molding, 518–19
See also Spray foam insulation
Polyvinyl acetate glue (Titebond II wood glue), 87, 88, 523, 630
Polyvinyl acetates (PVA), 473
Polyvinyl chloride. *See* PVC
Popping the grain, 603
Porcelain (Vitreous china), 414
Portland cement, 235
Post-and-beam construction, 194, 226–33
Post bases and caps, 86
Post caps, 225
Postformed plastic laminate countertops, 403, 405
Post jacks, 264
Posts
definition of, 630
foundation, 260
for jacking and shoring, 263–64, 266
for new beams, 224
newel, 16, 218–19, 629
replacement of, 269
Poured masonry chimney liner, 249
Powder-actuated tools, 66–67
Powder-post beetles, 215

Power (Volt-Amps), definition of, 287
Power layer, 302, 303
Power planers, 65, 124, 211
Power stretcher, 619, 624
Power tools, 56–58, 60–62, 305–306
Power washing, 567
Precased doors, 128
Prefinished flooring, 593
Prehung doors, 117, 119–31, 630
Prepaste activator, 572
Prepasted wallcovering, 574, 583
Presealers, 562
Preservatives, 630
Press-connect fittings, 357
Pressure-balancing valve, 357
Pressure-treated lumber, 70
Primer
    definition of, 539, 630
    for drywall & plaster, 551–52
    for exterior work, 567, 569
    latex vs. oil-based, 552
    PVA, 473
    for siding, 566–67
    tinted, 540
    for wallcoverings, 576
    when to use, 539
Probe testers, 294
Productivity, 54
Propylene gas torches, 354
Proud cut, 515, 527, 630
Pry bars, 64
PSL (Parallel strand lumber, Parallam), 73–74
Pull bars, 611, 613
Pump jacks, 65
Push-connect fittings, 357, 361
Putty Pads, 480–81
PVA (Polyvinyl acetates), 88, 473, 630
PVC (Polyvinyl chloride)
    bonder, 476–77
    cellular, 171
    definition of, 630
    flooring, 595
    pipe, 349, 362, 363, 364, 372–73

Q
Quarter-round, 517, 518, 522, 534, 630
Quartz composite countertops, 405

R
R&R (Resquared and rebutted) shingles, 178
Rabbet, 527, 528, 530, 537, 630
Radiation, 417
Radius-bullnose trim, 485
Rafters
    definition of, 630
    eaves trim on, 172–73
    inspection of, 15
    jack, 628
    valley, 232, 630
Rails, 117, 630
Rain gutters. See Gutters
Rain-screen walls, 180
Raised-panel wainscoting, 537

Rake, 94, 96, 103, 104–106, 630
Random-orbit floor sander, 598
Range hoods, 408
Rasps, 59, 457, 515
Ratchet-handle bit driver, 62
Rat-tail files, 59, 515
Razor knife, 575, 576
Reaming tool, 353
Rebar
    for cement block walls, 257
    cutting, 273
    for foundation walls, 260, 274–75
    installation of, 276
    for slabs, 283
Rebates, energy, 419
Receptacles
    AFCI, 18, 300, 311, 328–30, 626
    air-sealing, 427
    for bathrooms, 18, 310
    demolition and, 207
    duplex, 326, 327–31
    end of circuit, 327
    grounding, 296
    kitchen, 18, 310, 391
    midcircuit, 326
    polarized, 311
    requirements for, 300
    selection of, 310–11
    split-tab, 330–31
    tamper-resistant (TR), 311
    two-slot, 327
    wiring, 326, 327–31
    See also GFCI
Recessed lighting, 339–42, 425–26
Reciprocating saws, 56, 57, 210, 305, 353
Recycling, 46
Refrigerators, 390, 422
Refinishing wood floors, 595–605
Refinishing woodwork, 561–63
Refractory bricks (Firebricks), 251
Renewable energy devices, 419
Renovation, Repair, and Painting (RRP), 548
Replacement sashes, 148, 149–51
Repointing, 235, 241–42, 243, 630
Resilient channel, 480, 483
Resilient flooring, 594–95, 614–18
    characteristics of, 591
    installation of, 618–19
    template for, 616–18
    tile work over, 500
    transition to, 624
    underlayment panels for, 614–16, 632
Respirator mask, 53
    for blowing insulation, 449
    definition of, 630
    for fiber-cement, 183
    for lead paint removal, 156
    for masonry work, 235
    for painting, 541, 542, 544, 545, 547, 566
    for refinishing floors, 596, 599
Resquared and rebutted (R&R) shingles, 178

Retaining walls, 283
Reveal
    definition of, 630
    for doors, 122–23, 124, 125, 128, 130, 525
Reverse-trap toilets, 414–15
Ridge, 101, 110, 113, 630
Ridgeboard, 9, 11, 15, 113, 630
Ridge vents, 102–103, 110, 111, 113
Right-angle drills, 66, 305
Rigid foam insulation
    in attics, 426
    for basements, 284, 427–28
    for damp areas, 284
    definition of, 630
    EPS, 441, 628
    in headers, 200–201
    heat transfer and, 417
    overview of, 441
    Polyiso, 630
    R-value of, 440
    XPS, 279, 427, 441–42, 632
Rim joist, 266, 274, 630
Rimless sinks, 379
Rip/ripping, definition of, 630
Rise, definition of, 630
Riser, 216, 218, 220, 630
Rock (Mineral) wool insulation, 438
Roller pans & rollers, 543–44, 575
Roll roofing, 95, 630
Roofing jacks, 92, 107
Roofing nails, 94, 96, 104
Roof-mounted strap nailers, 191
Roofover, 109, 630
Roofs and roofing, 90–115
    built-up, 11, 626
    equipment for, 91–92
    flashing for, 90, 96–101
    flat, 11, 113–14
    inspection of, 9–11, 14, 15
    insulation over, 431
    insulation under, 430
    metal, 11
    nailing, 94, 96, 104
    over old shingles, 109, 630
    pitch of, 90, 92, 96–97, 115, 630
    reinforcing, 15, 40
    roll, 95, 630
    safety on, 90–91, 92, 247–48
    sheathing for, 93, 630
    shed, 101
    spray foam insulation on, 452–53
    stripping off old roofing, 92–93
    terminology of, 104
    tile and slate, 10, 11, 114–15
    underlayment for, 94–95, 96, 111
    unvented, 431, 447
    valleys in, 9, 97–98, 108–109, 632
    vented, 430–31
    ventilation for, 102–103, 111, 430–31, 432
    warranties on, 91
    See also Flashing; Shingles
Roof sheathing, 93, 630
Roof trusses, 72, 632

Room ratings, 25
Room temperature vulcanization (RTV), 479
Rotary hammer drills, 67
Rotted wood, 175, 215–17, 230, 231
Rough opening (RO)
    for bathtubs, 350, 351, 373
    definition of, 630
    for doors, 118, 120–21, 124, 131
    for DWV systems, 370–76
    framing of, 201–202
    for kitchen sinks, 350, 351
    for lavatories, 350
    layout of, 198–99
    for showers, 350, 351
    size of, 222
    for toilets, 373
Routers, 58
    drywall, 457–58, 468
    for finish carpentry, 515
    laminate, 58, 457–58
    plunge, 58
Router tables, 515
Row-running knife, 619
RRP (Renovation, Repair, and Painting), 548
RTV (Room temperature vulcanization), 479
Rubber-and-steel mounts, 480
Run, definition of, 630
R-value
    of building materials, 417, 441
    definition of, 630
    of insulation, 437, 440, 630
    recommendations for, 437
    of windows, 148, 417, 441

S
Saddle. See Cricket
Safety, 4, 196
    for blowing insulation, 449
    for electrical work, 52, 286–87, 288, 291, 312, 327
    equipment for, 51, 52, 53
    for jacking, 265–66
    for ladders, 52, 91, 563
    for lead paint, 548–51, 565–66
    for masonry work, 235, 236
    for mold removal, 436, 437
    for painting, 544–45, 551
    for plumbing, 354
    for refinishing floors, 596, 599
    on roofs, 90–91, 92, 247–48
    for spray painting, 547
    for stripping, 560
    for tile work, 486, 487
    for tool use, 51–53
Safety harness, 53
Salvage lumber, 71
Sanders, 515
    belt, 515, 597
    drum, 590, 597–98, 601
    edger, 598, 601–602
    orbital, 515
    palm, 515
    random-orbit, 598
Sand float stucco, 189

Sanding
    drywall, 459, 460, 541
    floors, 597–603
    joint compound, 460, 470, 472
    painting, 566
    wet, 459
    woodwork, 561
Sandpaper, 515, 516, 557, 599–600
Sash balance (Weights) & cord,
    153, 156, 157, 631
Sashes
    definition of, 631
    painting, 556
    replacement, 148, 149–51
Sawblades, 56, 514
Saws, 56–57, 58, 65, 514
Scaffolding, 64–65, 91
Scarf joints, 520–21
Scarifying tool, 577
Schipa, Gregory and Carolyn, 227–
    30
Scrapers
    Bahco by Snap-On carbide, 59
    paint, 560, 561, 562, 565–66
    for refinishing floors, 598
Screed, 280, 631
Screwdrivers, 61, 62
Screw guns, 458–59
Screw jacks, 221, 264
Screws
    coatings on, 82, 84
    corrosion-resistant, 497
    deck-head, 82
    drawer-front, 82
    drywall, 461, 465, 467, 468, 496
    framing, 82, 83
    for gutters, 192–93
    repositioning, 83
    SDS and SD, 82, 83–84
    stainless-steel, 398
    structural, 68, 82–84
    trim-head finish, 81, 82, 171–72
Screw threads, 82
Scribe pieces, 393, 395, 398
Scribing, 465, 488, 631
Scuff-sanding, 561
SDS and SD screws, 82, 83–84
Sealants, 86, 87, 88–89
    acoustic, 482, 483
    for air-sealing, 423–25
    for brick, 244
    for damp basements, 285
    definition of, 631
    for flooring, 603
    polyurethane foam, 424
    for sinks, 411–12
    See also Air-sealing; Caulk and
        caulking
Seam adhesive, 572
Seam rollers and irons, 619, 623
Second-floor additions, 45–47
Self-leveling compounds (SLCs),
    491, 497–98
Self-leveling lasers, 55
Self-rimming sink, 379, 409, 410,
    411
Semigloss paint, 539, 631

Separate-rim sinks, 409, 410
Septic system inspection, 19
Service entrance, 288
Service panel, 289–90
    description of, 288, 289–90
    inspection of, 18, 21, 295
    safety and, 287
    shutting off power from, 293–94
    size of, 299
Service pipe, 347
Setting a nail, 516, 518, 523, 631
Setting bed, 489, 490–91, 494, 496–
    500
Setting-type joint compounds, 462
Shakes, 10, 110–13, 631
Shears (Nibblers), 184, 575
Shear wall, 197, 631
Sheathing, 93, 174, 441, 631
Shed roofs, 101
Sheet-mounted tile, 485, 486, 506
Sheetrock. See Drywall
Shellac, 540, 563, 596, 603
Shims, 121, 123, 124, 129, 132
Shingles, 177–79
    architectural (laminated), 103,
        104, 108
    asphalt, 10, 101, 103–10
    cap, 110, 113
    cutting, 106
    inspection of, 10
    installation of, 105–10
    layout of, 104–105
    longevity of, 92
    over old shingles, 109
    removal of, 92–93
    repair of, 110, 113
    slate, 10
    terminology of, 104
    three-tab, 103, 104, 107
    weight of, 103
    See also Wood shingles
Shiplap siding, 181
Shoes and sweeps, 137–38
Shop vacuum, 460
Shoring
    definition of, 195
    a house, 263–69
    removal of, 268–69
    a wall, 206, 221
Shower pans, 493, 500
Showers
    drains for, 366
    hardware for, 412, 495
    installation of, 383
    pressure-balancing valve for, 357
    rough-in dimensions for, 350,
        351
    types of, 415
    windows in, 412
Shutoff valves, 347, 355, 357, 377,
    385, 495
Shy cut, 631
Side cutters, 63
Siding, 166, 176–90
    cedar, 182
    clapboards, 179–83
    definition of, 631

factory-finished, 182
    fiber-cement, 182, 183–85
    inspection of, 13, 14
    layout of, 176–77
    painting, 563–69
    shiplap, 181
    stucco, 14, 185–91, 631
    vinyl, 14
    See also Wood shingles
Silicone caulk, 88–89, 424
Sill pans, 124
Sill plate, 202, 226, 230, 630
Sills
    definition of, 631
    doorsills, 125, 126, 131
    mudsills, 267, 272, 273, 277–78,
        631
    weatherproofing, 169
    window, 167, 530
Simpson Strong-Tie connectors, 83,
    85–86
Single-line kitchens, 391, 392
Sinker nails, 78
Sinks
    acrylic, 409
    base cabinets for, 397–98
    cast iron, 411
    countertops and, 390, 406
    cutouts for, 406
    double, 380
    drains for, 369
    flush-mount, 409, 410
    hardware for, 410
    installation of, 378–80, 409, 411–
        12
    integral, 410
    kitchen, 350, 351, 369, 378–80,
        389–90, 406, 408–12
    pedestal, 350, 378
    removal of, 377, 495
    rimless, 379
    rough-in dimensions for, 350,
        351
    sealant for, 411–12
    solid-surface, 409
    stainless-steel, 409
    tile work around, 506, 508
    undermount, 379, 409, 410, 411,
        413
    See also Lavatories
Sistered joists, 213, 216
Site maps, 28–29
Sizing, 576
Skin care, 551
Skip-sheathing, 93, 110
Skylights, 116, 159–65
    in bathrooms, 412
    in bedroom remodels, 41
    ceiling openings for, 163
    flashing for, 101, 159, 163–65
    framing for, 160–62
    openable, 159, 163
Slab-on-grade foundation, 279–82
Slate roofs, 10, 114–15
SLCs (Self-leveling compounds),
    491, 497–98
Sleeper, definition of, 631

Slide-nut (Sliding-jaw) pliers, 352,
    353
Sliding compound-miter saw, 65
Slip-joint pliers, 63
Smoke pencil, 420
Snap cutters, 487, 488
Snap-in connectors, 309
Soapstone countertops, 405
Soffits, 172–73, 430, 631
Soffit-to-ridge ventilation, 432
Software, 33–34
Softwood, 70
Soil conditions, 261
Soil stack, 366, 367–68, 631
Soil tampers, 67
Soldering copper pipe, 354, 355,
    356–57
Sole (Sole plate), 201, 202, 211–12,
    214, 630
Solenoidal voltage testers, 304
Solid-surface countertops, 405
Solid-surface sinks, 409
Solid-wood flooring, 591–92, 605–
    10
Solvents, 545, 559
Sound-attenuating adhesives, 459
Sound isolation clips, GenieClip
    RST, 480–81, 482
Sound-mitigating drywall, 457
Soundproofing, 479–83
Sound studio, 482–83
Sound transmission coefficient
    (STC), 148, 631
Spade bits, 305
Span, 76, 631
Spanish stucco, 189
Speed Square, 54, 55, 56, 631
SPF (Spray polyurethane foam),
    631
    See also Spray foam insulation
Spike-and-ferrule hangers, 191
Spike nails, 78
Spirit levels, 55–56, 514, 575
Splash block, 631
Splashboards (Water table trim),
    173–74
Splay molding, 517
Split-tab receptacles, 330–31
Spot filler, 605
Spouts, faucet, 410, 415, 495
Sprayable caulk, 424–25
Spray-foam insulation
    for air-sealing, 423–26, 440
    for basements, 427–28
    for ceilings, 443–44
    closed-cell, 430, 442, 444, 452
    definition of, 631
    for door sealing, 129
    heat transfer and, 417
    open-cell, 442, 444, 445
    over a roof, 452–53
    overview of, 442–43
    R-value of, 440
    for window sealing, 153
Spray painting, 546–48, 556–58,
    568
Spring clamps, 62

Spun-glass lavatories, 414
Square (Unit of measurement), 631
Square cut (Butt joints), 523–24, 525–26, 527, 626
Squares (Tool), 54–55, 513, 525
Squeaky steps, 217
Stack venting, 366, 367–68, 371–72
Stain, 539, 540, 558, 562, 569
Stainless-steel chimney liner, 249–50
Stainless-steel sinks, 409
Stain-sealers, 603
Stair carriage, 219–20, 631
Stairs
    carpeting, 624–25
    inspection of, 16
    repair of, 217–20
    replacement of, 219
    terminology of, 217
    treads of, 16, 217, 218, 220, 632
Stair tool, 619
Standard lumber, 68–71
STC (Sound transmission coefficient), 148, 631
Steam rollers, 575–76
Steel doors, 126
Steel dowels, 260
Steel framing, 73, 74–75
Steel I-beams, 74, 75
    adding, 270, 271, 272
    boxing in, 535–36
    definition of, 628
    for jacking and shoring, 263–64
Steel reinforced foundations, 260, 274–75, 276
Step-flashing, 159, 164, 165, 631
Stiles, 117, 631
Stippled stucco, 189
Stock, definition of, 631
Stone countertops, 405, 411
Stone facade, 253
Stone floors, 591, 595
Stool, window, 528, 529–30, 631
Storm windows, 427
Story pole, 487, 500, 505, 510, 631
Straightedges, 487
Strap ties, 85, 225
Strap wrenches, 352
Strike plates, 127, 134, 135
Striking a joint. See Pointing mortar
Stringers, 216, 218, 219, 220, 631
Strip flooring, 591–92, 605–10, 631
Strippers, chemical, 544, 560–61, 567, 597
Stripping finishes, 544, 558–61, 567
Stripping tools, 305
Strong (Long cut), definition of, 631
Strongback, 161
Strong-ties, 85, 631
Structural and nonstructural panels, 75–78
Structural carpentry, 194–233
    damage to, 214–17
    demolition of, 206–11
    history of, 194

loads and, 195, 197
mapping, 27–28
nailing schedule for, 79, 197, 198
overview of, 194
permits for, 195
reinforcement and repair of, 211–16
See also Bearing walls
Structural framing lumber, 69
Structural screws, 68, 82–84
Stubby screwdriver, 62
Stub-outs, 360, 379, 380, 631
Stucco
    definition of, 631
    inspection of, 14, 187
    painting, 567, 569
    repair of, 185–91
    sheathing for, 274
    texturing, 189
Stud finder, 54, 55
Studs
    corner, 200
    definition of, 631
    edge protection for, 374
    end-wall, 202, 628
    gable-end, 204, 206
    layout of, 199, 200
    lumber for, 69
    metal, 213–14
    nailing, 201
    notching and drilling, 374
    removal of, 222
    standard length for, 199
    steel framing for, 74–75
    straightening and planing, 212–13
    terminology of, 199
    trimmer, 199, 201, 632
    See also Jack studs
Styrene-butadiene, 88
Subfloor, 606, 610, 612, 631
Submanifolds, 359
Subpanels, 289
Sump pumps, 284
Supply system. See Water supply system
Surface-bullnose trim, 485
Surform rasp, 457
Suspended ceiling, 631
Swanson Speed Square, 55
Sweep gaskets, 126–27
Swirled stucco, 189
Switches
    air-sealing, 424, 427
    demolition and, 207
    dimmers, 334, 335, 336–37
    drywall around, 467
    planning for, 302, 303
    requirements for, 300
    selection of, 310–11
    single-pole, 332–33, 335
    three-way, 344–45
    wireless, 337
    wiring, 332–37, 343–45
Switch loop, 332–34, 344
Synthetic mesh (Cedar Breather), 93, 111, 180
Synthetic-rubber caulks, 424

T
Tablesaw, 514
Tackless strips, 618–19, 620, 621, 624
Tack rags, 603, 632
Taillight warranty, 632
Tail pieces, 352, 378, 379, 382–83, 632
Tamper-resistant (TR) receptacles, 311
Tankless water heaters, 20, 385–88
Tannin bleed-through, 565
Tape
    for air-sealing, 429
    fiberglass mesh, 429, 474, 497, 499
    flashing, 168, 169–70
    foil-backed butyl, 429
    hot-melt carpet seam, 619, 621, 622, 623
    joint, 461
    painter's, 208
Tape measures, 55, 65, 487, 513
Taping compounds, 462, 471, 472
Taping knife, 459, 576
Tapping blocks, 611, 613
Task lighting, 392
T-bar applicators, 605
Tear-off (Stripping) shovel, 93
Teco nails, 80
Tee (Spread) foundation, 259
Temperature and pressure relief (TPR) valve
    permits for, 347
    purpose of, 20, 355
    water heater replacement and, 30, 383, 385, 387
Tempered glass, 159
Templates, 404, 519, 615, 616–18
Temporary walls, 208, 221
Termites, 215
Termite shield, 277, 632
Test plugs, 376
Texturing
    faux, 555
    plaster & drywall, 474
    stucco, 189
Theatrical smoke, 420
Thermal barrier, 417, 632
Thermal break, 147–48, 159
Thermal bridging
    definition of, 417, 632
    imaging of, 421
    insulation for, 426, 431
    unvented roofs and, 447
Thermal bypass, 632
Thermal bypass checklist, 632
Thermal envelope
    description of, 416, 417, 632
    energy audit of, 419–21
    heating system and, 419
    unvented roofs and, 431
    vapor barriers and, 429
Thermal imaging, 420–21, 440, 449
Thinset mortar/adhesive
    application of, 491, 497, 499, 500, 502, 503, 505

for countertops, 506, 507
description of, 489, 491, 632
epoxy, 491–92, 499, 500
for handmade tiles, 485
pros and cons of, 494
selection of, 486
for tub surrounds, 509, 510, 511
uses for, 490, 494, 498
Three-tab shingles, 103, 104, 107
Three-way switches, 344–45
Threshold
    for carpeting, 620
    definition of, 116, 632
    flooring under, 121, 496
    for French doors, 131
    gasketed, 138
    water-return, 137, 138–39
    weatherstripping, 126–27, 137–38
Tie-downs, 85, 632
Tile, 484–511
    adhesives for, 489, 491–92
    ceramic, 405
    cleanup for, 489
    countertops, 405, 495, 499, 505–508
    cutting, 487, 488, 501
    drywall under, 456, 490, 498–99
    estimating, 500–501
    floors, 502–505, 591, 594, 595
    grout for, 485, 488–89, 492
    handmade, 485–86, 489
    layout of, 487–88, 501–503, 504, 505, 510–11
    materials for, 489–93
    mission-barrel, 115
    mosaic, 506
    nonvitreous, 486
    offset joints for, 504–505
    prep work for, 493–96
    roof, 11, 114–15
    selection of, 484–86
    setting, 488–89
    setting beds for, 489, 490–91, 494, 496–500
    sheet-mounted, 485, 486, 506
    supporting, 510
    tiling over, 499–500
    tools for, 486–89
    trim, 485, 486, 506, 507
    tub surrounds, 413
    underlayment for, 114–15
    vitreous, 486, 487
    waterproofing membrane for, 489, 492–93, 494, 498, 507
    water resistance and durability of, 486
    See also Thinset mortar/adhesive
Tile-edge (Flush-mount) sinks, 409, 410
Tile nippers, 488
Tilt-and-turn windows, 145, 146, 151
Tin (Aviation) snips, 63
Titebond II wood glue (Polyvinyl acetate glue), 87, 523
Toekicks (Kickspaces), 393, 395–96, 398

Toenail, 79, 198, 201, 632
Toilets
    branch drain-stack connection
        for, 371–72
    dimensions needed for, 349–50
    drains for, 373–74
    flawed, 380
    inspection of, 16
    installation of, 381–82, 494
    planning for, 412
    removal of, 378, 494
    rough-in for, 373
    selection of, 414–15
    tile around, 494–95
    venting options for, 367–69
Tool belts, 56
Tools, 51–67
    Bazooka, 473
    for carpeting, 619
    cordless, 60–61, 305, 353, 458–
        59, 516
    for demolition, 207
    for drywall, 457–60, 473
    for electrical work, 304–306
    for finish carpentry, 512–16
    for flooring, 597–99, 607, 611
    hand, 55–56, 58, 59, 62–64, 304–
        305
    laser, 55, 65, 176
    layout, 54–55, 512
    lifting, 74, 458
    masonry, 235–36, 242
    measuring, 54–55, 512–14
    oscillating multitools, 57, 514,
        606–607
    for painting, 541–44
    for plumbing, 352–54, 360, 361
    powder-actuated, 66–67
    power, 56–58, 60–62, 305–306
    renting, 64–67
    safety for, 51–53
    for tile work, 486–89
    for wallcoverings, 575–76
Topping compounds, 462, 471, 472
Top plates
    air-sealing, 426
    attaching, 203, 212
    beams across, 222, 224–25
    definition of, 630
    gable-end, 204, 206
    headers and, 200, 202
    for knee walls, 204
    marking, 198–99
    nailing, 194, 197–98, 201
    in post and beam structures,
        227, 228, 229, 230
    single, 205
    for temporary walls, 220
    upper, 199
Torpedo levels, 56
TPR valve. See Temperature and
    pressure relief (TPR) valve
TR (Tamper-resistant) receptacles,
    311
Treads (Stair)
    carriage support of, 220
    definition of, 632

    inspection of, 16
    replacement of, 218
    squeaky steps and, 217
Trillium Dell Timberworks, 230–33
Trim
    definition of, 632
    inspection of, 13
    prefinished, 524
    removal of, 209
    See also Exterior trim; Interior
        trim
Trim-head finish screws, 81, 82,
    171–72
Trimmer studs, 199, 201, 632
Trim tile, 485, 486, 506, 507
Trisodium phosphate (TSP), 551,
    566
Trowel filler, 605
Trowels, 235
    curved, 459
    margin, 235, 488, 491, 511
    notched, 488
    pointing, 235
    techniques for using, 239–40
    for tile work, 488
    tuck-pointing, 236, 242
TrusJoists (I-joists), 72–74
Trusses, 15, 72, 233, 632
Try square, 55
TSP (Trisodium phosphate), 551,
    566
T-supports, 458, 469
Tubing benders, 357
Tubs. See Bathtubs
Tub-strainer wrenches, 352, 353
Tub surrounds, 16, 413, 494, 509–
    11
Tubular weatherstripping, 136, 424
Tuck-pointing. See Repointing
Tuck-pointing trowels & chisels,
    236, 242
Tung oil, 563, 603
Twist straps (Hurricane ties), 85
Two-slot receptacles, 327
Type G screws, 461
Type S screws, 461
Type W screws, 461

U
UBC (Uniform Building Code), 632
U-factor, 148, 632
Unconditioned space
    attics, 423, 425
    basements & crawlspaces, 427,
        449
    definition of, 632
    ventilation of, 418
Underlayment
    for flooring, 606, 612, 614–16
    for resilient flooring, 615, 616,
        632
    for roofing, 94–95, 96, 111
    for tile & slate, 114–15
Underlayment panels, 615, 616, 632
Undermount sink, 379, 409, 410,
    411, 413
Underused spaces, 32

Unified hydraulic jacking system,
    273
Uniform Building Code (UBC), 632
Uniform Plumbing Code (UPC),
    632
Unvented roofs, 431, 447
UPC (Uniform Plumbing Code),
    632
USG Sheetrock UltraLight Panels,
    457
U-shaped kitchens, 391, 396, 397,
    399
Utility knives, 59, 457, 466–67, 488,
    632

V
Valley rafters, 232, 630
Valleys, 108–109
    closed-cut, 97, 98, 109, 113
    definition of, 632
    flashing for, 97–98
    inspection of, 9
    open, 97, 112–13
    wood shingles and, 112–13
    woven, 98, 109
Vapor retarders (Vapor barriers)
    definition of, 632
    under flooring, 610
    plastic, 428, 429, 442, 451
    types of, 429
Variable-speed reversible (VSR)
    drill, 632
Varnish, 540, 596, 603
V-bronze (Metal tension strip), 136
V-cap trim, 485, 506, 507
Veneer grades, 76
Ventilation
    attic, 14
    bathroom, 16, 412
    at the eaves, 173, 447
    energy recovery, 435
    heat-recovery, 418, 435
    increasing controlled, 432–35
    kitchen, 16, 369, 408
    mechanical, 418, 432–35
    net-free ventilation area and, 430
    passive, 102
    roof, 102–103, 111, 430–31, 432
    soffit-to-ridge, 432
    of unconditioned space, 418
    See also HVAC
Venting pipes. See Drainage, waste,
    and vent (DWV) system
Vent-pipe flashing (Jack flashing),
    98–99
Vent runs, 375
Vents
    common, 369
    definition of, 632
    direct-vent pipe, 386, 387, 388
    intake, 102–103
    termination of, 370, 371
    wet, 366–67
    See also Drainage, waste, and
        vent (DWV) system;
        Ventilation
Vent stack, 370, 371, 375
Vermiculite, 449

Vinyl bead, 471
Vinyl flooring, 21, 594–95, 614–18
Vinyl siding, 14
Vinyl-to-vinyl adhesive, 572, 574
Vinyl wallcovering, 571–72, 573,
    574
Vise-Grip pliers, 63
Vitreous china (Porcelain), 414
Vitreous tile, 486, 487
Vix bit, 140
VOC (Volatile organic compound),
    21, 540, 632
Voltage, definition of, 287
Voltage testers
    Multimeters, 294, 304
    noncontact testers, 304
    plug-in circuit analyzers, 304
    solenoidal, 304
    types of, 294, 304
    uses for, 207–208, 287, 333
Volt-Amps (Power), definition of,
    287
VSR (Variable-speed reversible)
    drill, 632

W
Wago Wall-Nuts, 309
Wainscoting, 518, 537
Wall cabinets, 394, 395, 398–400
Wallcovering, 570–89
    adhesives/pastes for, 572, 573,
        574, 582, 583, 585
    amount of, 573
    for arches & alcoves, 589
    booking (folding), 585
    border, 572
    on ceilings, 585, 587, 588–89
    cork, 571, 572, 574
    on corners, 586–88
    custom, 572
    cutting, 583
    definition of, 572, 632
    dry-hanging, 585
    dye lot of, 573
    edges of, 573
    fabric, 571, 572, 574, 585
    foil, 571, 572, 574, 585
    grass, 571, 572, 585
    hanging, 570, 584, 585
    layout of, 580, 581
    lincrusta, 571, 572, 574
    mylar, 572, 574, 585
    natural textures, 572, 574
    over old wallcoverings, 577
    over paint, 577
    peelable, 578–79
    prepasted, 574, 583
    prep work for, 576–80
    primer-sealer for, 576
    repair of, 586, 589
    seams of, 582, 586
    sponging, 585
    stripping, 577–79
    techniques for, 581–89
    tools for, 575–76
    trimming and rolling, 586
    types of, 570–72, 574
    vinyl, 571–72, 573, 574

washable, 571
wood-veneer, 572, 574
Wall framing. *See* Framing
Wallpaper, 571
Wallpaper smoothers, 575
Walls
    air-sealing, 427
    bearing, 195, 206, 212, 221, 222–25, 626
    blocking in, 211–12
    building in place, 203–206
    concrete, 253
    cripple, 195, 222
    drywall on, 469–71
    faux-painted, 555
    flashing for, 101
    inspection of, 15
    insulation of, 418, 448–50
    kneewalls, 204, 206, 426–27
    layout of, 198–99
    load-bearing vs. nonbearing, 26, 27–28, 222
    nonbearing, 195, 211–16, 222, 629
    out-of-plumb, 119
    painting, 554
    plumbing and nailing, 203
    rain-screen, 180
    raising, 194, 202
    retaining, 283
    temporary, 208, 221
    wet, 28
    See also Exteriors; Framing
Wall-to-wall carpeting, 122, 595, 618–25
Warranties, 91, 632
Water-based caulks, 424
Water-based polyurethanes, 562, 603–604, 605
Water damage
    in damp basements, 284–85
    inspection of, 12, 13, 14–15, 16–18, 19
    moisture and, 432–34, 435–37
    to structural elements, 216–17
    See also Mold
Water-hammer arrestors, 377
Water heaters, 19–20
    permits for, 30
    replacement of, 383–88
    tankless, 20, 385–88
    TPR valves for, 20, 30, 347, 355, 383, 385, 387
Waterproofing membrane
    self-adhering, 88
    for tile work, 489, 492–93, 494, 498, 507
    See also Weather-resistant barriers
Waterproof shingle underlayment (WSU), 94–95, 97, 111
Water-resistant barriers. *See* Weather-resistant barriers
Water-resistant drywall. *See* Greenboard
Water-return threshold, 137, 138–39

Water-supply risers, 349
Water supply system
    air-sealing, 427
    copper pipe for, 347, 353, 354–55
    CPVC for, 347, 362
    galvanized steel pipe for, 362
    insulation of, 376–77
    overview of, 346, 347–48
    PEX for, 28, 347–48, 358–62
    rough-in for, 376–77
    slope for, 379
    supports for, 370, 377
Water table trim (Splashboards), 173–74, 176, 179, 182
Watts, definition of, 287
Wax crayon putty, 524
Wax-sealed floors, 596
Weather Hill Company, 227–30
Weathering grade, 238
Weather-resistant barriers (WRB)
    details for, 167, 168–70
    for doors, 124–25
    exterior walls for, 166
    role of, 166
    for roofing, 94, 96
    for tile work, 494
    for windows, 143
Weatherstripping
    for air-sealing, 424
    definition of, 632
    doors, 124–25, 126–27, 136–39
    kerf-in, 136
    sills, 169
    thresholds, 126–27, 137–38
    tubular, 136, 424
Weep screed, 187
Wet sanding, 459
Wet saws, 487, 488
Wet vents, 366–67
Wet walls, 28
Whole house exhaust fans, 418, 432, 435
Whole house renovation, case history of, 48–50
Wick, 179, 185, 436, 632
Wind baffle, 173, 632
Window inserts, 149, 151–54
Windows, 116, 142–59
    air-sealing, 427
    arched, 529, 532–33
    awning, 145, 146
    basement, 428
    casement, 145, 146
    casing for, 142, 150, 524, 528–30
    custom, 147
    double-hung, 144, 145, 146, 156–59, 556
    drywall around, 470
    energy efficiency of, 147–48, 149
    fixed, 145, 146
    flashing, 139, 142, 143–44, 167, 168–69
    frame materials for, 146–48
    glazing, 158–59
    hopper, 145, 146
    horizontal sliders, 145, 146
    inspection of, 13–14

installation of, 118, 154–56
    interior, 118
    interior trim for, 524, 526, 528–30
    new-construction, 149, 154–56
    painting, 556
    refurbishing, 156–59
    replacement of, 48
    replacement sashes for, 148, 149–51
    rough opening for, 222
    R-value of, 148, 417, 441
    selection of, 142, 144–49
    in showers, 412
    sizing, 150
    storm, 427
    terminology of, 147
    tilt-and-turn, 145, 146, 151
    types of, 144–46
    upgrading, 419
    wallcoverings around, 581
    weatherproofing, 143
Windowsills, 167, 530
Window stool, 528, 529–30, 631
Wire connectors, 309, 310, 334
Wire corners, 187
Wireless switches, 337
Wire nuts, 309, 310, 334
Wires, definition of, 289
Wiring
    aluminum, 297
    GFCI receptacles, 328–30
    guidelines for, 299
    inspection of, 18, 295, 296–97
    insulation around, 443, 447
    knob-and-tube, 17, 18, 295, 296, 297, 448
    lighting fixtures, 332–34, 343–45
    nail-protection plates for, 317, 318, 465, 629
    receptacles, 326, 327–31
    rough-in, 312–14, 315, 326
    soundproofing, 480
    splicing, 297, 309, 325, 334
    switches, 332–37, 343, 345
    types of, 309
    See also Electrical systems
Wiring trench, 323–24
Wood construction connectors, 84, 85–86
Wood countertops, 406
Wood doors, 126
Wooden gutters, 193
Wood-fiber composite, 171
Wood flooring, 591–93
    carpet transition to, 624
    installation of, 605–10
    refinishing, 595–605
    removal of, 210–11
Wood paneling, stripping, 558–61
Wood putty, 561
Wood shingles
    alignment of, 175
    corner boards and, 174
    cutting, 176
    replacement of, 179
    roof, 93, 109, 110–13

skip-sheathing for, 93
    valleys in, 109
    wall installation of, 177–79
    weaving corners with, 178–79
Woodstoves, 20, 247
Wood-veneer wallcovering, 572, 574
Woodwork
    refinishing, 561–63
    stripping, 558–61
Work area, isolation of, 208
Working drawings (Construction set), 34–35
Work triangle, 391
Woven valleys, 98, 109
WRB. *See* Weather-resistant barriers
WR board (Greenboard), 165, 456, 464, 628
Wrecking bars, 64
WSU (Waterproof shingle underlayment), 94–95, 97, 111
Wythe, 237, 632

X

XPS (Extruded polystyrene), 279, 427, 441–42, 632

Credits

Unless noted below, all photos appearing in *Renovation* are by Michael W. Litchfield and Ken Gutmaker; all drawings are by Vincent Babak.

Chapter 1

p. 9: Photo by Art Hartinger.

p. 9 (inset): Photo by Muffy Kibbey.

pp. 10–13, 15–16, 19: Photos by Roger Robinson, First Rate Property Inspection.

Chapter 2

p. 32: Floor plans courtesy of Steve Rynerson, www.rynersonobrien.com.

pp. 33–34: Illustrations courtesy of Russell Hamlet, www.studiohamlet.com.

p. 37 (left): Photo by Muffy Kibbey.

p. 37 (top and bottom right): Photo by David Glaser.

p. 37: Floor plans courtesy of Fran Halperin, www.halperinandchrist.com.

p. 38: Photos by Muffy Kibbey.

p. 39 (left): Photo by David Simone.

p. 39 (right): Photo by Beth Sumner.

p. 40 (top): Photo by David Simone.

p. 40 (bottom): Photo by Beth Sumner.

p. 41: Photos by David Simone.

p. 42 (bottom): Photo by David Milne.

p. 43–44: Photos by David Milne.

p. 45 (top): Photos by Russ Hamlet.

p. 45 (bottom): Photo by Mike Derzon.

pp. 46–47: Photos by Art Grice.

p. 48: Photos by Stephen Shoup.

pp. 49–50: Photos by Scott Hargis.

Chapter 4

p. 85: Illustration courtesy of Simpson Strong-Tie Company, Inc.

Chapter 5

p. 95: Photo by Roe A. Osborn, courtesy of *Fine Homebuilding*, © The Taunton Press, Inc.

p. 102: Photo by Mike Guertin, courtesy of *Fine Homebuilding*, © The Taunton Press, Inc.

Chapter 6

p. 126: Photo by Dan Thornton, courtesy of *Fine Homebuilding*, © The Taunton Press, Inc.

p. 146: Photo by Rodney Diaz, courtesy of *Fine Homebuilding*, © The Taunton Press, Inc.

p. 164: Drawing adapted from an original by Velux.

Chapter 7

p. 174: Photo by Roe A. Osborn, courtesy of *Fine Homebuilding*, © The Taunton Press, Inc.

p. 181: Photo by Charles Miller, courtesy of *Fine Homebuilding*, © The Taunton Press, Inc.

Chapter 8

p. 213: Photo courtesy of p. 221 (top): Photo by Roe A. Osborn, courtesy of *Fine Homebuilding*, © The Taunton Press, Inc.

p. 223: Photo by Tom O'Brien, courtesy of *Fine Homebuilding*, © The Taunton Press, Inc.

p. 225: Photo courtesy of Simpson Strong-Tie Co., Inc.

Chapter 9

p. 235: Photo courtesy of Bon Tool Co., www.bontool.com.

p. 240: Photos by Rosmarie Hausherr.

p. 241: Photo courtesy of Bon Tool Company.

p. 245: Photo courtesy of Roger Robinson, First Rate Property Inspection.

p. 247: Photos by Justin Fink, courtesy of *Fine Homebuilding*, © The Taunton Press, Inc.

p. 248: Photos by Justin Fink, courtesy of *Fine Homebuilding*, © The Taunton Press, Inc.

pp. 254–257: Photos courtesy of Quikrete Cement and Concrete Products.

Chapter 10

p. 260 (top): Photo courtesy of David Peterson Construction.

p. 263: Photo courtesy of Simpson Strong-Tie Company, Inc.

p. 270 (bottom left): Drawing adapted from Rob Thallon's Graphic Guide to Frame Construction, The Taunton Press.

Chapter 11

p. 288 (top): Photo by Roger Robinson, First Rate Property Inspection.

p. 289: Photo by Brian Pontolilo, courtesy of *Fine Homebuilding*, © The Taunton Press, Inc.

p. 295 (right): Photo by Roger Robinson, First Rate Property Inspection.

p. 296 (left): Photo by Roger Robinson, First Rate Property Inspection.

p. 296 (right): Photo by Brian Pontolilo, courtesy of *Fine Homebuilding*, © The Taunton Press, Inc.

p. 299: Photo by Brian Pontolilo, courtesy of *Fine Homebuilding*, © The Taunton Press, Inc.

p. 303: Photos by Muffy Kibbey.

p. 307 (top right and middle): Photos by Andy Engel, courtesy of *Fine Homebuilding*, © The Taunton Press, Inc.

p. 318 (top right): Photo by Brian Pontolilo, courtesy of *Fine Homebuilding*, © The Taunton Press, Inc.

Chapter 12

p. 349: Drawing adapted from Architectural Graphic Standards, Ramsey/Sleeper, 6th edition, ©1986, Reprinted with permission of John Wiley & Sons, Inc.

p. 359: Photo courtesy of Watts-Radiant, www.waterpex.com.

p. 361 (top 3): Photos by Daniel S. Morrison, courtesy of *Fine Homebuilding*, © The Taunton Press, Inc.

p. 361 (bottom): Photo by Krysta S. Doerfler, courtesy of *Fine Homebuilding*, © The Taunton Press, Inc.

p. 362: Photo courtesy of Watts-Radiant, www.waterpex.com.

p. 374: Photo courtesy of Simpson Strong-Tie Company, Inc.

Chapter 13

p. 400 (top): Photo courtesy of Hafele.

p. 400 (bottom): Photo courtesy of Rev-a-Shelf.

p. 403: Kitchen designed by Jean Revere. Tilework by Moncrieff & Brotman.

p. 413 (bottom left): Photo courtesy of Steve Rynerson, architect. Tile and stone work by Riley Doty.

Chapter 14

p. 416: Photo by GreenHomes America.

p. 417: Drawing adapted from Tom Barbour courtesy of Thermal Image UK, www.thermalimageuk.com.

p. 421 (top): Photo by Tom Barbour courtesy of Thermal Image UK, www.thermalimageuk.com.

p. 422: Photo by GreenHomes America.

p. 423 (top): Photo by CalCERTS Inc.

p. 423 (bottom): Photo by GreenHomes America.

p. 425 (top): Photo by John Larson.

p. 425 (bottom): Photo by GreenHomes America.

p. 427 (bottom): Photo by Tim Snyder.

p. 436: Photos by Roger Robinson.

p. 449: Photo by Tim Snyder.

p. 451: Photo by Charles Bickford, courtesy of *Fine Homebuilding*, © The Taunton Press, Inc.

Chapter 15

p. 459 (top left): Photo courtesy of SENCO.

p. 469 (center): Photo by Mark Kozlowski, courtesy of Myron Ferguson's Drywall, The Taunton Press.

p. 473 (bottom): Photo by Justin Fink, courtesy of *Fine Homebuilding*, © The Taunton Press, Inc.

Chapter 16

p. 506 (top): Photo by Rosmarie Hausherr.

Chapter 17

p. 518: Photos courtesy of Style Solutions/Balmer Architectural Molding.

p. 535 (right): Photo by Charles Bickford, courtesy of *Fine Homebuilding*, © The Taunton Press, Inc.

p. 537: Photo by James Kidd, courtesy of *Fine Homebuilding*, © The Taunton Press, Inc.

Chapter 18

p. 540: Photo courtesy of The Sherwin-Williams Company. Used with permission.

p. 548: Photo by Tom O'Brien, courtesy of *Fine Homebuilding*, © The Taunton Press, Inc.

p. 549: Photo by Rob Wotzak, courtesy of *Fine Homebuilding*, © The Taunton Press, Inc.

p. 550: Photos by Rob Wotzak, courtesy of *Fine Homebuilding*, © The Taunton Press, Inc.

p. 564: Photo courtesy of Behr/Rohm & Haas.

p. 566: Photo courtesy of American International Tool, Inc.

p. 568 (bottom): Photo by Tom O'Brien, courtesy of *Fine Homebuilding*, © The Taunton Press, Inc.

Chapter 20

p. 593: Photo by Albert Lewis, courtesy of Smith & Fong Plyboo.

p. 598 (left): Photo by Randy O'Rourke, courtesy of *Fine Homebuilding*, © The Taunton Press, Inc.

p. 604 (top right): Photo by Charles Bickford, courtesy of *Fine Homebuilding*, © The Taunton Press, Inc.